AF742414

Advances in Industrial Control

Series Editors

Michael J. Grimble, Industrial Control Centre, University of Strathclyde, Glasgow, UK
Antonella Ferrara, Department of Electrical, Computer and Biomedical Engineering, University of Pavia, Pavia, Italy

Editorial Board

Graham Goodwin, School of Electrical Engineering and Computing, University of Newcastle, Callaghan, NSW, Australia
Thomas J. Harris, Department of Chemical Engineering, Queen's University, Kingston, ON, Canada
Tong Heng Lee, Department of Electrical and Computer Engineering, National University of Singapore, Singapore, Singapore
Om P. Malik, Schulich School of Engineering, University of Calgary, Calgary, AB, Canada
Kim-Fung Man, City University Hong Kong, Kowloon, Hong Kong
Gustaf Olsson, Department of Industrial Electrical Engineering and Automation, Lund Institute of Technology, Lund, Sweden
Asok Ray, Department of Mechanical Engineering, Pennsylvania State University, University Park, PA, USA
Sebastian Engell, Lehrstuhl für Systemdynamik und Prozessführung, Technische Universität Dortmund, Dortmund, Germany
Ikuo Yamamoto, Graduate School of Engineering, University of Nagasaki, Nagasaki, Japan

Advances in Industrial Control is a series of monographs and contributed titles focusing on the applications of advanced and novel control methods within applied settings. This series has worldwide distribution to engineers, researchers and libraries.

The series promotes the exchange of information between academia and industry, to which end the books all demonstrate some theoretical aspect of an advanced or new control method and show how it can be applied either in a pilot plant or in some real industrial situation. The books are distinguished by the combination of the type of theory used and the type of application exemplified. Note that "industrial" here has a very broad interpretation; it applies not merely to the processes employed in industrial plants but to systems such as avionics and automotive brakes and drivetrain. This series complements the theoretical and more mathematical approach of Communications and Control Engineering.

Indexed by SCOPUS and Engineering Index.

Proposals for this series, composed of a proposal form downloaded from this page, a draft Contents, at least two sample chapters and an author cv (with a synopsis of the whole project, if possible) can be submitted to either of the:

Series Editors

Professor **Michael J. Grimble**
Department of Electronic and Electrical Engineering, Royal College Building, 204 George Street, Glasgow G1 1XW, United Kingdom
e-mail: m.j.grimble@strath.ac.uk

Professor **Antonella Ferrara**
Department of Electrical, Computer and Biomedical Engineering, University of Pavia, Via Ferrata 1, 27100 Pavia, Italy
e-mail: antonella.ferrara@unipv.it

or the

In-house Editor

Mr. **Oliver Jackson**
Springer London, 4 Crinan Street, London, N1 9XW, United Kingdom
e-mail: oliver.jackson@springer.com

Proposals are peer-reviewed.

Publishing Ethics

Researchers should conduct their research from research proposal to publication in line with best practices and codes of conduct of relevant professional bodies and/or national and international regulatory bodies. For more details on individual ethics matters please see: https://www.springer.com/gp/authors-editors/journal-author/journal-author-helpdesk/publishing-ethics/14214

More information about this series at http://www.springer.com/series/1412

Rubén Molina Llorente

Practical Control of Electric Machines

Model-Based Design and Simulation

Springer

Rubén Molina Llorente
BASc & MSC in Electronic Engineering
Universitat de Barcelona
Barcelona, Spain

ISSN 1430-9491 ISSN 2193-1577 (electronic)
Advances in Industrial Control
ISBN 978-3-030-34757-4 ISBN 978-3-030-34758-1 (eBook)
https://doi.org/10.1007/978-3-030-34758-1

© Springer Nature Switzerland AG 2020
This work is subject to copyright. All rights are reserved by the Publisher, whether the whole or part of the material is concerned, specifically the rights of translation, reprinting, reuse of illustrations, recitation, broadcasting, reproduction on microfilms or in any other physical way, and transmission or information storage and retrieval, electronic adaptation, computer software, or by similar or dissimilar methodology now known or hereafter developed.
The use of general descriptive names, registered names, trademarks, service marks, etc. in this publication does not imply, even in the absence of a specific statement, that such names are exempt from the relevant protective laws and regulations and therefore free for general use.
The publisher, the authors and the editors are safe to assume that the advice and information in this book are believed to be true and accurate at the date of publication. Neither the publisher nor the authors or the editors give a warranty, expressed or implied, with respect to the material contained herein or for any errors or omissions that may have been made. The publisher remains neutral with regard to jurisdictional claims in published maps and institutional affiliations.

This Springer imprint is published by the registered company Springer Nature Switzerland AG
The registered company address is: Gewerbestrasse 11, 6330 Cham, Switzerland

Dedicated to my family and friends in special to my wife Núria and my daughter Martina.
Rubén Molina Llorente

Series Editor's Foreword

Control system engineering is viewed very differently by researchers and those that must implement designs. The former group develops general algorithms with a strong underlying mathematical basis, while the latter have more local concerns over the limits of equipment, quality of control, and plant downtime. The series *Advances in Industrial Control* attempts to bridge this divide and hopes to encourage the adoption of more advanced control techniques when they are beneficial. The rapid development of new control theory and technology has an impact on all areas of control engineering and applications. This monograph series encourages the development of more targeted control theory that is driven by the needs and challenges of applications. A focus on applications is essential if the different aspects of the control design problem are to be explored with the same dedication that control synthesis problems have received. The series provides an opportunity for researchers to present an extended exposition of new work on advanced control, raising awareness of the substantial benefits, and exploring the challenges that can arise.

One of the unusual features of this monograph is that it deals with implementation problems very often neglected in more academic texts, drawing upon the author's very relevant application experience. Chapter 1 sets the tone, being concerned with embedded control systems. Problems in real-time implementation of control systems have become more onerous now that most advanced control system designs are "model-based." They have many advantages and allow for the multi-input multi-output nature of a process or machine explicitly, but implementation can be problematic. This first chapter is very wide-ranging, covering areas such as "hardware-in-the-loop testing, software tools, and system architectures."

Chapter 2 involves electrical machine control problems covering a range of classical and fuzzy control methods, and dealing with well-known problems in digital implementation. Various structures for control systems such as feedforward and feedback control systems and cascade control structures are considered. The problems of digital implementation are again considered from a practical viewpoint, describing the type of equipment involved. It is unusual for an introduction to classical PI or PID control to be extended into a discussion of real implementation issues. This should be particularly valuable to engineers in industry. The material on fuzzy control is also interesting and includes a useful electric machine speed control application.

Chapter 3 deals with three-phase electrical systems which are of course very common for use with motor control systems. In this case, the material is more basic electrical engineering but written from a control engineer's perspective. This provides an introduction to some of the material in Chap. 4 on the fundamentals of electrical machines. From a control viewpoint, it is important to consider the non-linearities mentioned in these systems. The general model information should also be valuable to those simulating such systems. This overview of electrical machine types will be particularly helpful to control engineers that often use very simplified models and do not need to cover the detailed electromagnetic characteristics of such machines. It is also a reminder that the viewpoint of an electrical engineer in terms of vector diagrams of machine currents is rather different to the usual control engineer's transient characteristic investigations.

Chapter 5 on modeling electrical machines returns to more familiar territory for the control engineer dealing with state-space systems and how they may be used to represent AC or DC machines. The discussions extend into simulation of these systems. Chapter 6 is concerned with measurements in electrical drive systems. This is also a topic which is often neglected in more academic texts but from a practical viewpoint is important since faults in systems can often be traced back to problems with measurements. There is a useful overview of the different types of sensors used and their characteristics.

Chapter 7 is concerned with microcontrollers for electric drive systems. Most texts avoid the details of technology since these change so rapidly; however, from a process or commissioning engineer's perspective it is one of the most important areas to understand. The different modules involved and timing problems are discussed, and aspects of analogue-to-digital conversion are covered. A family of microcontrollers is described, and the simulation of systems, including the machines and component parts, is considered.

Chapter 8 deals with three-phase voltage-source inverters for high-performance control of three-phase machines. This is another area where a good understanding of the power electronics is useful for both simulating the system and understanding the noise and uncertainties that are present. This is valuable when treating control problems and breakdowns. Chapter 9 is concerned with the topic of space vector modulation. Chapter 10 returns to more familiar territory dealing with the practical control of AC machines. It describes the speed and current loops using a DC machine as an introduction to AC machine vector control. The familiar tools of Bode diagrams and transient time responses are used. The block diagrams of various machine control systems are helpful, and topics such as sensorless control are explored.

Chapter 11 is concerned with model-in-the-loop development in vector control of induction machines. Control loop analysis is performed, and all aspects of simulation and implementation are discussed. The application to electric vehicle and electric aircraft propulsion control systems is particularly interesting and topical. This is of course a hugely important topic in the automotive and aircraft industry at present.

The book contains appendices that also cover practical material needed by design engineers and their commissioning colleagues. This text is therefore a valuable contribution to the *Advances in Industrial Control* series, bridging the gap between the electrical engineer and the control engineer, and going far into the application and equipment aspects of producing a real working control system.

Glasgow, UK
October 2019

Michael J. Grimble

Preface

It is well known that electric machines are widely used in numerous applications. Nowadays, recent applications such as electrified aircraft propulsion (EAP) use propulsors (propeller or fans) driven by electric machines. In the aviation sector, the electric machines and power converters should meet a power density 2–3 times state of the art in the MW power range, with efficiencies higher than 96 and 99%, respectively. In the last few years, applications such as drones and traction machines in electric vehicles are being a challenge because the demands in terms of efficiency and durability are also considerable. Renewable energies also consist of continuous improvement in terms of the efficiency of the energy developed.

The advanced design of the AC machines with finite element analysis (FEA) increasingly allows obtaining high-performance designs and high power density machines which together with sophisticated control systems and the appropriate hardware continue to optimize their operation in a wide range of speed. The result is that lower-performance machines such as brushed machines are being displaced in many applications.

The electric machine systems are a multidisciplinary area. The machine design, the mechanical systems, the electronic hardware composed by the power semiconductors, sensors, actuators, and the embedded systems are designed together with the collaboration of multidisciplinary engineering teams to guarantee success. Furthermore, the new engineering design process and new sophisticated co-simulation tools accelerate the time to market of the motor control units (MCUs) with high-quality results.

The speed/torque variation in electromechanical systems where the speed/torque is adapted according to the necessity of the system is the role of the electric drives. The control system of these drives and therefore the control of the machine are increasingly complex systems and usually consist of microprocessed embedded systems. During the last few years, the increase of the embedded system complexity in electrical machine control applications involves an increase of the model-based design (MBD) where the lines of code are mostly replaced by code generated on tested models in a personal computer environment. MBD provides a mathematical and visual approach to develop complex control systems. During the development process, models of the systems can be used for design, analysis, simulation, verification, and automatic code generation for the embedded systems. MBD is transforming the way of working of engineers and scientists since the design tasks of the laboratory and field are moved to a simulation environment in a desktop. The simulation and verification tools allow to test, refine, and retest the models without to build prototypes. Different test stages as model-in-the-loop (MiL) and software-in-the-loop (SiL) can be carried out in the MBD process. The MBD, together with appropriate software architecture patterns, guarantees success in the motor control units (MCUs). The automotive industry is one of the industries which applies this modern design methodology where the benefits can be observed today.

There is wonderful control of electric machine books based on the experience of their authors. However, in this book, the intention has been to dedicate as much as possible to the practical application of how the control of an electric machine could be carried out with the modern tools available today in an efficient manner.

The book consists of twelve chapters. Chapter 1 describes the modern design technics based on MBD, the V-model, the computation simulation software packages, and the software architecture patterns. Chapter 2 discusses the basic regulation based on classic controllers such as the proportional–integral–differential (PID), different control structures, digitalization methods for PIDs, aliasing, zero-order hold, quantifiers, and time delays, and with examples and simulations including a controller based on fuzzy logic. Chapter 3 discusses three-phase systems mostly used in AC machines. The three-phase systems with star and delta connections are analyzed. The power calculation with practical explanations is also analyzed, being the prelude to the in-depth analysis of the mathematical tools that facilitate the analysis of AC machines. Furthermore, practical graphics and examples of a digital implementation of mathematical axis transformations (Clarke and Park transformations), RMS, and electric power computations are represented.

In Chap. 4, is shown a classification of the most common electrical machines starting with the more traditional machines such as the DC-brushed and induction machines (IMs), and finishing with more sophisticated machines such as the permanent magnet-assisted synchronous reluctance machine (PMASynRM). Their primary structures, their mathematical expressions in a steady-state, in space vector, in dq_s transformations, and the electromagnetic torque expression for each machine are shown. Moreover, basic concepts of machine design, sections of different machines, and simulation results are introduced based on Altair, Flux™, and FluxMotor™ FEA software package. Chapter 5 is a continuation of Chap. 4 where it described the models with continuous state-space methods of the DC and the different AC machines such as IM, permanent magnet synchronous machine (PMSM), SynRM, and PMASynRM. As an example of the utility of the discrete models, the PMASynRM is described with its discrete model for its implementation in Simulink®, which can be used for C code generation, to be able to run it in real time in a DSP or FPGA. As a prelude of the next chapters, the closed-loop control of the current loops is introduced to verify the effect produced by the inherent cross-coupling of the AC machines, as well as a solution to the decoupling that optimizes its control.

Chapter 6 is reserved to treat a different subject as sensors and sensing circuits used in most of the DC and AC drives. Voltage, current, temperature, speed, and position are the basic measurements in the machine control applications for control algorithms and its protection in fail situations such as locked rotor or overtemperatures. In some applications, the control part should be isolated from the power stage, which is typically fed by high voltage. Then, as some of the sensors should be directly connected to the high-voltage side, some isolation mechanism should be used as it is shown in this chapter. In this chapter, the design of the above measurement variables with its hardware signal conditioning, software strategies, and experimental results is deeply analyzed for the proper design of high-performance machine control.

With the same purpose as Chap. 6, another subject about the knowledge of the standard microcontroller/DSP peripherals used in the implementation of the electric machine control such as I/O, timers, and A/D converters is reserved for Chap. 7. Specific features and functions such as smart high-resolution pulse width modulation (PWM) timers and Delta-Sigma A/D converters are also treated in detail. The high-resolution PWM signals help to generate smoother sinusoidal waveforms with high-frequency fine-tuning, while Delta-Sigma A/D converters allow measuring the machine phase current accurately. The microcontroller/DSP peripherals, as well as the CPU performance, are determinant for the microcontroller/DPS selection, but it should be in agreement with the application. The GTM Timer module from Bosch is an example of a smart high-resolution timer which can be found as an intellectual property (IP) integrated into different microcontroller manufacturers. In this chapter, A/D converters and specific peripherals for high-performance machine control are covered with simulations and real implementation examples for the AURIX™ family of Infineon and RX600 family of Renesas.

After describing the previous chapters, the reader can be more comfortable with Chap. 8, which discusses all necessary knowledge needed to design a voltage-source inverter (VSI). The voltage-source inverter (VSI) is a fundamental power electronic drive where high-performance control for three-phase electrical machines can be achieved. The continuous improvement of power devices that increasingly improve their performance, such as high electron mobility transistor (HEMT) devices, allows higher efficiencies and more and more wide range of use. The inverter not only is a three-phase bridge made by three half-bridge legs but also needs other elements for its correct operation. The stability of the voltage source required by the three-phase bridge is a key to optimizing its performance. Also, the inverter and machine protection elements allow having a safe behavior in the abnormal situations that prevent its destruction and other near components. These protection elements join the control logic and constitute the motor control unit (MCU). The analysis of the switching and conduction losses in the power devices is analyzed in detail, as well as the effects of capacitances, inductances, and parasitic resistances. The key elements, such as gate drivers, are also analyzed, even for devices in parallel. The effects of high dv/dt are also analyzed, especially when the length of the connection power cable between the inverter and the machine is considerable, providing different solutions such as the use of sinusoidal filters. In this chapter, all the necessary parts for the design of a VSI for control of three-phase machines are entered in detail, providing experimental results and simulations for better understanding, as well as a complete model of a VSI (power plant).

Once minimum knowledge of the VSI is acquired in Chap. 8, in Chap. 9 is discussed the space vector modulation (SVM), also known as SVPWM. SVM is increasingly replacing more traditional modulations such as six-step and sinusoidal modulation as it provides better use of the available DC-link voltage and a reduction in harmonic content. The model-based design is approached for SVM algorithm development which is tested to assure a correct operation before its implementation in real hardware. The theory of SVM in its two modes, continuous and discontinuous, deadtime compensation, model-based design using MATLAB and Simulink, simulations, and the experimental results are treated.

Chapter 10 comes back to the machine control by using high-performance control system as field-oriented control. The electric machine based on a control system with the machine model is not a simple task but requires necessary simulation tools to understand its basic operation. The complexity of their models suggests performing first simulations in both open and closed loops, without using an electric drive (VSI or DC servo stage). High-performance control, such as vector control for AC machines, can be achieved with excellent results by using suitable simulations. The chapter starts with the speed and current loop control theory of the rotation loads for DC machine, which is used as an introduction of AC machine vector control. Some simulation results with and without electric drive are illustrated. The chapter continues with the theoretical and practical part of vector control through simulations without electric drive, as well as the practical development of magnitude and position flux observers, and estimators for sensorless systems. The most relevant AC machines on the market are covered, as seen in Chap. 4.

Chapter 11, which discusses the model-based design (MBD) process in a field-oriented control for induction machines. Two induction machines are presented, one with 5 HP of power for industrial purposes and other with 110 HP of power for electric vehicle (EV) application or full electric aircraft propulsion propeller (one of my many passions), both with a three-phase VSI. The MBD is increasingly used in the field of electrical machine control because of the numerous advantages it offers, such as improving product quality and reducing development time. MBD is transforming the way of working of engineers and scientists since the design tasks of the laboratory and field are moved to a simulation environment in a desktop. The control system of an electrical machine can be rapidly prototyped using a simulation environment while in parallel the software architecture necessary for its implementation in a microcontroller or DSP can be discussed and designed. The success of these two steps guarantees a better performance of the model-in-the-loop (MiL), which is where the control of the machine on a simulated plant is evaluated, with a certain level of realism. If the control requirements are minimally met, where most of the errors are solved, the automatic code generation allows performing tests with the real plant, or on a processor-in-the-loop (PiL) or hardware-in-the-loop (HiL) scenarios. In EV/HEV automotive systems, the plant is the vehicle which is also modeled to evaluate the control of the electric machine.

In the Chap. 12 is treated a possible real-time model for emulation purpose of the machine and VSI. For faster development, the hardware composed by the machine, inverter, and controller can be replaced by a precise real-time model for emulation. The control systems increasingly use their verification and development through a digitized plant on an field-programmable gate array (FPGA) platform so that control algorithms can be evaluated without the need for real hardware, in this case an inverter and an electric machine. However, real-time simulation of electric machine models and the VSI can be especially complicated due to the rapid nature of the dynamics, that is, reduced time constants, especially on very low power machines. The switching of PWM signals of up to tens of kHz requires sampling rates of the order of several MHz to obtain reasonable accuracy, for example, to model the ripple produced by the PWM in the inductance of the machine. That is why FPGAs are the ideal platform for complex real-time simulations due to their ability to process data in parallel allowing sampling and execution rates up to the MHz range.

To arouse the curiosity of the readers, in the second part of the Appendix, it presented simulation detail of results performed with FluxMotor™ according to a similar machine analyzed in Chap. 4 in motor and generator mode. It corresponds to a 55 kW IPMSM machine with 48 stator slots, 8 poles, and permanent magnets mounted in V-pole configuration in the rotor. The construction details are deeply explained. To have an idea of the size, the machine has an external stator radius of 134.5 mm and a stack length of 84 mm. The total mass of the machine is 32.137 Kg, which means a power density of 1.7 kW/Kg. Due to this power density and other factors, the application of this machine can be for electric vehicles.

The book is intended for a wide variety of readers because during the explanations, my intention has always been to explain the theory as I would have liked it. The different readers can be academia and industry researchers, graduate students and their professors, engineers, and practitioners who are working in the field of the machine control systems for any industrial sector. The theory is always necessary, but I have intended to keep practical descriptions as much as possible. For students and newcomers, the main prerequisites are undergraduate courses on system control theory, basics in electric machines, and power electronics.

The simulations and experimental results in this book have been developed thanks to my 16 years of experience in power electronics and control systems, especially in machine control systems from which I decide to graduate with a final degree project based on field-oriented control for induction machine in 2003. In my modest opinion, this book is a book that I wish had during my years of research since it would have greatly facilitated my understanding of the control of electric machines thanks to its practical contribution.

Lastly, I give my special thanks and love to my wife, Núria, who will always be at my side, for her patience during these last 3 years which I have been very busy preparing this book.

Barcelona, Spain
October 2019

Rubén Molina Llorente

Acknowledgements I would like to particularly thank the following persons (in alphabetical order) for their contributions: Ph.D. Ramón Bargalló Perpiñà (Polytechnic University of Catalonia), Isabelle Feix (Infineon Technologies AG), Jasmin Hamp (AUTOSAR), Hua Jin (Powersim), Vincent Leconte (Altair Engineering Inc.), Amadeo Tierno (Altair Engineering Inc.), Bernd Westhoff (Renesas Electronics Corporation). In addition, I would like to thank MathWorks.

Trademark Acknowledgements

1. MATLAB®, Simulink®, Stateflow® and Simscape Electrical™ (formerly SimPowerSystems™ and SimElectronics®) are registered trademarks or trademarks of The MathWorks, Inc. For more information and a list of additional trademarks contact:
   The MathWorks, Inc., 1 Apple Hill Drive, Natick, MA 01760-2098, USA
   Web: mathworks.com/trademarks for a list of additional trademarks.
2. PSIM is a registered trademark of Powersim Inc. For more information, please contact:
   Powersim, Inc., 2275 Research Blvd, Suite 500, Rockville, MD 20850, USA
   E-mail: info@powersimtech.com, Web: www.powersimtech.com
3. Altair™ and FluxMotor™ is a registered trademark of Altair Engineering, Inc. For more information, please contact:
   Altair Engineering, Inc., 1820 E. Big Beaver Rd., Troy, MI 48083, USA
   Phone: +1 (248) 614-2400
   Fax: +1 (248) 614-2411
   Web: www.altair.com
4. TEKTRONIX is a registered trademarks of Tektronix, Inc. For more information, please contact:
   Tektronix, Inc., 14150 SW Karl Braun Drive, P.O. Box 500 Beaverton, OR 97077, USA
   Web: www.tek.com
5. Teledyne LeCroy is a registered trademark of Teledyne LeCroy, Inc. For more information, please contact:
   Teledyne LeCroy, Inc., 700 Chestnut Ridge Road, Chestnut Ridge, NY 10977-6499, USA
   Phone: +1 800-553-2769 or +1 845-425-2000
   Web: www.teledynelecroy.com

6. AUTOSAR is a trademark of AUTOSAR GbR. The word AUTOSAR and the AUTOSAR logo are registered trademarks. For more information, please contact:
   Frankfurter Ring 224, 80807 Munich, Germany
   Phone: +49 (89) 452450-395
7. AURIXTM and TriCore® are a trademark of Infineon Technologies AG. For more information, please contact:
   Infineon Technologies AG, IFAG C EC MR, Am Campeon 1-15, 85579 Neubiberg, Germany
8. AUDI is a trademark of AUDI AG. For more information, please contact:
   Ettinger Street, 85057 Ingolstadt, Bavaria, Germany
9. Renesas and the Renesas logo are trademarks of Renesas Electronics. For more information, please contact:
   TOYOSU FORESIA, 3-2-24 Toyosu, Koto-ku, Tokyo, 135-0061, Japan
   Web: www.renesas.com
10. BOSCH is a trademark of Robert Bosch GMBH. For more information, please contact:
    Robert Bosch GmbH, Postfach 13 42, 72703 Reutlingen, Germany
    Web: www.bosch.com
11. SIEMENS is a trademark of Siemens Aktiengesellschaft. For more information, please contact:
    Siemens Aktiengesellschaft, Werner-von-Siemens-Straße, 180333 Munich, Germany
    Web: www.siemens.com

Contents

Chapter 1
Embedded Control System Development Process: Model-Based Design and Architecture Basics

1.1 Introduction

In different markets, such as industry, appliances, automotive, marine, and avionics, rotating electrical machines are widely used. The adjustable speed drive (ASD) with DC machines had been used widely to control the torque and speed. However, the AC machine drive system driven by a variable-voltage/variable-frequency (VVVF) is widely used due to their high-performance control thanks to the improvements in the power electronics devices, in the machine efficiency, and in the performance of the microprocessors. In the literature, it is possible to find different nomenclatures to specify a variable speed AC drive such as variable-frequency drive (VFD), adjustable-frequency drive (AFD) where both provide a VVVF. The common part of these drives is the control of the speed/torque variation in electromechanical systems where the speed/torque is adapted according to the necessity of the system. The most used electronic power system (sometimes designed as AC drive) able to perform a VVVF is known in the literature as an inverter. As it will be discussed in Chap. 8, the inverter topology can be single-phase, or poly-phase, and basically consists in transform a DC source into a single- or poly-phase AC source.

The control system of the AC drive and therefore the control of the AC machine are increasingly complex systems and usually consist of embedded systems. The embedded system is referred to as an electronic system that is designed to perform a dedicated function by using a combination of computer hardware and software, which is often embedded within a larger system. A generic embedded system architecture is composed of a microprocessor, its memory, and the inputs and outputs. The embedded software is commonly stored in the non-volatile memory devices such as flash memory, read-only memory (ROM), or erasable programmable ROM (EPROM). The microprocessor uses the random-access memory (RAM) for its run-time computation. Once the embedded system is powered, the software code stored in the non-volatile memory is read to execute the instructions to process the input information and set the outputs according to the needs of the external control system. In the control systems, the inputs usually are sensors, while the outputs are actuators.

© Springer Nature Switzerland AG 2020

R. Molina Llorente, *Practical Control of Electric Machines*,
Advances in Industrial Control, https://doi.org/10.1007/978-3-030-34758-1_1

Both are managed by dedicated peripherals such as general-purpose input/output (I/O), timers, and analog-to-digital converter (ADC or A/D). When the microprocessor, memory, and peripherals are integrated together on a single chip, the device consists of a microcontroller. The microprocessor with an architecture optimized for digital signal processing is the digital signal processor (DSP). The DSP is an ideal processor choice for applications with intensive math computations in constrained environments. For example, the analog input signals, such as audio or video signals, are converted to digital with ADC, and then, it is manipulated digitally with sophisticated algorithms and finally converted back to analog form with a digital-to-analog converter (DAC). Nowadays, some DSPs have dedicated peripherals for a control system so that they compete with microcontrollers thanks to their increasingly affordable price and the tools improvements. New platforms based in programmable system-on-chip (SoC) which combine programmable logic, DSP, and microprocessor cores on the same chip are increasingly used in electric machine control. The advantage of using programmable logic or DSP is its high-performance computation where complex control algorithms can be implemented inside, while the microprocessor can be dedicated to other tasks such as communication interfaces. The field-programmable gate array (FPGA) contains programmable logic blocks such as AND and XOR with faster and parallelizable processing where the performance is higher than the microprocessor and DSPs. They are generally more expensive and more difficult to use, and its uses are limited for applications where the high performance is a requirement.

DSPs, microcontroller, SoC, and FPGAs are valid options to perform a machine control units (MCU), but as it will be seen in Chap. 7, the microcontroller option will be studied, in particular, the families RX63T of Renesas and Aurix™ Tricore of Infineon.

The embedded systems can combine any combination of a microprocessor, microcontrollers, DSPs, FPGA, and SoCs. The embedded systems have to manage millions of lines of code where their integration and debugging with other systems are increasingly more complex. To deal with the complexity of embedded software development in different markets, and especially in electronic control units (ECU) or more specifically, in MCUs of modern automobiles, the development is increasingly based in the model-based design. Hence, in the automotive sector, the traditional way of building embedded code, where lines of code are written by hand, has become obsolete. The model-based design (MBD) focuses on the models that describe the desired control and the behavior of the system under development. The MBD has been discussed for a few decades, but it is not until these recent years where it is being involved in the flow of system design. In the MBD, the engineer focuses on the functionality, the modeling of physical phenomena, the interface of modules, the general behavior of the system, and verification through their own simulations.

On the other hand, the definition of fundamental standard software architecture is mandatory if certain security, reliability, and interactive work methodology requirements between the designers, the client, and the quality regulators should be satisfied. For example, in automotive, in many occasions, the development of a project is executed between OEMs (Original Equipment Manufacturers) and their Tier1 and Tier2 suppliers. To allow the exchange of models and software, the standardization of a work platform is necessary. In this sense, the AUTOMotive Open System ARchitecture (AUTOSAR) partnership guarantees a process for the development of standard automotive software. As will be discussed in Sect. 1.4, AUTOSAR offers a framework guide for electronic automotive control systems which comes with layered software architecture.

Lastly, the MCU and ECUs are tested with selected methods in agreement with the standards, e.g., the International Electrotechnical Commission (IEC), to guarantee and minimum safety requirements. The objective is to ensure the safety of persons, to measure performances, and to ensure compatibility with other systems. For example, the IEC 61508 describes the basic functional safety standard applicable to all types of industry, while the International Organization for Standardization (ISO) 26262 specifies the functional safety for road vehicles which is derived from IEC 61508.

In this chapter, it basically introduces the MBD technique, the simulation tools used in this book, the software architecture of the ECUs and MCUs, and the different test systems for each of the development phases.

1.2 Model-Based Design

MBD is a mathematical and visual method that facilitates the complex designs of embedded systems. Instead of using extensive and complicated programming codes, designers can use MBD to define models with advanced functional features using continuous-time and discrete-time simulations. The main components of a model-based design are design and simulation at the system and component level, automatic code generation, and continuous testing and verification.

1.2.1 V-Model

The MBD can be considered a software development methodology based on the V-model. The V-model was first presented in 1991 (Forsberg and Mooz 1991), and it is a variation of the waterfall model in a V shape folded in half at the lowest level of decomposition. Figure 1.1 shows the V-model adapted for the software development process.

V-model is denoted as a linear refinement process that follows a top-down approach shown on the left side of the V, while validation and verification take a bottom-up approach that is shown on the right side of V. V-model demonstrates the relationships between each phase of the development life cycle and its associated testing phase. The horizontal and vertical axes represent the time or integrity of the project (from left to right) and the level of abstraction (the abstraction of coarser grain upwards), respectively.

The project starts with the system requirements where the system specification is derived. This defines a detailed specification where is specified the functionality

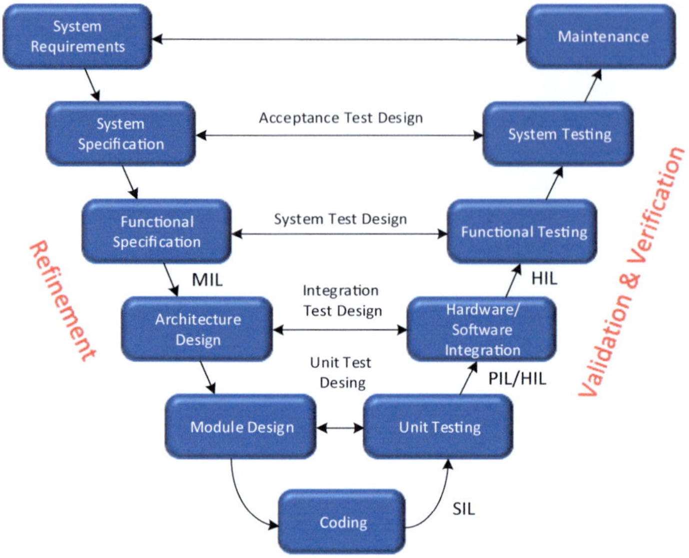

Fig. 1.1 V-model. The left side is denoted as a linear life-cycle process that follows a top-down approach, while the right side of the *V* is the validation and verification using a bottom-up approach

to meet and allows designing a software architecture. For example, it could define the speed control functionality of an electric machine. In this step is developed a basic MiL, which in the mentioned example, should consist of a speed control loop on a plant model. Taking into account the different functionalities, the architecture software platform can be designed. On this architecture, the design of the different modules or software components (SWC) is perfectly defined with their respective inputs and outputs. The code generation can be generated for each one of the SWCs designed to finally be tested by means of a unit test (UT). If the UT meets the requirements, a hardware/software integration test with the rest of the modules can be performed using a processor-in-the-loop (PiL) or hardware-in-the-loop (HiL). Otherwise, horizontally it would return to the design of the module to modify its features to rebuild the UT. If the integration test does not comply with the requirements, in particular, it is necessary to proceed with the redesign either of the architecture or of the different modules. If this is not the case, the functional tests are carried out. As in the previous cases, if the test requirements are not met, it is returned horizontally to the refinement.

1.2.2 Test Stage

As mentioned, it is possible to consider different test stages as illustrated in Fig. 1.1: model-in-the-loop (MiL), software-in-the-loop (SiL), PiL, and HiL. MiL stage is implemented during the refinement process in order to validate demodel according to the requirements specification. The rest of the test stage is usually implemented in the validation and verification of the model, as explained before. In the following section, each of them is explained separately.

1.2.2.1 Model-in-the-Loop (MiL)

The MiL scenario is a technique used to abstract the behavior of a system or subsystem so that this model can be used to test, simulate, and verify that model. For example, the control system of an electric machine based on a PID regulator with a power stage, it would be possible to adjust and test its correct operation on a modeled plant formed by the power stage and the electric machine. By using a chain of industry standard tools such as Simulink® to define the model, it is possible to test and refine that model within a personal computer (PC), which allows managing a complex system efficiently.

1.2.2.2 Software-in-the-Loop (SiL)

Unlike the MiL, the SiL scenario allows testing the code generated from the model also on a PC where it is possible to perform the simulation, but based on the model code. That is, on the same example of the previous control system, the PID controller based on a model, its code is generated and therefore tested in the same environment. This phase requires only the simulation model and is independent of the hardware, focusing on software interfaces and numerical results. The requirements and specifications of the software can be analyzed and verified here. The first phase of refinement of the requirements is carried out during the SiL simulations.

1.2.2.3 Processor-in-the-Loop (PiL)

The PiL scenario allows developing real-time control over a microprocessor target connected to a digital platform that emulates, in this case, the most complicated parts to obtain at the beginning of the project, such as the power stage and the electric machine. Unlike MiL and SiL, in this case, the real microprocessor of the ECU is tested, where the software not only consists of the application layer control algorithm but a part of the software architecture. Unlike the simulation in the SiL scenario, where the runtime metrics were not obvious, due to the higher calculation capacity of a PC, the PiL has the ability to detect insufficient hardware capabilities. For example, PiL provides real-time metrics, detects bottlenecks, adds used memory, supervises hardware and software interruptions, analysis of waveforms, thermal effects, and electromagnetic interferences, among others.

On the other hand, the digital platform that emulates the power stage and the machine can be a PC, a field-programmable gate array (FPGA), or even a DSP. The increase in the speed of processing of the DSPs and of their reduction in cost make that the option of using a DSP to emulate an electrical machine is one of the best alternatives.

1.2.2.4 Hardware-in-the-Loop (HiL)

The hardware-in-the-loop (HIL) scenario tends to become the standard electronic development tool for testing ECUs, MCUs, and, more particularly, its software for different OEMs. Especially during development, the individual software change tests (MBD changes) of the ECU can be tested using the HiL in real time. Increasing the complexity of control algorithms requires the use of this scenario, which advances the problems that can be encountered before testing in real conditions. In the case of an MCU, consisting of a power stage based on a three-phase inverter, and its control logic for an electrical machine, the HiL scenario can be used as a tool to develop and validate control strategies in all operating conditions, including extreme conditions, such as non-destructive failures in the machine itself. In this scenario, the machine is emulated, with a certain degree of fidelity, in a processor that acts on an electronic load that emulates its physical behavior. Additionally, it is possible to include a power supply emulator for the inverter where different power conditions can be tested. Then, the MCU is exposed to the different possible power conditions, and to the different behaviors of the electric machine. If the power supply of the MCU or inverter is by means of a battery as in electric vehicle (EV), the HiL scenario can emulate its operation in different states of charge (SOC), where the voltage is reduced as the battery energy is consumed. Otherwise, if the power is through the single-phase or three-phase electricity network grid, the voltage and frequency can be varied according to the voltages and frequencies and their corresponding limits set by each country. The most critical test cases can be evaluated, such as degradation and derating conditions. These are usually those where the power supply is low, and the demand for machine load is high, or when the temperature is above normal limits. In addition, if during the development phase changes the characteristics of the electric machine or even the type of machine, simply changing the model that controls the electronic load will suffice.

On the other hand, the different dynamic behaviors of the electric machine to be controlled, such as load variation, and inertia, can be tested by the electronic load. For example, in the case of the EV, one can simulate, among other things, the driving speed, the vehicle mass, the dynamics, the aerodynamics and the drag resistance, and the regenerative brake. In this case, during the iterative design phases, it is possible to find optimal parameters for the inverter + electrical machine set.

In this scenario, the MCU under test would be wholly tested and validated with the test cases created, providing the corresponding reports as output. If the results of the reports are satisfactory, it is possible to proceed to perform the test with the real machine on a test bench with a dynamometer. Dynamometers are widely used to test the torque and power of combustion engines and electric machines. It involves moving mechanical parts so that it can be dangerous if it is not carried out with all the necessary safety elements. The test bench consists of the real machine connected to a machine that acts as a brake (dynamometer) to simulate different pairs of dynamic loads. The torque and speed meter is monitored by means of instrumentation equipment connected to a PC.

It is essential to mention that the HiL scenario described above usually has a high cost, so it will not always be possible to have one during the development phase. That is why other more modest verification systems are often used as well as digitized plants. In this case, the hardware formed by the machine and the inverter + controller can be replaced by a precise model in real time for emulation, for faster development and if it is not expected to have a complete HiL scenario. It is true that control systems increasingly use their verification and development through a plant digitized on a field-programmable gate array (FPGA) platform so that control algorithms can be evaluated without the need for real hardware, in this case, an inverter and an electrical machine (Tavana and Dinavahi 2015). However, the real-time simulation of electric machine models and the inverter can be especially complicated due to the rapid nature of the dynamics, that is, reduced time constants especially in very low-power machines. The switching of PWM signals up to tens of kHz requires sampling rates of the order of several MHz to obtain a reasonable accuracy and model, for example, the ripple produced by the PWM in the inductance of the machine. That is why FPGAs are the ideal platform for complex simulations in real time due to their capacity to process data in parallel allowing sampling and execution rates up to the MHz range. The FPGA is a reconfigurable digital logic platform, which allows execution of millions of operations in parallel. As cited in (Le-Huy et al. 2006), research has advanced considerably in modeling and real-time simulation of different power systems that use the FPGA as computational devices. The control algorithm designed in this case for the control of an electric machine is loaded into a card where it will be tested with the FPGA modeling the inverter and the machine. Different test cases can be verified quickly by advancing many of the possible problems, for example, on the stability of the control.

Finally, there will be occasions when, due to cost, although not as high as that of the HiL scenario, these plant modeling systems will not be available. In this case, the tests are carried out directly on the actual plant or on a test bench, typically being the longest and most complicated development, since numerous software and hardware errors will be found in the real system or in the test bench according to the case.

1.2.3 MBD Process

The MBD begins with the MiL scenario (Lamberg et al. 2004), which consists of developing models submitted to simulated test environments at the beginning of the design. The models are then refined and transformed into software. This software can be tested in the SiL scenario, or in the microprocessor, PiL scenario. Finally, the HiL scenario (Hanselmann 1993) contains the real hardware and software (MCU), integrated into a simulated environment. The plant (electric machine) controlled by

MCU is developed during the functional specification phase, which can be used by the HiL scenario, defining the simulated environment.

The model-based design allows systems to be developed with a model-centered approach, since the basic idea is to develop the model without the need for a physical prototype, in a simulation-based verification environment. The model includes all the relevant components for the behavior of the system: algorithms, control logic, physical components, and the environment. Once the model has been developed and verified that it works according to the requirements, code of the control logic can be generated in the chosen programming language, for its later implementation in a microcontroller, FPGA, or DSP. The programming language will vary depending on the chosen hardware, being typically C/C++ language for the microcontroller and DSP, while VHSIC (very high speed integrated circuit) hardware description language (VHDL) is reserved for the FPGA.

The model development environment also allows generating reports and other types of documentation that are very beneficial for the developers and the customer. MBD has the ability to develop a functional version of the system from the beginning, with the plant, sensors, actuators, etc. For example, in electric machine control, it is possible to develop the speed control system in an environment where the plant will be an AC machine model, together with a power stage and the necessary sensing stages. That is to say, when the hardware is not available, some of the problems that could only be encountered with the physical prototype can be progressed in a very effective way. The results of the simulation can be shared with the customer instead of the results of the hardware tests and can be used to measure progress, verify that the system meets the requirements, and perform automatic test reports with coverage results of the control system. In addition, as the model is developed, it can be used to perform a MiL, generate code for SiL/PiL-type tests, and finally for HiL tests. Code generation is automatic in a way that reduces any manual implementation and therefore reduces development times. Figure 1.2 shows the structure of the MBD where two parts are differentiated. On the one hand, the design, simulation, and verification based on a personal computer (PC) (MiL and SiL), and on the other hand the verification on real hardware (PiL and HiL). The hardware can be modeled with a reasonable degree of precision in order to verify the design results of the models with an adequate degree. Different input test cases can be implemented in order to test accurately the models where the results should be compared with the expected outputs.

The change of requirements or conditions that affect one or several models can be modified and verified independently without waiting for the other models to be finalized. Through the simulations, it is verified that the changes do not cause an involuntary behavior of the system by providing regression tests.

In the previous example, a control engineering team is developing software for a control system of an AC electric machine without a speed sensor. The system bases the machine speed estimation by means of an adaptive model by measuring the current and voltage that circulates through the machine terminals. In the first phase, each sub-team (or group of engineers) models their respective sub-systems, using a model in simulation software shared at the system level to coordinate their work.

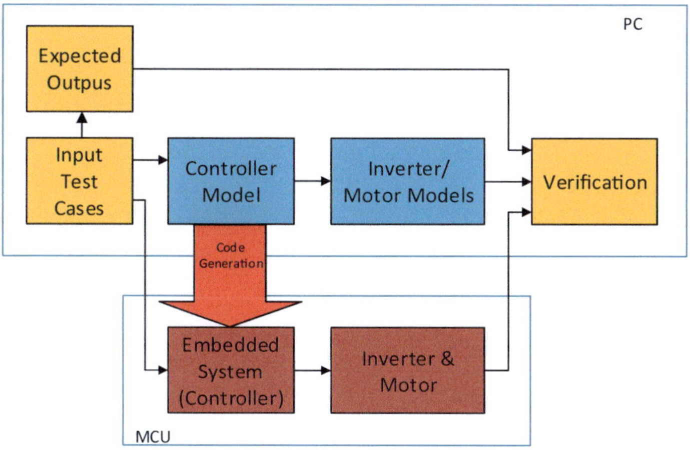

Fig. 1.2 MBD setup application in an MCU

Because the system will only control the machine, the application layer of the SW architecture will be composed of the control model. In this first phase, simulations can be run to see how the control behaves under various conditions. Each of the sub-systems can be separately debugged (MiL), where the parameters to be optimized are identified, and the performance metrics are displayed without generating a single line of code.

The first test version will be the one where all the MiL sub-systems are integrated to perform the main task of the system where it is possible to extract performance metrics. At this point, the teams can adjust or improve the models according to the needs of the customers as, for example, the need to extend the maximum speed of the machine at speed higher than the initial one of the project. The code generated by the models is a part of the system code so that there will be an integration process with the rest of the code. The rest of the code basically consists of sub-systems such as drivers for the ADC converter, driver for the I/O control, and Timers configuration. In the next phase, the models can be verified by SiL tests that are more rigorous than the MiL to have a more advanced level of verification. Here, the compiled model for its simulation is no longer verified, but it is verified as the generated code performs the same functions as the compiled model.

The last step consists of the most rigorous checks in real time such as the PiL and HiL test to ensure that the design meets the customer's requirements. At this point, there has been time for the hardware team to design and develop the real hardware. It is here, where the control of the machine can be evaluated on a real plant, that is, with an inverter and an electric machine, or in its absence, on an advanced simulation system of the plant. In addition, they also verify that the established standards and guidelines of the models and code comply, use static analysis and formal methods to prove the absence of critical errors at runtime, and produce reports and other artifacts in preparation for standards certification. As the project progresses, the needs of customers may change. For example, the customer can request a machine position sensing to make the system more robust in the face of any adversity in the control. Due to the fact that a system model is being used in this project, the team will only have to create a new model for the processing of the rotor position of the machine and compare it with the estimated position to activate the corresponding alarms in case of particular deviation. The remaining models remain unchanged so that the team must re-run simulations for MiL and SiL verification and share them with the client before executing the PiL and the HiL. If it works as expected, it is possible to proceed to perform HiL tests for the final verification of the new functionality.

1.3 Computer Simulations

In Engineering, there are three methods to resolve a problem: analytical method, numerical method, and experimental method. The analytical method is the classical approach, with 100% accurate results. However, it is applicable only for simple problems. The numerical method consists of a mathematical representation where some approximation and assumptions are made, and the results cannot be believed in 100%. The last method is the experimental method which consists of actual measurement and only applicable if the physical prototype is available.

Computer simulations, in general terms, use numerical methods to mimic the operations of real-world processes according to appropriately developed assumptions taking the form of logical, statistical or mathematical relationships that have been developed and formed in a model (McHaney 2009). These techniques for imitating real-world process operations are generally called systems, and assumptions about their functioning are generally made. These assumptions, which usually take the form of logical or mathematical relationships, constitute the model used to try to understand how the system acts (Law 2006). A system is defined as a collection of entities (such as people or machines), which act and interact together to achieve a logical end. In practice, the definition of "the system" depends on the objectives of a particular study. The collection of entities that comprise a system for a study may be only a subset of the totality of another system. The state of a system is defined as a collection of variables necessary to describe a system at a particular time, relative to the objectives of a study. The systems are categorized into two types, discrete and continuous. A discrete system is one for which state variables change instantaneously at separate points of time. A continuous system is one for which state variables change continuously over time. In practice, few systems are entirely discrete or entirely continuous. But since a type of change predominates in most systems, it is generally possible to classify a system as discrete or continuous. The models can be simulated in a computer either in continuous or discrete time, depending on the system, in increasingly sophisticated simulation environments that facilitate their development and understanding of their behavior.

In general, most simulation packages use numerical methods to perform the output results following a discrete approach. The discretization of a continuous system with infinite degrees of freedom (DOF) to a finite degree of freedom is known as meshing (nodes and elements). Figure 1.3 illustrates the meshing concept for an AC machine where phase winding is not meshing. As it can be observed, there are a numerous number of nodes and elements, in concrete, 39,669 and 88,582, respectively. The finite element (FE) is a numerical technique used to determine the approximated solution for a partial differential equation on a defined domain (Altair University 2019). FE is more focused to solve problems of engineering and mathematical physics as electromechanical phenomena, coupled effects, electromagnetic effects, mechanical structures, etc.

The FE has an excellent performance to solve partial differential equations over complex domains that can vary with time. It only makes calculations at a limited (finite) number of points and then interpolates the results for the entire domain (surface or volume). It is possible to define "Finite" as the reduction of the degrees of freedom from infinite to finite with the help of discretization or meshing (nodes and elements), and the "Element" as the entity joining nodes and forming a specific shape such as quadrilateral or triangular as shown in Fig. 1.3b. Hence, all of the calculations are made at a limited number of points known as nodes. To get the value of a variable anywhere in between the calculation points, an interpolation function is used as commented before.

On the other hand, simulation packages, which are not based in FE, use solvers. A solver applies a numerical method to solve the set of ordinary differential equations that represent the model [MathWorks]. The continuous solvers use the numerical integration to calculate continuous states of a model at the current time step based on the states at previous time steps and the state derivatives. Continuous solvers depend on individual blocks to calculate the values of the discrete states of the model at each time step. Discrete solvers are mainly for solving purely discrete models. Simulink provides two main types of solvers, fixed-step where it fixes step sizes from the beginning to the end of the simulation, and variable-step solvers, where solvers vary the step size during the simulation. In general, a smaller step size increases the accuracy of the results but increases the time required to simulate the system. The advantage of the variable-step is that it is possible to reduce the simulation time required since the size step is automatically chosen according to the actual results.

In power electronics, thanks to the simulation packages, the complicated algebra in power circuits is sometimes avoided, such as root mean square (RMS) of the currents, transients, and power losses for example. The precision of the power circuit can be fine-tuned with the knowledge of the layout where stray components such as inductance, resistance, and capacitance can be included in the simulation. It improves the results in such a way it is possible to perform accurate harmonic content simulations which can prevent future hardware modifications.

In the market, there are a large number of simulation packages for device-level power electronics which offers accurate and precision results of the behavior of the power stage. These packages can generate the dynamic equations of the system, which can be simplified for a computationally efficient. The device-level model such as power semiconductors device, resistors, capacitors, inductors, voltage, and current sources are used to perform the power circuit where often are solved by using

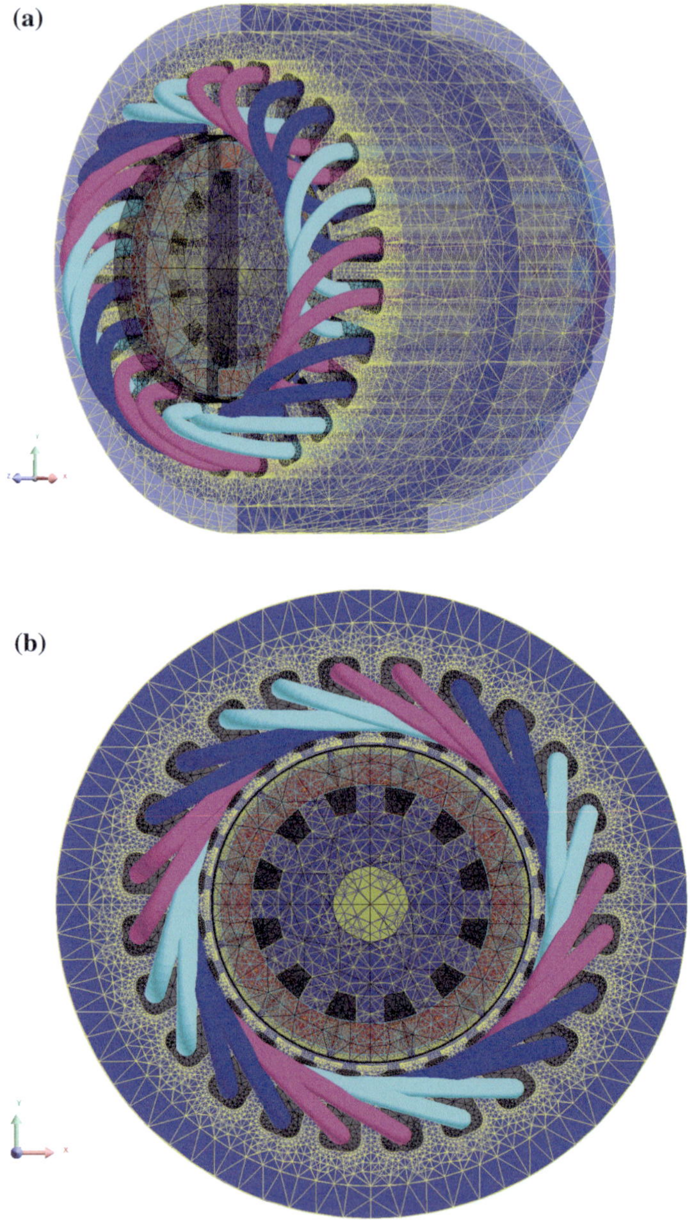

Fig. 1.3 Discretization of a continuous model by using nodes and elements. **a** Three dimensions, **b** two dimensions

non-linear differential-algebraic equations. The high accuracy is, in general, due to the high precision device model and the different numerical integration algorithms such as Newton-Raphson (solver) with a very low step size. The great detail of the waveforms results allows to analyze precisely, for example, the transients. However, the simulation execution speed is relatively slow, which can derive to bottleneck during the project development. For this reason, different simulations areas such as system level as MATLAB/Simulink® and device level such as PSIM® can be used during the development to exploit the modeling strengths for every area. Moreover, analog, digital, and mixed-signal simulations are often used in the same environment.

On the other hand, the co-simulation is becoming more prevalent in the power electronic circuits simulations. The co-simulation allows interconnection between several simulation tools such as system level and device level, where the advantages of both systems are emphasized. In this section, the simulation MATLAB/Simulink®, PSIM®, and Altair Flux™ finite element analysis (FEA) are introduced.

1.3.1 MATLAB/Simulink

Simulink® developed by MathWorks is a block diagram environment for multi-domain simulation and model-based design. It is compatible with system-level design, simulation, automatic code generation, and continuous testing and verification of integrated systems. Simulink® provides a graphics editor, customizable block libraries, and solvers for modeling and simulating dynamic systems. It is integrated with MATLAB®, allowing the incorporation of MATLAB® algorithms in models and the export of simulation results to MATLAB® for further analysis. These factors make Simulink® a powerful engineering tool. Simulink® control toolbox allows to design and simulate the behavior of the plant control in the same environment with algebraic differential equations. The control theory is shown graphically as plots such as Bode diagram and root locus, where stability analysis can be performed in a friendly way. Figure 1.4 shows the Simulink® environment.

On the other hand, Simscape Electrical Specialized Power Systems library of Simulink® contains blocks that use their own, *specialized* electrical domain. It provides a unique environment for modeling and simulating physical systems that span mechanical, electrical, and other physical domains. It provides fundamental building blocks of these domains that can be assembled into models of physical components, such as electrical machines, hydraulic valves, and other mechanisms. Simscape models can be used to develop system-level control and performance systems. It is possible to expand the libraries using the Simscape language based on MATLAB®, which enables the text-based creation of components, domains, and physical modeling libraries. Using the Simscape language, users can control accurately what effects are captured in their models. Figure 1.5 represents an example of a DC machine control with a power stage based in two isolated gate bipolar transistor (IGBT) power semiconductor devices made with Simscape Electrical Specialized Power Systems library.

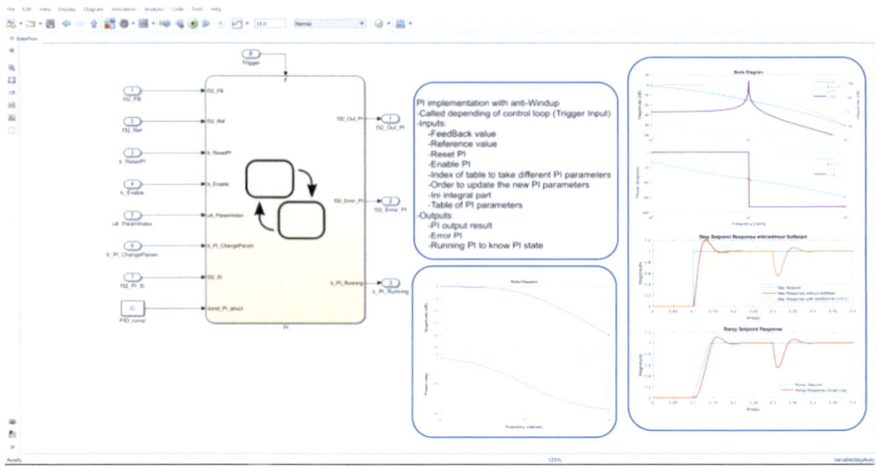

Fig. 1.4 Simulink environment software package

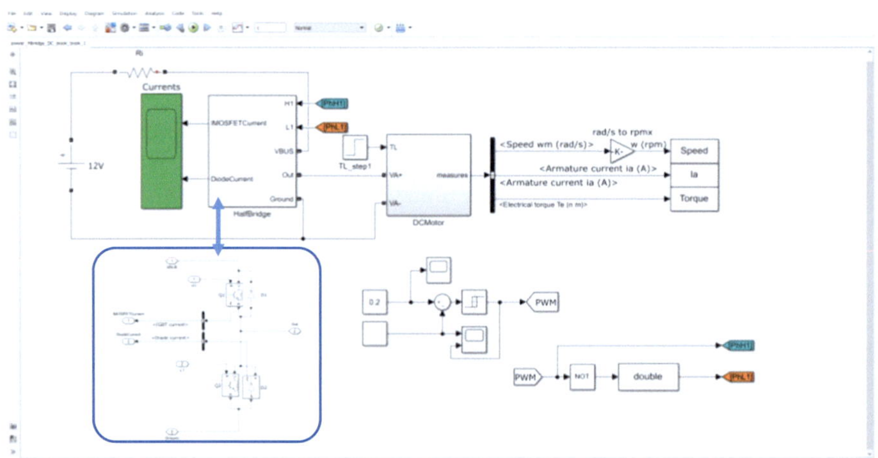

Fig. 1.5 DC machine control system made with Simscape electrical specialized power systems library

With Simscape, users build a system model in the same way that they would assemble a system. Simscape employs a physical network approach, also known as acausal modeling, building models: Components (blocks) corresponding to physical elements, such as pumps, machines, and operational amplifiers are joined by lines corresponding to physical connections that transmit power. This approach allows users to describe the physical structure of a system instead of the underlying mathematics. From the model, which looks a lot like a scheme, Simscape automatically

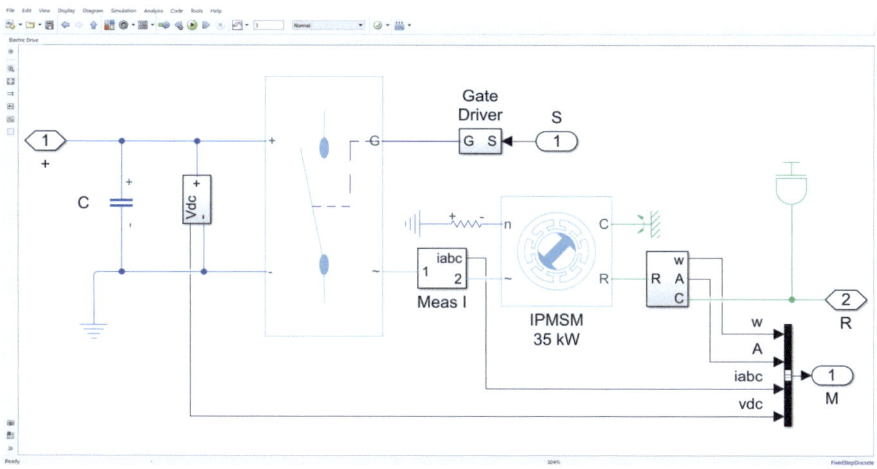

Fig. 1.6 AC machine control system made with Simscape electrical™ library

constructs the algebraic differential equations that characterize the system's behavior. These equations are integrated with the rest of the Simulink model, and the differential equations are solved directly. The variables for the components in the different physical domains are solved simultaneously, avoiding problems with algebraic loops. This is the main advantage of using the physical system of Simscape modeling. In addition, Simscape Electrical™ extends Simulink with other component libraries with analysis tools to model and simulate electronic, mechatronic, and electrical power systems. It includes semiconductor and electric machine models, as well as components for applications such as electromechanical drive, and renewable energy systems. Figure 1.6 represents as an example of a machine control system made with the Simscape Electrical™ library. In these systems, the harmonic analysis, calculation of the total harmonic distortion, and other critical analyses of the electric power system are automated.

1.3.2 PSIM®

PSIM® is a simulation software package (Fig. 1.7) for electronic power circuits and simulations of electrical machines. PSIM is used by the industry and educational institutions for research and product development. Developed by Powersim, PSIM uses nodal analysis and the integration of trapezoidal rules as the basis of its simulation algorithm. PSIM provides a schematic capture interface and a Simview waveform viewer. PSIM has several modules that extend their functionality to specific areas of simulation and circuit design, which include, among others: control theory, electrical machines, photovoltaics, and wind turbines.

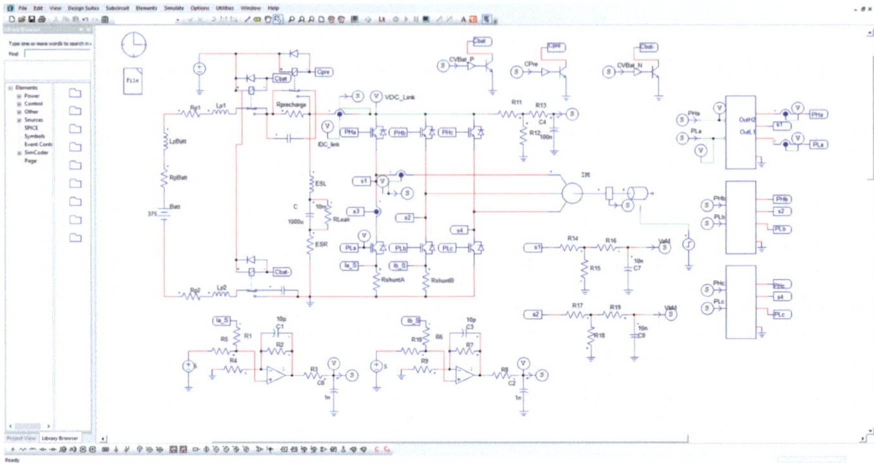

Fig. 1.7 PSIM environment software package

The PSIM modules are agile, easy to implement, and integrate with other popular platforms, such as Simulink® to perform a co-simulation. PSIM has a module called SimCoupler (SimCoupler Tutorial 2009) that allows an interface between PSIM and MATLAB/Simulink® environments. For example, the simulation of an adjustable speed drive based in an electric machine, the power stage, and the machine can be implemented in PSIM environment, whereas the control system is implemented in MATLAB/Simulink® environment. Figure 1.8 represents an example of the co-simulation between PSIM and MATLAB/Simulink.

PSIM package also offers a PiL module that allows to quickly verify the operation of the digital control algorithm in a DSP or microcontroller. In a typical PSIM simulation, the PSIM engine simulates the power stage and the control algorithm. In the PSIM PiL simulation, the power stage is simulated by the PSIM engine, while the control algorithm is executed by the DSP. The DSP communicates with PSIM through a USB/JTAG link. Changing the control to the DSP, while the power stage remains in the simulation gives the flexibility to iterate the design of the power stage without the need to build a prototype, test the effects of the changing operating conditions of the power stage, and easily configure the intermediate DSP variables. The code in the DSP is not executed in real time; this allows to have a very complex PSIM power stage.

The SPICE module of PSIM, powered by the CoolSPICE engine from CoolCAD Electronics, provides the ability to run the SPICE simulation in the PSIM environment. Version 11.1 of the PSIM includes the ability to use the LTspice engine (copyright by Linear Technology Co.) to run models that are not compatible with the SPICE engine native to PSIM. The subtle differences of syntax in the main variants of SPICE are covered mainly by the execution of PSIM-SPICE and the execution of LTspice from PSIM.

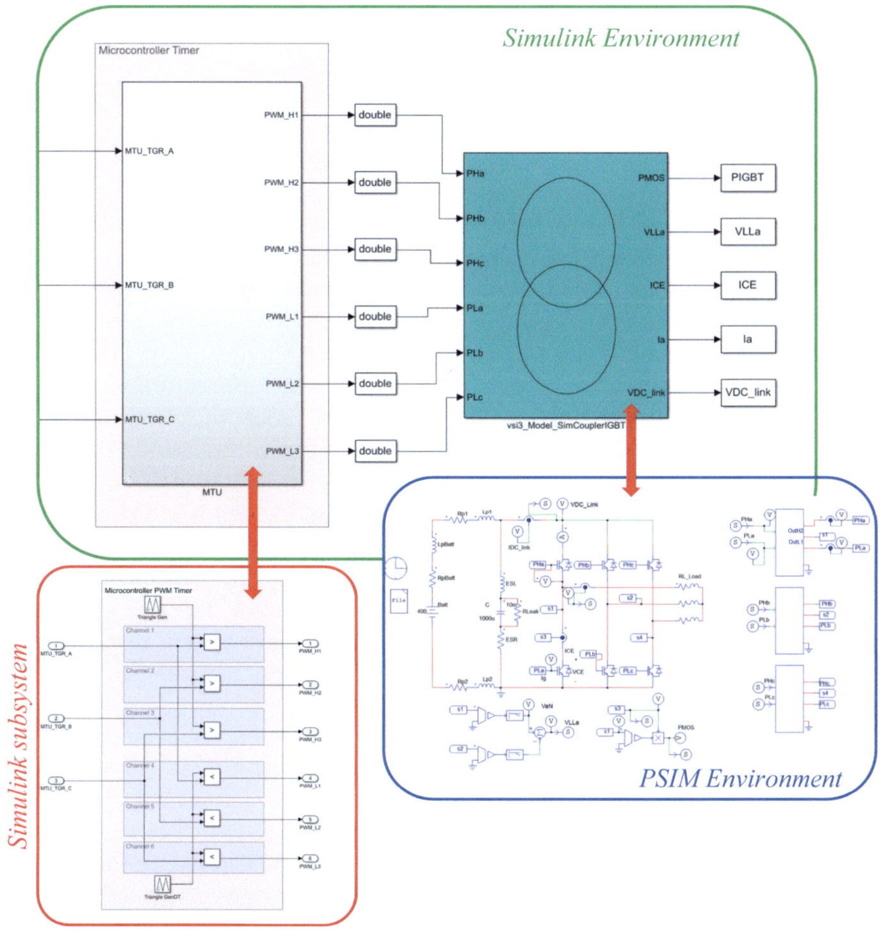

Fig. 1.8 PSIM environment software package

PSIM also offers code generation through SimCoder. It generates high-quality, consistent C code from a PSIM control schematic automatically. On the other hand, PSIM offers some design suite as Motor Control Design Suite which gives a toolbox to design to motor drive system with greater efficiency and less effort, and Hybrid Electric Vehicle (HEV) Design Suite, which offers a complete design solution for complex HEV powertrain systems.

1.3.3 Finite Element in Electric Machines

There is a different high-performance interactive software package that uses FEA to solve electrical or magnetic problems. As commented before, FE refers to the method from which the solution is numerically obtained from an arbitrary geometry by dividing it into simple pieces called finite elements. It is possible to affirm that FEA is the industry standard for simulating electromechanical phenomena, coupled effects, electromagnetic effects, mechanical structures, etc.

Rotating electrical machines are susceptible to phenomena that can only be estimated numerically and cannot be measured directly in the range of interest. Magnetic flux density, acoustic and magnetic noise, velocity, torque, power losses converted to heat, stator current, mechanical stresses, vibration spectra, and temperature influences can be analyzed by FEA. There are a large variety of software packages on the market specialized in the analysis of electrical machines using FEA. Altair Flux™ (Fig. 1.9) is an FEA software for electromagnetic and thermal simulations. It has embedded multi-parametric analysis capabilities; its open interface deals with different simulation domains and is well suited for multi-physics couplings.

On the other hand, Altair FluxMotor™ (Fig. 1.9a) is a flexible standalone software tool focusing on the pre-design of electric rotating machines. The user can design and create new machines from standard or customized parts, as well as to naturally add windings and materials to perform a selection of tests and compare machine behavior.

Nowadays, FEA software packages support to import the simulation data into other environments, as Simulink® for realistic closed-loop control system simulation. In this situation, the electrical machine model can contain, for example, non-linear characteristics due to saturation, enabling to build a realistic closed-loop simulation before it is manufactured. This is because the machine parameters as the inductances are no constant, but it is a non-linear mapping which relates the current, temperature, speed, and torque.

1.4 Software Architecture Patterns

1.4.1 Introduction

An architecture based on software layers in a microprocessed ECU/MCU has many benefits such as define reusable software components, improve the efficiency during its development, define test cases, apply safety standards, and generate documentation increasingly demanded by the domestic, industrial sectors, and especially in the automotive sector. The software inside of ECUs or MCUs is becoming more and more complex systems where minimum safety levels defined by the standard IEC 61508 regulation must be guaranteed, which describes the basic functional safety standard applicable to all types of industry. Different safety standards are derived

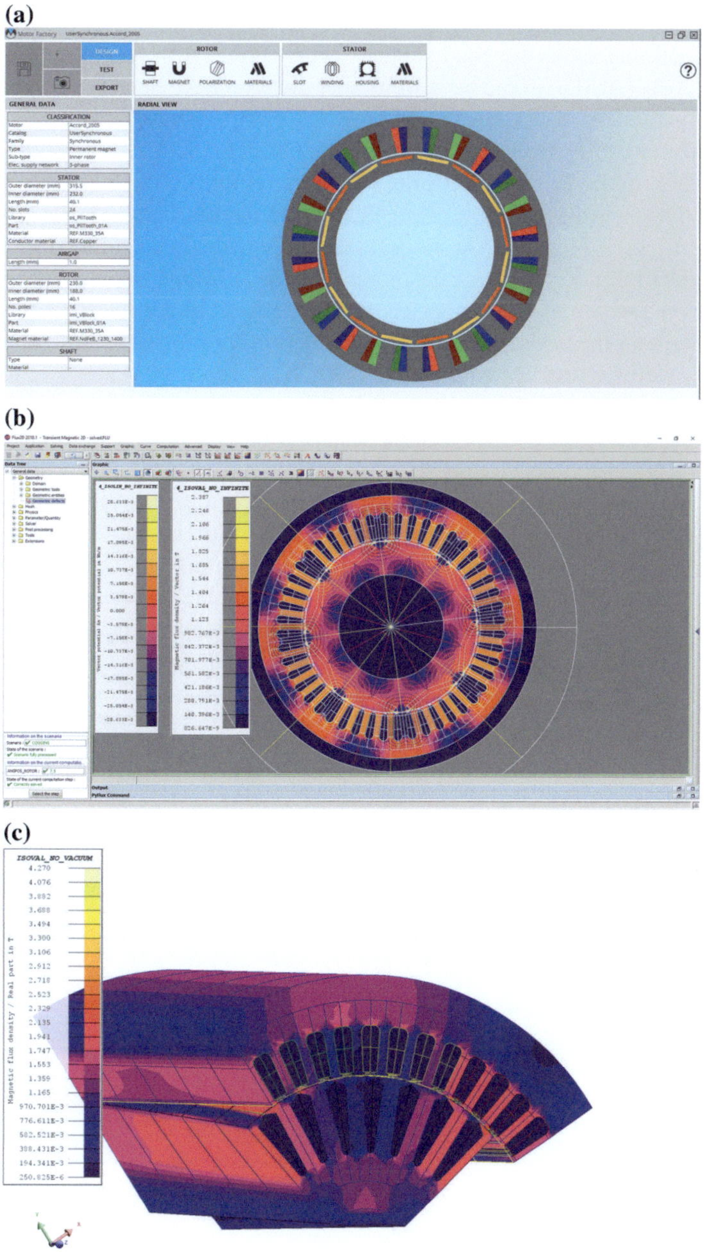

Fig. 1.9 **a** Altair FluxMotor^{TM} environment software package. Image developed using FluxMotor^{TM} provided courtesy of Altair Engineering, Inc. **b** Altair Flux2D^{TM} environment software package. Image developed using Flux2D^{TM} provided courtesy of Altair Engineering, Inc. **c** Induction machine FEA simulation with Altair Flux3D^{TM}. Image developed using Flux3D^{TM} provided courtesy of Altair Engineering, Inc.

from IEC 61508 for other more specific sectors such as ISO 26262, which specifies the functional safety for road vehicles. This defines different automotive safety integrity levels (ASIL). Some parts of its specification have been welcomed in other transport systems such as railway (IEC 62279), marine, or avionics.

The different software layers define the functions of each of the modules and to abstract from the DSP/microcontroller hardware. If during the development phase, the DSP/microcontroller must be changed, the modules that belong to the upper layers will not be affected. However, the modules that refer to the DSP/microcontroller hardware must be modified. These are the modules of the DSP/microcontroller abstraction layer (MCAL) which makes higher layers independent of the microcontroller. This layer is where all the peripherals and other functions of the microcontroller are configured according to the needs of the ECU/MCU. In addition, it is where the drivers of direct access to the DSP/microcontroller as memory driver, communication, and I/O drivers are located. It is essential to mention that each peripheral must have its associated driver and the missing hardware features compensated by suitable software modules at this point. Above this layer, there is a layer (middle layer) which makes higher levels independent of ECU/MCU hardware layout. It has the interface functionality for internal and external peripherals, such as I/O, memory, watchdog, and communications. It is possible to find an optional layer above the middle layer called service layer. It gives services to the application layer where one of the functionalities is to manage the tasks through a kernel or an operating system. The service layer is mostly independent of the hardware. Finally, the application layer will be as its name indicates the layer where the functionality of the ECU/MCU is performed. In this layer, there can be different software components (SWC) each one is its function, which communicates between them and toward the lower layer through a data interface called data flow. Figure 1.10 depicts the minimum layers that form the architecture pattern described above.

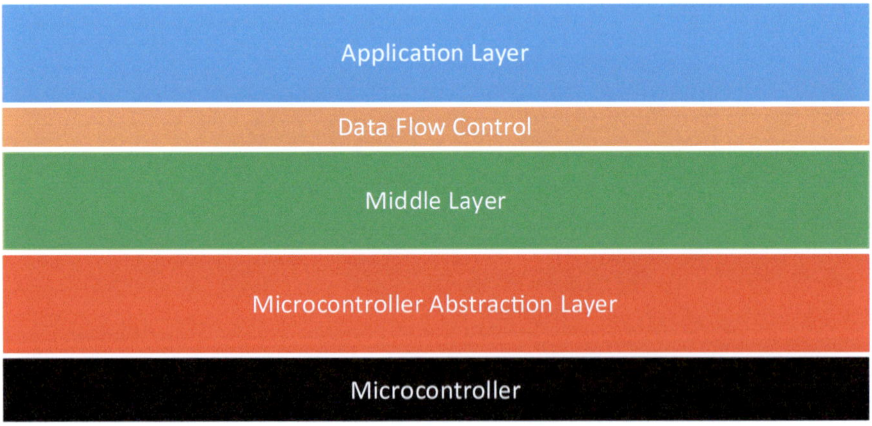

Fig. 1.10 Minimum layers for proper software architecture design

1.4.2 Automotive Open System Architecture (AUTOSAR)

The AUTOSAR development partnership was initiated in 2003 with the goal of establishing an open standard architecture for electrical/electronic in the automotive industry. It is a worldwide development partnership of automotive original equipment manufacturers (OEMs), suppliers and other companies in the software, semiconductor, and electronics industries. The cooperation of such companies created an open, standardized software architecture designed as AUTOSAR. AUTOSAR offers a framework guide for electronic automotive control systems which comes with layered software architecture.

In any vehicle, it is possible to find more and more multiple complex ECUs and MCUs where their methodology, application interfaces, and basic architecture must be well defined to guarantee compliance with ISO 26262, and a standard and methodical follow-up of work. The methodology refers to the exchange of formats or descriptive templates to allow a natural process of configuration of the stack of basic modules and smooth integration of the application software in the ECU/MCUs. AUTOSAR includes a methodology to use this framework. The application interface specifies the automotive application interfaces typical of all domains in terms of syntax and semantics that should act as standard for application software. For example, the ARXML files (AUTOSAR XML) provide the description of the SWC that contains specifications for the application, communication interfaces, access to the sensor signals available in the ECU/MCU, and many other things. All this will simplify the integration of the software of the functions since it is based on the compatibility provided by AUTOSAR. The definitions of these components, through ARXML files, are found in manufacturing, supply chains, and the management of the product's useful life in the automotive business. The basic architecture, for example, AUTOSAR architecture, has a layered structure similar to the one discussed in the previous section. The main difference is that it only applies to the electronic components sector of the automotive sector. AUTOSAR offers open and documented standards of how its architecture should be constructed in a detailed and more sophisticated way than previously seen. The objectives are to offer a serviceability over the entire product life cycle as software updates and upgrades, an abstraction of the hardware of the software, a standard configuration of code generation and model tools, to improve the quality of the standardizing software, a reusability of the modules and software components, a standardization of elementary functions of the system, integration of multiple suppliers, and portability between vehicles and platforms.

The AUTOSAR architecture can be divided into four layers, the application layer, the runtime environment (RTE) layer, the basic software layer (BSW), and the ECU-hardware layer of the microcontroller. Figure 1.11 represents the basic AUTOSAR architecture.

The application software layer is mostly hardware independent and consists of an interconnection of different SWCs. An AUTOSAR SWC is an atomic piece of software that has to be deployed on one ECU or MCU. The communication between the SWC and the access to the BSW is through RTE, this being the complete interface

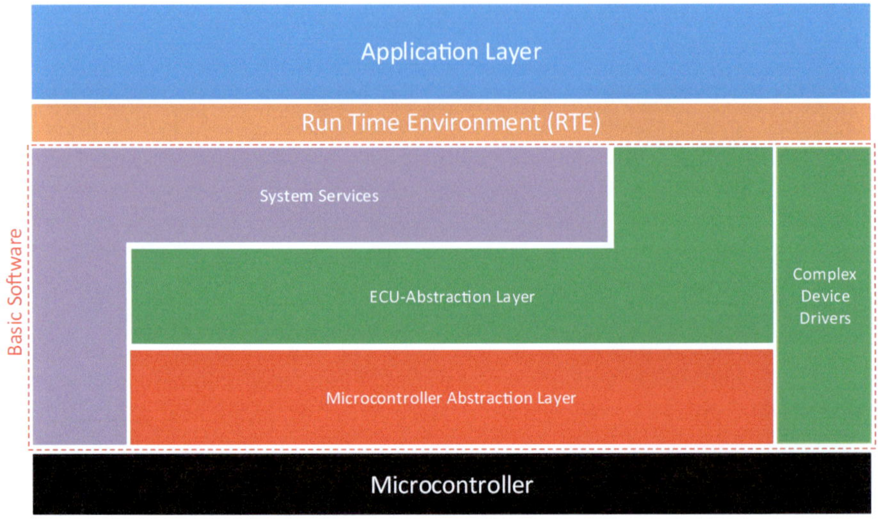

Fig. 1.11 Basic AUTOSAR software architecture

for applications. The RTE implements the virtual functional bus (VFB) being the sum of the communication mechanisms and the BSW interfaces (Naumann 2009). The BSW is divided into three main layers, the services layer, the ECU abstraction layer (ECUAL), and the microcontroller abstraction layer (MCAL). The services are also divided into functional groups that represent the infrastructure for system, memory, and communication services. More information about AUTOSAR is given in (www.autosar.org).

Figure 1.12 represents the basic software layer which is divided into different stacks corresponding to the general functionality.

In the design of MCUs, it is possible that the OEM/client requires that it be designed according to the architecture, methodology, and interfaces of the AUTOSAR application. In this case, all the layers described above are used in the design of the microcontroller software architecture. In particular, the minimum necessary is the I/O stack consisting of the I/O drivers, the hardware abstraction I/O, the RTE, and the application. The I/O stack provides standardized access to sensors, actuators, and the other ECU/MCUs on-board peripherals. The I/O Hardware Abstraction Layer communicates with the application layer by means of the RTE. However, in some MCUs is needed the direct microcontroller access for a particular control for sensors and actuators. It usually needs special timing conditions or the integration of non-AUTOSAR technologies. To support these particular conditions, the complex device drivers (CDD) can be used. The CDD has direct access to resources for critical applications such as injection control, battery management system (BMS), and electric machine control. The CDD is not provided by AUTOSAR, but each manufacturer who needs those drivers has to implement them themselves.

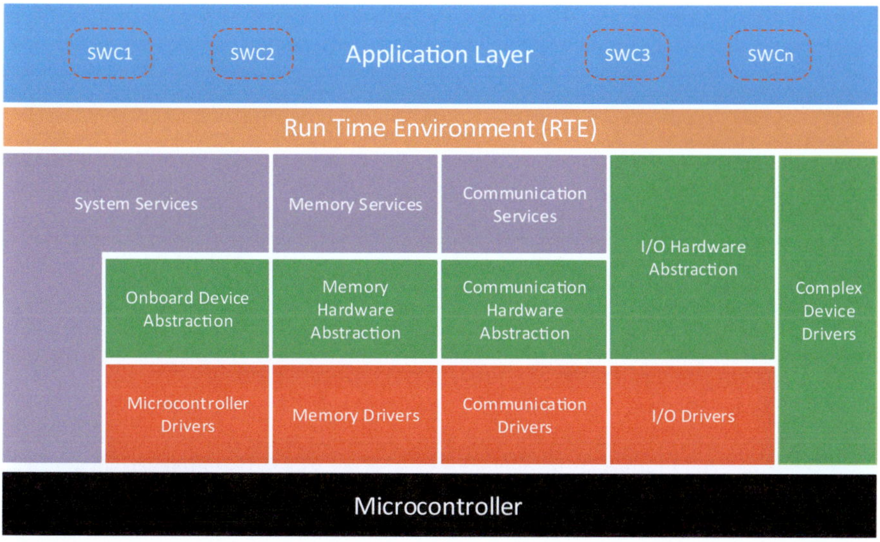

Fig. 1.12 AUTOSAR software architecture with the stacks of functionality in the basic software layer

The time-critical current loops of the electric machine control can be handled in an interrupt context in the application layer or in a CDD as it will be shown in Sect. 1.5.

However, if an interface for the management of the MCU is needed, for example, for its control and monitoring by a personal computer (PC), or for its possible control by another ECU, it will be necessary to add a communication interface based, for example, on the standard CANbus interface communication for automotive. The necessary modules are the communication service, the communication hardware abstraction, and the communication driver. Taking into account what has been described above, it is possible to rewrite the architecture part of the interest, as shown in Fig. 1.13.

The necessary hardware or peripherals of the microcontroller for the implementation of the control of the machine are usually timers for the management of the PWM outputs and digital inputs for the capture of sensing signals, an A/D converter, and an input/output interface of the ports for the activation/deactivation of auxiliary systems. The modules needed for the CAN interface are also shown in Fig. 1.13 in a very generic way.

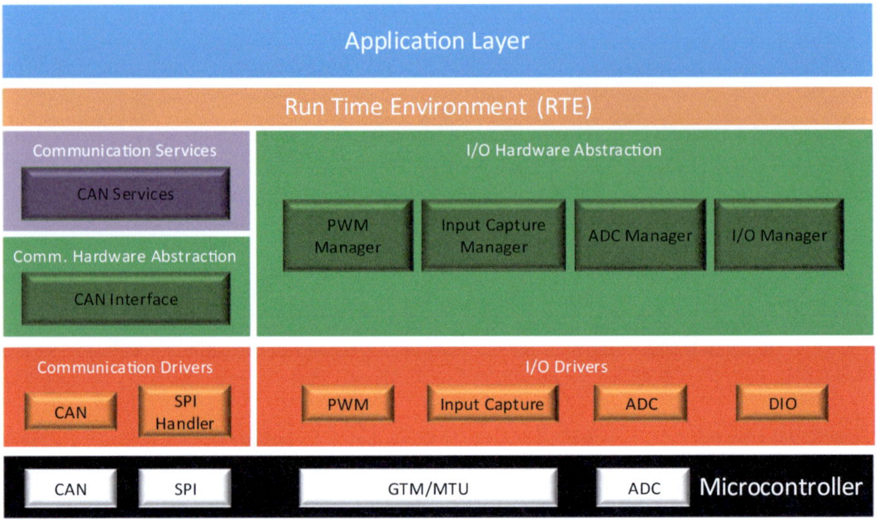

Fig. 1.13 AUTOSAR classic platform architecture applied to machine control with a power stage based on an inverter

1.5 Discrete-Time Electric Machine Control System Overview

Like most control systems, the electric machine control consists of sensing, control, and actuation. The electric machine is the controlled system which is a continuous-time physical process. As expected, the inputs and outputs of the machine are continuous-time signals. For instance, the voltage and current inputs are continuous-time signals. The analog controllers are avoided in most machine control systems because of its complexity and its functionality. The digital implementation of the controllers makes it possible to perform advanced algorithms controls as field-oriented control (FOC). The analog signals from the physical process (electric machine) should be converted to digital employing an analog-to-digital converter (ADC). The specific control algorithm produces the control commands according to the input signals. Hence, by using a digital-to-analog converter (DAC), the commands can be applied to the electric machine.

In digital machine control systems is assumed asynchronous process with a constant rate of sampling time. The ADC samples the analog signal periodically with a sampling time T. Due to the conversion time of ADC, the analog signal should be held during a certain time. This is done by a circuit known as sample-and-hold circuitry. The sampling time should be shorter compared with the time constant of the system for effective digital control and to treat the system as a continuous system. The sampling time and the processing delay (e.g., ADC and signal conditioning) introduce phase shift terms in the discrete-time systems which affect its behavior. The

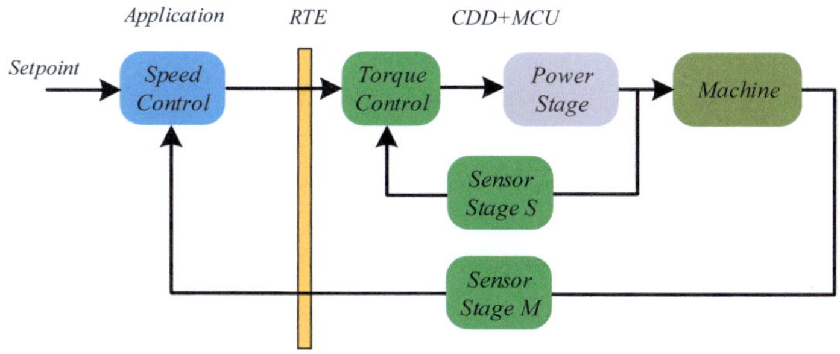

Fig. 1.14 Typical control system for electric machine based in current and speed control loops. The fast loop corresponds to the torque control, which is implemented in the CDD, while the slow loop is the speed control implemented in the application layer

processing delay should be minimized as much as possible, while the sampling time should be selected in agreement with the time constant of the system as commented before.

Figure 1.14 shows a simplified cascade control structure for a generic electric machine. As can be observed, there is a power stage, sometimes called electric drive in the literature, which sets the necessary voltage and current to the machine for its proper control according to the control stage. The control stage consists of two closed-loop digital controllers connected in cascade as mentioned. Due to this structure, it is possible to identify two algorithm components, the speed and torque algorithm with its corresponding sensor feedback signal. The speed feedback comes from a sensor mounted in the rotor (axe) of the machine, while the feedback of the torque should be the current throughout the machine because, as it will see in Chap. 5, controlling the current the torque can be controlled. The sensor stages S and M are composed basically by a sample-and-hold and ADC, one per each sensor. The sample-and-hold and ADC will be seen in Chaps. 2 and 7, while the sensors and its signal conditioning will be seen in Chap. 6. The command output of the torque controller is a digital voltage signal which is converted to analog signal inside the power stage.

As it will see in Chap. 2, the bandwidth of the inner torque control loop should be wider than the outer speed control loop when a cascade control structure is used. In other words, the inner loop control should be faster (rates of a few kHz) than the outer loop control (rates of a few hundreds of Hz) in order to respond rapidly to a possible disturbance in the machine without affecting the outer loop. For a digital implementation, it is possible to differentiate two critical tasks in this control structure: the torque control with its sensor processing S, and the speed control also with its sensor processing M. The mentioned tasks should be executed periodically, for example, every sampling time T of the sensor signal processing. The algorithms components requiring faster rates above 20 kHz should be implemented on the DSP, FPGA, or programmable logic of an SoC.

In a microcontroller implementation (below 20 kHz), the periodic tasks can be called by using an internal timer which can be programmed to launch an interrupt service routine (ISR) every fixed periodic time (e.g., $T = 0.1$ ms or 10 kHz). The ISR can execute the torque control task with the actual discrete acquisition sensor signal S, and update its output. The speed control task can be executed for example every 10 times slower, i.e., 1 ms, either using another timer with another ISR or using the same ISR but with a slow loop inside. In both cases, the output of the speed control task will be updated every 1 ms with the actual discrete acquisition sensor signal M. In Fig. 1.14, below the RTE the faster periodic task (torque control loop) is found in the CDD, while in the application layer the speed control loop task is found.

On the other hand, it is possible to introduce the execution time of a task, often called a job. It is the amount of time required to fully complete its execution without interruptions (Fan 2015). That is, there is no other task competing for resources. In the present example, the torque loop task is called every 0.1 ms. It means that the computational time of this task in all cases should be lower than 0.1 ms to avoid losing the synchronization with the actual discrete sample. It is important to comment that the computational time can vary from one sampling period to the next. For example, inside of the task, it is possible to find different paths with different lengths due to if-then-else statements, which can produce different execution times.

Moreover, if during the computation of the task is interrupted by software or hardware interruption, the execution time will also vary. As it can be observed, the torque loop task is a critical task which should be designed with the best possible code optimization. For this reason, the implementation in CDD will be the right choice if AUTOSAR architecture is used.

References

Fan X (2015) Real-time embedded systems. Design principles and engineering practices. 2015 Elsevier Inc

Forsberg K, Mooz H (1991) The relationship of system engineering to the project cycle. In: Proceedings of the National Council for Systems Engineering (NCOSE) Conference. Chattanooga, Tennessee, pp 57–65

Hanselmann H (1993) Hardware-in-the-loop simulation as a standard approach for development, customization, and production test. In: International congress and exposition, Mar 1993

Lamberg K, Beine M, Eschmann M, Otterbach R (2004) Model based testing of embedded automotive software using Mtest. In: SAE 2004 world congress and exhibition, Mar 2004

Law AM (2006) Simulation modeling and analysis. 4th edn. McGraw-Hill

Le-Huy P, Guerette S, Dessaint LA, Le-Huy H (2006) Real-time simulation of power electronics in power systems using an FPGA. In: Canadian conference on electrical and computer engineering, pp 873–877. May (2006)

McHaney R (2009) Understanding computer simulation

Naumann N (2009) AUTOSAR Runtime Environment and Virtual Function Bus. Technical report, Hasso-Plattner-Institut f¨ur Softwaresystemtechnik

Practical aspects of finite element simulation. A study guide. 5th edition released. Altair University (2019)

Tavana NR, Dinavahi V (2015) A general framework for FPGA-based real-time emulation of electric machines for HIL applications. IEEE Trans Ind Electron 62(4):2041–2053

Tutorial on how to use the SimCoupler Module (2009). Powersim Inc.

Chapter 2
Electric Machine Control Technics

2.1 Control Theory Overview

In electrical machine control systems, the control system structure is based on a compensator, a power converter, the electrical machine (plant), and the feedback signals. Moreover, disturbances of the system, such as disturbances in the load torque or in the supply voltage of the power converter stage, are usually included. The control system must also be designed to be robust in terms of the variations between different electrical machines, power converter, and its measurements, as well as the variations due to their aging. The disturbance is usually between the power stage and the electrical machine so that the system will not react to the disturbance until it affects the output. As will be seen later, the reaction to the disturbance can be compensated with feedforward and cascade-type control structures.

In Fig. 2.1 is shown the basic closed-loop speed control system for an electric machine according to the blocks discussed above. The controller, such as proportional–integral–differential (PID), is responsible for minimizing the error $E(s)$ between the reference/command and the feedback signal to zero. The machine is the plant of the system which produces the system response. It also included the coupling elements to its load. As the electrical machine, in general, is a passive system, the power stage is needed to supply energy (voltage and current) to the machine. In electronics systems, the power is often delivered through pulse width modulation (PWM) which switches transistors at high frequency. The sensor stage produces the feedback signal to the controller if the closed-loop control strategy is used. Some electrical machines can be controlled in open loop, but its responses to load changes will be difficult to predict.

For a better understanding of the operation of the closed-loop control of an electric machine, it is possible to mention, as an easy example, the speed control of a wiper washer of a vehicle. The electric machine of the wiper consists of a DC machine designed to operate with the battery of the vehicle that is 12 V. The voltage variation of the electric machine causes the torque to vary and therefore the speed as it will be shown in Chap. 4. The voltage variation is made by the power converter, which

© Springer Nature Switzerland AG 2020

R. Molina Llorente, *Practical Control of Electric Machines*,
Advances in Industrial Control, https://doi.org/10.1007/978-3-030-34758-1_2

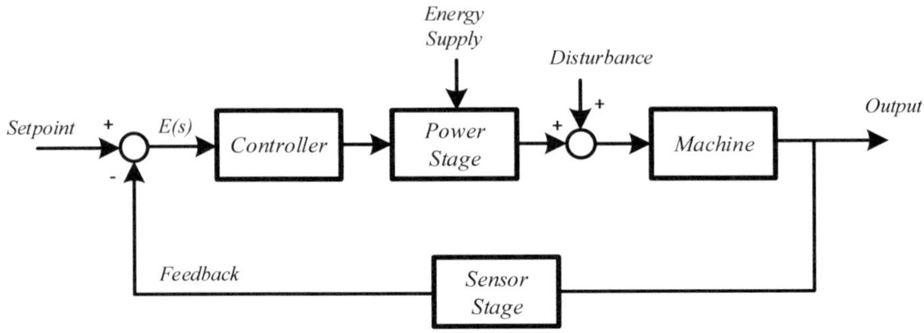

Fig. 2.1 Block diagram of closed-loop control for an electrical machine with disturbance

modifies the supply voltage of the electrical machine to adjust, for example, the speed of the rotor for a given load torque. To vary the voltage from 0 to 12 V, it is possible to use a power stage which consists of a metal–oxide–semiconductor field-effect transistor (MOSFET) device controlled by a PWM. The PWM is a digital signal which controls, using its duty cycle, the time that the MOSFET remains active, that is, the time it allows the 12 V of the battery to pass, in order to vary its average value. For instance, with a 50% of duty cycle, the average voltage value obtained is 6 V. The duty cycle is controlled, for example, by a PID compensator where its input is the error between the speed reference and speed measured in the axe of the machine. The speed measured is the feedback signal of the closed-loop control structure. The ideal speed sensor measures the speed with perfect accuracy and without delay, but it is not the real situation. Feedback stage and its sensors inject some delay between the actual signal and the measurement as it will be discussed in Chap. 6.

On the other hand, the disturbances that affect the correct operation of the control can be, for example, the battery voltage variation and the load torque variation. For the first case, the voltage of the battery will not always be 12 V but may vary depending on the state of charge (SOC) and the power that is being demanded of it. The voltage fluctuation of the battery voltage affects the speed of the DC machine directly. Hence, the speed controller should compensate with the variation of the duty cycle of the PWM. For the second case, the variation of the load torque could produce such as activation in dry, or wet, and/or with a high speed of the vehicle. The disturbance in the load torque will affect the speed variation so that the speed controller should also compensate with the duty cycle variation. If the PID controller is well tuned within a reasonable change (minimum and maximum operating definition for the battery voltage and the load torque), the control system should operate correctly for most cases.

On the contrary, in an open-loop control where the speed feedback is removed, it will be complicated to maintain the command speed according to the voltage and torque variations. The open-loop control is usually used in low-cost applications that do not require excellent control of speed and/or torque, with a lower control dynamic.

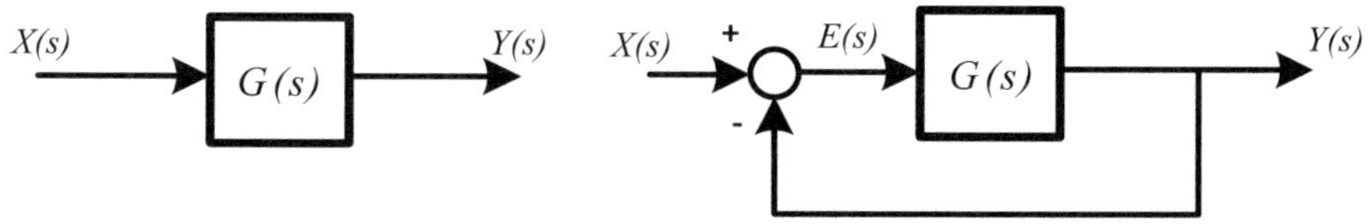

Fig. 2.2 Open- and closed-loop systems

For these cases, it will be necessary to protect the electric machine and the power converter with other mechanisms such as overcurrent protection, especially in the case of rotor locked.

The simplified block diagram of an open- and closed-loop control where the transfer functions are represented in (2.1) and (2.2), respectively, is depicted in Fig. 2.2. The transfer function of the negative feedback path of the closed-loop control is unity.

As can be observed in the transfer functions, the gain between the output $Y(s)$ and the input $X(s)$ in case of open loop (2.1) is much higher than for closed-loop control (2.2) which can generate instability. For example, in (2.1) any variation in the input $X(s)$ is amplified by the gain $G(s)$, what does not happen with (2.2) where the gain is approximate to the unity if $G(s) \gg 1$.

$$\frac{Y(s)}{X(s)} = G(s) \tag{2.1}$$

$$\frac{Y(s)}{X(s)} = \frac{G(s)}{1 + G(s)} \tag{2.2}$$

2.1.1 Stability Analysis of Second-Order Systems

Stability describes how predictably a system follows the command or reference. Control systems with negative feedback tend to stabilize the system in most of its frequencies and increase its bandwidth, and it is less sensitive to the variation of its internal parameters. The instability of these systems for a unitary loop gain can be due to the existence of some frequencies where the phase lag accumulates at 180°, making the feedback then positive instead of negative. Each stage of the control system contributes to varying the gain and delay, and even if a delay of 180° is accumulated if the gain is less than unity the system will produce sustained oscillations (marginally stable). Otherwise, the change of sign produced by the accumulation of delays and with a gain equal to zero, the system will become unstable.

2.1.1.1 Time Domain

In the time domain, stability is most commonly measured from the step response in the reference. The system output will react to the step with a typical feature called overshoot. It is possible to add that the step is easily applicable in any control system where the response to it is measurable and provides a complete view of the stability of the system, although in some real cases it is not convenient to apply as described below.

The unit amplitude step $u(t)$, also called the Heaviside step, takes the value of the unit for time t greater than a, as can be seen in Fig. 2.3. This is the generic way of representing the step function where a indicates the delay with respect to $t = 0$. As it is the system input, it is defined that $x(t - a) = u(t - a)$.

The second-order transfer function system can be represented according to the natural pulsation of the system ω_n with (2.3) or by the time constant τ according to (2.4), where the time constant is $\tau = 1/\omega_n$. The gain of both transfer functions is K.

$$G(s) = \frac{K\omega_n^2}{s^2 + 2\xi\omega_n s + \omega_n^2} \tag{2.3}$$

$$G(s) = \frac{K}{\tau^2 s^2 + 2\xi\tau \cdot s + 1} \tag{2.4}$$

As previously mentioned, the second-order system can be analyzed with its open-loop response to a step of amplitude A where the transfer function, according to the Laplace transform, is $X(s) = A/s$. So,

$$Y(s) = X(s) \cdot G(s) = \frac{A}{s} \cdot \frac{K\omega_n^2}{s^2 + 2\xi\omega_n s + \omega_n^2} \tag{2.5}$$

The denominator of (2.5) corresponds to a second-order equation. Depending on the value taken by the damping coefficient ξ, it will have real roots and/or imaginary, in the left or right half plane. In an easy way, it is possible to get the roots of the second order (2.6) and perform Table 2.1.

$$s = \omega_n\left(-\xi \pm \sqrt{\xi^2 - 1}\right) \tag{2.6}$$

The equations of the time-domain response are also represented in Table 2.1 except for an unstable system case.

$$x(t-a) = u(t-a)$$

$$u(t-a) = \begin{cases} 1 & t > a \\ 0 & t \le a \end{cases}$$

Fig. 2.3 Unitary step in the time domain

Table 2.1 Pole situation and its effects according to damping ratio

Damping ratio	Effect and description	s-plane (pole situation)
$\xi > 1$	– Overdamped – Stable system – Two real roots ($s_1 = -\mathrm{Re}1$ and $s_2 = -\mathrm{Re}2$) $y(t) = \left(1 + \frac{\omega_n^2}{s_1(s_1-s_2)}e^{s_1 t} + \frac{\omega_n^2}{s_2(s_2-s_1)}e^{s_2 t}\right) \cdot u(t)$	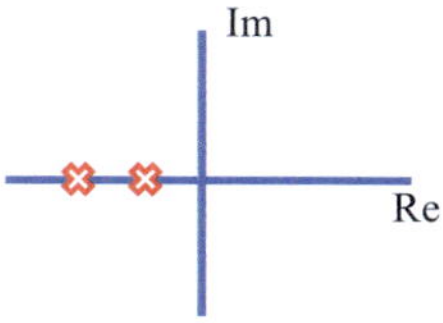
$0 < \xi < 1$	– Underdamped – Stable system – Two conjugated with imaginary part and negative real part roots (−Re±Im) $y(t) =$ $A \cdot K\left[1 + \frac{1}{\sqrt{1-\xi^2}}e^{-\xi\omega_n t}\sin\left(\omega_n\sqrt{1-\xi^2}t - \alpha\right)\right] \cdot u(t)$	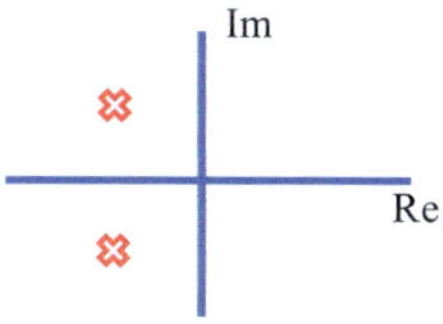
$\xi = 0$	– Underdamped – Marginally stable system – Two imaginary conjugated roots (±Im) $y(t) = A \cdot K[1 - \cos(\omega_n t)] \cdot u(t)$	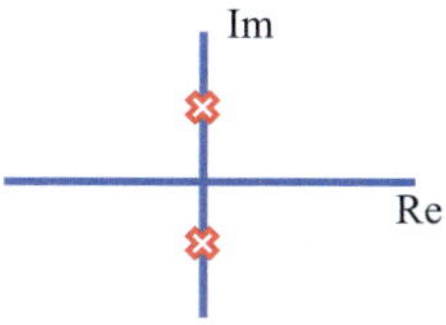
$\xi = 1$	– Critically damped – Stable system – Two double real roots (2Re) $y(t) = A \cdot K\left[1 - (\omega_n t + 1)e^{-\omega_n t}\right] \cdot u(t)$	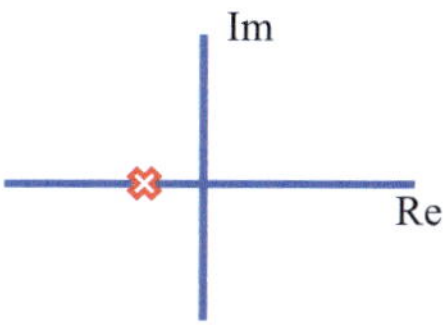
$\xi < 0$	– Unstable system – Two conjugated with imaginary part and positive real part roots (+Re±Im)	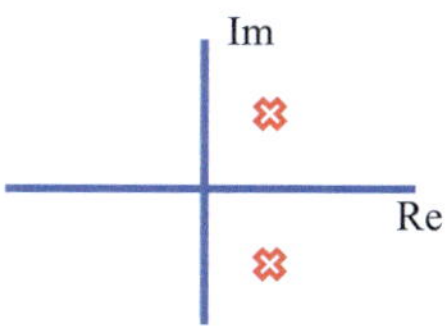

The second-order system will react differently to the unit step input according to the value taken by the damping coefficient ξ of the $G(s)$ system. The effect that it produces can be observed using the time-domain response, according to Fig. 2.4, where a unit step has been applied. When ξ takes the value zero, the system is marginally stable: The step produces in the output a constant oscillation in amplitude and frequency ω_n. A small variation in the system can make it unstable since its roots are on the imaginary axis and can easily pass to the right side of the s plane. The other output responses stabilize with more or less time to the value of the unit step. From

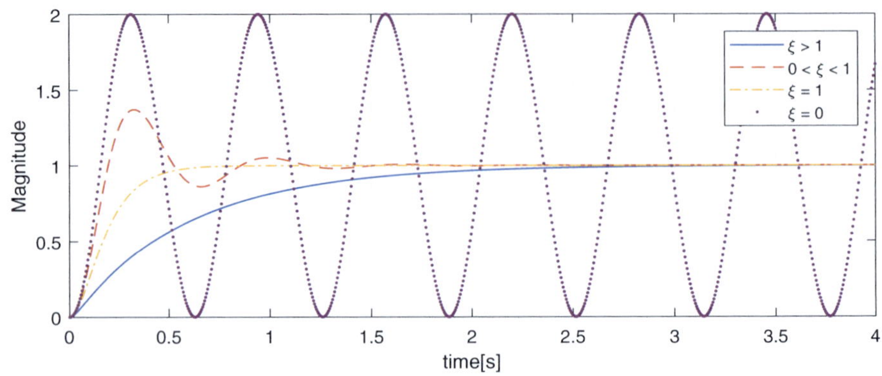

Fig. 2.4 Unitary step response for different damping ratios

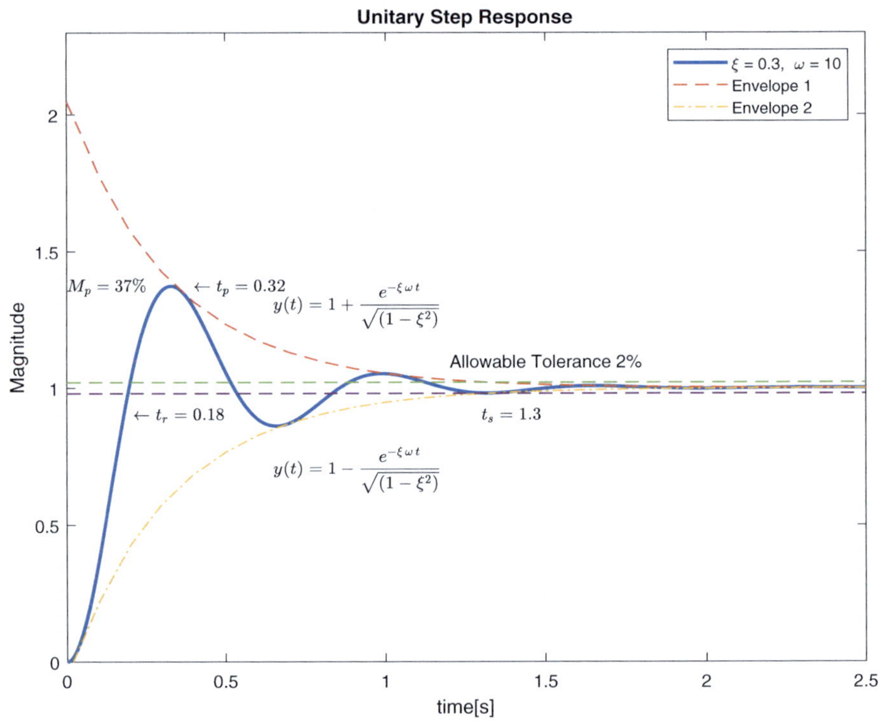

Fig. 2.5 Underdamped response system when a unitary step is set

the response where damping coefficient ξ is between zero and one, it is possible to extract a series of parameters that characterize the response as illustrated in Fig. 2.5.

In Fig. 2.5 is shown the damped control response to a unit step of a second-order system with roots conjugated with imaginary part and negative real part ($-\text{Re}\pm\text{Im}$).

As commented before, the damping coefficient term ξ takes values between zero and one. It is possible to observe different parameters that characterize the response, which provides important information about the behavior. In a closed-loop system, the response to a step may have similar behavior as represented in the figure. Hence, the following parameters are valid for analyzing the responses of open-loop and closed-loop systems in the time domain. It is possible that some of these parameters are specified as control requirements so that the controller, in the case of a closed loop, must be designed to meet these requirements.

The maximum output value exceeded above the reference is the overshoot M_p, which is expressed as a percentage. The response time t_r corresponds to the time required when the output passes from 10 to 90% of the reference. The time required from the application of the step until the first impulse is reached at the output is called peak time t_p. The settling time t_s is the time required for the system output to fall within 5 or 2%, depending on the case, of the steady-state value for a step input. The allowed tolerance is the error between the reference and the output that is allowed in steady-state, usually less than 1%. Lastly, the natural frequency of the system is ω_n, as can be observed.

In Fig. 2.5, the overshoot M_p corresponds to 37%, which is very excessive value. Ideally, it should be less than 10%, and sometimes it is complicated to reduce it to zero. In an electric machine practical example, if the signal in Fig. 2.5 is the speed of the electric machine, the speed exceeds the reference with a high percentage during a certain time. This could not be desirable because, in closed-loop control, the controller will attempt to correct the exceeded speed applying negative torque. The negative torque means braking the machine, and the power flow is back into the power stage. The power stage maybe is not able to return this energy to the AC power line or battery. In this case, for low energy regeneration, the energy can be absorbed by the DC-link capacitors and returned to the machine during the next motoring period. On the other hand, if the energy regeneration is large enough, then some protection elements as a resistor are typically used to dissipate the energy in a controlled manner.

The ideal response is to achieve an almost instantaneous response time without overshoot and minimal settling time. In electrical machine control, this is very difficult to achieve, especially in machines with large inertias and high time constant so that the requirements of the specification must be appropriate for each case. It is important to mention that speed steps in real applications will not always be welcome as an instantaneous change of reference since it can generate a pronounced speed overshoot, as well as high current transient peaks. The soft-start process should avoid overshoots and high current transients. The soft start can be implemented either using a command ramp or command step with lower control dynamic only during the start-up of the machine. In Fig. 2.6 is shown the soft-start process for both cases commented. In Fig. 2.6a, it used lower dynamic PI gain parameters during the start-up, and then it is changed to the normal steady-state values. During the start-up, slower response to the command step is expected. For example, at time instant $t = 0.15$, the parameters are changed to the same ones used in the PI controller without soft start. The reduction of the overshoot from 17 to 7.5% can be clearly observed. However,

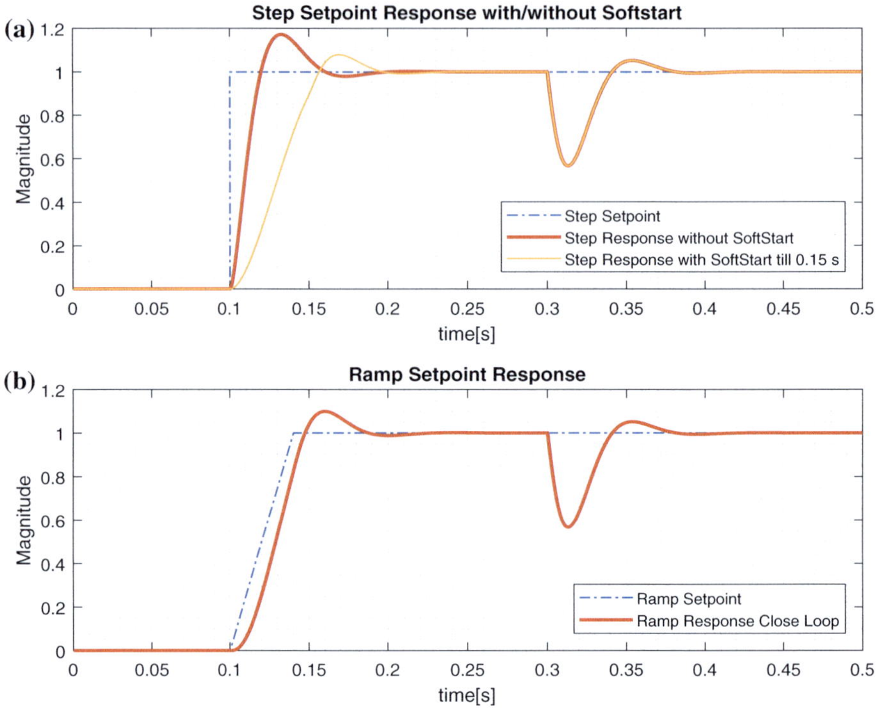

Fig. 2.6 Closed-loop response for step and ramp commands. Step command is with and without soft start

the response time t_r is higher, but the reduction of the overshoot may become more relevant. It is also possible to observe that the settling time is almost unaffected so that it can be a solution for most electric machine control systems. In order to corroborate that the PI gain comes to the normal steady-state values, at time instant $t = 0.3$ a disturbance is applied. As expected, the two responses with and without soft start act precisely the same. It is essential to mention that the change in PI gain parameters should be carried out smoothly. For example, if the proportionality constant K_P in soft start takes the value of 5 and in the steady-state, it takes the value of 15, the change from 5 to 15 is better to do it smoothly in small increments in a given time. As a reference, the time it takes to change from 5 to 15 should be 10 times slower than the periodicity of execution of the digital controller. In this way, the digital controller is executed 10 times with parameters that are changing progressively, avoiding small instabilities and unexpected spikes in the current of the electric machine.

On the other hand, if a command ramp is used, an output response similar to the soft start can be achieved without changing the PI gain parameters of the controller, as can be seen in Fig. 2.6b.

In the cases in which the machine remains in steady-state, and the speed should increase, the concept of soft start could also be applied, although it must pay attention to not generate instabilities that at higher speeds may be more challenging to control.

2.1.1.2 Frequency Domain

The response of a system can also be measured in the frequency domain by inspecting gain and phase in the Bode diagram. The transfer function of the second-order system has a behavior similar to low-pass filter where for low frequency it has a constant gain and a much lower gain at high frequencies. The bandwidth of the system defines the response of the closed-loop control. It is defined as the frequency in which gain drops -3 dB ($20\log_{10}(1/\sqrt{2})$). In Fig. 2.7, the Bode diagram of the previous second-order system can be observed for the different damping coefficients. In case of damping coefficient $\xi = 0$, the output of the system is amplified just at the natural frequency ω_n, and the phase changes abruptly from zero to $-180°$, causing the output to have a constant oscillation. In the case of $0 < \xi < 1$, the output is also amplified near the natural frequency ω_n, which makes the response faster, as shown in Fig. 2.7. For lower frequencies than natural frequency, the gain is unitary, and the response to the step will be stable and fast, while for $\xi > 1$ it is attenuated and slower. On the contrary, at frequencies higher than the natural frequency, none of the previous cases will respond.

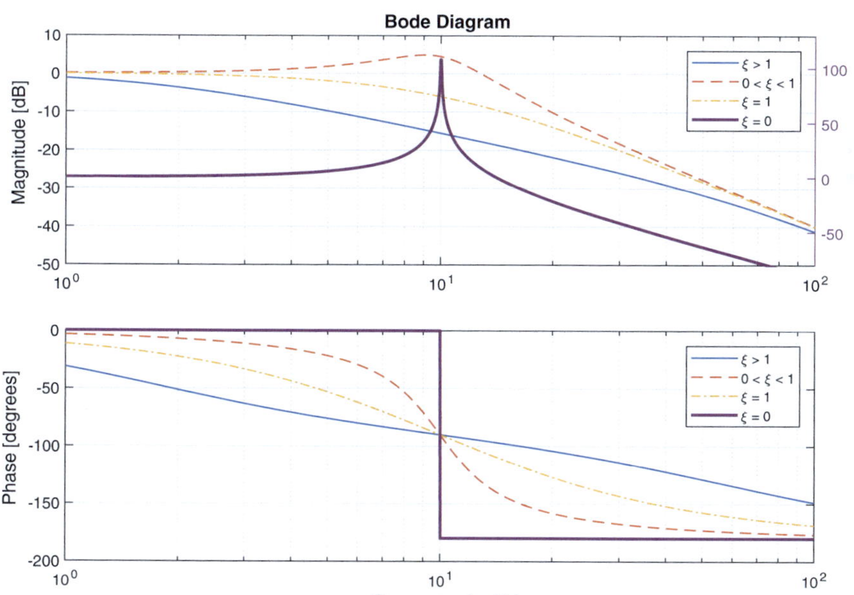

Fig. 2.7 Bode diagram for different damping ratios

Most control systems demonstrate a good response of the system at low frequencies but fail to respond at high frequencies. At low frequencies, the controller is fast enough to control the system. As the frequency increases, the controller cannot maintain the same dynamic regulation. The settling time with the different cases of Fig. 2.4, according to Fig. 2.7, is closely related to the bandwidth of the system.

The higher the bandwidth, the shorter the settling time and vice versa. The time constant of the system can be calculated approximately as the inverse of the bandwidth frequency in rad/s. With a simple approximation, the settling time to 5% can be calculated as $3/\omega_{BW}$ and for 2% $4/\omega_{BW}$.

2.2 Control Structures

In general, in the control systems, including the control of electrical machines, the first step to analyze is the objectives to achieve with the control, as well as the variables measurable through sensors and the variables that can be manipulated through the actuators. Once this step has been carried out, the design of the control structure can be performed that is to establish which are the most appropriate regulation loops. Note that the conventional feedback structure seen in the previous section is not the only possible control structure, but there are other alternatives that in certain situations are more appropriate or can also add some improvement to the feedback structure such as feedforward or cascade control. The implementation of these control structures requires a better knowledge of the system, as well as additional elements such as sensors and further processing. However, this increase can be justified in many cases by the improvements obtained with these structures as there are cases in which the conventional feedback structure can be a weak solution, such as:

- The corrective action is not started until a deviation of the controlled variable from the desired value is detected
- It does not contemplate a predictive control action to compensate for the effect of measurable and known disturbances.

In the following section, it will be seen that these two examples can be corrected with the cascade and feedforward structures, respectively.

2.2.1 Feedforward Control

The feedforward control basically tries to compensate for the disturbance before it affects the variable that is being controlled. The disturbance, as soon as it occurs, it is necessary to measure it to start acting on the controlled variable, so that the effect of this action compensates the disturbance, getting negligible the overall effect on the output. On the other hand, the conventional feedback structure does not start corrective action until the disturbance has affected the process sufficiently to cause

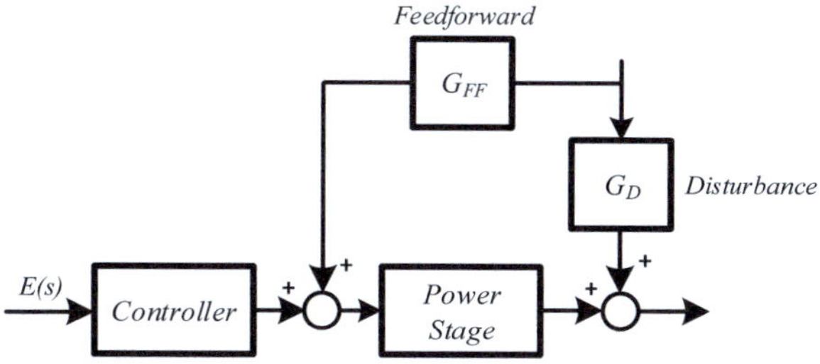

Fig. 2.8 Feedforward compensation

an error between the output and the reference. The block diagram of the feedforward structure can be seen in Fig. 2.8 where two new blocks are introduced, a feedforward compensator $G_{FF}(s)$ and a block $G_D(s)$ that models how the disturbance impacts in the system.

For adequate compensation, the $G_D(s)$ dynamic must be slower than the $G(s)$ to ensure that the disturbance is transmitted more slowly than the control signal in the process. The advantage concerning the conventional feedback structure is quite evident in theory, but it does not suppose a definitive solution for all the cases since the feedforward control will not always be applied according to the following points:

- Online measurement of disturbances is necessary
- The model of the process must be perfectly known
- The feedforward controller $G_{FF}(s)$ is not always physically feasible, so an approximation to the ideal feedforward controller is usually made.

In practical applications, feedforward control is usually applied together with feedback control, as can be seen in Fig. 2.9.

In this case, the purpose of the feedforward control is to reduce the effects of measurable disturbances and the feedback control for reducing the effects of modeling

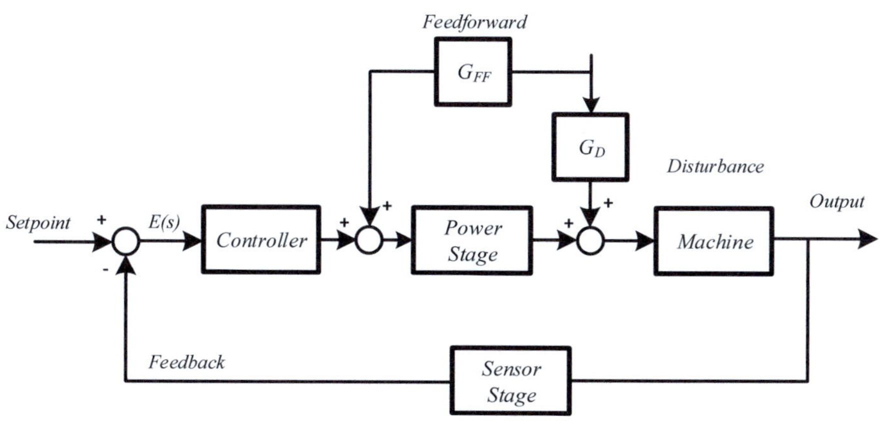

Fig. 2.9 Feedforward structure in a disturbance with feedback control allowing an improvement in the bandwidth and rapid response at disturbances

errors, the variation of parameters, non-measurable disturbances, etc. The feedforward control is an open-loop control, while the feedback control is closed loop so that both do not have interactions between them. That is, the feedforward control does not affect the dynamics of the closed loop, as can be seen in Eq. (2.7).

$$Y(s) = \frac{G_c(s) \cdot G(s)}{1 + G_c(s) \cdot G(s) \cdot H(s)} E(s) + \frac{G_D(s) - G_{FF}(s) \cdot G(s)}{1 + G_c(s) \cdot G(s) \cdot H(s)} D(s) \tag{2.7}$$

Observing (2.7), to ultimately reduce the effect of the disturbance $D(s)$ on the output $Y(s)$, the feedforward compensator must meet $G_{FF}(s) = G_D(s)/G(s)$. In this case, the feedforward compensator is called perfect cancelation or ideal compensator, but it depends on $G(s)$. This dependence causes the perfect cancelation compensator not to be used since it can also be non-causal. For example, if $G_D(s) = 1$ y $G(s) = 1/s$ the compensator $G_{FF}(s)$ would be the inverse of the plant $G(s)$ that is to say the derivative ($G_{FF}(s) = s$). As will be seen later, the ideal derivative is non-causal and has a capacity to amplify noise that can negatively influence the system. A practical solution that usually works for most systems is to use an approximation to the ideal compensator based on a lead–lag compensator as can be seen in (2.8):

$$G_{FF}(s) = Kf \frac{\tau_1 s + 1}{\tau_2 s + 1} \tag{2.8}$$

There are techniques for tuning the three parameters of the lead–lag compensator when used for feedforward control as described in Franklin et al. (1994).

The feedforward control can also be applied in systems where a fast reaction to the change of reference is necessary, as can be seen in Fig. 2.10. The change of reference is detected by the feedforward controller, which quickly modifies the input of the power stage to arrive as soon as possible at the new reference, advancing the feedback controller. As in the previous case, the additional loop does not affect the closed-loop control of the system and the lead–lag compensator seen above can also be used for the controller design.

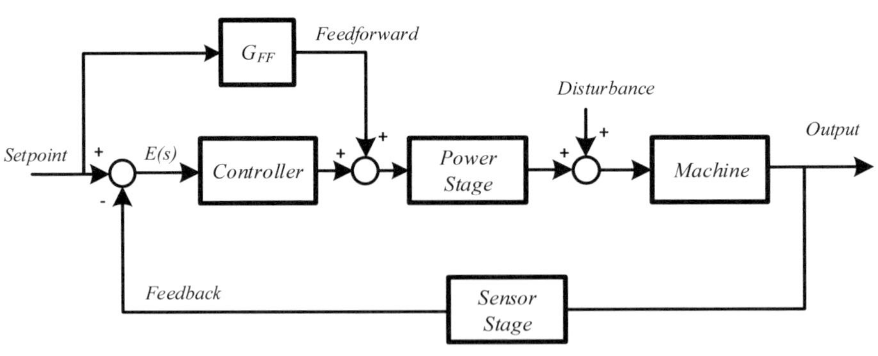

Fig. 2.10 Feedforward structure in the command plus feedback control allowing an improvement of the bandwidth and rapid response at command change

2.2.2 Cascade Control Structure

The cascade control structure is a variation of the feedback control structure since it is composed of several simultaneous feedback loops. The use of cascade control has been increasingly extended in different control systems for different markets such as household appliances, industrial, and automotive thanks to the digital implementation of controllers in the microcontrollers or DSPs. These digital controllers, basically PID, allow in a simple way the connection in cascade mode, which consists of the output of one controller being the input of the next controller. This characteristic allows detecting disturbances at the input and correcting its effect before it propagates at the output. The objective is to avoid that the effect accumulates and affects the output. For example, the closed-loop speed control of a DC machine regulates its speed without taking into account the torque (current of the DC machine) so that for some speeds the maximum current that the DC machine can withstand can be exceeded. The control cascade is able to regulate the current and the speed avoiding to exceed the maximum speed and the maximum current of the DC machine. Another interesting point of cascade control is the treatment it performs in external disturbances as the following explained. A disturbance at the input of the plant modifies the level of control action expected by the controller, as can be seen in Fig. 2.11. In this example, the disturbance could be in the supply voltage of the power stage.

The effect of a disturbance at the input manifests itself in the output after a certain time. The disturbance $D(s)$ affects the system for a period of time without any action being taken. It is usually easy to measure the disturbances that may affect the output of the actuators. Cascade control avoids this phenomenon, that is, acts on the disturbance before it manifests itself on the output. The value of the action that is actually applied is measured or estimated. In Fig. 2.12 is represented the block diagram of the cascade control structure where the perturbation $D(s)$ affects the internal feedback loop (inner loop, or secondary loop, or slave) which is corrected by its controller $Cs(s)$ immediately. The condition is that the internal loop must be more responsive (higher bandwidth) than the outer loop (outer loop, primary loop, or master) since this will compensate for the disturbances in the input before they affect

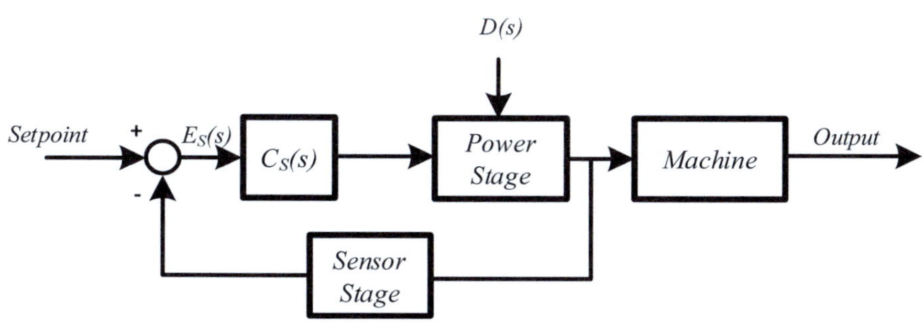

Fig. 2.11 Closed-loop control system with disturbance in the input of the plant

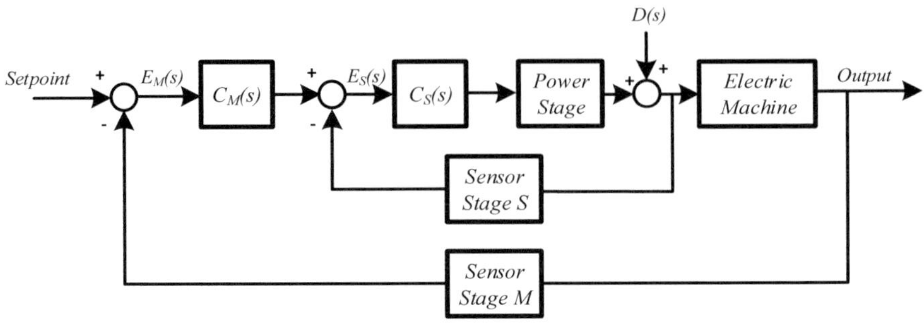

Fig. 2.12 Cascade control structure with master (outer loop), slave (inner loop), and the disturbance which can be corrected before affecting the output

the primary variable or output. A general rule is that the inner loop time constant should be smaller than four to ten times the time constant of the outer loop. In contrary case, cascade control should be avoided because the system will tend to be unstable.

In general, it is possible to affirm that cascade control eliminates the effect of some disturbances by making the regulation response of the system more stable and faster and improves the dynamics of the control loop.

In electrical machines, the control variables are usually well defined. For example, to control a DC machine, the control variable of the slave loop would be the current, and for the master, it would be the speed. Any disturbance in the current of the machine will be corrected by the slave loop without the output (speed) being significantly affected. The block diagram in Fig. 2.13 presents the structure of cascade control for a generic electric machine system with two disturbances. The cascaded structure of up to three control loops, the current or torque control loop, the speed control loop, and the position control loop, is illustrated.

The current controller sets the command voltage or current in the power stage to regulate the current measured in the electric machine terminals. The current through the machine develops an electromagnetic torque where the mechanical system composed of a rotor and load inertia transforms the difference between the developed

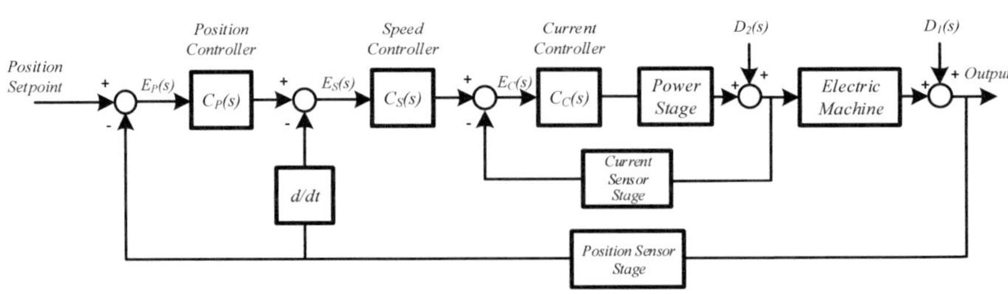

Fig. 2.13 Block diagram of cascade control structure for electric machine control with two disturbances

and the load torques into speed. The speed of the electric machine is the integral of the torque divided by the total inertia J_T as it will be seen in Chap. 4. In the same way, the position is obtained by integrating the speed.

The inner current control loop will ensure that any disturbance $D_2(s)$ in the electrical system does not affect too much the torque output. The disturbances in the load $D_1(s)$ (e.g., load torque) will be corrected by the speed control without the position loop having almost to contribute. For the cascade structure to operate as described above, the current loop must be designed with a bandwidth higher than the speed loop. At the same time, the speed loop must also be designed to have a bandwidth higher than that of the position loop. The way to fine-tune the cascade control systems is first to tune internal loops and then tune the outer loops. Once the internal loop has been tuned, it will behave as a low-pass filter for the outer loop. The outer loop will have its compensator and then the equivalent to a low-pass filter in which the parameters of the external compensator can be designed in a more straightforward way.

2.3 Classical PID Controllers

The PID controller or also called compensator is composed by some of the combinations corresponding to a proportional part (P), proportional plus derivative (PD), proportional plus integrative (PI), and proportional plus derivative and integrative (PID). The input of the controller is the error $E(s)$ between the command and the measured variable. The output is the control signal $U(s)$. About the amount of error, the controller will try to act on the system to get the error to be zero. For this, the controller must be perfectly tuned to the plant, and the measurement sensors must give measurement with an acceptable delay resolution. The gain parameters of the controller are the proportional gain K_P, the integral time T_I, and the derivative time T_D. For most systems with a PI compensator, it may be sufficient since it offers stability and good response at high frequency, achieving zero error. Moreover, it only presents two configuration parameters. However, the response to the step can produce a high overshoot if the derivative term is not used. Various PID structures can be found (Fig. 2.14) such as ideal or standard form given by (2.9), the series or cascade form is given by (2.10), and finally the parallel form is given by (2.11). The parallel form has the advantage that the parameters K_P, K_I, and K_D are decoupled, the opposite of the ideal formal and series that if the proportional parameter K_P or K is modified, it affects the integral and derivative actions of the controller.

$$\frac{U(s)}{E(s)} = K_P\left(1 + \frac{1}{T_I s} + T_D s\right) \tag{2.9}$$

$$\frac{U(s)}{E(s)} = K'_P\left(1 + \frac{1}{T'_I s}\right)(1 + T'_D s) \tag{2.10}$$

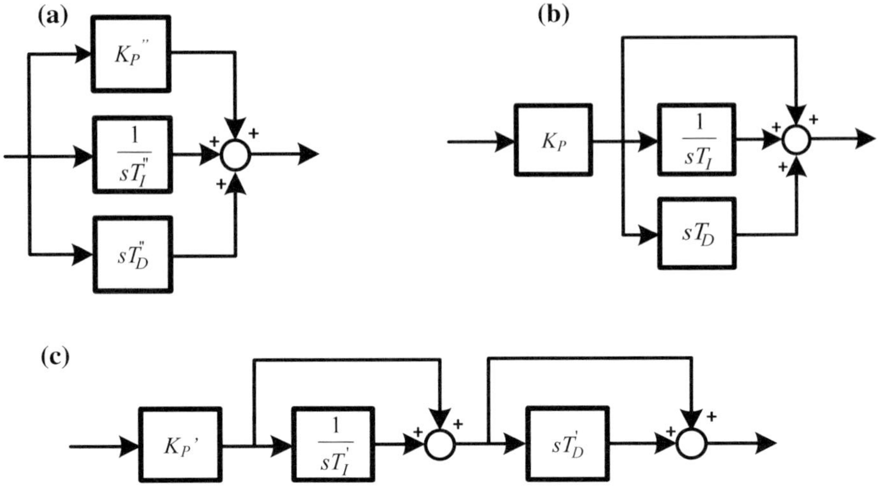

Fig. 2.14 PID structures: **a** parallel, **b** ideal or standard, **c** series

$$\frac{U(s)}{E(s)} = K_P'' + \frac{1}{T_I'' s} + T_D'' s \tag{2.11}$$

It is important to mention that, for a given structure with defined parameters, if the structure of the controller should be changed, the controller parameters must be adjusted to have the same response. For example, if the structure should be changed from series to parallel, the equivalence will be:

$$\begin{aligned} K_P'' &= K_P'\left(1 + \frac{T_D'}{T_I'}\right) \\ K_I'' &= \frac{K_P'}{T_I'} \\ K_D'' &= K_P' T_D' \end{aligned} \tag{2.12}$$

In the previous transfer functions, it is observed that all the PID controllers introduce a pole at the origin due to the integral action and two zeros that can be real or complex in the cases of the standard and parallel structure, but that are real in the structure series. The zeros depend solely on the integral and derivative time constants T_I and T_D. In some control systems, the controller's mission is to cancel the pole or dominant poles of the process. One of the zeros of the controller can be used to cancel the dominant pole of the process, and the other can be chosen to improve the stability. In the cases of the standard and serial PID, proportional gain is available as the third parameter, which places the poles of the system in closed loop in suitable positions. In the parallel case, the procedure is not so easy since the choice of zeros imposes some restrictions on the three parameters. That is, the subsequent modification of any of them would force the modification of the others.

The transfer functions of the different PID structures are difficult to perform in practice since it is not true that the denominator $E(s)$ has order greater or equal to the numerator $U(s)$. This, as previously mentioned, is due to the derivative part, which was not causal and very susceptible to amplify the noise. The solution is to add a pole in the derivative part so that it behaves like a low-pass filter. Hence, the numerator and denominator will have the same order. Therefore, if the parallel PID structure is considered Eq. (2.11) can be rewritten as

$$\frac{U(s)}{E(s)} = K_P + K_I\frac{1}{s} + K_D\frac{s}{\alpha \cdot s + 1} = \frac{s^2(K_P\alpha + K_D) + s(K_P + K_I\alpha) + K_I}{s^2\alpha + s} \tag{2.13}$$

If α is zero, Eq. (2.11) is obtained.

2.3.1 PD Controller

The ideal PD controller has a constant gain up to the frequency $\omega_{PD} = 1/T_D$, and from there the gain is increased with a positive slope of 20 dB/dec. The phase is lead by a maximum of 90° so that it is a lead compensation where its behavior is as a high-pass filter. The transfer function of the PD controller in the "ideal" form is shown in (2.14), where its root remains in the left half plane at $s = -1/T_D$, while the PD with low-pass filter is shown at (2.15).

$$\begin{aligned} G_c(s) &= K_p(1 + T_D s) = K_P T_D\left(s + \frac{1}{T_D}\right) \\ \omega_{PD} &= \frac{1}{T_D} \end{aligned} \tag{2.14}$$

$$G_c(s) = K_p\left(1 + T_D\frac{s}{\alpha \cdot s + 1}\right) \tag{2.15}$$

The PD controller offers an increase in the response speed since it increases the bandwidth thanks to the zero introduced by the derivative time T_D. The action of the derivative can be interpreted as providing prediction by linear extrapolation over time T_D. Damping increases with increasing derivative time but decreases again when the derived time is too large. The action of the derivative does not help if the prediction time T_D is too large. The asymptotic Bode diagram of the PD controller can be seen in Fig. 2.15.

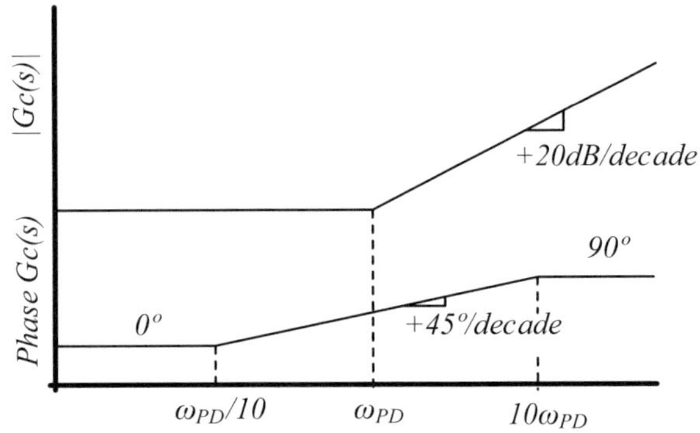

Fig. 2.15 Asymptotic Bode diagram of the PD controller

2.3.2 PI Controller

In a PI controller, the proportional gain K_P affects how the controller reacts to the magnitude of the error. A higher value makes the proportional component faster to reach the command, and a lower value makes it less fast. The proportional part by itself is not capable of reducing the error to zero in steady-state. The error will decrease with increasing K_P gain, but the tendency to oscillate will also increase. To reduce the error to zero, it is necessary to add the integral action. The drawback of adding integral action is that it can have a destabilizing influence on the response of the system if the integral time T_I is not selected appropriately. That is, the tendency to oscillation also increases with the reduction in T_I.

The PI controller has a pole at the origin that makes the gain fall with a negative slope of 20 dB/dec until it reaches the zero introduced by the integrative part leaving the gain constant. The zero, according to (2.16), sets the bandwidth of the controller. From the point of view of the frequency domain, if zero is set suitably, it is accomplished that the gain of the controller is enough to reduce the control error to zero and decrease the oscillations caused by the excess gain of the proportional part. The PI controller adds an integral part adjustable phase lag that decreases with increasing frequency in the system. K_P is designed to have a desired gain and bandwidth, while ω_{PI} is selected to have an adequate phase margin.

$$G_c(s) = K_P\left(1 + \frac{1}{T_I s}\right) = \frac{s + \frac{1}{T_I}}{\frac{s}{K_P}}$$

$$\omega_{PI} = \frac{1}{T_I} \tag{2.16}$$

The conclusion is that a PI regulator adds to the system an adjustable phase lag with the T_I parameter that decreases when increasing frequency. The phase lag variation is from $-90°$ to $0°$. Its behavior reminds of a low-pass filter. At the same time, the proportional parameter K_P has a direct influence on the cut-off frequency in open

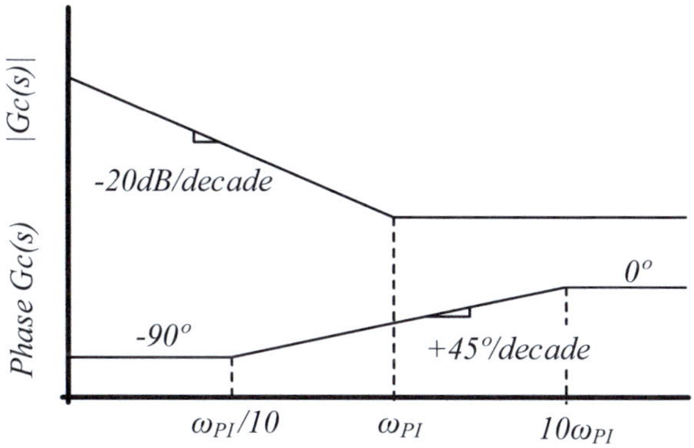

Fig. 2.16 Asymptotic Bode diagram of PI controller

loop, since the gain can raise or fall according to K_P. Because it is a regulator that adds phase lag, it also is known as a lag compensator.

The asymptotic Bode diagram of the magnitude and phase of the PI controller is illustrated in Fig. 2.16. The gain falls with a negative slope of 20 dB/dec because the pole is at the origin and stabilizes at frequency ω_{PI}. If the error varies with a frequency higher than the bandwidth of the PI controller, the error will be significantly attenuated. The phase at low frequencies has an offset of $-90°$ that decreases to $0°$ as the error frequency increases.

2.3.3 *PID Controller*

The PID controller combines the best features of the PD and PI controller into a single controller so that it is also known as a lead–lag compensator. The tuning of the controller is more complicated because it has a third additional variable and its behavior is similar to that of a band-pass filter. The PID has two zeros caused by the integrative and derivative part, and a pole in origin also caused by the integrative part according to (2.17). The zero of the integrative part is located in $\omega_I = 1/T_I$ as in the case of the PI controller, while the zero of the derivative part is located in $\omega_D = 1/T_D$. The zero of the derivative part could amplify the high frequencies as it happened with the derivators, so it would be advisable to perform the derivative part with a filter as shown in Sect. 2.3.1. The transfer function of the PID controller is shown in Eq. (2.17).

$$G_c(s) = K_P\left(1 + \frac{1}{T_I s} + T_D s\right) = \frac{s^2 + \frac{s}{T_D} + \frac{1}{T_D T_I}}{\frac{s}{T_D K_P}} = \frac{K_P T_D}{T_I}\frac{\left(s + \frac{1}{T_I}\right)\left(s + \frac{1}{T_D}\right)}{s} \tag{2.17}$$

2.3.4 Anti-windup

One of the main undesirable effects after saturation limits have been reached by the actuators is that the PI or PID controllers will continue integrating the error. Thus, the state of the integrator in question can reach excessive values, which will deteriorate the transitory response of the system, generally producing large overshoots. This effect is called a windup integrator. There are different alternatives to avoid the windup of the integrators. One of the most common is simply to limit the integral action with a limiter on its output so that it is bounded between a maximum and minimum value as represented in Fig. 2.17.

An alternative to this and perhaps more elegant method is to feed back the saturation error produced by the difference between the unrestricted output of the PID controller and the limited output, as can be seen in Fig. 2.18. When the output of the PID controller $U_r(s)$ is higher than the output $U(s)$ because the saturation error has been saturated, it subtracts integrative action. The correction gain of the integrator K_{aw} is a new parameter to fine-tune as it happens with the maximum and minimum limits of the limiter.

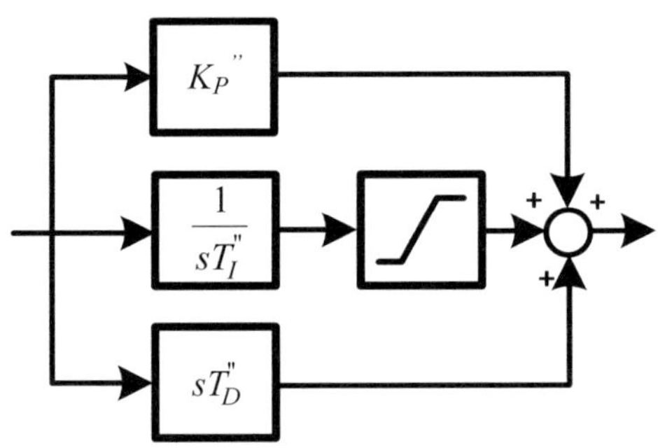

Fig. 2.17 Anti-windup with saturation in the integral part

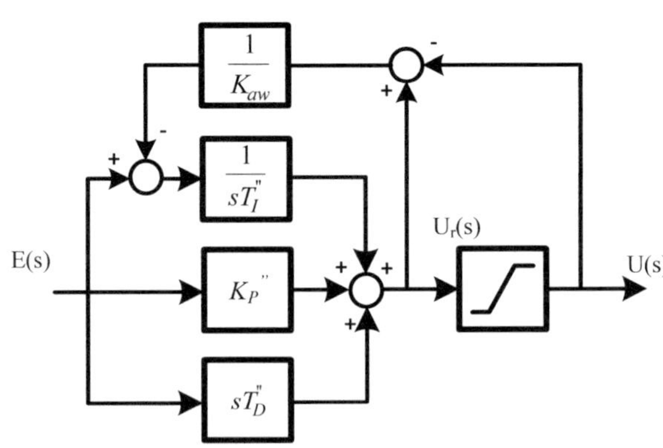

Fig. 2.18 Anti-windup with control of saturation of the integral part with a new parameter K_{aw}

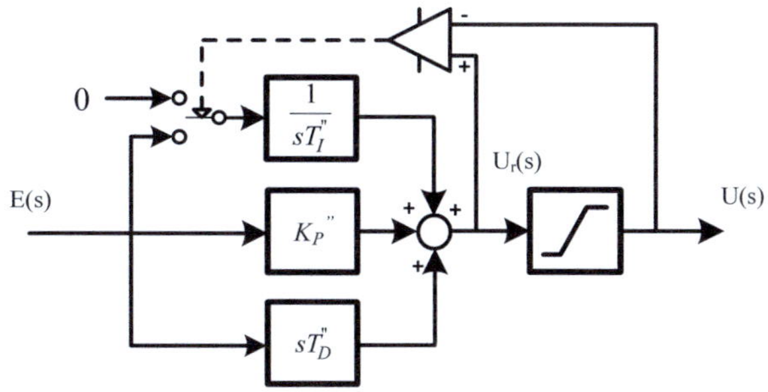

Fig. 2.19 Anti-windup with a comparator to select zero or the error signal according to if the output is saturated

On the other hand, an alternative to the previous method, where the K_{aw} parameter is not needed, is the conditional method structure, as shown in Fig. 2.19. As can be observed, when saturation error is positive, it disconnects the integrating action by setting zero in its input.

2.4 Digital Control

The digital control can use a microcontroller, microprocessor, or DSPs and present some advantages with respect to the analog control. For example, the gain parameters of the digital controllers persist forever in the tuned value while analog controllers depend on some passive components such as resistances and capacitances which they can undergo a drift with environmental variations and aging. Also, the gain parameters can be changed easily without any alteration in the hardware, and digital control is less sensitive to noise. Another advantage is that it can perform advanced controls, e.g., non-linear and adaptive control algorithms. The most common disadvantage of the digital control is the limited bandwidth due to the sampling loop, the digital execution delay which is inevitable due to the nature of the serial execution of the software, the numerical problems due to quantization and rounding, and the limited resolution of analog-to-digital converters and the digital PWM. However, the digital control is often used in the electric machine control due to the flexibility, because some sophisticated algorithms, as vector control, are not suitable in analog control.

The digital control is executed with a finite periodicity called sampling period T_s. In practice, in electric machine control systems, there are different sampling periods in the different feedback paths, e.g., current, and speed paths in cascade control. The sampling period selection usually is small, and it can be approximated by 10 times smaller than the inverse of the bandwidth frequency of the closed-loop system. For example, if the bandwidth of the closed-loop system is 1000 Hz, the estimate sampling period is 100 μs. In fact, in the control systems is very common to analyze first the system in continuous mode. Then, it is possible to add the sample-data system elements with the above sampling period estimation, digitalize the control laws, and finally check the performance using discrete models. The z-domain is used to digitalize the control laws, and the z-transform is a mathematical tool which converts a discrete time-domain signals into a complex variable frequency domain. The term $1/z$ is defined as e^{-sT}, and it is the delay operation. The effects of delays introduced by the digital controllers are easily represented in the z-domain.

The analog signals need to be converted to discrete-time signals. The discrete-time signal is characterized by its values that take a numerable set of time instants spaced x called instants of sampling. In other words, the discretization of an analog signal collects samples with a constant periodicity T_s. The ideal sampler is like a switch (Fig. 2.20) activated periodically every sampling time, which lets the signal pass at that instant to the next stage, usually a zero-order hold (ZOH). The zero-order hold, as it will be shown later, holds the sample during the sample period. The mathematical expression that describes a discrete signal sampled ideally can be expressed as the sequence of unit impulses multiplied by the analog signal according to (2.18).

$$x_s(t) = x(t)x_\delta(t) \tag{2.18}$$

The sequence of unit impulses is expressed according to (2.19).

$$x_\delta(t) = \sum_{k=-\infty}^{k=\infty} \delta(t - kT_s) \tag{2.19}$$

where $\delta(t)$ is the Dirac delta function or the impulse function.

So,

$$x_s(kT_s) = x(t) \sum_{k=-\infty}^{k=\infty} \delta(t - kT_s) \tag{2.20}$$

Once the analog signal is sampled and holds, it is converted into digital form. Most microcontroller/DSPs has an on-chip analog-to-digital converter with a conversion time less than 1 μs with 12-bit resolution. The conversion time is the time it takes to perform an n-bit conversion that usually is specified in the converter specification.

The sampling period of the control system will be limited either by the speed of the CPU or by the conversion time of the A/D converter. The processing speed of the CPU is usually higher than the frequency at which a new sample of the A/D converter is obtained. In this case, it is not useful to execute the digital control at the CPU frequency since it will use the same samples until the new conversion of the A/D is performed. Therefore, in this case, the maximum frequency at which

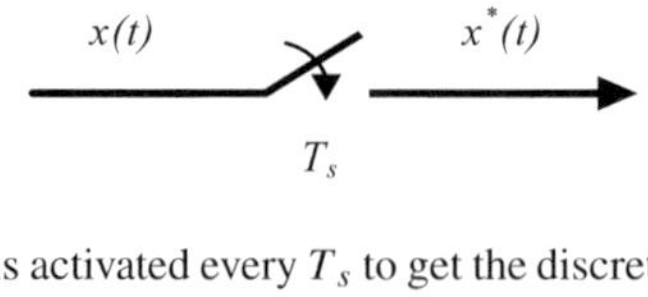

Fig. 2.20 Ideal switch which is activated every T_s to get the discrete value of the continuous signal

the digital control could be executed would be fixed by the conversion time of the A/D converter. For example, if the conversion time is 3 μs with 12-bit resolution when the controller task is executed with a periodicity of 1 μs, this would be, in the worst case, with 3 identical samples. This is called oversampling. However, if the controller task is executed every 4 μs, it will probably always process it with a new sample since A/D converters are hardware processes independent of CPU processing. In this case, the 3 μs A/D conversion time is very close to the 4 μs controller task, and there will be cases in which the controller will not process any of the samples. The main reason is that the 4 μs periodicity task is a synchronous process launched by an ISR as explained in Chap. 1, where its periodicity is not always assured and will vary depending on the additional tasks that the microcontroller has to perform. To minimize this risk, either the controller's task becomes slower than the A/D conversion time, or the controller execution task is given maximum priority. Another perhaps more elegant technique would be to synchronize the A/D converter with the execution of the controller. That is, every 4 μs periodically the controller is executed with a new sample of the analog signal. When the sample is converted in k instant, an automatic ISR of the A/D is triggered to perform the controller processing. Inside of the ISR, the first task is to disable the A/D and then executed the controller task, and finally, at the end of the ISR, the A/D converter should be enabled so that it can perform the next acquisition in $k + 1$. Figure 2.21 shows the flowchart of the controller processing synchronized with the A/D measurement according to the above explanation.

As commented before, the cascade control in electric machine control with inner current loop and outer speed loop is usually used. If the rotor speed sensor consists of a DC tachometer which gives a voltage proportional to the speed, an extra A/D channel will be needed. Then, two analog signals should be digitalized, the current in the machine terminals and the rotor speed. If the simultaneous conversion is not

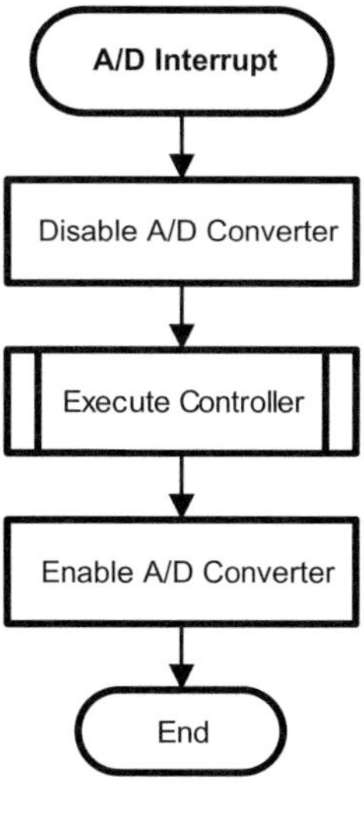

Fig. 2.21 Digital controller synchronized with the A/D converter

possible because the microcontroller/DSP has N analog-to-digital channels, but only one A/D converter, the A/D converter should be shared. As the sampling period is different for the two paths, it will not be an issue.

In the previous examples, the limits of the microcontroller have been played by imposing the minimum possible sampling period without taking into account how the output of the digital control is translated into the analog world of the electric machine. It can be said that between the digital control and the plant (electrical machine) there will be a block that will perform the task of converting the signal from digital-to-analog (D/A). The D/A converter will have a specific conversion time and resolution. As previously mentioned, the variation of the speed of a continuous electrical machine can be done by varying its average voltage. The DC voltage can be digitally varied by a PWM-type signal where the duty cycle will set the amount of RMS value of the voltage. The digital PWM signal is generated by the microcontroller itself in one of its outputs. This output has a limited resolution due to the clock of the microcontroller itself and the bits used by the timer to set the period and the duty cycle, as will be seen in Chap. 7. Most electrical machines can be governed by digital PWM signals through a power stage where the voltage is regulated. The voltage applied to the machine is a pulsating voltage according to the PWM which is filtered basically by a low-pass filter with L/R constant, giving a non-pulsating current. In some way, it can be said that the microcontroller, the power stage, and the electric machine behave like a D/A converter.

In Fig. 2.22 is shown the block diagram of the digital cascade control of current $C_c(z)$ and speed $C_s(z)$ controllers for an electric machine. The digital control consists of two channels CH1 and CH2 A/D converters, the samplers, two digital controllers $C_c(z)$ and $C_s(z)$, and a timer to generate the PWM. In the analog part, it is possible to find the power stage, the current and speed signal conditioner, and finally, the electric machine. The RMS voltage of the electric machine will be varied by the duty cycle of the PWM, which regulates the current and, then, the motor torque. The variation of the torque of the electric machine thanks to the total system rotational inertia is translated into a variation of speed that will be measured by the outer control loop. In Chap. 7, the operation of the A/D converter and the PWM generator for the Renesas RX63T microprocessor will be seen.

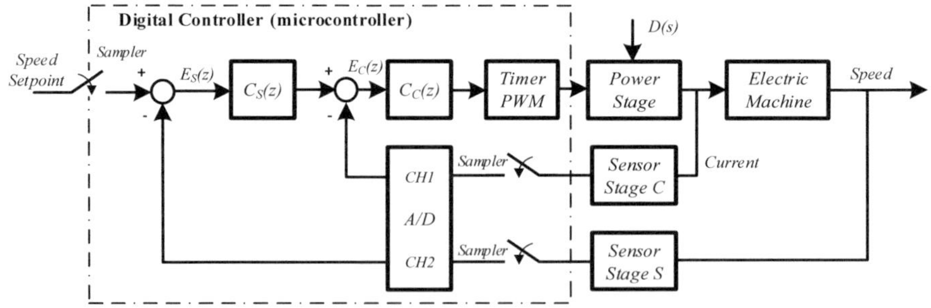

Fig. 2.22 Digital cascade control structure for an electric machine with torque and speed controllers

2.4.1 Aliasing

If the Fourier transform is performed in Eq. (2.20), the frequency spectrum of the sampled signal has the same shape as that of the unsampled signal, attenuated by a factor of $1/T_s$ and repeated on the frequency every $\Delta\omega = 2\pi/T_s$ radians per second. It is possible to observe that the spectra will be further away when the sampling frequency is larger. Figure 2.23 shows the spectrum of a discretized signal.

The reconstruction of a sampled signal consists in obtaining the original continuous signal from the sampled signal. The idea would be to filter one of the spectra of the original signal and from there to perform the inverse Fourier transform. This will be possible if the spectra are not overlapping, that is, if $\omega_s \geq 2\omega_o$ is met.

The sampling frequency must be at least twice as fast as the maximum frequency of the analog signal, according to Shannon's theorem. If this condition is met, the analog signal can be replicated by the sampled signal without having the effect of overlapping or aliasing. The effect of the aliasing appears when the sampling frequency is less than twice the maximum frequency of the analog signal causing its spectra to overlap, distorting the resulting discretized signal. Therefore, the maximum sampling time would be according to (2.21) although, in practice, it is usually much smaller.

$$T_s \leq \frac{\pi}{\omega_o} \tag{2.21}$$

Low-pass anti-aliasing filters are usually used whose cut-off frequency is fixed to twice the sampling frequency. There are cases in which the anti-aliasing filter cannot be used, or its cut-off frequency must be fixed to a higher value than twice the maximum signal frequency. For this last case, it is essential to avoid excessive delay. For example, in the control of three-phase electric machines with PWM, the voltage measurement of the machine must be filtered since it contains infinite harmonics produced by the sinusoidal PWM voltage that is applied. The filter should be tuned in such a way that it does not excessively delay the voltage but that it also fulfills filtering adapted to assure the best possible control performance. In a four-pole machine and at a maximum mechanical speed of 7200 RPM, the electric frequency is twice the mechanical frequency so that the frequency of the voltage and current is 240 Hz. According to Shannon's theorem, it is possible to sample the voltage and current as a minimum at 480 Hz, and therefore, the anti-aliasing filter should have a minimum

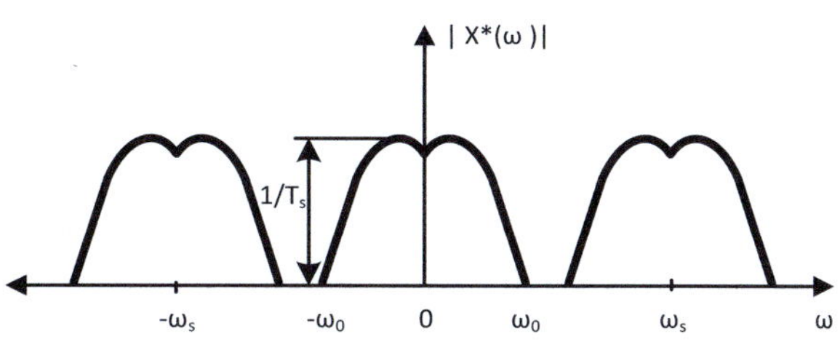

Fig. 2.23 Spectrum of the discrete-time signal when sampled at $\omega_s \geq 2\omega_o$

cut-off frequency of 480 Hz. At this cut-off frequency, there is a delay of 45° and at 240 Hz of 26.46°. That is when the machine turns at the maximum speed of 240 Hz, the voltage and current will have a delay of 307.5 μs, which may be excessive. The primary criterion to follow in this case would be to take a cut-off frequency of 10 times the maximum frequency of the signal, which for this case will be 4800 Hz. The delay that will occur in the signal to be measured will then be 2.86° (33.1 μs). It is important to say that the low-pass filter will not behave as anti-aliasing, but taking a sampling frequency of 4800 Hz, the aliasing effect is avoided filter, and the selection of the sampling frequency should be selected carefully. In high dynamic control, such as vector control, the sampling frequency is usually the PWM frequency where it is usually between 2 and 20 kHz. It can be seen that in this range the aliasing is nonexistent, even if it samples at half the frequency of the PWM. For this same case, it does not mean that an oversampling is being done, but, in order to have the best possible control dynamics, it is essential to execute the cycle-to-cycle control with current samples.

2.4.1.1 Zero-Order Hold (ZOH)

The reconstruction of signals using ideal filters is not possible since this is non-causal, that is, unrealizable. The alternative is to use zero- or higher-order holds. The idea of the zero-order hold is to maintain the sample constant until the next sample. With higher order, more precision in the reconstruction of the analog signal is achieved, although it can produce stability problems. The most common holders are zero and one order. From the impulse response of the zero-order hold, its transfer function can be extracted as a sum of two-step signals of opposite sign and one of them delayed by T seconds (2.22):

$$G_{ho}(s) = \frac{1}{s} + \frac{-1}{s}e^{-sT} = \frac{1 - e^{-sT}}{s} \tag{2.22}$$

Figure 2.24 shows the block diagram of the zero-order hold.

The frequency response of the zero-order hold can be achieved by substituting Laplace s for $j\omega$ in (2.22). By developing the steps, it is possible to find the expression (2.23), where the frequency response is illustrated in Fig. 2.25. In the zero-order hold, to minimize the effect on the sampled signal, it must be fulfilled that the maximum frequency of the signal to be sampled is below half the sampling frequency, i.e., $\omega_o \ll \omega_s/2$. Otherwise, the repetitions of the spectrum centered on $1.5\omega_s$ and $2.5\omega_s$ of the holder will distort the reconstruction of the signal.

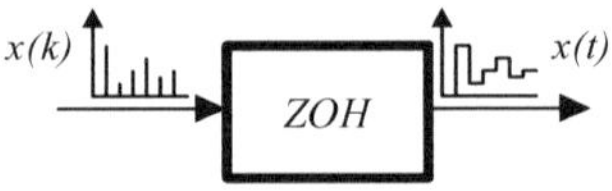

Fig. 2.24 Block diagram of a zero-order hold

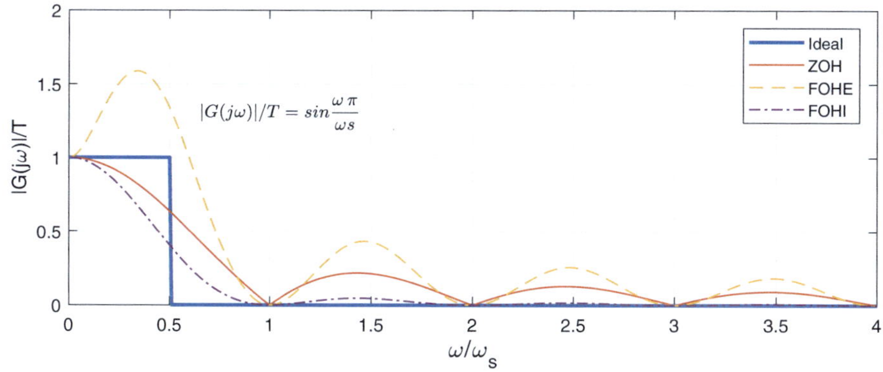

Fig. 2.25 Frequency response for ideal, zero-order hold, first-order hold extrapolation, and first-order hold interpolation

$$G_{ho}(j\omega) = T\frac{\sin\left(\frac{\pi\omega}{\omega_s}\right)}{\frac{\pi\omega}{\omega_s}}e^{-j\frac{\pi\omega}{\omega_s}}$$

$$\left|\frac{G_{ho}(j\omega)}{T}\right| = \frac{\sin\left(\frac{\pi\omega}{\omega_s}\right)}{\frac{\pi\omega}{\omega_s}}$$

$$\angle G_{ho}(j\omega) = -\frac{\pi\omega}{\omega_s} \tag{2.23}$$

In the case of the interpolated first-order hold (Ogata 1996), the filtering is improved since the reduction is more abrupt, and the repetitions have less amplitude. However, the difference is not very significant. On the contrary, the extrapolated first-order hold is closer to the ideal up to 0.5 with an even steeper drop than in the

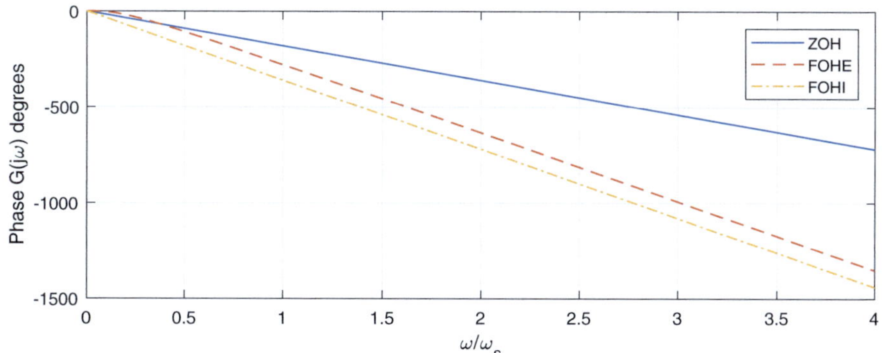

Fig. 2.26 Phase response in degrees for zero-order hold, first-order hold extrapolation, and first-order hold interpolation

previous case, but the repeats have a higher amplitude than even the zero-order hold (Fig. 2.25). Considering the phase, the zero-order hold seems to be better since it shifts the signal less in the whole frequency range as can be seen in Fig. 2.26.

For electric machine control applications, it will be sufficient in most cases to use a zero-order hold.

2.4.2 *Quantifier*

Once the sampling and hold process has been completed, it is necessary to quantify the signal. Before encoding the signal, it is necessary to quantify the signal. The quantifier assigns a finite set of possible values; that is, the value of the sample is approximated at the quantizer output to the nearest value. The distance between two adjacent levels of the quantifier is called the quantization interval Δ. When this value is constant, it is said that the quantifier is uniform. The difference between the input and output of the quantizer is called quantization noise, which is always less than half the quantization interval $\Delta/2$. The uniform quantifier is a non-linear system, with a characteristic between the input and the output, as shown in Fig. 2.27. If the zero is assigned as a quantization level, the quantifier is called midtread, whereas if the zero is a decision level, the quantifier takes the name of midriser. Preferably, it is convenient to use a midtread quantifier if the zero must be quantified since it will avoid fluctuations in the decision due to the presence of noise.

In general, the number of quantizer output Q levels is a power of 2. This is due to the fact that the levels are going to be addressed with an integer number of bits. That is, if n is the number of encoder bits (per sample), the expression $Q = 2^n$ is fulfilled. Usually, half of the quantizer values ($Q/2$) are positive and half negative. $X_{\max}$ is the highest quantizable value without saturation so that the dynamic range of the quantifier corresponds to the range of values that can be quantized without saturation $[-X_{\max}, X_{\max}]$.

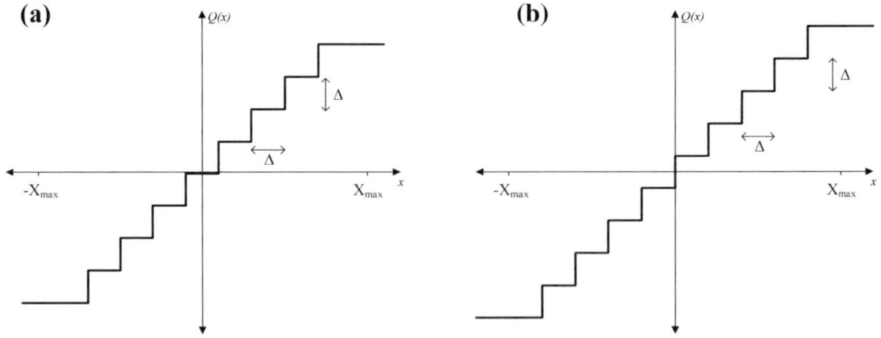

Fig. 2.27 Input/output characteristic of a quantifier. **a** Midtread and **b** midriser

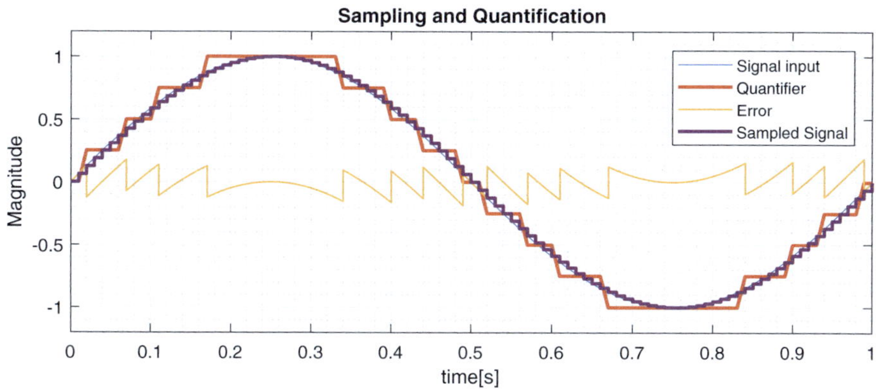

Fig. 2.28 Sinusoidal signal sampled at $T_s = 0.01$ and quantizer of 8 bits. It also shows the quantifier error

It is possible to define two types of errors when quantifying:

- Saturation noise: Appears when a value that exceeds the dynamic range is quantified
- Granular noise: Appears when quantifying a value within the dynamic range.

The signal-to-noise ratio (SNR) is the parameter that will provide information on how much the quantized signal is being distorted. The SNR is the quotient between the power of the original signal and the power of the quantization error. With a series of calculations, it is possible to reach (2.24). From (2.24), it is possible to see that if the number of bits increases one unit, the SNR will increase by 6 dB. On the other hand, if $X_{\max}$ is much higher than the power of the signal, there will be a lower SNR. In this case, the entire dynamic range is not used because they are only using the lowest quantization levels. To avoid this problem, non-uniform quantifiers are usually used, where smaller values use a lower level of quantification than that used for large levels.

$$SNR(\text{dB}) = 10\log\left(\frac{\sigma_x^2}{\sigma_Q^2}\right) = \left(4.77 + 6n - 10\log\left(\frac{X_{\max}^2}{\sigma_x^2}\right)\right)\text{dB} \tag{2.24}$$

The effect of a uniform 8-bit quantizer for a sinusoidal signal can be seen in Fig. 2.28. The signal retained by the zero-order hold is also shown where the quantization error between the two signals can be seen.

2.4.3 *Time Delays*

Time delays in control systems are very common due to the nature of the actuators, sensors, and their processing. In order to achieve an accurate model, it is essential

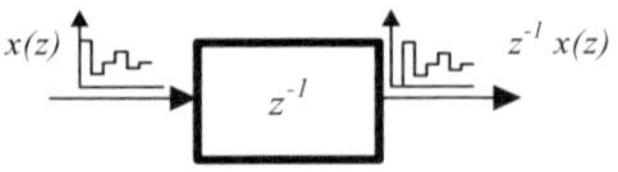

Fig. 2.29 Digital sampling period delay block diagram representation

to quantify the delays that may exist in each of the models of the control system. In digital control, the sample-and-hold (S/H) applies a delay which is increased with the frequency of the signal, and digital controllers have a delay caused by the time required to execute the control.

On the other hand, the sensors also have a time delay sometimes specified in its specification, but on the contrary case, it is possible to measure it experimentally to model it. In general, the delay times become more relevant in the analog and digital filters that are used in the control system, e.g., anti-aliasing filters. Thanks to the modeling of the filters, it is possible to determine their response in frequency and therefore the delay time for each frequency. The time delay of all the blocks that form the model of the system is accumulated and can generate instability in the system. Therefore, it is essential to quantify them in order to add them to the model.

The ideal delay is when it delays the same amount of time for any input signal regardless of its frequency and magnitude. For a continuous signal $x(t)$ if a time T is delayed, the result is the previous signal delayed T according to $x(t - T)$. In the s-domain, it remains according to (2.25). However, in the z-domain, it can be represented according to (2.26).

$$L\{x(t - T)\} \rightarrow e^{-sT} X(s) \tag{2.25}$$

$$Z\{x(t - kT_s)\} = z^{-k} X(z) \tag{2.26}$$

Figure 2.29 shows the digital sampling period delay block diagram representation.

2.4.4 Integrators

Analog integrators have a frequency response similar to that of a low-pass filter where the higher frequencies will be attenuated considerably. The single pole located at the origin causes the slope of the gain of its frequency response to falling to −20 dB/dec. The ideal and continuous analog integrator in s-domain is defined according to (2.27).

$$H_{Int\,ideal}(s) = \frac{1}{s} \tag{2.27}$$

The analog integrator constructed with an operational amplifier as depicted in Fig. 2.30 presents an implicit non-ideality, making the pole not just at the origin but at a higher frequency, as can be seen in (2.28). The operational amplifier has the characteristic that if the signal input enters through its negative input, its output will

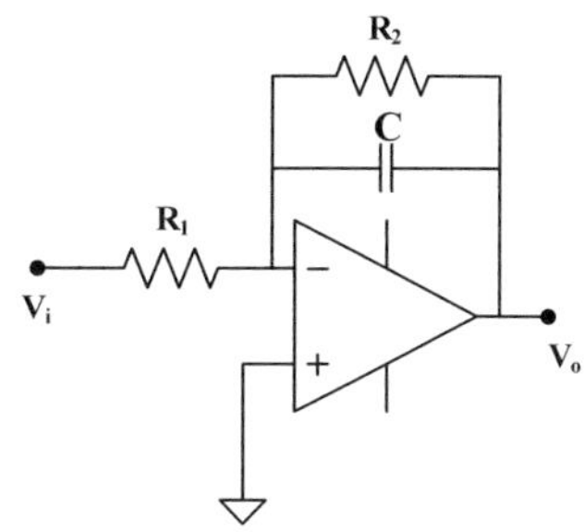

Fig. 2.30 Analog integrator implemented with an operational amplifier

be inverted 180° as it is represented with a negative sign in (2.28). Since the expected output offset of the integrator is $-90°$, in the case of the analog integrator, a unity gain inverter operational amplifier can be used at the output to invert again to have a desired total phase shift of $-90°$.

$$H_{Int}(s) = \frac{M}{s+N}$$
$$M = -\frac{1}{R_1 \cdot C}$$
$$N = \frac{1}{R_2 \cdot C} \tag{2.28}$$

The analog integrator of (2.28) has a pole at $s = -N$ so that the low frequencies will not be so attenuated while the ideal integrator reduces the low frequencies considerably (Fig. 2.31).

The discretization of an analog integrator can be implemented with the following approximation methods (Table 2.2): rectangular forward approach (forward Euler), rectangular approach backward (backward Euler), and bilinear approach (also known as trapezoidal approach or Tustin) (Kuo 1995).

The bilinear method combines the other two methods to achieve a better approach to an ideal integrator by calculating trapezoids each T_s.

The frequency response of the ideal analog integrator and the discretized integrators according to the methods previously seen is shown in Fig. 2.32. The frequency and magnitude are linear, where the frequency has been normalized.

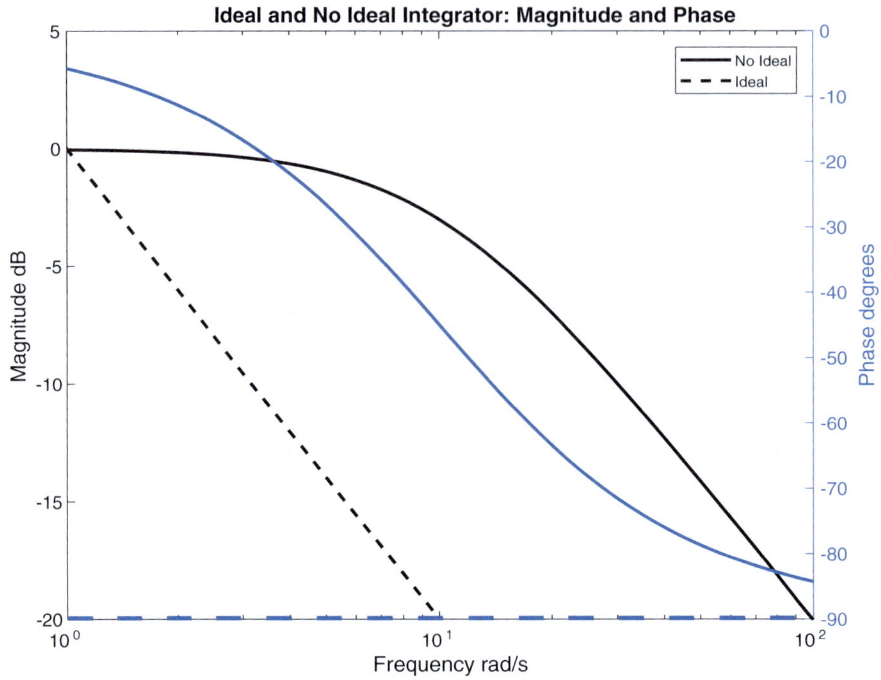

Fig. 2.31 Bode diagram of ideal and non-ideal analog integrator

Table 2.2 Discretization methods to implement a digital integrator with the z-transform and discrete equations

Discretization method	z-transform	Difference equation
Forward Euler	$H_{FR}(z) = T_s\left(\frac{z^{-1}}{1-z^{-1}}\right)$	$y[n] = y[n-1] + T_s \cdot x[n-1]$
Backward Euler	$H_{BR}(z) = T_s\left(\frac{1}{1-z^{-1}}\right)$	$y[n] = y[n-1] + T_s \cdot x[n]$
Trapezoidal or Tustin	$H_{TR}(z) = \frac{T_s}{2}\left(\frac{1+z^{-1}}{1-z^{-1}}\right)$	$y[n] = y[n-1] + \frac{T_s}{2} \cdot (x[n] + x[n-1])$

2.4.5 Derivative

The derivators have a behavior similar to that of a high-pass filter where the highest frequencies are amplified and the lowest attenuated. Its use is very interesting when suppression of the continuous DC value of a signal is desired (DC suppressors). However, the ideal derivator is non-causal and has the ability to amplify the higher-frequency electrical noise that is always not suitable for control. In order to solve this problem, the derivator can be constructed by placing one or several zeros at a certain frequency so that at high frequency the signal is not amplified, but rather attenuated. In this way, the derivator will behave as such for the bandwidth of interest designed,

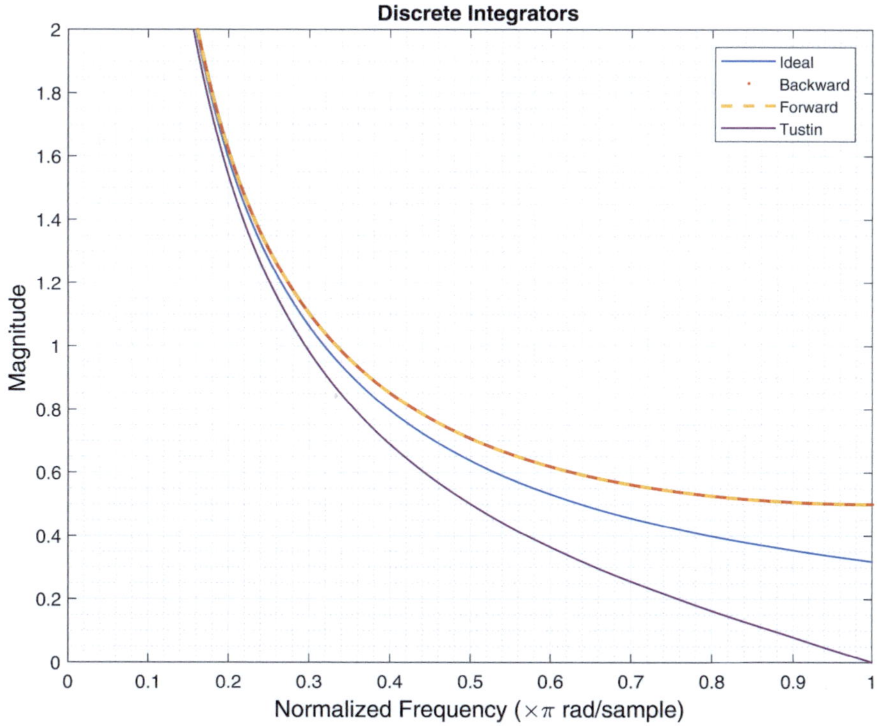

Fig. 2.32 Frequency response of ideal analog integrator with different discrete integrators for backward, forward, and Tustin methods

rejecting the higher frequencies. In addition, the relationship between the input and output will be causal. The analog derivative of Fig. 2.33 made with an operational amplifier, as in the case of the integrator, has an inherent non-ideal characteristic that causes the gain in high frequencies to be attenuated. However, for the implementation of a discrete derivative, it is necessary to apply a non-ideality to attenuate the high frequency as if it were a high-pass filter. The ideal derivative is defined according to (2.29), while a non-ideal derivative with attenuation of the high frequencies is defined according to (2.30).

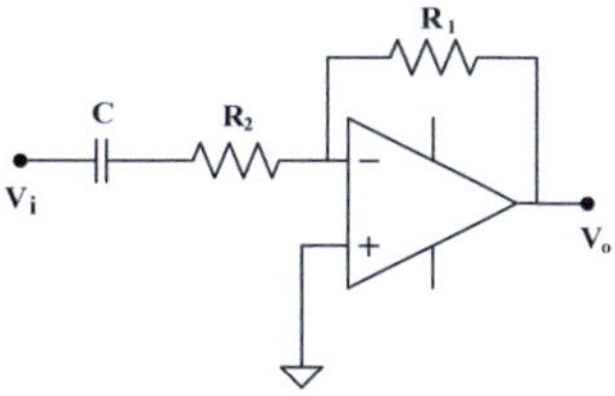

Fig. 2.33 Analog derivator implemented with an operational amplifier

$$H_{Dev}(s) = s \tag{2.29}$$

$$\begin{aligned} H_{DFilt}(s) &= \frac{Ms}{s+N} \\ M &= -\frac{R_1}{R_2} \\ N &= \frac{1}{R_2 \cdot C} \end{aligned} \tag{2.30}$$

The analog derivator of (2.30) has a pole at $s = -N$ so that the high frequencies will be attenuated. This does not occur with the ideal derivator, where it will be amplified since there is no pole.

As in the case of integrators, different methods of approximation can be used to discretize a derivator. The ideal derivative should not be used as discussed above, so it is advisable to use a filter-based derivative like that in (2.30) that sets the degree of attenuation at high frequencies. The most common case where a derivator is used is in the position sensor, which gives us the instantaneous angle of the rotor position. The slope of the angle position at a specific time determines the speed of the rotor. The higher the speed, the higher the slope for the same moment of time, so that the

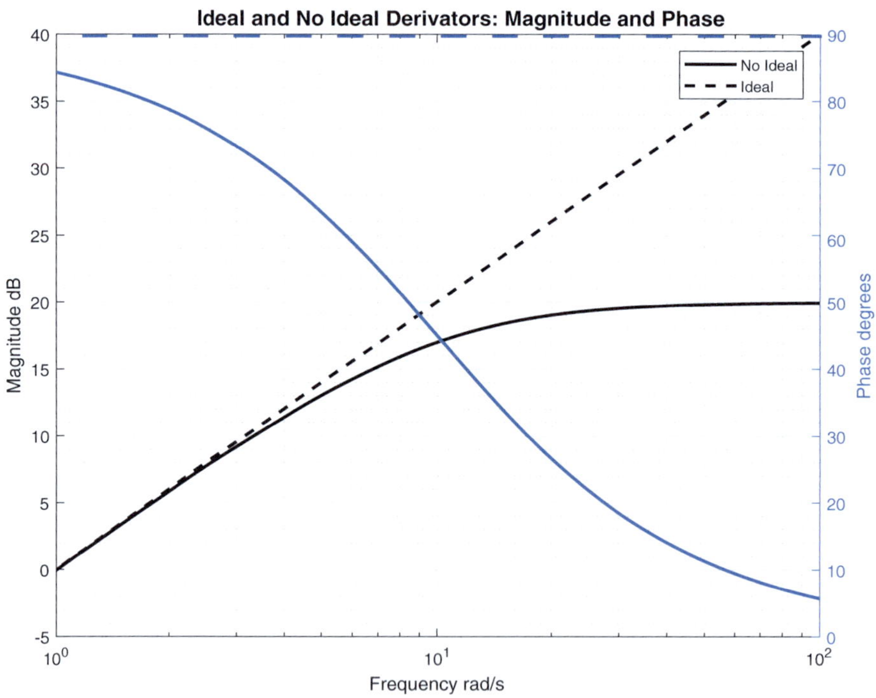

Fig. 2.34 Bode diagram of ideal non-ideal analog derivator

Table 2.3 Discretization methods to implement a digital differentiator with the z-transform and discrete equations

Discretization method	z-transform	Difference equation
Backward Euler	$H_{BR}(z) = \frac{1-z^{-1}}{T_s}$	$y[n] = \frac{1}{T_s} \cdot (x[n] - x[n-1])$
Forward Euler	$H_{FR}(z) = \frac{z-1}{T_s}$	$y[n] = \frac{1}{T_s} \cdot (x[n+1] - x[n])$
Central difference	$H_{CD}(z) = \frac{1-z^{-2}}{2T_s}$	$y[n] = \frac{1}{2T_s} \cdot (x[n] - x[n-2])$

derivative directly provides the value of the slope, that is, the speed. If the signal contains high-frequency noise, the noise will be amplified so that it will affect the accuracy of the obtained speed. As previously mentioned, if the derivator is based on a filter, the position angle noise will be attenuated so that a better speed accuracy is achieved. In Fig. 2.34 is represented the Bode diagram of no-ideal and ideal analog derivator.

The discretization of an analog derivative can be done with the following approximation methods (Table 2.3): rectangular approach backward (backward Euler), forward Euler, and approximation of central difference.

The response of an ideal, discrete backward, and central derivators is shown in Fig. 2.35. As before, the frequency and magnitude are linear where the frequency is normalized.

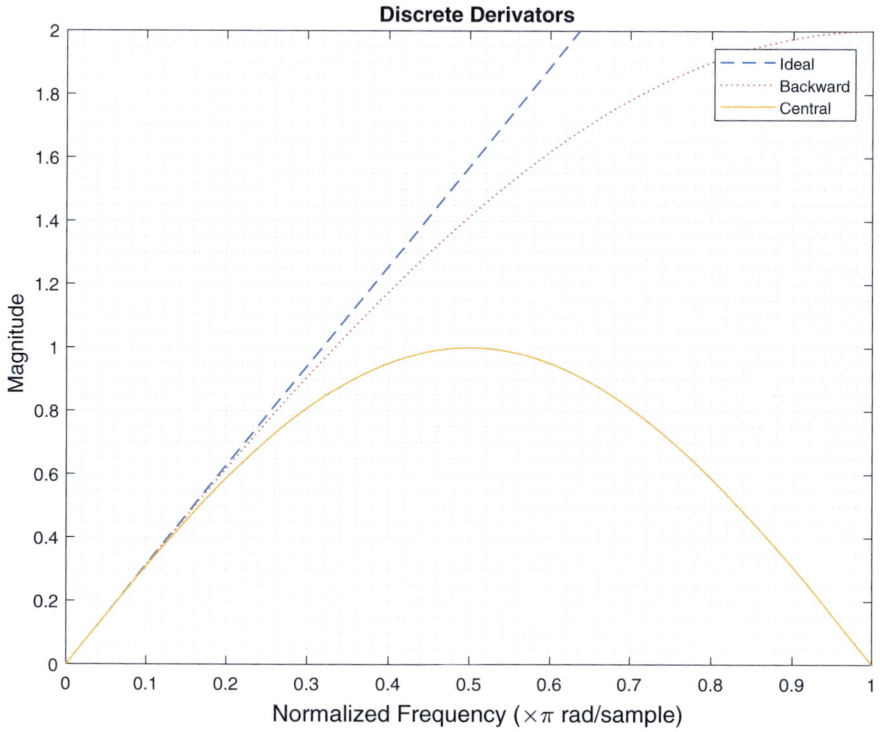

Fig. 2.35 Frequency response of ideal analog derivator with backward and central discrete methods

2.5 Digital PID Implementation

2.5.1 Discrete PI

In this section, the PI controller will be discretized instead of the PID because it is the most used in the control of electrical machines. In the case of the discretization of a PID controller, the forward approximation cannot be used because the controller will not be causal; that is, the future value is needed for the calculation of the derivative. In the PI controller, the derivative action does not exist so that any of the previously seen methods can be used for the discretization of the integrative action.

2.5.1.1 Trapezoidal Discrete PI Controller

As mentioned, most of the control systems with a PI controller may be sufficient, so that the discretization of the PI controller according to the Tustin or trapezoidal method for the integral part will be obtained. For this, the bilinear transformation of (2.31) is applied; that is, the expressions in the s plane can be transformed into the z plane. The bilinear transformation is a first-order approximation of the natural logarithmic function.

$$s \approx \frac{2}{T_s}\left(\frac{z-1}{z+1}\right) \tag{2.31}$$

Starting from the equation of the continuous PI according to (2.16), if the bilinear transformation is applied, it is easy to arrive at the expression in the z plane of (2.32).

$$G_c(z) = K_P + \frac{K_I \cdot T_s}{2}\left(\frac{z+1}{z-1}\right) = \frac{(K_P + \frac{K_I T_s}{2})z + \frac{K_I T_s}{2} - K_P}{z-1} \tag{2.32}$$

If Eq. (2.32) multiplies the numerator and denominates by z^{-1}, Eq. (2.33) is obtained:

$$G_c(z) = \frac{U(z)}{E(z)} = \frac{(K_P + \frac{K_I T_s}{2}) + (\frac{K_I T_s}{2} - K_P)z^{-1}}{1 - z^{-1}} \tag{2.33}$$

Then, the discrete difference equation according to the Tustin method can be written as in (2.34):

$$U[n] = U[n-1] + \left(K_P + \frac{K_I T_s}{2}\right)E[n] + \left(\frac{K_I T_s}{2} - K_P\right)E[n-1] \tag{2.34}$$

2.5.1.2 Backward and Forward Discrete PI Controller

If the integral part is calculated using the differential equation, a more straightforward expression can be obtained as good as the previous one. The integrative part can be obtained by differentiating both sides applying the backward approach method such as:

$$y(t) = K_I \int x(t)\mathrm{d}t \rightarrow \frac{\mathrm{d}Y}{\mathrm{d}t} = K_I x(t) \tag{2.35}$$

The derivative term is approached with the backward approximation method:

$$\frac{Y[k] - Y[k-1]}{T_s} = K_I X[k] \tag{2.36}$$

Then, the recursive Eq. (2.37) for the integrative action is obtained:

$$Y[k] = Y[k-1] + K_I T_s \cdot X[k] \tag{2.37}$$

In the PI controller, the input is $E[k]$, and the output is $U[k]$. The output of the controller is formed by the proportional action and the integrative action, which is calculated as:

$$I[k] = I[k-1] + K_I T_s \cdot E[k] \tag{2.38}$$

where $I[k-1]$ is the accumulation of the integral part which must be memorized according to (2.38). The output of the controller will then be:

$$U[k] = K_P \cdot E[k] + I[k] \tag{2.39}$$

If the derivative term is approached with the forward approximation method:

$$\frac{Y[k+1] - Y[k]}{T_s} = K_I X[k] \tag{2.40}$$

Then, the recursive Eq. (2.41) for the integrative action is obtained:

$$Y[k+1] = Y[k] + K_I T_s \cdot X[k] \tag{2.41}$$

The future integrative action is calculated with the current value of the memorized integrative action plus the new value that depends on the error:

$$I[k+1] = I[k] + K_I T_s \cdot E[k] \tag{2.42}$$

where $I[k]$ is the accumulation of the integral part which must be memorized. The output of the PI controller remains as the proportional action plus the future integrative action. In this case, the order of execution would first be to execute (2.42) and then the equation of integrative action (2.43).

$$U[k] = K_P \cdot E[k] + I[k+1] \tag{2.43}$$

In Fig. 2.36 is represented the Bode diagram of a PI controller for ideal analog, real analog, real discrete forward, and real discrete Tustin methods. It is possible to observe in the magnitude that real analog PI controller matches both discrete implementation methods at low frequency, but when the frequency is approaching the one half of the sampling frequency, as Nyquist said, the systems no longer match. In case of phase, at frequency increases phase lag due to sampling delay starts to appear.

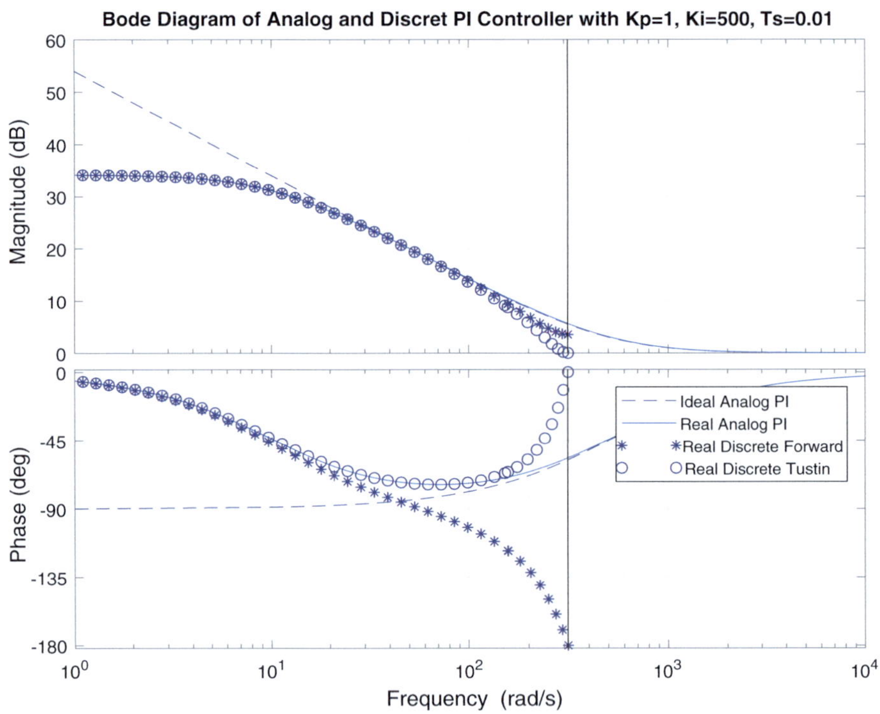

Fig. 2.36 Bode diagram of analog and different discrete PI controllers as forward and Tustin methods

2.5.2 *Digital PI Implementation*

The Simulink® simulation environment of MATLAB® has a MATLAB function block, which is possible to write a MATLAB® function for use in a Simulink® model. It is possible to configure different inputs and outputs, as well as the interaction between them. It allows performing custom functions, controlling its implementation at all times. On the other hand, MATLAB® also has Stateflow® that works with the Simulink® software. Stateflow® is an environment for modeling and simulating combinatorial and sequential decision logic based on state machines and flowcharts. It includes an animation of state machines and static and runtime checks to test the consistency and integrity of the design before implementation. It is important to mention that both the MATLAB® function of Simulink® and Stateflow® allow generating code for its possible implementation in a microcontroller or DSP with the Simulink Coder™.

This section will show the implementation of a custom digital PI controller in the Simulink® simulation environment by using MATLAB® function and Stateflow® that will be used later in the following chapters.

2.5.2.1 MATLAB Function Implementation

In Fig. 2.37 is shown the Simulink® and MATLAB function of the custom discrete PI controller. The MATLAB function implements the algorithm shown below in MATLAB® code with the backward derivation method. The inputs are the set point or reference signal, the max, and min saturation values, the feedback signal, the PI gain parameters, the previous integral value, and the sampling time T_s. On the other hand, the outputs are the PI result and the integral part value.

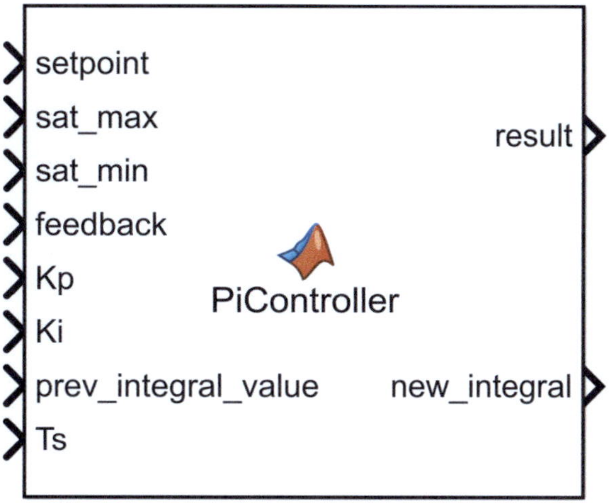

Fig. 2.37 MATLAB function block of PI controller to be used in Simulink®

The MATLAB® function, being discrete, must be executed periodically every sampling time T_s. The anti-windup method of saturation of the integral part has been added, as can be seen in lines 7–16. If there is a saturation of the integrative action, the output of the PI controller will be directly the saturation value *sat_max* or *sat_min*. However, if the integrative action is not saturated, the proportional action will be added to it. Then, it will be rechecked if the sum is above the saturation value or not, to proceed with its limitation if necessary. It is important to mention that the new integral action must be memorized in each execution of the PI controller as seen in line 33 so that in the next input cycle it can calculate the new integrative action from the previous value. If at any time, the integrative action must be reset or initialized to a specific value, the algorithm allows it, and it will be enough to modify the variable *new_integral*. This is very useful when the electric machine goes to the stopped state and is restarted again so that it always starts from the same initial conditions. The start-up response time can be improved with the adjustment of the initialization of the integrative action at instant time $k-1$. So the first execution step of the PI controller at instant time k, the output can react more rapidly. It is necessary to pay special attention to the possible overshoot that can generate an incorrect initialization of the integrative action.

```
%% PI Backward Method

function [result, new__integral] = PiController(setpoint, sat_max, ...
sat_min, feedback, Kp, Ki, prev_integral_value, Ts)

% Local variables
clamping = false;
integral_value_buffer = single(0);


1 prev_integral_value= new_integral
2 error = setpoint - feedback;
3
4 % New integral term calculation
5 integral_value_buffer = prev_integral_value + error*Ki*Ts;
7 % Integral clamp (no update if clamping)
8 if integral_value_buffer > sat_max
9    result = sat_max;
10   clamping = true;
10 elseif integral_value_buffer < sat_min
11    result = sat_min;
12    clamping = true;
13 else
14    clamping = false;
15    result = 0;
16 end
```

```

% only continue if there is no clamping
if clamping==false
    % Sum of proportional term
    result = integral_value_buffer+error*Kp;
    % final clamp and saturation control
    if result > sat_max
        result = sat_max;
    elseif result < sat_min
        result = sat_min;
        error = -error;
    else
        error = 0;
    end
    % Integral update (only if output is not saturated
    % or if output and error are opposite in sign
    if error <= 0
        new_integral = integral_value_buffer;
    end
end
end
```

2.5.2.2 Stateflow® Implementation

The implementation of a custom PI controller can also be implemented using Stateflow®, as previously mentioned. Although the PI controller is not the best example to see the potentiality of Stateflow®, here it will be seen as an introduction since in later chapters, it will be used deeply.

In Fig. 2.38a, the PI Simulink® block created in Stateflow® is illustrated, while in Fig. 2.38b is represented the content of the block. The content of the Simulink® block corresponds to Stateflow® block, which is indicated by the symbol located in the center of it. In both, the most common controller inputs and outputs such as reference, feedback, and output can be seen. Also, additional inputs can be observed such as the initialization of the integrative action, controller enabling, reset of the integrative action, parameter table of K_P, K_I, max, and min, index of the table that points to the desired parameters, command to change parameters, and finally a "Trigger" trigger input. This last input corresponds to a trigger signal which sets how often the controller must be executed, that is, the sampling period T_s. As additional outputs, the error and the status of the PI controller are seen.

The parameter table is shown below in MATLAB® code, where each index that varies between 0 and 5 causes certain parameters to be loaded.

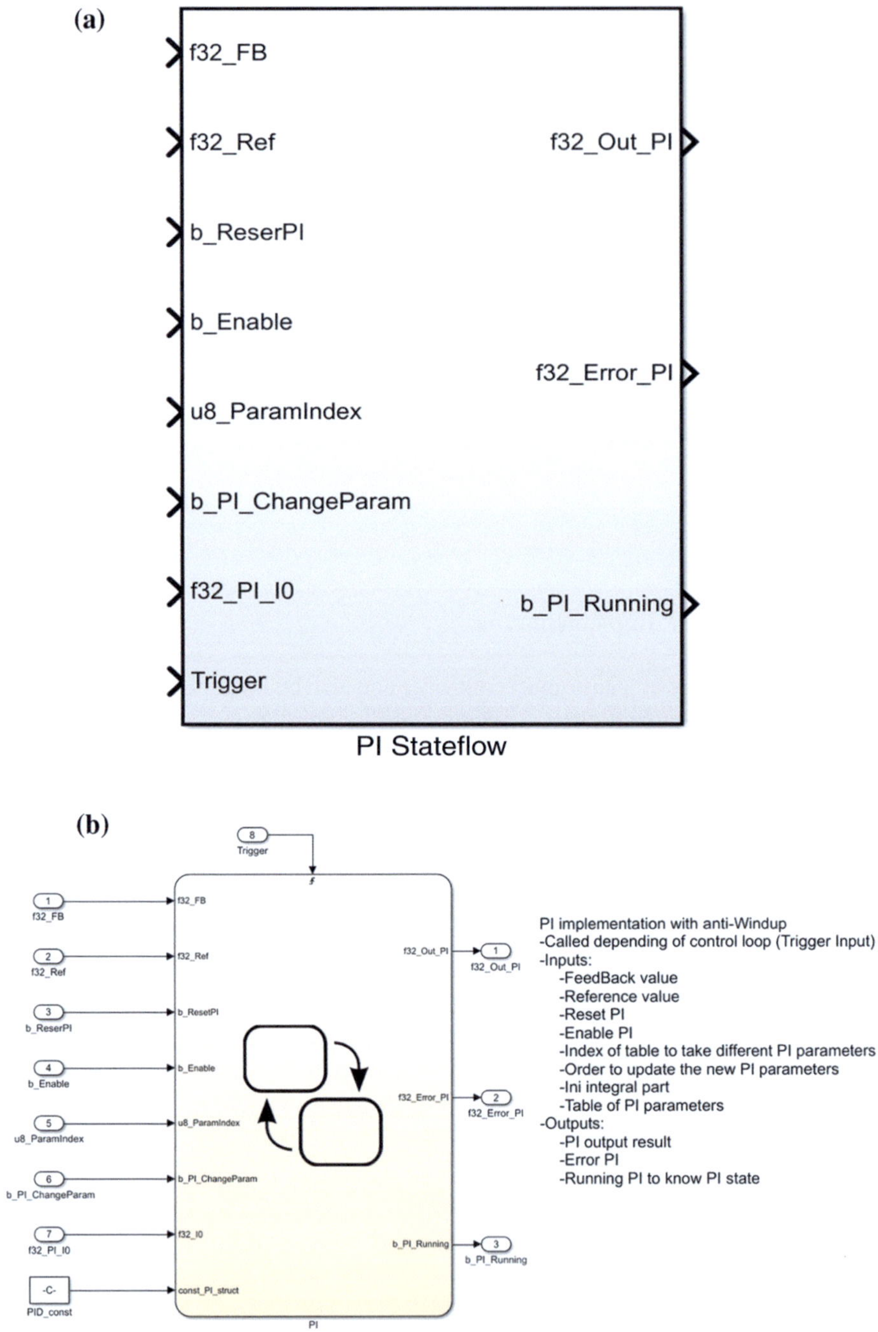

Fig. 2.38 Stateflow® block diagram for a custom PI controller

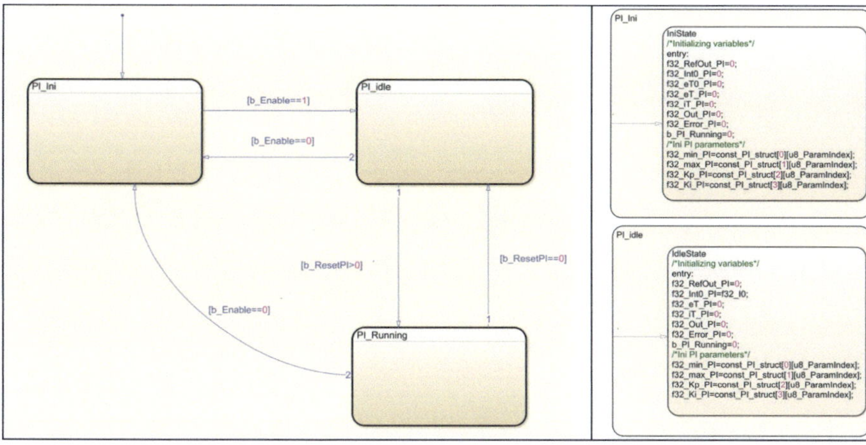

Fig. 2.39 PI controller states implemented in Stateflow®

```
f32_PI_TableParam = [-100,  -200,  0,    0,    0,    0;     %Min
                      100,   200,  40,   40,   40,   40;    %Max
                      2,     10,   12,   13,   14,   20;    %Kp
                      0.01,  0.2,  0.3,  0.4,  0.5,  0.6];  %Ki
```

Inside the Stateflow® block, three sub-blocks or states called PI_Ini, PI_idle, and PI_Runnig are shown, as can be seen in Fig. 2.39. The first corresponds to an initialization state. It will always remain in this state if the PI is not enabled. In this state, the values of the controller parameters prefixed by the aforementioned index are loaded. If the PI controller is enabled, the state is changed to *PI_idle* state, which is where the initialization value of the integrating action is loaded. If in this state the variable *Reset* is true, it will change to the state of *PI_Running*. The *PI_Running* state is when the PI controller starts to operate. If in this state the input variable *Reset* becomes false or the controller is disabled, the status will change to *PI_idle* or *PI_Ini*, respectively.

In Fig. 2.39, the content of the *PI_Ini* and *PI_idle* states can be observed. A part of the *PI_Running* state structure can be seen in Fig. 2.40. The implementation in this state is more complex since this is where the PI controller is implemented. Basically, connective junctions are shown here where the transitions between these junctions are labeled with conditions marked inside the brackets "[]". If any of the conditions are true, it will pass to the next union and execute (if it exists) the code that is closed between the braces { }. On the contrary, if any condition is true, no transition occurs. The conditions are evaluated with the priority marked by the number taken in the union, where 1 means the highest priority or the one that is evaluated first. If it is true, it passes to the next union, and condition 2 is not evaluated.

It is possible to observe that the inputs, outputs, and variables take different prefixes that designate the type of variable as shown in Table 2.4, "b", "f32", "u8".

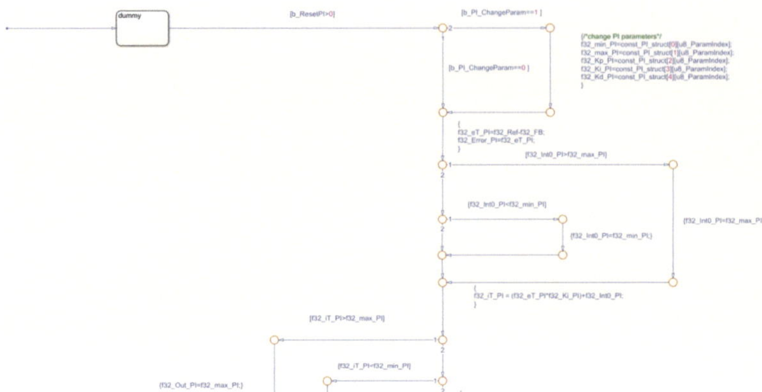

Fig. 2.40 PI_Running state

Table 2.4 Variable types with MATLAB® interpretation

Prefix	Type	MATLAB® interpretation
b	Boolean	Boolean
u8	Unsigned char	8-bit unsigned integer
s8	Signed char	8-bit signed integer
u16	Unsigned integer	16-bit unsigned integer
s16	Signed integer	16-bit signed integer
f32	Float 32 bits	Single precision

Stateflow® allows inserting MATLAB® functions in a straightforward way. The previously explained MATLAB® function of the PI controller can be integrated into Stateflow® as shown in Fig. 2.41. The function is called as shown in the figure, that is, in the middle of a transition. Each time it passes through this transition, the PI function will be called, so that this transition, for the correct functioning of the digital PI, should be called periodically every T_s.

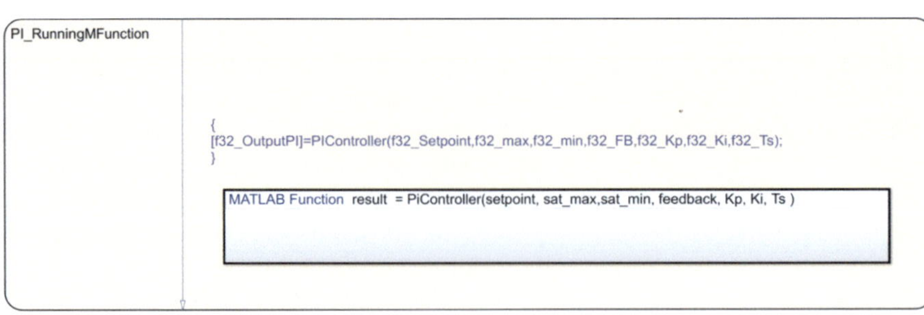

Fig. 2.41 M-function inserted in Stateflow®

2.6 Fuzzy Logic as Controllers

2.6.1 Overview

In the previous section, it was seen that the classic PID controller presents a good response and zero error in steady-state being sufficient for the control of most electrical machines. However, for applications that require more control dynamics with almost zero overshoot, it is necessary to apply a non-classical control such as a controller based on fuzzy logic. It is expected that the controller based on fuzzy logic is non-linear to improve the performance that a linear controller cannot satisfy. Fuzzy logic has quickly become one of the most successful technologies today to develop control systems. With their help, very complex requirements can be implemented in a simple, cheap, and easy to maintain control. The fuzzy logic controller makes the decisions regarding the adjustment of the process by directly interconnecting continuous analog variables to discrete systems.

Fuzzy logic was developed for applications of complex non-linear control systems where a basic controller would not be able to control it adequately and where it is not necessary to know the plant accurately but the operator's experience. This is not the case for the control of electrical machines, where the system is not complex, and a basic controller could control it without difficulties, but given its non-linear characteristic makes it possible to improve the characteristic of a PI controller for example. As it will be seen later, the combination that allows implementing a better control dynamics with almost zero overshoot and zero error in a steady-state is that of a fuzzy controller plus a PI.

Fuzzy logic does not need that the control rules are not known with precision, but they can be satisfied to a certain extent. This property of fuzzy logic is what characterizes approximate reasoning. This means the rule cannot be controlled completely. For example, to answer some questions in certain surveys, the answers could be "Not Satisfied" or "Fairly Satisfied," which are confusing or ambiguous answers.

Fuzzy inference systems are algorithms based on fuzzy logic and reasoning approximation. Diffuse inference systems are basically composed of two elements, the inference mechanism and the knowledge base. However, depending on the application, it may be necessary to add additional elements such as fuzzifier and defuzzifier to connect the fuzzy inference system to the real world. Figure 2.42 shows the block diagram of the control of an electric machine based on a fuzzy controller where the interaction of the elements described above can be appreciated.

The control rule is the core of the fuzzy inference system and is related to the human experience. Lotfali Askar Zadeh (Zadeh 1973) is considered the founder of fuzzy mathematics, fuzzy set theory, and fuzzy logic in 1965 where he published "fuzzy sets, information, and control article." The article mentions the idea of creating a new type of mathematics that allows expressing human knowledge. It was not recognized until 1974 when Seto Assilian and E. H. Mamdani established the first milestone in the industrial application of fuzzy control (Mamdani and Assilian 1975).

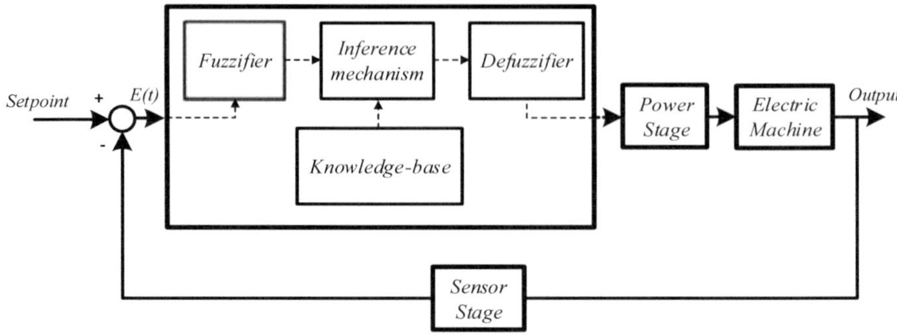

Fig. 2.42 Fuzzy control applied in an electric machine control

By means of a system based on linguistic rules, it was possible to control a steam generator according to the experience of the plant operator. The steam generator was not possible to control by conventional techniques. They demonstrated the power of fuzzy theory in the actual control of the machine. Since then, the control scheme mostly in use is the Mamdani–Assilian model or more frequently like the Mamdani model.

2.6.2 *Fuzzy Logic System*

Recently, fuzzy logic control (FLC) has more and more applications due to robust, adaptive performance, for linear and non-linear systems with a parameter variation. It has the ability to control non-linear systems, even systems with mathematical models that cannot be obtained due to their complexity, such as nuclear reactors. These systems cannot be analyzed with conventional control methods, but it is possible to take advantage of the good experience of the plant operator to implement a control system based on fuzzy logic (FL). By using FL, it is not necessary to know the plant (mathematical model of the plant) but only with the experience of the operator of the plant is enought to establish the rules. In other words, the fuzzy controller automates the way a human expert performs the task of controlling the system. The general reasoning methodology in FL uses the rules "IF … THEN …". Diffuse operators, AND and OR, are not used in the YES (antecedent) that is part of the rule. On the contrary, THEN is the consequent according to the evaluation of the antecedent YES, which is the implication from the antecedent to the consequent. Fuzzy inference can be defined as the process of obtaining an output value for an input value using fuzzy set theory. There are different types of diffuse inference in the literature, such as the Mamdani and TSK types (Takagi, Sugeno, and Kang) (Takagi and Sugeno 1985). The basic difference between these two rules is in their implication part or THEN, where Mamdani's style has a fuzzy style, but TSK has a linear equation. In general, Mamdani's method is used more widely because it appeared earlier and because the expert knowledge is well represented. It allows us to describe expert knowledge in an intuitive way. The main drawback is its high computational cost, which is why the TSK inference method is used more frequently for control applications and optimization problems.

In general, the fuzzy system consists of four functional blocks: fuzzifier, defuzzifier, knowledge base, and inference engine. Next, each part of the components is briefly explained.

2.6.2.1 Fuzzifier

The fuzzifier takes input signals by applying scaling factors for proper fuzzification. Fuzzification takes clear numerical values to transform to diffuse quantities or linguistic variables. This transformation is done through the use of member or membership functions. A member function is a mathematical function that defines how the values of a fuzzy variable in a given region are assigned to a membership value μ (or degree of membership) between 0 and 1. There are many mathematical functions that can represent a member function as triangular, trapezoidal, polynomial, Gaussian, sigmoid functions, etc., but the triangular member function, shown in Fig. 2.43, should be sufficient to solve most problems. The user initially determines the number of member functions and the form.

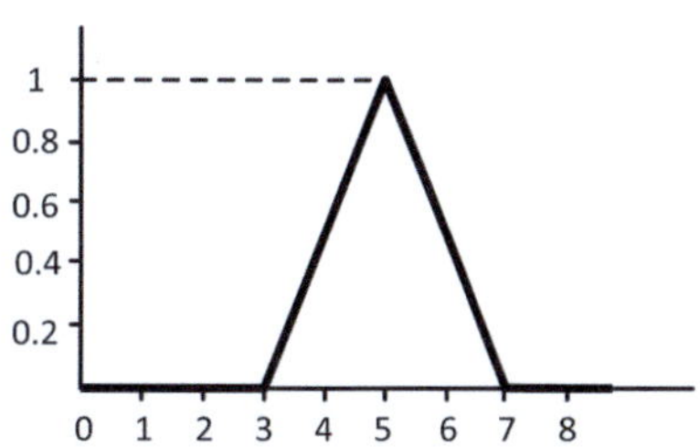

$$\mu_{triangle}(x, a, b, c) = \begin{cases} 0 & x \le a \\ \dfrac{x-a}{b-a} & a \le x \le b \\ \dfrac{c-x}{c-b} & b \le x \le c \\ 0 & c \le x \end{cases}$$

Fig. 2.43 Triangular membership for $a = 3$, $b = 5$, and $c = 7$

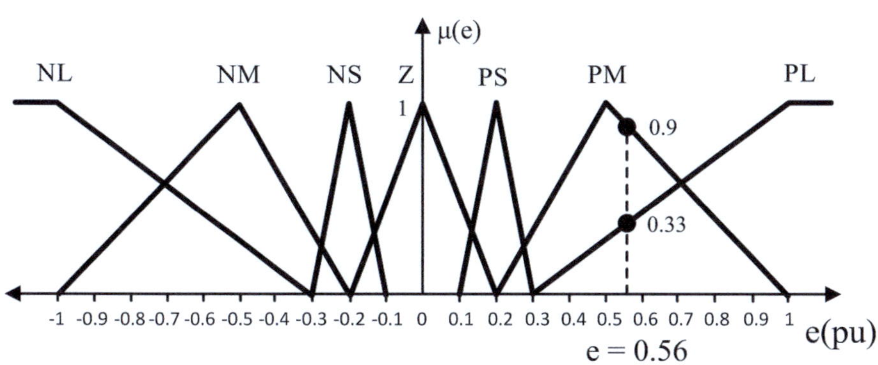

Fig. 2.44 Triangular membership with different linguistic variables: Z → zero, PS → positive small, PM → positive medium, PL → positive large, NS → negative small, NM → negative medium, NL → negative large

As an example of fuzzification, Fig. 2.44 is considered, where it has the following linguistic variables: Z (zero), PS (small positive), PM (medium positive), PL (large positive), NS (small negative), NM (medium negative), NL (big negative). When the input variable has been scaled $e(t)$, which is the error, if it takes a value of 0.56, the membership value $\mu(0.56)$ is equal to 0.9 for PM and 0.33 for PL. It means that there is a 90% certainty of PM and 33% of PL. Conversely, for $e(t) = 0$, $\mu(0) = 1.0$, which is an absolute certainty of 100% that $e(t) = 0$ is what it means by Z (zero).

2.6.2.2 Knowledge Base

The knowledge base contains rules of linguistic control if-then. As discussed earlier, the rules can be established using the experience and knowledge of an expert control system. The previous inputs, that is, the output of the fuzzifier, are applied to the background of the fuzzy rules. If a rule has multiple antecedents, the AND or OR operators are used to obtain a single number that represents the result of the evaluation. This number applies to the consequent.

2.6.2.3 Inference Mechanism

The inference mechanism performs the inference operation of the rules and has the ability to simulate human decisions based on fuzzy concepts. It evaluates, after the fuzzification, the set of if-then rules (given in the knowledge base), and the result is also a linguistic value where it will be transformed into a real output by defuzzifier.

2.6.2.4 Defuzzifier

Defuzzifier performs appropriate scale mapping and defuzzification. There are different methods for defuzzification. Perhaps the most used and known, although with a certain degree of complexity of calculation, is the center of gravity or also called center of area. This takes the weighted average of the contributions of the control action of each of the different fuzzy rules according to (2.44).

$$COA = \frac{\sum_{i=1}^{n} x_i \cdot \mu(x_i)}{\sum_{i=1}^{n} \mu(x_i)} \tag{2.44}$$

2.6.3 *Fuzzy Logic Control*

As mentioned above, the mathematical knowledge of the plant is not strictly necessary to perform control based on fuzzy logic. It is not the case of the control of electrical machines since the mathematical model of the system can be defined appropriately but for specific parameters. If these parameters vary due to, for example, the temperature of the electric machine or cannot be defined accurately, the conventional control will not adapt to these variations, losing accuracy in the control. To solve this problem, adaptive control is a very common solution, and FLC has the basic characteristic of being an adaptive and non-linear control that provides robustness before variations or inaccuracies of plant parameters.

Using the fuzzy logic and the approximate reasoning theory introduced by Zadeh, it is possible to create a controller based on this logic. The traditional form of the control laws with self-regulation (feedback) seen above is based on the fact that the output $u(k)$ is a function of the current error input $e(k)$ and previous $e(k-1)$ as well as the output previous $u(k-1)$:

$$u(k) = F(e(k), e(k-1), \ldots, e(k-v), u(k-1), u(k-2), \ldots, u(k-v))$$

The fuzzy logic controller uses control laws consisting of logical rules if-then in conjunction with fuzzy membership functions to control a process and minimize the error. Sets and fuzzy operators are the subjects and predicates of fuzzy logic. The if-then logical rules are used to formulate conditional expressions that use fuzzy logic.

The FLC uses the error $e(k)$ and the error change $\Delta e(k)$ to produce changes in the controller output function that can be $\mu(k)$ or $\Delta\mu(k)$:

$$\begin{aligned} e(k) &= x(k) - y(k) \\ \Delta e(k) &= e(k) - e(k-1) \\ \mu(k) &= F(e(k), \Delta e(k)) \\ \Delta\mu(k) &= F(e(k), \Delta e(k)) \end{aligned}$$

e(k) is defined as the reference minus the output:

$$\begin{aligned} &\text{If } e(k) > 0 \text{ then } x(k) > y(k) \\ &\text{If } \Delta e(k) > 0 \text{ then } e(k) > e(k-1) \end{aligned}$$

In Fig. 2.45, the block diagram of the FLC with all its elements is illustrated. The inputs of the FLC are the error e and the derivative of the error Δe or Ce (error change), modulated by the constants K_0 and K_1, respectively. The output is the derivative of the manipulated variable that, after modulation with a K_u constant, is integrated before acting. Inside of the fuzzy logic system block are the sub-blocks discussed above as fuzzifier, defuzzifier, knowledge base, and inference engine.

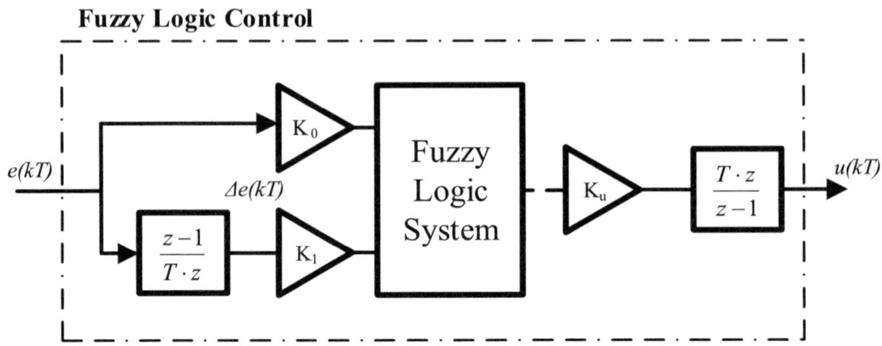

Fig. 2.45 Fuzzy logic control structure composed with the fuzzy logic system, fine-tuning gains, digital derivator, and integrator

2.6.3.1 Electric Machine Speed Control Application

The fuzzy logic controller has the advantage that the algorithm inputs are simply the error and its derivative. Its setting is based on a procedure like PID controllers. In particular, the constants of a PI (K_P, T_I) must be fine-tuned and, through them and with the help of (2.45) and (2.46), the K_u and K_1 are obtained (choosing a certain K_0).

$$T_I = \frac{K_1}{K_0} \tag{2.45}$$

$$K_P = K_u K_1 \tag{2.46}$$

The constants obtained are an approximation, but they must be fine-tuned to optimize the control according to the desired requirements. In addition, these constants scale the entries to operate within the per unit range (pu).

In the case of speed control of an electric machine, as in the case of the PID controller, the input will be the speed error. The error, the change of error, and change in output membership functions for fuzzy speed control are represented in Fig. 2.46. The linguistic variables that could be defined for the speed control of an electric machine based on the knowledge of the system to be controlled are the following:

Z	Zero
PS	Positive small
PM	Positive medium
PL	Positive large
NS	Negative small
NM	Negative medium
NL	Negative large

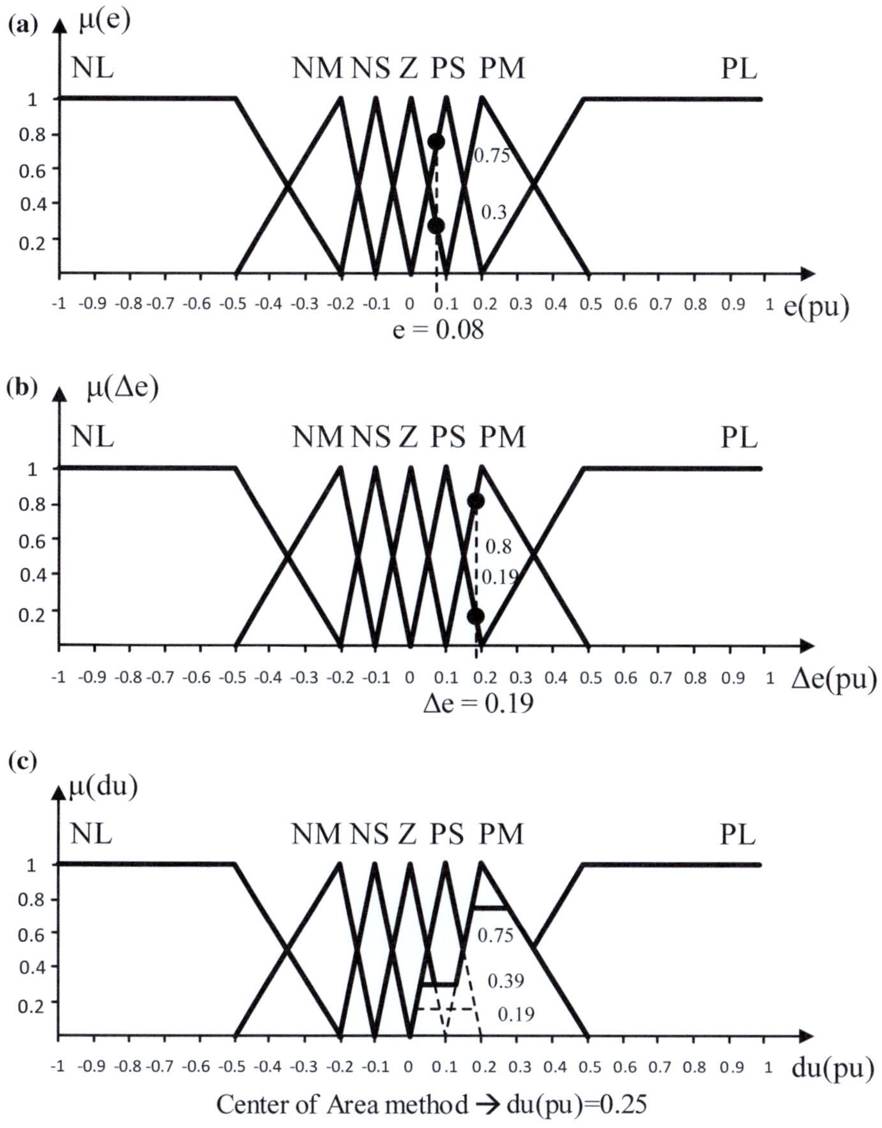

Fig. 2.46 Membership functions for fuzzy speed control. **a** Error, **b** change of error, **c** change in output control

The error in the steady-state control will be close to the origin, that is, the zero linguistic variables, Z. This is where the precision should be better so that more linguistic variables can be added to achieve it, such as PVS (positive very small) and NVS (negative very small). Another factor that helps the precision of the control in steady-state is the asymmetry of member functions as can be seen in Fig. 2.44. Based on the previous linguistic variables and the knowledge of the system to be controlled, a table of the rules for the fuzzy speed controller can be made, where the two inputs are reflected, error and error change, and the output. The interpretation, according to the Mamdani method of Table 2.5 for any of the possible combinations, is as follows:

IF error = Z AND change of error = Z THEN output du = Z
IF error = PL AND change of error = PM THEN output du = PL.

According to the rules established in Table 2.5 based on the knowledge of the speed control rules of an electric machine, there will only be two overlapping member functions. There will be no more than four rules at the same time. This simplifies the calculations in each sampling time of the algorithm since once the intervening member functions are undefined, it is possible to apply the four correct rules.

The basic algorithm for fuzzy speed control is shown below, which must be executed every sampling time T_s (Bose 2002).

1. Sample the reference speed $\omega_r^*(kT)$ and current speed $\omega_r(kT)$.
2. Calculate $E(kT)$ and CE(kT).

$$E(kT) = \omega_r^*(kT) - \omega_r(kT)$$

$$\Delta E(kT) = E(kT) - E(k-1)T$$

   where $k = 0, 1, 2, \ldots$.
3. Apply fine-tuning gain factors to error and change of error.

$$e(kT) = E(kT) \cdot K_0$$

Table 2.5 Rule table for speed control based on the knowledge of the system

		Error						
		NL	**NM**	**NS**	**Z**	**PS**	**PM**	**PL**
Change of error	**NL**	NL	NL	NL	NM	NS	NS	Z
	NM	NL	NL	NM	NS	NS	Z	PS
	NS	NL	NM	NS	NS	Z	PS	PS
	Z	NM	NS	NS	Z	PS	PS	PM
	PS	NS	NS	Z	PS	PS	PM	PL
	PM	NS	Z	PS	PS	PM	PL	PL
	PL	Z	PS	PS	PM	PL	PL	PL

$$\Delta e(kT) = \Delta E(kT) \cdot K_1$$

4. Identify in the table the position for inputs $e(kT)$ and $\Delta e(kT)$. For the present example, the $e(kT) = 0.08$ and $\Delta e(kT) = 0.19$. Then, the inputs belong to fuzzy sets PS and PM.
5. For given inputs $e(kT)$ and ce(kT), calculate the degree of membership according to previous identification, $\mu_Z(e) = 0.3$, $\mu_{PS}(\Delta e) = 0.19$, $\mu_{PS}(e) = 0.75$, $\mu_{PM}(\Delta e) = 0.8$.
6. Identify the four valid rules from the previous table and apply, for example, AND or min operator for each rule combination.

   Rule 1: IF error = Z AND change of error = PS THEN output du = PS

   $$\min\{\mu_Z(e) = 0.3, \mu_{PS}(\Delta e) = 0.19\} = 0.19$$

   Rule 2: IF error = Z AND change of error = PM THEN output du = PS

   $$\min\{\mu_Z(e) = 0.3, \mu_{PM}(\Delta e) = 0.8\} = 0.3$$

   Rule 3: IF error = PS AND change of error = PS THEN output du = PS

   $$\min\{\mu_{PS}(e) = 0.75, \mu_{PS}(\Delta e) = 0.19\} = 0.19$$

   Rule 4: IF error = PS AND change of error = PM THEN output du = PM

   $$\min\{\mu_{PS}(e) = 0.75, \mu_{PM}(\Delta e) = 0.8\} = 0.75$$

7. Apply defuzzification method selected as center of area for example.
8. Apply fine-tuning gain factor to the output.

$$u(kT) = du(kT) \cdot K_u$$

2.6.4 Adaptive Fuzzy PI

Once the adjustment of the proportional gain parameters K_P and the integrating action K_I of the PI controller for the control system has been made, these parameters remain constant. Thanks to the digital implementation of the PI controller, it was seen that a soft start could be made to improve the starting of the electric machine by modifying the K_P and K_I parameters in a particular condition. The adaptive PI controller consists in that its parameters K_P and K_I are not constant, but vary according to some rules. Using fuzzy logic, it is possible to perform an adaptive PI controller where its output is the parameters of the PI controller. These will be adapted according to the error

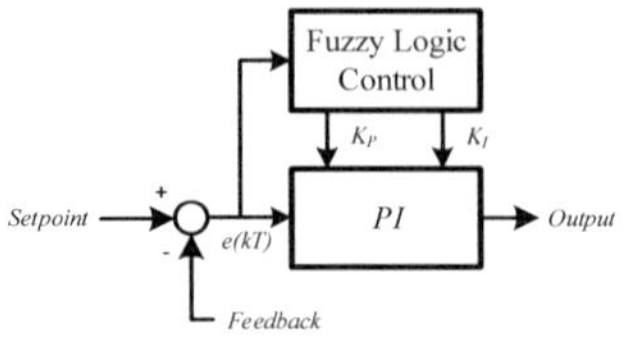

Fig. 2.47 Adaptive fuzzy PI where the output of fuzzy logic control is the PI parameters

and the error change that the fuzzy controller interprets. As in the previous case, there will be a series of rules and membership functions. For example, the fuzzy controller will make the K_P gain larger when the error between the reference and the output is too large to arrive at that reference as soon as possible. As the error decreases, the gain K_P will decrease thanks to the fuzzy controller so as not to have a high overshoot. Figure 2.47 shows the adaptive PI controller based on fuzzy logic.

In this case, the linguistic variables defined for the output are different from those of the speed control seen previously, since the following definition is closer to the case to be studied. For example, for the parameters of the PI controller, it does not make sense to define negative values but a range from very small to a large value. However, the inputs being the speed error, which can be positive or negative the linguistic variable takes the same name.

VS	Very small
SM	Small medium
M	Medium
ML	Medium large
VL	Very large

Since the behavior of the K_P and K_I parameters is not similar, it is preferable to define two outputs for the fuzzy controller with different rule tables. The matrix of fuzzy controller rules for the adaptive PI controller can be seen in Tables 2.6 and 2.7.

Table 2.6 Rule table for K_P parameter base based on the knowledge of the system

		Error				
		NL	**N**	**Z**	**P**	**PL**
Change of error	**NL**	VL	VL	ML	ML	M
	N	VL	ML	ML	M	SM
	Z	VL	ML	M	SM	VS
	P	ML	M	SM	SM	VS
	PL	M	SM	SM	VS	VS

Table 2.7 Rule table for K_I parameter base based on the knowledge of the system

		Error				
		NL	**N**	**Z**	**P**	**PL**
Change of error	**NL**	VS	VS	SM	SM	M
	N	VS	SM	SM	M	ML
	Z	VS	SM	M	ML	VL
	P	SM	M	ML	ML	VL
	PL	M	ML	ML	VL	VL

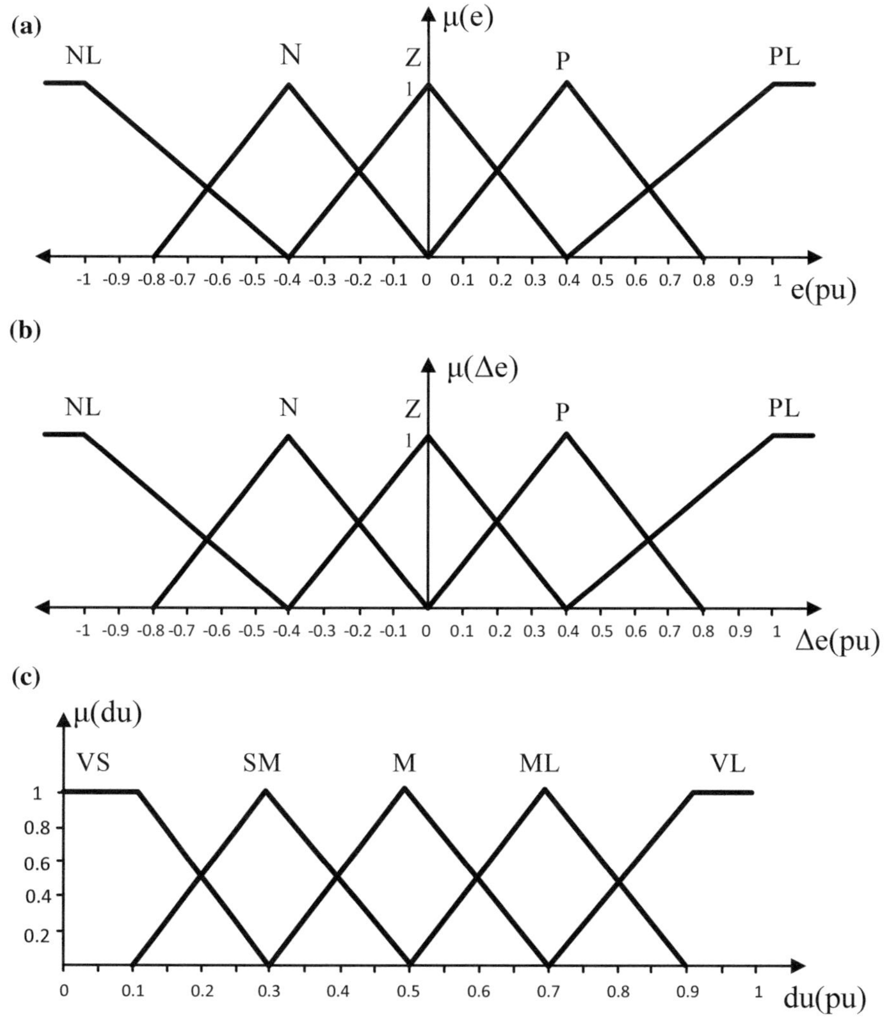

Fig. 2.48 Membership functions for adaptive PI control. **a** Error, **b** change of error, **c** change in output control for K_P and K_I

The membership function for the error input, error change, and output is shown in Fig. 2.48. Unlike the previous case, fewer linguistic variables have been used to speed up the design task. In this case, the precision in the parameters of the PI controller is not so relevant. Then, the algorithm processing by the CPU is less.

2.6.5 *Fuzzy + PI*

With the combination of an FLC followed by a PI controller, a more dynamic response can be obtained without overshoot. The output of the FLC attacks the input of the PI controller so that the FLC acts as an amplifier or error attenuator with an adaptive speed depending on the error between the reference and the output of the plant. For example, for a step-type reference where the output of the plant to be

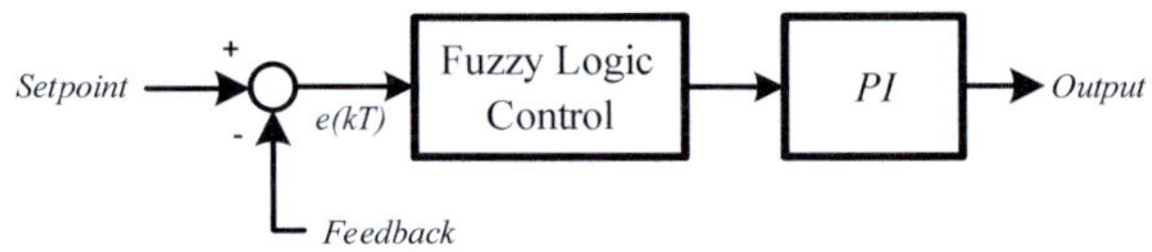

Fig. 2.49 Fuzzy logic control plus PI

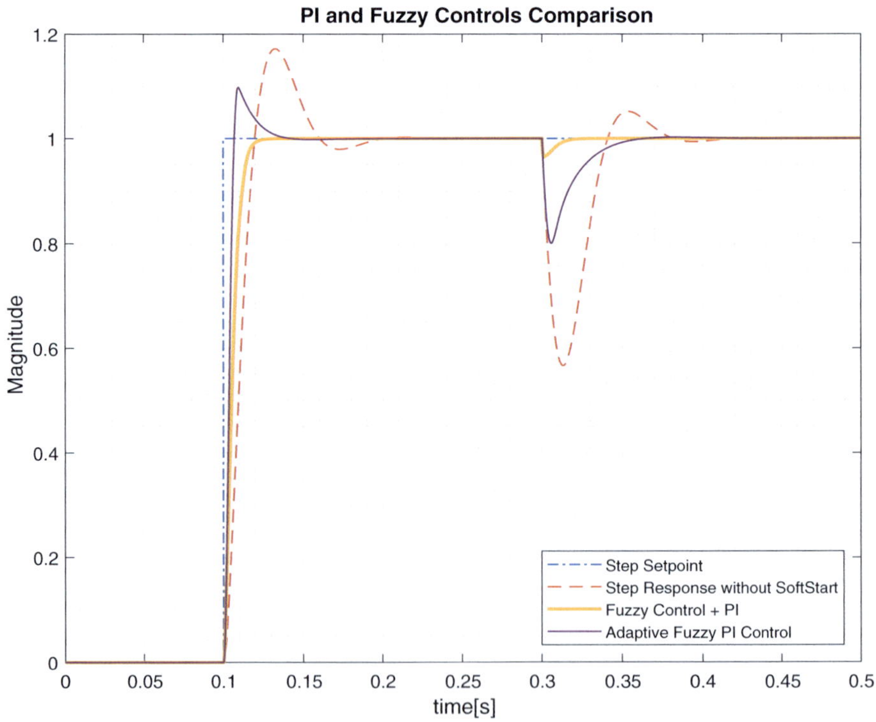

Fig. 2.50 Step response comparative for different control methods studied in this chapter as PI control without soft start, fuzzy control plus PI, and adaptive fuzzy PI

controlled is zero, the error is maximum so that the FLC will act by giving its output a large value at the input of the PI. Then, the PI will immediately begin to act to get the output of the plant to reach the reference. As the output approaches the reference, the FLC begins to decrease its output adequately to avoid having overshoots. The member functions and the rules for the FLC + PI controller could be the same as for the speed control seen above. The block diagram of the control system based on fuzzy logic plus PI controller can be seen in Fig. 2.49.

Figure 2.50 shows the different responses of the PI, PI controllers with soft start, and FLC + PI for the same plant. For this plant, a response without overshoot and with a very fast control dynamic before the step and the perturbation at $t = 0.3$ is through the fuzzy controller plus a PI controller (FLC + PI).

References

Åström KJ, Hägglund T (1988) Automatic tuning of PID controllers. Instrument Society of America

Åström KJ, Hägglund T (1995) PID controllers: theory, design and tuning. Instrument Society of America

Åström KJ, Hägglund T (2005) Advanced PID control. ISA—The Instrumentation, Systems, and Automation Society, Research Triangle Park, NC

Bose BK (2002) Modern power electronics and AC drives. Prentice Hall

Chen G, Pham TT (2001) Introduction to fuzzy sets, fuzzy logic, and fuzzy control systems. CRC Press

Clevelaizd WB (1976) First-order-hold interpolation. Digital-to-analog converter with application to aircraft simulation. NASA

Erenoglu I, Eksin I, Yesil E, Guzelkaya M (2006) An intelligent hybrid fuzzy PID controller. In: Proceeding of the 20th European, modeling and simulation (ECMS'06). Wolfgang Borutzky, Alessandra Orsoni, Richard Zobel, pp 1–5

Franklin GF, Powell JD, Emami-Naeimi A (1994) Feedback control of dynamic systems, 3rd edn. Addison-Wesley

Gao M, He S (2008) Self-adapting fuzzy-PID control of variable universe in the non-linear system. Int Conf Intell Comput Technol Autom ICICTA 1:473–478

Kuo BC (1995) Automatic control systems, 7th edn. Prentice Hall

Kiencke U, Nielsen L (2000) Automotive control systems: for engine, driveline, and vehicle. Springer, Berlin

Mamdani EH, Assilian S (1975) An experiment in linguistic synthesis with a fuzzy logic controller. Int J Man Mach Stud 7(1):1–13

Mendel JM (1995) Fuzzy logic systems for engineering: a tutorial. Proc IEEE 83:345–377

Ogata K (1996) Discrete time control systems. Prentice Hall

Ogata K (1998) Modern control engineering. Prentice Hall

Piña AJB (2009) Síntesis de sistemas de control borroso estables por diseño. Escuela Politécnica Superior, Universidad de Huelva

Passino KM, USA (1998) Adaptive fuzzy control

Shinskey FG (1996) Process-control systems. Application, design, and tuning, 4th edn. McGraw-Hill, New York

Takagi T, Sugeno M (1985) Fuzzy identification of systems and its applications to modeling and control. IEEE Trans Systems Man Cybern 14(1):116–132

Zadeh LA (1973) Outline of a new approach to the analysis of complex systems and decision processes. IEEE Trans Syst Man Cybern 3(1):28–44

Chapter 3
Three-Phase Electrical Systems

3.1 Introduction

Because grid voltage can be single-phase and three-phase, AC machines, especially induction machines (IM), have been adapted to these grid combinations by including the low- and high-voltage possibilities for three-phase systems. This allows a direct connection of the machine to the grid without using any electric drive; however, the optimal working output of the machine is lost and, as a result, the efficiency will be poor. The maximum speed of the machine will be conditioned to the frequency of the voltage grid: 50 Hz for Europe and 60 Hz for the USA if the poles of the machine do not change. Modern electric drives control AC machines most efficiently, and speed range can vary from 0 to 1500 Hz. For a two-pole machine, this means that the speed of 90,000 RPM could be reached instead of 3000 RPM in the case of a direct connection to the 50 Hz grid. It should be noted that the machine may not be designed to reach 90,000 RPM, as its bearings may undergo extra forces for which it has perhaps not been designed. This is only an example of reaching higher speeds than those permitted by the AC frequency grid.

Three-phase induction machines can be connected to a single-phase grid. These machines have a capacitor installed, which is connected between phases to supply delayed voltage to the remaining phase that is alone. When these machines use the single-phase grid, the torque of the machine will be reduced as compared to when the same machine is connected to the three-phase grid. Most of the water pumps used in the domestic sector are built by three-phase induction machines with a capacitor installed to operate in a single-phase grid. As comented before, the capacitor supply voltage to the third phase allowing also that the machine starts-up in the correct direction of water flow. In a domestic water pump plumbing system, when the tap is opened, the pressure in the pipe falls. This is detected by a pressure control switch which then closes contacts to supply energy to the pump. Subsequently, the pump starts, and the pressure of the pipes starts to increase until the pressure detected reaches a pre-defined pressure, causing the contacts to open. This is an example of the three-phase machine's direct connection to a single-phase grid through a switch

© Springer Nature Switzerland AG 2020

R. Molina Llorente, *Practical Control of Electric Machines*,
Advances in Industrial Control, https://doi.org/10.1007/978-3-030-34758-1_3

controlled by water pressure. The disadvantage of this system is the fact that there are continuous sparks in the contacts when the pump switches OFF to ON and viceversa.

Furthermore, the current spikes during the start-up of the pump when switching ON are large, and the efficiency will be reduced. There is a more expensive solution for this water pump system in which, between the pump and the grid, there is an electric drive which controls and protect the electric machine. By using a water pressure sensor, it can switch the pump ON using a soft start to avoid current spikes where the pump speed is gradually increased. Likewise, sparks are also avoided since this method uses electronics devices as power semiconductors. Although the speed is not controlled, the electric drive improves the efficiency of the machine and can monitor the current to detect several failures, such as a locked rotor and overheating. In addition, the capacitor of the single-phase machine connection is avoided as the electric drive—as will be shown in Chap. 8—can generate three-phase voltages using single-phase grid voltage.

Delta and star connections for supply source and load are the sole possibility in three-phase systems in which they each have their advantages and disadvantages. For example, delta connections for machines are used for low-voltage, three-phase grids like 230 VAC, while star connections are used for high-voltage, three-phase grids like 400 VAC—both for Europe. Since the power of the machine is the same for both connections, the current will be higher in delta connections than in star connections, with a relationship of 400/230, that is to say, $\sqrt{3}$.

In the analysis of a three-phase system for AC machine control, all currents and voltages for each phase are sinusoidal and 120° apart, and phase differences between voltages and currents for every phase are considered to be the same. It means that the three-phase equivalent circuit of the machine is balanced and the neutral point for the star connection is not connected electrically to any part. This facilitates the analysis and the obtaining of the machine model.

Due to the nature of electric machines, the loading effect of the energy supplier is inductive. For inductance losses, a resistor is considered in series with the inductance, which is directly responsible for heating the machine. The leakage effect between the windings and the interaction between the stator and rotor of the machine are considered to be leakage inductance.

In accordance with three-phase analysis, the physical variables of the machine are a time-variant-system differential equation, while the rotor of the machine rotates. The inductances of AC machines vary in accordance with the rotor position. In order to get time-invariant differential equations, reference frame theory techniques are applied. This makes it easy to obtain a machine model in which all physical variables such as voltage, current, and flux are transformed, as will be analyzed in this chapter. Furthermore, three-phase systems, mathematical tools, and reference frame theory techniques and implementation will likewise be analyzed in this chapter.

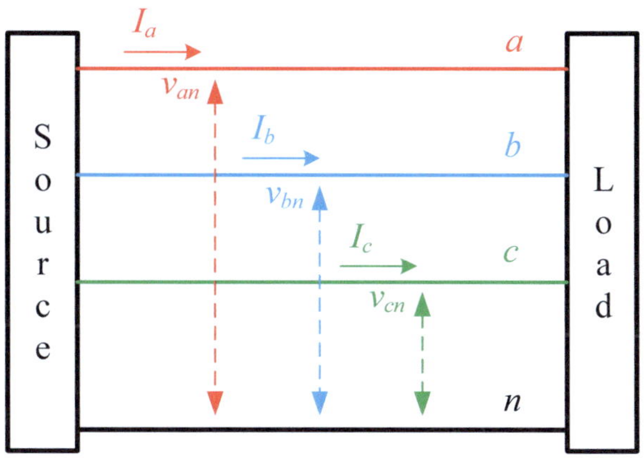

Fig. 3.1 Source and load interconnection with the line-to-line voltages and phase currents

3.2 Three-Phase Balanced Linear Load

It is common to assume that three-phase machines are constructed in such a way that their phases are balanced, with all three impedances being identical, and isolated neutral point in the star connection, thus allowing for proper rotational movement and simplifying machine analysis. For example, the calculations are not much more complicated than for the case of single-phase since two-thirds of the work is eliminated. A fault internally in the machine can cause an unbalanced operation which is considered an abnormality and can be detected by measuring the machine impedance.

Figure 3.1 represents in a general way a three-phase load powered by the three-phase source. Between the source and the load, there are only three wires since the neutral connection is avoided as commented before. The source can be the three-phase supply network (voltage source) or an electric drive which can supply three-phase voltage or current, while the load is an AC machine. This section will be focused on the load, where two types of load connection circuits will be analyzed—star and delta—for three-phase balanced systems.

3.2.1 Star (Wye) Connection

As its name suggests, star connection means that the load impedances are connected like a star, with a common point called neutral. Figure 3.2 shows a star three-phase source connected to a three-phase star load. Current I_a from source V_{AN} is the same as i_a of impedance A. The same occurs in other phases. The goal is to analyze the load, i.e., the circuit from a, b, c points to the right.

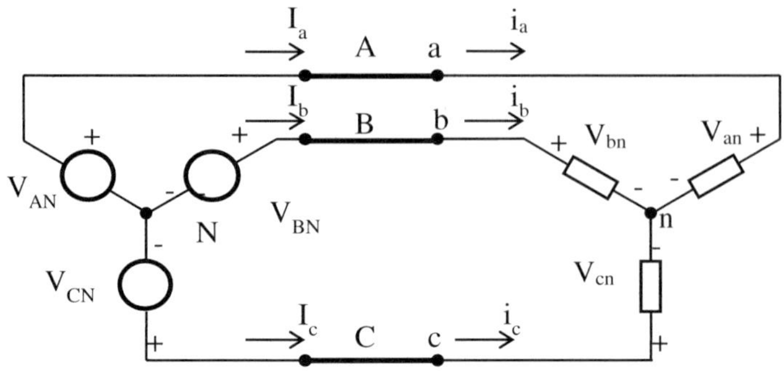

Fig. 3.2 Voltage source in a star connection with a star balanced load

The above system follows the equations below, following Kirchhoff's Current and Voltage Laws:

$$
\begin{aligned}
& i_a + i_b + i_c = 0 \rightarrow KCL \\
& V_{ab} + V_{bc} + V_{ca} = 0 \rightarrow KVL
\end{aligned}
\tag{3.1}
$$

It is possible to define three-phase instantaneous voltage as:

$$
\begin{aligned}
V_{an}(t) &= \sqrt{2} \cdot V \cdot \cos(\omega t) \\
V_{bn}(t) &= \sqrt{2} \cdot V \cdot \cos(\omega t - 120^\circ) \\
V_{cn}(t) &= \sqrt{2} \cdot V \cdot \cos(\omega t + 120^\circ)
\end{aligned}
\tag{3.2}
$$

and as a phasor:

$$
\begin{aligned}
V_{an} &= V\angle 0^\circ \\
V_{bn} &= V\angle -120^\circ \\
V_{cn} &= V\angle 120^\circ
\end{aligned}
\tag{3.3}
$$

For balanced linear loads with identical impedance, the line-to-line voltages are:

$$
\begin{aligned}
& V_{ab} = V_{an} - V_{bn} = V\angle 0^\circ - V\angle -120^\circ = \sqrt{3} \cdot V\angle 30^\circ \\
& V_{bc} = V_{bn} - V_{cn} = V\angle -120^\circ - V\angle +120^\circ = \sqrt{3} \cdot V\angle -90^\circ \\
& V_{ca} = V_{cn} - V_{an} = V\angle 120^\circ - V\angle 0^\circ = \sqrt{3} \cdot V\angle 150^\circ \\
& V_{ab} + V_{bc} + V_{ca} = 0 \\
& V_{an} + V_{bn} + V_{cn} = 0
\end{aligned}
\tag{3.4}
$$

Voltages referenced to a neutral point (phase voltage in the load) can be calculated as:

$$V_{an} = \frac{1}{3}(V_{ab} - V_{ca})$$
$$V_{bn} = \frac{1}{3}(V_{bc} - V_{ab})$$
$$V_{cn} = \frac{1}{3}(V_{ca} - V_{bc}) \tag{3.5}$$

and neutral point potential is:

$$V_n = \frac{1}{3}(V_a + V_b + V_c) \tag{3.6}$$

Lastly, it is easy to represent phase voltages as a function of V_a, V_b, V_c:

$$V_{an} = \frac{2}{3}V_a - \frac{1}{3}V_b - \frac{1}{3}V_c$$
$$V_{bn} = \frac{2}{3}V_b - \frac{1}{3}V_c - \frac{1}{3}V_a$$
$$V_{cn} = \frac{2}{3}V_c - \frac{1}{3}V_a - \frac{1}{3}V_b \tag{3.7}$$

The neutral point *N* at the source side and the neutral point *n* at the load side may not be connected. This will make the phase voltage at the source side be different from the phase voltage at the load side. Figure 3.3 shows the vector representation of the three-phase system analyzed. Dashes represent the line-to-line voltages, 30° apart, with a higher magnitude. On the contrary, regular lines represent the phase

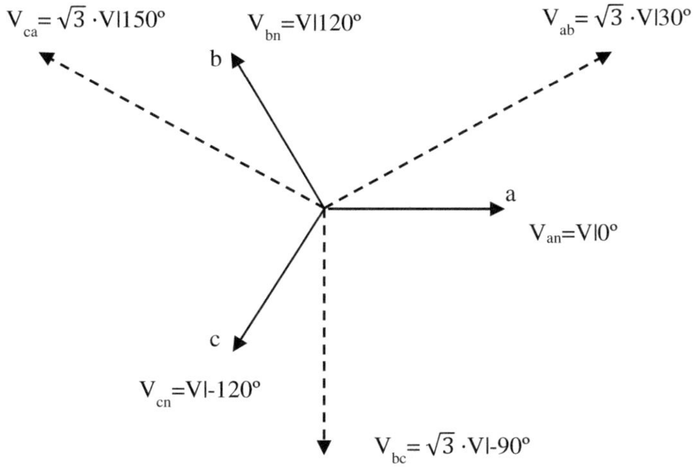

Fig. 3.3 Voltage phasor diagram in a balanced three-phase system

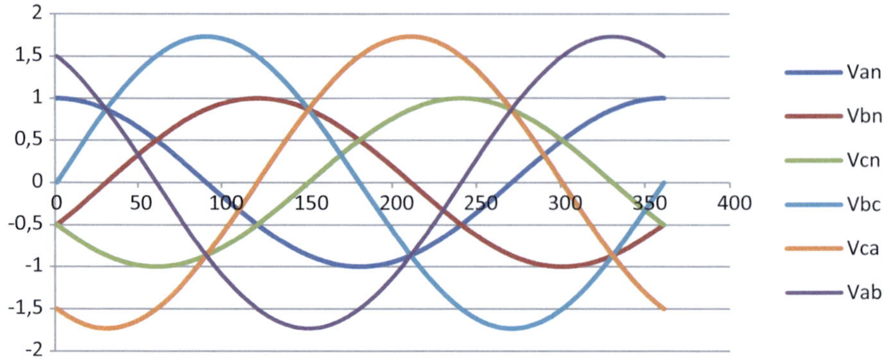

Fig. 3.4 Instantaneous voltage in three-phase systems

voltage—that is to say, the drop in voltage in loads *A*, *B*, *C*—with lower magnitudes when compared with line-to-line voltages.

To help understand this important concept, Fig. 3.4 shows the six instantaneous voltages from 0° to 360° in which the magnitude of phase voltage is unity. As discussed before, it can be observed that magnitude and angle lag for line-to-line voltages change with regard to phase voltages.

3.2.2 Delta Connection

In a delta connection, there is no common or neutral point since load impedances are connected like a triangle. This can be seen in Fig. 3.5, in which line-to-line voltage V_{ab} drops in impedance *A*, so current i_a is higher than phase current i_{ab}.

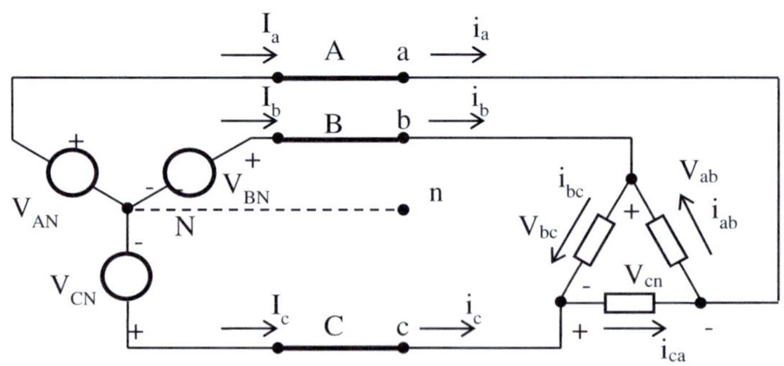

Fig. 3.5 Voltage source in star connection with delta balanced load

Since impedance is the same, the power absorbed by the load in both connections—star and delta—will be the same. From the point of view of machine control, the internal machine connection does not matter.

The above system follows the equations below, in accordance with Kirchhoff's Current and Voltage Laws:

$$\begin{aligned} i_a &= i_{ab} - i_{ca} \rightarrow KCL \\ i_b &= i_{bc} - i_{ab} \\ i_c &= i_{ca} - i_{bc} \\ V_{ab} &+ V_{bc} + V_{ca} = 0 \rightarrow KVL \end{aligned} \tag{3.8}$$

The line-to-line voltages for balanced linear loads with identical impedance are:

$$\begin{aligned} V_{ab} &= V_{an} - V_{bn} \\ V_{bc} &= V_{bn} - V_{cn} \\ V_{ca} &= V_{cn} - V_{an} \\ V_{ab} &+ V_{bc} + V_{ca} = 0 \\ V_{an} &+ V_{bn} + V_{cn} = 0 \end{aligned} \tag{3.9}$$

Like the star connection, the line voltage is ultimate:

$$\begin{aligned} V_{ab} &= V_{an} - V_{bn} = V\angle 0^\circ - V\angle{-120^\circ} = \sqrt{3} \cdot V\angle 30^\circ \\ V_{bc} &= V_{bn} - V_{cn} = V\angle{-120^\circ} - V\angle{+120^\circ} = \sqrt{3} \cdot V\angle{-90^\circ} \\ V_{ca} &= V_{cn} - V_{an} = V\angle 120^\circ - V\angle 0^\circ = \sqrt{3} \cdot V\angle 150^\circ \end{aligned} \tag{3.10}$$

where the above equations correspond to the phase voltage for the load.

Note that impedances A, B, C, see $\sqrt{3}$ more voltage than in the star connection, but the current for every impedance is $\sqrt{3}$ less, maintaining the same power.

3.2.3 *Low- and High-Voltage AC Machine Connection*

Historically, the induction machines (IM) have been directly connected to the supply network (electric grid), which has led to adapted designs, according to the National Electrical Manufacturers Association (NEMA), for the different voltage and frequencies available in the electrical grid. For example, in the USA, induction machines are designed to use the line voltage there: 120, 208, 230/460 V, with a voltage relation of 1:2 and a frequency of 60 Hz, while in northern and central Europe, they are designed to use 230/400 V and 50 Hz with a 1:$\sqrt{3}$ voltage relation. Low-voltage or delta connections use 230 V for both countries while high-voltage or star connections use 460 V for the USA and 400 V for Europe. In any case, in accordance with NEMA design, the rated voltage of the machine is defined with a 10% tolerance for voltages above and below the rated value.

Even if voltages could be similar for 230/460 and 230/400 V, induction machines are optimized to use the frequency of the country's line voltage. Obviously, this is mandatory in order to avoid using induction machines whose designed frequency differs from the line voltage. If this is the case, it is necessary to use an electric drive which controls voltage and frequency, and it is possible to find different voltages combinations. In agreement with the above explanation, several AC machines, such as the permanent magnet synchronous machine (PMSM) and synchronous reluctance machine (SynRM), also have the possibility of using low or high voltage. Hereinafter, AC machines will be used to designate the IM, PMSM, and SynRM.

In AC machines with both voltage possibilities, there is a connection box where wiring is numerically identified for the proper selection of high and low voltage. This selection can be made automatically by contactor devices; for example, when starting low-voltage high-power IM connected directly to the grid, where high voltage (star) is selected first and then low voltage (delta) in order to reduce overcurrent peaks. On the other hand, if an electric drive is used, it must be selected correctly for the available grid voltage.

The increasing use of electric drives, variable frequency drives (VFD) often called adjustable speed drives (ASD), has led to the design of exclusive IM for this purpose labeled for many manufacturers as "inverter duty." This label should fulfill the standard NEMA MG1 Part 30–31, where among other points, voltage stress is mentioned. Due to the rapid rise time and frequency of drive switching transients, and depending on cable length and physical characteristics, the voltage seen at the machine terminals when supplied by an ASD can be upwards of 2–4$\times$ rated supply voltage. NEMA MG1 Part 31 requires machine insulation systems for 460 V rated machines to be capable of withstanding 1600 volts' peak, at a rise time of 0.1 μs.

On the other hand, if the voltage source is a DC battery instead of the electrical grid, it is necessary to reconstruct a three-phase voltage with 120° phase difference with one another from a DC voltage. This electric drive is popularly known as a three-phase inverter. The three-phase inverter, as will be seen in Chap. 8, is capable of modifying the voltage and frequency of the phases independently with a phase shift of 120° between phases. In this case, the selection of low or high voltage will depend on the available voltage of the battery.

3.2.3.1 Delta/Star Connection with Six-Lead Terminal Wiring

As shown earlier, for northern and central Europe, the line voltage is 230/400 V with a 1:$\sqrt{3}$ voltage relation, meaning that voltage selection is made based upon a simple star and delta connections with three windings and six terminals. Figure 3.6 depicts as an example of a simple voltage connection for induction machines, with the wiring numerically identified.

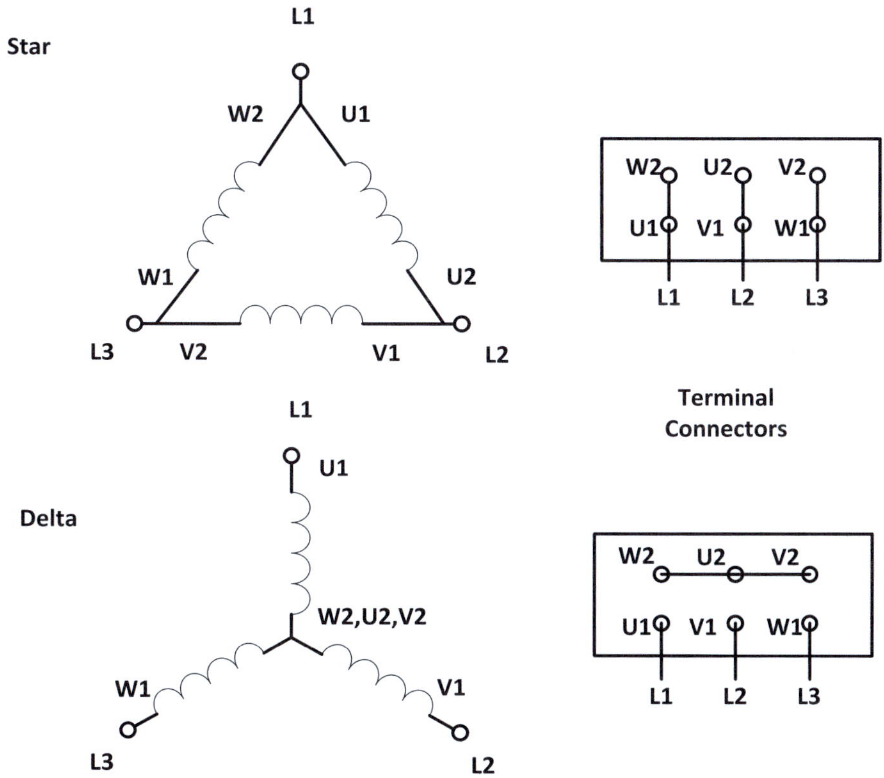

Fig. 3.6 Star and delta connection for an induction machine

For example, the same motor with a delta connection can be used in a three-phase 120 V grid, or a three-phase 208 V grid for the USA with a star connection. In this case, the voltage yielded by the stator windings will be the same since, $\sqrt{3} \cdot 120 = 208$ V, while the phase current yielded by the delta connection in the stator winding will be $\sqrt{3}$ times smaller in order to maintain the same power.

The machine's parameters will change with the selected voltage. Table 3.1 shows how machine parameters change with the delta/star voltage selected. It is possible to observe the conversion factor K between delta/star.

3.2.3.2 Low and High Voltage with Nine-Lead Terminal Wiring

In the USA, where the voltage relationship is 1:2—that is to say, 230/460 V—voltage selection is made based upon complex star and delta connections with six windings and nine terminals. This combination allows for having half (or for doubling) the voltage and current. For example, for a low-voltage (i.e., 230 V) operation, the

Table 3.1 Machine parameters in both voltage combinations

Parameter	Unit	K	High voltage (star)	Low voltage (delta)
Stator resistance R_s	Ω	3	1.7	5.1
Stator inductance L_{ls}	mH	3	1.55	4.65
Rotor resistance R_r	Ω	3	0.85	2.55
Rotor inductance L_{lr}	mH	3	1.42	4.26
Magnetizing inductance L_m	mH	3	52.3	156.9
Nominal voltage	V	$1/\sqrt{3}$	200	115
Nominal current	A	$\sqrt{3}$	4	7

windings are connected in parallel (half voltage, double current), while for 460 V, they are connected in series (double voltage, half current).

3.2.3.2.1 Star Connection

In Fig. 3.7, the wiring diagram for high-voltage setups with a star connection is shown. Terminals 1, 2, and 3 are connected to voltage phases L1, L2, and L3—or R, S, T. The terminals that are individually joined are 4-7, 5-8, and 6-9, respectively.

In Fig. 3.8, the wiring diagram for low-voltage setups with a star connection is shown. Terminals 4, 5, and 6 are connected together. The terminals that are individually joined are 1-7, 3-9, and 2-8, respectively, with voltage phases L1, L2, and L3—or R, S, T.

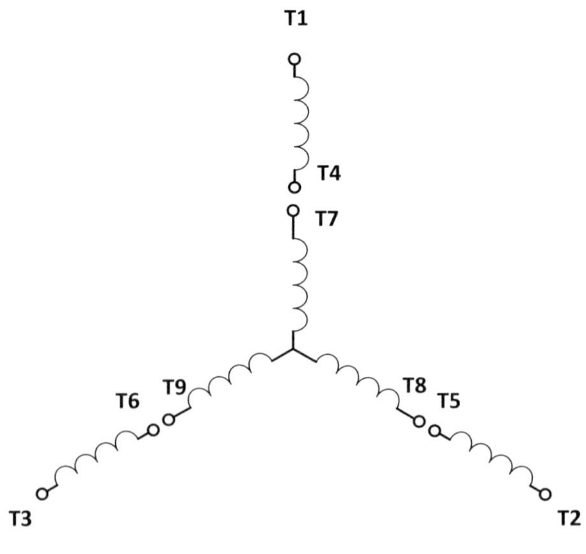

Fig. 3.7 Nine-lead terminal wiring for high-voltage setups with a star connection

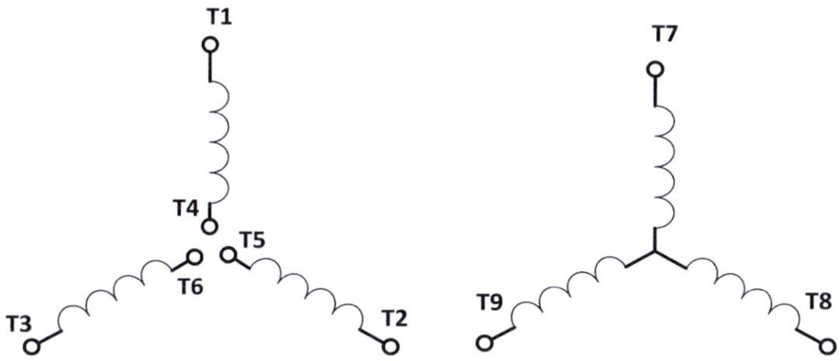

Fig. 3.8 Nine-lead terminal wiring for low-voltage setups with a star connection

Table 3.2 Machine parameters in both voltage combinations with nine-lead terminal wiring for star connections

Parameter	Unit	K	High voltage (star)	Low voltage (delta)
Stator resistance R_s	Ω	1/4	1.20	0.30
Stator inductance L_{ls}	mH	1/4	5.00	1.25
Rotor resistance R_r	Ω	1/4	0.84	0.21
Rotor inductance L_{lr}	mH	1/4	7.66	1.92
Magnetizing inductance L_m	mH	1/4	141.4	35.35
Nominal voltage	V	1/2	460	230
Nominal current	A	2	7.5	15

Table 3.2 shows how machine parameters are affected by both voltage combinations.

3.2.3.2.2 Delta Connection

In Fig. 3.9, the wiring diagram for a high-voltage setup with a delta connection is shown. Terminals 1, 2, and 3 are connected to voltage phases L1, L2, and L3—or R, S, T. The terminals which are individually joined are 4-7, 5-8, and 6-9, respectively.

The wiring diagram for a low-voltage setup with a delta connection is shown in Fig. 3.10. The terminals that are individually joined are 1-6-7, 2-4-8, and 3-5-9, respectively, with voltage phases L1, L2, and L3—or R, S, T.

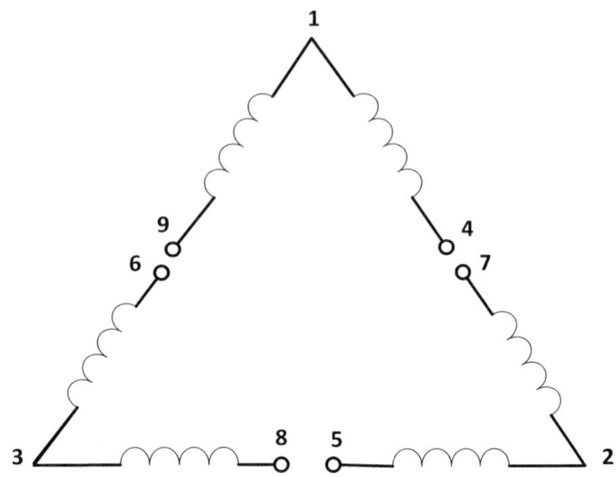

Fig. 3.9 Nine-lead terminal wiring for high-voltage setups with a delta connection

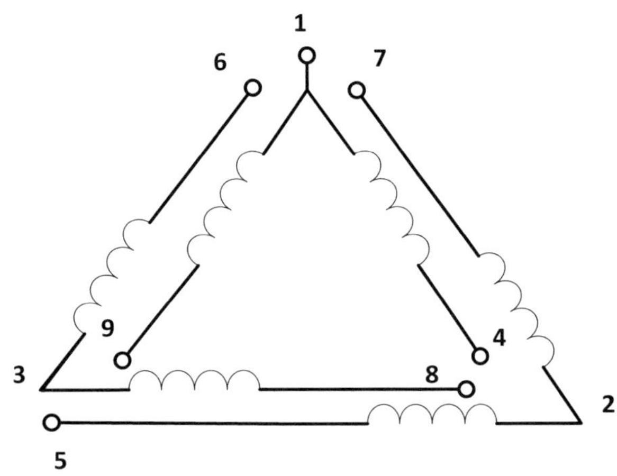

Fig. 3.10 Nine-lead terminal wiring for low-voltage setups with a delta connection

3.3 Power in Three-Phase Systems

In AC systems, the power factor relationship plays an important role. Power factor is a measure of how efficiently the load current is being converted into useful work output. In other words, if all the power delivered by the supply is sent to the load. In AC systems with load resistance, the power factor will be 1, which means that all the power delivered is useful to the load. This is the desired value for the inductive and capacitive load. As will be shown in this section, a power factor equal to 1 for inductive and capacitive loads is not possible. There are some electronic circuits (power factor correctors) which control the power factor so that it is 1 for inductive and capacitive loads, but this type of circuit is not within the scope of this book. In PMSM, there is a control method called "unity power factor," in which the power factor can be controlled to be 1 in a wide range of machine speeds, thus avoiding the use of power factor correctors. In this section, how the power of three-phase systems can be calculated will be analyzed, as well as the behavior thereof in accordance with the power factor.

The average power that is sent to one-phase can be calculated according to the following equation:

$$P = V \cdot I \cdot \cos\theta \rightarrow [\mathrm{W}] \tag{3.11}$$

where angle θ is the phase angle between voltage and current. The cosines of this angle represent the power factor of the system.

$$PF = \cos\theta \tag{3.12}$$

Total average power in a three-phase balanced system is the sum of the three powers; that is to say, three times the aforementioned equation:

$$P_T = 3 \cdot V \cdot I \cdot \cos\theta \rightarrow [\mathrm{W}] \tag{3.13}$$

The reactive power is the power which goes to the supply. For one-phase, it can be calculated as:

$$Q = V \cdot I \cdot \sin\theta \rightarrow [\mathrm{VAr}] \tag{3.14}$$

In Fig. 3.11, one-phase analysis is shown. Voltage, current, instant power, average power, and reactive power are represented. In this case, there is an inductive load with a 72° delay angle between voltage and current. It can be observed that power

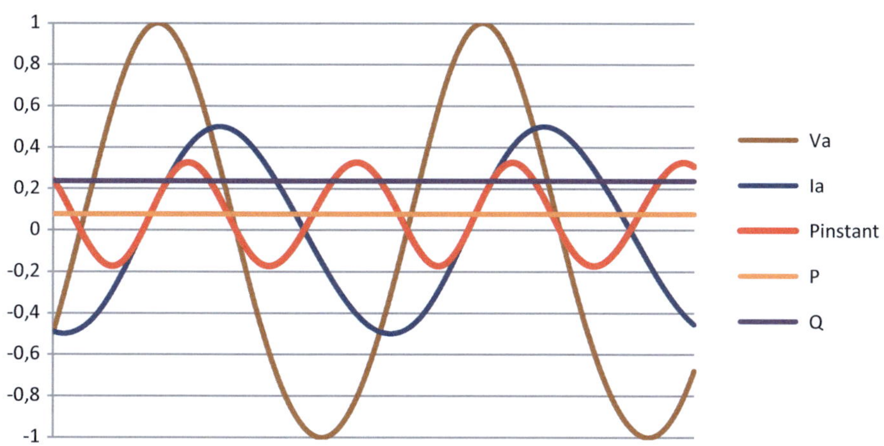

Fig. 3.11 Representation of one-phase voltage, current, instant power, average power (P = 0.077 W) and reactive power (Q = 0.237 VAr) for an inductive load angle of 72°

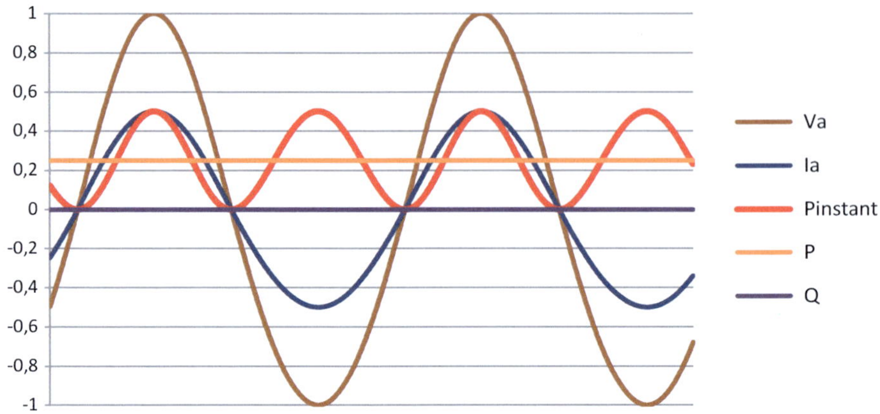

Fig. 3.12 Representation of one-phase voltage, current, instant power, average power ($P = 0.25$ W) and reactive power ($Q = 0$ VAr) for a resistive load

flows in both directions periodically, and the average power absorbed by the load is 0.077 W. Reactive power is 0.237 VAr.

$$Q = V \cdot I \cdot \sin\theta \to \frac{1}{\sqrt{2}} \cdot \frac{0.5}{\sqrt{2}} \cdot \sin 72^\circ = 0.237 \to [\text{VAr}] \quad (3.15)$$

$$P = V \cdot I \cdot \cos\theta \to \frac{1}{\sqrt{2}} \cdot \frac{0.5}{\sqrt{2}} \cdot \cos 72^\circ = 0.077 \to [\text{W}] \quad (3.16)$$

In Fig. 3.12, the same example is shown but for a resistive load. It is possible to observe that power flows in one direction to the load. The average power absorbed by the load is 0.25 W, and this is the maximum average power. Reactive power is zero.

$$Q = V \cdot I \cdot \sin\theta \to \frac{1}{\sqrt{2}} \cdot \frac{0.5}{\sqrt{2}} \cdot \sin 0^\circ = 0.0 \to [\text{VAr}] \quad (3.17)$$

$$P = V \cdot I \cdot \cos\theta \to \frac{1}{\sqrt{2}} \cdot \frac{0.5}{\sqrt{2}} \cdot \cos 0^\circ = 0.25 \to [\text{W}] \quad (3.18)$$

Total reactive power in a three-phase balanced system is three times the phase reactive power, as the following equation illustrates:

$$Q_T = 3 \cdot V \cdot I \cdot \sin\theta \to [\text{VAr}] \quad (3.19)$$

The apparent power for every phase is:

$$S = V \cdot I \to [\text{VA}] \quad (3.20)$$

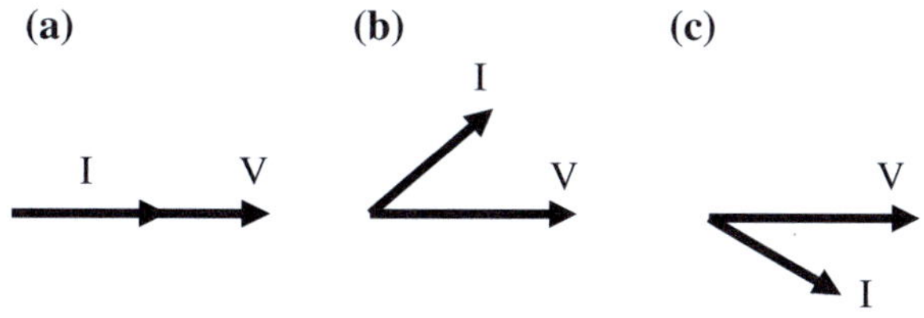

Fig. 3.13 Representation of different power factors. **a** Unity, **b** leading, and **c** lagging

And the total apparent power of the three-phase load is:

$$S_T = 3 \cdot V \cdot I \rightarrow [\mathrm{VA}] \tag{3.21}$$

The power factor of the system can be calculated by dividing the average power by the apparent power:

$$PF_{\mathrm{T}} = \frac{P_{\mathrm{T}}}{S_{\mathrm{T}}} = \cos\theta \tag{3.22}$$

It should be noted that voltage and current in the previous equations are the phase voltage and current seen by the load. For example, in a star connection, the phase voltage is the line voltage divided by $\sqrt{3}$, while the phase current matches the line current.

The different power factor combinations are illustrated in Fig. 3.13. They are leading when current is advanced with respect to voltage (capacitive load) and lagging when it is delayed (inductive load). As commented, the power factor is unity when the phase angle between current and voltage is zero, meaning that all the average power is supplied to the load; for example, when the load is a resistive load without inductors or capacitors.

In Fig. 3.14, the three-phase voltages and currents are represented when the load is a three-phase inductive load; thus, the current is lagging—in this case—72° with respect to voltage. The power factor can be calculated using the following equation:

$$PF_T = \cos\theta \rightarrow \cos 72^\circ = 0.309 \tag{3.23}$$

Figure 3.15 shows the instantaneous power for every phase for the above example, with average and reactive power being three times greater:

$$Q = 3 \cdot V \cdot I \cdot \sin\theta \rightarrow 3 \cdot \frac{1}{\sqrt{2}} \cdot \frac{0.5}{\sqrt{2}} \cdot \sin 72^\circ = 0.711 \rightarrow [\mathrm{VAr}] \tag{3.24}$$

$$P = 3 \cdot V \cdot I \cdot \cos\theta \rightarrow 3 \cdot \frac{1}{\sqrt{2}} \cdot \frac{0.5}{\sqrt{2}} \cdot \cos 72^\circ = 0.231 \rightarrow [\mathrm{W}] \tag{3.25}$$

Fig. 3.14 Representation of three-phase voltage and current when the current is 72° lagging. This is a three-phase inductive load

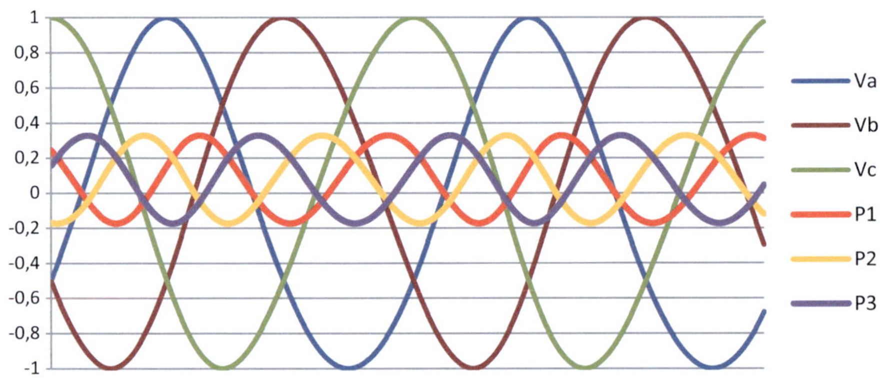

Fig. 3.15 Representation of phase voltage and instant power for every phase, in accordance with the above example

3.4 Vector Representation in Three-Phase Systems

The definition of the space vector by three-phase components are represented in (3.26):

$$\boldsymbol{f}_{abc} = f_a + \boldsymbol{\alpha} \cdot f_b + \boldsymbol{\alpha}^2 \cdot f_c \tag{3.26}$$

where $\alpha = e^{j\frac{2}{3}\pi}$ and $\alpha^2 = e^{j\frac{4}{3}\pi}$ represent the spatial operators, f_a, f_b, and f_c are the variables at each a, b, and c phase, respectively. In the three-phase AC machines, the variable f can be voltage, current, and flux linkages for each phase spatially 120° apart, and $\boldsymbol{f}_{abc}$ stands for the total voltage, current, or flux linkages, which may rotate.

In many cases, in the analysis of three-phase electric machine models (synchronous or asynchronous machines), descriptions with vectors of two-axis theory are commonly used. The description in two components consists on complex coordinate axes defined as $S = s_d + js_q$, where s_d is Re(S) (real part) and s_q is $I_m(S)$ (imaginary part) as represented in (3.27):

$$\boldsymbol{f}_{abc} = f_d + jf_q \tag{3.27}$$

The three-phase voltages, currents, and fluxes of AC machines can be analyzed in complex vector spaces using the two-axis theory with the orthogonal components. For example, the stator current of the machine can be defined as follows. Assuming that $i_{sa}(t)$, $i_{sb}(t)$, and $i_{sc}(t)$ are instantaneous currents in the stator phases, then the current vector (also called space phasor) is defined as:

$$\boldsymbol{i}_s = i_{sa}(t) + \boldsymbol{\alpha} \cdot i_{sb}(t) + \boldsymbol{\alpha}^2 \cdot i_{sc}(t) \tag{3.28}$$

The magnitude of the space vector of the stator current equals 3/2 of the peak value of the sinusoidal current. The current vector $\boldsymbol{i}_s$ can be found in the literature, reduced by the factor 2/3, as in the following equation:

$$\boldsymbol{i}_s = \frac{2}{3}\left(i_{sa}(t) + \boldsymbol{\alpha} \cdot i_{sb}(t) + \boldsymbol{\alpha}^2 \cdot i_{sc}(t)\right) \tag{3.29}$$

This facilitates the use of space vectors since the parameters which align with the real equivalent circuit of the machine can be employed. In complex coordinate axes, the stator current can be expressed as:

$$\boldsymbol{i}_s = i_{sd}(t) + ji_{sq}(t) \tag{3.30}$$

where the subscripts s_d and s_q represents the instantaneous value of the direct-axis stator current component, and the quadrature-axis stator current component expressed in the stationary reference frame fixed to the stator.

If the stator windings are supplied with three-phase instantaneous currents $i_{sa}(t)$, $i_{sb}(t)$, and $i_{sc}(t)$ and the neutral point is isolated, the zero-sequence stator instantaneous current $i_{s0}(t)$ is zero. That is:

$$\boldsymbol{i}_{s0}(t) = i_{sa}(t) + i_{sb}(t) + i_{sc}(t) = 0 \tag{3.31}$$

If (3.31) is fulfilled, as it will be shown in the next section, the Clarke transformation is reduced to two-axis.

The following diagram shows an example of the stator current vector as the vector space in three-phase and two-phase components (complex coordinate axes) for an electrical machine. Vector $\boldsymbol{i}_s$ rotates at the speed of the phase current, i.e., 50 Hz (Fig. 3.16):

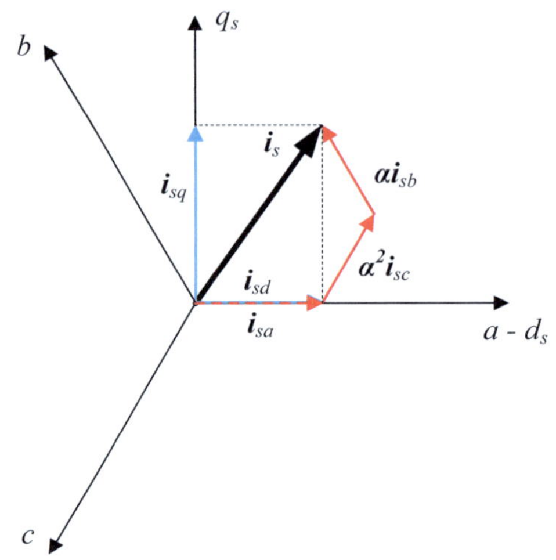

Fig. 3.16 Stator current vector representation with two and three axes in a three-phase AC machine

Vector $\boldsymbol{i}_s$ is given by *abc* or *dq* components. These components are on the same plane, where *abc* components represent the phase current *a*, *b*, and *c*. The s_d and s_q components can be found in the literature as α and β, respectively.

Figure 3.17 shows the three current components in a machine for two instant speeds, $\omega t = 0$ and $\omega t = \pi/3$. In the case of Fig. 3.17a, the $\boldsymbol{i}_s$ vector has an angle equal to zero because of the sum of the three vectors, while in Fig. 3.17b, the position for current vector is $\pi/3$:

Three-phase stator phase current can be represented in Fig. 3.18 from zero to 360° according to the equations below in which amplitude I_m is fixed to unity. The

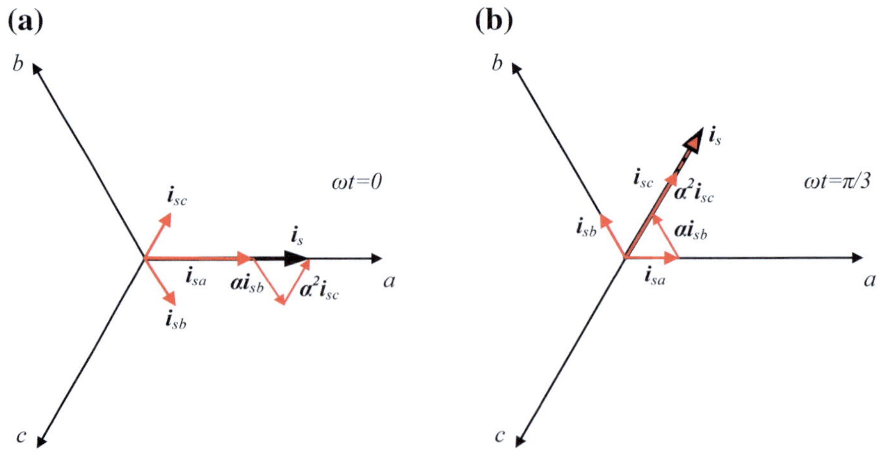

Fig. 3.17 Representation of different $\boldsymbol{i}_s$ vectors at two observation times

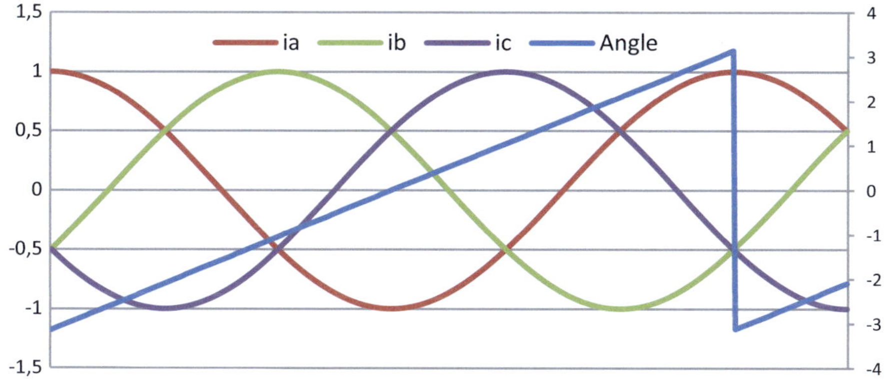

Fig. 3.18 Representation of currents i_a, i_b, i_c, and angle in radians between $-\pi$ and π

components of the above example at observation moments $\omega t = 0$ and $\omega t = \pi/3$ can be observed graphically. For instance, for $\omega t = \pi/3$, or 60°, the i_a and i_b components take the value of 0.5, while the i_c component takes the value of -1.

$$
\begin{aligned}
i_a &= I_m \cdot \cos(\omega t) \\
i_b &= I_m \cdot \cos\left(\omega t - \frac{2\pi}{3}\right) \\
i_c &= I_m \cdot \cos\left(\omega t + \frac{2\pi}{3}\right)
\end{aligned}
\tag{3.32}
$$

The above graph is drawn with Excel so that anyone can interpret it through the implementation below. Cell A—degrees—must reach, at least, 360 to see an entire period. In Table 3.4, several values, calculated from 0° to 10°, are shown, with the formulas in Table 3.3 (3.32) having been used (Table 3.4).

Table 3.3 Machine parameters in both voltage combinations

Row	A	B	C	D	E
0	Angle degrees	Angle	i_a	i_b	i_c
1	0	=Radians(A1)	=COS(B5)	=COS(B5−2*PI()/3)	=COS(B5+2*PI()/3)

Table 3.4 Values calculated according to the aforementioned implementation in Excel, up to 10°

Angle degrees	Angle	i_a	i_b	i_c
0	0	1	−0.5	−0.5
1	0.01745329	0.9998477	−0.48480962	−0.51503807
2	0.03490659	0.99939083	−0.46947156	−0.52991926
3	0.05235988	0.99862953	−0.4539905	−0.54463904
4	0.06981317	0.99756405	−0.43837115	−0.5591929
5	0.08726646	0.9961947	−0.42261826	−0.57357644
6	0.10471976	0.9945219	−0.40673664	−0.58778525
7	0.12217305	0.99254615	−0.39073113	−0.60181502
8	0.13962634	0.99026807	−0.37460659	−0.61566148
9	0.15707963	0.98768834	−0.35836795	−0.62932039
10	0.17453293	0.98480775	−0.34202014	−0.64278761

3.5 Mathematical Transformation for AC Machine Analysis

The electrical variables of voltage, current, and flux in a, b, c three-phase systems can be transformed into $dq0$ variables (direct, quadrature, 0-component) on an orthogonal axis by using reference frame theory techniques, and vice versa. The 0-component shall be avoided as electromechanical power conversion is the only concern that will be considered in the analysis of electric machine (Sul 2011). According to notation, direct d_s, and quadrature q_s, means stationary reference frame (with s). Direct d_e and quadrature q_e mean synchronously rotating reference frame (with e), while d_ω and q_ω mean arbitrary rotating reference frame. In other literature, as commented before, it is also possible to find α and β, respectively, for direct d_s, and quadrature q_s; while direct d_e, and quadrature q_e may be found as $dq0$.

In this section, different mathematical transformation tools applied to space vectors of time-domain signals (for example, voltage, current, and flux) which simplify the analysis of three-phase systems will be seen. Said tools are Clarke's [Edith Clarke, 1883–1959] and Park's [Robert H. Park 1902–1994] transformation theories. Furthermore, the pseudocode implementation of every transformation, taking into account a floating-point microcontroller will be seen.

3.5.1 *The Clarke and Concordia Transformation*

The Clarke transformation converts from three-phase quantities to two-axis orthogonal stationary reference frame quantities:

$$\left[f_{\alpha\beta0}\right] = \left[\boldsymbol{T}_{\alpha\beta0}\right] \cdot \left[f_{abc}\right] \tag{3.33}$$

alternatively, with another notation:

$$\left[\boldsymbol{f}^{s}_{dq0}\right] = \left[\boldsymbol{T}^{s}_{dq0}\right] \cdot \left[\boldsymbol{f}_{abc}\right] \tag{3.34}$$

with the transformation matrix:

$$\left[\boldsymbol{T}_{\alpha\beta 0}\right] = \frac{2}{3}\begin{bmatrix} 1 & -\frac{1}{2} & -\frac{1}{2} \\ 0 & \frac{\sqrt{3}}{2} & -\frac{\sqrt{3}}{2} \\ \frac{1}{2} & \frac{1}{2} & \frac{1}{2} \end{bmatrix} \tag{3.35}$$

The inverse of the Clarke transformation converts from two-axis orthogonal stationary reference frame quantities to three-phase reference frame quantities:

$$\left[\boldsymbol{f}_{abc}\right] = \left[\boldsymbol{T}^{s}_{dq0}\right]^{-1} \cdot \left[\boldsymbol{f}_{dq0}\right] \tag{3.36}$$

where

$$\left[\boldsymbol{T}^{s}_{dq0}\right]^{-1} = \begin{bmatrix} 1 & 0 & 1 \\ -\frac{1}{2} & \frac{\sqrt{3}}{2} & 1 \\ -\frac{1}{2} & -\frac{\sqrt{3}}{2} & 1 \end{bmatrix} \tag{3.37}$$

An example of the current is representing below. The 0-component current i_{s0} is zero for a balanced load and isolated neutral as commented before.

$$\begin{bmatrix} i_{sds}(t) \\ i_{sqs}(t) \\ i_{s0}(t) \end{bmatrix} = \frac{2}{3}\begin{bmatrix} 1 & -\frac{1}{2} & -\frac{1}{2} \\ 0 & \frac{\sqrt{3}}{2} & -\frac{\sqrt{3}}{2} \\ \frac{1}{2} & \frac{1}{2} & \frac{1}{2} \end{bmatrix}\begin{bmatrix} i_{sa}(t) \\ i_{sb}(t) \\ i_{sc}(t) \end{bmatrix}; \begin{bmatrix} i_{sa}(t) \\ i_{sb}(t) \\ i_{sc}(t) \end{bmatrix} = \begin{bmatrix} 1 & 0 & 1 \\ -\frac{1}{2} & \frac{\sqrt{3}}{2} & 1 \\ -\frac{1}{2} & -\frac{\sqrt{3}}{2} & 1 \end{bmatrix}\begin{bmatrix} i_{sds}(t) \\ i_{sqs}(t) \\ i_{s0}(t) \end{bmatrix} \tag{3.38}$$

The 2/3 factor comes from so-called classical, non-power-invariant form of the phase transformation from three-phase to quadrature-phase components. If voltage and current variables are transformed with (3.38), when the instantaneous power and torque is calculated, it is larger than that of corresponding variables in a three-phase axis, so that the 3/2 factor has to be applied as will be seen in Sect. 3.6. However, for the power-invariant form, i.e., to maintain the same instantaneous power and torque in the transformation, the Concordia transformation (Robyns et al. 2012) is chosen which use the $\sqrt{3/2}$ factor as represented below:

$$[C] = \sqrt{\frac{2}{3}}\begin{bmatrix} 1 & -\frac{1}{2} & -\frac{1}{2} \\ 0 & \frac{\sqrt{3}}{2} & -\frac{\sqrt{3}}{2} \\ \frac{1}{\sqrt{2}} & \frac{1}{\sqrt{2}} & \frac{1}{\sqrt{2}} \end{bmatrix} \tag{3.39}$$

The Concordia inverse matrix is orthogonal and thus equals its transpose:

$$[C]^{-1} = \sqrt{\frac{2}{3}} \begin{bmatrix} 1 & 0 & \frac{\sqrt{2}}{2} \\ -\frac{1}{2} & \frac{\sqrt{3}}{2} & \frac{\sqrt{2}}{2} \\ -\frac{1}{2} & -\frac{\sqrt{3}}{2} & \frac{\sqrt{2}}{2} \end{bmatrix} = [C]^T \tag{3.40}$$

Then, with the same example for the phase current:

$$\begin{aligned} \begin{bmatrix} i_{sds}(t) \\ i_{sqs}(t) \\ i_{s0}(t) \end{bmatrix} &= \sqrt{\frac{2}{3}} \begin{bmatrix} 1 & -\frac{1}{2} & -\frac{1}{2} \\ 0 & \frac{\sqrt{3}}{2} & -\frac{\sqrt{3}}{2} \\ \frac{1}{\sqrt{2}} & \frac{1}{\sqrt{2}} & \frac{1}{\sqrt{2}} \end{bmatrix} \begin{bmatrix} i_{sa}(t) \\ i_{sb}(t) \\ i_{sc}(t) \end{bmatrix}; \begin{bmatrix} i_{sa}(t) \\ i_{sb}(t) \\ i_{sc}(t) \end{bmatrix} \\ &= \sqrt{\frac{2}{3}} \begin{bmatrix} 1 & 0 & \frac{\sqrt{2}}{2} \\ -\frac{1}{2} & \frac{\sqrt{3}}{2} & \frac{\sqrt{2}}{2} \\ -\frac{1}{2} & -\frac{\sqrt{3}}{2} & \frac{\sqrt{2}}{2} \end{bmatrix} \begin{bmatrix} i_{sds}(t) \\ i_{sqs}(t) \\ i_{s0}(t) \end{bmatrix} \end{aligned} \tag{3.41}$$

The mathematically simplest way is to avoid using and representing the 0-component since it is considered that three-phase systems and machines are always balanced with isolated neutral point. Then, with this consideration in mind, the unitary gain transformation for any variable a, b, c to dq_s is resumed as represented in (3.42) when the Clarke transformation is used.

$$\begin{aligned} d_s &= \frac{2}{3}a - \frac{1}{3}(b - c) \\ q_s &= \frac{2}{\sqrt{3}}(b - c) \end{aligned} \tag{3.42}$$

When d is superposed with a, then (3.42) is more straightforward as:

$$\begin{aligned} d_s &= a \\ q_s &= \frac{(a + 2 \cdot b)}{\sqrt{3}} \end{aligned} \tag{3.43}$$

The digital pseudocode implementation for the Clarke transformation is shown below. The function can be called every time conversion from three-phase reference to two-axis reference should be performed. It should be noted that phase c is redundant since, as commented, the system is balanced.

```
#define INVSQRT3 0.577350269
Function abc_to_dq (inputs: a, b, outputs: ds, qs)
    ds=u
    qs=(a+2*b)* INVSQRT3
End Function
```

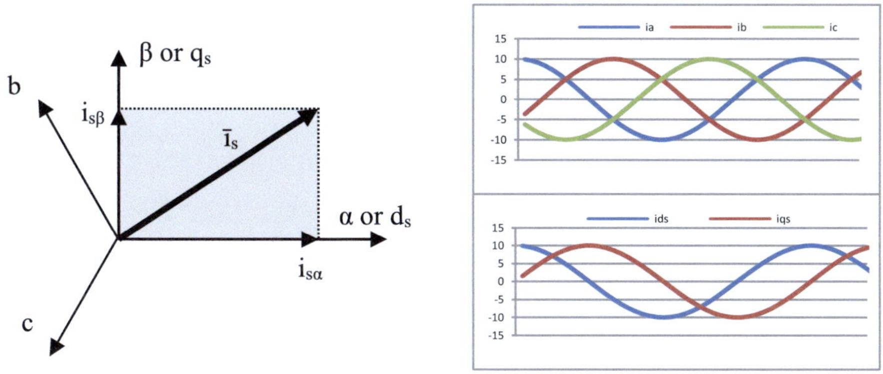

Fig. 3.19 Clarke transformation example for current

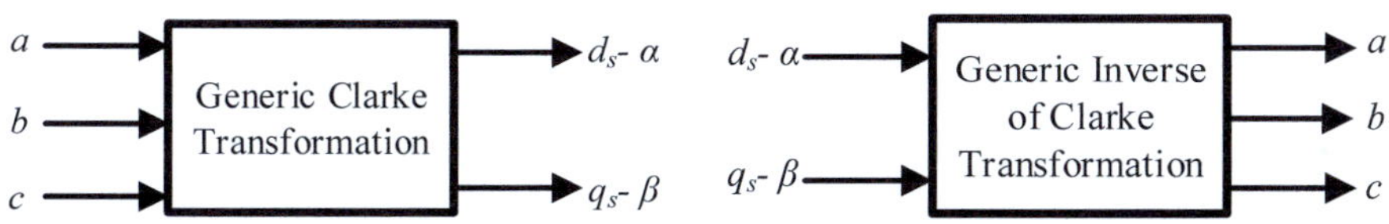

Fig. 3.20 Generic Clarke and inverse transformation block diagram, where a, b, and c variables can be voltage, current, or flux linkage

The vector $\boldsymbol{i}_s$ is entirely defined by a, b, c stationary axes, like (3.28), or by stationary dq_s axes as illustrated in Fig. 3.19.

As before, if the 0-component is null, the unitary gain transformation dq_s to abc for the inverse of the Clarke transformation is:

$$
\begin{aligned}
a &= d_s \\
b &= \frac{(d_s \cdot \sqrt{3} - q_s)}{2} \\
c &= \frac{(-d_s \cdot \sqrt{3} - q_s)}{2}
\end{aligned}
\tag{3.44}
$$

Figure 3.20 represents the block diagram for generic Clarke transformation and its inverse.

The digital pseudocode implementation is shown below for the inverse of the Clarke transformation.

```
#define SQRT3 1.7320508
Function dq_to_abc (inputs: ds, qs, outputs: a, b, c)
     a=ds
     b=((SQRT3 *ds)-qs)*0.5
     c=(-(SQRT3 *ds)-qs)*0.5
End Function
```

3.5.2 The Rotation Transformation

The rotation matrix converts from two-axis orthogonal stationary reference frame quantities to two-axis in arbitrary rotating reference frame quantities. Since the new reference frame is rotating, we need the angle θ as a third input. The rotation matrix is defined as:

$$\begin{bmatrix} \boldsymbol{f}_{qe} \\ \boldsymbol{f}_{de} \end{bmatrix} = [\boldsymbol{R}(\theta)] \cdot \begin{bmatrix} \boldsymbol{f}_{qs} \\ \boldsymbol{f}_{ds} \end{bmatrix} \tag{3.45}$$

where

$$[\boldsymbol{R}(\theta)] = \begin{bmatrix} \cos\theta & \sin\theta & 0 \\ -\sin\theta & \cos\theta & 0 \\ 0 & 0 & 1 \end{bmatrix} \tag{3.46}$$

where the angular displacement, θ, of the arbitrary reference.

The unitary gain transformation dq_s to synchronously rotating reference frame dq_e is:

$$\begin{aligned} d_e &= d_s \cdot \cos(\theta) + q_s \cdot \sin(\theta) \\ q_e &= q_s \cdot \cos(\theta) - d_s \cdot \sin(\theta) \end{aligned} \tag{3.47}$$

With the same example for the vector $\boldsymbol{i}_s$, it can be defined by a new axis which rotates and the same synchronous speed as represented in Fig. 3.21.

The angle can be defined between $-\pi$ and $+\pi$, but it depends on the mathematical functions sine and cosine for the microcontroller. The digital pseudocode implementation for the rotation matrix in the synchronously rotating reference frame is shown below:

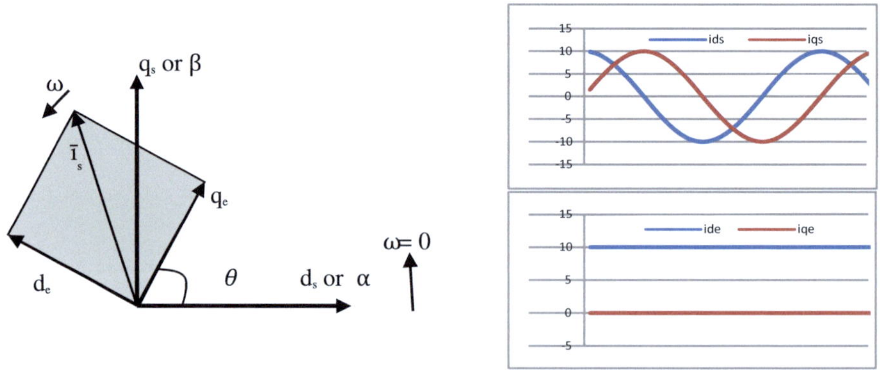

Fig. 3.21 Vector $\boldsymbol{i}_s$ is defined with new two-axis which rotates at the same synchronous speed

```
Function dqs_to_dqe (inputs: ds, qs, angle, outputs: de, qe)
      de=ds*cos(angle)+qs*sin(angle)
      qe=qs*cos(angle)-ds*sin(angle)
End Function
```

The inverse of a rotation matrix is its transpose and converts from two-axis arbitrary rotating reference frame quantities to two-axis orthogonal stationary reference frame quantities:

$$\begin{bmatrix} f_{qs} \\ f_{ds} \end{bmatrix} = [\boldsymbol{R}(\theta)]^{-1} \cdot \begin{bmatrix} f_{qe} \\ f_{de} \end{bmatrix} \tag{3.48}$$

where

$$[\mathrm{R}(\theta)]^{-1} = \begin{bmatrix} \cos\theta & \sin\theta & 0 \\ -\sin\theta & \cos\theta & 0 \\ 0 & 0 & 1 \end{bmatrix} \tag{3.49}$$

The unitary gain transformation from synchronously rotating reference frame dq_e to stationary reference frame dq_s is:

$$\begin{aligned} d_s &= d_e \cdot \cos(\theta) - q_e \cdot \sin(\theta) \\ q_s &= d_e \cdot \sin(\theta) + q_e \cdot \cos(\theta) \end{aligned} \tag{3.50}$$

Figure 3.22 represents the block diagram for generic Rotation transformation and its inverse.

The digital pseudocode implementation for the inverse of a rotation matrix with synchronously rotating reference frame notation is shown below. Like before, the angle can be defined between $-\pi$ and $+\pi$, but it depends on the mathematical functions sine and cosine for the microcontroller.

```
Function dqe_to_dqs (inputs: de, qe, angle, outputs: ds, qs)
      ds=de*cos(angle)-qe*sin(angle)
      qs=de*sin(angle)+qe*cos(angle)
End Function
```

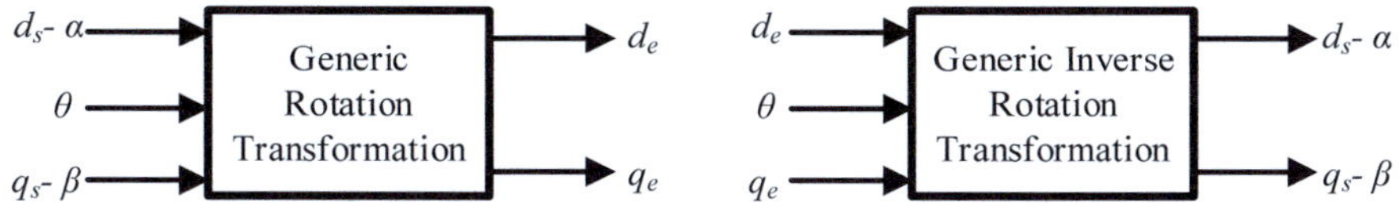

Fig. 3.22 Rotation and inverse of rotation transformation. *dq* variables can be voltage, current, or flux linkages. Synchronously, rotating reference frame notation is used

The Concordia transformation and a rotation matrix sometimes are used directly to convert from three axes to two axes rotating in arbitrary speed $dq\omega$ with the power-invariant form. It is the product of the two frame changes:

$$\left[\boldsymbol{f}_{dq\omega}\right] = [\boldsymbol{R}(\theta)][\boldsymbol{C}]\left[\boldsymbol{f}_{abc}\right] \tag{3.51}$$

where the angular displacement, θ, of the arbitrary reference.

On the other hand, the Park transformation establishes the arbitrary speed rotating dq_e transformation with the non-power-invariant form. It is also the product of the two frame changes:

$$\left[\boldsymbol{f}_{dqe}\right] = [\boldsymbol{R}(\theta)][\boldsymbol{T}]\left[\boldsymbol{f}_{abc}\right] = [\boldsymbol{K}(\theta)]\left[\boldsymbol{f}_{abc}\right] \tag{3.52}$$

where $\boldsymbol{K}(\theta)= [\boldsymbol{R}(\theta)][\boldsymbol{T}]$, then

$$\boldsymbol{K}(\theta) = \sqrt{\frac{2}{3}}\begin{bmatrix} \cos(\theta) & \cos\left(\theta - \frac{2\pi}{3}\right) & \cos\left(\theta + \frac{2\pi}{3}\right) \\ -\sin(\theta) & -\sin\left(\theta - \frac{2\pi}{3}\right) & -\sin\left(\theta + \frac{2\pi}{3}\right) \\ \frac{1}{\sqrt{2}} & \frac{1}{\sqrt{2}} & \frac{1}{\sqrt{2}} \end{bmatrix} \tag{3.53}$$

And its inverse is represented as:

$$\boldsymbol{K}(\theta)^{-1} = \sqrt{\frac{2}{3}}\begin{bmatrix} \cos(\theta) & -\sin(\theta) & \frac{\sqrt{2}}{2} \\ \cos\left(\theta - \frac{2\pi}{3}\right) & -\sin\left(\theta - \frac{2\pi}{3}\right) & \frac{\sqrt{2}}{2} \\ \cos\left(\theta + \frac{2\pi}{3}\right) & -\sin\left(\theta + \frac{2\pi}{3}\right) & \frac{\sqrt{2}}{2} \end{bmatrix} \tag{3.54}$$

Then, by using the current as the same example is possible to get (3.55) for power-invariant form (Concordia) and (3.56) for non-power-invariant form (Park):

$$\begin{bmatrix} i_{sd\omega} \\ i_{sq\omega} \\ i_{s0} \end{bmatrix} = \sqrt{\frac{2}{3}}\begin{bmatrix} \cos(\theta) & \cos\left(\theta - \frac{2\pi}{3}\right) & \cos\left(\theta + \frac{2\pi}{3}\right) \\ -\sin(\theta) & -\sin\left(\theta - \frac{2\pi}{3}\right) & -\sin\left(\theta + \frac{2\pi}{3}\right) \\ \frac{1}{\sqrt{2}} & \frac{1}{\sqrt{2}} & \frac{1}{\sqrt{2}} \end{bmatrix}\begin{bmatrix} i_{sa}(t) \\ i_{sb}(t) \\ i_{sc}(t) \end{bmatrix}$$

$$\begin{bmatrix} i_{sa}(t) \\ i_{sb}(t) \\ i_{sc}(t) \end{bmatrix} = \sqrt{\frac{2}{3}}\begin{bmatrix} \cos(\theta) & -\sin(\theta) & \frac{\sqrt{2}}{2} \\ \cos\left(\theta - \frac{2\pi}{3}\right) & -\sin\left(\theta - \frac{2\pi}{3}\right) & \frac{\sqrt{2}}{2} \\ \cos\left(\theta + \frac{2\pi}{3}\right) & -\sin\left(\theta + \frac{2\pi}{3}\right) & \frac{\sqrt{2}}{2} \end{bmatrix}\begin{bmatrix} i_{sd\omega} \\ i_{sq\omega} \\ i_{s0} \end{bmatrix} \tag{3.55}$$

$$\begin{bmatrix} i_{sd\omega} \\ i_{sq\omega} \\ i_{s0} \end{bmatrix} = \frac{2}{3}\begin{bmatrix} \cos(\theta) & \cos\left(\theta - \frac{2\pi}{3}\right) & \cos\left(\theta + \frac{2\pi}{3}\right) \\ -\sin(\theta) & -\sin\left(\theta - \frac{2\pi}{3}\right) & -\sin\left(\theta + \frac{2\pi}{3}\right) \\ \frac{1}{2} & \frac{1}{2} & \frac{1}{2} \end{bmatrix}\begin{bmatrix} i_{sa}(t) \\ i_{sb}(t) \\ i_{sc}(t) \end{bmatrix}$$

$$\begin{bmatrix} i_{sa}(t) \\ i_{sb}(t) \\ i_{sc}(t) \end{bmatrix} = \begin{bmatrix} \cos(\theta) & -\sin(\theta) & 1 \\ \cos\left(\theta - \frac{2\pi}{3}\right) & -\sin\left(\theta - \frac{2\pi}{3}\right) & 1 \\ \cos\left(\theta + \frac{2\pi}{3}\right) & -\sin\left(\theta + \frac{2\pi}{3}\right) & 1 \end{bmatrix}\begin{bmatrix} i_{sd\omega} \\ i_{sq\omega} \\ i_{s0} \end{bmatrix} \tag{3.56}$$

Equations (3.55) and (3.56) can be expressed in synchronous rotating reference frame if arbitrary angle θ is a synchronous angle θ_e and notation should be changed from ω *to e*.

In practice, it is more flexible and perhaps more efficient to use the Clarke or Concordia transform along with the rotation matrix, instead of the $dq\omega$ or Park transformation. For example, in the implementation of a microcontroller, it is more interesting to have both transformations separately because they are used separately in the vector control systems.

In order to understand the importance of the aforementioned mathematical tools, a machine rotor has been analyzed as an example of Fig. 3.23. The rotor is, in this case, a permanent magnet with two poles: one north and one south. The stator is not shown, but it is a three-phase system, 120° apart with magnetic field rotating at synchronous speed ω_e. The rotor is rotating at $\omega_m = \omega_e$ in rad/s. A three-phase reference frame (*A*, *B*, *C* axes) and orthogonal stationary reference frame (dq_s axes) are fixed and on the same plane (left part of Fig. 3.23). The flux linkage can be expressed more easily by using the stationary two-axis instead of the three-axis.

In the right part of Fig. 3.23, the orthogonal synchronous rotating reference frame dq_e, in which d_e is at angle θ_m (rotation angle) with respect to axis *A*, is attached to the rotor. Thus, it is rotating at the same (ω_m) speed, and any stator (*a*, *b*, *c*) complex equation can be referenced to the rotating axis (dq_e) rotor, thus reducing its complexity. This is called rotor reference frame, in which stator quantities are converted to the rotor reference frame. For instance, the expression for stator flux, created by the three-phase current windings on the *a*, *b*, *c* axes, is hard to obtain; however, if it is referenced to the rotor by using the rotating axes, it is easier to calculate, as it will see in Chap. 4.

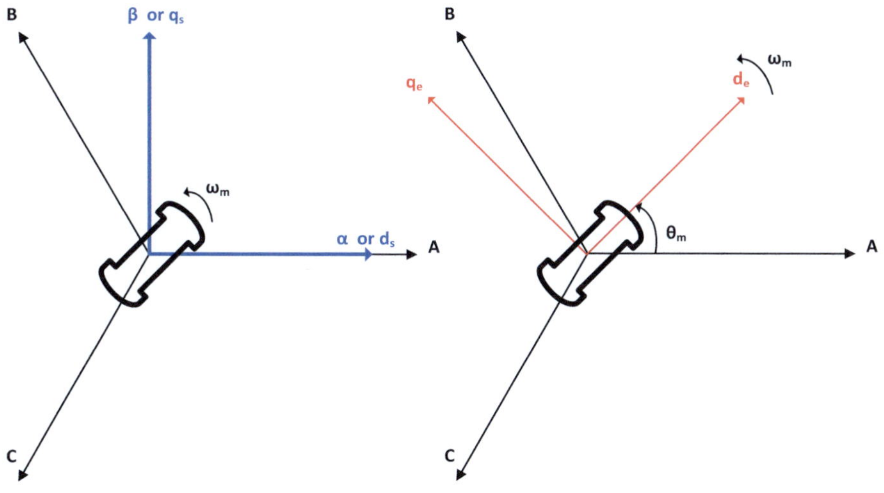

Fig. 3.23 Clarke and Park transformation saw in a permanent-magnet rotor

3.6 Instantaneous Power in Three-Phase Systems

In a symmetrical three-phase system, the total input of instantaneous power is equal to the sum of the instantaneous powers produced by each of the three phases:

$$p(t) = v_a(t)i_a(t) + v_b(t)i_b(t) + v_c(t)i_c(t) \tag{3.57}$$

where $v_a(t)$, $v_b(t)$, $v_c(t)$, and $i_a(t)$, $i_b(t)$, $i_c(t)$ are the phase voltages and currents, respectively, in the three phases. They can have arbitrary variation in time. As shown before, the definitions of the space phasor for the phase voltages and currents were expressed as:

$$\boldsymbol{v}_s = v_{sa}(t) + \boldsymbol{\alpha} \cdot v_{sb}(t) + \boldsymbol{\alpha}^2 \cdot v_{sc}(t) \tag{3.58}$$

$$\boldsymbol{i}_s = i_{sa}(t) + \boldsymbol{\alpha} \cdot i_{sb}(t) + \boldsymbol{\alpha}^2 \cdot i_{sc}(t) \tag{3.59}$$

Neglecting the zero-sequence (0-component) voltages $v_{s0}(t)$ and currents $i_{s0}(t)$, instantaneous power can be expressed in terms of voltage and current space phasors in the following way:

$$p = \frac{3}{2}\mathrm{Re}(\boldsymbol{v}_s \boldsymbol{i}_s *) \tag{3.60}$$

Thus, reactive power can be expressed as:

$$q = \frac{3}{2}\mathrm{Im}(\boldsymbol{v}_s \boldsymbol{i}_s *) \tag{3.61}$$

where the space vector $\boldsymbol{v}_s$ and is are expressed as

$$\boldsymbol{v}_s = v_{sds} + jv_{sqs} \tag{3.62}$$

And $\boldsymbol{i}_s$ as

$$\boldsymbol{i}_s = i_{sds} + ji_{sqs} \tag{3.63}$$

As commented before, the 3/2 factor comes from non-power-invariant form of the phase transformation from three-phase to quadrature-phase components.

For simplicity's sake, it is useful to use the two-axis components for both magnitudes (voltage and current) in the stationary reference frame to calculate active and reactive power in a microcontroller. This method keeps us from measuring three instantaneous voltages and three instantaneous currents. Equations (3.64) and (3.65) show that active and reactive power can be monitored by employing a two-axis reference frame, first applying the Clarke transformation to *a*, *b*, and *c* magnitudes:

$$p = \frac{3}{2}\text{Re}\left[\left(v_{sds} + jv_{sqs}\right)\left(i_{sds} + ji_{sqs}\right)\right] = \frac{3}{2}\left(v_{sds}i_{sds} + v_{sqs}i_{sqs}\right) \rightarrow [\text{W}] \quad (3.64)$$

$$q = \frac{3}{2}\text{Re}\left[\left(v_{sds} + jv_{sqs}\right)\left(i_{sds} - ji_{sqs}\right)\right] = \frac{3}{2}\left(v_{sqs}i_{sds} - v_{sds}i_{sqs}\right) \rightarrow [\text{VAr}] \quad (3.65)$$

The active and reactive power can also be calculated by using the synchronously rotating reference—i.e., the Park transformation—as follows:

$$p = \frac{3}{2}\left(v_{sde}i_{sde} + v_{sqe}i_{sqe}\right) \rightarrow [\text{W}] \quad (3.66)$$

$$q = \frac{3}{2}\left(v_{sqe}i_{sde} - v_{sde}i_{sqe}\right) \rightarrow [\text{VAr}] \quad (3.67)$$

3.6.1 Instantaneous Power Computation

When instantaneous power is calculated online with a microcontroller or DSP, it is possible to monitor the electrical machine operation. For example, it is possible to detect short circuits in the stator winding, asymmetries, air-gap eccentricity, and in case of a squirrel-cage induction machine, it is possible to detect failures in the rotor bars (Vas 1993).

For microcontroller/DSP implementation, the digital method to obtain instantaneous power is as follows:

$$p(k) = \frac{3}{2}\left(v_{sds}(k)i_{sds}(k) + v_{sqs}(k)i_{sqs}(k)\right) \rightarrow [\text{W}] \quad (3.68)$$

$$q(k) = \frac{3}{2}\left(v_{sqs}(k)i_{sds}(k) - v_{sds}(k)i_{sqs}(k)\right) \rightarrow [\text{VAr}] \quad (3.69)$$

The stationary reference frame is used, where $v_{sds}(k)$, $v_{sqs}(k)$, and $i_{sds}(k)$, $i_{sqs}(k)$ are the sampled values of the voltage and current. It should be noted that it only needs to detect two voltages and two currents.

If voltages and currents are symmetric (no DC offset values), there is no noise during acquisition, and it is purely sinusoidal. Instantaneous power remains constant at different sampling times ($k=0, k=1$, etc.) for fixed voltage and current amplitudes, as we can see in Fig. 3.24. This example shows active power p of 2.59 W, reactive power q of 1.5 VAr. Voltage amplitude is 2 V_{peak}, the current is delayed 30° as regards voltage, and amplitude is 1 A_{peak}. The orthogonal function, or 90° phase, between $v_{sds}(k)$, $v_{sqs}(k)$ and $i_{sds}(k)$, $i_{sqs}(k)$ is noticed.

In previous examples, an ideal discrete analysis is shown for a better understanding of how instantaneous power can be calculated. Even if ADC acquisition is previously

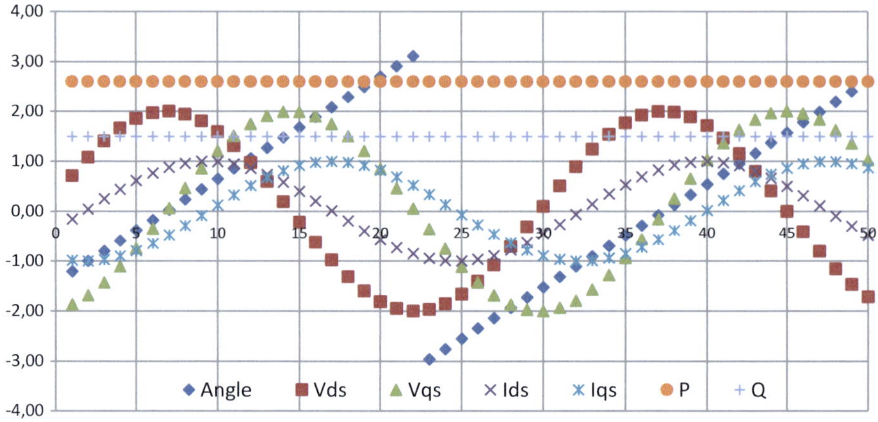

Fig. 3.24 Discrete values for $v_{sds}(k)$, $v_{sqs}(k)$ and $i_{sds}(k)$, $i_{sqs}(k)$, as well as instantaneous calculation, every sampling time, of the active $p(k)$ and reactive power $q(k)$. Angle magnitude is also shown

filtered externally through a low-pass filter, some noise can be captured by the microcontroller/DSP, which will directly affect the instantaneous power. Because of this, it is a good idea to filter or average the power calculation, by taking N samples, for example. We can use the same prior example to give instantaneous power every 50 samples instead of every sampled value, that is to say, to take a window of 50 samples.

If the voltage or current is not purely sinusoidal or they have some DC offset, instantaneous powers will be affected, as can be seen in Fig. 3.25, in which current

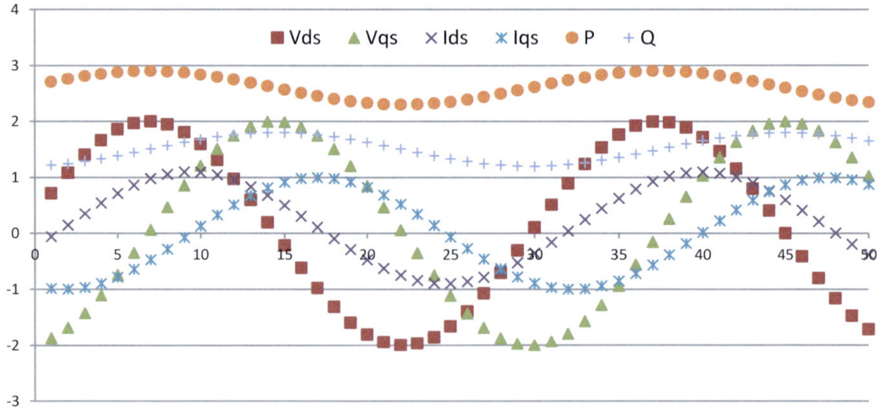

Fig. 3.25 Digital values for $v_{sds}(k)$, $v_{sqs}(k)$ and $i_{sds}(k)$, $i_{sqs}(k)$, as well as the instantaneous calculation, every sampling time , of the active $p(k)$ and reactive power $q(k)$. Current $i_{sds}(k)$ has 0.1 A of DC offset to see how it affects instantaneous power

$i_{sds}(k)$ has a DC offset of 0.1 A. It should be noted that both powers take sinusoidal shape and have the same frequency as the voltage and current.

The average of N voltage and current samples can be calculated as:

$$p = \frac{1}{N}\frac{3}{2}\sum_{k=1}^{N}\left[v_{sds}(k)i_{sds}(k) + v_{sqs}(k)i_{sqs}(k)\right] \rightarrow [\mathrm{W}] \tag{3.70}$$

$$q = \frac{1}{N}\frac{3}{2}\sum_{k=1}^{N}\left[v_{sqs}(k)i_{sds}(k) - v_{sds}(k)i_{sqs}(k)\right] \rightarrow [\mathrm{VAr}] \tag{3.71}$$

In the previous example, the new values for powers p and q are 2.68 W and 1.57 VAr, respectively, with $N = 50$ average samples.

Instead of using the above-average method, another valid method consists of filtering the instantaneous power calculation by using a first-order digital filter. It is possible to use the IIR (infinite impulse response) first-order low-pass filter type due to its low computation. The difference equation is:

$$y[n] = y[n-1] + k(x[n] - y[n-1]) \tag{3.72}$$

The IIR first-order low-pass filter transfer function in z-domain is:

$$\frac{Y(z)}{X(z)} = \frac{1 - \mathrm{e}^{-\omega_c T_s}}{1 - z^{-1}\mathrm{e}^{-\omega_c T_s}} \tag{3.73}$$

where the constant k can be calculated according to the sampling time T_s and the cut-off frequency ω_c as follows:

$$k = \frac{T_s}{T_s + \left(\frac{1}{\omega_c}\right)} \tag{3.74}$$

Following, it is shown the digital pseudocode implementation:

```
Function power_calculation(inputs: Uqs_s, Uds_s, Iqs_s, Ids_s,
Power_kt, outputs: Power, QPower)
    Power=Power+((3.0/2.0)*(Uqs_s*Iqs_s+Uds_s*Ids_s)-Power)*k
    QPower=QPower+((3.0/2.0)*(Uqs_s*Ids_s-Uds_s*Iqs_s)-QPower)*k
End Function
```

The aforementioned function should be called periodically every sampling time preferable between 10 and 40 ms.

3.7 RMS Computation

In power systems, the power quality can be analyzed by using the measurement of the root mean square (RMS) value of voltage and current (Styvaktakis et al. 2002). In the electric machines, control system is not an exception, and sometimes, the microcontroller/DSP should be ready to compute the RMS value of the voltage and current.

The RMS value of an electrical magnitude is calculated mathematically by the expression (3.75), where for the case of voltage is:

$$V_{RMS} = \sqrt{\frac{1}{T}\int_{0}^{T} v^2(t)\mathrm{d}t} \tag{3.75}$$

where T is the period of the signal. In a pure sinusoidal voltage $V_p\ \sin(\omega t)$ with a period equal to 2π, it is possible to demonstrate that the RMS value is the peak value V_p divided by $\sqrt{2}$. During a transition, the RMS calculation does not give the correct value until the new period of the signal is detected.

To compute (3.75) in a microcontroller or DSP, the discretization expression is:

$$V_{RMS} = \sqrt{\frac{1}{N}\sum_{n=1}^{N} v_n^2} \tag{3.76}$$

It is possible to observe that to compute the voltage RMS, it is needed some N samples window of the voltage. Instead, to use wait to have the N samples to update the RMS value, it is possible to use a moving window where the values are continuously updated every new sample.

On the other hand, considering a symmetric three-phase sinusoidal currents (e.g., three-phase electric machine stator phase currents), it is possible to demonstrate that RMS calculation can be calculated with the current instants values of phase a and c as (Alfredo et al. 1998):

$$I_{sRMS} = \sqrt{\frac{2}{3}}\sqrt{i_{as}(i_{as} + i_{cs}) + i_{cs}^2} \tag{3.77}$$

The discretization of (3.77) expression is:

$$I_{sRMS} = \sqrt{\frac{2}{3}}\sqrt{i_{as}(k)(i_{as}(k) + i_{cs}(k)) + i_{cs}^2(k)} \tag{3.78}$$

The real part of the stator current is the RMS current multiplied by a constant, and the power factor $\cos(\theta)$:

$$I_{s(\mathrm{Re})} = \frac{3}{\sqrt{2}} I_{sRMS} \cos(\theta) \tag{3.79}$$

The real part of the current magnitude can be used for several control strategies, such as scalar control. Since the power factor can be challenging to compute/calculate, it is also possible to calculate the real part for the current by using the instant values, as shown below:

$$\begin{aligned} I_{sRMS} \cos(\theta) = {} & i_{as}\left(\cos(\theta_e) - \cos\left(\theta_e - \frac{2\pi}{3}\right)\right) \\ & + i_{cs}\left(\cos\left(\theta_e + \frac{2\pi}{3}\right) - \cos\left(\theta_e - \frac{2\pi}{3}\right)\right) \end{aligned} \tag{3.80}$$

where angle θ_e is the angle of voltage magnitudes. The instant value of phase currents a and c with their angle, multiplied by the cosines of the voltage angle, gives the real part of the current.

The discretization of (3.80) expression is:

$$I_{s(\mathrm{Re})} = \frac{3}{\sqrt{2}} \begin{bmatrix} i_{as}(k)\left(\cos(\theta_e(k)) - \cos\left(\theta_e(k) - \frac{2\pi}{3}\right)\right) + \\ + i_{cs}(k)\left(\cos\left(\theta_e(k) + \frac{2\pi}{3}\right) - \cos\left(\theta_e(k) - \frac{2\pi}{3}\right)\right) \end{bmatrix} \tag{3.81}$$

Every new sample of the current of phase a, c, and the angle, the real part of the current I_s can be computed in a microcontroller or DPS. For accurate results, the average of N samples can be used (3.82):

$$I_{s(\mathrm{Re})} = \frac{1}{N} \frac{3}{\sqrt{2}} \sum_{k=1}^{N} \begin{bmatrix} i_{as}(k)\left(\cos(\theta_e(k)) - \cos\left(\theta_e(k) - \frac{2\pi}{3}\right)\right) + \\ + i_{cs}(k)\left(\cos\left(\theta_e(k) + \frac{2\pi}{3}\right) - \cos\left(\theta_e(k) - \frac{2\pi}{3}\right)\right) \end{bmatrix} \tag{3.82}$$

As before, instant currents must be symmetrical and sinusoidal for accurate results. Figure 3.26 shows an example of discrete three-phase stator current and voltage, with 5 A_{peak} of amplitude and 15 V_{peak}. It is also shown the RMS value (3.54 A) and a real part value (1.21 A) which is calculated every k instant.

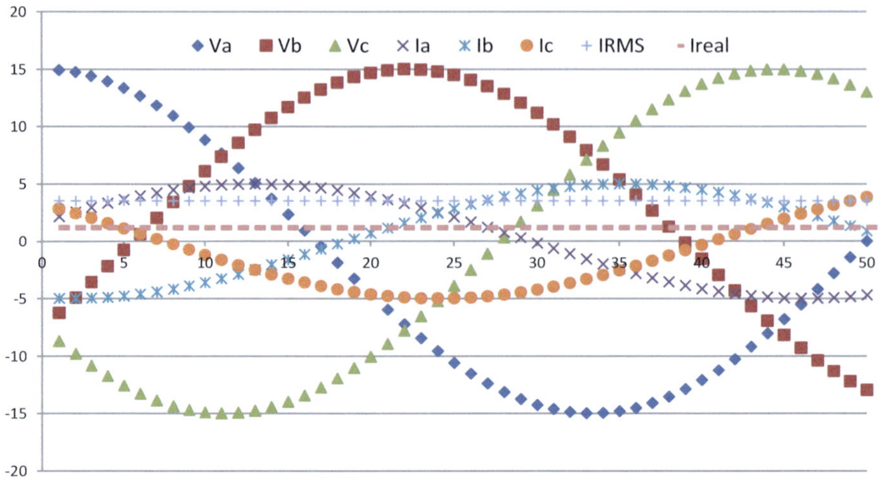

Fig. 3.26 Digital values for $v_a(k)$, $v_b(k)$, $v_c(k)$, $i_a(k)$, $i_b(k)$, $i_c(k)$ and $i_{\mathrm{RMS}}(k)$, $i_{\mathrm{real}}(k)$ calculated in accordance with the prior equations, at every instant k

References

Alfredo MG, Thomas AL, Donald WN (1998) A new induction motor V/f control method capable of high-performance regulation at low speeds. IEEE Trans Ind Appl 34(4):813–820 July–August

Robyns B et al (2012) Vector control of induction machines, power systems. Springer, London

Sul S-K (2011) Control of electric machine drive systems. Wiley-IEEE Press

Styvaktakis E, Bollen MHJ, Gu IYH (2002) Automatic classification of power system events using RMS voltage measurements. In: Proceedings of the IEEE power engineering society summer meeting, Chicago, IL, USA, 21–25 July 2002, pp 824–829. ISBN 0-7803-7518-1

Vas P (1993) Parameter estimation, condition monitoring, and diagnosis of electrical machines. Clarendo Press, Oxford

Chapter 4
Fundamentals of Electric Machines

4.1 Introduction

The electrical machines' configuration can be performed mechanically in different ways. The most common machine is the radial flux machine where stator and rotor are cylindrical, and flux is produced radially along the sideways of the rotor. The axial flux machines consist of a disk-shaped rotor where the flux is produced axially along the axis of the rotor with a more complicated flux path (Pyrhönen et al. 2016). The axial flux principle was used in Faraday's disk generator in 1831, but it was no popular for a long time due to manufacturing technologies. However, the use of the axial machines is increasing more and more in direct-drive applications such as HEV and EV, thanks to their compact structure, and the technologies improvements. Lastly, the linear machine is another electrical machine configuration used in some applications such as ultrasonic cutting machine. Here, the focus will be on radial flux machines unless otherwise specified.

High-performance electric machines' systems are becoming more and more present in the different sectors such as domestic appliance, industry, automotive, aviation, and electric locomotive markets. The challenge requirements are light, efficient, and high power density machines, especially in the recent developing of the electrified aircraft propulsion (EAP). It uses propulsors such as propeller or fans driven by electric machines. In like manner, generators which convert shaft power to electricity also requires light, efficient, and high power density in the hybrid-electric and turbo-electric aircraft architectures.

In the case of the domestic appliance market, the electric machines are in some home appliances such as refrigerators, washing machines, dishwashers, and coffee machines. Even though these appliances have included electric machines for many years, the tendency today is to improve their efficiency and power density so that the appliances are better ranked in terms of energy use—for example, A and A++ classification. For example, in a refrigerator with a compressor moved by an electric machine, an efficiency improvement consists of substituting the ON/OFF operation of the compressor with a soft start-up and pressure regulation by using power stage

© Springer Nature Switzerland AG 2020

R. Molina Llorente, *Practical Control of Electric Machines*,
Advances in Industrial Control, https://doi.org/10.1007/978-3-030-34758-1_4

with semiconductor switches' devices. Without soft start-up, every time the compressor switches to ON, current spikes, and overshoots are produced, reducing the whole efficiency. With a soft start-up, current spikes and overshoots are avoided. Furthermore, in the ON/OFF operation, there is no way to regulate the pressure that controls the freezing ability of the refrigerator in accordance with its necessity.

In the automotive industry, historically the economic DC-brushed machines (DC machines) have been the preferred choice for numerous applications such as wiper washers, fans, and starters. However, due to new, strict regulations to reduce CO_2 emissions, the electrification of vehicles is increasing. For example, some of the mechanical loads as the air-conditioning compressors are substituting for electrical compressors supplied by an auxiliary battery. The efficiency of these machines is an essential role in reducing the CO_2, so DC-brushed machines are avoided. The higher efficiency of the AC machines makes it preferred for low-voltage applications such as air-conditioning compressors, electrical turbo compressors, start-stop, electrical powered steering (EPS), and high-voltage application such as traction machines in EV/HEV. In low-voltage applications, the 12 V battery system does not have enough power for these applications, and new higher-voltage batteries' system is required.

For this reason, some vehicles have 48 V systems to supply the necessary power to the new electrical loads. However, for low-power electronics or smaller size machines, 12 V is used, and energy transfer between batteries should be possible by using high-efficiency DC-DC converters. The electric vehicle is a reality in which electric traction machines can provide the required torque/speed characteristics without a complex gear system. These machines operate in a high-voltage range, between 200 and 800 V, to increase the efficiency to reduce the amount of copper. Furthermore, traction operation systems operate in the four quadrants in a torque/speed plane, in contrast to the internal combustion engine.

In industrial markets, traditionally the electric machines have been applied in a wide range of applications from the last century. As happen with above markets, the efficiency and power density of the electrical machines are also a requirement for the industry and electrical locomotive markets to reduce also the CO_2 emissions and the expended energy.

On the other hand, the simulation of electrical machines, power electronic devices, and their control are the design process basis where the output's quality depends on the simulation model accuracy. The simulation of electric machines are usually expensive software packages that need high computing and is one of the slowest parts in the development process. This will require a compromise between speed and precision to optimize the design process. The finite element analysis (FEA) is a detailed approach to model the behavior of electrical machines. Its ability to perform a refined electromagnetic analysis has been the motivation for its use as a tool for machine design problems. The magnetic properties of the linear and non-linear magnetic material, its dependence on temperature, its saturation, deformations of the geometry, and the details of the windings can be modeled using FEA. The current FEA simulation software packages have the option to use lumped-parameter models, such as a *dq* model to increase the speed of the simulation. The *dq* model, as discussed in this chapter, is usually a pure electric circuit, and useful for determining electrical magnitudes, flux, torque, and speed in steady-state or in its transitions. However, this loses the possibility of performing magnetic analysis and does not present details, such as the slotting effects, the skewing effects, and magnetic saturation. For this reason, the design packages often implement a combination of the FEA and *dq* models where they benefit from the accuracy of the FE solutions and the velocity of the *dq* model in the same development environment tool.

In the first part of this chapter, machine classification and a thorough analysis of brushed machines (universal machines and DC machines) and its structure will be undertaken. It will be the key to understanding other machine types and their control strategy—something that is usually more complex. In the second part of this chapter, the three-phase brushless machine will be analyzed deeply showing different analysis strategies as space vector, phasors, steady-state, and dynamic equations. Some FEA simulations are presented for the different machines. Machine model, control loop strategy, simulations, and implementation will be analyzed for brushed and brushless machines in the following chapter.

4.2 Electric Machine Classification

Electric machines can be classified into two big groups: brushed and brushless machines. The difference between both machines is related to the use (or lack of use) of a mechanical commutator to power the machine's rotor. According to the National Electrical Manufacturer's Association (NEMA) standard, "Motion/Position Control Motors and Controls" defines the commutator as follows: "set of mechanical contacts arranged angularly or linearly, contacted by the brushes to provide the electrical path from the power source to the armature. The purpose of the commutator is to control the relative electrical phase angle between the fields of the stationary element and the moving element of a motor."

According to this, electricity is supplied to the commutator through the brush, generally made of graphite, which is in contact with the rotating rotor part (commutator). Because of the said contact, and depending on the current, some spikes are usually produced (fire arc), which must be avoided in explosive environments. Moreover, the maintenance of brushed motors is more frequent when compared with brushless machines because of the mechanical parts, which are subject to wear.

The typical brushed machine is the DC machine in which the stator produces a fixed magnetic flux using permanent magnets (PMs), and the rotor (rotating part) varies its flux polarity through automatic electric polarity in its coils, thanks to the mechanical commutator's final production of the torque.

Brushless machines do not need a mechanical commutator and brushes to power the rotor due to the fact that they are built in a different way. They are typically constructed with three-phase coils in the stator which produce a rotating synchronous magnetic flux when supplied with three-phase AC voltage 120° apart. For example, the magnets of a permanent magnet AC (PMAC) machine are in the rotor (moving part), and the stator generates a synchronous rotational magnetic flux in which the magnets of the rotor try to continue running likewise at the same, synchronous

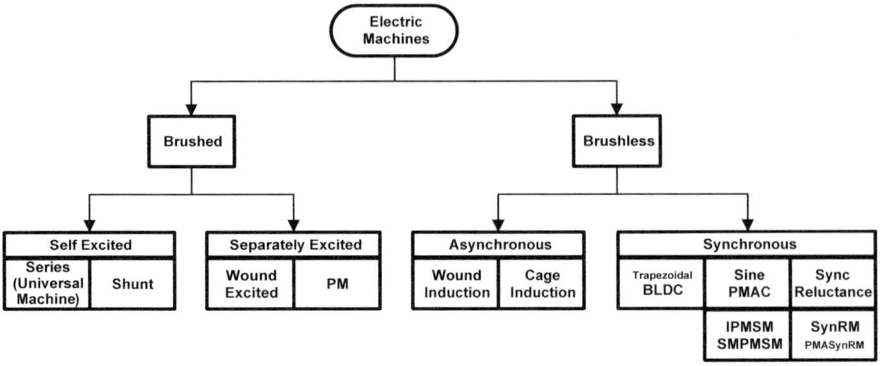

Fig. 4.1 Classification of electric machines based on their brushing mechanism

speed. Sudden load torque variations in the rotor can cause this synchronism to be lost, producing overcurrent and stopping the machine. For this reason, depending on the application, a system is needed to sense or to estimate the position of the rotor in brushless PMAC machines.

On the other hand, induction machines (IM) use the induction principle to induce a current in the shortcut squirrel cage, which produces a pair of forces to move the rotor. This induced current is produced by the same principle as the PMAC: Three-phase AC voltage produces a synchronous rotating magnetic flux at the same speed as the AC voltage. The relative difference speeds between stator and rotor produce the torque to move the rotor (asynchronous machine).

In Fig. 4.1, a classification scheme for electric machines into two big groups is shown, brushed and brushless machines. For brushed machines, the most common is the universal machine and the DC machine. It is classified in self-excited or separately excited machines. Brushless machines are classified in if the rotor speed is synchronous or asynchronous. The induction machine is the asynchronous machine which can be constructed with a wound or squirrel-cage rotor. For synchronous machines, it is possible to find more types as PMAC, brushless DC (BLDC), and synchronous reluctance machine (SynRM).

4.3 Brushed Machine

In brushed machines, it is very common to talk about the armature reaction, which is a resulting magnetic field distortion that arises as a result of interactions between the stator and rotor magnetic fields. The stator magnetic field is produced directly by the magnets or field winding current. The rotor magnetic field is produced by the armature current through the rotor winding itself. Armature reaction gets worse as armature current gets larger in magnitude. A compensation winding and a compensation pole are usually installed on the machine to cancel out the armature reaction. In this section, it is assumed that the magnetic structure does not saturate and the armature reaction is canceled by compensation winding.

4.3.1 Universal Machine

A universal machine is a self-excited machine designed to be used in most applications since it can operate with AC or DC. Some of its applications in home appliances are in vacuum cleaners, washing machines, and blenders where high speed, power, and small size are required. The terminal connections of some universal machine types allow connections in series and in parallel (separately excited), but in this section self-excited universal machine will be discussed.

Due to improvements in brushless machines and cost as well as power electronics, the applications for these machines are decreasing more and more. In applications where high torque is required at low speed, as well as low torque for high speed, this is still an interesting machine due to the cost of power electronics (typically TRIAC circuit) and the machine is less expensive than the brushless machine with its electric drive (three-phase inverter). For example, low-cost washing machines are still using universal machines connected in series and controlled through a TRIAC circuit, with two relays to change the direction of the motor. All these devices can be controlled by an eight-bit microcontroller in order to implement closed-loop speed control, usually using a speed sensor (tachometer). In the case of washing machines, universal machines are specially designed by the manufactures to include a tachometer generator on the machine shaft which, with low-cost electronic drive, allows competitive solutions when compared with other machines' types. Most low-cost washing machines use universal motors, as brushless machines offer higher costs because of the electric drive (although their efficiency is higher than universal machines). Brushless machines are used in high-end washing machines where noise and efficiency are more relevant.

In Fig. 4.2, the wiring of a series–shunt universal machine is shown. It can be observed that the stator current is the same as the rotor current if the machine is connected in series by joining T2 and T4. In this case, voltage is applied between one of the Lines *L* and neutral *N*. Line L2, in the middle of the coil, is the half field

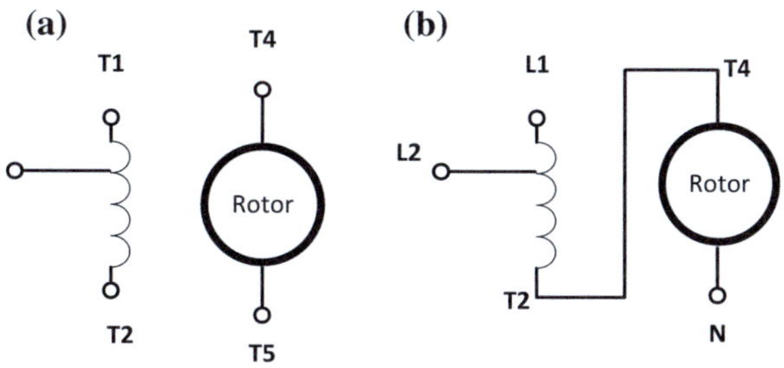

Fig. 4.2 **a** Universal machine connection wiring. **b** Universal machine connected in series

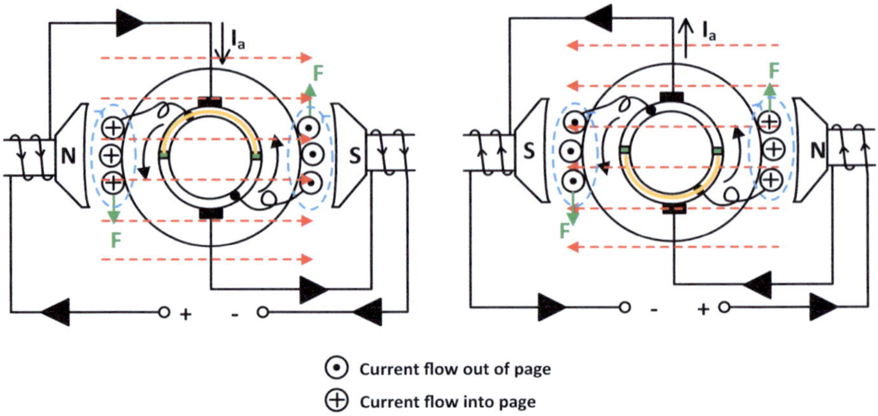

Fig. 4.3 Current flow and the magnetic field generated in a universal machine

or flux weakening control, used to increment the torque and the speed. To achieve higher torque, inductance is reduced (half field) and current is increased. In this case, the supply voltage is between L2 and *N*.

A mechanical commutator is used to change the current flow automatically from one rotor coil to the next as represented in Fig. 4.3. The commutator (usually in the rotor) is in contact with two brushes to connect the current. When the rotor is attracted by a magnetic field created in the stator, the commutator changes the current to the next pole to continue the rotation, thus maintaining synchronism. In Fig. 4.3, although the polarity of the voltage and current changes, the machine continues running in the same direction, since the armature current I_a and the polarity of the electromagnet change.

In Fig. 4.3 is represented two-pole machine with armature and field winding created by the same current I_a. The armature winding is connected through a commutator and a brush. The stator field, in horizontal, is created by an electromagnet according to the armature current, while the rotor flux lines are created by the armature winding, with three copper conductors with the same armature current itself. The interaction with stator filed and the armature current in the conductors creates a force vector F in the conductors as represented. As can be observed, the electromagnet flux (horizontal dashes lines) created by the stator is perpendicular or in quadrature to the rotor flux created by the armature current. This is the condition to get the maximum torque available of the machine which is performed thanks to the correct commutation of the armature current with the electromagnetic flux of the stator. In the figure, the distortion of the flux produced by the armature current has been omitted. When armature flux superimposes with the stator field flux, the spatial flux distribution is distorted as represented in Fig. 4.4. This effect is known as armature reaction in universal and DC machines.

The armature reaction debilitates the main flux and shift the position of the magnetic neutral axis, as shown in the figure, where the stator flux lines are perpendicular.

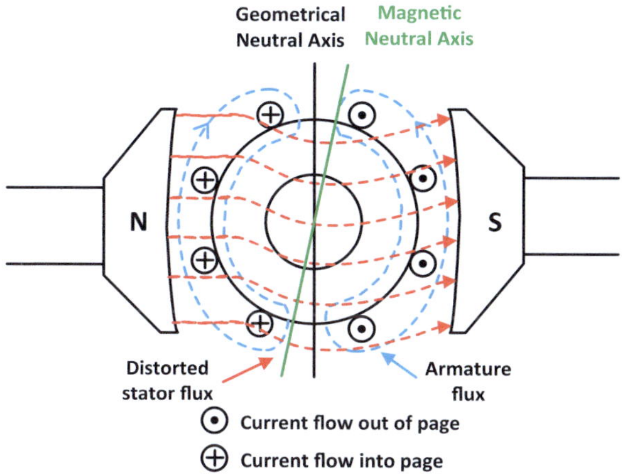

Fig. 4.4 Armature reaction when both fluxes are superimposed. As a consequence, the position of the magnetic neutral axis is shifted

Is in this axe where the brushed should be placed; otherwise, it will lead to sparking at the brush face. It is hard to determine the exact position of the magnetic neutral axis; thus, a compensation winding and/or interpoles can be used to reduce this effect. The compensation winding is placed in series with armature winding, to carry the same current but in the opposite direction to cancel the armature reaction effect. The interpoles are small auxiliary poles placed between the main field (stator) poles which are connected in series with the armature current. In the case of large machines, the compensation winding and the interpoles are indispensable.

A cross-sectional view of a DC machine for simulation purposes where the stator flux is performed by using permanent magnets is represented in Fig. 4.5. As can be observed, the stator flux is created by four permanent magnets, which indicates a four poles DC machine. The armature rotor consists of 12 slots with its corresponding

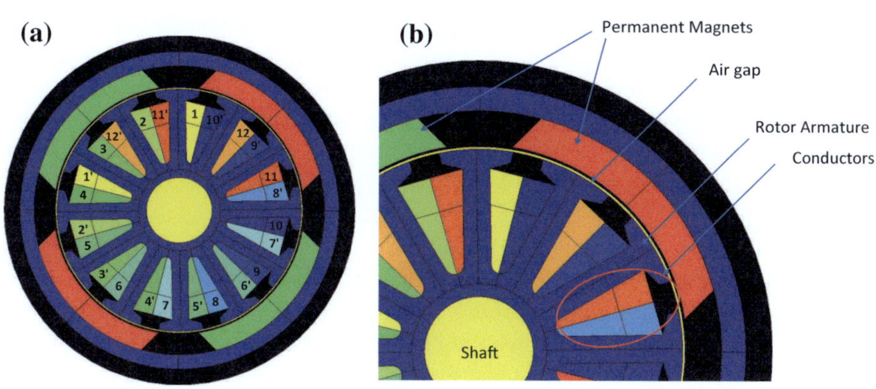

Fig. 4.5 Cross-sectional view of the 4 poles DC machine with rotor armature with 12 slots and stator flux created by permanent magnets. Image developed using Flux2D provided courtesy of Altair Engineering, Inc.

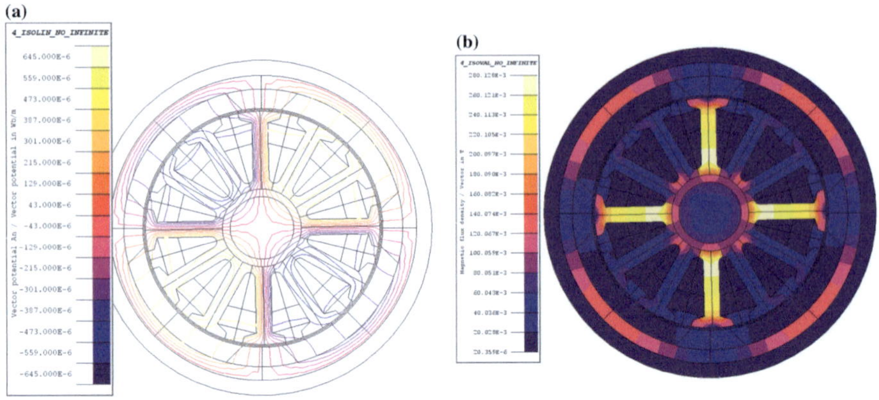

Fig. 4.6 DC machine FEA simulation with no magnets in the stator with current flowing through the armature winding. **a** Rotor flux lines are perpendicular to magnets flux lines to get the maximum torque available. **b** Magnetic rotor flux density in Tesla units. Image developed using Flux2D provided courtesy of Altair Engineering, Inc.

winding and 4 brushes equal to the number of poles. The numbers in Fig. 4.5a indicate the windings interconnection of the stator.

To get the maximum torque available, the rotor flux lines should be perpendicular or quadrature with the magnet's lines. If the permanent magnets of the stator are removed, the flux line in the armature is due only to the current through the armature windings. In Fig. 4.6a is represented this situation, an FEA simulation without permanent magnets, but with nonzero armature current at the best brush angle position to get this optimal point. In Fig. 4.6b is represented the magnetic rotor flux density in Tesla units in the same situation.

As mentioned before, in-series machines without permanent magnets, the stator and the rotor are in series, so the same current is responsible for producing the magnetic field in the stator and for turning the rotor. The more voltage that is applied to the series circuit (rotor-stator), the more the current flows. The current creates the magnetic field so that it is stronger as the current increases. Since torque is proportional to the magnetic field, the higher the magnetic field is, the more torque there is produced in the rotor. This torque is opposed to the load torque, and only when these two forces are compensated is speed stabilized. Depending on the load torque, the magnetic field will be strong or weak; that is, the current may be high or low, or the applied voltage may be high or low.

DC voltage yields a little more efficiency than AC voltage since iron or core losses are smaller. DC voltages imply working in one quadrant of the hysteresis curve of Fig. 4.7, while AC voltages imply working in two quadrants. Each time the magnetic field is reversed, a small amount of energy is lost as a result of hysteresis within the core from friction. With DC voltage, this friction is avoided. In applications where

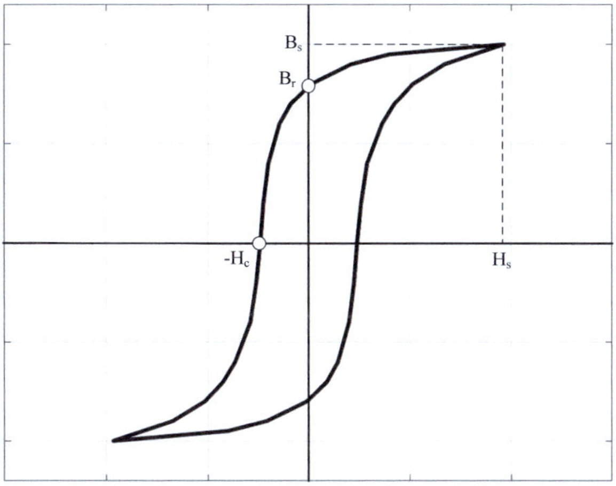

Fig. 4.7 Typical hysteresis curve loop for ferromagnetic materials

efficiency is relevant, and the available source is AC voltage, it will be necessary to convert to DC using a diode bridge rectifier, although this implies a more expensive solution.

A typical hysteresis curve for a ferromagnetic material is shown in Fig. 4.7. In the figure, the saturation point of that material is when a high magnetizing force H_s is applied and does not cause a useful increase in B. The saturation flux density B_s is produced at the magnetizing force H_s as shown in Fig. 4.7. The remanence B_r flux is the polarized flux remaining in the core after the excitation has been removed. The magnetizing force, $-H_c$, is called coercivity. It is the amount of magnetizing force required to bring the remanence flux density back to zero. The *B–H* curve of a magnetic material is non-linear and is highly dependent on its operating condition, such as frequency, magnetic intensity, and temperature.

The core loss is due to hysteresis loss and the eddy current loss. Hysteresis loss is proportional to the area of the *B–H* curve of the material, while the eddy current loss is caused when the lines of flux pass through the core, inducing electrical currents in it. The eddy currents produce heat in the core and can be minimized with a core with high electrical resistance. The typical method to increase the electrical resistance is by using thin sheets of laminations of the magnetic material. For example, in the stator core of the electric machines is usually not built as a single solid core, but stacked together with thin (0.1–1 mm) lamination sheets. These lamination sheets have a very thin (some μm) insulation layer to prevent eddy currents flowing in the axial direction of the core.

In the designing magnetic components or electric machines, the core loss is a significant design factor which can be controlled by selecting the right material and thickness. It will prevent overheating that could result in damage to the wire insulation and/or the potting compound.

4.3.1.1 Torque Variation

The machine is driven by a voltage which is proportional to the developed torque. In Fig. 4.8, how voltage can be modified depending upon the firing angle of the triode AC (TRIAC) switch controller is shown. The TRIAC controller, in this case, acts like an AC-AC converter, since it converts from one AC voltage to another AC voltage with a lesser or equal root mean square (RMS) value. The AC-AC converter, when it is used to control a universal machine, is also known as an electric drive.

The corresponding formula for a continuous function that can be used to calculate the RMS is as follows:

$$f_{RMS} = \sqrt{\frac{1}{T_2 - T_1} \int_{T_1}^{T_2} [f(t)]^2 \mathrm{d}t} \tag{4.1}$$

By applying this formula to the sinusoidal signal based upon the firing angle, it is possible to obtain:

$$V_{o\_RMS} = \sqrt{\frac{1}{T/2 - \alpha} \int_{\alpha}^{\frac{T}{2}} [V_i \sin \omega t]^2 \mathrm{d}t + \frac{1}{T - \alpha} \int_{\alpha}^{T} [V_i \sin \omega t]^2 \mathrm{d}t} \tag{4.2}$$

Lastly, by performing some operations, the output RMS value of the AC-AC converter, which depends on the firing angle, can be calculated as:

$$V_{o\_RMS} = \frac{V_{i\_peak}}{\sqrt{2}} \sqrt{\frac{1}{\pi}\left(\pi - \alpha + \frac{1}{2} \sin 2\alpha\right)} \tag{4.3}$$

If the firing angle α increases, less RMS voltage is applied to the machine (low torque or low speed). On the contrary, if the firing angle α decreases close to 0°, more RMS voltage is applied to the machine (high torque or high speed). In Fig. 4.8, output RMS variation is shown in accordance with the firing angle in degrees. It can

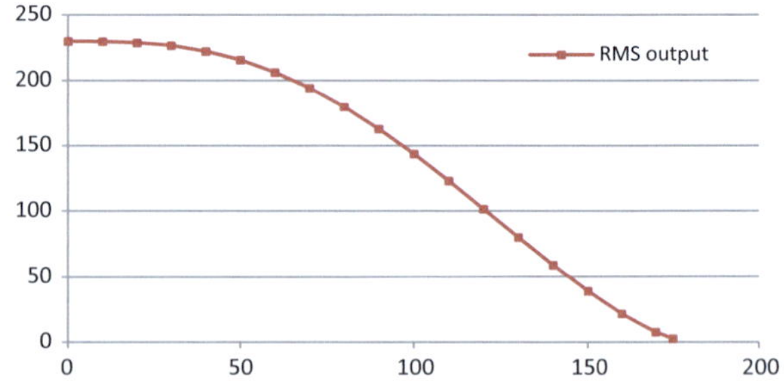

Fig. 4.8 Voltage output variation in accordance with the firing angle on the *x*-axis in degrees. For 110°, the voltage output is 143.63 V

be observed that voltage output variation is not linear as expected, due to the fact that portions of the sinusoidal signal are taken.

In Fig. 4.9, an example firing angle of 110° is shown. The shaded area represents the RMS voltage in the output of the AC-AC converter. Zero-crossing points for the input line voltage are also shown and must be detected to allow for synchronization with the firing angle. With an easy trimmer circuit, the sinusoidal signal can be converted into a digital signal so that both edges may be read by a microcontroller or DSP by using an input capture timer. This timer always monitors the current period of line voltages of 50 or 60 Hz and related variations. Once the zero crossing is detected at number 1 of the figure, a timer starts and waits the necessary time according to the desired firing angle (number 2). In that case, if there is a 50 Hz voltage line frequency, the half period is 10 ms and, for a 110° firing angle, the wait time must be 6.1 ms. When this time has elapsed, activation pulses must be produced in the gate of the TRIAC controller to switch ON and to give the desired voltage at the output of the AC-AC converter (number 3). Here, two pulses (burst pulses) are shown; however, there can be more than two, and this depends on the inductive load of the machine. Instead of using short pulses, it is possible to use a long pulse till next half period. Finally, number 4 shows the wait for the next zero crosses.

In the event that DC voltage is used to supply the machine, the AC signal must be converted to DC. If a full-wave diode rectifier is used, a negative semiperiod will appear on the positive side, as can be seen in Fig. 4.9b. The same above explanation is valid for the machine supplied by DC voltage.

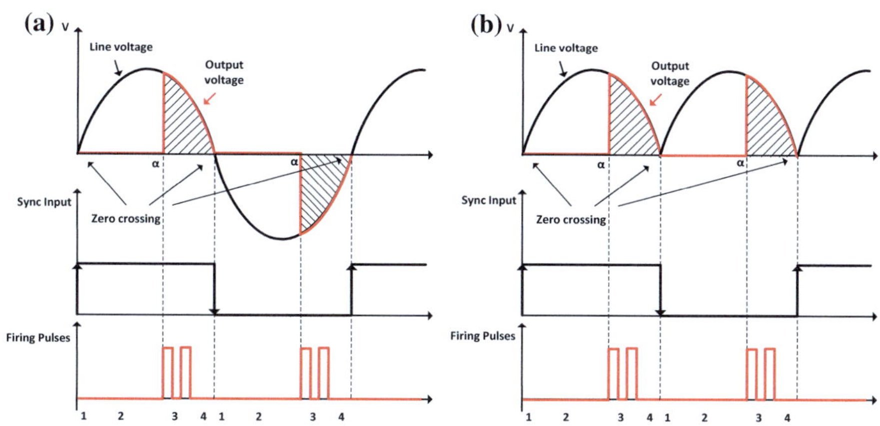

Fig. 4.9 Voltage variation in accordance with the firing angle. **a** For AC universal machine supply. **b** For DC universal machine supply

4.3.2 Self-Excited and Separately Excited Torque Expression

In order to facilitate the following analysis, it is considered that the machine is supplied by DC current and the armature reaction is entirely canceled out with the interpoles and the compensation winding.

In a general way, it is common to describe the DC machine model with separately excited terminals:

$$V_f = R_f i_f + \frac{d\lambda_f}{dt} \tag{4.4}$$

$$\lambda_f = L_f i_f \tag{4.5}$$

where V_f is the voltage applied to the separately excited field winding terminals. A current i_f is created in the coil L_f (equivalent field winding inductance), producing a flux λ_f. R_f is the equivalent field winding resistance. The flux linkage λ_f can also be provided by a permanent magnet. Thus, the flux can be considered constant.

In Fig. 4.10 is represented the equivalent circuit and the machine structure showing the commutator, brush, and how it is connected to the voltage supply. As commented before, in order to get the maximum torque available, the field and armature fluxes should be in quadrature.

The back electromotive force (back-EMF) is the voltage induced in stator terminals when the rotor is in motion and it is proportional to the speed and flux produced by separately exciting voltage. Here, K_e stands for the back-EMF constant in V/(Wb turn rad/s), while machine speed ω_m is in rad/s. It is considered one pair pole so mechanical and electrical matches. λ_f is the linkage flux to the armature current in Wb turn, produced by the field winding current. The back-EMF voltage depends on the machine speed ω_m, the back-EMF constant K_e, and the flux linkage λ_f as shown in (4.6):

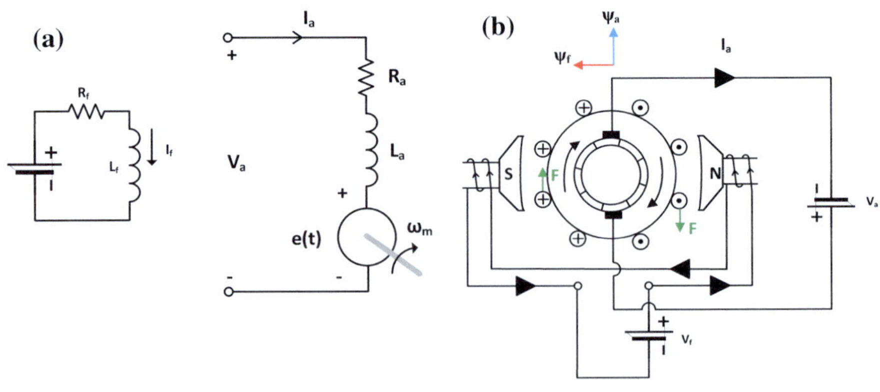

Fig. 4.10 Separately excited machine. **a** Equivalent circuit. **b** Commutator, brush, and DC voltage supplies

$$e(t) = K_e \lambda_f \omega_m(t) \tag{4.6}$$

Thus, V_a is the armature voltage that is made up of the voltage drop in the equivalent armature winding resistance, the voltage in the equivalent armature winding inductance, and the back-EMF (4.7). Additional terms may be added in (4.7) for brush voltage drop; however, here, this idea is neglected:

$$V_a(t) = R_a i_a(t) + L_a \frac{\mathrm{d}i_a}{\mathrm{d}t} + e(t) \tag{4.7}$$

The torque produced by the DC motor is proportional to the armature current I_a, flux linkage λ_f, and torque constant K_T as described in Eq. (4.8). Units for K_T constant are Nm/(Wb turn A):

$$T_e(t) = K_T \lambda_f i_a(t) \tag{4.8}$$

According to (4.8), if the flux linkage λ_f is considered constant, either by considering permanent magnets or field winding is excited by a constant current i_f, the DC machine differential equations can be considered as a linear system.

It is possible to get some torque/speed curves for better understanding. Equations (4.6)–(4.8) show that separately excited shunt machines have the advantage that currents, armature, and field winding can be controlled independently, obtaining lineal versatile torque–speed curves' characteristics when provided field flux remains constant. In steady-state, (4.7) is simplified as

$$V_a(t) = R_a i_a(t) + e(t) \tag{4.9}$$

where derivative term $\mathrm{d}i_a(t)/\mathrm{d}t$ is zero in steady-state. Then, by using Eqs. (4.6), (4.8), and (4.9), eliminating $i_a(t)$ and $e(t)$, Eq. (4.10) is deduced.

$$T_e = K_T \lambda_f \frac{V_a(t)}{R_a} - \frac{K_T K_e \lambda_f^2}{R_a} \omega_m \tag{4.10}$$

In Fig. 4.11, the torque versus speed curves is shown in accordance with (4.10). In order to calculate these curves, three voltages have been used, with the nominal parameters for a 600 W machine shown in Table 4.1.

The cubic load torque, typical in water pump systems, is shown in the figure below, with three working points *A*, *B*, and *C*. If the machine is working at point *A*, and the speed should be incremented to point *B*, the voltage of the machine must also be incremented to reach the $V_a = 100$ V curve. Between these two points, there is infinite torque versus speed curves. The same occurs if speed must be incremented to point *C*, where the final voltage will be 150 V. In the case of speed reduction (e.g., from point *B* to *A*), the voltage must be reduced until it reaches 50 V. It can also be seen that Eq. (4.10) and Fig. 4.11 lead to a stable behavior of the machine. As

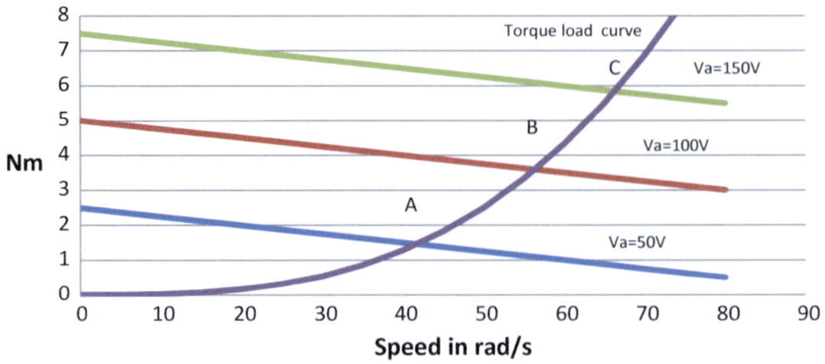

Fig. 4.11 Torque–speed curves characteristic for separately excited DC machine 220 V–600 W

Table 4.1 Parameters used to calculate the torque versus speed curves

Parameter description	Symbol	Value	Unit
Equivalent field winding resistance	R_f	0.01	Ohms
Back-EMF constant	K_e	10	V/(Wb turn rad/s)
Torque constant	K_T	0.01	Nm/(Wb turn A)
Equivalent field winding inductance	L_f	0.001	H
Coil resistance	R_a	0.01	Ohms
Magnet flux	λ_f	0.05	Wb turn
Friction torque	T_f	0	Nm

the speed increases, the torque decreases, thus reducing the power supplied to the system and, consequently, reducing the speed as well.

On the other hand, closed-loop control with speed feedback allows a proper load torque control by regulating the machine voltage. If the voltage is fixed (e.g., 50 V in an open loop) and the load torque curve changes with more torque, the machine's speed will be reduced.

According to Eq. (4.10), torque development is constant even though speed is increased since the negative term of the equation is increased by the speed but compensated by the increase in V_a voltage. This is true until the base voltage of V_a is reached. At this point, since more voltage is not permitted, if speed is increased, field flux should be reduced. If field flux is reduced, speed can be extended; however, in order to yield the same torque, in accordance with (4.10), armature current should be increased in the same proportion. Since armature current is limited to nominal value to keep from overheating the machine (copper losses), torque will start to decrease proportionally to the reduction of field flux as depicted in Fig. 4.12. This is the flux weakening principle.

In a series excited DC motor, the armature current is the same as the field current, and characteristic torque–speed curves (4.11) vary in a quadratic manner, as shown in Fig. 4.13. High torque is present at low speed, and low torque is present at high speed. In the first part of the curve of Fig. 4.12, the torque varies linearly with the current, back-EMF is small, and flux is constant. When speed is increased, torque undergoes a quadratic decrement because of the field weakening. Half-field techniques are used to increase the current and torque at these speeds.

$$T_e = K_T L_f \frac{V_a^2}{\left(R_a + R_f + \omega_m K_e L_f\right)^2} \tag{4.11}$$

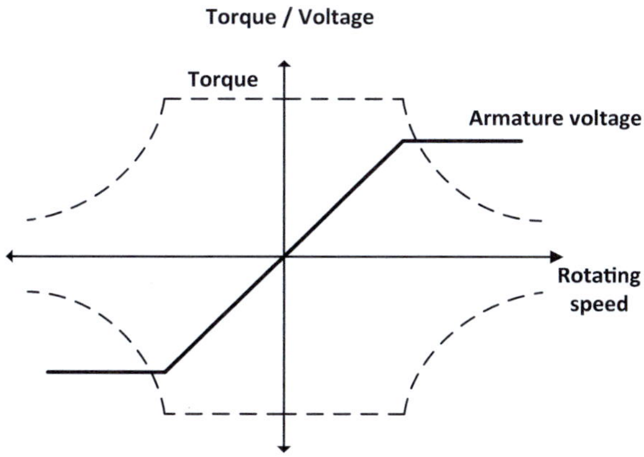

Fig. 4.12 Torque–speed operating area

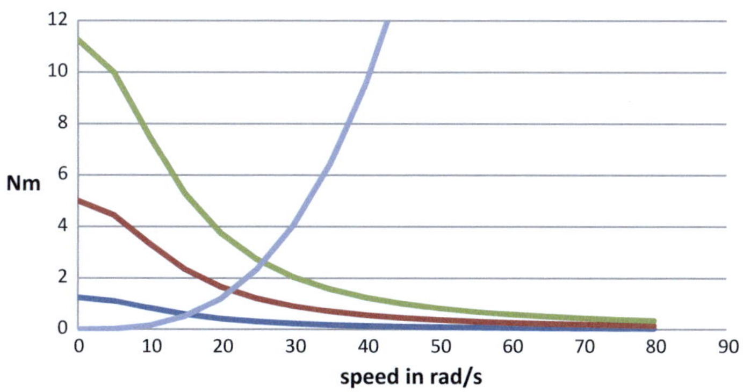

Fig. 4.13 Torque–speed curves which are characteristic for series–shunt DC machines

4.3.3 Brushed Machine Operation

The brushed machine is working in motoring mode when it takes energy from the supply source; i.e., the motor develops electromagnetic torque to move a connected load. The same behavior is also valid in the reverse direction when current and voltage are negative, but the power (which is the product of voltage and current) is still positive. When torque developed is opposite to the machine's rotational speed, the machine is working in regenerative braking. Depending on the direction, forward or reverse current, and voltage will be either negative or positive. In regenerative braking mode, the resulting power is negative; thus, the machine is working as a generator and energy flows from the machine to the supply source. The motoring and regenerative mode example of an in-series brushed machine can be observed in Fig. 4.14. Rotational speed direction ω_m, electromagnetic torque $T_{em,}$ and load torque T_L are shown in the figure.

The operation modes explained previously can be placed on a Cartesian coordinate system where the *y*-axis represents the machine speed, and the *x-axis* represents the torque. There are four quadrants where the brushed machine can operate. The first and third quadrants are when the machine is operating in the motoring mode; i.e., there is power flow from the supply source to the machine. On the contrary, in the second and fourth quadrants, the machine is operating in regenerative braking mode, in which power flows from the machine to the supply source. The regenerative operation is only possible if the electric drive can reverse the energy (e.g., bidirectional converters).

The above concept is valid for other machines, such as induction machines (IM), PMAC, and SynRM. In Fig. 4.15, the four operations' modes on the speed/torque axes are shown for the brushed machine. In the first quadrant, the machine rotates in a clockwise direction thanks to the torque, which is also going in the same direction. The machine is generating torque; induced EMF, voltage, current, and power are positive. Similar operations occur when the machine is rotating anticlockwise in the third quadrant; however, in this case, induced EMF, voltage, and current are

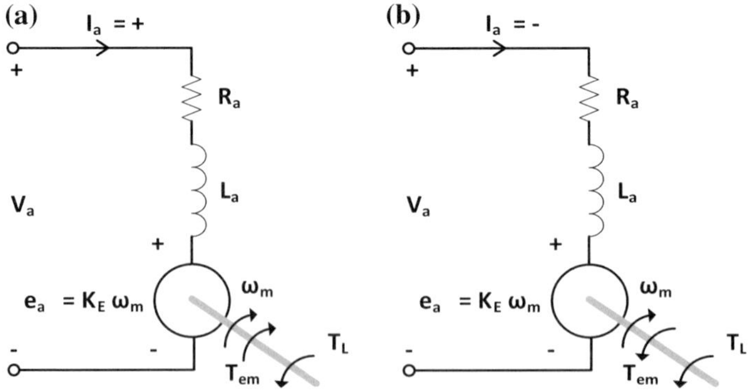

Fig. 4.14 **a** Universal machine in motoring stage. **b** Universal machine in regenerative braking stage

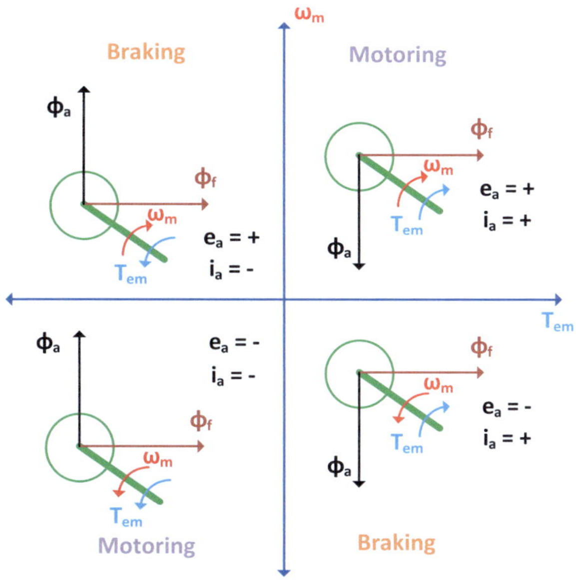

Fig. 4.15 Brushed machine operations in accordance with the direction of power flow

negative, and power is still positive. It should be noted that armature flux ϕ_a changes its direction to change the direction of the machine, and it follows the same direction as the current, as expected. In regenerating braking mode, the energy is fed back while braking, since torque and speed are in opposite directions. Current and torque are in a reversed direction.

4.4 Three-Phase Brushless AC Machine

The three-phase brushless AC machines as the asynchronous induction machine (IM), the synchronous permanent magnet AC machine (PMAC), and the synchronous reluctance machine (SynRM) do not need brushes to be able to operate, unlike the DC machine with brushes seen in Sect. 4.2. They share a similar structure in the stator formed by a three-phase winding to form a rotating magnetic field (Fig. 4.16), while the rotor may or may not have permanent magnets depending on the machine type. The three-phase windings can be connected into either star connection or delta connection.

The brushless AC machines are replacing the DC machines in numerous applications due to its higher efficiency, high robustness, low maintenance, and even low cost, for the case of the induction machine. The induction machine is still the most popular machine in the industry for its simplicity, its low cost, and because its simple manufacturing. Moreover, the induction machine can operate directly with the electrical

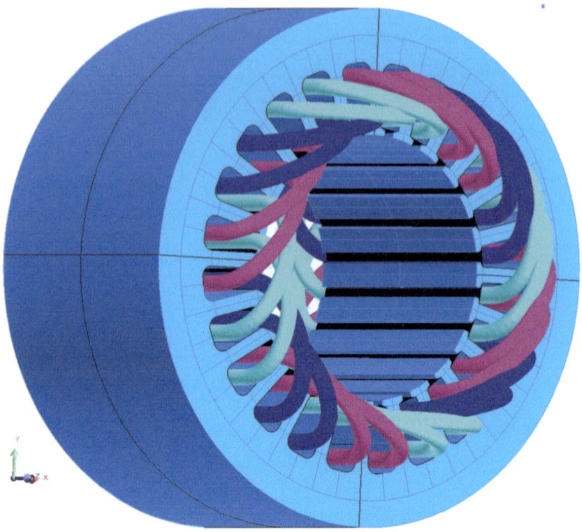

Fig. 4.16 Stator with 4 poles, 24 slots, and three-phase distributed windings. Image developed using Flux3D provided courtesy of Altair Engineering, Inc.

network. The use of the three-phase inverter with closed-loop control allows obtaining similar performances to those of the DC machine with brushes. The advance in the design of induction machine through FEA, the advance of the power electronics, and the high microcontrollers/DSPs processing have allowed maximizing its performance in terms of maximum torque per ampere and the optimal extension of the maximum speed. However, it is a fact that the PMAC machines and the SynRM are increasingly replacing the induction machine in numerous applications because they increase power density and efficiency (Weh et al. 1990) and can operate with a unity power factor. They usually have better use of energy so that it is opening a large number of new possibilities in the industrial, domestic, and automotive sectors, increasingly demanding with CO_2 emissions. However, the PMAC machine and the SynRM cannot be connected directly to the power grid as in the case of the induction machine, but require an electric drive (inverter).

As in the case of the DC machine, the three-phase brushless AC machines also have a steady-state equivalent circuit that allows calculating, in different points of operation of the machine, power, efficiency, torque, etc. The parameters of the equivalent circuits that are explained in this section should have sufficient precision to obtain representative results (close to reality) and to optimize the control as it will be seen in Chaps. 10 and 11. The electrical machines' manufacturers can provide these parameters either by means of specific tests, as a no-load test, locked-rotor test, and DC test, carried out in a laboratory, or by means of software based on FEA simulation. Otherwise, they can be obtained ourselves by means of this specific test which requires expensive laboratory instruments to perform tests without load and with a locked rotor. In most cases, these tests are impracticable since inverters and machines are usually offered separately, and because the machine in a test must rotate without load and another test must be mechanically locked. To avoid these inconveniences, the inverter should have mechanisms either to insert parameters manually or to measure or estimate some of the machine parameters by means of voltage and current injection in the machine terminals in a standstill and without blocking its rotor. In (Quang et al. 2015), it is described how to estimate the necessary parameters for the equivalent circuit and for its control through the nominal characteristics of an induction machine (nameplate), or through voltage injection in the terminals of the machine (standstill test).

On the other hand, the mathematical modeling of the physical relations in three-phase AC machines is composed of first-order differential equations, one for each winding. These differential equations are coupled to each other through mutual inductances between the windings. The coupling between the stator and rotor is a function of the rotor position. When the rotor rotates, the couplings make it vary in time. The transformation of axes of Chap. 3 basically uncouples variables, in a system, of complex equations variant in time, in order to obtain a more straightforward system of equations' invariant in time. These transformations serve for studies of the AC machines in some transitory conditions of their behavior and to obtain their dynamic model for simulation. In steady-state condition, where derivative terms are approximated to zero, or it has small variations, in three-phase power distribution systems per phase terms' analysis has been used to simplify to a single-phase system. For example, an induction machine has been traditionally analyzed by using the per-phase equivalent circuit, which is the representation of the single-axis equation as it will be seen in this chapter.

On the other hand, the theory of field orientation control (FOC) allows that brushless AC machines can be modeled as a magnetized DC machine separately through the axis transformations discussed above. With the model of the AC machine transformed, control methods similar to those of the DC machine can be applied, as will be seen in Chap. 5. The FOC control or also known as vector control can obtain optimal performance in the transients, a higher maximum speed extension, and torque control without oscillations from zero speed, displacing more and more the use of the universal machine and the DC machine.

4.4.1 AC Induction Machine

The induction machine is widely used in industry since it can easily be supplied directly from the AC mains. However, when it is directly connected to the AC mains, traditional induction acquires the high performance only at speed near the synchronous speed due to its working principle. The most rotor structure used in the induction machine is squirrel-cage rotors made of copper or die-cast in aluminum, which contain conductor bars and end rings, as shown in Fig. 4.17a. The rotor cage is a closed conductor as the conductor bars are all short-circuited by the end rings (Fig. 4.17b) acting as a three-phase sinusoidally distributed short-circuited winding

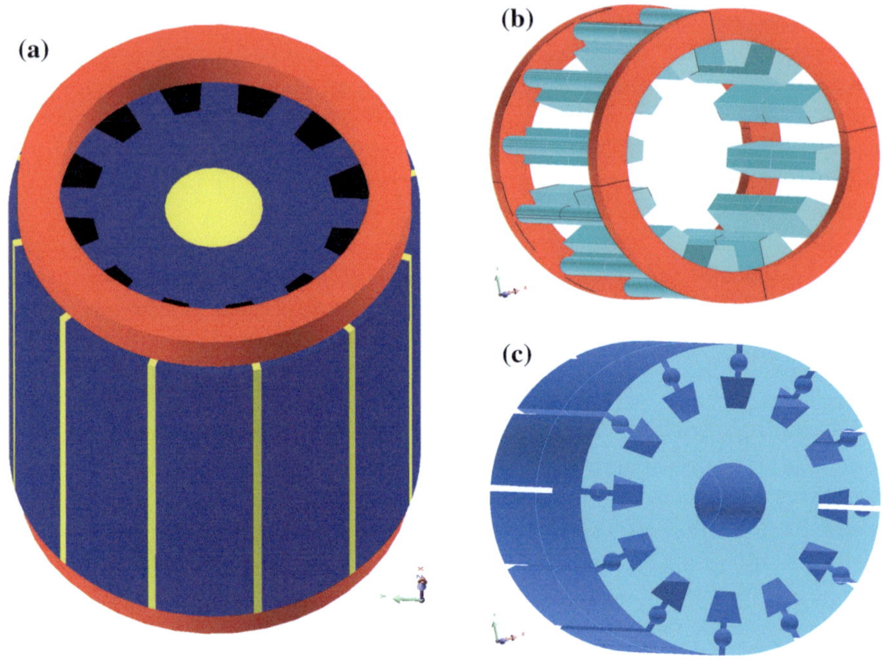

Fig. 4.17 **a** Complete squirrel-cage rotor of an induction machine (no skewing rotor). **b** Squirrel cage with conductor bars and end rings. **c** Die-cast. Image developed using Flux3D provided courtesy of Altair Engineering, Inc.

with finite rotor resistance (Hughes 1994). In the figure is shown the rotor cage without skewing the conductor's bars. Skew of conductor bars is used to reduce cogging torque and harmonic torque components.

The rotor is mounted on the shaft with two bearings, and one shaft ends are used for driving the load while the other one for mounting the shaft position or the speed measurement devices. The other less used rotors are wound rotor which uses brushes/slip rings to access to the three-phase windings of wound rotor. The advantage of this type of induction machine is that additional external resistance can be connected to modify the rotor resistance changing the torque/speed characteristics.

As seen in Fig. 4.16, the stator can be made up with several pole pairs. Each pole pair has three windings placed symmetrically in space which are supplied with three-phase voltages. The three-phase sinusoidal voltages with 120° phase shift in the three-phase stator winding generate a sinusoidal rotating magnetic field in the air gap. The relative speed between stator rotating magnetic field and rotor conductors causes an induced EMF in the rotor conductors, according to Faraday's law of electromagnetic induction. The EMF induces a current in the rotor since the squirrel-cage model is composed of resistance and inductance. The induced current in the rotor will also produce alternating flux around it. This rotor flux lags behind the stator flux. The direction of induced rotor current, according to Lenz's law, is such that it will tend to oppose the cause of its production resulting in a mechanical pair of opposing forces that cause the rotor to rotate. There is no electromagnetic torque developed if the rotor speed matches the speed of the magnetic field since no rotor current is induced if there is no relative movement between rotor bars and the magnetic field.

In Fig. 4.18a is illustrated a cross section of a two-pole induction machine made with 36 stator slots and 30 rotor slots. Every phase has 12 slots, 6 for the positive phase, and others 6 for the negative phase. In this case, the machine has distributed winding as can be observed. There is another type of winding distribution called concentrated distribution winding, which will be seen in Sect. 4.4.2. In Fig. 4.18b is represented the flux lines FEA simulation result with no load in one instant time where two poles have been created with the three windings supplied by a three-phase voltage.

In Fig. 4.18c, d are represented another induction machine operating at the rated load where flux lines are represented for two-time instants. It is possible to observe

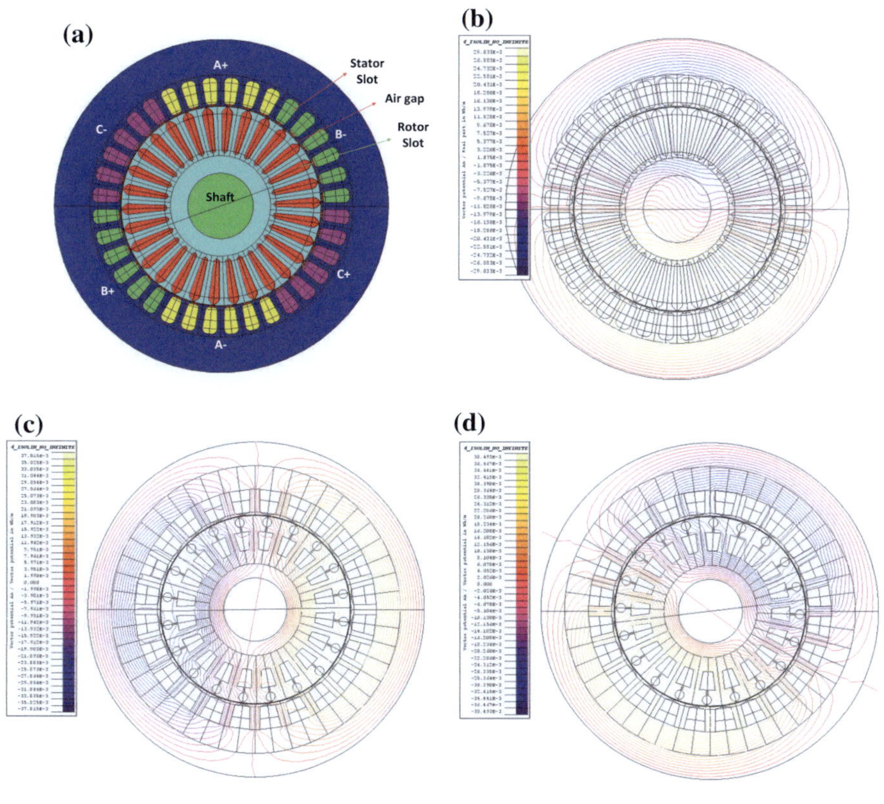

Fig. 4.18 **a** Cross-sectional view of the induction machine with 36 stator slots, 30 rotor slots, and two poles. **b** FEA simulation of flux lines where two poles have been crated (no load). **c** Flux lines of an induction machine operating at rated load. **d** Another instantaneous time. Image developed using Flux2D provided courtesy of Altair Engineering, Inc.

that most of the flux lines cross the air gap diagonally, trying to rotate the rotor in counterclockwise. It is possible to say that the flux lines act like rubber bands which try to contract (Pyrhönen et al. 2016). The flux lines in Wb/m are shown in a color scale.

The rotor must always rotate at speed lower than that of the stator so that the induction machine can operate since it is strongly related to the slip angle. In that case, the magnetic field vector will surpass the vector of the surface, since it rotates at a higher speed. In steady-state, the slip is constant, while the angle (integral of the speed) has a straight line with slope $\omega_e - \omega$.

The relation between the mechanical ω_{sm} and electrical speed ω_e sawn is:

$$\omega_{sm} = \frac{\omega_e}{P}. \tag{4.12}$$

The synchronous speed in RPM is:

$$N_s = \frac{60 \cdot \omega_{sm}}{2\pi} = \frac{60 f_e}{P} \tag{4.13}$$

where f_e stands for the synchronous frequency in Hz and P the number of pair poles. The slip s is expressed in (4.14) as the difference between speeds or sometimes in a normalized manner (per unit):

$$s = \omega_{sm} - \omega_{rm}$$

or

$$s = \frac{\omega_{sm} - \omega_{rm}}{\omega_{sm}} = \frac{\omega_e - \omega_r}{\omega_e} \tag{4.14}$$

4.4.1.1 Space Vector Theory in Induction Machine

In the induction machine, only the stator windings are fed by a voltage or current source by means an inverter or directly by the three-phase grid as discussed. The stator and rotor are linked only by the flux interaction as expected in the brushless machine. The space vector theory of Chap. 3 is usually applied in the analysis of the three-phase AC machines since simplifies the machine representation considerably. The equations for each winding can be a hard task which is avoided with the space vector theory. In addition, the equivalent circuit can be obtained in a simple way.

On the other hand, an alternative frequently used is the representation in the two-phase method based on the orthogonal system also of Chap. 3, which simplifies, even more, the equations as it will be shown in Sect. 4.4.1.2.

To describe the general equations of the induction machine in space vector theory, the assumptions are exposed below (Krause 1986):

- Constant air gap
- Magnetic linear circuit
- Three-phase stator windings are identic and symmetric. Only one spatial sinusoidal distributed magnetomotive force (MMF) is generated with sinusoidal distributed and balance currents and voltages
- Winding or cage rotor with a spatial sinusoidal distribution of MMF forces with the same number of poles of the stator
- Stator and rotor permeability is assumed to be infinite
- Effects of anisotropy, magnetic saturation, iron losses, and eddy currents are neglected
- Coil resistance and reactance are taken to be constant.

The stator voltage equations of induction machine in the stator reference frame system are expressed with sub-index s for a, b, and c as follows:

$$\boldsymbol{V}_{abcs} = R_s \boldsymbol{i}_{abcs} + \frac{\mathrm{d}\boldsymbol{\Psi}_{abcs}}{\mathrm{d}t} \tag{4.15}$$

where R_s is the stator winding resistance and $\boldsymbol{\Psi}_{abcs}$ is the total flux composer by the sum of the stator and rotor flux.

According to vector theory, the stator vector voltage in the stator reference frame is given in the familiar form:

$$\boldsymbol{V}_s^s = R_s \boldsymbol{i}_s^s + \frac{\mathrm{d}\boldsymbol{\Psi}_s^s}{\mathrm{d}t} \tag{4.16}$$

Since squirrel cage can be modeled as three-phase windings, the rotor vector voltage can be easily obtained as

$$\boldsymbol{V}_r^r = R_r \boldsymbol{i}_r^r + \frac{\mathrm{d}\boldsymbol{\Psi}_r^r}{\mathrm{d}t} \tag{4.17}$$

where R_r is the rotor resistance. Equations (4.16) and (4.17) are only one vectorial equation which describes the machine behavior instead of three-phase equations where zero-sequence component has been neglected.

The stator flux linkage $\boldsymbol{\Psi}_s$ and the stator current $\boldsymbol{i}_s$ can be described from the rotor reference frame with

$$\begin{aligned} \boldsymbol{\Psi}_s^s &= \boldsymbol{\Psi}_s^r e^{j\theta_r} \\ \boldsymbol{i}_s^s &= \boldsymbol{i}_s^r e^{j\theta_r} \end{aligned} \tag{4.18}$$

where the angle θ_r is the angle between the stator and rotor axes. $\boldsymbol{\Psi}_s^r$ and $\boldsymbol{i}_s^r$ are the space vector of the flux linkage and current, respectively, in the reference frame fixed to the rotor.

Then, Eq. (4.16) can be expressed in the stator reference frame, but using the rotor reference frame quantities as

$$\boldsymbol{V}_s^s = R_s \boldsymbol{i}_s^r e^{j\theta_r} + \frac{\mathrm{d}}{\mathrm{d}t}\left(\boldsymbol{\Psi}_s^r e^{j\theta_r}\right) = R_s \boldsymbol{i}_s^r e^{j\theta_r} + \frac{\mathrm{d}\boldsymbol{\Psi}_s^r}{\mathrm{d}t} e^{j\theta_r} + j\omega_r \boldsymbol{\Psi}_s^r e^{j\theta_r} \tag{4.19}$$

If both sides of the previous equation are multiplied by $e^{-j\theta r}$, Eq. (4.20) is derived which corresponds to the stator voltage equation in the rotor reference frame.

$$\boldsymbol{V}_s^s e^{-j\theta_r} = \boldsymbol{V}_s^r = R_s \boldsymbol{i}_s^r + \frac{\mathrm{d}\boldsymbol{\Psi}_s^r}{\mathrm{d}t} + j\omega_r \boldsymbol{\Psi}_s^r \tag{4.20}$$

In a common rotating coordinate system designed by c symbol with angular speed ω_c, Eq. (4.20) can be rewritten as

$$\boldsymbol{V}_s^c = R_s \boldsymbol{i}_s^c + \frac{\mathrm{d}\boldsymbol{\Psi}_s^c}{\mathrm{d}t} + j\omega_c \boldsymbol{\Psi}_s^c \tag{4.21}$$

The rotor voltage equation can be derived in the same form as shown for the stator voltage but with the consideration that induced voltage term is due to the difference of speeds:

$$\boldsymbol{V}_r^c = R_r \boldsymbol{i}_r^c + \frac{\mathrm{d}\boldsymbol{\Psi}_r^c}{\mathrm{d}t} + j(\omega_c - \omega_r) \boldsymbol{\Psi}_r^c \tag{4.22}$$

where ω_r is the rotor electrical speed.

The stator and rotor flux linkages can be written as

$$\begin{aligned} \boldsymbol{\Psi}_s^c &= L_s \boldsymbol{i}_s^c + L_m \boldsymbol{i}_r^c \\ \boldsymbol{\Psi}_r^c &= L_r \boldsymbol{i}_r^c + L_m \boldsymbol{i}_s^c \end{aligned} \tag{4.23}$$

In Eq. (4.23), L_m is the magnetizing inductance, $L_s = L_m + L_{ls}$ is the total inductance of stator, and $L_r = L_m + L_{lr}$ is the total inductance of the rotor. L_{ls} and L_{lr} are the leakage inductance of the stator and the rotor, respectively.

With Eqs. (4.21)–(4.23) is possible to describe the T equivalent circuit in Fig. 4.19 based on the space vector theory. This equivalent circuit is also valid for transient states where iron losses are neglected.

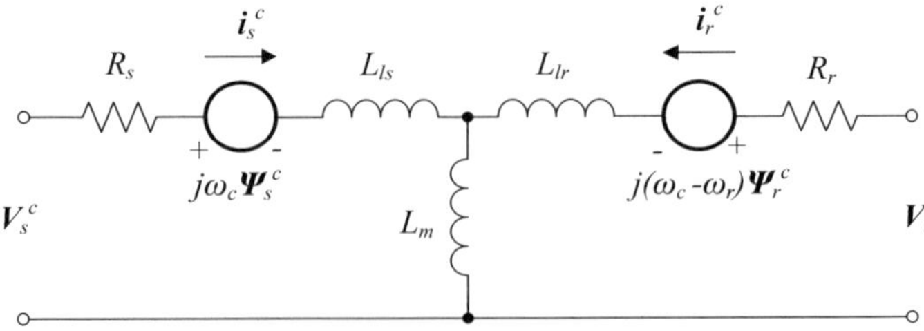

Fig. 4.19 Equivalent circuit of an induction machine in general frame according to the space vector theory

Equations (4.21)–(4.23) describe the induction machine in the general frame or arbitrary reference. Hence, the reference frame fixed to the stator is when $\omega_c = 0$ (stationary reference frame), and reference frame fixed to the rotor is when $\omega_c = \omega_r$. Lastly, the reference frame is rotating at synchronous speed when $\omega_c = \omega_e$ (synchronously rotating reference frame). Sometimes called the reference frame fixed to the magnetic field. For example, in a reference frame fixed to the rotor ($\omega_c = \omega_r$), Eqs. (4.21)–(4.23) can be rewritten as

$$\boldsymbol{V}_s^r = R_s \boldsymbol{i}_s^r + \frac{\mathrm{d}\boldsymbol{\Psi}_s^r}{\mathrm{d}t} + j\omega_r \boldsymbol{\Psi}_s^r \tag{4.24}$$

$$\boldsymbol{V}_r^r = R_r \boldsymbol{i}_r^r + \frac{\mathrm{d}\boldsymbol{\Psi}_r^r}{\mathrm{d}t} \tag{4.25}$$

$$\boldsymbol{\Psi}_s^r = L_s \boldsymbol{i}_s^r + L_m \boldsymbol{i}_r^r$$

$$\boldsymbol{\Psi}_r^r = L_r \boldsymbol{i}_r^r + L_m \boldsymbol{i}_s^r \tag{4.26}$$

On the other hand, the equivalent circuit in T is not completely identifiable by the no-load and locked-rotor test, so that $L_{lr} = L_{ls}$ is assumed. There are two possible equivalent circuits called the equivalent circuit Γ and their inverse (Quang et al. 2015). The first is particularly suitable for the treatment of stator flux orientated control method while the second is more suitable for the treatment of rotor flux orientated control method.

4.4.1.2 Two-Axis Model

From Eqs. (4.16), (4.17) and (4.23), the voltage equations can be represented in terms of stator and rotor current with the space vector variables as

$$\boldsymbol{V}_{abcs} = R_s \boldsymbol{i}_{abcs} + L_s \frac{\mathrm{d}\boldsymbol{i}_{abcs}}{\mathrm{d}t} + L_m \frac{\mathrm{d}}{\mathrm{d}t}\left(\boldsymbol{i}_{abcr} e^{j\theta_r}\right) \tag{4.27}$$

$$\boldsymbol{V}_{abcr} = R_r \boldsymbol{i}_{abcr} + L_r \frac{\mathrm{d}\boldsymbol{i}_{abcr}}{\mathrm{d}t} + L_m \frac{\mathrm{d}}{\mathrm{d}t}\left(\boldsymbol{i}_{abcs} e^{-j\theta_r}\right) \tag{4.28}$$

By using $\boldsymbol{V}_{dqs}^c = \boldsymbol{V}_{abcs} \cdot \boldsymbol{e}^{-j\theta}$ and $\boldsymbol{i}_{dqr}^c = \boldsymbol{i}_{abcr} \cdot \boldsymbol{e}^{-j(\theta-\theta r)}$, Eqs. (4.27) and (4.28) can be rewritten as Eqs. (4.29) and (4.30) in terms of the space vector variables at dq axes rotating in an arbitrary reference frame. The 0-axis equations are not considered since the impedance of the stator winding is balanced, and the neutral point is isolated.

$$\boldsymbol{V}_{dqs}^\omega = R_s \boldsymbol{i}_{dqs}^\omega + (L_{ls} + L_m)\frac{\mathrm{d}\boldsymbol{i}_{dqs}^\omega}{\mathrm{d}t} + L_m \frac{\mathrm{d}\boldsymbol{i}_{dqr}^\omega}{\mathrm{d}t} + j\omega\left[(L_{ls} + L_m)\boldsymbol{i}_{dqs}^\omega + L_m \boldsymbol{i}_{dqr}^\omega\right] \tag{4.29}$$

$$\boldsymbol{V}^{\omega}_{dqr} = R_r\boldsymbol{i}^{\omega}_{dqr} + (L_{lr} + L_m)\frac{\mathrm{d}\boldsymbol{i}^{\omega}_{dqr}}{\mathrm{d}t} + L_m\frac{\mathrm{d}\boldsymbol{i}^{\omega}_{dqs}}{\mathrm{d}t} + j(\omega - \omega_r)\left[(L_{lr} + L_m)\boldsymbol{i}^{\omega}_{dqr} + L_m\boldsymbol{i}^{\omega}_{dqs}\right] \tag{4.30}$$

Then, separating the dq terms, stator and rotor voltages in d and q quantities are:

$$v^{\omega}_{ds} = R_s i^{\omega}_{ds} + (L_{ls} + L_m)\frac{\mathrm{d}i^{\omega}_{ds}}{\mathrm{d}t} + L_m\frac{\mathrm{d}i^{\omega}_{dr}}{\mathrm{d}t} - \omega\left[(L_{ls} + L_m)i^{\omega}_{qs} + L_m i^{\omega}_{qr}\right] \tag{4.31}$$

$$v^{\omega}_{qs} = R_s i^{\omega}_{qs} + (L_{ls} + L_m)\frac{\mathrm{d}i^{\omega}_{qs}}{\mathrm{d}t} + L_m\frac{\mathrm{d}i^{\omega}_{qr}}{\mathrm{d}t} + \omega\left[(L_{ls} + L_m)i^{\omega}_{ds} + L_m i^{\omega}_{dr}\right] \tag{4.32}$$

$$v^{\omega}_{dr} = R_r i^{\omega}_{dr} + (L_{lr} + L_m)\frac{\mathrm{d}i^{\omega}_{dr}}{\mathrm{d}t} + L_m\frac{\mathrm{d}i^{\omega}_{ds}}{\mathrm{d}t} - (\omega - \omega_r)\left[(L_{lr} + L_m)i^{\omega}_{qr} + L_m i^{\omega}_{qs}\right] \tag{4.33}$$

$$v^{\omega}_{qr} = R_r i^{\omega}_{qr} + (L_{lr} + L_m)\frac{\mathrm{d}i^{\omega}_{qr}}{\mathrm{d}t} + L_m\frac{\mathrm{d}i^{\omega}_{qs}}{\mathrm{d}t} + (\omega - \omega_r)\left[(L_{lr} + L_m)i^{\omega}_{dr} + L_m i^{\omega}_{ds}\right] \tag{4.34}$$

The stator and rotor flux linkage can be written as:

$$\lambda^{\omega}_{ds} = L_{ls}i^{\omega}_{ds} + L_m\left(i^{\omega}_{ds} + i^{\omega}_{dr}\right) \tag{4.35}$$

$$\lambda^{\omega}_{qs} = L_{ls}i^{\omega}_{qs} + L_m\left(i^{\omega}_{qs} + i^{\omega}_{qr}\right) \tag{4.36}$$

$$\lambda^{\omega}_{dr} = L_{lr}i^{\omega}_{dr} + L_m\left(i^{\omega}_{ds} + i^{\omega}_{dr}\right) \tag{4.37}$$

$$\lambda^{\omega}_{qr} = L_{lr}i^{\omega}_{qr} + L_m\left(i^{\omega}_{qs} + i^{\omega}_{qr}\right) \tag{4.38}$$

Then, Eqs. (4.31)–(4.34) can be expressed as:

$$v^{\omega}_{ds} = R_s i^{\omega}_{ds} + \frac{\mathrm{d}\lambda^{\omega}_{ds}}{\mathrm{d}t} - \omega\lambda^{\omega}_{qs} \tag{4.39}$$

$$v^{\omega}_{qs} = R_s i^{\omega}_{qs} + \frac{\mathrm{d}\lambda^{\omega}_{qs}}{\mathrm{d}t} + \omega\lambda^{\omega}_{ds} \tag{4.40}$$

$$v^{\omega}_{dr} = R_r i^{\omega}_{dr} + \frac{\mathrm{d}\lambda^{\omega}_{dr}}{\mathrm{d}t} - (\omega - \omega_r)\lambda^{\omega}_{qr} \tag{4.41}$$

$$v^{\omega}_{qr} = R_r i^{\omega}_{qr} + \frac{\mathrm{d}\lambda^{\omega}_{qr}}{\mathrm{d}t} + (\omega - \omega_r)\lambda^{\omega}_{dr} \tag{4.42}$$

The T equivalent circuit in the arbitrary reference frame according to Eqs. (4.39)–(4.42) is represented in Fig. 4.20 where 0-component has been omitted.

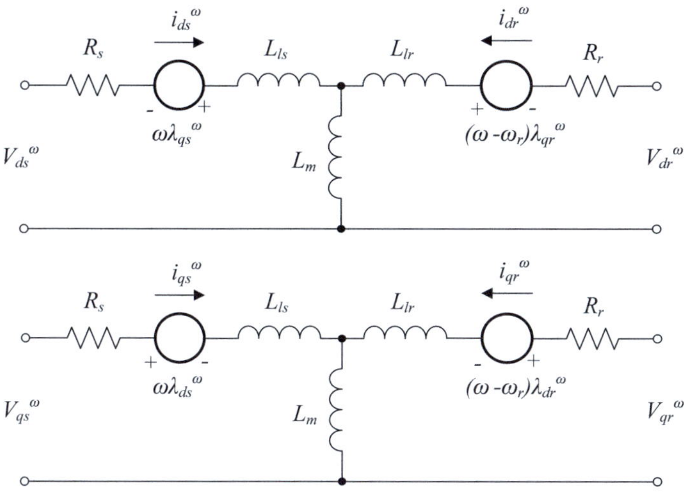

Fig. 4.20 *T* equivalent circuit of an induction machine at *dq* axes rotating in arbitrary speed

Fig. 4.21 Stator current i_s representation in a stationary frame with three and two axes. Also, it is represented with two-axis arbitrary rotating frame

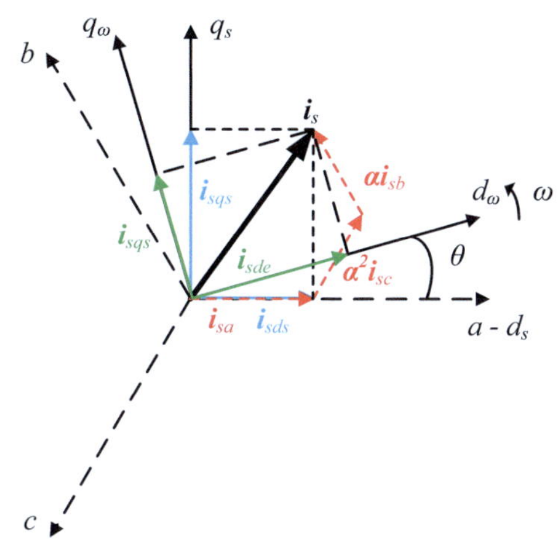

Figure 4.21 illustrates the vector diagram of stator current vector of an induction machine in the *abc*, dq_s, and dq_ω reference frames in motoring operation rotating in counterclockwise.

4.4.1.2.1 Torque Equation

The sum of the instantaneous input powers of the six stator and rotor windings is (note that the squirrel-cage rotor is usually modeled as three windings):

$$p_{in} = \underbrace{v_{sa}(t) \cdot i_{sa}(t) + v_{sb}(t) \cdot i_{sb}(t) + v_{sc}(t) \cdot i_{sc}(t)}_{Stator\_winding} + \underbrace{v_{ra}(t) \cdot i_{ra}(t) + v_{rb}(t) \cdot i_{rb}(t) + v_{rc}(t) \cdot i_{rc}(t)}_{Rotor\_winding} \quad (4.43)$$

In case of a squirrel-cage rotor induction machine, the rotor is shorted by end ring, and the rotor voltage is zero as commented before. Then, it is possible to neglect the rotor winding part of (4.43). The instantaneous power can also be represented in terms of two-axis rotating at arbitrary speed by using the Park transformation matrix $K(\theta)$ derived in Chap. 3:

$$p_{in} = \boldsymbol{V}_{abc}^{T} \boldsymbol{I}_{abc} = \left[\boldsymbol{K}(\theta)^{-1} \boldsymbol{V}_{dq0}^{\omega}\right]^{T} \left[\boldsymbol{K}(\theta)^{-1} \boldsymbol{I}_{dq0}^{\omega}\right] = \frac{3}{2} \boldsymbol{V}_{dq0}^{\omega} \cdot \boldsymbol{I}_{dq0}^{\omega} \quad (4.44)$$

Thus, the instantaneous power is:

$$p_{in} = \frac{3}{2}\left(v_{qs}^{\omega} \cdot i_{qs}^{\omega} + v_{ds}^{\omega} \cdot i_{ds}^{\omega} + 2v_{0s}^{\omega} \cdot i_{0s}^{\omega}\right) \quad (4.45)$$

By substituting (4.39) and (4.40) into (4.45), the instantaneous power can be expressed as:

$$p_{in} = \underbrace{\frac{3}{2} R_s\left(i_{ds}^{2\omega} + i_{qs}^{2\omega}\right)}_{Copper\_loss\_Power} + \underbrace{\frac{3}{2}\left(i_{ds}^{\omega} \frac{\mathrm{d}\lambda_{ds}^{\omega}}{\mathrm{d}t} + i_{qs}^{\omega} \frac{\mathrm{d}\lambda_{qs}^{\omega}}{\mathrm{d}t}\right)}_{Magnetic\_Power} + \underbrace{\frac{3}{2} P \cdot \omega_r\left(\lambda_{ds}^{\omega} \cdot i_{qs}^{\omega} - \lambda_{qs}^{\omega} \cdot i_{ds}^{\omega}\right)}_{Mechanical\_Power} \quad (4.46)$$

where 0-axis component once again is supposed zero since the impedance of the stator winding is balanced and the neutral point is isolated. Also, the electrical speed is P (pole pairs of the machine) times the mechanical speed, as shown below:

$$\omega_r = P \cdot \omega_{rm} \quad (4.47)$$

In Eq. (4.46), it is possible to observe three terms: due to copper loss, due to magnetic power in windings, and mechanical power. Therefore, the electromagnetic torque which is the mechanical power divided by the mechanical speed can be represented as

$$T_e = \frac{3}{2} P\left(\lambda_{ds}^{\omega} \cdot i_{qs}^{\omega} - \lambda_{qs}^{\omega} \cdot i_{ds}^{\omega}\right) \quad (4.48)$$

Also, by using the stator and rotor flux linkage Eqs. (4.36) and (4.37) the torque can be expressed as a function of the stator and rotor currents as

$$T_e = \frac{3}{2}P \cdot L_m\left(i_{qs}^{\omega} \cdot i_{dr}^{\omega} - i_{ds}^{\omega} \cdot i_{qr}^{\omega}\right) \tag{4.49}$$

Equation (4.49) suggests that the torque is proportional to magnetizing inductance L_m, the pole pair, and the difference of the cross components of stator and rotor currents.

4.4.1.3 Steady-State Equivalent Circuit

The steady-state equivalent circuit can be obtained when $\omega_c = \omega_{e,}$ that is, when the reference frame is rotating at synchronous speed, to refer the rotor variables to the stator. Hence, the XL_{ls}, XL_{lr}, and XL_m reactance depends on the frequency of the stator, and only the resistance of the rotor depends on the slip s. In the literature, it is possible to find the variables of the rotor with the prime symbol ($'$) or (s) to indicate that it is referenced to the stator, but here has been omitted. The derivative terms of (4.21) and (4.22) are zero since flux in steady-state is constant. Then, (4.21) and (4.22) can be rewritten in the synchronously rotating reference frame but without symbol e as

$$\boldsymbol{V}_s = R_s\boldsymbol{I}_s + j\omega_e\boldsymbol{\Psi}_s \tag{4.50}$$

$$\boldsymbol{V}_r = R_r\boldsymbol{I}_r + j(\omega_e - \omega_r)\boldsymbol{\Psi}_r \tag{4.51}$$

And the stator and rotor flux linkages

$$\boldsymbol{\Psi}_s = L_s\boldsymbol{I}_s + L_m\boldsymbol{I}_r$$

$$\boldsymbol{\Psi}_r = L_r\boldsymbol{I}_r + L_m\boldsymbol{I}_s \tag{4.52}$$

The magnetizing current $\boldsymbol{I}_m$ is the sum of the stator and rotor currents:

$$\boldsymbol{I}_m = \boldsymbol{I}_s + \boldsymbol{I}_r \tag{4.53}$$

Substituting (4.53) into (4.52) and then into (4.50) and (4.51), Eqs. (4.54) and (4.55) are obtained.

$$\boldsymbol{V}_s = R_s\boldsymbol{I}_s + j\omega_e L_{ls}\boldsymbol{I}_s + j\omega_e L_m\boldsymbol{I}_m \tag{4.54}$$

$$\frac{\boldsymbol{V}_s}{s} = \frac{R_r}{s}\boldsymbol{I}_r + j\omega_e L_{lr}\boldsymbol{I}_r + j\omega_e L_m\boldsymbol{I}_m \tag{4.55}$$

where the slip was represented in (4.14).

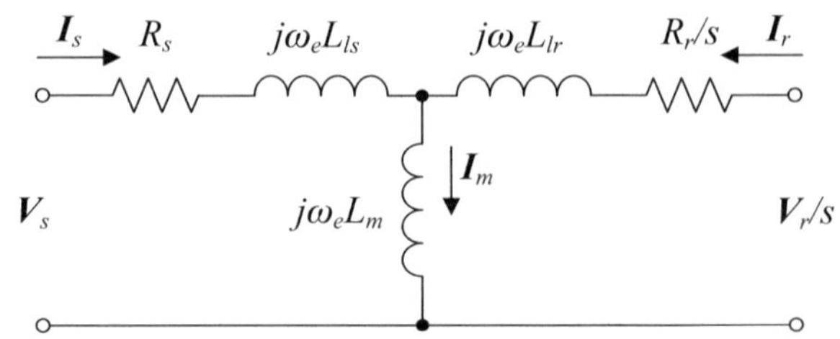

Fig. 4.22 T equivalent phase circuit of an induction machine where rotor variables are expressed at the stator side without the prime symbol (′)

Table 4.2 Usual nameplate data of induction machine

Nameplate	Symbol	Unit
Nominal power	P_N	kW
Line-to-line nominal voltage	V_N	V
Nominal current	I_N	A
Nominal frequency	f_N	Hz
Nominal speed	N_N	RPM
Nominal power factor	cos φ	
Service factor	SF	

The circuit which describes Eq. (4.54) is represented in Fig. 4.22. In total, the equivalent circuit in *T* contains five parameters. Using the steady-state equivalent circuit, the characteristics of the induction machine can be investigated using the common phasor analysis method of the AC circuits.

Table 4.2 shows the most common nameplate data of the induction machine, while Table 4.3 shows the necessary parameters for the *T* equivalent circuit, plus the mechanical parameters as rotor inertia *J* and damping coefficient *B* (motion parameters).

Table 4.3 Induction machine *T* equivalent circuit and motion parameters

Parameters for machine model	Symbol	Unit
Leakage stator inductance	L_{ls}	H
Leakage rotor inductance	L_{lr}	H
Magnetizing inductance	L_m	H
Stator resistance	R_s	Ohms
Rotor resistance	R_r	Ohms
Rotor inertia	J	$kg \cdot m^2$
Damping coefficient	B	$kg \cdot m^2/s$

4.4.1.4 Power Flow

The equivalent steady-state circuit of the previous section allows calculating the different powers involved in the induction machine in order to calculate, for example, the efficiency. The efficiency is the relationship between the input electrical power and the mechanical output power of the induction machine which is shown in Fig. 4.23. The input power P_{in} is in the form of three-phase voltages and currents and is calculated according to the real part of the complex number of the space vector theory and the complex conjugate of the current according to (4.56).

$$P_{in} = 3R_e\left[\boldsymbol{V}_s\boldsymbol{I}_s^*\right] \tag{4.56}$$

The difference in phase between V_s and I_s is the power factor angle φ.

The first losses found in the induction machine are the copper losses in the stator windings. Then, a certain amount of power is lost as by hysteresis losses and by Eddy currents in the stator (P_{core}). The remaining power at this point is transferred to the rotor of the machine through the air gap between the stator and rotor. This power is called air-gap power P_{GAP}. After the energy is transferred to the rotor, part of it is lost as rotor copper losses, and the rest is converted from electricity to mechanical form (P_{conv}). Finally, the losses by friction and windage are subtracted, leaving the mechanical power P_{out} at the output of the machine.

The core losses do not always appear in the power flow diagram. The losses of the core of an induction machine come partly from the stator circuit and partly from the rotor circuit. As an induction machine normally operates at speed close to the synchronous, the relative motion of the magnetic fields on the surface of the rotor is quite slow, and the losses of the rotor core are minimal compared to the losses of the stator core. Since the most significant fraction of core losses comes from the stator circuit, all core losses are grouped together at that point in the diagram (P_{core}). These losses are represented in the equivalent circuit of the induction machine by a resistor R_m in parallel to the inductance L_m. It is often ignored since the effects tend to be minimal, although, for the evaluation of the efficiency of the machine, it should be considered. The higher the speed of the machine (up to the nominal velocity), the lower the core losses. On the other hand, higher friction and higher parasitic losses, especially if the machine has a fan for its own cooling (windage losses), are given at higher speeds.

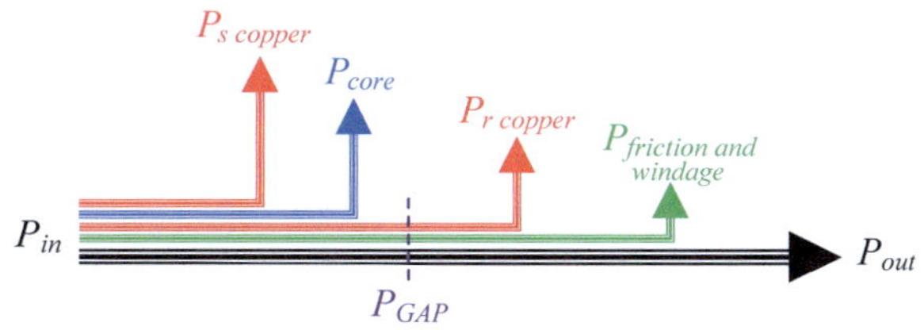

Fig. 4.23 Power flow in the induction machine

The copper losses in the stator and the rotor can be determined by (4.57) and (4.58), respectively.

$$P_{s\_copper} = 3R_s\boldsymbol{I}_s^2 \tag{4.57}$$

$$P_{r\_copper} = 3R_r\boldsymbol{I}_r^2 = 3 \cdot s \cdot \frac{R_r}{s}\boldsymbol{I}_r^2 \tag{4.58}$$

where $\boldsymbol{I}_s$, $\boldsymbol{I}_r$ are the phasors of stator and rotor currents in the steady-state.

Developing (4.43), the active input power P_{in} of the induction machine is

$$P_{in} = 3\boldsymbol{V}_s\boldsymbol{I}_s\cos\varphi \tag{4.59}$$

The power in the air gap P_{GAP} is the input power minus the power dissipated in the stator copper and the losses in the core:

$$P_{GAP} = P_{in} - P_{s\_copper} - P_{core} \tag{4.60}$$

The power of the core can be calculated by knowing the voltage drop V_m in extremes of the resistance R_m of the core:

$$P_{core} = 3\frac{\boldsymbol{V}_m^2}{R_m} \tag{4.61}$$

Finally, the mechanical power P_{mec} is the power developed by the shaft and corresponds to the difference between the air-gap power and rotor copper losses (4.62):

$$\begin{aligned} P_{mec} &= P_{GAP} - P_{r\_copper} \\ P_{mec} &= 3\frac{R_r}{s}(1-s)\boldsymbol{I}_r^2 \end{aligned} \tag{4.62}$$

The efficiency of the induction machine can be calculated as the quotient between the useful power (mechanical power minus the friction power) and the electric power absorbed:

$$\eta = \frac{P_{mec} - P_f}{P_{in}} \tag{4.63}$$

The electromagnetic torque can be written as the quotient between the mechanical power and the mechanical speed of the machine, as shown in (4.64):

$$T_{em} = \frac{P_{mec}}{\omega_{rm}} = P\frac{(1-s)P_{GAP}}{(1-s)\omega_e} = P\frac{P_{GAP}}{\omega_e} = 3P\frac{R_r}{s\omega_e}\boldsymbol{I}_r^2 \tag{4.64}$$

where P was the number of pair poles.

4.4.1.5 Speed Variation in an Induction Machine

According to (4.64), to vary the torque and therefore, the speed in the induction machine can be performed in three ways:

- Variation of the number of pairs of *P* poles (Dahlander connection)
- Slip *s* variation
- Variation of the frequency ω_e.

The Dahlander connection limits the number of speeds allowed. Depending on the connection made, the machine will rotate at a certain speed. Usually, the induction machine that uses this method has more cables to be able to make the connections through contactors or electronic devices, and therefore, the control method will be straightforward.

The variation of the slip is achieved either by varying the stator voltage V_s or by varying the rotor resistance R_r. In the latter case, it will only be possible in winding rotors. The first case is usually used for low-performance fan speed variation. In case of rotor resistance variation, the slip also varies, and therefore, the torque remains constant. With this method, it is able to move large loads with small machines, but the main disadvantage is the low performance that it presents due to the use of resistances.

The last method to regulate the speed consists of the frequency variation of the stator voltages. This method is the most efficient compared with the other methods discussed above and is achieved by a three-phase inverter where the voltage and frequency can be varied independently. The inverter allows the induction machine to operate in the regions of constant torque, constant power, and field weakening. The first case, in order to achieve a constant torque, a proportion of constant frequency and voltage variation or also known as "constant air gap flux operation" must be maintained. That is, if the electric frequency increases, the voltage must be increased so that the torque is kept constant. This control of the voltage–frequency ratio is called scalar control or variable voltage and variable-frequency (VVVF) control. On the contrary, for constant power, naturally, it must maintain constant current and voltage. This is only possible when the rated voltage is delivered to the induction machine.

4.4.1.6 Capability Curve of an Induction Machine

From the circuit of Fig. 4.22, the air-gap voltage V_m can be expressed as (the suffix "*e*" has been omitted)

$$\boldsymbol{V}_m = \left(\frac{R_r}{s} + jXL_{lr}\right)\boldsymbol{I}_r \tag{4.65}$$

And the power factor of the air gap can be represented by

$$\cos\phi_r = \frac{\frac{R_r}{s}}{\sqrt{\left(\frac{R_r}{s}\right)^2 + (\omega_e L_{lr})^2}} \tag{4.66}$$

The electromagnetic torque can also be expressed as a function of the rotor current I_r, the air-gap power factor $\cos(\varphi_r)$, and the air-gap voltage V_m.

$$T_{em} = 3P\frac{|\boldsymbol{V}_m|}{\omega_e}|\boldsymbol{I}_r|\cos\varphi_r \tag{4.67}$$

The air-gap flux λ_m can be defined as the quotient between the voltage V_m and the stator frequency ω_e:

$$\boldsymbol{\lambda}_m = \frac{\boldsymbol{V}_m}{j\omega_e} \tag{4.68}$$

So that the electromagnetic torque can be expressed as a function of the air-gap flux $\boldsymbol{\lambda}_m$ and the rotor current $\boldsymbol{I}_r$ as

$$T_{em} = 3P \cdot |\boldsymbol{\lambda}_m| \cdot |\boldsymbol{I}_r|\cos\varphi_r \tag{4.69}$$

where the rotor current is

$$|\boldsymbol{I}_r| = \frac{|\boldsymbol{V}_m|}{\sqrt{\left(\frac{R_r}{s}\right)^2 + (\omega_e L_{lr})^2}} \tag{4.70}$$

If the air-gap voltage is much higher than the voltage drop in the stator impedances formed by its resistance R_s and its reactance XL_{ls} as indicated in (4.71), then $\boldsymbol{V}_s \cong \boldsymbol{V}_m$.

$$\boldsymbol{V}_m \gg |(R_s + jXL_{ls})\boldsymbol{I}_s| \tag{4.71}$$

From (4.70), if the air-gap power factor $\cos(\varphi_r)$ is close to the unit which means that $R_r/s \gg XL_{lr}$, the expression of the electromagnetic torque (4.69) is similar to that of the DC machine according. From (4.69) to (4.70), the torque–speed characteristic can be extracted in different regions delimited by the limitations of voltage, current, and maximum slip. The torque–speed characteristic of the induction machine can be divided into the three regions, as shown in Fig. 4.24.

The first region is called constant torque region where the speed increment is compensated with the voltage V_s increment. That is, the air-gap flux $\boldsymbol{\lambda}_m$ remains constant. In this region, the slip is constant, and the stator current is regulated by the stator voltage to obtain the constant torque. When the voltage V_s reaches its limit value (at nominal speed), the torque can no longer be constant so that it begins to fall

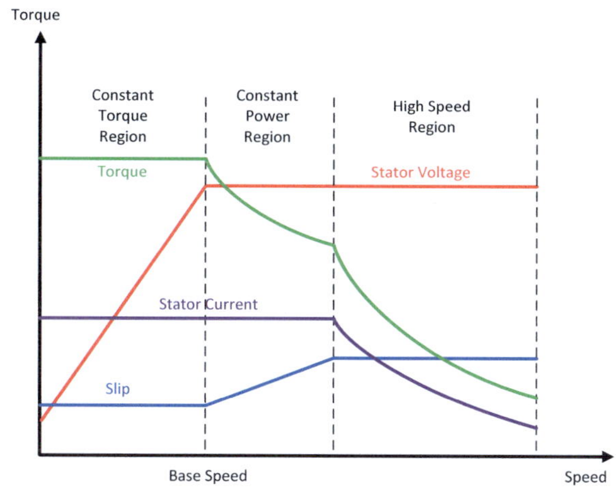

Fig. 4.24 Capability curve of an induction machine in the first quadrant of torque–speed curve

as described in the second region. In the second region, called the constant power region, the slip increases to a maximum value corresponding to the point where the pull-out torque occurs, so that the stator current remains constant and the rotor can maintain its capacity of power. The stator voltage in this region is the nominal of the machine, and it is in this region where the nominal mechanical power is delivered. The increase in speed causes the torque to fall at a rate of $1/\omega_e$ according to (4.67), but the power remains constant because the speed of the machine compensates for the factor $1/\omega_e$, ($P_{mec} = T_{em}\omega_{rm}$). It is in this region where the weakening of the field begins since the flux also falls at the $1/\omega_e$ ratio. Finally, in the third region, the slip remains constant while the stator and rotor current decreases (4.70) since the rotor reactance magnitude $\omega_e L_{lr}$ increases with the frequency. The torque decays with the square of the speed ω_e since the current I_r and the air-gap flux $\boldsymbol{\lambda}_m$ decay by $1/\omega_e$. Thus, the output power drops at $1/\omega_e$, as can be observed in Fig. 4.24.

4.4.1.6.1 Maximum Power

Using the same steady-state equivalent circuit of Fig. 4.22 of the induction machine, it is possible to obtain with a simple analysis of the maximum power condition. The equivalent impedance seen from the equivalent interaction between the rotor and the stator is

$$Z_{TH} = \frac{1}{\frac{1}{R_s + jXL_{ls}} + \frac{1}{jXL_m}} \tag{4.72}$$

And Thevenin voltage is

$$\boldsymbol{V}_{TH} = \frac{jXL_m}{R_s + j(XL_{ls} + XL_m)} \cdot \boldsymbol{V}_s \tag{4.73}$$

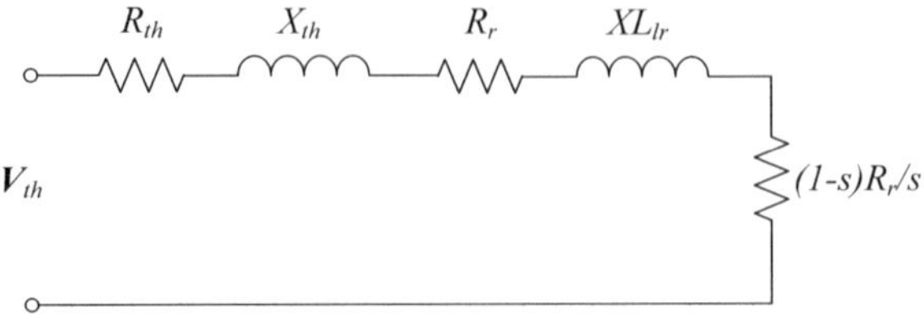

Fig. 4.25 Equivalent Thevenin circuit of the induction machine

To have maximum power, the load of this Thevenin equivalent circuit must be the conjugate of the calculated impedance. Observing the circuit of Fig. 4.25, it is easy to deduce the previous condition so that it is possible to establish the slip for which the power is maximized according to (4.74).

$$(1-s)\frac{R_r}{s} = \sqrt{(R_{th}+R_r)^2 + (X_{th}+XL_{lr})^2} \tag{4.74}$$

Finally, the slip to achieve a maximum power transfer can be calculated according to the expression (4.75):

$$s_{Pmax} = \frac{R_r}{\sqrt{(R_{th}+R_r)^2 + (X_{th}+XL_{lr})^2} + R_r} \tag{4.75}$$

4.4.1.6.2 Maximum Torque (Pull-out Torque)

Using the torque analytical expression, it is possible to determine the maximum torque ratio. Due to the linear relationship between the torque and the delivered power, the maximum torque condition corresponds to the peak of the delivered power. The maximum active power transfer is when the equivalent impedance module formed by the impedance R_{th}, X_{th}, and XL_{lr} is equal to the load resistance. That is, the condition for maximum torque is:

$$R_r + (1-s)\frac{R_r}{s} = \frac{R_r}{s} \tag{4.76}$$

and occurs when

$$\frac{R_r}{s} = \sqrt{R_{th}^2 + (X_{th}+XL_{lr})^2} \tag{4.77}$$

The maximum or critical slip to achieve maximum torque (pull-out torque) can be obtained by clearing the slip s:

$$s_{Tmax} = \frac{R_r}{\sqrt{R_{th}^2 + (X_{th}+XL_{lr})^2}} \tag{4.78}$$

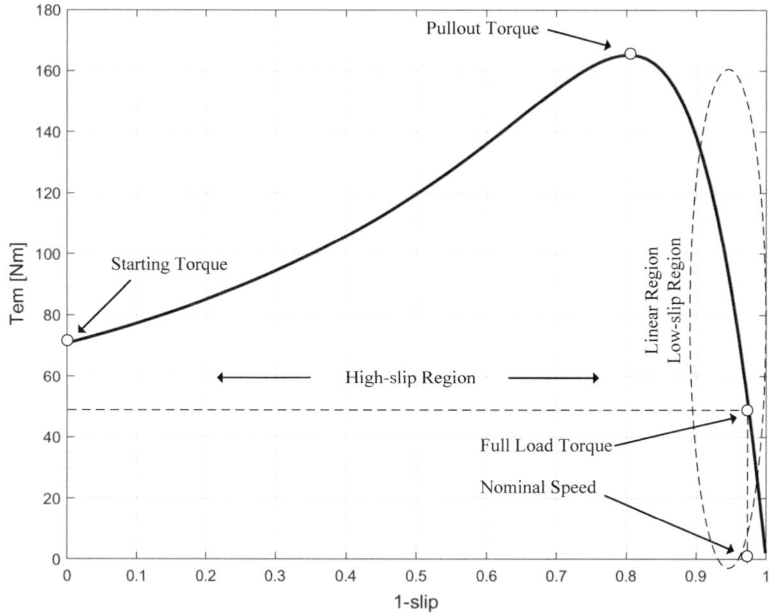

Fig. 4.26 Torque–speed curve where is shown the starting torque, pull-out torque, and full-load torque. The maximum slip to have the pull-out torque is 0.2 approx. The linear region is the region where the induction machine is more stable

Figure 4.26 shows the torque-speed characteristic of the induction machine for a specific supply voltage. There will be different similar curves for different supply voltages. It is possible to observe different points such as the starting torque, the pull-out torque, and the full load torque. The linear region indicated in the figure is the stable region of the induction machine, where it should operate.

4.4.1.7 Induction Machine NEMA Classification

NEMA classifies the induction machine as A, B, C, and D, depending on the torque–speed curve, as shown in Fig. 4.28. These different characteristics are achieved by varying the rotor resistance R_r, which, in the case of the squirrel-cage rotor, is achieved by varying its structure. Generally, the farther away the rotor bar of the stator, the higher the XL_{lr} reactance of the rotor since the lower percentage of the bar flux will reach the stator. Thus, if the bars of the rotor cage are closer to the rotor surface, they will have a lower leakage flux, and the reactance XL_{lr} of the rotor will be lower. Figure 4.27 shows the different rotor sections, where it is possible to observe the different geometries of the rotor bars.

The most used induction machine is type B since it has a high torque at zero speed of up to 150% and a pull-out torque of up to 225% of rated torque. However, if the machine is controlled with a VVVF control by an inverter, type A is usually a better

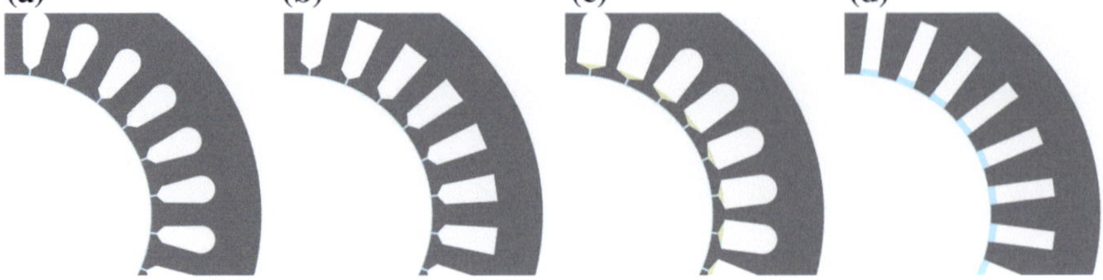

Fig. 4.27 Cross section of different rotors and rotor bars. Image developed using FluxMotor provided courtesy of Altair Engineering, Inc.

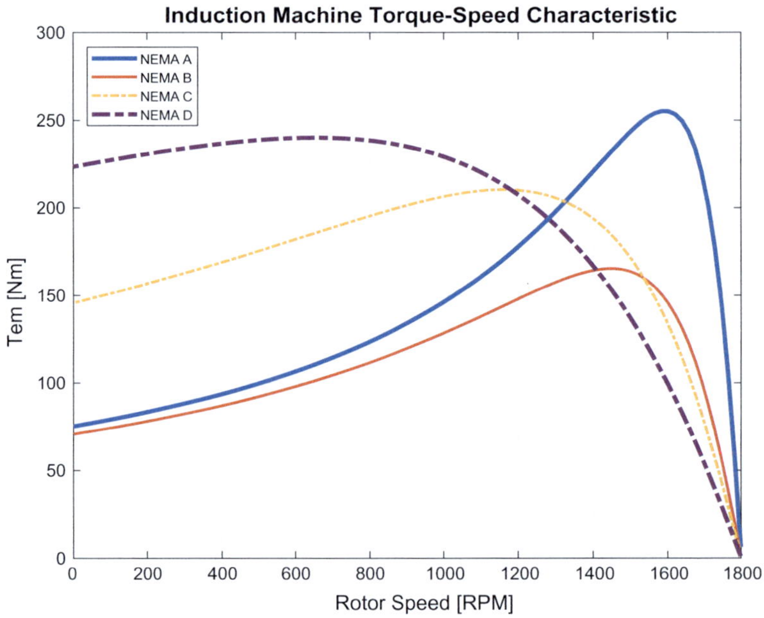

Fig. 4.28 Typical torque–speed curves for different rotor designs according to NEMA classification for two pair-pole machine. NEMA A: $R_r = 0.166$, $XL_{lr} = 0.232$. NEMA B: $R_r = 0.498$, $XL_{lr} = 1.392$. NEMA C: $R_r = 0.664$, $XL_{lr} = 0.696$. NEMA D: $R_r = 0.996$, $XL_{lr} = 0.37$

option because it offers better efficiency due to smaller rotor resistance. This offers a torque at zero speed of up to 150% and pull-out torque of up to 275% of the rated torque.

In the case of *C* classification, the machine has a higher starting torque between 240 and 275% of the rated torque, while *D* classification is between 275 and 300%. The induction machines with a high slip that operate at speed between 85 and 95% of the rated speed are the classification *D*. It is usually used to start large loads with high inertias at a lower starting current close to its nominal. However, its efficiency is poor, especially when the slip is large.

4.4.1.8 Induction Machine Operation

As in the case of the DC machine, the induction machine can also operate in the four quadrants, as shown in Fig. 4.29. The first and third quadrants are when the induction machine operates in motor mode, i.e., when the slip s is between zero and one ($0 > s \leqq 1$), while in the second and fourth it operates in braking mode and regenerative braking mode. As discussed, regenerative braking mode is the process by which the energy from the induction machine returns to the battery or electrical grid during the braking process.

In the induction machine, it is possible to distinguish two types of brakes: the braking mode (plugging $s > 1$) and the regenerative braking mode ($s < 0$) as shown in Fig. 4.30. In the braking mode, the rotor is forced to rotate against the stator field, which causes a considerable amount of back-EMF voltage and induced currents in the rotor conductors. This mode can easily be imposed on the induction machine by inverting the magnetic field, by exchanging the sequence of the supply voltages. However, the braking torque is low, so this deceleration method of the machine is not very effective. In addition, both the kinetic energy yielded by the load and the electrical energy supplied to the machine are dissipated in the stator winding and especially in the rotor. Therefore, energy is not recovered, and it is converted in heat, causing a possible machine overheat. The slip in this mode becomes more significant than the unit, and the torque developed tries to force the machine to rotate in the opposite direction. The plugging mode can be used in low inertia loads with high rotor resistance induction machines in order to have higher slips value.

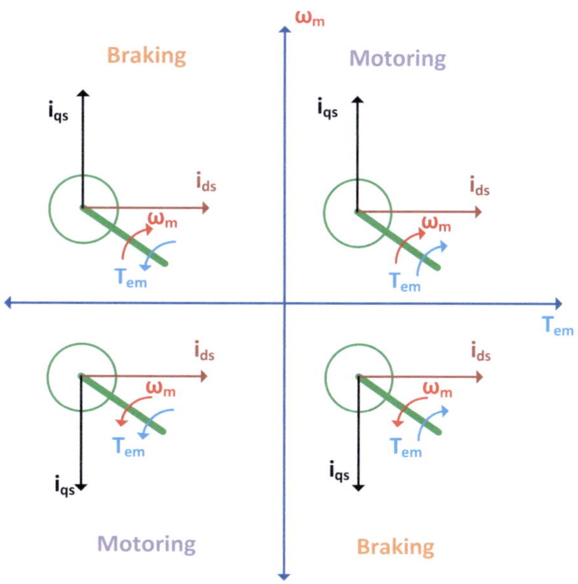

Fig. 4.29 Induction machine operations in accordance with the direction of power flow

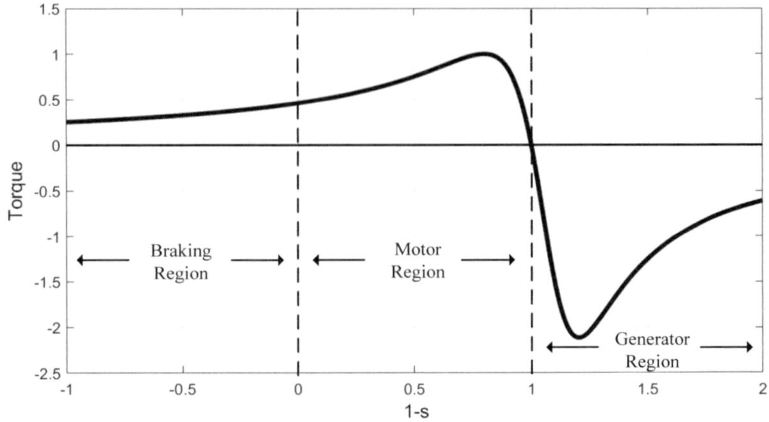

Fig. 4.30 Torque–speed curve of an induction machine in the three working regions: braking, motor, and generator regions

Regenerative braking is more efficient since it forces the machine to operate in generation mode, which requires the rotor to rotate faster than the stator magnetic field. This is achieved by reducing the field speed so that it rotates slower than the rotor. By means of an inverter, as it will see in Chap. 8, the regenerative brake can be applied easily, keeping track the rotor speed and reducing the feeding frequency accordingly. Regenerative braking is preferable for EV/EHV due to its higher efficiency and its ability to return power to the battery.

4.4.2 PMAC and BLDC Machine

The permanent magnet machines can be classified in radial flux permanent magnet machines (RFPMs) and axial flux permanent magnet machines (AFPM). Note that the previous machines seen such as induction and brushed machines were radial flux machines. The AFPM principle was used in Faraday's disk generator in 1831, but it was no popular for a long time due to manufacturing technologies. In the AFPM, the flux is produced axially along the axis of the rotor, while in RFPM the flux is produced radially along the sideways of the rotor. Thanks to the new generation of permanent magnets and the evolution of power electronics, new AFPM designs have been developed with higher performance than RFPM for some applications. For example, the AFPM is preferred in direct-drive traction applications due to their wide speed range operation capacity, compact structure (Mebarki et al. 2014), and higher power density. Thus, due to their compact structure, the AFPM can be integrated into internal combustion (IC) engines for HEV or can be used in EV wheel for traction. In Fig. 4.31 is represented the structure for a three-phase 8 poles AFPM.

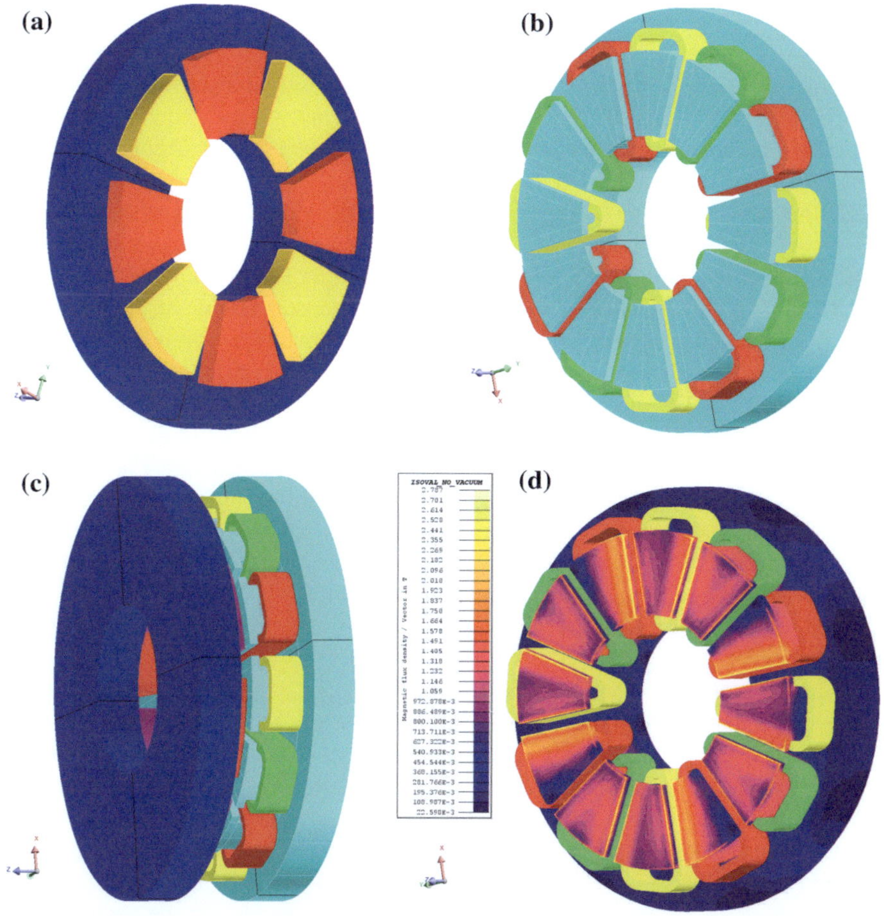

Fig. 4.31 3D model of an axial flux permanent magnet machine. **a** Rotor core with 8 permanent magnets (8 poles). **b** Stator core with three-phase winding, 12 slots, and 8 poles. **c** Complete AFPM. **d** 3D FEA simulations where stator and winding are shown. Image developed using Flux3D provided courtesy of Altair Engineering, Inc.

As can be observed, the stator core is composed of 12 slots with 8 poles, while the rotor cores have 8 permanent magnets (8 poles). In Fig. 4.31d is represented the FEA simulation result for the magnetic flux density at machine start-up.

Once a brief introduction to the AFPM has been explained, the following analysis is considered for the RFPM, although most of the equations are entirely valid.

The PMAC and BLDC machines are electric machines with permanent magnets usually in the rotor which does not need magnetizing currents like induction machine. Less weight per volume, smaller rotors, the high relationship between Torque/Inertia are some of the advantages compared with induction machines due mainly to the absence of field coil losses. On the contrary, PMAC and BLDC machines need to synchronize the stator voltages with the position of the rotor to produce high-performance torque. In the literature, it is possible to find the PMAC machine as PM synchronous machine (PMSM). The PMSM has a sinusoidal distribution of the flux density, operates with 120° phase-shift sinusoidal currents and induces sinusoidal back-EMF phase voltage. The PMSM requires sinusoidal stator currents to produce constant torque with low ripple.

On the other hand, the BLDC machine has a square distribution of the flux density, trapezoidal back-EMF phase voltage, and square-wave stator current since the stator winding is distributed in a square manner. The BLDC machine has the advantage of being a simple machine with higher power density, simple discrete position sensors, and simple control compared to a sinusoidal machine (Pillay and Krishnan 1991). According to NEMA standard, "Motion/Position Control Motors and Controls" defines a BLDC machine as follows: "rotating self-synchronous machine with a permanent magnet rotor and with known rotor shaft positions for electronic commutation. A motor meets this definition whether the drive electronics are integral with the motor or separate from it." The electronic commutation emulates the mechanical commutator in the brushed DC machine thanks to the shaft position sensors which dictates the absolute position of the rotor. Once the position of the rotor is known, an appropriate commutation sequence can be set for the next commutation cycle. The position sensor is usually either a three Hall effect sensor placed 120° apart or an optical encoder.

Although both machines have constant torque, the BLDC machine has higher torque ripple than PMSM. One of the main reasons is due to the switching on and off of the phases at 60° intervals as it will be seen in Chap. 8. This torque ripple in BLDC machines is called commutation torque ripple and has a frequency of six times the electrical frequency. There are some techniques to minimize the commutation torque ripple as Kim et al. (2006) suggest in BLDC machines. Other reasons for the torque ripple in BLDC machines are the cogging torque which will be discussed later and the distortions in the trapezoidal back-EMF. However, in PMSM the torque ripple contribution is mainly due to the reluctance torque which appears in the saliency poles machines, the cogging torque, and the electromagnetic torque developed. In any case, the torque ripple is undesirable due to cause noise and vibration in the load (Islam et al. 2009).

The BLDC machines are widely used in the applications without requirements of high motion precision with a low cost-effective due to its simple control. In theory, the PMSM and BLDC machines can be driven by either sinusoidal or square current.

The rotor of BLDC machines is usually constructed with the PM on the surface as represented in Fig. 4.32a. However, based on the structure of the PMSM, there are two possible combinations, the surface-magnet (SPMSM) rotor, and interior-magnet (IPMSM) rotor. The cross-sectional view of the SPMSM machine is illustrated in Fig. 4.32a. As can be observed, the machine has 24 stator slots and 4 poles with a three-phase concentrated winding.

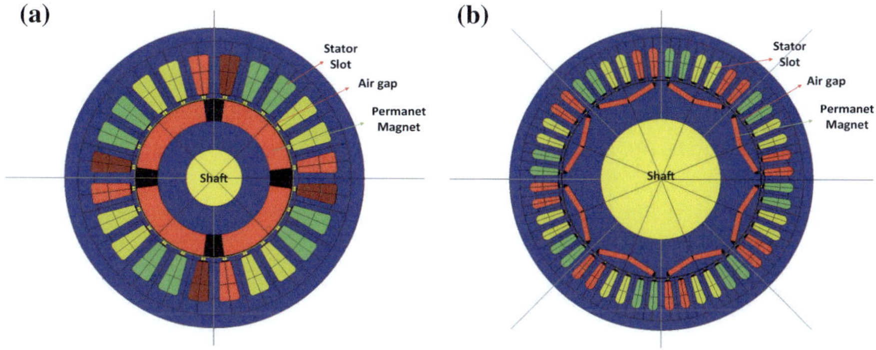

Fig. 4.32 **a** Cross-sectional view of the surface-magnet SPMSM machine with 24 stator slots and 4 poles. **b** A cross-sectional view of the internal-magnet 50 kW IPMSM with 48 stator slots and 8 poles. Image developed using Flux2D provided courtesy of Altair Engineering, Inc.

On the other hand, in Fig. 4.32b is shown the cross-sectional view of 50 kW IPMSM machine with 48 stator slots and 8 poles, also with three-phase concentrated winding. As can be observed, the permanent magnets of the rotor are mounted in V-pole configuration.

The distribution of the internal magnets of the IPMSM machine allows obtaining better performance, especially at high speed during field weakening, although with greater complexity in its control. The interior-magnet rotor has radially magnetized and alternately poled magnets as Fig. 4.32b shows. The internal magnet construction protects the magnet against centrifugal forces allowing an upper limit of rotation speed. On the other hand, surface-magnet machines are magnetized radially or in some cases circumferentially. An external high conductivity non-ferromagnetic cylinder is sometimes used to protect the PMs against the demagnetizing action of armature reaction and centrifugal forces. The SPMSM requires a more straightforward control since the expression of the electromagnetic torque is simpler than the IPMSM machine, as will be seen in Sect. 4.4.2.2.1. Note that for small machines, it is common to have rotors made with the permanent magnet itself where the steel shaft is inserted. However, the PM can be segmented to reduce the area of the induced eddy currents to reduce the losses, incrementing the manufacturing time and cost.

On the other hand, the stator is made up of several pole pairs, where each pole pair has three windings placed symmetrically in space as an induction machine. As commented early, in Fig. 4.32, the distribution winding is concentrated.

In Fig. 4.32, the mechanical angle between d and q axes is $\pi/(2P)$ where P is the number of pole pairs. If the machine is fed with a three-phase sinusoidal voltages at a constant frequency, in the stator windings will flow sinusoidal currents, which create a rotating magnetic field. The permanent magnets in the rotor tend to stay aligned with the rotating field, so the rotor rotates at synchronous speed. Also, the flux of the PM induces in the stator three sinusoidal voltages 120° apart. The main challenge in the control of these machines is to know the rotor position in real-time, so mainly implementation is using a position sensor or position estimator.

4.4.2.1 IPMSM Machine Analysis Overview with FEA

The 3D model of the 50 kW IPMSM machine is represented in Fig. 4.33a, while the machine efficiency in a torque–speed area is represented in Fig. 4.33b. In Fig. 4.33a, it is possible to observe the stator, the rotor, and the three-phase concentrated winding. The machine has an external stator radius of 141 mm and a stack length of 75 mm. The shaft radius is 56 mm, and rotor external radius is 96 mm. The stator and rotor are made with a lamination-type M270-35A, that is, electrical steel stacks of 0.35 mm of thickness. The main characteristics of this material are high permeability to let higher amount of flux through the core, low electrical conductivity to reduce the eddy currents and hence the Foucault losses, and a mass density of 7650 kg/m^3. The rotor permanent magnets' type is neodymium iron boron (NdFeB), and the dimensions are 24 × 5 × 54 mm, with a mass density of 7700 kg/m^3. The PM has a magnetic flux remanence B_r equal to 1.3 T and coercivity field strength H_c equal to 1000 kA/m at 20 °C.

Regarding the machine efficiency, the machine achieves the maximum efficiency of 94.86% in a wide area as can be observed in Fig. 4.33b. The area is between 1347–3320 RPM of rotor speed while torque varies between 25 and 127 Nm. However, at a maximum rotor speed of 6000 RPM, the maximum torque is 54 Nm with an efficiency of 90.93%. These characteristics, together with the size, weight, and output power, are ideal for automotive applications such as HEV and EV. The winding layout is represented in Fig. 4.33c. As can be observed, it is spatially distributed in the stator slots so that the stator currents are distributed as sinusoidally as possible. Moreover, it is possible to see the slots, the phases, and the slots per pole.

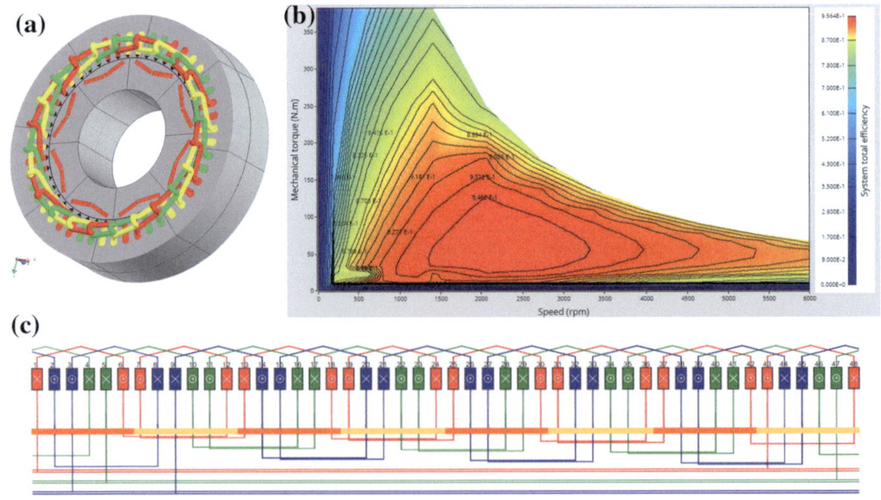

Fig. 4.33 **a** 3D model of the 50 kW IPMSM machine. Image developed using Flux3D provided courtesy of Altair Engineering, Inc. **b** Machine efficiency in torque–speed area. **c** Winding layout. Image developed using FluxMotor provided courtesy of Altair Engineering, Inc.

In Appendix 12.2 is provided more information with detailed test results of a similar machine working in motor and generator modes.

4.4.2.1.1 Cogging Torque

As discussed early, the cogging torque is one of the causes of the torque ripple in the BLDC machines but also in the PMSM. The cogging torque is the torque due to the interaction between the permanent magnets of the rotor and the stator slots. When an axis of a BLDC/PMSM is rotated without power, it is possible to feel it, since it is perceived as a jump from the current point and settles in another. The torque experienced in these jumps is the torque required to separate from the current rotor/stator alignment and move on to the next rotor/stator alignment. The cogging torque is a function of the angular position and is periodic with a frequency proportional to the pole pitch. The pole pitch is defined as the distance between two adjacent poles; in this case, it is 360°/8 equal to 45°. The cogging torque usually is higher for IPMSM machines since the air gap is lower than for SPMSM machines. The improvements in the geometry of the machine can help to reduce the cogging torque such the V-pole configuration, variation of slot opening, special slots' shapes, and maybe the most effective technique by steeping and skewing the rotor.

For medium power machines, the cogging torque can be equivalent to the rated torque of a 1 HP machine, that is, between 3 and 5 Nm. For example, in Fig. 4.34 is represented the FEA simulation computation cogging torque with the 50 kW IPMSM

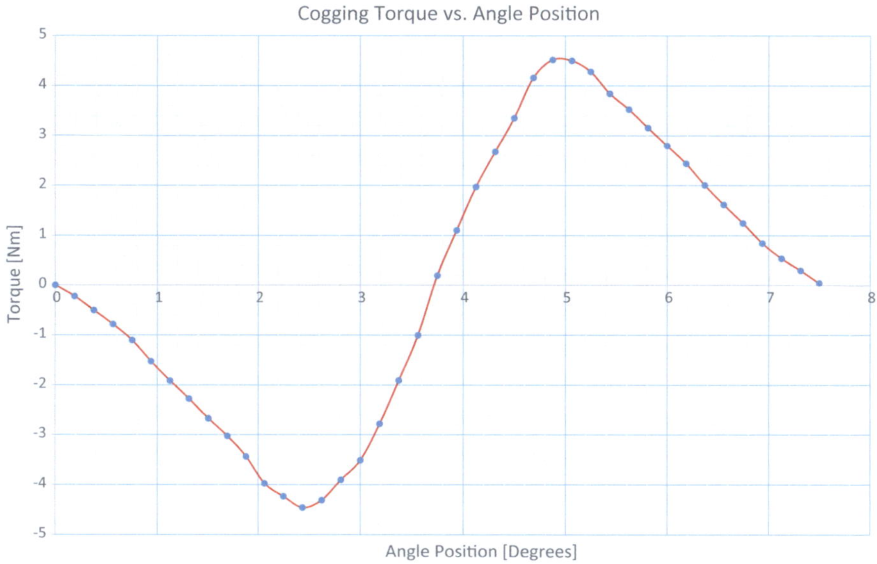

Fig. 4.34 Cogging torque computation result from 0 to 7.5° of a 50 kW IPMSM machine. Image developed using Flux2D data provided courtesy of Altair Engineering, Inc.

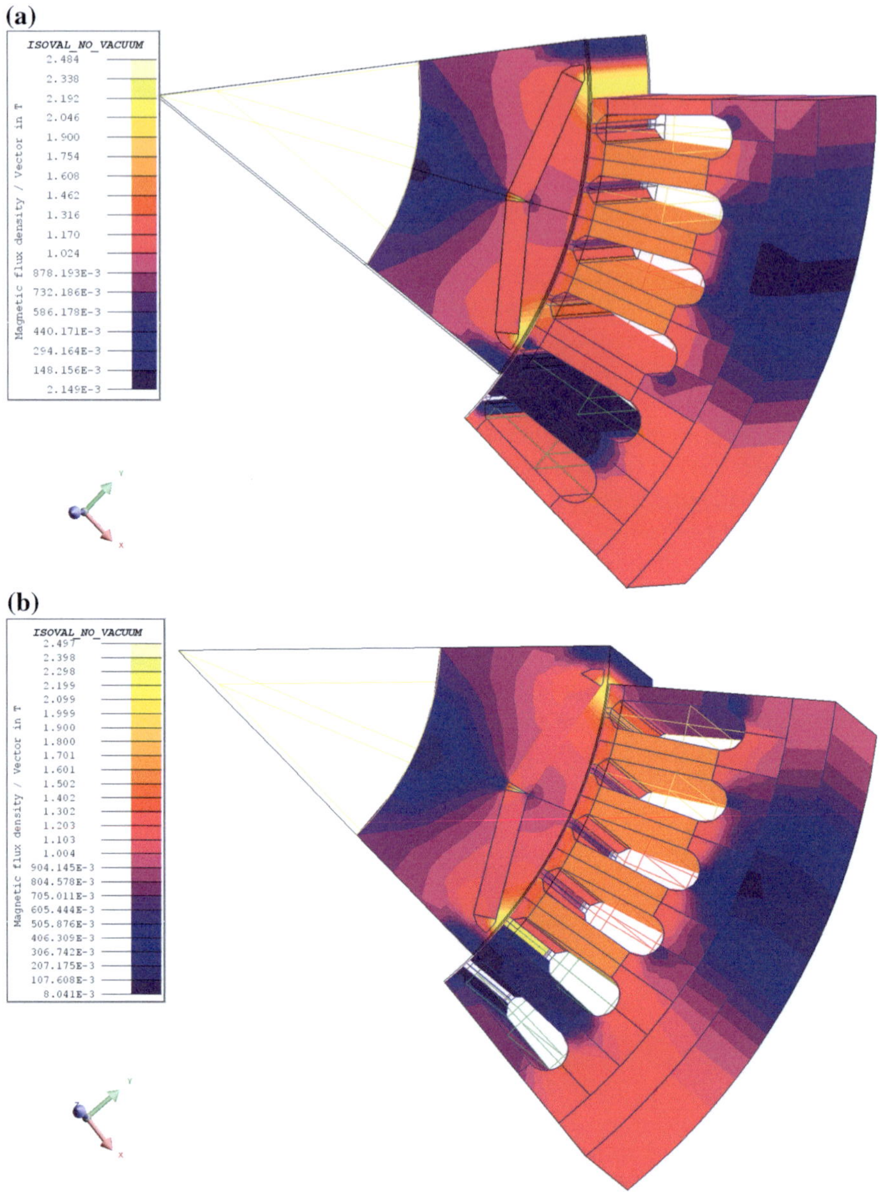

Fig. 4.35 3D FEA magnetic flux density provided by the PM at no-load conditions with a portion of the machine. **a** An angle equal to 7.5°. **B** An angle equal to 5°. Image developed using Flux3D provided courtesy of Altair Engineering, Inc.

machine of Figs. 4.32b and 4.33. This machine can develop a peak electromagnetic torque of 400 Nm, and as it can be observed in the figure, the maximum cogging torque is approximately 4.5 Nm. The stable positions are in 0° and 3.75°. In the stable position, if the rotor rotates in a positive direction, the torque is negative and tends to go back to the initial position. However, if the rotor rotates in a negative direction, the torque is positive and also tends to go back to the initial position. The period of the cogging torque is 360/48 equal to 7.5°.

The magnetic flux density provided by the PM at no-load with 3D FEA for the angular position of 7.5° and 5° is represented in Fig. 4.35.

In Fig. 4.35 is possible to observe the flux distribution around the stator and rotor cores. The simulation confirms that there are no saturation parts and the flux lines go from a north pole to a south pole of the magnets. The most critical parts of the machine are the iron bridges, where more flux is concentrated, as can be seen in orange with a value around 1.6 T. The width of the iron bridges is determinant to reduce the flux leakage from the magnets. The stator teeth see a flux density lower than in the iron bridges, that is around 1.3 T due to its higher area. Lastly, the flux density takes values around 2.5 T in two specific areas, as shown in the figure. The reason is that the flux path is closed by the magnet itself at this point.

4.4.2.1.2 Back-EMF

Without powering the machine, by an external force, the rotor can rotate at a constant speed in order to produce a constant back-EMF voltage at machine terminals. It can be simulated by the FEA simulation, as shown in Fig. 4.36 where the three-phase voltages at machine terminals are represented for 1000 RPM rotor speed. As discussed, the induced voltage is proportional to the magnetic flux density. It is possible to observe the slot-opening effect on the top of the waves where four peaks are performed. The waveform of the back-EMF is made with the sum of all harmonic components. It is possible to reduce the harmonic amplitudes to obtain a more sinusoidal waveform by reducing the slot opening, also reducing the cogging torque (Bianchini et al. 2012). However, the amplitude of the fundamental component and the saliency ratio could be reduced. For this case, the fundamental component has an amplitude of 153.65 V. Hence, it is possible to deduce the permanent magnet flux linkage Ψ_m, being 0.2117 V/rad/s.

4.4.2.1.3 Torque Developed

The torque developed by a PMSM depends on the angle between the permanent magnet magnetization axis and the applied current vector. The electromagnetic torque can be analyzed at different positions maintaining the current vector in steady-state conditions. It is possible to represent the simulation results for developed torque versus the angle position for different currents, as shown in Fig. 4.37.

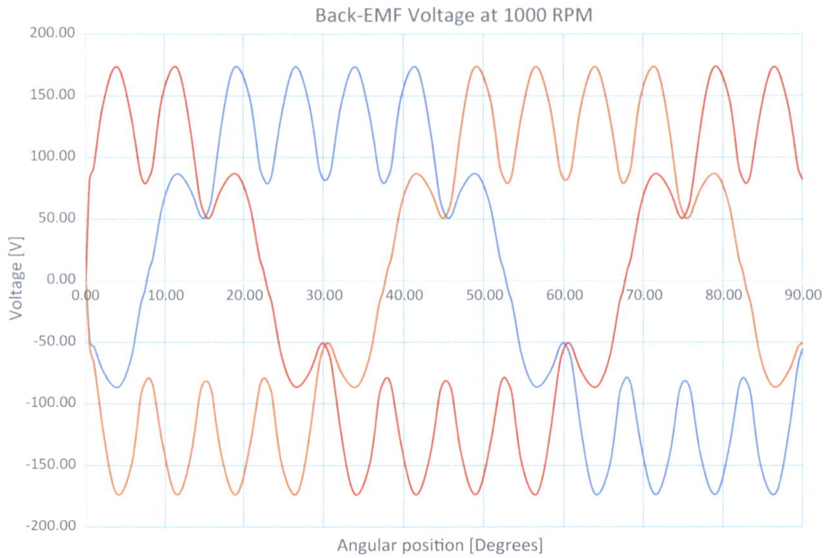

Fig. 4.36 Induced phase voltage at no-load condition (back-EMF voltage) from angular position 0° to 90°. The rotor speed is 1000 RPM. Image developed using Flux2D provided courtesy of Altair Engineering, Inc.

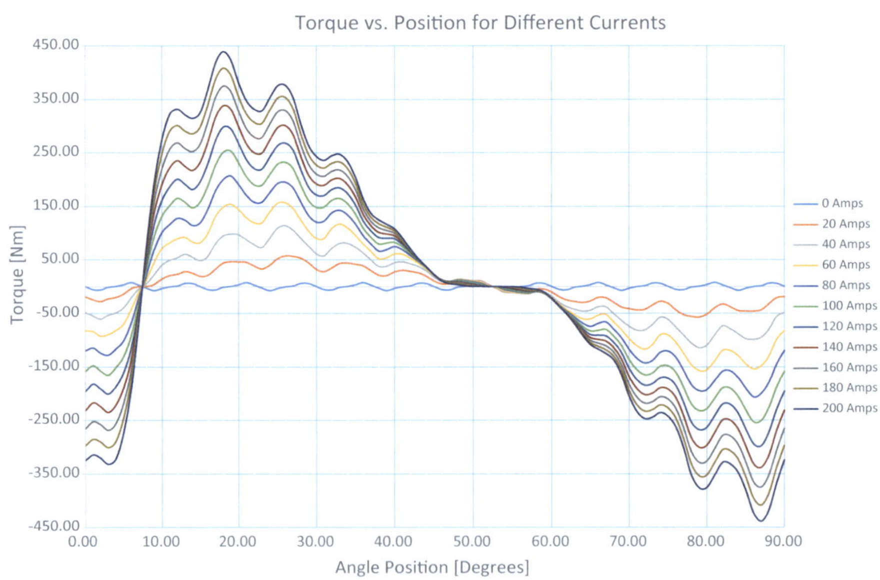

Fig. 4.37 Torque versus angle position for different currents where the current vector is maintained in the same position. Image developed using Flux2D data provided courtesy of Altair Engineering, Inc.

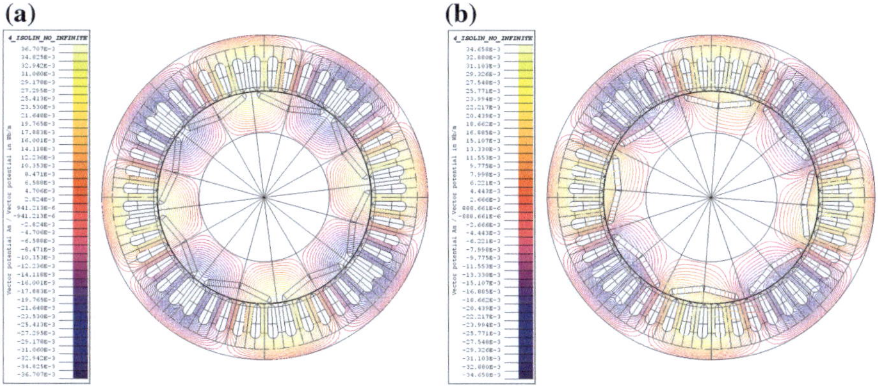

Fig. 4.38 Magnetic vector potential when current is 200 A. **a** Zero torque developed at rotor angle position equal to 52°. **b** Maximum torque developed at rotor angle position equal to 18°. Image developed using Flux2D provided courtesy of Altair Engineering, Inc.

As can be observed, the maximum positive torque is achieved between 16° and 28°, while it is close to zero between 46° and 58°. It is also possible to observe that the maximum peak torque is achieved at 200 Amps at 18° angle position, which corresponds to 439 Nm. In Fig. 4.38 is shown the magnetic vector potential when current is 200 A, for 52° and 18° rotor angles positions. In Fig. 4.38b is possible to observe that most of the flux lines cross the air gap diagonally, trying to rotate the rotor in counterclockwise. As seen for the induction machine, it is possible to say that the flux lines act like rubber bands which try to contract.

4.4.2.1.4 Torque Ripple

When the machine is rotating in a steady-state condition, the torque ripple can be analyzed with the FEA simulation. As commented before, the torque ripple is undesirable since it creates noise, vibrations, mechanical stress.

In Fig. 4.39 is illustrated the torque ripple result versus the angular position between 0 and 90°, that is, a quarter of a mechanical turn. The torque ripple impacts in the instantaneous input and output power. The instantaneous input power of the machine varies according to the instantaneous output power, which varies with the instantaneous torque ripple. In the figure, the torque ripple is approximately 28.02%, the average is 389.6 Nm, and the RMS value is 391.24 Nm. As discussed early, the torque ripple can be improved, for example, by skewing the rotor core and reducing the slot opening.

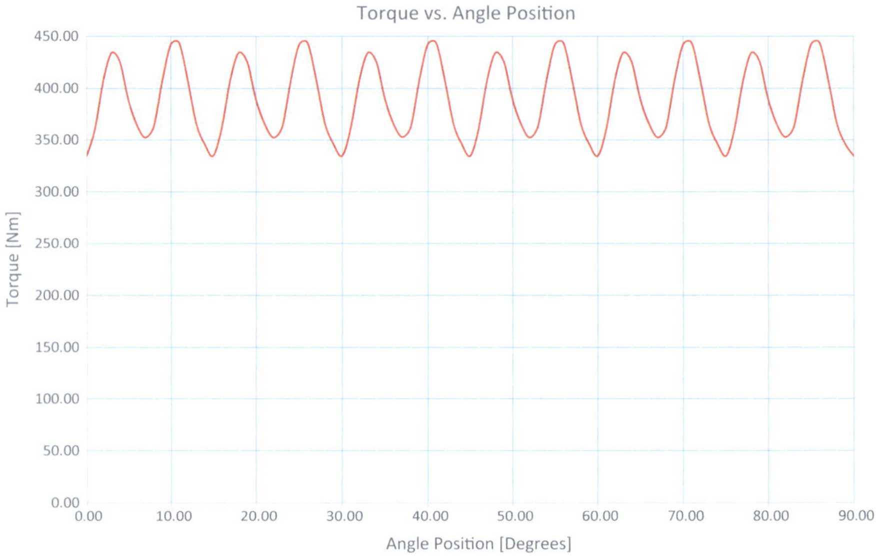

Fig. 4.39 Torque ripple versus angular position. Image developed using Flux2D data provided courtesy of Altair Engineering, Inc.

4.4.2.2 Space Vector Theory in PMSM

The mathematical analysis of PMSM in this section is also valid for BLDC machines. As it was shown in Sect. 4.4.1.1, the space vector theory of Chap. 3 can be applied in the analysis of the PMSM, to simplify the machine.

To describe the general equations of the PMSM in space vector theory, similar assumptions as shown for induction machine are still valid, but the following are added:

- Permanent magnet has a spatial sinusoidal distribution of MMF forces with the same number of poles of the stator
- No damper winding.

The stator voltage (4.79) in vector form of the induction machine is still valid for PMSM:

$$\boldsymbol{V}_{abcs} = R_s \boldsymbol{i}_{abcs} + \frac{\mathrm{d}\boldsymbol{\Psi}_{abcs}}{\mathrm{d}t} \tag{4.79}$$

where R_s was the stator winding resistance and $\boldsymbol{\Psi}_{abcs}$ is the total flux composer by the sum of the stator and rotor flux:

$$\boldsymbol{\Psi}_{abcs} = \boldsymbol{\Psi}_{abcs(s)} + \boldsymbol{\Psi}_{abcs(r)} \tag{4.80}$$

The stator flux $\boldsymbol{\Psi}_{abcs}$ depends on the phase current, the self-inductance of the diagonal elements of the matrix of (4.81) like L_{aas}, L_{bbs}, L_{ccs}, and the mutual inductance L_{abs}, L_{acs}, L_{cas} and its reciprocal. The matrix is symmetric because the flux coupling between the two windings is equal in both directions.

$$\boldsymbol{\Psi}_{abcs(s)} = \begin{bmatrix} L_{aas} \; L_{abs} \; L_{acs} \\ L_{bas} \; L_{bbs} \; L_{bcs} \\ L_{cas} \; L_{cbs} \; L_{ccs} \end{bmatrix} \cdot \boldsymbol{i}_{abcs} \tag{4.81}$$

On the other hand, the rotor flux depends only on flux linkage Ψ_m of the permanent magnet and its position, as shown in (4.82).

$$\boldsymbol{\Psi}_{abcs(r)} = \Psi_m \begin{bmatrix} \sin(\theta) \\ \sin\left(\theta - \frac{2\pi}{3}\right) \\ \sin\left(\theta - \frac{4\pi}{3}\right) \end{bmatrix} \tag{4.82}$$

The mutual and self-inductances of each winding are defined in (4.83).

$$\begin{aligned}
L_{aas} &= L_{ls} + L_A - L_B \cdot \cos(2\theta) \\
L_{bbs} &= L_{ls} + L_A - L_B \cdot \cos\left(2\theta + \frac{2\pi}{3}\right) \\
L_{ccs} &= L_{ls} + L_A - L_B \cdot \cos\left(2\theta - \frac{2\pi}{3}\right) \\
L_{abs} &= L_{bas} = -\frac{1}{2}L_A - L_B \cdot \cos\left(2\theta - \frac{2\pi}{3}\right) \\
L_{acs} &= L_{cas} = -\frac{1}{2}L_A - L_B \cdot \cos\left(2\theta + \frac{2\pi}{3}\right) \\
L_{bcs} &= L_{cbs} = -\frac{1}{2}L_A - L_B \cdot \cos(2\theta)
\end{aligned} \tag{4.83}$$

where L_{ls} stands for the leakage inductance of the stator winding, L_A stands for the inductance independent of the rotation of the rotor, and L_B is the maximum value for the inductance varying with the position. Both inductances depend on the rotor dimension, radius r, length l, air gap, max and min distance between rotor and stator $g_{\max}$ and $g_{\min}$, the number of turns of a stator winding N, and the permeability μ_o as represented in (4.84).

$$\begin{aligned}
L_A &= \left(\frac{N}{2}\right)^2 \pi \mu_o r l \frac{1}{2}\left(\frac{1}{g_{\min}} + \frac{1}{g_{\max}}\right) \\
L_B &= \left(\frac{N}{2}\right)^2 \pi \mu_o r l \frac{1}{2}\left(\frac{1}{g_{\min}} - \frac{1}{g_{\max}}\right)
\end{aligned} \tag{4.84}$$

In the mutual inductance, the coefficient $-1/2$ comes due to the fact that stator phases are displaced 120°, and $\cos(120) = -1/2$.

The stator winding flux linkage of each phase is the sum of the flux linkages related to the stator current and the mutual flux linkage resulting from the permanent magnet. The flux linkage space vector can be expressed as:

$$\boldsymbol{\Psi}_{abcs} = \left(L_{ls} + \frac{3}{2}L_A\right)\boldsymbol{i}_{abc} - \frac{3}{2}L_A\boldsymbol{i}^*_{abc}e^{j2\theta_r} + \Psi_m e^{j(\theta_r - \pi/2)} \quad (4.85)$$

where current space vector $\boldsymbol{i}^*$ is the conjugate of $\boldsymbol{i}$.

The magnetic field vector $\boldsymbol{\Psi}_m$, whose position in respect to the stator, is determined by the angle θ between the vector direction and the stator reference frame.

Finally, (4.79) can be rewritten as:

$$\begin{bmatrix} V_{as} \\ V_{bs} \\ V_{cs} \end{bmatrix} = \begin{bmatrix} R_s & 0 & 0 \\ 0 & R_s & 0 \\ 0 & 0 & R_s \end{bmatrix} \cdot \begin{bmatrix} i_{as} \\ i_{bs} \\ i_{cs} \end{bmatrix} + \frac{\mathrm{d}}{\mathrm{d}t}\left(\begin{bmatrix} L_{aas} & L_{abs} & L_{acs} \\ L_{bas} & L_{bbs} & L_{bcs} \\ L_{cas} & L_{cbs} & L_{ccs} \end{bmatrix} \cdot \begin{bmatrix} i_{as} \\ i_{bs} \\ i_{cs} \end{bmatrix} + \Psi_m \begin{bmatrix} \sin(\theta) \\ \sin\left(\theta - \frac{2\pi}{3}\right) \\ \sin\left(\theta - \frac{4\pi}{3}\right) \end{bmatrix}\right) \quad (4.86)$$

Equation (4.86) demonstrates the complexity of the PMSM analysis when the space vector theory with the three components is applied. In case of induction machine, the equation gets more complexity due to the relative speed difference of the rotor respect to the stator and to the squirrel cage of the rotor. In Wach (2011), the induction machine equations are derived which it is possible to observe that the inductances related to the mutual stator-rotor inductances have time-varying inductances. In both cases, it is difficult to solve, and it presents some significant difficulties, in terms of solution, that should be avoided. That is, it is difficult to find an expression of the torque. As it was performed for the induction machine, the space vector theory in a single equation can be derived for PMSM, but it will not represent the saliency of the rotor. That is, the magnetic asymmetry of the salient pole of IPMSM machine is not represented by a single equation. Therefore, the two-axis model is preferred to have the two components of the magnetic axis of the machine. It is possible to use the Clarke transformation to pass from the three-phase system to stationary two-axis orthogonal system. It does not modify the module of the vectors and depends still on the angle. To remove this dependence, rotation matrix transformation can be applied which pass from the stationary orthogonal system to another orthogonal system which rotates in synchronism. According to this, it is considered:

- d is preferred axis for pole flux orientation of Ψ_m
- q is orthogonal to d.

Using the Clarke and rotation matrix transformation, the three-phase voltage, current, and flux linkage space vectors in stationary *abc*-axis reference frame can be transformed to the *dqe* space vectors in rotating reference (synchronous rotating coordinate system, designed by *e* symbol with angular speed $\omega_e = \omega_r$) frames as follows:

$$\begin{aligned} \boldsymbol{V}_{abc} &= \boldsymbol{V}_{dqe} e^{j\theta_r} \\ \boldsymbol{i}_{abc} &= \boldsymbol{i}_{dqe} e^{j\theta_r} \\ \boldsymbol{\Psi}_{abc} &= \boldsymbol{\Psi}_{dqe} e^{j\theta_r} \end{aligned} \tag{4.87}$$

where the angle θ_r is the angle between the stator fixed axes and the synchronous rotating axes. Equation (4.79) can be transformed as follows:

$$\boldsymbol{V}_{abcs} = R_s \boldsymbol{i}_{abcs} + \frac{\mathrm{d}\boldsymbol{\Psi}_{abcs}}{\mathrm{d}t} \rightarrow \boldsymbol{V}_{dqe} e^{j\theta_r} = R_s \boldsymbol{i}_{dqe} e^{j\theta_r} + \frac{\mathrm{d}}{\mathrm{d}t}\left(\boldsymbol{\Psi}_{dqe} e^{j\theta_r}\right) \tag{4.88}$$

Hence, the voltage equation can be expressed in rotor *dqe* reference frame as:

$$\boldsymbol{V}_{dq} = R_s \boldsymbol{i}_{dqe} + \frac{\mathrm{d}\boldsymbol{\Psi}_{dqe}}{\mathrm{d}t} + j\omega_r \boldsymbol{\Psi}_{dqe} \tag{4.89}$$

Separating the real and imaginary components in (4.89) gives:

$$V_{ds}^e = R_s \boldsymbol{i}_{ds}^e + \frac{\mathrm{d}\Psi_{ds}^e}{\mathrm{d}t} - \omega_r \Psi_{qs}^e \tag{4.90}$$

$$V_{qs}^e = R_s \boldsymbol{i}_{qs}^e + \frac{\mathrm{d}\Psi_{qs}^e}{\mathrm{d}t} + \omega_r \Psi_{ds}^e \tag{4.91}$$

The flux linkage space vector of (4.85) can be solved as:

$$\begin{aligned} \boldsymbol{\Psi}_{dqe} = \boldsymbol{\Psi}_{abcs} e^{-j\theta_r} &= \left(L_{ls} + \frac{3}{2}L_A\right)\boldsymbol{i}_{abc} e^{-j\theta_r} - \frac{3}{2}L_B \boldsymbol{i}_{abc}^* e^{j2\theta_r} e^{-j\theta_r} \\ &+ \Psi_m e^{j(\theta_r - \pi/2)} e^{-j\theta_r} = \left(L_{ls} + \frac{3}{2}L_A\right)\boldsymbol{i}_{dq} - \frac{3}{2}L_B \boldsymbol{i}_{dq}^* + \Psi_m e^{-j\pi/2} \end{aligned} \tag{4.92}$$

It is possible to observe that in (4.92), the inductance coefficients are no longer time-dependent, compared with (4.85).

The *dq* magnetizing inductance is defined as:

$$L_{md} = \frac{3}{2}(L_A + L_B) \tag{4.93}$$

$$L_{mq} = \frac{3}{2}(L_A - L_B) \tag{4.94}$$

Solving to obtain L_A and L_B gives:

$$L_A = \frac{L_{md} + L_{mq}}{3} \tag{4.95}$$

$$L_B = \frac{L_{md} - L_{mq}}{3} \tag{4.96}$$

Substituting in (4.92) gives:

$$\boldsymbol{\Psi}_{dqe} = \left(L_{ls} + \frac{L_{md} + L_{mq}}{2}\right)\boldsymbol{i}_{dq} - \left(\frac{L_{md} - L_{mq}}{2}\right)\boldsymbol{i}^{*}_{dq} + \Psi_m e^{-j\pi/2} \tag{4.97}$$

The inductance in the *dqe* axes is defined as L_d and L_q. It is the sum of the leakage inductance and the corresponding magnetizing inductance of the axis:

$$L_d = L_{ls} + L_{md} \tag{4.98}$$

$$L_q = L_{ls} + L_{mq} \tag{4.99}$$

Separating the real and imaginary components in (4.97) and using Eqs. (4.98) and (4.99) give the following *dqe* fluxes:

$$\Psi^e_{ds} = L_d i^e_{ds} + \Psi_m \tag{4.100}$$

$$\Psi^e_{qs} = L_q i^e_{qs} \tag{4.101}$$

Finally, (4.90) and (4.91) can be rewritten as:

$$V^e_{ds} = R_s \boldsymbol{i}^e_{ds} + L_d \frac{\mathrm{d}i^e_{ds}}{\mathrm{d}t} - \omega_r \Psi^e_{qs} \tag{4.102}$$

$$V^e_{qs} = R_s \boldsymbol{i}^e_{qs} + L_q \frac{\mathrm{d}i^e_{qs}}{\mathrm{d}t} + \omega_r \Psi^e_{ds} \tag{4.103}$$

Equations (4.102) and (4.103) describe a circuit one for each as represented in Fig. 4.40.

The terms which depend on the angle has disappeared. Voltage and currents are now constant values. This is a linear equation system where the electromagnetic torque can be calculated in a simple way. In Fig. 4.41 is shown the vector diagram of a salient-pole and non-salient-pole PMSM, where the main difference is that for non-salient-pole PMSM the direct current is zero, and $L_d = L_q = L_s$.

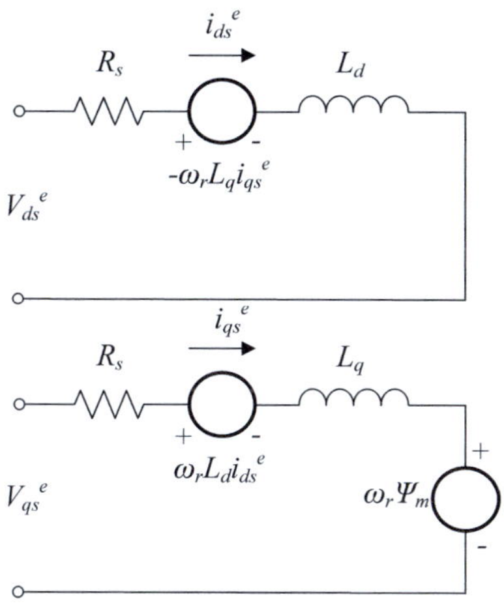

Fig. 4.40 Equivalent circuit of a PMSM in *dqe* axes

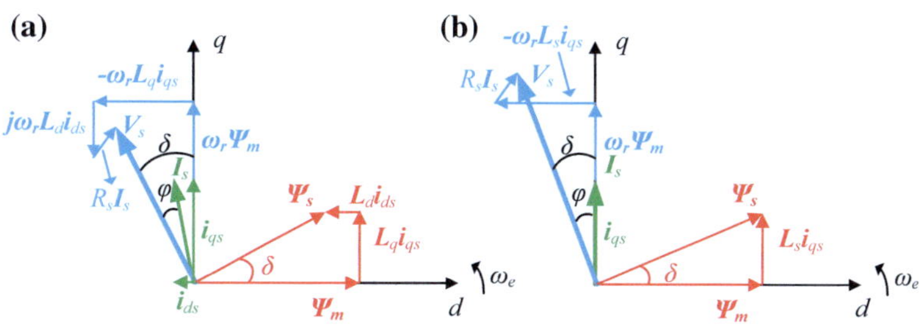

Fig. 4.41 Vector diagram of a PMSM in synchronously rotating reference frame. **a** Salient-pole machine. **b** Non-salient-pole machine

4.4.2.2.1 Torque Expression

The instantaneous power discussed for induction machine is also valid for PMSM, but here is defined in the synchronously rotating reference frame:

$$p_{in} = \frac{3}{2}\left(v_{qs}^e \cdot i_{qs}^e + v_{ds}^e \cdot i_{ds}^e + 2v_{0s}^e \cdot i_{0s}^e\right) \tag{4.104}$$

By using (4.102) and (4.103) and neglecting the 0-axis component, the instantaneous power can be expressed as:

$$p_{in} = \underbrace{\frac{3}{2} R_s \left(i_{ds}^{2e} + i_{qs}^{2e} \right)}_{Copper\_loss\_Power} + \underbrace{\frac{3}{2} \left(L_d i_{ds}^e \frac{di_{ds}^e}{dt} + L_q i_{qs}^{\omega} \frac{di_{qs}^e}{dt} \right)}_{Magnetic\_Power} + \underbrace{\frac{3}{2} P \cdot \omega_r \left(\Psi_{ds}^e \cdot i_{qs}^e - \Psi_{qs}^e \cdot i_{ds}^e \right)}_{Mechanical\_Power} \tag{4.105}$$

As mentioned, the mechanical power is the mechanical speed multiplied by the electromagnetic torque, which can be expressed as:

$$p_{em} = \omega_{mec} T_e = \frac{3}{2} \omega_r \left(\Psi_{ds}^e \cdot i_{qs}^e - \Psi_{qs}^e \cdot i_{ds}^e \right) = \frac{3}{2} \omega_r \left[\Psi_m i_{qs}^e + (L_d - L_q) i_{ds}^e i_{qs}^e \right] \tag{4.106}$$

Electrical speed is P (pole pairs of the machine) times the mechanical speed, as shown below:

$$\omega = P \cdot \omega_{mec} \tag{4.107}$$

Finally, the electromagnetic torque is:

$$T_e = \frac{3}{2} P \left[\Psi_m i_{qs}^e + (L_d - L_q) i_{ds}^e i_{qs}^e \right] \tag{4.108}$$

Equation (4.108) suggests that the torque is linear and proportional to three machine parameters: the pole pair, the flux linkage of the magnet, and to the difference of the synchronous inductances $L_d - L_q$. It is also proportional to both currents when L_d is not equal to L_q. Then, in this case, there are two types of torque, one due to flux linkage of the magnet and the other due to reluctance difference between L_d and L_q. Note that for IPMSM machine, $L_d < L_q$, and i_{dse} current should be negative under positive i_{qse} current to get positive reluctance torque. In the case of SPMSM machine, L_d and L_q are nearly identical, and the contribution of the electromagnetic torque is usually due to i_{qse} current control. The instantaneous electromagnetic torque reacts directly to the changes in the instantaneous current i_{qse}. This is the principle of vector control: the instantaneous stator current is controlled to be aligned with q-axis while the pole flux is orientated to d-axes to get the field-oriented control (FOC).

It should be noted that the flux linkage of the magnet has negative temperature sensitivity, such that with the rise of the rotor temperature, the magnet flux linkage decreases. The temperature coefficient of a ferrite magnet is typical −0.2%/°C. So, the flux linkage decreases by 20% according to a temperature increment of 100 °C in the magnet. If the rotor temperature is monitored, it is possible to compensate this flux linkage for the variations of the rotor temperature to avoid the degradation of the electromagnetic torque. The inductances L_d and L_q also vary with the temperature but in a minor way compared with the stator current magnitude level. Depending on the PMSM type, the inductances L_d and L_q vary several times according to the magnitude of the stator current, that is the operating condition. Typically, the inductances L_d

and L_q decrease with the increase of the stator current magnitude. The variation of L_d and L_q should be considered in the design of the high-performance PMSM drive system. For the high-performance control system, with a look-up table stored in the microcontroller/DSP memory is possible to have the profile for the flux linkage of the magnet variations, and for L_d, L_q variation to be compensated according to temperature measurement and stator current level, respectively.

4.4.2.3 Particularity for SMPMSM

For non-salient-pole machine (SMPMSM) $L_B = 0$ in accordance with (4.84) since both air gap distances $g_{\min}$ and $g_{\max}$ are equal. Then, there is only one synchronous inductance L, as expressed in (4.109).

$$\begin{aligned} L_d &= L_q = L \\ L &= L_{ls} + \frac{3}{2} L_A \end{aligned} \tag{4.109}$$

4.4.2.4 Particularity for IPMSM Machine

For salient-pole machine (IPMSM), $L_B \neq 0 \rightarrow L_q > L_d$. The synchronous inductances L_d and L_q are expressed in (4.110).

$$\begin{aligned} L_d &= L_{ls} + \frac{3}{2}(L_A + L_B) \\ L_q &= L_{ls} + \frac{3}{2}(L_A - L_B) \end{aligned} \tag{4.110}$$

4.4.2.5 Steady-State Equations of PMSM

As described for the induction machine, the steady-state circuit of the PMSM can be derived for the rotating synchronous reference frame. That is when the rotor rotates at the same speed as the angular speed of the voltage vector (taking into account the number of pole pairs). Note that the voltage vector is stationary in the rotor reference. As represented in Fig. 4.42, the *dqe* voltage components can be written in terms of the V_s voltage and load angle δ (Chandana Perera 2002).

The load angle δ represents the angle difference between the stator electrical applied voltage and the back-EMF e_m generated by the rotor magnets when rotating. Note that the upper index "*e*" has been omitted in the following equations.

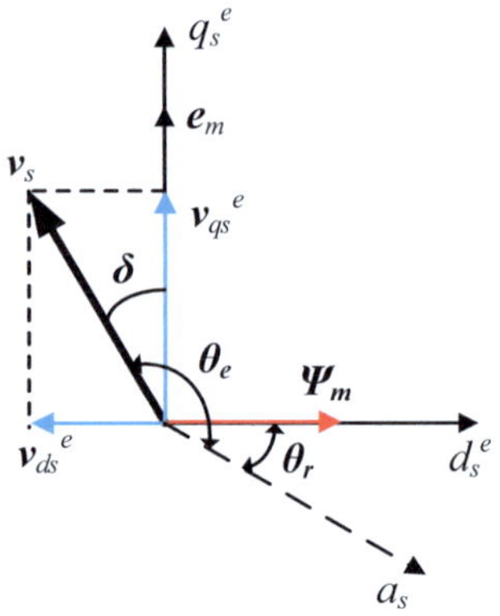

Fig. 4.42 Stator voltage V_s representation in the synchronously rotating reference frame with its components V_{ds} and V_{qs}

Thus,

$$\begin{aligned} V_{ds} &= -\boldsymbol{V}_s \sin\delta \\ V_{qs} &= \boldsymbol{V}_s \cos\delta \end{aligned} \tag{4.111}$$

The description of Eqs. (4.102) and (4.103) using the two axes, direct and quadrature in steady-state, is:

$$V_{ds} = R_s i_{ds} + \frac{\mathrm{d}\Psi_{ds}}{\mathrm{d}t} - \omega_r \Psi_{qs} \tag{4.112}$$

$$V_{qs} = R_s i_{qs} + \frac{\mathrm{d}\Psi_{qs}}{\mathrm{d}t} + \omega_r \Psi_{ds} \tag{4.113}$$

In the steady-state, the term derivative can be eliminated since the variables are constant. Therefore, it is possible to create a single phasor equation from the steady-state equations by multiplying Eq. (4.113) by j and adding it to Eq. (4.112):

$$V_{ds} + jV_{qs} = R_s\left(i_{ds} + ji_{qs}\right) + j\omega_r(L_d i_{ds} + \Psi_m) - \omega_r L_q i_{qs} \tag{4.114}$$

Thus,

$$\boldsymbol{V}_s = R_s \boldsymbol{I}_s - \omega_r L_q i_{qs} + j\omega_r \Psi_m + j\omega_r L_d i_{ds} \tag{4.115}$$

In Fig. 4.43, the phasor diagram of the PMSM is shown according to Eq. (4.115).

On the other hand, knowing that the derivative term disappears in steady-state, it is possible to determine the currents i_{ds} and i_{qs} in terms of the steady-state voltage from (4.112) and (4.113) such as:

$$i_{ds} = \frac{R_s V_{ds} + \omega_r L_q V_{qs} - \omega_r^2 \Psi_m L_q}{R_s^2 + \omega_r^2 L_d L_q} \tag{4.116}$$

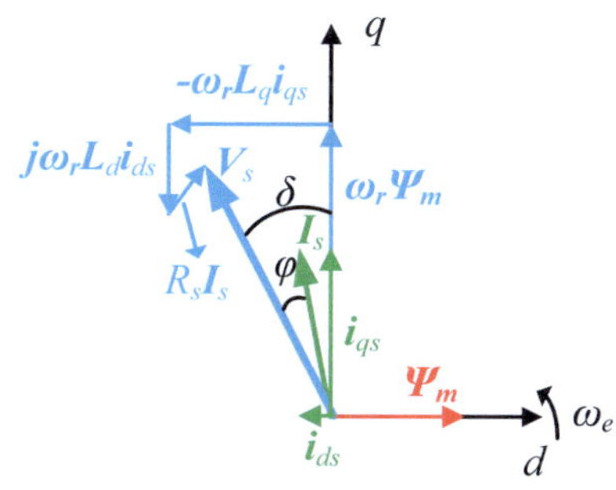

Fig. 4.43 Phasor diagram of a salient-pole PMSM

$$i_{qs} = \frac{R_s V_{qs} - \omega_r L_d V_{ds} - \omega_r \Psi_m R_s}{R_s^2 + \omega_r^2 L_d L_q} \tag{4.117}$$

If the value of the resistance of the stator R_s is neglected, the following simplified equations are obtained:

$$i_{ds} = \frac{V_{qs} - \omega_r \Psi_m}{\omega_r L_d} \tag{4.118}$$

$$i_{qs} = -\frac{V_{ds}}{\omega_r L_q} \tag{4.119}$$

Therefore, the electromagnetic torque (4.108) can be expressed as

$$\begin{aligned} T_e &= \frac{3}{2} P \left[\Psi_m i_{qs} + (L_d - L_q) i_{ds} i_{qs} \right] \\ &\approx \frac{3}{2} P \left[\left(\frac{L_d - L_q}{L_d L_q} \right) \left(\frac{V_{qs} V_{ds}}{\omega_r^2} - \frac{\omega_r \Psi_m V_{ds}}{\omega_r^2} \right) - \Psi_m \frac{V_{ds}}{\omega_r L_q} \right] \end{aligned} \tag{4.120}$$

where stator resistance has been neglected. Note that V_{ds} voltage is negative in the motoring mode, since $V_{ds} = -\boldsymbol{V}_s \sin\delta$. The above equation is valid for frequencies above zero. If SMPMSM $L_d = L_q$, thus, (4.120) can be rewritten as:

$$T_e \approx \frac{3}{2} P \cdot \Psi_m \frac{-V_{ds}}{\omega_r L_q} \approx \frac{3}{2} P \cdot \Psi_m \frac{\boldsymbol{V}_s}{\omega_r L_q} \sin\delta \tag{4.121}$$

The maximum torque is achieved when load angle δ is equal to 90°. Equation (4.121) suggests a capability curve of torque/speed of the SMPMSM similar to the induction machine. When V_{ds} voltage reaches its maximum value, the constant torque region ends. From this point, if the speed in incremented the torque will decrease at $1/\omega_r$.

On the other hand, if $V_{qs} V_{ds} \gg \omega_r \varPsi_m V_{ds}$ in (4.120), it can be simplified for IPMSM machine as:

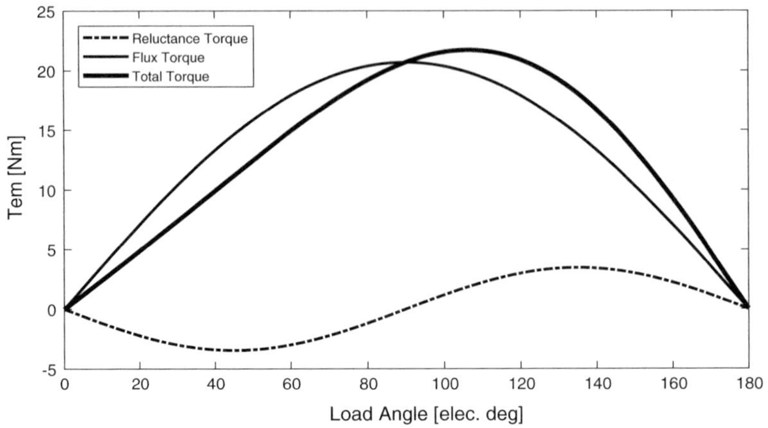

Fig. 4.44 Developed torque components of an IPMSM machine. It is shown the torque produced by the reluctance, and flux components. The total torque is the sum of both. The maximum torque is at 106° approx

$$T_e \approx \frac{3}{2}P\left[\left(\frac{L_d - L_q}{L_d L_q}\right)\left(\frac{V_{qs}V_{ds}}{\omega_r^2}\right) - \Psi_m \frac{V_{ds}}{\omega_r L_q}\right] \tag{4.122}$$

And in function of load angle δ:

$$T_e \approx \frac{3}{2}P\left[-\left(\frac{L_d - L_q}{L_d L_q}\right)\frac{\boldsymbol{V}_s^2}{\omega_r^2}\frac{1}{2}\sin 2\delta + \Psi_m \frac{\boldsymbol{V}_s}{\omega_r L_q}\sin\delta\right] \tag{4.123}$$

In this case, the maximum torque produced by the machine is reached at load angle above 90°, as shown in Fig. 4.44, since the inductance L_q is often slightly higher than L_d. Due to this low difference, the reluctance torque contribution will be small percentage respect the torque produced by the magnets in the IPMSM machines.

In case of SMPMSM, where $L_d = L_q$, the reluctance torque will be null, and the maximum torque will be produced at a load angle of 90°.

Table 4.4 Usual nameplate data of PMSM

Nameplate	Symbol	Unit
Nominal power	P_N	kW
Line-to-line nominal voltage	V_N	V
Nominal current	I_N	A
Nominal frequency	f_N	Hz
Nominal speed	N_N	RPM
Maximum speed	$N_{N\max}$	RPM
Nominal torque	T	Nm

Table 4.5 PMSM equivalent circuit and motion parameters

Parameters for machine model	Symbol	Unit
Direct inductance	L_d	H
Quadrature inductance	L_q	H
Stator resistance	R_s	Ohms
Permanent magnet flux	Ψ_m	Wb/rad
Rotor inertia	J	$kg \cdot m^2$
Viscous coefficient	B	$kg \cdot m^2/s$

As in the case of the induction machine, by means of the equivalent circuit, the characteristics of the SPMSM machine can be investigated using the common phasor analysis method of the AC circuits.

Table 4.4 shows the most common nameplate data of the PMSM, while Table 4.5 shows the necessary parameters for the equivalent circuit plus the mechanical parameters (motion parameters).

4.4.2.6 PMSM Operation

Figure 4.45 illustrates the four operations' modes on the speed/torque axes for PMSM. In the first quadrant, the PMSM is rotating in a clockwise direction. The machine is generating torque; quadrature current I_{qs} is positive while I_{ds} current is

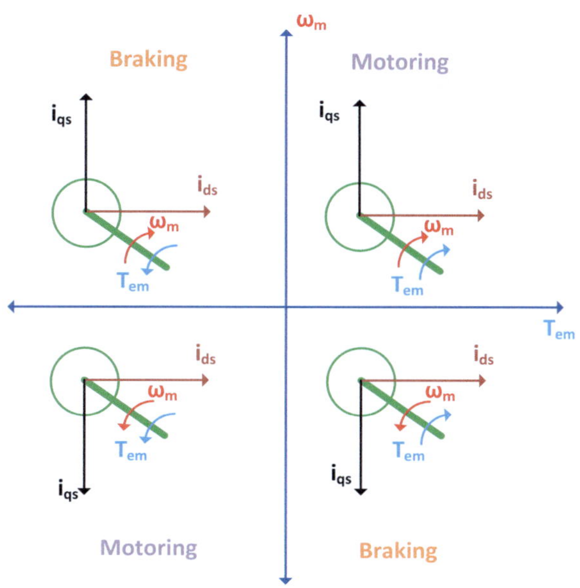

Fig. 4.45 PMSM operations in accordance with the direction of power flow

set to zero for SPMSM machines. The flux linkage Ψ_m of the magnet is in the d-axis direction. When the machine is rotating anticlockwise in the third quadrant, the quadrature current I_{qs} is negative. In regenerating braking mode, the energy is fed back while braking, since torque and speed are in opposite directions. Quadrature current and torque are in a reversed direction.

4.4.3 Synchronous Reluctance Machine

The synchronous reluctance machine (SynRM) is a synchronous machine in general without permanent magnets, where the torque is produced with the reluctance principle. The SynRM has a structure simple and rugged as an induction machine, but it is not necessary to compute the slip and does not have any rotor copper loss. However, it has lower average torque and larger torque pulsation which can be improved by using different methods proposed by Park et al. (2006) and Hofmann et al. (2004).

The operation principle is based on this reluctance variation where the rotor tends to be aligned with the rotating magnetic field of the stator at the lower reluctance path in a synchronous way. This machine differs from the variable reluctance machine in that the stator is constructed from a cylindrical structure as an induction machine and only the rotor has salient poles. The complexity of the machine resides in the rotor design to get the most efficiency reluctance variation and to reduce the torque ripple. The electromagnetic torque produced varies with the position of the rotor and depends on the saliency ratio, which is defined as L_d/L_q. The saliency ratio should reach values above 6 to be comparable to the torque density of induction machines (Lipo 1991). The SynRM usually is less expensive than a PMSM, and there are no restrictions to prevent demagnetization of the permanents' magnets. However, there are rotors designs which mount permanent magnets in the rotor flux barriers (Pellegrino et al. 2016) to improve the power factor and to get higher efficiency than induction machine and comparable with PMSM. These machines are known as permanent magnet-assisted SynRM (PMASynRM). The magnets are magnetized so as to force the permanent magnet flux into the quadrature axis. It is important to mention that the amount of flux which is needed is not significant since the machine is basically magnetized by the d-axis component of stator current as in an induction machine. In Fig. 4.46 is represented as a cross-sectional view of the SynRM with 24 slots and 4 rotor poles with distributed winding. This is the most common rotor structure of the SynRM.

The synchronous reluctance machine (SynRM) uses the concept of the variation of its reluctance for the production of torque. This is achieved thanks to the rotor geometry under a sinusoidal MMF generated by the traditional stator, similar to induction or PMAC machines.

In Fig. 4.47b is shown an object with an anisotropic geometry under a magnetic field ψ. If the object has a different angle δ of zero between the d-axis and the field lines, a pair of forces F will be produced that will move the object to its equilibrium position ($\delta = 0$). To produce a continuity of the movement of the object, the direction

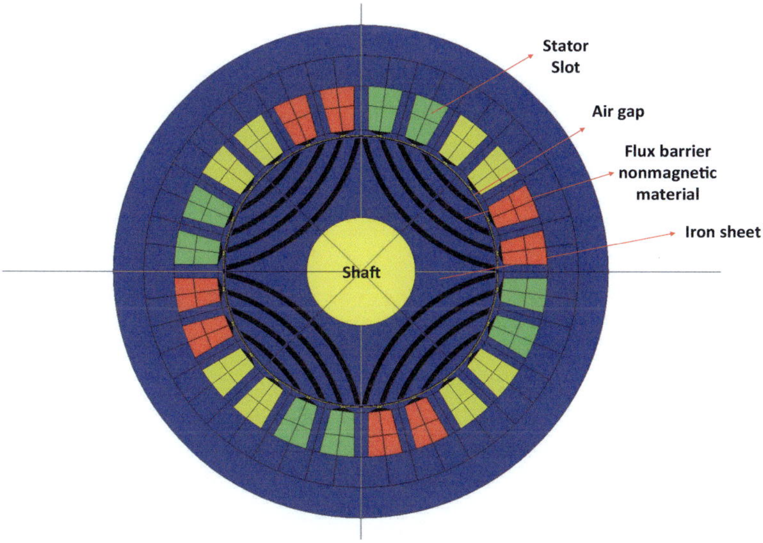

Fig. 4.46 Cross-sectional view of the SynRM with 24 slots and 4 rotor poles. Image developed using Flux2D provided courtesy of Altair Engineering, Inc.

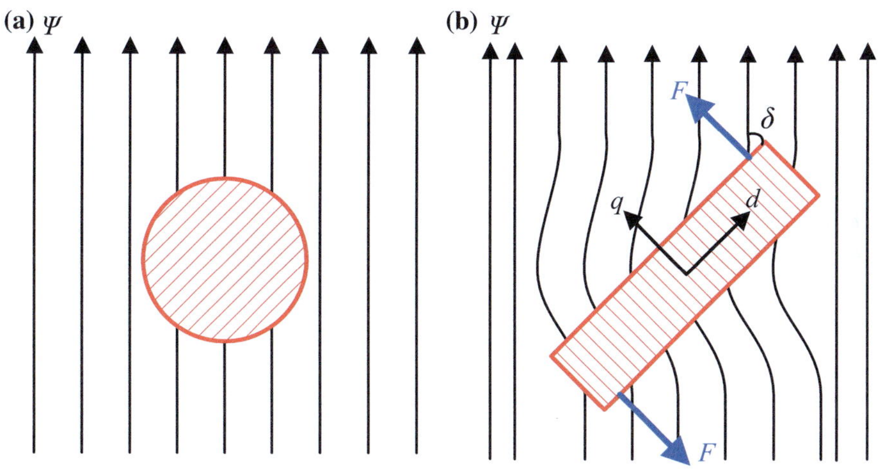

Fig. 4.47 Magnetic field ψ through two object types. **a** An object with isotropic geometry. **b** An object with anisotropic geometry where torque is produced

of the magnetic field must be modified to have again an angle δ different from zero. That is, if the object is placed inside a conventional stator with a winding distributed sinusoidally fed with three three-phase voltages; this will produce a sinusoidal MMF producing a continuous rotation of the object. It is evident that if the d-axis of the object is not aligned with the field, it will introduce a field distortion in the main field,

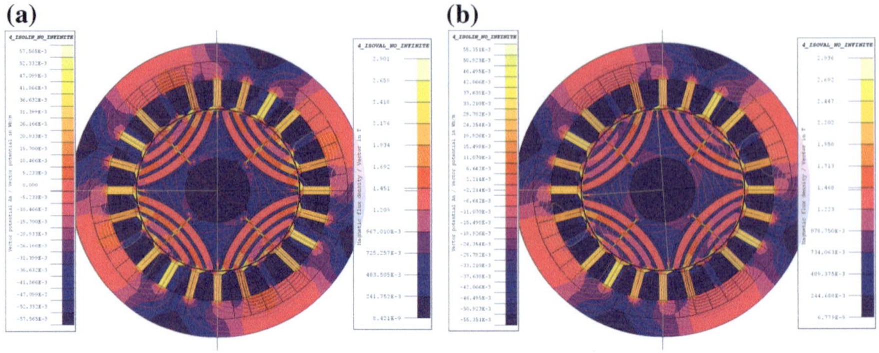

Fig. 4.48 Cross-sectional view of SynRM. **a** The rotor is aligned with the stator magnetic field. Angle $\delta = 0°$ and no torque is produced. **b** The rotor is not aligned with the stator magnetic field. Angle $\delta = 4°$ and torque is produced in a clockwise direction. Image developed using Flux2D provided courtesy of Altair Engineering, Inc.

its direction being solidary with the *q*-axis as shown in Fig. 4.47b. In this situation, there will always be a torque that acts to reduce the distortion field on the *q*-axis by making the angle δ always tends to zero. If the angle δ is kept constant because the load applies a load torque, the electromagnetic energy will continuously be converted into mechanical energy.

In Fig. 4.48 is represented an FEA simulation of a SynRM when the rotor is aligned with the stator magnetic field where no torque is produced, and when is no aligned producing torque. The produced torque tends to align the rotor with the stator magnetic field to get a stable point. As commented before, when rotor gets this stable point, the stator magnetic field should be changed to continue with the movement of the rotor to the next stable point.

As the induction machine, the stator current is responsible for both the magnetization and the production torque that tries to reduce field distortion. As described in (Xu et al. 1991), this can be done by controlling the angle θ, which is the angle between the current vector of the stator winding and the rotor *d*-axis in the synchronous reference frame as represented in Fig. 4.49.

As represented in Fig. 4.50, there are two paths for the flux. The first path for the flux is commonly referred to as the *d*-axis path. It has high permeability, where the flux lines flow in rotor iron paths, as shown in Fig. 4.50a. On the other hand, the second path for the flux has a low permeability, where flux lines cross the rotor flux barriers (commonly referred to as the *q*-axis path) as represented in Fig. 4.50b.

In Fig. 4.51 is represented the flux linkage in the quadrature and direct axes as a function of the current. In Fig. 4.50a, the higher flux linkage is along *d*-axis as commented before. The iron saturation at higher currents limits the flux. The *q*-axis flux linkage is lower, limited by the rotor flux barriers, and it is linear with the current.

The SynRM runs at a poorer power factor than the induction machine (Lipo et al. 1992). A solution to this is to insert permanent magnet in the flux barriers as

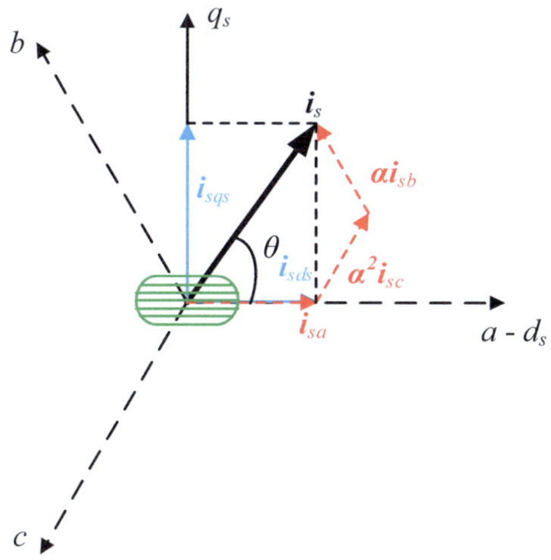

Fig. 4.49 Vector diagram of SynRM

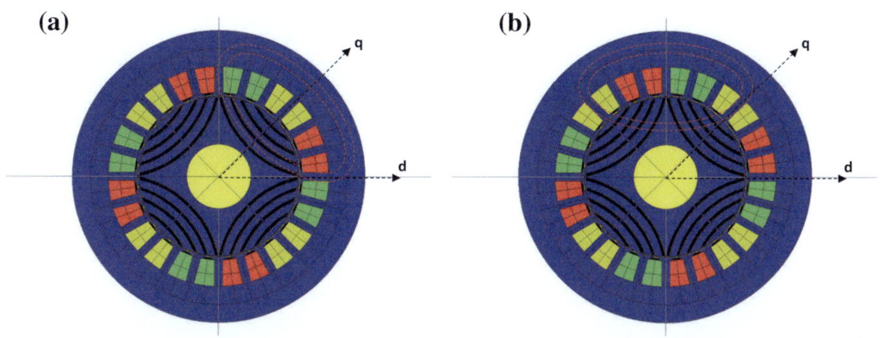

Fig. 4.50 Two different rotor paths for the flux. **a** high permeability path, where the flux lines flowing in rotor iron paths. **b** Low permeability, where the flux lines cross the rotor flux barriers. Image developed using Flux2D provided courtesy of Altair Engineering, Inc.

commented before. If a permanent magnet is inserted in the flux barriers, it produces a negative flux linkage along the q-axis which is represented in Fig. 4.52b. It is the PMASynRM, which adds two main advantages to the SynRM. The first advantage is the increase of the power factor of the machine since the voltage vector is rotated toward the current vector. Thus, PMASynRM requires a lower Volt-Ampere rating for given nominal mechanical power. The second advantage is the little increment of the electromagnetic torque of the machine since the permanent magnet adds a torque term in the reluctance torque. Moreover, a part of the flux of the permanent magnet tends to saturate the iron bridges, which implies a reduction of the L_q inductance. Then, according to the electromagnetic torque expression, the torque is increased.

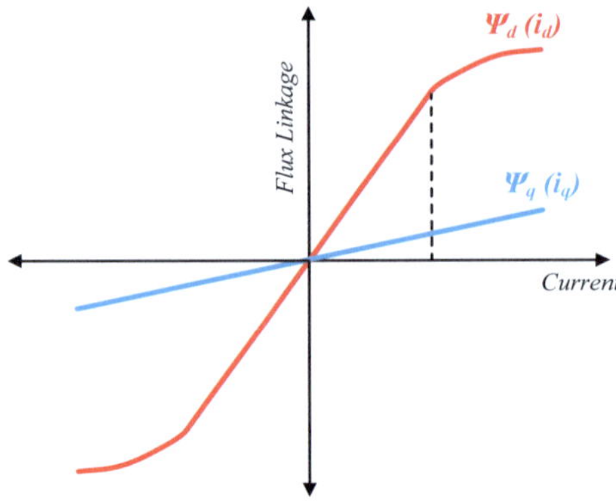

Fig. 4.51 Flux linkage in the quadrature and direct axes as a function of their respective currents

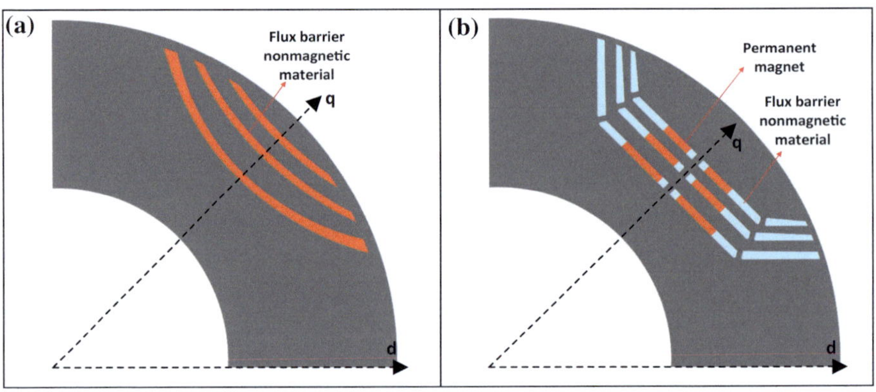

Fig. 4.52 Cross-sectional view of the four-pole rotor. **a** SynRM. **b** PMASynRM. Image developed using FluxMotor provided courtesy of Altair Engineering, Inc.

The cross-sectional view of a typical four-pole SynRM is shown in Fig. 4.52a for SynRM and Fig. 4.52b for PMASynRM.

As IPMSM machines, the L_d inductance varies according to the magnitude of i_{dse} and i_{qse} doing the control a complicated task. Moreover, the saliency ratio decreases rapidly, as shown in Fig. 4.53, as the magnitude of the current increase. It is due to the magnetic saturation and the consequence is that the electromagnetic torque drops down severely.

As saliency ratio depends on the air gap, it should be small enough to get higher saliency ratio in SynRM. This is a complicated task for higher-power machines such as one over several hundred kilowatts, where there are mechanical limitations due to bearing problems, eccentrically rotor, and centrifugal forces.

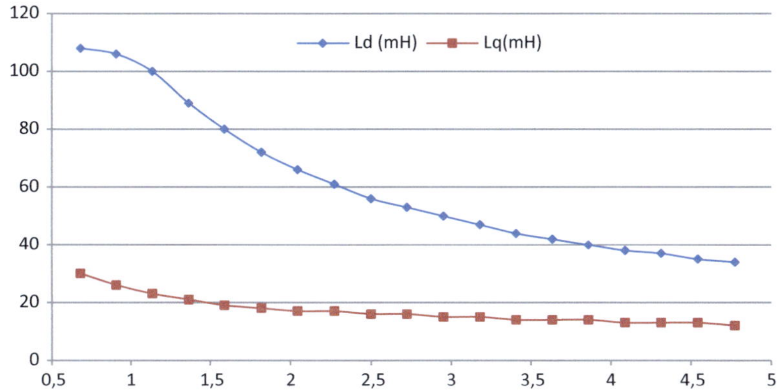

Fig. 4.53 FEA simulation results of L_d and L_q variation in a SynRM when current is increased

4.4.3.1 Space Vector Theory in SynRM and PMASynRM

The PSynRM and PMASynRM can be analyzed in a similar way as shown for PMSM and induction machine considering the same assumptions. The PMSM and both SynRM share many similar expressions of their equations, which it is not necessary to repeat again. In a synchronous rotating coordinate system, designed by *e* symbol with angular speed ω_e, the dependence of the angle θ is eliminated as shown for PMSM:

$$\boldsymbol{V}_s^e = R_s \boldsymbol{i}_s^e + \frac{\mathrm{d}\boldsymbol{\Psi}_s^e}{\mathrm{d}t} + j\omega_e \boldsymbol{\Psi}_s^e \tag{4.124}$$

And the stator flux linkage can be written as

$$\boldsymbol{\Psi}_s^e = L_s \boldsymbol{i}_s^e + \boldsymbol{\Psi}_m^e \tag{4.125}$$

As a PMSM, the single equation does not describe the magnetic rotor asymmetry of the reluctance paths of SynRM. Therefore, the two-axis model is preferred to have the two components of the magnetic axis of the machine. Using the Clarke and rotation matrix transformation, the three-phase voltage, current, and flux linkage space vectors in stationary *abc*-axis reference frame can be transformed into the *dqe* space vectors in the rotating reference frame.

According to this, it is considered:

- q is the preferred axis for magnet flux orientation of Ψ_m of PMASynRM.

The stator voltage *dqe* equations for SynRM are identical to the PMSM, but with the difference that SynRM does no has flux magnet Ψ_m. However, the PMASynRM with the magnets inserted in the q flux path implies that a new term is added to the quadrature flux equations as represented in (4.129). Equations (4.124)–(4.129) represent the general equations for SynRM with and without magnets inserted, where flux magnet Ψ_m is zero for SynRM.

$$V_{ds}^e = R_s i_{ds}^e + L_d \frac{\mathrm{d}i_{ds}^e}{\mathrm{d}t} - \omega_r \Psi_{qs}^e \tag{4.126}$$

$$V_{qs}^e = R_s i_{qs}^e + L_q \frac{\mathrm{d}i_{qs}^e}{\mathrm{d}t} + \omega_r \Psi_{ds}^e \tag{4.127}$$

where linkage fluxes are

$$\Psi_{ds}^e = L_{ls} i_{ds}^e + L_{md} i_{ds}^e = L_d i_{ds}^e \tag{4.128}$$

$$\Psi_{qs}^e = L_{ls} i_{qs}^e + L_{md} i_{qs}^e + \Psi_{mq} = L_q i_{qs}^e - \Psi_{mq} = L_q i_{qs}^e - \Psi_m \tag{4.129}$$

As a PMSM, the inductance in the *dqe* axes is defined as L_d and L_q. It is also the sum of the leakage inductance and the corresponding magnetizing inductance of the axis:

$$L_d = L_{ls} + L_{md} \tag{4.130}$$

$$L_q = L_{ls} + L_{mq} \tag{4.131}$$

The equivalent circuit is depicted in Fig. 4.54 with Eqs. (4.126)–(4.129).

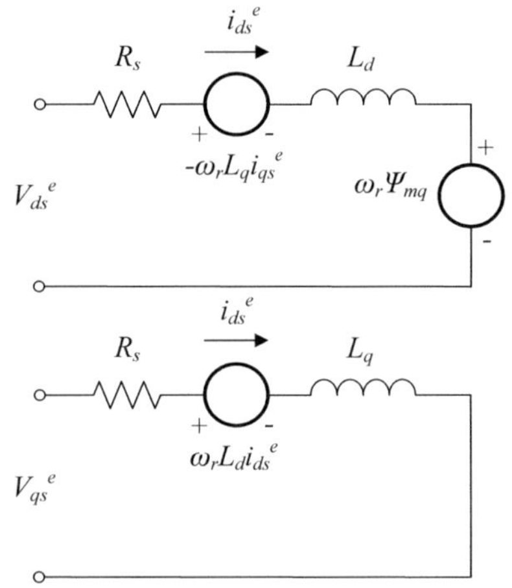

Fig. 4.54 Equivalent circuit of the SynRM and PMASynRM in *dqe* axes

4.4.3.1.1 Torque Expression

As discussed by PMSM, the mechanical power, which is the mechanical speed multiply by the electromagnetic torque, can be expressed as:

$$p_{em} = \omega_{mec} T_e = \frac{3}{2}\omega_r\left(\Psi^e_{ds} \cdot i^e_{qs} - \Psi^e_{qs} \cdot i^e_{ds}\right) = \frac{3}{2}\omega_r\left[(L_d - L_q) i^e_{ds} i^e_{qs} + \Psi_m i^e_{ds}\right] \tag{4.132}$$

And the electromagnetic torque for SynRM and PMASynRM is:

$$T_e = \frac{3}{2}P\left[(L_d - L_q) i^e_{ds} i^e_{qs} + \Psi_m i^e_{ds}\right] \tag{4.133}$$

The electromagnetic torque for PMASynRM of (4.133) is expressed similar to electromagnetic torque expression of IPMSM machine, but considering that the magnets are included in the q-axis flux path and $L_d > L_q$. Thus, to produce positive torque in motoring mode, i_{dse} current should be positive under positive i_{qse} current. This is the contrary seen in the IPMSM machine, where i_{dse} should be negative.

In addition, the above expression suggests that to get the maximum torque for a SynRM is necessary to set $L_q = 0$. This possibility can be achieved by the use of the second term of the electromagnetic torque expression (4.133) of PMASynRM. Equation (4.133) can be rewritten as:

$$T_e = \frac{3}{2}P\left[L_d i^e_{ds} i^e_{qs} - \left(L_q i^e_{qs} - \Psi_m\right) \cdot i^e_{ds}\right] \tag{4.134}$$

Hence, the second negative term can be made as small as necessary by placing magnets in the q-axis of the rotor. Note that the saliency ratio of the PMASynRM has been increased in high proportion since L_q contribution is close to zero where torque density can be compared with PMSM, as mentioned above, but with less permanent magnet and as consequence lower price. In Fig. 4.55 is represented the

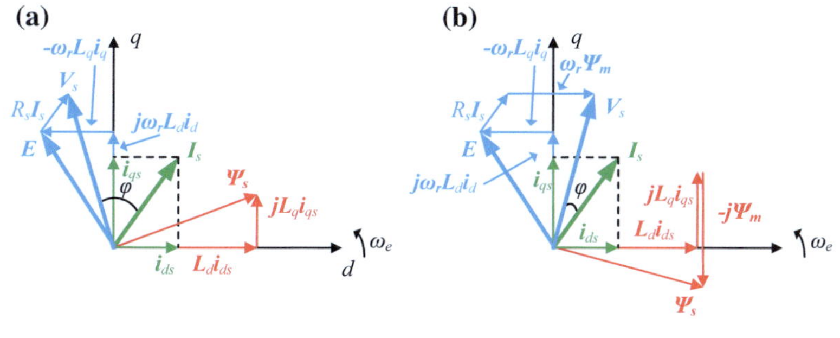

Fig. 4.55 Vector diagram of a synchronous reluctance machine. **a** Without magnets in the q-axis. **b** With magnets in the q-axis to improve the power factor (cos φ) and to reduce the L_q inductance

vector diagram of SynRM and PMASynRM according to Eqs. (4.126)–(4.129) in the synchronously rotating reference frame where suffix "*e*" has been omitted.

Graphically, it is possible to see how PMASynRM can improve the power factor is the magnets are correctly chosen and inserted along the *q*-axis of the rotor of the machine.

4.4.3.2 Steady-State Equations of SynRM

As described for the PMSM, the circuit in steady-state of the SynRM and PMASynRM can also be derived for the rotating synchronous reference frame.

The description using the two direct and quadrature axes in steady-state is:

$$V_{ds} = R_s i_{ds} + \frac{\mathrm{d}\Psi_{ds}}{\mathrm{d}t} - \omega_r \Psi_{qs} \tag{4.135}$$

$$V_{qs} = R_s i_{qs} + \frac{\mathrm{d}\Psi_{qs}}{\mathrm{d}t} + \omega_r \Psi_{ds} \tag{4.136}$$

Note that the upper index "*e*" has been omitted in above and below equations.

In the stationary state, the term derivative can be eliminated since the variables are constant. By performing the same steps that were already done for the PMSM, it is possible to obtain in a phasor equation:

$$\boldsymbol{V}_s = R_s \boldsymbol{I}_s - \omega_r L_q i_{qs} + \omega_r \Psi_m + j\omega_r L_d i_{ds} \tag{4.137}$$

In the same way as the PMSM, the currents i_{ds} and i_{qs} can be calculated in terms of the steady-state voltage as (Lipo et al. 1992):

$$i_{ds} = \frac{R_s V_{ds} + \omega_r L_q V_{qs} + \omega_r \Psi_m R_s}{R_s^2 + \omega_r^2 L_d L_q} \tag{4.138}$$

$$i_{qs} = \frac{R_s V_{qs} - \omega_r L_d V_{ds} - \omega_r^2 \Psi_m L_d}{R_s^2 + \omega_r^2 L_d L_q} \tag{4.139}$$

If the value of the resistance of the stator R_s is neglected, the following simplified equations are obtained:

$$i_{ds} = \frac{V_{qs}}{\omega_r L_d} \tag{4.140}$$

$$i_{qs} = \frac{-V_{ds} - \omega_r \Psi_m}{\omega_r L_q} \tag{4.141}$$

Therefore, the electromagnetic torque Eq. (4.133) can be expressed as

$$\begin{aligned} T_e &= \frac{3}{2}P\left[\Psi_m i_{qs} + (L_d - L_q) i_{ds} i_{qs}\right] \\ &\approx \frac{3}{2}P\left[\left(\frac{L_d - L_q}{L_d L_q}\right)\left(\frac{V_{qs}V_{ds} - \omega_r \Psi_m V_{qs}}{\omega_r^2}\right) - \Psi_m\left(\frac{V_{ds} + \omega_r \Psi_m}{\omega_r L_q}\right)\right] \end{aligned} \tag{4.142}$$

where stator resistance has been neglected. The above equation is valid for frequencies above zero. If SynRM, $\Psi_m = 0$, thus, Eq. (4.142) can be rewritten as:

$$T_e \approx \frac{3}{2}P\left(\frac{L_d - L_q}{L_d L_q}\right)\left(\frac{V_{ds}}{\omega_r}\right)\left(\frac{V_{qs}}{\omega_r}\right) \tag{4.143}$$

Substituting (4.111) into (4.143), the torque in terms of the Volt per Hertz and torque angle δ is obtained as:

$$T_e \approx \frac{3}{2}P\left(\frac{1}{L_q} - \frac{1}{L_d}\right)\left(\frac{\boldsymbol{V}_s}{\omega_r}\right)^2 \frac{\sin 2\delta}{2} \tag{4.144}$$

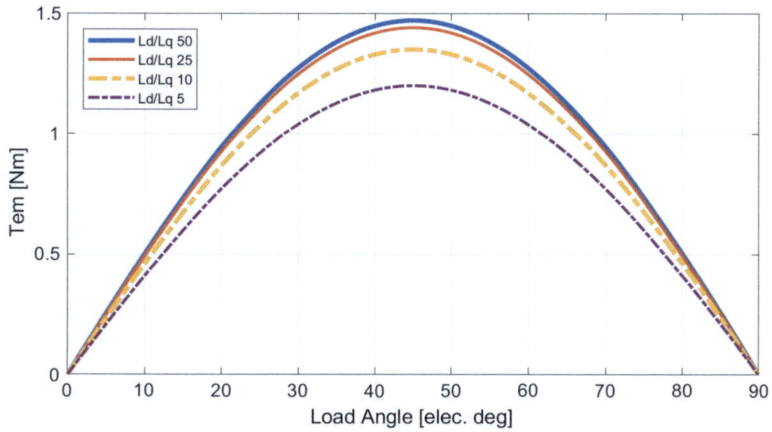

Fig. 4.56 Developed torque of a SynRM at difference saliency ratios. The maximum torque is at 45°

As it can be seen, the torque varies as the square of the Volt per Hertz and as the sine of twice of the angle δ. When the Volt per Hertz is fixed, the maximum torque is clearly reached when $\delta = 45°$. In Fig. 4.56 is illustrated the torque/load angle curves for different saliency ratios.

The torque Eq. (4.144) is still valid for PMASynRM since contribution torque of magnet flux Ψ_m is very low. As commented before, the main advantages of inserting magnets in the q-axis are to improve the power factor and to reduce the L_q inductance, which increases the developed torque.

The most common nameplate data of the PMASynRM which describes the nominal parameters, the equivalent circuit parameters, and the motion parameters are usually the same ones that were seen for PMSM.

References

Bianchini C, Immovilli F, Lorenzani E, Bellini A, Davoli M (2012) Review of design solutions for internal permanent-magnet machines cogging torque reduction. IEEE Trans Magn 48(10):2685–2693

Chandana Perera PD (2002) Sensorless control of permanent-magnet synchronous motor drives. Ph.D. dissertation, Institute of energy technology. Aalborg university

Hofmann HF, Sanders SR, Antably A (2004) Stator-flux-oriented vector control of synchronous reluctance machines with maximized efficiency. IEEE Trans Ind Electron 51(5):1066–1072

Hughes A (1994) Electric motors and drives. Newnes

Islam R, Husain I, Fardoun A, McLaughlin K (2009) Permanent-magnet synchronous motor magnet designs with skewing for torque ripple and cogging torque reduction. IEEE Trans Ind Appl 45(1):152–160

Kim DK, Lee KW, Kwon B-I (2006) Commutation torque ripple reduction in a position sensorless brushless DC motor drive. IEEE Trans Power Electron 21(6):1762–1768

Krause PC (1986) Analysis of electric machinery. McGraw-Hill

Lipo TA (1991) Synchronous reluctance machines, a variable alternative for AC drives? University of Wisconsin-Madison

Lipo TA, Matsuo T (1992) Performance of synchronous reluctance motors, part of: synchronous reluctance motors and drives—a new alternative, tutorial held 4 Oct 1992, Annual Meeting IEEE-IAS, pp 1–56

Mebarki A, Gerada D, Brown NL (2014) Analysis of an axial PM machine with field weakening capability for engine integration. In: Proceedings of 7th IET international conference on power electronics, machines and drives, vol 1, pp 1–6

Park JM, Kim S, Hong JP, Lee JH (2006) Rotor design on torque ripple reduction for a synchronous reluctance motor with concentrated winding using response surface methodology. IEEE Trans Magnet 42(10):3479–3481

Pellegrino G, Jahns ThM, Bianchi N, Soong W, Cupertino F (2016) The rediscovery of synchronous reluctance and ferrite permanent magnet motors. Springer, Berlin

Pillay P, Krishnan R (1991) Application characteristics of permanent magnet synchronous and brushless DC motors for servo drives. IEEE Trans Ind Appl 27(5):986–996

Pyrhönen J, Hrabovcova V, Scott Semken R (2016) Mobipocket, "electrical machine drives control: an introduction". Wiley

Quang NP, Dittrich JA (2015) Vector control of three-phase AC machines. Springer, Berlin

Wach P (2011) Dynamics and control of electrical drives. Springer, Berlin, Heidelberg

Weh H, May H, Shalaby M (1990) Highly effective magnetic circuits for permanent magnet excited synchronous machines. In: International conference on electrical machines, Cambridge, MA, 13–15 August 1990, pp 1040–1045

Xu L, Xu X, Lipo TA, Novotny DW (1991) Vector control of a synchronous reluctance motor including saturation and iron loss. IEEE Trans Ind Appl 27(5):977–987

Chapter 5
Modeling Electric Machines

5.1 Mechanical Motion Model (Newton's Laws of Motion)

The rotating load can be handled by employing different transmission methods, i.e., timing belt or gear. It is also possible to find some applications where the machine is connected to the load directly, without any mechanical transmission. Of course, this depends on the application, but it is useful to use a timing belt or gearbox in applications in which the high torque at very low speed is needed, or in which the speed range over the speeds of the machine must be extended. For example, for induction machines, the cost is correlated with the machine's torque rating, and if high torque is required at a low speed, the cost of the machine increases. Thus, via mechanical transmission through belt or gears, torque can be increased at low speed without using an induction machine with high torque. Additionally, the inertia seen by the machine (reflected inertia) is reduced. In Fig. 5.1, a machine connected to a rotating load using a timing belt is shown. As can be observed, the load and the machine have different pulley diameters D_m and D_L, and different inertias J_m, and J_L.

The ratio N is the relationship between the diameters of the pulley of motor D_m and the pulley of load D_L:

$$N = \frac{D_L}{D_m} \tag{5.1}$$

The torque seen by the electric machine is T_L/N. The total inertia seen from the machine is:

$$J_T = J_m + \frac{1}{N^2} J_L \tag{5.2}$$

The total friction coefficient B_T is composed by the friction coefficient of the machine B_m plus the friction coefficient of the load divided by the square of ratio N:

© Springer Nature Switzerland AG 2020
R. Molina Llorente, *Practical Control of Electric Machines*,
Advances in Industrial Control, https://doi.org/10.1007/978-3-030-34758-1_5

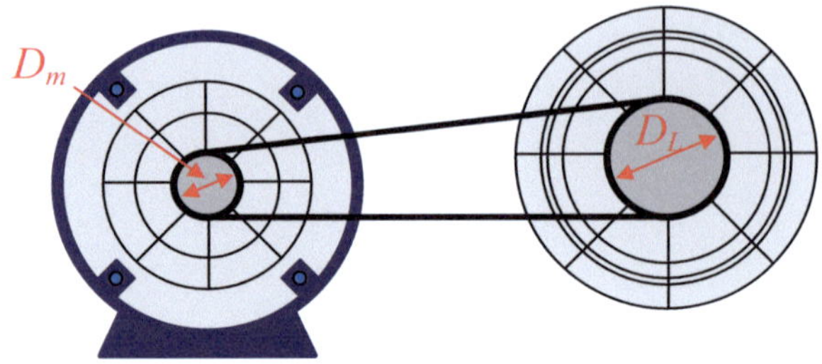

Fig. 5.1 Machine connected to a rotating load with a timing belt

$$B_T = B_m + \frac{1}{N^2} B_L \tag{5.3}$$

When a machine lifts an object (Fig. 5.2a) of mass m by means of a pulley with inertia J_m, the overall reflected inertia at the machine is:

$$J_T = J_m + \frac{1}{4} m \cdot D_m^2 \tag{5.4}$$

When an object is driven by a conveyor belt, as shown in Fig. 5.2b the reflected inertia at the machine is:

$$J_T = 2J_m + \frac{1}{4} m \cdot D_m^2 \tag{5.5}$$

On the other hand, the equation below shows the application of Newton's second law to the rotating system:

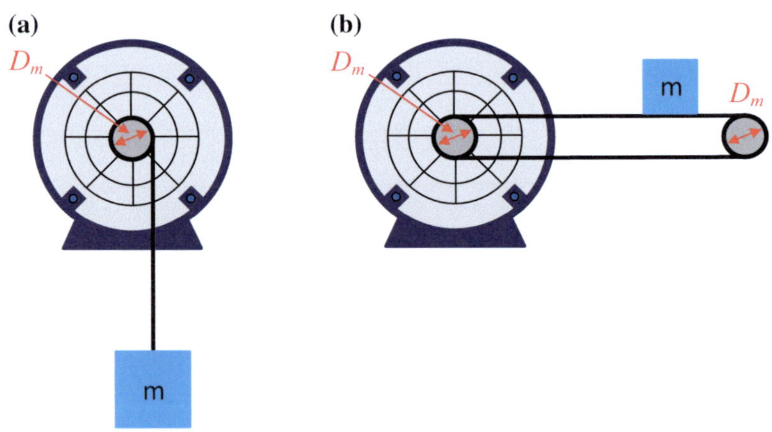

Fig. 5.2 **a** Lifting an object by means of a belt. **b** Linear movement of an object by means of a conveyor belt

$$T_e - \frac{T_L}{N} - B_T \omega_m = J_T \frac{\mathrm{d}\omega_m}{\mathrm{d}t} \tag{5.6}$$

In case of timing belt load:

$$T_e - \frac{T_L}{N} - \left(B_m + \frac{1}{N^2} B_L\right)\omega_m = \left(J_m + \frac{1}{N^2} J_L\right)\frac{\mathrm{d}\omega_m}{\mathrm{d}t} \tag{5.7}$$

where T_e is the electromagnetic torque developed by an electric machine.

According to Fig. 5.1, the electromagnetic torque T_e saw by the machine is the load torque divided by the timing belt ratio *N*, as expected when the diameter of the machine pulley is smaller than the diameter of the load. In the case of the inertia and friction coefficient, as commented before, the inertia and friction coefficient seen in the machine are divided by the square of the timing belt ratio. It is also possible to observe that, at a constant speed (derivative terms disappears), both torques T_L and T_e (neglecting B_T) are identical and, in order to accelerate the load, the electromagnetic torque developed by the machine must be higher than the load torque.

The most common and general expression of the machine motion is, according to (5.8). As discussed early, T_L stands for load torque, and T_f stands for the friction torque. Total inertia J_T of the rotating part in $\mathrm{kg} \cdot \mathrm{m}^2$ is composed by the rotor and load inertia. B_T stands for the total friction coefficient of the machine and the load in Nm/rad/s. The additional friction torque produced by the friction coefficient is increased with speed as Eq. (5.8) shows. The friction coefficient is typically considered to be a constant, but it can change over time due to inadequate or failing lubrication and normal wear. There could also be a small change due to a change in operating temperature. If the load is belt-driven, the load will cause a side force on the bearings called "overhung load." That could cause a change in the force due to friction to change due to load changes. If the shaft supports a pump impeller or a fan, the load could cause a thrust force on the bearings that would change with speed and load. In addition to bearing friction, machines have aerodynamic drag or "windage" that is a type of friction between the moving parts of the machine and air. Most machines have some windage that is deliberately added by incorporating a fan or rotor fins for cooling purposes. Brushes machines have brush friction in addition to bearing friction which can be different according to the direction of rotation.

$$T_e(t) = J_T \frac{\mathrm{d}\omega_{rm}}{\mathrm{d}t} + B_T \omega_{rm}(t) + T_L(t) + T_f \tag{5.8}$$

Neglecting T_f, the block diagram, which relates the previous variables in Laplace form, is represented in Fig. 5.3.

Where the rotor speed can be expressed as a function of angle rotor position θ_{rm}:

$$\omega_{rm}(t) = \frac{\mathrm{d}\theta_{rm}}{\mathrm{d}t} \tag{5.9}$$

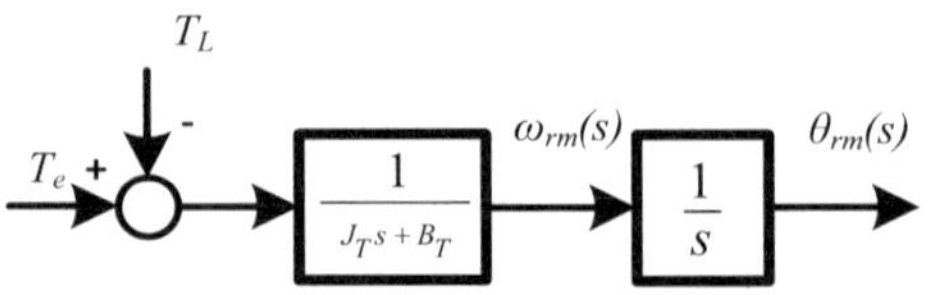

Fig. 5.3 Block diagram of mechanical motion

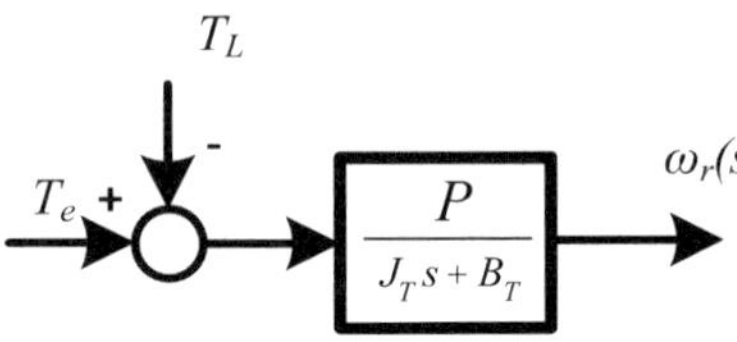

Fig. 5.4 Block diagram of mechanical motion with machine pole pairs included

The electromechanical power can be expressed as a function of the mechanical speed and electromagnetic torque T_e:

$$p_{em} = \omega_{rm} T_e \tag{5.10}$$

Electrical speed ω_r is P (pole pairs of the machine) times the mechanical speed, as shown below:

$$\omega_r = P \cdot \omega_{rm} \tag{5.11}$$

Then, Eq. (5.8) can be rewritten with pole pair dependence and electrical speed instead of mechanical speed as:

$$T_e(t) = \frac{J_T}{P}\frac{\mathrm{d}\omega_r}{\mathrm{d}t} + \frac{B_T}{P}\omega_r(t) + T_L(t) + T_f \tag{5.12}$$

Equation (5.12) will be useful for the state-space model of AC machines as shown in Sect. 5.3 and ahead (Fig. 5.4).

5.2 State-Space Overview

In order to develop and evaluate the performance of a control algorithm that acts on a dynamic system, such as an electrical machine, an intimate knowledge of the system is desirable. The description of these dynamic systems through their transfer function presents some limitations. It does not provide information about the physical structure of the system and is only valid for time-invariant linear systems with one input and one output. In addition, it does not provide information on what goes on inside the system and requires that the initial conditions of the system be null. However, many systems are analyzed through their transfer function since the system is usually linearized at a point of interest to apply the advantages offered by the Laplace transform. This is not the case of AC electric machines, because it is usually described by time-variant non-linear systems. Representation in the state-space allows the analysis of systems with more than one input or more than one output to systems that are variant or invariant over time. State-space equations provide information on what happens within the system (observability) and apply to linear and non-linear systems. The concept of observability (Kalman 1960) is related to the determination of the state of a system based on the knowledge of input–output data. That is, it allows to know what happens inside a system observing its outputs.

In conclusion, with the representation in state-space, the internal dynamics of a system and its response can be obtained.

The state-space system is described in continuous-time form according to Eqs. (5.13) and (5.14). Equation (5.13) is the state equation, and Eq. (5.14) is the output equation. $\boldsymbol{y}$ is the output vector which is expressed as a linear combination of the state vector $\boldsymbol{x}$ and the input vector $\boldsymbol{u}$.

$$\dot{\boldsymbol{x}}(t) = \boldsymbol{A}\boldsymbol{x}(t) + \boldsymbol{B}\boldsymbol{u}(t) \tag{5.13}$$

$$\boldsymbol{y}(t) = \boldsymbol{C}\boldsymbol{x}(t) + \boldsymbol{D}\boldsymbol{u}(t) \tag{5.14}$$

where

$\boldsymbol{x}$ State vector
$\boldsymbol{y}$ Output vector
$\boldsymbol{u}$ Input vector
$\boldsymbol{A}$ State coefficient matrix
$\boldsymbol{B}$ Source coefficient matrix
$\boldsymbol{C}$ Output coefficient matrix
$\boldsymbol{D}$ Output source coefficient matrix.

In Fig. 5.5 is represented the block diagram corresponding to Eqs. (5.13) and (5.14).

Equations (5.13) and (5.14) are a particular case for linear systems with constant parameters A, B, C, and D, but as it will be seen later, the machine models except the DC machine are a linear system with time-variant parameters. In this case, Eqs. (5.13) and (5.14) take the form:

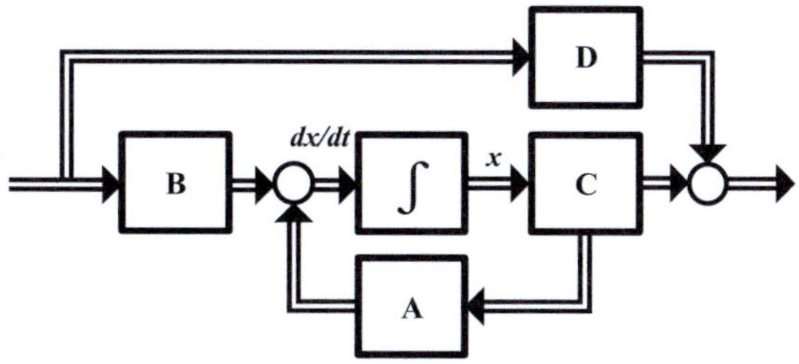

Fig. 5.5 Continuous state-space block diagram

$$\begin{aligned}\dot{\boldsymbol{x}} &= \boldsymbol{A}(t)\boldsymbol{x}(t) + \boldsymbol{B}(t)\boldsymbol{u}(t)\\ \boldsymbol{y}(t) &= \boldsymbol{C}(t)\boldsymbol{x}(t) + \boldsymbol{D}(t)\boldsymbol{u}(t)\end{aligned} \tag{5.15}$$

The discrete-time form is

$$\begin{aligned}\boldsymbol{x}(k+1) &= \boldsymbol{A}_d(k)\boldsymbol{x}(k) + \boldsymbol{B}_d(k)\boldsymbol{u}(k)\\ \boldsymbol{y}(k) &= \boldsymbol{C}_d(k)\boldsymbol{x}(k) + \boldsymbol{D}_d(k)\boldsymbol{u}(k)\end{aligned} \tag{5.16}$$

where

$$\boldsymbol{A}_d(k) = e^{\mathbf{A}(kT)T} \tag{5.17}$$

And T is de sampling period which is presumed constant.

Equation (5.18) corresponds to the series expansion of the matrix exponential function e^{AT} performed considering the linear terms. It is possible since, in the sampling interval, matrices A can be considered to be constant during the sample period.

$$e^{AT} = \boldsymbol{I} + AT + \frac{(AT)^2}{2} + \cdots = \sum_{\nu=0}^{\infty} \frac{(AT)^\nu}{\nu!} \approx \boldsymbol{I} + AT \tag{5.18}$$

where $\boldsymbol{I}$ is an identity matrix.

Similarly:

$$\begin{aligned}\boldsymbol{B}_d(k) &\approx \boldsymbol{B}T\\ \boldsymbol{C}_d(k) &= \boldsymbol{C}\end{aligned} \tag{5.19}$$

It should be noted that Eqs. (5.18) and (5.19) lost some accuracy which can be improved if second-order series expansion is used. However, it increases the computational time. In addition, as commented before, the sampling time should be as small as possible to give a stable and accurate discretized model.

The stability and natural response characteristics of the system can be studied from the eigenvalues of the matrix $\boldsymbol{A}$ or $\boldsymbol{A}_d$. The transfer function can always be derived from the state-space model and vice versa. However, only the completely controllable and observable part of the system is described in the state-space model.

On the other hand, for a system to be considered linear, in terms of its cause–effect relationship (excitement/response), it is necessary to comply with the principle of superposition. A system complies with the superposition principle if, with the input excitation $x_1(t)$, the system responds with $y_1(t)$, and for an excitation $x_2(t)$, its output is $y_2(t)$; then for an input of linear combination of the previous excitations, their response responds in a combination linear as

$$x(t) = \alpha \cdot x_1(t) + \beta \cdot x_2(t) \Rightarrow y(t) = \alpha \cdot y_1(t) + \beta \cdot y_2(t) \tag{5.20}$$

If the system is non-linear, it should be linearized with the standard techniques of linear control system theory. In case of the machine model, it is possible to contemplate two methods (Krause et al. 1995): the Taylor expansion method, and the substitution of the state-space variables by and steady-state which fix the operating point (DC term) x_{io} plus a perturbance variable (AC term) Δx_i as

$$x_i = x_{io} + \Delta x_i \tag{5.21}$$

Both methods lose accuracy, but they facilitate the application of classical control techniques.

The Taylor expansion linearization method can be employed in non-linear systems that usually operate around an operating point stable (equilibrium point), being small the alterations to which it can be yielded. Any machine variable f_i can be written in terms of a Taylor expansion about its fixed value, f_{io} in a linear approximation way as

$$g(f_i) \cong g(f_{io}) + \left[\frac{\mathrm{d}g(f_{io})}{\mathrm{d}f}\right]_0 (f_i - f_{io}) \tag{5.22}$$

where the terms higher than the first order have been neglected, and

$$f_i = f_{io} + \Delta f_i \tag{5.23}$$

The linearized form of the state-space equation definition can be written as

$$\Delta \dot{\boldsymbol{x}} = \boldsymbol{A}(X)\Delta \boldsymbol{x} + \boldsymbol{B}(X)\Delta \boldsymbol{u} \tag{5.24}$$

where Δx is the perturbation matrix for state variable x, $A(X)$ is state transition matrix, Δu is the input perturbation matrix, and $B(X)$ is the input matrix. It should be considered that each machine state variable is composed of the steady-state component x_{io} and a small perturbation Δx_i:

$$x_i = x_{io} + \Delta x_i \tag{5.25}$$

where x_i is the ith state variable.

The transfer function from the state-space can be obtained, taking the Laplace transform of Eqs. (5.13) and (5.14):

$$\begin{aligned}
&L\{\dot{\boldsymbol{x}}(t) = \boldsymbol{Ax}(t) + \boldsymbol{Bu}(t)\} \\
&sX(s) - x(0^-) = AX(s) + BU(s) \\
&L\{\boldsymbol{y}(t) = \boldsymbol{Cx}(t) + \boldsymbol{Du}(t)\} \\
&Y(s) = CX(s) + DU(s)
\end{aligned} \tag{5.26}$$

which gives for $X(s)$

$$
\begin{aligned}
&(sI - A)X(s) = BU(s) + x(0^-) \rightarrow \\
&X(s) = (sI - A)^{-1}BU(s) + (sI - A)^{-1}x(0^-)
\end{aligned}
\tag{5.27}
$$

And for $Y(s)$

$$
Y(s) = C(sI - A)^{-1}BU(s) + DU(s) + C(sI - A)^{-1}x(0^-) \tag{5.28}
$$

The transfer function $G(s)$ is defined as:

$$
G(s) = \frac{Y(s)}{U(s)} = C(sI - A)^{-1}B + D \tag{5.29}
$$

And the initial condition response is:

$$
C(sI - A)^{-1}x(0^-) \tag{5.30}
$$

The roots of the determinant $sI - A$ are the system transfer function poles, and Eq. (5.30) is part of the response, but not part of the transfer function.

In conclusion, the state-space models give a powerful method for the dynamic modeling of numerous systems. It is advantageous to analyze the performance of the system influenced by external disturbances and, also, for applications of non-linear control methodology such as fuzzy logic control. In particular, for electric machines, the machine model with steady-state and dynamic quantities can be described. The state variables are associated directly with storage elements because the averaged values observed are usually currents, voltages, speed, and flux. In case of non-linear systems, it can be easily linearized in a steady-state with fix operating point (DC term) plus a perturbance variable (AC term).

5.3 Modeling DC Machine

In this section, the separately excited DC machine is used as a reference to obtain the state-space representation of the machine. It is assumed in the DC machine that the magnetic structure does not saturate and the armature reaction is canceled by compensation winding.

With the help of Eqs. (4.6), (4.7), and (4.8) of Chap. 4, the motion Eq. (5.12) is possible to represent a block diagram of the DC machine as shown in Fig. 5.6.

Where T_f is negligible and has been omitted, and:

$$
\begin{aligned}
K_T' &= K_T \lambda_f \\
K_e' &= K_e \lambda_f
\end{aligned}
\tag{5.31}
$$

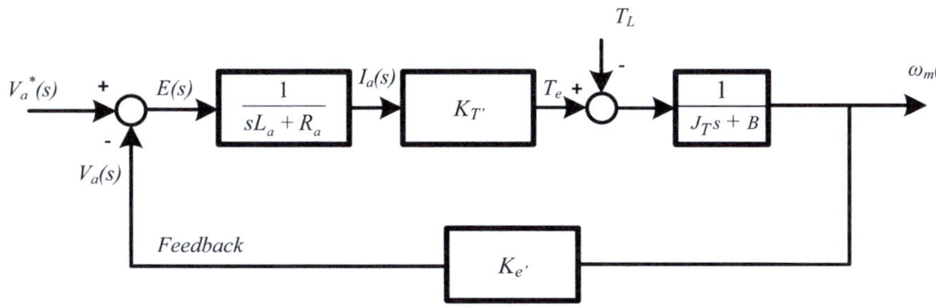

Fig. 5.6 Block diagram of DC machine

If the flux linkage λ_f is considered constant, the DC machine model is described by linear time-invariant (LTI) system which can be analyzed by using the conventional control techniques.

In Fig. 5.6, at time instant $t = 0$ where machine is stopped, if $V_a^*(s)$ is incremented, the error $E(s)$ is also increased since $V_a(s)$ is still zero. This increment produces and armature current increment which produces an electromagnetic torque T_e. If it is higher than load torque T_L, the rotor of the machine will start to move producing a back-EMF voltage $V_a(s)$, reducing the error $E(s)$ and starting current. On the contrary, if the rotor of the machine is blocked by the load, the armature current will be incremented to try to increment the electromagnetic torque. If the maximum value of current is reached, the start-up of the machine should be aborted to avoid to damage the machine. The structure of the block diagram of the DC machine is similar to a typical closed-loop control structure with a PI controller.

5.3.1 *Continuous State-Space*

As discussed early, the state-space technique gives a DC machine model with steady-state and dynamic quantities. Each state is defined with the state-space representation. In the DC machine, the input vector u is the voltage V_a, and the load and friction torque T_L and T_f, respectively. The state vector x is the armature current I_a and the machine speed ω_m. Then, the state-space matrix can be represented as:

$$\begin{bmatrix} \frac{di_a}{dt} \\ \frac{d\omega_m}{dt} \end{bmatrix} = \begin{bmatrix} \frac{-R_a}{L_a} & -\frac{K_e'}{L_a} \\ \frac{K_T'}{J_T} & \frac{-B}{J_T} \end{bmatrix} \begin{bmatrix} i_a(t) \\ \omega_m(t) \end{bmatrix} + \begin{bmatrix} \frac{1}{L_a} & 0 & 0 \\ 0 & \frac{-1}{J_T} & \frac{-1}{J_T} \end{bmatrix} \begin{bmatrix} V_a \\ T_L \\ T_f \end{bmatrix} \tag{5.32}$$

where

$$
\begin{aligned}
\boldsymbol{A} &= \begin{bmatrix} \frac{-R_a}{L_a} & -\frac{K'_e}{L_a} \\ \frac{K'_T}{J_T} & \frac{-B}{J_T} \end{bmatrix} \\
\boldsymbol{B} &= \begin{bmatrix} \frac{1}{L_a} & 0 & 0 \\ 0 & \frac{-1}{J_T} & \frac{-1}{J_T} \end{bmatrix} \\
\boldsymbol{C} &= \begin{bmatrix} 1 & 0 \\ 0 & 1 \end{bmatrix} \\
\boldsymbol{D} &= \begin{bmatrix} 0 & 0 \\ 0 & 0 \end{bmatrix}
\end{aligned} \tag{5.33}
$$

The electromagnetic torque is:

$$
T_e(t) = K'_T i_a(t) \tag{5.34}
$$

The differential equations of the separately DC machine, in general are non-linear equations. In order to analyze the stability, the equations can be considered linear in an operating point under assumption of the small disturbance. Thus, the state-space model of the DC machine can be considered as a linear system with constant parameters if the linearization is performed using a fix operating point (DC term) plus a perturbance variable (AC term). This is also known as small-signal AC technique, where all variables are consisting of a DC term imposing the operating point, and an AC modulation signal as disturbance:

$$
\begin{aligned}
\frac{\mathrm{d}(I_a + i_a)}{\mathrm{d}t} &= \frac{-R_a}{L_a}(I_a + i_a) - \frac{K'_e}{L_a}(\Omega_m + \omega_m) + \frac{1}{L_a}(V_a + v_a) \\
\frac{\mathrm{d}(\Omega_m + \omega_m)}{\mathrm{d}t} &= \frac{K'_T}{J_T}(I_a + i_a) - \frac{B}{J_T}(\Omega_m + \omega_m) - \frac{1}{J_T}(T_L + t_L) - \frac{1}{J_T}\left(T_f + t_f\right)
\end{aligned} \tag{5.35}
$$

where capital letter denotes the value of a quantity at operating point and in lower case denotes its variation around that value.

In steady-state, the derivative terms and the disturbances are zero:

$$
\begin{aligned}
0 &= \frac{-R_a}{L_a} I_a - \frac{K'_e}{L_a}\omega_m + \frac{1}{L_a} V_a \\
0 &= \frac{K'_T}{J_T} I_a - \frac{B}{J_T}\omega_m - \frac{1}{J_T} T_L - \frac{1}{J_T} T_f
\end{aligned} \tag{5.36}
$$

Equation (5.36) can be rewritten to calculate the armature current I_a in steady-state as:

$$
I_a = \frac{V_a - K'_e \omega_m}{R_a} \tag{5.37}
$$

And the electromagnetic torque is:

$$T_e = K'_T I_a \tag{5.38}$$

Substituting (5.37) in (5.38), and introducing (5.31), is possible to find the known torque expression of Sect. 4.3.2:

$$T_e = K_T \lambda_f \frac{V_a}{R_a} - \frac{K_T K_e \lambda_f^2}{R_a} \omega_m \tag{5.39}$$

According to (5.39), it is possible to extract the transfer function from the previous state-space model as following steps.

The denominator or the roots of the transfer function is:

$$|sI - A| = \frac{1}{\left(s + \frac{R_a}{L_a}\right)\left(s + \frac{B}{J_T}\right) + \frac{K^2}{L_a J_T}} \tag{5.40}$$

And the transfer function which relates the speed and the armature voltage V_a is:

$$\frac{\omega_m}{V_a} = \frac{\frac{K}{L_a J_T}}{\left(s + \frac{R_a}{L_a}\right)\left(s + \frac{B}{J_T}\right) + \frac{K^2}{L_a J_T}} = \frac{K}{(sL_a + R_a)(sJ_T + B) + K^2} \tag{5.41}$$

where $K = K'_T = K'_e$.

According to the transfer function (5.41), when armature resistance R_a and damping coefficient B become zero, the system is marginally stable with an oscillation angular frequency of:

$$\omega = \frac{K}{\sqrt{J_T L_a}} \tag{5.42}$$

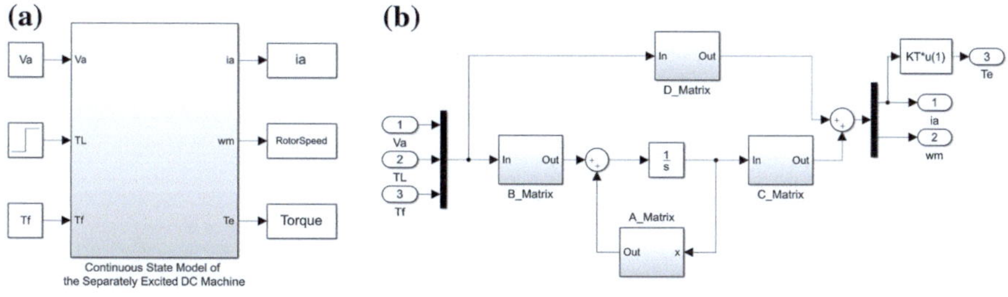

Fig. 5.7 Continuous state-space DC machine model implemented in Simulink. **a** DC machine block. **b** Internal view of the DC machine model block

Table 5.1 DC machine parameters

Parameters for DC machine	Symbol	Value	Unit
Armature inductance	L_a	0.276	mH
Armature resistance	R_a	0.57	mH
Back-EMF constant	K_e	0.0175	V/(Wb turn rad/s)
Torque constant	K_T	0.0175	Nm/(Wb turn A)
Friction torque	T_f	0.005	Nm
Rotor inertia	J	5.5e−6	$kg \cdot m^2$
Damping coefficient	B	0.00002	$kg \cdot m^2/s$

In the real machine, these parameters are no zero which degrades the efficiency of the machine, but they help to enhance the stability of the system.

In Fig. 5.7 is represented the block diagram of the continuous state-space of the DC machine with an internal view (Fig. 5.7b) showing the state-space matrix. As inputs, there is the V_a voltage, the load torque T_L, and friction torque T_f. As outputs, there is the armature current i_a, rotor speed in rad/s, and developed torque T_e.

By using the parameters of the DC machine of Table 5.1 is possible to simulate the model in a step response of the armature voltage V_a. The parameters correspond to a servo DC machine of 12 V of rate voltage and 50 W of rate power.

In Fig. 5.8 is represented the open-loop step response when armature voltage V_a is set to 6 V at $t = 0$, and the load torque is set to 0.1 Nm. In this case, the system stabilizes rapidly without oscillations. At time $t = 0.25$, 0.15 Nm load torque step is applied. It can see the reaction of the current and the mechanical speed when the step of load torque is applied. Since there is no control of the mechanical speed, the speed is reduced to 250 RPM as it can be observed. In a closed-loop speed control system, this load step is compensated by increasing the armature voltage to increase the current to keep the command speed.

In the case of damping coefficient and armature resistance are zero as commented before, the system would oscillate without decaying as represented in Fig. 5.9. As can be observed, the system is marginally stable with angular frequency according to Eq. (5.42).

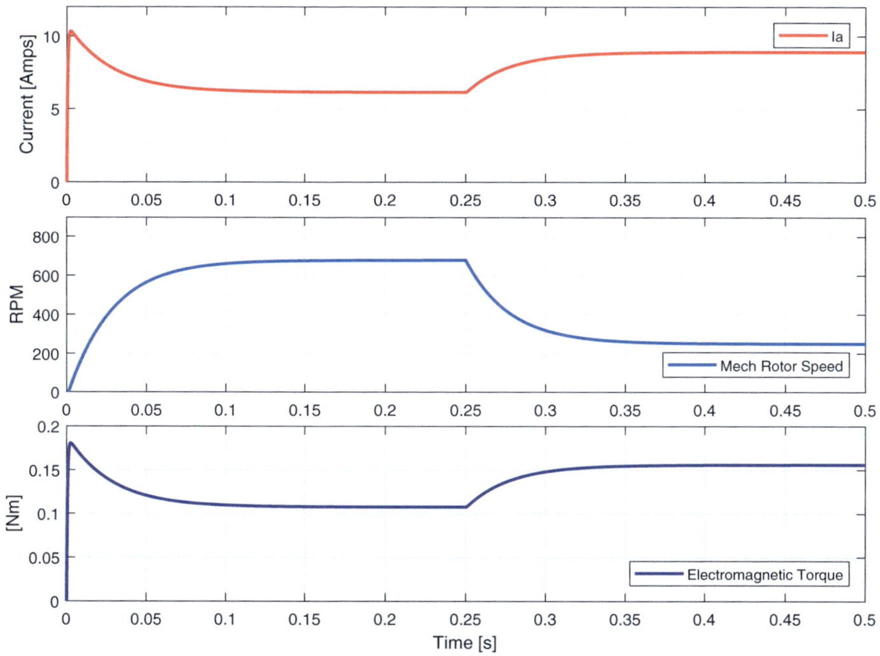

Fig. 5.8 Run-up of armature current, mechanical speed, and torque in open loop of loaded DC machine. Load torque is 0.1 Nm, and friction torque T_f is 0.005 Nm till 0.25 s. At 0.25 s, 0.15 Nm load torque is applied. Voltage step V_a of 6 V is applied at $t = 0$. Simulation variable-step size. Minimum step $T_{sp} = 10$ µs, and automatic solver selection

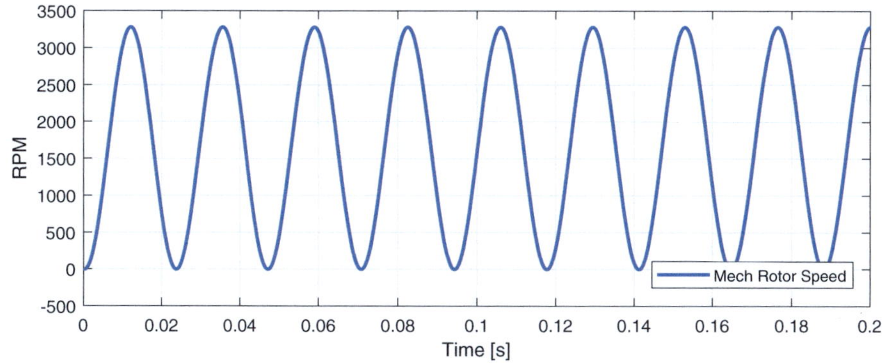

Fig. 5.9 Run-up of mechanical speed in open loop of loaded DC machine. Load torque is 0.1 Nm, and friction torque T_f is 0.005 Nm. Damping coefficient B and armature resistance R_a are zero. Voltage step V_a of 6 V is applied at $t = 0$. Simulation variable-step size. Minimum step $T_{sp} = 10$ µs, and automatic solver selection

5.4 Three-Phase Brushless AC Machine Model

5.4.1 Induction Machine

5.4.1.1 Continuous State-Space Model of Induction Machine

Because the currents and flux linkages are not independent variables, the machine equations can be written with currents or flux linkages, or both, as state variables, keeping as inputs the stator voltages and the load torque, and as outputs the electromagnetic torque and rotor speed or angular speed. The choice is generally determined by the application. The state-space variables selected in this section to describe the mathematical model of the IM are a combination between the stator current and rotor flux linkage. This is preferable for rotor flux orientation in a vector control strategy.

Equations (4.39)–(4.42) describe the induction machine in the arbitrary reference frame. These equations can be described in stationary reference frame where ω should be substituted by zero. Since rotor currents are not measurable, it can be eliminated from the equation system. The stator flux can also be eliminated since it is not interesting for the model. Then, from the flux Eqs. (4.35)–(4.38) it follows:

$$i_{dr}^{s} = \frac{1}{L_r}\left(\lambda_{dr}^{s} - L_m i_{ds}^{s}\right) \tag{5.43}$$

$$i_{qr}^{s} = \frac{1}{L_r}\left(\lambda_{qr}^{s} - L_m i_{qs}^{s}\right) \tag{5.44}$$

The stator flux linkage is expressed as:

$$\lambda_{ds}^{s} = L_s i_{ds}^{s} + \frac{L_m}{L_r}\left(\lambda_{dr}^{s} - L_m i_{ds}^{s}\right) \tag{5.45}$$

$$\lambda_{qs}^{s} = L_s i_{qs}^{s} + \frac{L_m}{L_r}\left(\lambda_{qr}^{s} - L_m i_{qs}^{s}\right) \tag{5.46}$$

If Eqs. (5.45) and (5.46) are substituted into the voltage Eqs. (4.39)–(4.42), new expression are obtained with no dependence of rotor current and stator flux variables:

$$\begin{cases} v_{ds}^{s} = R_s i_{ds}^{s} + \sigma L_s \frac{\mathrm{d}i_{ds}^{s}}{\mathrm{d}t} + \frac{L_m}{L_r}\frac{\mathrm{d}\lambda_{dr}^{s}}{\mathrm{d}t} \\ v_{qs}^{s} = R_s i_{qs}^{s} + \sigma L_s \frac{\mathrm{d}i_{qs}^{s}}{\mathrm{d}t} + \frac{L_m}{L_r}\frac{\mathrm{d}\lambda_{qr}^{s}}{\mathrm{d}t} \\ 0 = -\frac{L_m}{T_r} i_{ds}^{s} + \frac{1}{T_r}\lambda_{dr}^{s} + \omega_r \lambda_{qr}^{s} + \frac{\mathrm{d}\lambda_{dr}^{s}}{\mathrm{d}t} \\ 0 = -\frac{L_m}{T_r} i_{qs}^{s} + \frac{1}{T_r}\lambda_{qr}^{s} - \omega_r \lambda_{dr}^{s} + \frac{\mathrm{d}\lambda_{qr}^{s}}{\mathrm{d}t} \end{cases} \tag{5.47}$$

With

$$\sigma = 1 - \frac{L_m^2}{L_r L_s} \tag{5.48}$$

$$T_s = \frac{L_s}{R_s} \tag{5.49}$$

$$T_r = \frac{L_r}{R_r} \tag{5.50}$$

where T_s is the stator time constant, and T_r is the rotor time constant.

Then, if the derivative part is separated, and the rotor flux derivative part of stator voltage equations are substituted by using the rotor voltage equations, Eq. (5.51) is obtained:

$$\begin{cases} \frac{\mathrm{d}i_{ds}^s}{\mathrm{d}t} = -\left(\frac{1}{\sigma T_s} + \frac{1-\sigma}{\sigma T_r}\right) i_{ds}^s + \frac{1-\sigma}{\sigma T_r}\frac{\lambda_{dr}^s}{L_m} + \frac{1-\sigma}{\sigma}\frac{\omega_r \lambda_{qr}^s}{L_m} + \frac{v_{ds}^s}{\sigma L_s} \\ \frac{\mathrm{d}i_{qs}^s}{\mathrm{d}t} = -\left(\frac{1}{\sigma T_s} + \frac{1-\sigma}{\sigma T_r}\right) i_{qs}^s + \frac{1-\sigma}{\sigma T_r}\frac{\lambda_{qr}^s}{L_m} - \frac{1-\sigma}{\sigma}\frac{\omega_r \lambda_{dr}^s}{L_m} + \frac{v_{qs}^s}{\sigma L_s} \\ \frac{\mathrm{d}\lambda_{dr}^s}{\mathrm{d}t} = \frac{L_m}{T_r} i_{ds}^s - \frac{1}{T_r}\lambda_{dr}^s - \omega_r \lambda_{qr}^s \\ \frac{\mathrm{d}\lambda_{qr}^s}{\mathrm{d}t} = \frac{L_m}{T_r} i_{qs}^s - \frac{1}{T_r}\lambda_{qr}^s + \omega_r \lambda_{dr}^s \end{cases} \tag{5.51}$$

The expression (5.51) represents the most used model for rotor flux-oriented control strategies since comprises the stator current and rotor flux linkage, which can be used as state-space variables.

On the other hand, the relation between electromagnetic torque produced, the load, and the speed with the pole pair P dependence similarly as (5.12) is:

$$T_e = \frac{J_T}{P}\frac{\mathrm{d}\omega_{rm}}{\mathrm{d}t} + \frac{B}{P}\omega_{rm} + T_L \tag{5.52}$$

where J_T was the total inertia (machine and load) and B the viscous coefficient or friction.

The torque expression can be rewritten using the equation of rotor current (5.43) and (5.44) and substituting in the torque Eq. (4.49):

$$T_e = \frac{3}{2}P \cdot \frac{L_m}{L_r}\left(i_{qs}^s \cdot \lambda_{dr}^s - i_{ds}^s \cdot \lambda_{qr}^s\right) \tag{5.53}$$

According to the following state-space definition in the stationary reference frame with s symbol:

$$\begin{aligned} \dot{\boldsymbol{x}}^s &= \boldsymbol{A}^s \boldsymbol{x}^s(t) + \boldsymbol{B}^s \boldsymbol{u}^s(t) \\ \boldsymbol{y}^s(t) &= \boldsymbol{C}^s \boldsymbol{x}^s(t) + \boldsymbol{D}^s \boldsymbol{u}^s(t) \end{aligned} \tag{5.54}$$

Equation (5.51) can be rewritten in the matrix form as

$$\begin{bmatrix} \frac{di_{ds}^s}{dt} \\ \frac{di_{qs}^s}{dt} \\ \frac{d\lambda_{dr}^s}{dt} \\ \frac{d\lambda_{qr}^s}{dt} \end{bmatrix} = \begin{bmatrix} -\left(\frac{1}{\sigma T_s} + \frac{1-\sigma}{\sigma T_r}\right) & 0 & \frac{1-\sigma}{\sigma T_r}\frac{1}{L_m} & \frac{1-\sigma}{\sigma}\frac{\omega_r}{L_m} \\ 0 & -\left(\frac{1}{\sigma T_s} + \frac{1-\sigma}{\sigma T_r}\right) & -\frac{1-\sigma}{\sigma}\frac{\omega_r}{L_m} & \frac{1-\sigma}{\sigma T_r}\frac{1}{L_m} \\ \frac{L_m}{T_r} & 0 & -\frac{1}{T_r} & -\omega_r \\ 0 & \frac{L_m}{T_r} & \omega_r & -\frac{1}{T_r} \end{bmatrix} \begin{bmatrix} i_{ds}^s \\ i_{qs}^s \\ \lambda_{dr}^s \\ \lambda_{qr}^s \end{bmatrix} + \begin{bmatrix} \frac{1}{\sigma L_s} & 0 \\ 0 & \frac{1}{\sigma L_s} \\ 0 & 0 \\ 0 & 0 \end{bmatrix} \begin{bmatrix} v_{ds}^s \\ v_{qs}^s \end{bmatrix} \tag{5.55}$$

where the system and input matrix are

$$A^s = \begin{bmatrix} -\left(\frac{1}{\sigma T_s} + \frac{1-\sigma}{\sigma T_r}\right) & 0 & \frac{1-\sigma}{\sigma T_r}\frac{1}{L_m} & \frac{1-\sigma}{\sigma}\frac{\omega_r}{L_m} \\ 0 & -\left(\frac{1}{\sigma T_s} + \frac{1-\sigma}{\sigma T_r}\right) & -\frac{1-\sigma}{\sigma}\frac{\omega_r}{L_m} & \frac{1-\sigma}{\sigma T_r}\frac{1}{L_m} \\ \frac{L_m}{T_r} & 0 & -\frac{1}{T_r} & -\omega_r \\ 0 & \frac{L_m}{T_r} & \omega_r & -\frac{1}{T_r} \end{bmatrix}$$

$$B^s = \begin{bmatrix} \frac{1}{\sigma L_s} & 0 \\ 0 & \frac{1}{\sigma L_s} \\ 0 & 0 \\ 0 & 0 \end{bmatrix} \quad C^s = \begin{bmatrix} 1 & 0 & 0 & 0 \\ 0 & 1 & 0 & 0 \end{bmatrix} \quad D^s = \begin{bmatrix} 0 & 0 \\ 0 & 0 \end{bmatrix} \tag{5.56}$$

Both matrices $\boldsymbol{A}^s$ and $\boldsymbol{B}^s$ depend on the induction machine parameters. In $\boldsymbol{A}^s$ matrix, there is a dependence of the electrical rotor speed ω_r, which is a time-variant parameter. The input vector $\boldsymbol{u}^s$ is the sinusoidal machine voltage in the stationary reference frame with the two components *dqs*, while the output $\boldsymbol{y}^s$ is the machine sinusoidal current also in the stationary reference frame. In addition, in the state vector $\boldsymbol{x}^s$ the rotor flux and current in the stationary reference frame can be obtained to be used to electromagnetic torque calculation. The continuous state model of the induction machine is based on (5.53), (5.55), and (5.56). These equations can be implemented in Simulink® following the block diagram of state-space seen in Sect. 5.3 for DC machine. In Fig. 5.10a is represented the block diagram of the continuous state-space model of the induction machine in the stationary reference frame. The detail of Fig. 5.10a is represented in Fig. 5.10b. As expected, the inputs are the phase voltage v_a, v_b, v_c, and load torque T_L, while the outputs are the electromagnetic torque T_{em}, the rotor electrical speed ω_r, the phases' currents i_a, i_b, i_c and rotor flux linkage *flux_r*$_d$, *flux_r*$_q$ in the two-axis *dqs* components.

In order to corroborate the continuous state-space model of the induction machine, several simulations can be performed in open loop. The induction machine selected corresponds to a squirrel-cage rotor, of 2-pair poles, with 2 HP rated power and nine-lead terminal wiring (high-voltage star and low-voltage delta). The rated torque

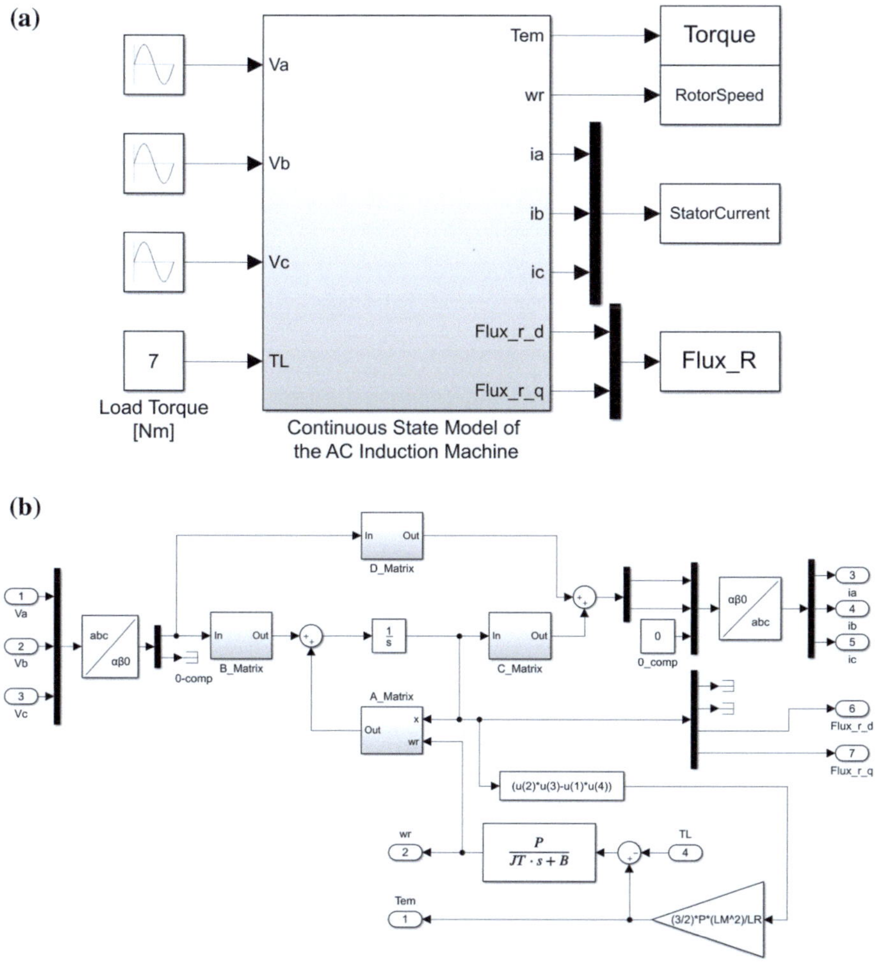

Fig. 5.10 **a** Continuous state-space model of the induction machine in the stationary reference frame. **b** Detail block diagram with the state-space system, torque, motion, and axes' transformation

Table 5.2 Nameplate data of induction machine for two voltage connections. High-voltage star connection (460 V) and low-voltage delta connection (230 V)

Nameplate	Symbol	460/230 V	Unit
Nominal power	P_N	2	HP
Line-to-line nominal voltage	V_N (RMS)	460/230	V
Nominal current	I_N (RMS)	3.1/6.2	A
Nominal frequency	f_N	60	Hz
Nominal speed	N_N	1740	RPM
Nominal power factor	$\cos\varphi$	0.755	
Service factor	SF	1.15	

Table 5.3 Induction machine parameters for two voltage connections. High-voltage star connection (460 V) and low-voltage delta connection (230 V)

Parameters for machine model	Symbol	460 V	230 V	Unit
Leakage stator inductance	L_{ls}	16.79	4.2	mH
Leakage rotor inductance	L_{lr}	14.83	3.7	mH
Magnetizing inductance	L_m	347.5	86.9	mH
Stator resistance	R_s	5.33	1.33	Ω
Rotor resistance	R_r	3.14	0.785	Ω
Rotor inertia	J	0.0038	0.0038	$kg \cdot m^2$
Damping coefficient	B	0	0	$kg \cdot m^2/s$

is about 8.1 Nm while the rated speed is 1740 RPM. The nameplate of the induction machine is shown in Table 5.2, and machine parameters are shown in Table 5.3.

In the simulation, the machine is powered directly by the three-phase grid, where the line-to-line voltage is 230 V, and the frequency is 60 Hz. Hence, the machine is wired in delta connection. The simulation starts with unloading condition till steady-state speed, and then, rated torque is applied at the load. The simulation results are

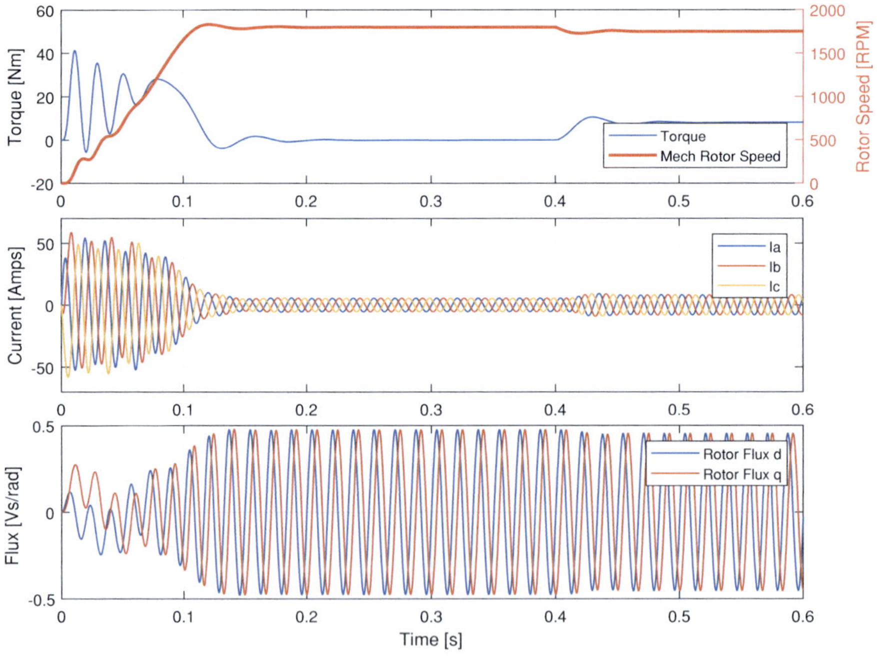

Fig. 5.11 Run-up speed, torque, phase currents, and rotor flux *dqs* components of unloaded induction machine till 0.4 s. At 0.4 s, rated torque is applied in the load. Simulation variable-step size. Minimum step $T_{sp} = 10$ μs, and automatic solver selection

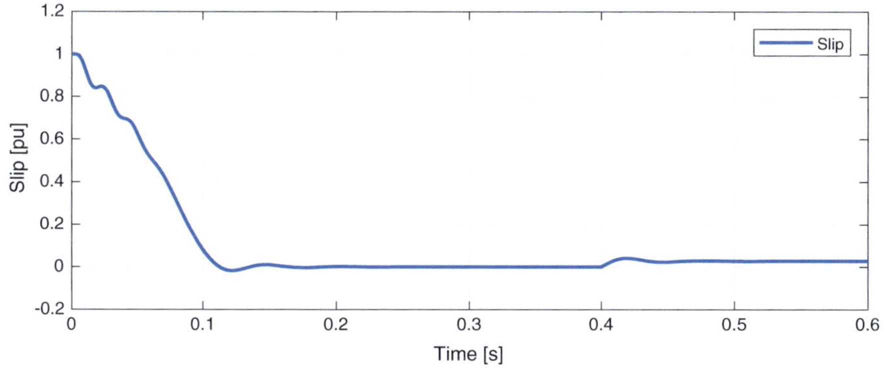

Fig. 5.12 Run-up slip in per unit of unloaded induction machine till 0.4 s. From 0.4 s, rated torque is applied in the load. Simulation variable-step size. Minimum step $T_{sp} = 10\ \mu s$, and automatic solver selection

represented in Figs. 5.11 and 5.12. In Fig. 5.11 is represented the rotor speed in RPM, the electromagnetic torque, the three-phase stator currents, and the rotor flux linkage in *dqs* components. As it can be observed, at the start-up with an unloaded machine, the phase currents have significant magnitude due to the slip (in per unit) which is the maximum as shown in Fig. 5.12. The rotor flux is low since rotor speed is also low. When the speed is incremented, the rotor flux is also incremented, and as a consequence, the current is decremented.

When the steady-state condition is reached, the current is relaxed and developed torque goes to zero (slip = 0). At time instant $t = 0.4$ s, rated torque is applied in the machine. Due to the maximum voltage of the machine applied, the rotor speed is decreased till 1740 RPM approx (rated speed). At this point, the machine is operating in rated conditions and in the constant power region.

5.4.2 PMAC Machine

5.4.2.1 PMSM Model

5.4.2.1.1 Machine Model in *dqe* Axes (Synchronous Rotating Reference Frame)

Due to complex equations of Sect. 4.4.2.2 of PMSM, it is possible to use the Clarke transformation to pass from the three-phase system to stationary two-axis orthogonal system. It does not modify the module of the vectors and depends still on the angle. To remove this dependence, rotation matrix transformation can be applied which pass from stationary orthogonal system to another orthogonal system which turns in synchronism as discussed in Chap. 3.

It is considered:

- d is preferred axis for pole flux orientation of Ψ_m
- q is orthogonal to d.

After applying the Clarke and rotation matrix transformation in (4.86), the following equation is obtained:

$$\begin{bmatrix} V_{os}^e \\ V_{ds}^e \\ V_{qs}^e \end{bmatrix} = \begin{bmatrix} R_s & 0 & 0 \\ 0 & R_s & 0 \\ 0 & 0 & R_s \end{bmatrix} \cdot \begin{bmatrix} i_{os}^e \\ i_{ds}^e \\ i_{qs}^e \end{bmatrix} + \frac{\mathrm{d}}{\mathrm{d}t} \begin{bmatrix} L_{ls} & 0 & 0 \\ 0 & L_d & 0 \\ 0 & 0 & L_q \end{bmatrix} \cdot \begin{bmatrix} i_{os}^e \\ i_{ds}^e \\ i_{qs}^e \end{bmatrix} + \begin{bmatrix} 0 & 0 & 0 \\ 0 & 0 & -\omega_r L_q \\ 0 & \omega_r L_d & 0 \end{bmatrix} \cdot \begin{bmatrix} i_{os}^e \\ i_{ds}^e \\ i_{qs}^e \end{bmatrix} + \begin{bmatrix} 0 \\ 0 \\ \omega_r \Psi_m \end{bmatrix} \tag{5.57}$$

It is possible to observe that Eq. (5.57) does not depend on the angle as expected. Then, removing the matrix form:

$$v_{os}^e = R_s i_{os}^e + L_{ls}\frac{\mathrm{d}i_{os}^e}{\mathrm{d}t} = 0$$

Equal to zero since it is a balance system, $i_a + i_b + i_c = 0$

$$\begin{aligned} v_{ds}^e &= R_s i_{ds}^e + L_d \frac{\mathrm{d}i_{ds}^e}{\mathrm{d}t} - \omega_r \lambda_{qs}^e \\ v_{qs}^e &= R_s i_{qs}^e + L_q \frac{\mathrm{d}i_{qs}^e}{\mathrm{d}t} + \omega_r \lambda_{ds}^e \end{aligned} \tag{5.58}$$

where the new inductance depends on the leakage stator inductance L_{ls}, and L_A and L_B:

$$L_d = L_{ls} + \frac{3}{2}(L_A + L_B) \tag{5.59}$$

$$L_q = L_{ls} + \frac{3}{2}(L_A - L_B) \tag{5.60}$$

And stator flux linkages equations are:

$$\begin{aligned} \lambda_{qs}^e &= L_q i_{qs}^e \\ \lambda_{ds}^e &= L_d i_{ds}^e + \Psi_m \end{aligned} \tag{5.61}$$

The torque expression was:

$$T_e = \frac{3}{2}P\left[\Psi_m i_{qs}^e + \left(L_d - L_q\right) i_{ds}^e i_{qs}^e\right] \tag{5.62}$$

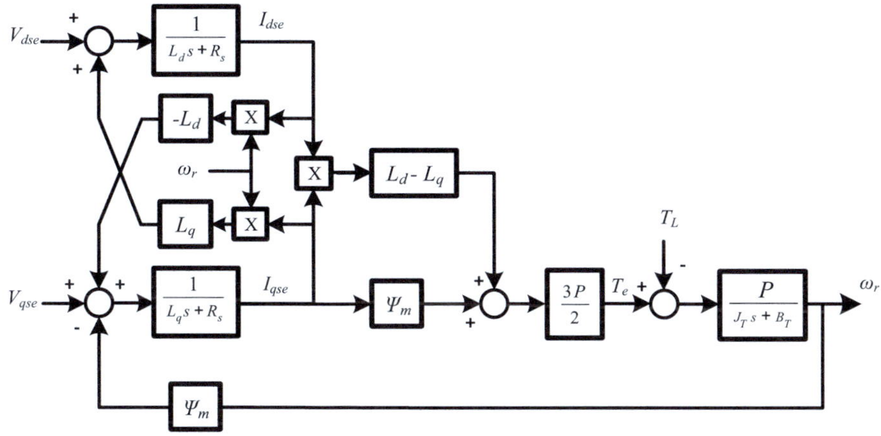

Fig. 5.13 Complete PMSM model to be used in the simulation

Finally, to obtain the PMSM model, it is necessary to use the relation between torque produced, the load, and then speed which was shown in (5.12), with the pole pair P dependence.

Figure 5.13 illustrates the complete model for PMSM, which uses Eqs. (5.12), (5.58), (5.61), and (5.62). This model can be used to perform simulations with Simulink® or PSIM environments.

The blocks that depend on the inductances L_d and L_q and the flux Ψ_m can be fixed and linear parameters. However, for more precise models these parameters are non-linear and dependent on current and temperature. For example, the value of the inductances L_d and L_q depends on the dq current magnitude components, as seen in Sect. 4.4.2.2.1. On the other hand, the temperature has an effect on the permanent magnet that causes its flux density to decrease with the increase in temperature; that is, they have a negative temperature coefficient. The above non-linearities can be modeled; even look-up tables can be used with a current and temperature dependence.

The resume of the parameters needed in the PMSM model is represented in Table 5.4.

Table 5.4 PMSM model parameters

Parameter	Description	Units
L_d	Direct synchronous inductance	H
L_q	Quadrature synchronous inductance	H
R_s	Stator resistance	Ω
Ψ_m	Permanent magnet flux	Wb/rad or Vs/rad
J_r	Inertia of rotor	$kg \cdot m^2$
B	Viscous coefficient	(Nm)/(rad/s)
P	Pole pair	–

5.4.2.1.2 Machine Model in *dqs* Axes (Stationary Reference Frame)

In the stationary reference frame, the *dqs* axes are stationary. It means that the equations have angle dependence, but can be useful to implement a rotor angle estimator as will be shown in Chap. 10. If Eq. (4.79) is decomposed into their *dqs* axes of the stationary reference frame, Eq. (5.63) is obtained.

$$\begin{aligned} V_{ds}^{s} &= R_s i_{ds}^{s} + \frac{\mathrm{d}\lambda_{ds}^{s}}{\mathrm{d}t} \\ V_{qs}^{s} &= R_s i_{qs}^{s} + \frac{\mathrm{d}\lambda_{qs}^{s}}{\mathrm{d}t} \end{aligned} \tag{5.63}$$

where direct and quadrature fluxes are expressed in (5.64) and (5.65). As it can be observed, it is composed of two elements, the stator flux component and the magnet flux component.

$$\lambda_{ds}^{s} = L_d i_{ds}^{s} + \lambda_{dsm}^{s} = L_d i_{ds}^{s} + \Psi_m^{s} \cos(\theta) \tag{5.64}$$

$$\lambda_{qs}^{s} = L_q i_{qs}^{s} + \lambda_{qsm}^{s} = L_q i_{qs}^{s} + \Psi_m \sin(\theta) \tag{5.65}$$

If the derivative is applied in (5.64) and (5.65):

$$\begin{aligned} \frac{\mathrm{d}\lambda_{ds}^{s}}{\mathrm{d}t} &= L_d \frac{\mathrm{d}i_{ds}^{s}}{\mathrm{d}t} - \omega_r \Psi_m \sin(\theta) = L_d \frac{\mathrm{d}i_{ds}^{s}}{\mathrm{d}t} + \omega_r \lambda_{qsm}^{s} \\ \frac{\mathrm{d}\lambda_{qs}^{s}}{\mathrm{d}t} &= L_q \frac{\mathrm{d}i_{qs}^{s}}{\mathrm{d}t} + \omega_r \Psi_m \cos(\theta) = L_q \frac{\mathrm{d}i_{qs}^{s}}{\mathrm{d}t} + \omega_r \lambda_{dsm}^{s} \end{aligned} \tag{5.66}$$

Then, Eq. (5.63) can be expressed as:

$$v_{ds}^{s} = R_s i_{ds}^{s} + L_d \frac{\mathrm{d}i_{ds}^{s}}{\mathrm{d}t} - \omega_r \lambda_{qsm}^{s} \tag{5.67}$$

$$v_{qs}^{s} = R_s i_{qs}^{s} + L_q \frac{\mathrm{d}i_{qs}^{s}}{\mathrm{d}t} + \omega_r \lambda_{dsm}^{s} \tag{5.68}$$

where $\omega_r \lambda_{qsm}$ and $\omega_r \lambda_{dsm}$ are the back-EMF. In Chap. 10, (5.67) and (5.68) will be used to perform a speed estimator for PMSM.

5.4.2.1.3 Continuous State-Space Model of PMSM

In the previous section, both references' frames which describe the mathematical model of the PMSM were shown. In comparison with IM, the stator and rotor speed match, where the pole position of PMSM can always be clearly identified. For this reason, the synchronously rotating reference frame is preferred in PMSM. However, both references, synchronous rotating and stationary references, help, for example, to study different control strategies and different speed estimators. In this section, the synchronously rotating reference frame is used as state-space variables to describe the PMSM model.

The state-space variables in the synchronously rotating reference frame are the stator current, the speed, and the load angle δ. As shown in Sect. 4.4.2.5, the *dqs* voltage components can be written in terms of the V_s voltage and load angle δ according to (5.69).

$$\begin{aligned} v_{ds}^e &= -v_s \sin\delta \\ v_{qs}^e &= v_s \cos\delta \end{aligned} \tag{5.69}$$

where the load angle δ represented the angle difference between the stator electrical applied voltage and the back-EMF generated by the rotor magnets when rotating. Thus, the load angle derivative term corresponds to the difference between the stator and rotor speed as represented in (5.70):

$$\frac{\mathrm{d}\delta}{\mathrm{d}t} = \omega_e - \omega_r \tag{5.70}$$

If Eqs. (4.100) and (4.101) of Sect. 4.4.2.2 are substituted into the voltage Eqs. (4.102) and (4.103) new expression are obtained with no dependence of rotor current and stator flux:

$$\begin{aligned} v_{ds}^e &= R_s i_{ds}^e + L_d \frac{\mathrm{d}i_{ds}^e}{\mathrm{d}t} - \omega_r L_q i_{qs}^e \\ v_{qs}^e &= R_s i_{qs}^e + L_q \frac{\mathrm{d}i_{qs}^e}{\mathrm{d}t} + \omega_r L_d i_{ds}^e + \omega_r \Psi_m \end{aligned} \tag{5.71}$$

Then, if the derivative part is separated:

$$\begin{cases} \frac{\mathrm{d}i_{ds}^e}{\mathrm{d}t} = -\frac{1}{T_{sd}} i_{ds}^e + \omega_r \frac{L_q}{L_d} i_{qs}^e + \frac{1}{L_d} v_{ds}^e \\ \frac{\mathrm{d}i_{qs}^e}{\mathrm{d}t} = -\frac{1}{T_{sq}} i_{qs}^e - \omega_r \frac{L_d}{L_q} i_{ds}^e - \omega_r \frac{\psi_m}{L_q} + \frac{1}{L_q} v_{qs}^e \end{cases} \tag{5.72}$$

where the time constant is:

$$\begin{aligned} T_{sd} &= \frac{L_d}{R_s} \\ T_{sq} &= \frac{L_q}{R_s} \end{aligned} \tag{5.73}$$

Then, expression (5.72) can be rewritten by using (5.69) as

$$\begin{cases} \frac{\mathrm{d}i_{ds}^e}{\mathrm{d}t} = -\frac{1}{T_{sd}} i_{ds}^e + \omega_r \frac{L_q}{L_d} i_{qs}^e - \frac{1}{L_d} v_s \sin\delta \\ \frac{\mathrm{d}i_{qs}^e}{\mathrm{d}t} = -\frac{1}{T_{sq}} i_{qs}^e - \omega_r \frac{L_d}{L_q} i_{ds}^e - \omega_r \frac{\psi_m}{L_q} + \frac{1}{L_q} v_s \cos\delta \end{cases} \tag{5.74}$$

The torque expression was:

$$T_e = \frac{3}{2} P \left[\Psi_m i_{qs}^e + \left(L_d - L_q\right) i_{ds}^e i_{qs}^e \right] \tag{5.75}$$

Using the motion Eq. (5.12) and torque expression (5.75), the derivative term of the electrical speed ω_r is

$$\frac{\mathrm{d}\omega_r}{\mathrm{d}t} = \frac{3}{2} \frac{P^2}{J_T} \left[\Psi_m i_{qs}^e + \left(L_d - L_q\right) i_{ds}^e i_{qs}^e \right] - \frac{1}{J_T} B\omega_r - \frac{P}{J_T} T_L \tag{5.76}$$

The state-space definition in the stationary reference frame with e symbol is:

$$\begin{aligned} \dot{\boldsymbol{x}}^e &= \boldsymbol{A}^e \boldsymbol{x}^e(t) + \boldsymbol{B}^e \boldsymbol{u}^e(t) \\ \boldsymbol{y}^e(t) &= \boldsymbol{C}^e \boldsymbol{x}^e(t) + \boldsymbol{D}^e \boldsymbol{u}^e(t) \end{aligned} \tag{5.77}$$

Equations (5.70), (5.74), and (5.76) can be rewritten in the matrix form as

$$\begin{aligned} \begin{bmatrix} \frac{\mathrm{d}i_{ds}^e}{\mathrm{d}t} \\ \frac{\mathrm{d}i_{qs}^e}{\mathrm{d}t} \\ \frac{\mathrm{d}\omega_r}{\mathrm{d}t} \\ \frac{\mathrm{d}\delta}{\mathrm{d}t} \end{bmatrix} &= \begin{bmatrix} -\frac{1}{T_{sd}} & \omega_r \frac{L_q}{L_d} & 0 & 0 \\ -\omega_r \frac{L_d}{L_q} & -\frac{1}{T_{sq}} & -\frac{\psi_m}{L_q} & 0 \\ \frac{3}{2} P^2 \frac{1}{J_T} \left(L_d - L_q\right) i_{qs}^e & \frac{3}{2} P^2 \frac{1}{J_T} \psi_m & -\frac{B}{J_T} & 0 \\ 0 & 0 & -1 & 0 \end{bmatrix} \begin{bmatrix} i_{ds}^e \\ i_{qs}^e \\ \omega_r \\ \delta \end{bmatrix} \\ &+ \begin{bmatrix} -\frac{\sin\delta}{L_d} & 0 & 0 \\ \frac{\cos\delta}{L_q} & 0 & 0 \\ 0 & 0 & -\frac{P}{J_T} \\ 0 & 1 & 0 \end{bmatrix} \begin{bmatrix} v_s \\ \omega_e \\ T_L \end{bmatrix} \end{aligned} \tag{5.78}$$

with

$$\begin{bmatrix} i_{ds}^e \\ i_{qs}^e \\ \omega_r \\ \delta \end{bmatrix} = \begin{bmatrix} 1 & 0 & 0 & 0 \\ 0 & 1 & 0 & 0 \\ 0 & 0 & 1 & 0 \\ 0 & 0 & 0 & 1 \end{bmatrix} \begin{bmatrix} i_{ds}^e \\ i_{qs}^e \\ \omega_r \\ \delta \end{bmatrix} + \begin{bmatrix} 0 & 0 & 0 \\ 0 & 0 & 0 \\ 0 & 0 & 0 \end{bmatrix} \begin{bmatrix} v_s \\ \omega_e \\ T_L \end{bmatrix} \tag{5.79}$$

where the system and input matrix are

$$
\boldsymbol{A}^e = \begin{bmatrix} -\frac{1}{T_{sd}} & \omega_r \frac{L_q}{L_d} & 0 & 0 \\ -\omega_r \frac{L_d}{L_q} & -\frac{1}{T_{sq}} & -\frac{\psi_m}{L_q} & 0 \\ \frac{3}{2}P^2 \frac{1}{J_T}\left(L_d - L_q\right) i_{qs}^e & \frac{3}{2}P^2 \frac{1}{J_T}\psi_m & -\frac{B}{J_T} & 0 \\ 0 & 0 & -1 & 0 \end{bmatrix}
$$

$$
\boldsymbol{B}^e = \begin{bmatrix} -\frac{\sin\delta}{L_d} & 0 & 0 \\ \frac{\cos\delta}{L_q} & 0 & 0 \\ 0 & 0 & -\frac{P}{J_T} \\ 0 & 1 & 0 \end{bmatrix} \quad \boldsymbol{C}^e = \begin{bmatrix} 1 & 0 & 0 & 0 \\ 0 & 1 & 0 & 0 \\ 0 & 0 & 1 & 0 \\ 0 & 0 & 0 & 1 \end{bmatrix} \quad \boldsymbol{D}^e = \begin{bmatrix} 0 & 0 & 0 \\ 0 & 0 & 0 \\ 0 & 0 & 0 \end{bmatrix} \tag{5.80}
$$

Both matrices $\boldsymbol{A}^s$ and $\boldsymbol{B}^s$ depend on the PMSM parameters as IM. In the $\boldsymbol{A}^s$ matrix, there is a dependence of the rotor speed ω_r, which is a time-variant parameter. Also, the derivative term of the rotor speed depends on the two *dqe* current components where the principle of the superposition is not accomplished.

The input vector $\boldsymbol{u}^s$ is the machine voltage quantities in the synchronously rotating reference frame with the two components *dqe*, while the output $\boldsymbol{y}^s$ is the DC quantities of current, rotor electrical speed, and the load angle. Equations (5.78), (5.79), and (5.80) can also be implemented in Simulink® as shown for the IM. Figure 5.14a reperesents the block diagram of the continuous state-space model of the PMSM in the synchronously rotating reference frame. The detail of Fig. 5.14a is represented in Fig. 5.14b.

The PMSM model equations are non-linear since the superposition method is no accomplished as commented before. The PMSM compared with IM should no be controlled in open loop because the machine has unstable zones. As discussed, the rotor should be synchronized with stator in all instants, where in open-loop control methods, it is not guaranteed. In order to find the unstable zones, the PMSM model can be analyzed. The first step is to linearize the model equations in order to apply the standard techniques of linear control system theory. As shown in Sect. 5.2, two methods can be considered to linearize the PMSM model (Krause et al. 1995). One is Taylor's expansion method. The other method consists in the substitution of the variables of the state-space system by and steady-state, which fix the operating point (DC term), plus a perturbance variable (AC term) as

$$
x_i = X + \Delta_{xi} \tag{5.81}
$$

Here, it will be applied the second method.

The linearized form of the state-space equation definition can be written as

$$
\Delta\dot{\boldsymbol{x}} = \boldsymbol{A}(X)\Delta\boldsymbol{x} + \boldsymbol{B}(X)\Delta\boldsymbol{u} \tag{5.82}
$$

where Δx is the perturbation matrix for state variable x, $A(X)$ is state transition matrix, Δu is the input perturbation matrix, and $B(X)$ is the input matrix. It should be considered that each machine state variable is composed of the steady-state component X and a small perturbation Δx_i:

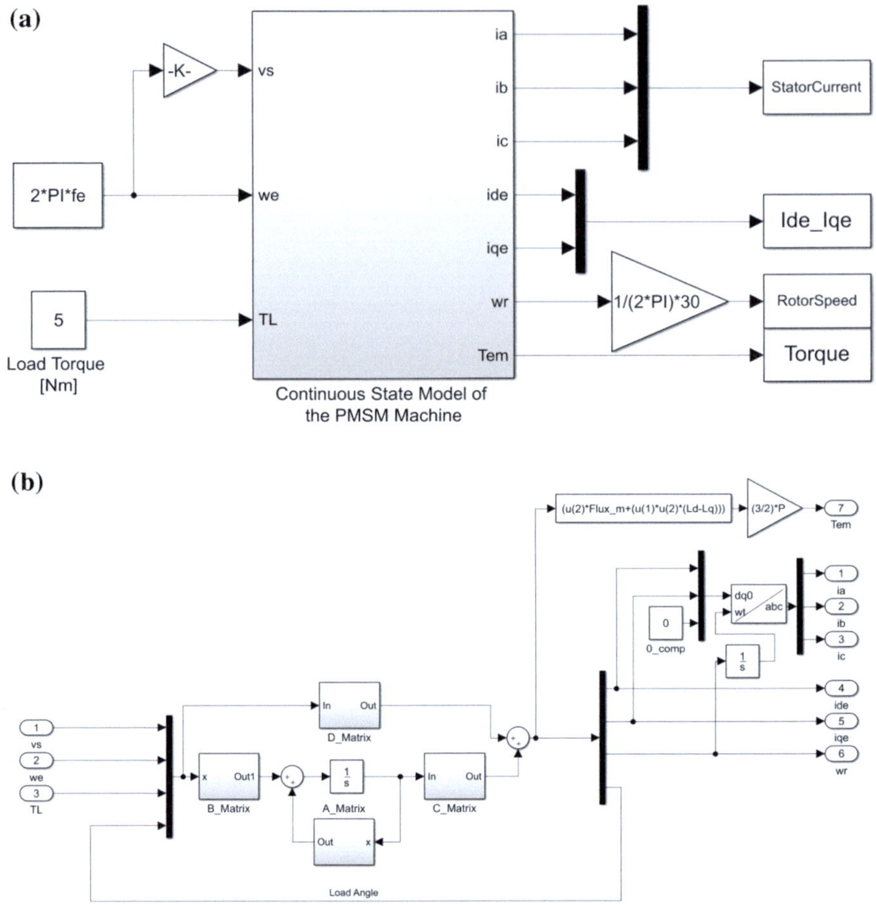

Fig. 5.14 **a** Continuous state-space model of the PMSM in the synchronously rotating reference frame. **b** Detail block diagram with the state-space system, torque, motion, and axes' transformation

$$x_i = X + \Delta_{xi} \tag{5.83}$$

where x_i is the ith state variable.

For example, by using Eq. (5.84)

$$\frac{\mathrm{d}i_{ds}^e}{\mathrm{d}t} = -\frac{1}{T_{sd}} i_{ds}^e + \omega_r \frac{L_q}{L_d} i_{qs}^e - \frac{1}{L_d} v_s \sin\delta \tag{5.84}$$

It is possible to obtain (5.85) applying (5.83):

$$\frac{\mathrm{d}\left(i_{ds0}^e + \Delta i_{ds}^e\right)}{\mathrm{d}t} = -\frac{1}{T_{sd}}\left(i_{dso}^e + \Delta i_{ds}^e\right) + (\omega_{r0} + \Delta\omega_r)\frac{L_q}{L_d}\left(i_{qs0}^e + \Delta i_{qs}^e\right)$$

$$-\frac{1}{L_d}(v_{s0}+\Delta v_s)\sin(\delta_0+\Delta\delta) \tag{5.85}$$

Developing (5.85):

$$\begin{aligned}\frac{\mathrm{d}\left(i^e_{ds0}+\Delta i^e_{ds}\right)}{\mathrm{d}t} &= -\frac{1}{T_{sd}}\left(i^e_{dso}+\Delta i^e_{ds}\right)+\omega_{r0}\frac{L_q}{L_d}i^e_{qs0}+\omega_{r0}\frac{L_q}{L_d}\Delta i^e_{qs}\\ &+\Delta\omega_r\frac{L_q}{L_d}i^e_{qs0}+\Delta\omega_r\frac{L_q}{L_d}\Delta i^e_{qs}\\ &-\frac{1}{L_d}(v_{s0}+\Delta v_s)(\sin\Delta\delta\cos\delta_0+\cos\Delta\delta\sin\delta_0)\end{aligned} \tag{5.86}$$

Now, if the multiplication terms $\Delta f_1 \Delta f_2$ are neglected, $\sin\Delta\delta=\Delta\delta$, $\cos\Delta\delta = 1$, and the steady-state expression are canceled from both sides of the equation, Eq. (5.86) can be rewritten as

$$\begin{aligned}\Delta\dot{i}^e_{ds} &= -\frac{1}{T_{sd}}\Delta i^e_{ds}+\omega_{r0}\frac{L_q}{L_d}\Delta i^e_{qs}+\Delta\omega_r\frac{L_q}{L_d}i^e_{qs0}\\ &-\frac{1}{L_d}(v_{s0}\Delta\delta\cos\delta_0+\Delta v_s\sin\delta_0)\end{aligned} \tag{5.87}$$

Applying this linearization method to the state-space model of the PMSM, the linearized model is obtained as

$$\begin{aligned}\begin{bmatrix}\Delta\dot{i}^e_{ds}\\ \Delta\dot{i}^e_{qs}\\ \Delta\dot{\omega}\\ \Delta\dot{\delta}\end{bmatrix} &= \begin{bmatrix}-\frac{1}{T_{sd}} & \omega_0\frac{L_q}{L_d} & \frac{L_q}{L_d}i^e_{qs0} & -v_{s0}\frac{\cos\delta_0}{L_d}\\ -\omega_0\frac{L_d}{L_q} & -\frac{1}{T_{sq}} & -\frac{(\psi_m+L_d i^e_{ds0})}{L_q} & -v_{s0}\frac{\sin\delta_0}{L_q}\\ \frac{3}{2}P^2\frac{1}{J_T}(L_d-L_q)i^e_{qs0} & \frac{3}{2}P^2\frac{1}{J_T}\left(\psi_m+(L_d-L_q)i^e_{ds0}\right) & -\frac{B}{J_T} & 0\\ 0 & 0 & -1 & 0\end{bmatrix}\begin{bmatrix}\Delta i^e_{ds}\\ \Delta i^e_{qs}\\ \Delta\omega\\ \Delta\delta\end{bmatrix}\\ &+\begin{bmatrix}-\frac{\sin\delta_0}{L_d} & 0 & 0\\ \frac{\cos\delta_0}{L_q} & 0 & 0\\ 0 & 0 & -\frac{P}{J_T}\\ 0 & 1 & 0\end{bmatrix}\begin{bmatrix}\Delta v_s\\ \Delta\omega_e\\ \Delta T_L\end{bmatrix}\end{aligned} \tag{5.88}$$

where the added subscript 0 denotes steady-state quantities, and steady-state terms have been canceled. The stability of the system is determined by the eigenvalues of the state matrix *A*(*X*). In no-load condition, the machine does not produce torque. Thus, the voltage applied to the machine is only to compensate for the back-emf voltage in the *q*-axis. In Eq. (5.89) is represented the conditions of no load to find the roots of the system.

$$\begin{aligned}v_s &= \omega_r\Psi_m\\ i^e_{ds} &= i^e_{qs}=0\\ \delta &= 0\end{aligned} \tag{5.89}$$

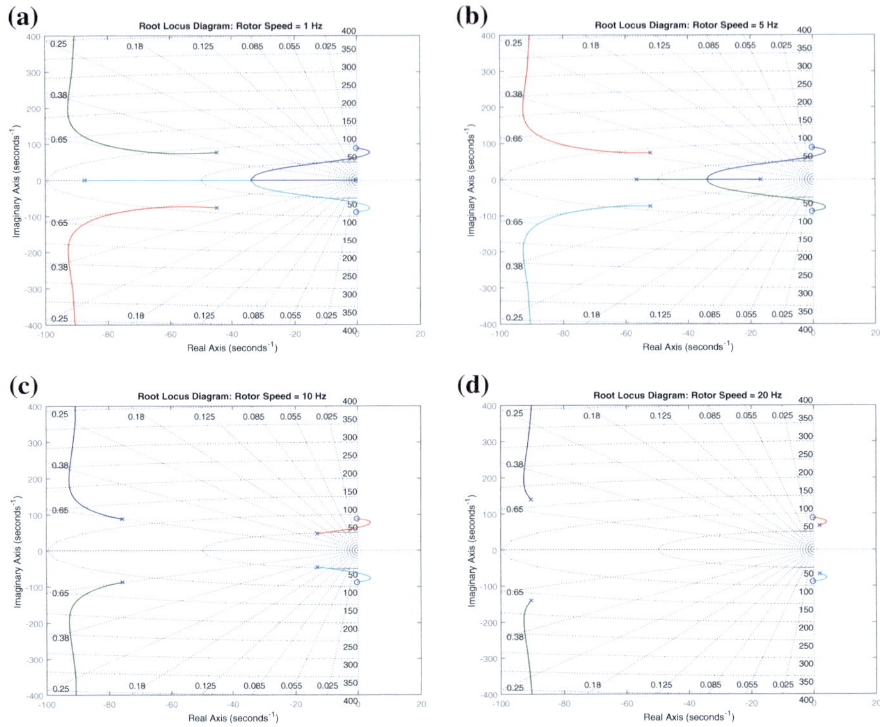

Fig. 5.15 Root locus of the PMSM operating at no load for different electrical command speeds when it is controlled in open-loop control strategy. **a** Electrical speed 1 Hz. **b** Electrical speed 5 Hz. **c** Electrical speed 10 Hz. **d** Electrical speed 20 Hz

If (5.88) is computed, the roots of the system for different rotor speeds can be found (Chandana Perera 2002; Chandana Perera et al. 2003). In Fig. 5.15 is represented the root locus diagram of the PMSM at no-load with above conditions as a function of the rotor speed. The root locus diagram is represented for four electrical speeds, 1, 5, 10, and 20 Hz.

The machine parameters used in the simulation are shown in Table 5.5. It corresponds to a 5 HP SMPMSM with a rated speed of 1800 RPM.

Table 5.5 Nameplate data of 5 HP PMSM

Nameplate	Symbol	Value	Unit
Nominal power	P_N	5	HP
Line-to-line nominal voltage	V_N (RMS)	208	V
Nominal current	I_N (RMS)	11.8	A
Nominal frequency	f_N	60	Hz
Nominal speed	N_N	1800	RPM
Number of poles	P	4	

Table 5.6 PMSM parameters

Parameters for machine model	Symbol	Value	Unit
Direct inductance	L_d	4.75	mH
Quadrature inductance	L_q	4.75	mH
Stator resistance	R_s	0.42	Ω
Permanent magnet flux	Ψ_m	0.43	Wb/rad
Rotor inertia	J	0.018	$kg \cdot m^2$
Viscous coefficient	B	0.001	$kg \cdot m^2/s$

The SMPMSM parameters are shown in Table 5.6.

It is possible to observe two different types of symmetrical poles: the stator and rotor poles. The stator poles in the left part represent the fast electrical stator dynamics, and the rotor poles in the right part represent the slow mechanical dynamics (Krause et al. 2002; Verghese 1986). At speed close to 17.5 Hz, the pole is placed just in the marginal stable zone. Speeds less than 17.5 Hz the PMSM is stable since all poles are in the left part of the diagram. For higher speeds, like 20 Hz, there is a symmetrical pole in the unstable region (right part of the diagram) as shown in Fig. 5.15d.

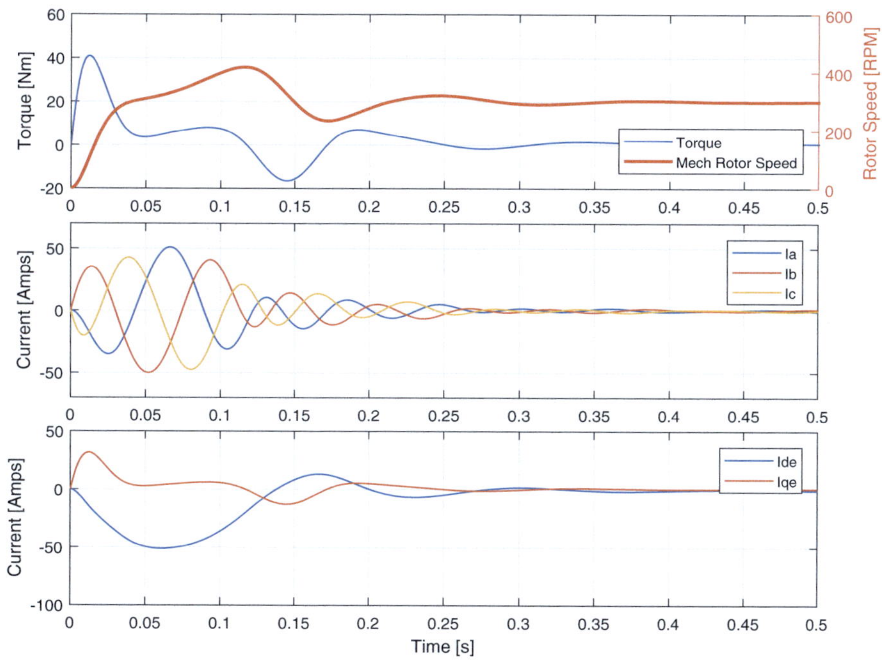

Fig. 5.16 Run-up speed, torque, phase currents, and i_{de} and i_{qe} components of unloaded PMSM at electrical set point speed of 10 Hz (only friction torque as load torque). Simulation variable-step size. Minimum step $T_{sp} = 10\ \mu s$, and automatic solver selection

In Fig. 5.16 is represented the run-up simulation of the PMSM at an electrical speed of 10 Hz in open-loop control. It represents the rotor speed, the development torque, the phases' current, and the i_{de} and i_{qe} current components. It is possible to observe that the PMSM reaches a stable situation where steady-state can be observed in Fig. 5.17. The machine has to develop the necessary torque to compensate only the friction torque. The stator V_s voltage applied is the electrical speed multiplied by the flux linkage.

From electrical speeds of 15 Hz, the PMSM starts to have unstable points as represented in Fig. 5.18. In this case, the machine reaches a stable situation in 1.5 s, but increasing the frequency a little bit, the machine takes the marginally stable situation as demonstrate with the root locus diagram.

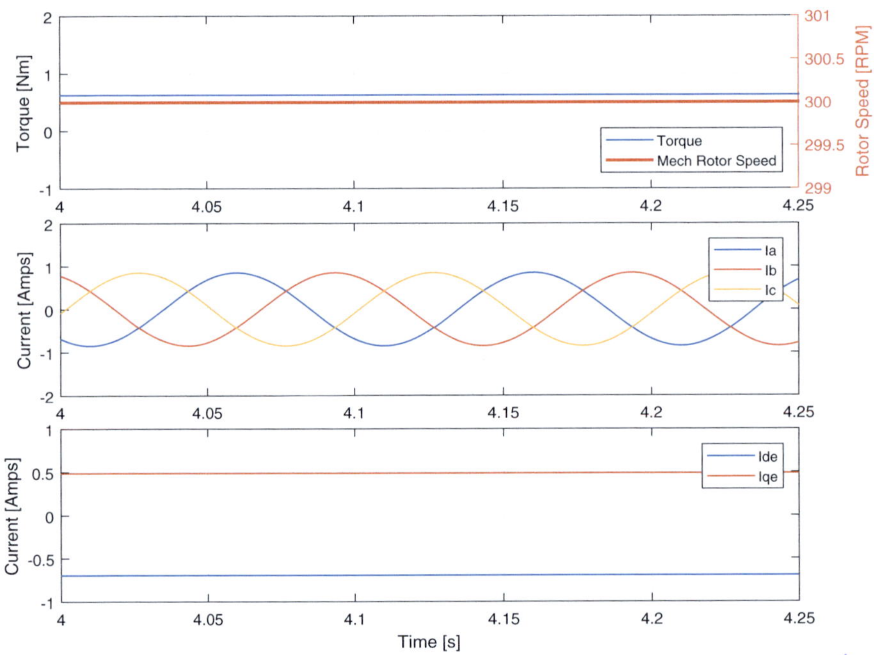

Fig. 5.17 Steady-state of the previous simulation at electrical set point speed of 10 Hz (only friction coefficient as load torque). Simulation variable-step size. Minimum step $T_{sp} = 10\ \mu s$, and automatic solver selection

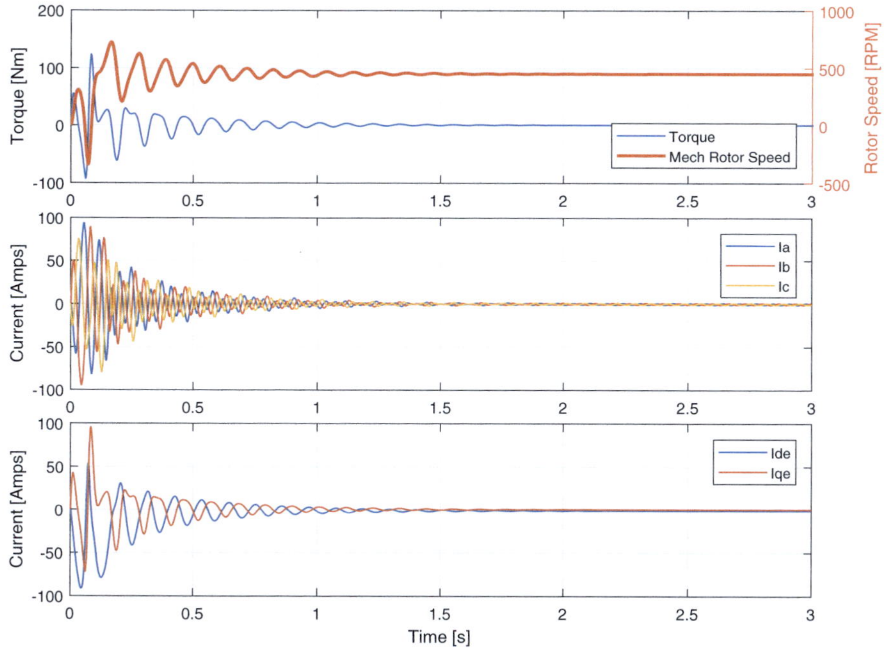

Fig. 5.18 Run-up speed, torque, phase currents, and i_{de} and i_{qe} components of unloaded PMSM at electrical set point speed of 15 Hz (only friction coefficient as load torque). Simulation variable-step size. Minimum step $T_{sp} = 10\ \mu s$, and automatic solver selection

5.4.2.2 Synchronous Reluctance Machine

5.4.2.2.1 Continuous State-Space Model of SynRM

In Chap. 4 was discussed the SynRM where mathematical equations were very similar to the PMSM. The synchronous rotating reference frame is usually used to describe the SynRM model as shown for PMSM. SynRM can be described in general form with permanent magnets; that is, the PMASynRM model is developed in this section.

As before, the derivative parts are separated:

$$\begin{cases} \frac{\mathrm{d}i_{ds}^e}{\mathrm{d}t} = -\frac{1}{T_{sd}} i_{ds}^e + \omega_r \frac{L_q}{L_d} i_{qs}^e - \omega_r \frac{\psi_m}{L_d} + \frac{1}{L_d} v_{ds}^e \\ \frac{\mathrm{d}i_{qs}^e}{\mathrm{d}t} = -\frac{1}{T_{sq}} i_{qs}^e - \omega_r \frac{L_d}{L_q} i_{ds}^e + \frac{1}{L_q} v_{qs}^e \end{cases} \tag{5.90}$$

$$\frac{\mathrm{d}\omega_r}{\mathrm{d}t} = \frac{3}{2} \frac{P^2}{J_T} \left[(L_d - L_q) i_{ds}^e i_{qs}^e + \Psi_m i_{qs}^e \right] - \frac{1}{J_T} B \omega_r - \frac{P}{J_T} T_L \tag{5.91}$$

$$T_{sd} = \frac{L_d}{R_s}$$

$$T_{sq} = \frac{L_q}{R_s} \tag{5.92}$$

The torque expression was similar to the IPMSM machine, but the flux is multiplicated by the direct current instead of the quadrature current:

$$T_e = \frac{3}{2}P\left[\left(L_d - L_q\right)i^e_{ds}i^e_{qs} + \Psi_m i^e_{ds}\right] \tag{5.93}$$

In matrix form:

$$\begin{bmatrix} \frac{di^e_{ds}}{dt} \\ \frac{di^e_{qs}}{dt} \\ \frac{d\omega_r}{dt} \\ \frac{d\delta}{dt} \end{bmatrix} = \begin{bmatrix} -\frac{1}{T_{sd}} & \omega_r\frac{L_q}{L_d} & -\frac{\psi_m}{L_d} & 0 \\ -\omega_r\frac{L_d}{L_q} & -\frac{1}{T_{sq}} & 0 & 0 \\ \frac{3}{2}P^2\frac{1}{J_T}\left[\left(L_d - L_q\right)i^e_{qs} + \psi_m\right] & 0 & -\frac{B}{J_T} & 0 \\ 0 & 0 & -1 & 0 \end{bmatrix} \begin{bmatrix} i^e_{ds} \\ i^e_{qs} \\ \omega_r \\ \delta \end{bmatrix} + \begin{bmatrix} -\frac{\sin\delta}{L_d} & 0 & 0 \\ \frac{\cos\delta}{L_q} & 0 & 0 \\ 0 & 0 & -\frac{P}{J_T} \\ 0 & 1 & 0 \end{bmatrix} \begin{bmatrix} v_s \\ \omega_e \\ T_L \end{bmatrix} \tag{5.94}$$

$$\begin{bmatrix} i^e_{ds} \\ i^e_{qs} \\ \omega_r \\ \delta \end{bmatrix} = \begin{bmatrix} 1 & 0 & 0 & 0 \\ 0 & 1 & 0 & 0 \\ 0 & 0 & 1 & 0 \\ 0 & 0 & 0 & 1 \end{bmatrix} \begin{bmatrix} i^e_{ds} \\ i^e_{qs} \\ \omega_r \\ \delta \end{bmatrix} + \begin{bmatrix} 0 & 0 & 0 \\ 0 & 0 & 0 \\ 0 & 0 & 0 \end{bmatrix} \begin{bmatrix} v_s \\ \omega_e \\ T_L \end{bmatrix} \tag{5.95}$$

where the system and input matrix are

$$\boldsymbol{A}^e = \begin{bmatrix} -\frac{1}{T_{sd}} & \omega_r\frac{L_q}{L_d} & -\frac{\psi_m}{L_q} & 0 \\ -\omega_r\frac{L_d}{L_q} & -\frac{1}{T_{sq}} & 0 & 0 \\ \frac{3}{2}P^2\frac{1}{J_T}\left[\left(L_d - L_q\right)i^e_{qs} + \psi_m\right] & 0 & -\frac{B}{J_T} & 0 \\ 0 & 0 & -1 & 0 \end{bmatrix}$$

$$\boldsymbol{B}^e = \begin{bmatrix} -\frac{\sin\delta}{L_d} & 0 & 0 \\ \frac{\cos\delta}{L_q} & 0 & 0 \\ 0 & 0 & -\frac{P}{J_T} \\ 0 & 1 & 0 \end{bmatrix} \quad \boldsymbol{C}^e = \begin{bmatrix} 1 & 0 & 0 & 0 \\ 0 & 1 & 0 & 0 \\ 0 & 0 & 1 & 0 \\ 0 & 0 & 0 & 1 \end{bmatrix} \quad \boldsymbol{D}^e = \begin{bmatrix} 0 & 0 & 0 \\ 0 & 0 & 0 \\ 0 & 0 & 0 \end{bmatrix} \tag{5.96}$$

In Fig. 5.19a is represented the block diagram of the continuous state-space model of the PMASynRM in the synchronously rotating reference frame. The detail is represented in Fig. 5.19b.

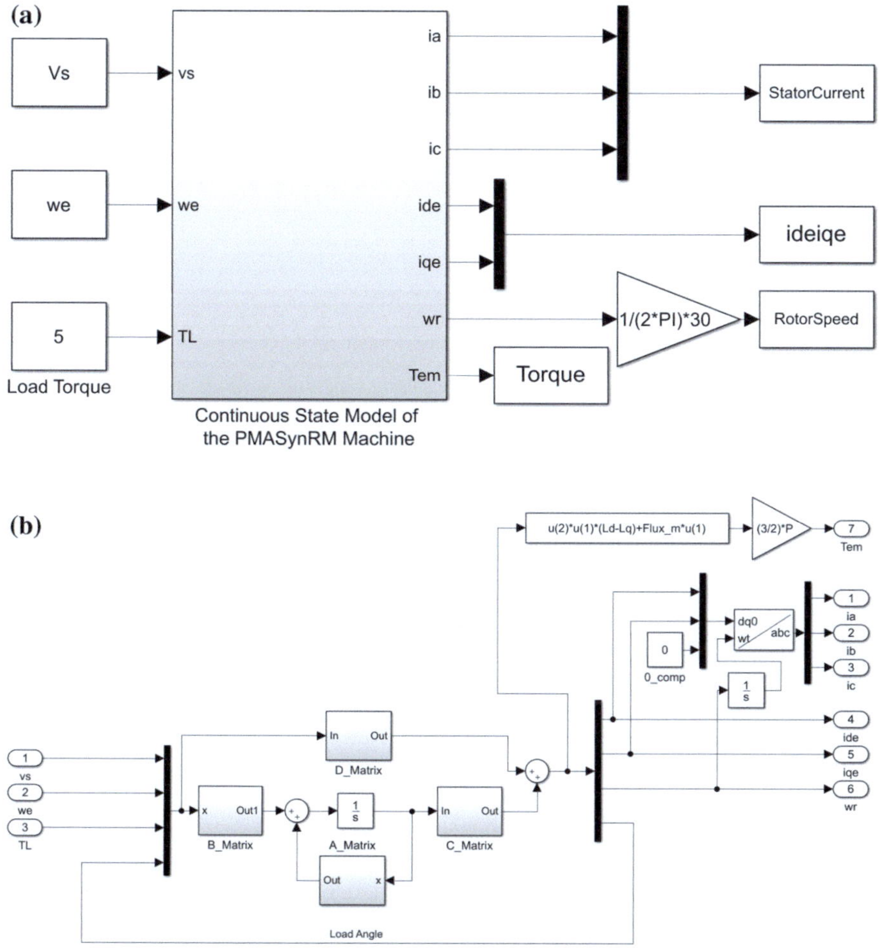

Fig. 5.19 **a** Continuous state-space model of the PMASynRM in the synchronously rotating reference frame. **b** Detail block diagram with the state-space system, torque, motion, and axes' transformation

In Fig. 5.20 is illustrated the run-up simulation of the SynRM at an electrical speed of 8 Hz in open-loop control. As the PMSM, it is possible to observe that the SynRM reaches a stable situation where steady-state can be observed in Fig. 5.21. As a PMSM, the machine has to develop the necessary torque to be equal to the friction torque. Also, the stator V_s voltage applied is the electrical speed multiplied by the flux.

The machine parameters used in the simulation are shown in Table 5.7. It corresponds to a 5 HP SynRM with a rated speed of 1800 RPM.

The SynRM parameters are shown in Table 5.8.

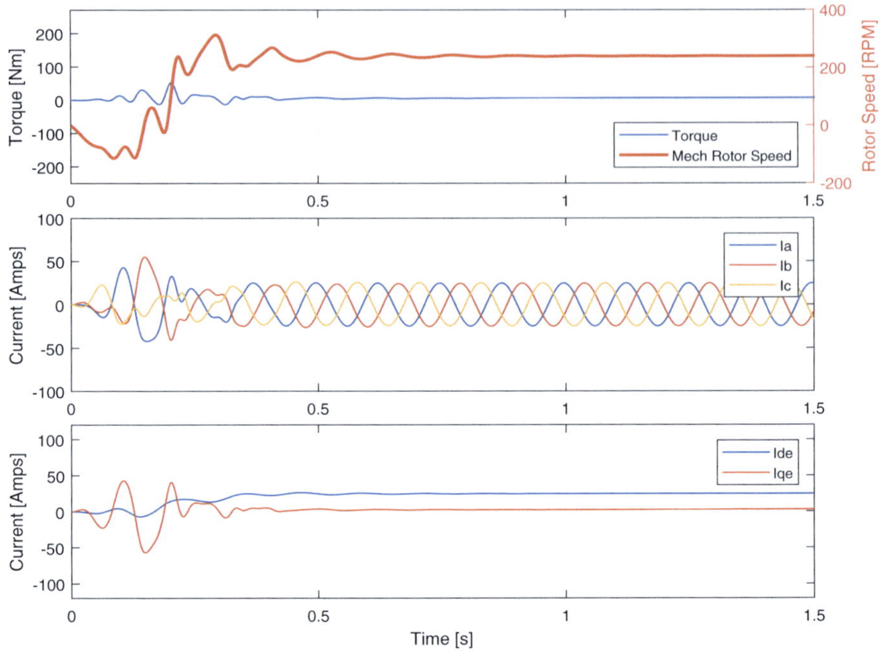

Fig. 5.20 Run-up speed, torque, phase currents, and i_{de} and i_{qe} components of loaded SynRM at electrical set point speed of 8 Hz. Load torque is 5 Nm. Simulation variable-step size. Minimum step $T_{sp} = 100$ ns and automatic solver selection

5.4.2.2.2 PMASynRM Discrete Model

In the state-space models obtained from different machines appear terms that depend on the state and that are time-variant. In order to find an approximate discrete model of the machine it is assumed that this dependence is constant during an interval of sampling period T. For this, the sampling time should be chosen sufficiently small, preferable less to 1 ms, so that the state-space system of (5.97) is linear and variant in time for each sampling period (piecewise linear system). Therefore, the methods of discretization of the conventional continuous model can be applied in the same way as it is applied in continuous linear systems and time-invariant.

The time-discrete state-space model of the PMASynRM can be obtained from (5.16), that is:

$$\begin{aligned} \boldsymbol{x}(k+1) &= \boldsymbol{A}_d(k)\boldsymbol{x}(k) + \boldsymbol{B}_d(k)\boldsymbol{u}(k) \\ \boldsymbol{y}(k) &= \boldsymbol{C}_d(k)\boldsymbol{x}(k) \end{aligned} \tag{5.97}$$

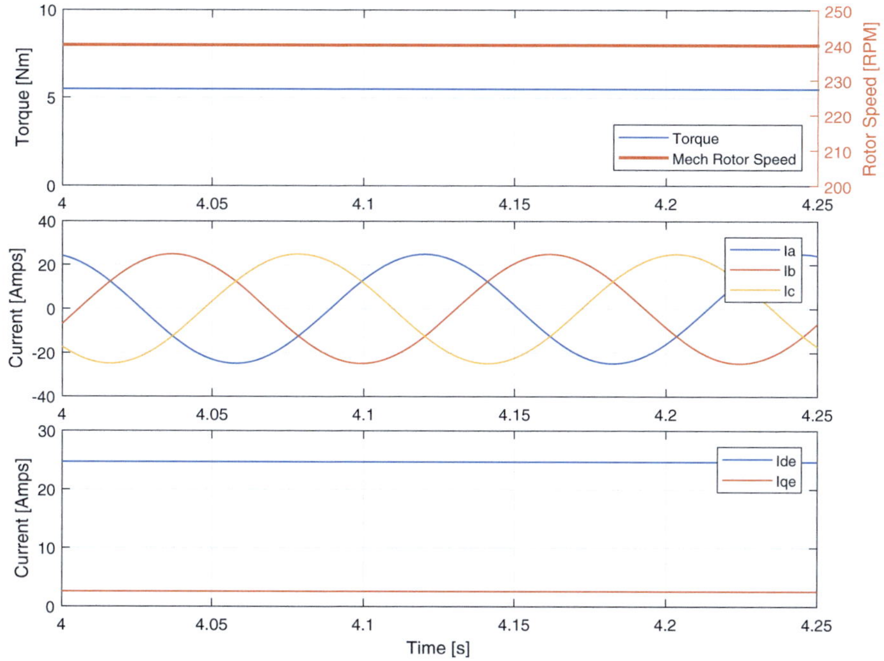

Fig. 5.21 Steady-state of the previous simulation at electrical set point speed of 8 Hz and 5 Nm of load torque. Simulation variable-step size. Minimum step $T_{sp} = 100$ ns and automatic solver selection

Table 5.7 Nameplate data of 5 HP SynRM

Nameplate	Symbol	230 V	Unit
Nominal power	P_N	5	HP
Line-to-line nominal Voltage	V_N (RMS)	170	V
Nominal current	I_N (RMS)	13.3	A
Nominal frequency	f_N	60	Hz
Nominal speed	N_N	1800	RPM
Number of poles	P	4	

Table 5.8 Parameters of 5 HP SynRM

Parameters for machine model	Symbol	170 V	Unit
Direct inductance	L_d	31.4	mH
Quadrature inductance	L_q	2.5	mH
Stator resistance	R_s	0.24	Ω
Permanent magnet flux	Ψ_m	0	Wb/rad
Rotor inertia	J	0.015	$\text{kg} \cdot \text{m}^2$
Viscous coefficient	B	0.001	$\text{kg} \cdot \text{m}^2/\text{s}$

where

$$\begin{aligned} \boldsymbol{A}_d(k) &= \mathrm{e}^{\boldsymbol{A}(kT)T} \\ \mathrm{e}^{\boldsymbol{A}T} &\approx \boldsymbol{I} + \boldsymbol{A}T \\ \boldsymbol{B}_d(k) &\approx \boldsymbol{B}T \\ \boldsymbol{C}_d(k) &= \boldsymbol{C} \end{aligned} \tag{5.98}$$

And T was de sampling period which is presumed constant. The sampling time should be as small as possible to give a stable and accurate discretized model. As a reference, the sampling time should be smaller than the time constants of the machine.

In Eqs. (5.99) and (5.100) is shown the discrete state-space model for PMASynRM according to (5.94) and (5.95).

$$\begin{bmatrix} i_{ds}^e(k+1) \\ i_{qs}^e(k+1) \\ \omega_r(k+1) \\ \delta(k+1) \end{bmatrix} = \begin{bmatrix} 1-\frac{T}{T_{sd}} & \omega_r \frac{L_q}{L_d} T & -\frac{\psi_m}{L_d} T & 0 \\ -\omega_r \frac{L_d}{L_q} T & 1-\frac{T}{T_{sq}} & 0 & 0 \\ \frac{3}{2} P^2 \frac{T}{J_T} \left[\left(L_d - L_q \right) i_{qs}^e + \psi_m \right] & 0 & 1-\frac{B}{J_T} T & 0 \\ 0 & 0 & -T & 0 \end{bmatrix} \begin{bmatrix} i_{ds}^e(k) \\ i_{qs}^e(k) \\ \omega_r(k) \\ \delta(k) \end{bmatrix} + \begin{bmatrix} -\frac{\sin\delta}{L_d} T & 0 & 0 \\ \frac{\cos\delta}{L_q} T & 0 & 0 \\ 0 & 0 & -\frac{PT}{J_T} \\ 0 & T & 0 \end{bmatrix} \begin{bmatrix} v_s \\ \omega_e \\ T_L \end{bmatrix} \tag{5.99}$$

$$\begin{bmatrix} i_{ds}^e(k) \\ i_{qs}^e(k) \\ \omega_r(k) \\ \delta(k) \end{bmatrix} = \begin{bmatrix} 1&0&0&0 \\ 0&1&0&0 \\ 0&0&1&0 \\ 0&0&0&1 \end{bmatrix} \begin{bmatrix} i_{ds}^e(k) \\ i_{qs}^e(k) \\ \omega_r(k) \\ \delta(k) \end{bmatrix} + \begin{bmatrix} 0&0&0 \\ 0&0&0 \\ 0&0&0 \end{bmatrix} \begin{bmatrix} v_s \\ \omega_e \\ T_L \end{bmatrix} \tag{5.100}$$

With the same previous restrictions of maintaining a constant and small sampling time, (5.94)–(5.95) equations can be discretized in a more intuitive way using the Euler forward discretization method seen in Chap. 2.

In Eq. (5.101), it is possible to observe the discretized equations of both currents i_{dse} and i_{qse} considering the method mentioned above where T is the sampling period.

$$\begin{aligned} \frac{i_{ds}^e(k+1) - i_{ds}^e(k)}{T} &= -\frac{1}{T_{sd}} i_{ds}^e(k) + \omega_r(k) \frac{L_q}{L_d} i_{qs}^e(k) - \omega_r(k) \frac{\psi_m}{L_d} + \frac{1}{L_d} v_{ds}^e(k) \\ \frac{i_{qs}^e(k+1) - i_{qs}^e(k)}{T} &= -\frac{1}{T_{sq}} i_{qs}^e(k) - \omega_r(k) \frac{L_d}{L_q} i_{ds}^e(k) + \frac{1}{L_q} v_{qs}^e(k) \end{aligned} \tag{5.101}$$

By developing both equations, it is possible to arrive at Eqs. (5.102) and (5.103):

$$i_{ds}^e(k+1) = \left(1 - \frac{T}{T_{sd}} \right) i_{ds}^e(k) + \frac{L_q}{L_d} T \omega_r(k) i_{qs}^e(k) - \frac{\psi_m}{L_d} T \omega_r(k) + \frac{T}{L_d} v_{ds}^e(k) \tag{5.102}$$

$$i_{qs}^{e}(k+1)=\left(1-\frac{T}{T_{sq}}\right)i_{qs}^{e}(k)-\frac{L_d}{L_q}T\omega_r(k)i_{ds}^{e}(k)+\frac{T}{L_q}v_{qs}^{e}(k) \tag{5.103}$$

It is possible to observe that Eqs. (5.102) and (5.103) are identical to those obtained according to the discrete state-space model of Eq. (5.99).

Considering the following constants:

$$\begin{array}{ll} k_1=\left(1-\frac{T}{T_{sd}}\right) & k_5=\left(1-\frac{T}{T_{sq}}\right) \\ k_2=\frac{L_q}{L_d}T & k_6=\frac{L_d}{L_q}T \\ k_3=\frac{\Psi_m}{L_d}T & k_7=\frac{T}{L_q} \\ k_4=\frac{T}{L_d} & \end{array} \tag{5.104}$$

Equations (5.102) and (5.103) can be expressed as:

$$i_{ds}^{e}(k+1)=k_1 i_{ds}^{e}(k)+k_2\omega_r(k)i_{qs}^{e}(k)-k_3\omega_r(k)+k_4 v_{ds}^{e}(k) \tag{5.105}$$

$$i_{qs}^{e}(k+1)=k_5 i_{qs}^{e}(k)-k_6\omega_r(k)i_{ds}^{e}(k)+k_7 v_{qs}^{e}(k) \tag{5.106}$$

In the previous expressions (5.105) and (5.106), it is possible to see the cross-coupling that exists between the d-axis and q-axis, precisely in the terms $k_2\omega_r(k)i_{qse}(k)$ and $-k_6\omega_r(k)i_{dse}(k)$. That is, the expression of the current i_{dse} also depends on the current i_{qse} and vice versa. This coupling of equations also exists in the expressions in continuous mode as can be seen in Eqs. (5.94) and (5.95). This is not exclusive to the SynRM but also occurs in the PMSM and in the induction machine. The independent current control is not possible, but depending on each other, degrading the performance of the control. In Sect. 5.4.2.2.3, it will be demonstrated the coupling, as well as a mechanism that allows independent current control.

For the case of the rotor speed, the procedure is exactly like the one carried out before:

$$\begin{aligned}\frac{\omega_r(k+1)-\omega_r(k)}{T} &= \frac{3}{2}\frac{P^2}{J_T}\left[(L_d-L_q)i_{ds}^{e}(k)i_{qs}^{e}(k)+\Psi_m i_{ds}^{e}(k)\right] \\ &\quad -\frac{1}{J_T}B\omega_r(k)-\frac{P}{J_T}T_L(k)\end{aligned} \tag{5.107}$$

$$\omega_r(k+1)=\left(1-\frac{T}{J_T}B\right)\omega_r(k)+\frac{3}{2}\frac{P^2}{J_T}T\left[\begin{array}{l}(L_d-L_q)i_{ds}^{e}(k)i_{qs}^{e}(k)\\ +\Psi_m i_{ds}^{e}(k)\end{array}\right]-\frac{P\cdot T}{J_T}T_L(k) \tag{5.108}$$

$$\omega_r(k+1)=k_8\omega_r(k)+k_9\Big((L_d-L_q)i_{ds}^{e}(k)i_{qs}^{e}(k)+\Psi_m i_{ds}^{e}(k)\Big)-k_{10}T_L(k) \tag{5.109}$$

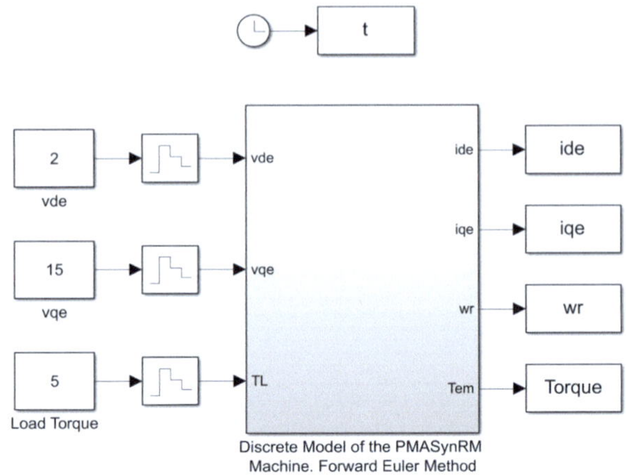

Fig. 5.22 Forward Euler discrete method model of the PMASynRM

where the constants are:

$$k_8 = \left(1 - \frac{T}{J_T}B\right)$$
$$k_9 = \frac{3}{2}\frac{P^2}{J_T}T$$
$$k_{10} = \left(\frac{P \cdot T}{J_T}\right) \tag{5.110}$$

In this case, the result obtained for the equation of the speed is also precisely the same as the discrete model of the state-space (5.99).

Figure 5.22 shows the Simulink® model of the discrete PMASynRM according to (5.105), (5.106), and (5.109) equations of the forward Euler method.

If PMASynRM model is excited with a voltage v_{de} and v_{qe} as shown in the figure (open loop) with a load torque T_L of 5 Nm, the responses of Fig. 5.23 are obtained. It is represented the rotor speed and the electromagnetic torque. The parameters of

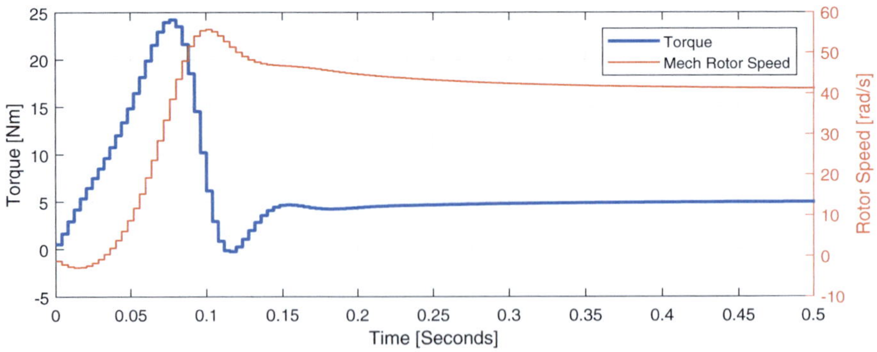

Fig. 5.23 Run-up speed and torque of loaded PMASynRM in open loop. Load torque is 5 Nm, and steady-state mechanical speed is 42 rad/s. Simulation fixed-step size auto and discrete solver

the PMASynRM are those of Table 5.8, and the sampling time T is 4 ms for better visualization of the discrete system.

5.4.2.2.3 Cross-Coupling Effect

As discussed above, the expressions describing the direct i_{dse} and quadrature current i_{qse} are cross-coupled so that it is difficult to control the current loop efficiently without affecting the control performance. The cross-coupling effect causes a non-zero error between the command and the actual value when trying to regulate a current loop since it affected by the other loop. In order to demonstrate this effect, it is possible to perform a discrete-time simulation in closed loop with the model of the previous machine and a PI regulator for each of the i_{dse} and i_{qse} currents. Both PI regulators have the function of regulating the voltages v_{de} and v_{qe} so that the i_{dse} and i_{qse} current should match the i_{dse}* and i_{qse}* commands, respectively. In (5.111), the discrete PI regulator can be observed according to the bilinear transformation method seen in Chap. 2, which is applied to regulate the direct current and quadrature loops. The error of the PI compensator is formed by the difference between the command and the measured signal (actual value).

$$\begin{aligned} v_{dsePI}(z) &= \left[K_P + \frac{K_I \cdot T}{2}\left(\frac{z+1}{z-1}\right)\right]\left(i_{dse}^*(z) - i_{dse}(z)\right) \\ v_{qsePI}(z) &= \left[K_P + \frac{K_I \cdot T}{2}\left(\frac{z+1}{z-1}\right)\right]\left(i_{qse}^*(z) - i_{qse}(z)\right) \end{aligned} \tag{5.111}$$

Figure 5.24 shows the block diagram of the closed-loop control system of both

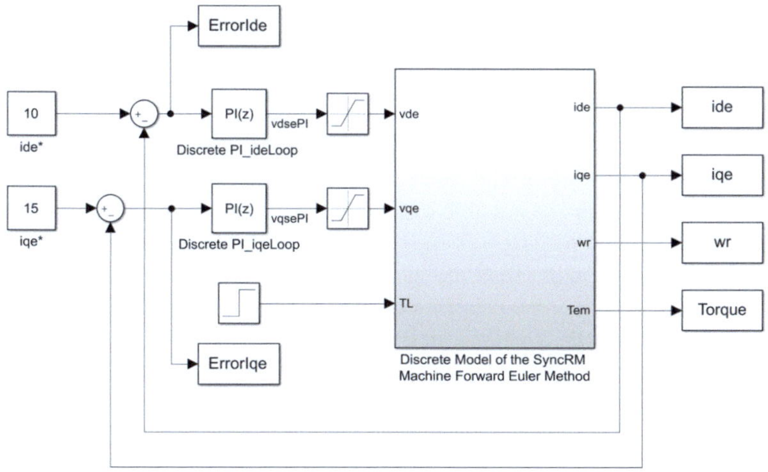

Fig. 5.24 Block diagram of discrete closed-loop current control of the PMASynRM

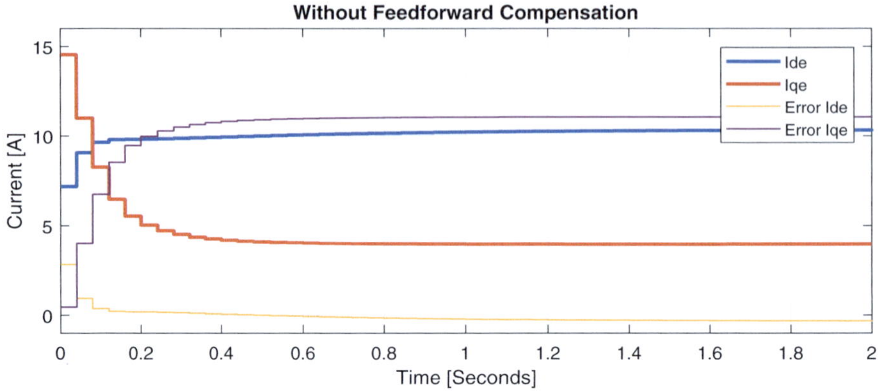

Fig. 5.25 Simulation result of direct and quadrature outputs currents in the discrete model PMASynRM without feedforward compensation. It also shows the error of direct and quadrature currents. Sampling time is 40 ms for best visualization of the discrete response. Simulation fixed-step size auto and discrete solver

Fig. 5.26 Block diagram of feedforward compensation. It is represented in the discrete mode

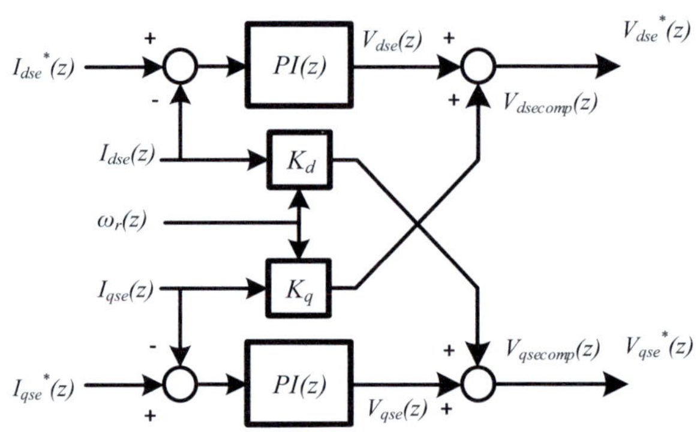

currents for the discretized model of the PMASynRM.

The simulation is performed with a sampling time T of 40 ms for a better appreciation of the discrete system. The result obtained is represented in Fig. 5.25. As can be seen, both controllers are not able to regulate the command in this case of $i_{dse} =$ 10 A and $i_{qse} = 15$ A by the effect of the cross-coupling. There is a steady-state error that is not null and that will affect the dynamics of the control.

To avoid the inherent cross-coupling of the AC machines, it is typical to apply a feedforward compensator, as shown in Fig. 5.26.

The value of K_d and K_q will depend on the machine type. In the case of the PMASynRM, Eqs. (5.105) and (5.106) can be rewritten as follows:

$$i_{ds}^{e}(k+1) = k_1 i_{ds}^{e}(k) + k_2 \omega_r(k) i_{qs}^{e}(k) - k_3 \omega_r(k) + k_4 \left(v_{ds}^{e}(k) + v_{dscomp}^{e}(k) \right) \tag{5.112}$$

Table 5.9 K_d and K_q parameters for the different AC machines

	Induction machine	PMSM	SynRM	PMASynRM
K_d	$\sigma L_s \omega_e i_{ds}^e + \omega_r \frac{L_m^2}{L_r} i_{ds}^e$	$L_d \omega_r i_{ds}^e + \psi_m \omega_r$	$L_d \omega_r i_{ds}^e$	$L_d \omega_r i_{ds}^e$
K_q	$-\sigma L_s \omega_e i_{qs}^e - R_r \frac{L_m^2}{L_r^2} i_{ds}^e$	$-L_q \omega_r i_{qs}^e$	$-L_q \omega_r i_{qs}^e$	$-L_q \omega_r i_{qs}^e + \psi_m \omega_r$

$$i_{qs}^e(k+1) = k_5 i_{qs}^e(k) - k_6 \omega_r(k) i_{ds}^e(k) + k_7 \left(v_{qs}^e(k) + v_{qscomp}^e(k) \right) \tag{5.113}$$

If the compensated voltages take the following expressions:

$$\begin{aligned} &k_4 v_{dscomp}^e(k) = -k_2 \omega_r(k) i_{qs}^e(k) + k_3 \omega_r(k) \rightarrow \\ &v_{dscomp}^e(k) = -L_q \omega_r(k) i_{qs}^e(k) + \psi_m \omega_r(k) \\ &K_q = -L_q \omega_r(k) i_{qs}^e(k) + \psi_m \omega_r(k) \end{aligned} \tag{5.114}$$

$$\begin{aligned} &k_7 v_{qscomp}^e(k) = k_6 \omega_r(k) i_{ds}^e(k) \rightarrow v_{qscomp}^e(k) = L_d \omega_r(k) i_{ds}^e(k) \\ &K_d = L_d \omega_r(k) i_{ds}^e(k) \end{aligned} \tag{5.115}$$

Table 5.9 shows the values that K_d and K_q take for the different AC machine types studied.

These previous terms are considered as disturbances that are canceled using the decoupling method that uses non-linear feedback of the coupled voltages.

Therefore, if the coupling is perfect and ideal, Eqs. (5.112) and (5.113) remain as a linear system of the first order:

$$i_{ds}^e(k+1) = k_1 i_{ds}^e(k) + k_4 v_{ds}^e(k) \tag{5.116}$$

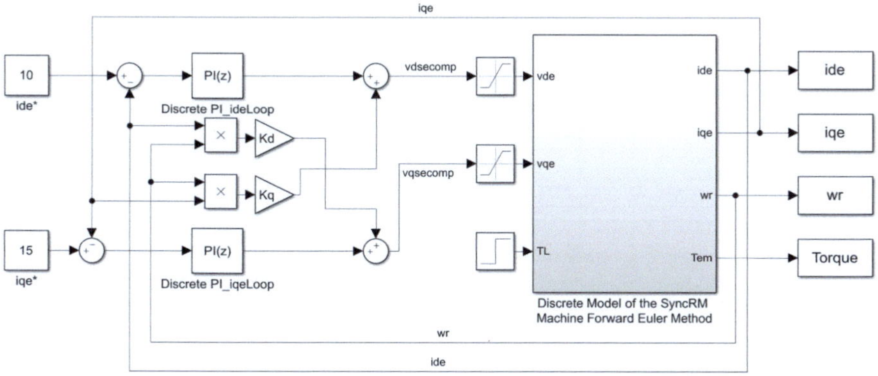

Fig. 5.27 Block diagram of discrete closed-loop current control of the SynRM with the feedforward compensation

$$i_{qs}^{e}(k+1) = k_5 i_{qs}^{e}(k) + k_7 v_{qs}^{e}(k) \tag{5.117}$$

As it can see, this compensation tries to counteract the effect produced by the opposite current of each of the expressions of the current. Figure 5.27 shows the current control system with the feedforward compensation in this case for the SynRM where the magnet flux Ψ_m is zero.

By performing this decoupling compensation, it is possible to cancel the crossover coupling in a steady-state, as shown in Fig. 5.28. In this case, both currents are stabilized at their respective commands, increasing the control performance of both current loops. However, the feedforward compensation needs to know the parameters of the machine L_d and L_q. These parameters, among other factors, vary with temperature and current so that the compensation will not be perfect if nominal values are taken.

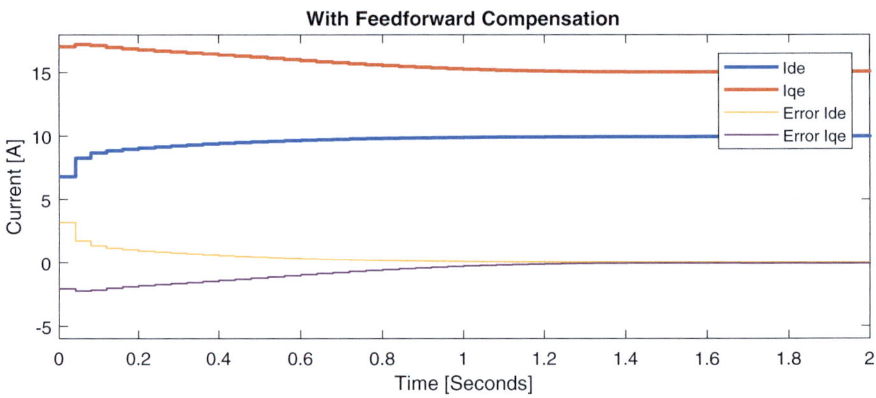

Fig. 5.28 Simulation result of direct and quadrature outputs currents in the discrete model SynRM with feedforward compensation. It is also shown the error of direct and quadrature currents. Sampling time is 40 ms for best visualization of the discrete response. Simulation fixed-step size auto and discrete solver

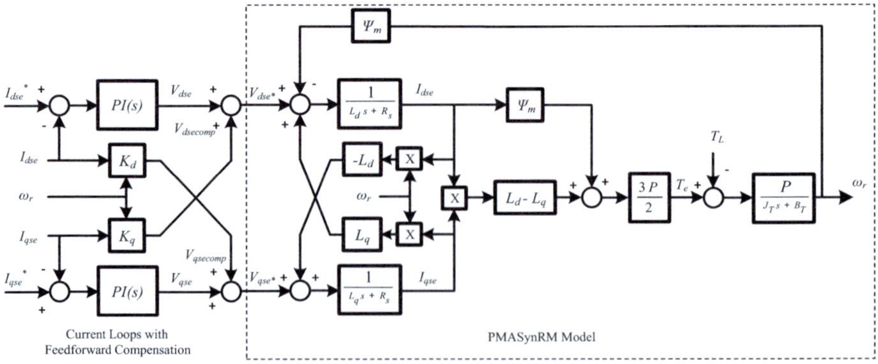

Fig. 5.29 Current loops with feedforward compensation connected to the PMASynRM

Finally, Fig. 5.29 shows the current control loops with the feedforward compensation together with the PMASynRM model in the s plane. It is possible to appreciate how the compensation decouples the cross-coupling of the machine model.

References

Chandana Perera PD (2002) Sensorless control of permanent-magnet synchronous motor drives. Ph.D. dissertation, Institute of Energy Technology, Aalborg University

Chandana Perera PD, Blaabjerg F, Pedersen JK, Thøgersen P (2003) A sensorless, stable *V/f* control method for permanent-magnet synchronous motor drives. IEEE Trans Ind Appl 39(3):783–791

Kalman RE (1960) Control system analysis and design via the second method of Liapunov: I. Continuos-time systems; II. Discrete-time systems

Krause PC, Wasynczuk O, Sudhoff SD (1995) Analysis of electric machinery. IEEE Press

Krause PC, Wasynczuk O, Sudhoff SD (2002) Analysis of electrical machinery and drive systems, 2nd edn. IEEE Press

Verghese GC, Lang JH, Casey LF (1986) Analysis of instability in electrical machines. IEEE Trans Ind Appl IA-22(5):853–864

Chapter 6
Measurement in Electric Drives

6.1 Introduction

Any process control system needs to use sensors and actuators. The sensors take measurements in the environment and pass that information to the controller. The controller compares the measured values with a defined value, taking action when the two are different. The actuator can influence the environment by closing the loop. In Fig. 6.1, the aforementioned explanation is shown graphically—the loop can be seen in an anti-clockwise direction.

In our specific case, the actuator is the electric machine, the environment is the load on the shaft, and the sensor will provide the feedback signals needed by the controller—things such as currents, voltages, and speed.

Physical connections between sensors and conductors will be necessary, for example, to measure the stator current in the motor. These connections can be via direct contact or magnetic or optical coupling. This means that the measurement may be sensitive to the measurement noise from a high dv/dt or di/dt value in the conductor, due to the electric drive's (power stage) switching. The accuracy and sensitivity to noise are the main concerns in the measurements of the variables. For best accuracy, the tolerance of the components should be selected according to the requirements, e.g., the resistance components used should be 1% or less of tolerance—especially for measurements with A/D conversion requirements, such as voltage, current, and temperature. Also, the low-pass filters and thorough design of the layout in the printed circuit board (PCB) will minimize the noise in the measurements.

On the other hand, in the automotive market, the components used should meet the automotive electronics council (AEC) standards. For example, the AEC Q200 establishes a qualification requirement for passive components as resistance and capacitors, while the AEC Q101 establishes the qualification requirement for discrete semiconductors as MOSFET and IGBT.

© Springer Nature Switzerland AG 2020
R. Molina Llorente, *Practical Control of Electric Machines*,
Advances in Industrial Control, https://doi.org/10.1007/978-3-030-34758-1_6

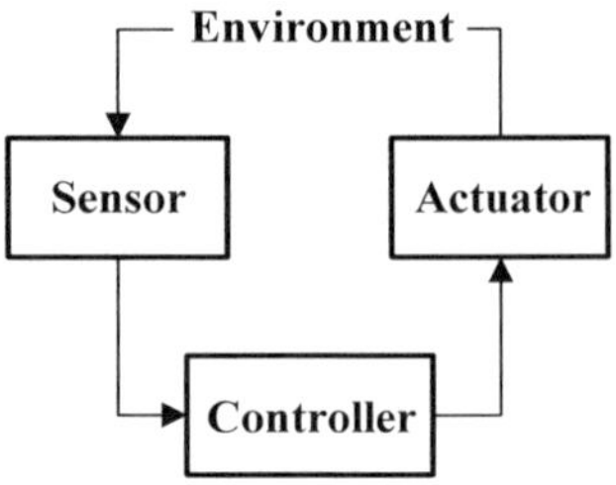

Fig. 6.1 Wiring for a process control system

In this section, several sensors and circuits for different measurements will be described. Voltage, current, temperature, speed, and rotational position sensors will be analyzed in detail. Some of the circuits described are very simple and also contain active elements. All are practice circuits used in industry, home appliances, and automotive.

6.2 Voltage Measurement

Sometimes, it is necessary to measure the voltage level for control monitoring and use. The non-isolated voltage sensor can only be made using high-precision resistor dividers, which reduces the voltage to the desired value in accordance with the A/D conversion range. It is recommendable to use a low-pass filter close to the A/D module to avoid introducing electrical noise in the measurement. An accurate cut-off frequency selection will be useful for digital signal processing.

6.2.1 Non-isolated Voltage Measurement

When the AC line is rectified by using a full-bridge rectifier and the output is filtered using a capacitor, the voltage between terminals is called DC BUS voltage—also known as DC-link in other literature. This is an AC–DC converter which converts the AC line voltage to constant DC voltage that has the maximum value of the AC input peak. The converter is typically used as the prior stage before the inverter. The inverter, which will be seen in Chap. 8, is a DC–AC converter which converts the DC voltage to AC voltage. If it is used for three-phase machine control, the output of the inverter must be three phase. The DC BUS voltage should be monitored in order to detect any failures (for example, an undervoltage, and an overvoltage caused by machine regeneration) and to know the actual voltage available for machine control. The sensor for DC BUS voltage consists of a voltage divider made with resistance where the reference is the ground. It is a non-isolated sensor when sensor shares the same ground between high-and low-voltage side.

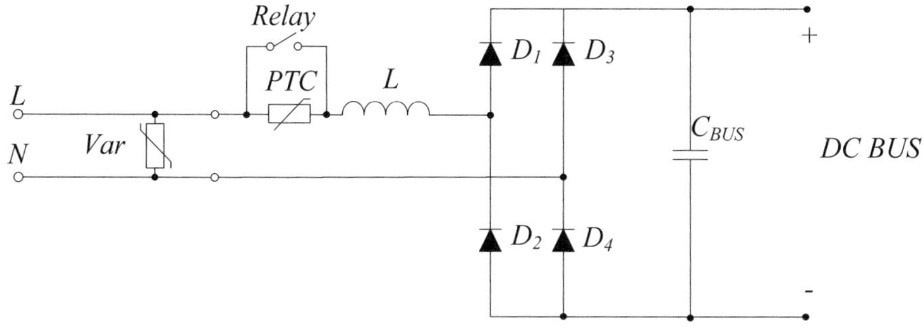

Fig. 6.2 Circuit diagram of an AC–DC converter in which output is a DC voltage, called DC BUS or DC-link

An AC–DC converter made with a full-bridge rectifier with some hardware protections such as a varistor and PTC is shown in Fig. 6.2.

For a single-phase grid with nominal voltage V_{AC} of 240 VAC, it is common to choose the bottom tolerance of 25% (180 VAC) and upper tolerance of 10% (265 VAC) of the nominal voltage in order to cover the input voltage variation. As a result, in the DC BUS, the variation will be from 254.56 to 374.77 V where $V_{AC}\sqrt{2}$ has been applied. Then, the rated voltage of C_{BUS} capacitor should be 400 V. It is recommended to design the voltage sensor to be able to measure voltages higher than 400 V in order to detect the overvoltage situation and thus protect the capacitor and machine. For instance, 430 V as a maximum voltage value for measurement is a good choice.

High voltage is scaled to a lower range with the voltage divider to be connected directly to the microcontroller/DSP analog-to-digital converter. Sometimes, it is necessary to install external diodes in order to protect the microcontroller/DSP port, even if it has internal diode protection. Internal microcontroller/DSP diodes have a maximum current at which, if exceeded, the diodes can be damaged. If in the worst-case scenario, the current which flows through microcontroller/DSP port diode is higher than the maximum current limit, external diodes should be installed. In this case, two external diodes are used as can be seen in Fig. 6.3, where its drop voltage is V_D. In the figure, the V_{sensed} is the DC BUS voltage.

Equation (6.1) is the transfer function and calculates the output voltage in accordance with relation $R_3/(R_1 + R_2 + R_3)$, DC BUS voltage, and voltage drop in the diodes. The leakage current I_{Leak} of the A/D converter is considered negligible.

$$V_0 = (V_{sensed} - 2 \cdot V_D)\frac{R_3}{R_1 + R_2 + R_3} \tag{6.1}$$

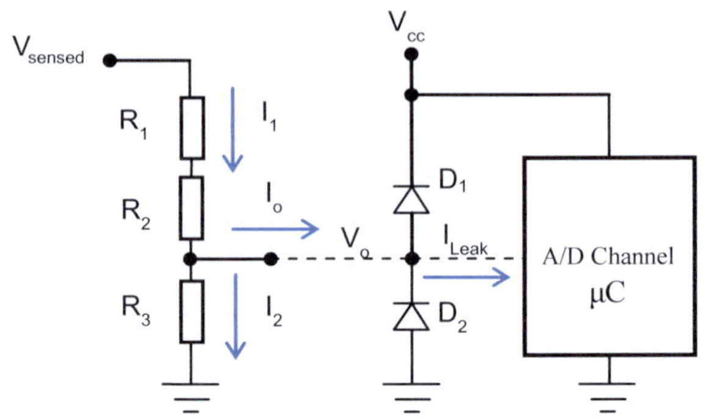

Fig. 6.3 Voltage sensor with scale reduction

After setting $R_1 = R_2$ to the standard value of 332 kΩ and R_3 to 5.11 kΩ, the new voltage range will be 0–432 VDC. Tolerance for this resistance should be 1% in order to have an accurate voltage measurement. Also, the power dissipation of the resistance must be considered because of the high-voltage range. Since the input pin of the microcontroller/DSP has a high impedance current, I_0 will be zero, thus $I_1 = I_2$. Then, the power dissipation in the worst-case scenario (i.e., at a maximum voltage of 432 V) can be calculated as follows:

$$P_{Total} = \frac{V_{sensed}^2}{R_1 + R_2 + R_3} = \frac{(432 - 2 \cdot 1.05)^2}{332K + 332K + 5K11} = 276 \text{ mW} \tag{6.2}$$

Now, the power dissipation for every resistor will be

$$\begin{aligned} I_1 = I_2 &= \frac{V_{sensed}}{R_1 + R_2 + R_3} = \frac{432 - 2 \cdot 1.05}{332K + 332K + 5K11} = 642.49 \ \mu\text{A} \\ P_1 &= R_1 I_1^2 = 137.04 \text{ mW} \\ P_2 &= R_2 I_1^2 = 137.04 \text{ mW} \\ P_3 &= R_3 I_1^2 = 2.1 \ \mu\text{W} \end{aligned} \tag{6.3}$$

Since drop voltage in R_1 and R_2 is high, power dissipation will also be high. Instead of using two resistors, if it is used only one resistor equal to 662 kΩ, power dissipation will be approximately 274 mW. For a standard thick film SMD resistor whose size is 1206, maximum power dissipation is 250 mW. This value is higher than nominal, and it is the reason that two resistors are being used instead of one.

In Table 6.1, the voltage variation and resistance variation are shown in accordance with the tolerance of the components and voltage.

Table 6.1 Maximum, minimum, and nominal values for the components and voltages in accordance with tolerance

Item	Min	Nominal	Max	Units	Tolerance
V_{cc} voltage	3.20	3.30	3.40	V	3%
Nominal VAC	180.00	240.00	265.00	V	25–10%
Nominal VDC	254.56	339.41	374.77	V	25–10%
R_1	328,680	332,000	335,320	Ω	1%
R_2	328,680	332,000	335,320	Ω	1%
R_3	5058.9	5110	5161.1	Ω	1%
Maximum input voltage uC			3.7000	V	–
Diode forward voltage	–	1.050	–	V	–

6.2.2 *Adding a Low-Pass Filter (LPF)*

A variation of the previous circuit is to add a low-pass filter after the voltage divider to eliminate the high-frequency electrical noise in the microcontroller/DSP pin. The resistance and capacitor of the low-pass filter should be placed close to the analog microcontroller/DPS pin. The cut-off frequency is in the order of 100 kHz, but it should be selected experimentally with the final design board since it depends on the PCB layout and input impedance of the A/D converter.

In three-phase machine control applications with PWM control, it could be necessary to measure the machine voltage in accordance with the control strategy. If this is the case, in Fig. 6.4, the electric drive and the voltage sensors are shown with a single low-pass filter for each phase. The electric drive is composed of an AC–DC converter plus an inverter (DC–AC converter). The last converter converts the three-phase voltage needed for the three-phase machine, as will be seen in Chap. 8. The low-pass filter is used to get the first harmonic of sinusoidal PWM modulation. The PWM frequency in the application is between 8 and 20 kHz. Thus, the cut-off

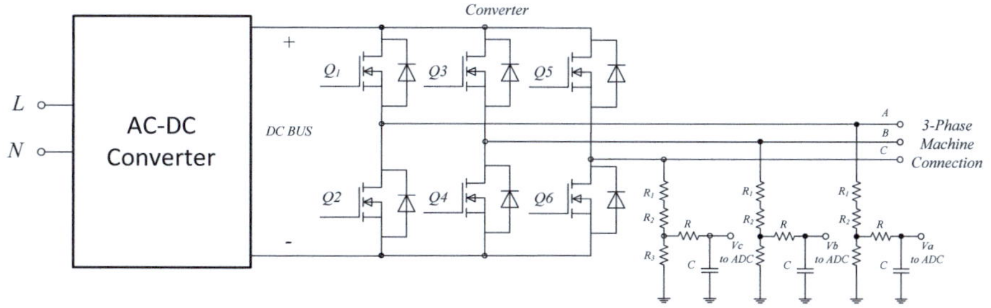

Fig. 6.4 Three-phase AC machine phase voltage monitoring through a voltage divider and low-pass filter scheme with an inverter

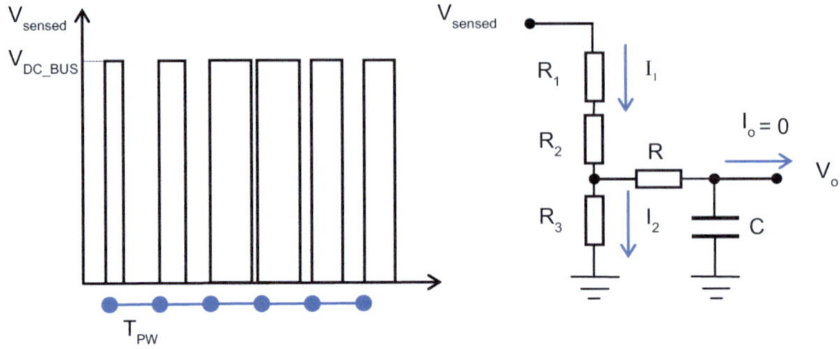

Fig. 6.5 Voltage divide with a low-pass filter

frequency of the low-pass filter should be lower. In this case, the high-frequency electrical noise and higher harmonics, which are above fundamental frequency, should be suppressed.

In Fig. 6.5, the shape of the voltage is shown with regard to the ground, where its amplitude is V_{DC_BUS}. The voltage is switching between DC BUS and zero in a PWM and sinusoidal waveform.

To obtain the line-to-line voltage V_{ab}, it is sufficient to execute the following equation in the microcontroller/DSP: $V_{ab} = V_a - V_b$. Furthermore, the voltage with regard to machine neutral can also be calculated inside the microcontroller/DSP by using the following equations, which were seen in Chap. 3:

$$\begin{aligned} V_{an} &= \frac{2}{3}V_a - \frac{1}{3}V_b - \frac{1}{3}V_c \\ V_{bn} &= \frac{2}{3}V_b - \frac{1}{3}V_c - \frac{1}{3}V_a \\ V_{cn} &= \frac{2}{3}V_c - \frac{1}{3}V_a - \frac{1}{3}V_b \end{aligned} \tag{6.4}$$

The equations that define the low-pass filter are (6.6) and (6.7). They are for the cut-off frequency and the angle delay:

$$f_c = \frac{1}{2\pi \cdot R \cdot C} \tag{6.6}$$

$$\varphi = -\arctan(2\pi \cdot f \cdot R \cdot C) \tag{6.7}$$

Cut-off frequency f_c should be ten times below PWM frequency in order to accept the first harmonic of the PWM voltage signal. For example, if PWM frequency is 16 kHz, ten times less would be 1.6 kHz. After subsequently setting the capacitor value to the standard value of 1 nF, resistance would be 99,471 kΩ. Taking the standard value of 100 kΩ for the resistance value, the new cut-off frequency would be 1591 Hz. For a signal with a frequency of f_c, the amplitude would attenuate 3 dB,

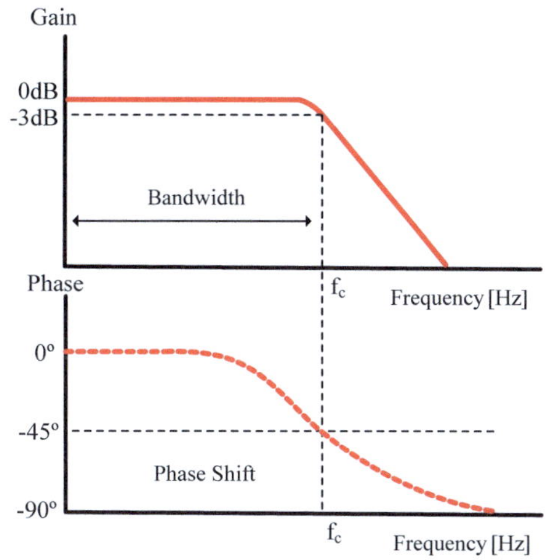

Fig. 6.6 First-order filter response

while phase shift would be approx. −45°. For higher frequencies, the signal would be attenuated by 20 dB/decade, and phase shift would increase to −90°, as shown in Fig. 6.6. For example, a machine with 12 poles that runs at 16,000 RPM has a voltage frequency of 1600 Hz. Thus, the designed filter is not valid since it attenuates 3 dB of the measured voltage. In this case, the cut-off frequency must be selected with a higher value. As mentioned, the cut-off frequency is selected by the PWM frequency and with the maximum speed of the machine.

It should be noted that it is essential to consider the phase shift of the low-pass filter because it adds a delay between the actual signal and the output signal which depends on the first harmonic frequency (6.7). According to Fig. 6.6, phase shift increases with frequency. In some applications, it is essential to compensate this delay since the electrical frequency of the first PWM voltage harmonic can be close to the cut-off frequency. In a high-performance microcontroller, this can be done with software due to the fact that electrical frequency and constant RC are known. As it will be seen in Chaps. 7 and 8, there is another technique for measuring the voltage of the machine without any delay by measuring the PWM duty cycle and reconstructing the machine voltage through software.

Figure 6.7 shows an experimental result of a machine phase voltage (V_{sense}) when DC PWM voltage is applied to the machine. It also shows scaling and filtering signals V_0 in the microcontroller pin. The PWM frequency is 8 kHz, and the max peak voltage is 336 V. The filtered and scaled V_0 signal voltage is 1.379 V of average. The analog-to-digital converter operates between 0 and 3.3 V.

The sinusoidal PWM phase voltage experimental result is shown in Fig. 6.8 when the machine is running at its maximum voltage. When the signal has been scaled and filtered (V_0), phase voltage with respect to the ground waveform reminds us of

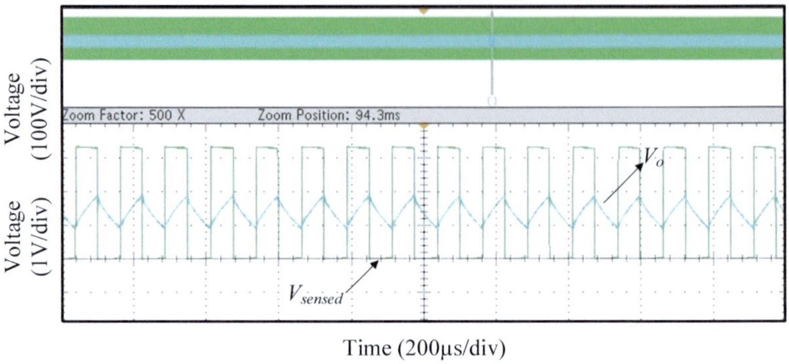

Fig. 6.7 Machine phase voltage and scaling and filtering signals

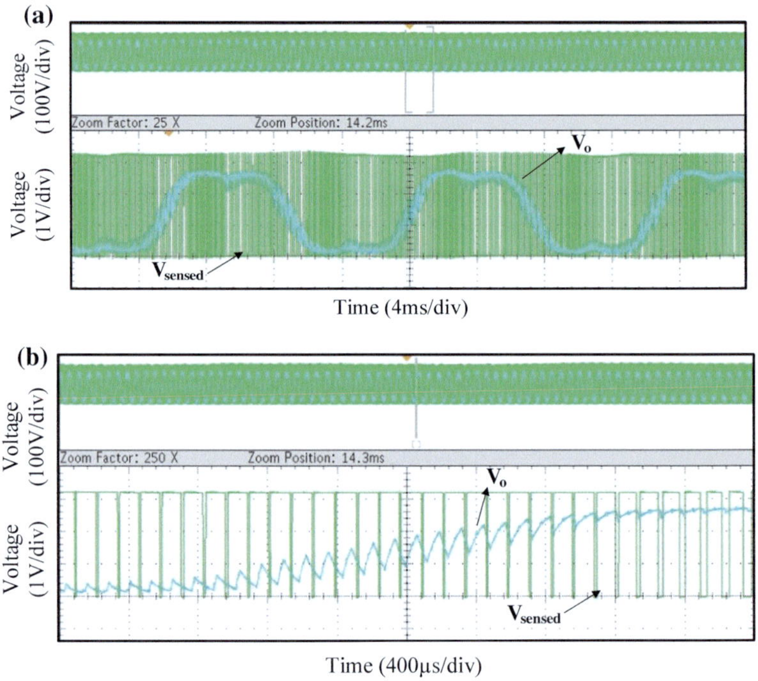

Fig. 6.8 **a** Phase voltage when the machine is running, with space vector PWM at maximum voltage, with voltage scaled and filtered. **b** Zoom to be able to see the sinusoidal space vector PWM and filtered voltage

space vector PWM, which it will be seen in Chap. 9. Phase shift should be evaluated to see if it can affect the regulation of the machine's control loop.

As can be observed, to minimize the ripple effect of the V_0 signal, it is important to synchronize the A/D conversion with the PWM in order to take the sample always in the same instant. Lower filter cut-off frequency allows the voltage ripple reduction, but the phase shift delay will be increased, and the signal can be attenuated at maximum machine speed as discussed early.

6.3 Temperature Measurement

For electronic devices, temperature plays an important role. It can limit the performance of the device, and sometimes, the device can be damaged if the maximum junction temperature is reached. For example, the power semiconductor devices as IGBT and MOSFETs have a safe operation area (SOA) where the device can be expected to operate without self-damage. The operation detection outside the SOA limits can be performed with the temperature measurement. The temperature measurement is used to keep the junction temperature within the SOA under the worst-case condition of maximum power dissipation and maximum ambient temperature.

In electric drives, where there is a couple of power semiconductor devices, the average temperature should be enough if they are placed together. For example, in a three-phase bridge (inverter), there are at least six power devices for which temperature monitoring is relevant, and the temperature sensor can be mounted on a heat sink. If discrete power devices are used, the temperature sensor should be installed close to the devices, creating an average temperature of the six devices. Therefore, the temperature sensor can be used to implement derating algorithms to reduce the power delivered to the load in order to control the temperature.

There are different temperature sensors, such as thermistors and thermocouples. The thermocouples are mostly used in ECU or MCU temperature test validation where a couple of sensors is placed, for example, in different electronic components such a MOSFET, capacitor, inductor, or inside the stator winding to monitor the machine temperature. Basically, the thermocouple consists of two metal wires in contact where a voltage difference is generated in the terminals thanks to the union between the two different metals. This principle is based on the Seebeck effect and needs special equipment to obtain a high accuracy temperature in Celsius degrees. On the other hand, the thermistors are mostly used for microcontroller-based systems such as ECUs and MCUs to monitor internal temperatures for protection purposes as derating algorithms. Figure 6.9 illustrates a power module (PM) composed with six IGBTs with a thermistor mounted close to the devices.

Fig. 6.9 Power module composed of six IGBTs and three thermistors

6.3.1 The Thermistor for Temperature Measurement

A thermistor is a resistive sensor in which resistance varies with temperature. There are two types of thermistors, depending on if the temperature coefficient (TC) is positive (PTC) or negative (NTC). In the first case, when the temperature increases, its resistance also increases: it is dealing with a conductor. On the contrary, for an NTC, if temperature increases, the resistance decreases. In this case, it is dealing with a semiconductor. Both thermistors have a non-linear resistance vs. temperature function.

In some applications, the PTC is used as a current limit device for circuit protection. If the current is large enough to generate more heat than the device can dissipate, the device heats up, causing its resistance to increase. A PTC thermistor has the ability to raise resistance suddenly at a specific critical temperature (switching point) as illustrated in Fig. 6.10. On the other hand, an NTC thermistor can be used as an in-rush current limiter in power supply circuits. In this case, it presents a higher resistance initially, which prevents large currents when the power supply circuit is turned on. Rapid heat up does reduce the resistance to allow higher current flow under normal operation. It is possible to shortcut an NTC thermistor by using a relay or contactor under normal operating conditions in order to switch the current from the NTC thermistor to the relay if a higher current is needed.

On the other hand, typically, NTC or PTC thermistors do not increase or decrease in resistance and temperature in a linear way as commented before. Figure 6.10 shows typical non-linear resistance value as a temperature function for NTC and PTC thermistors.

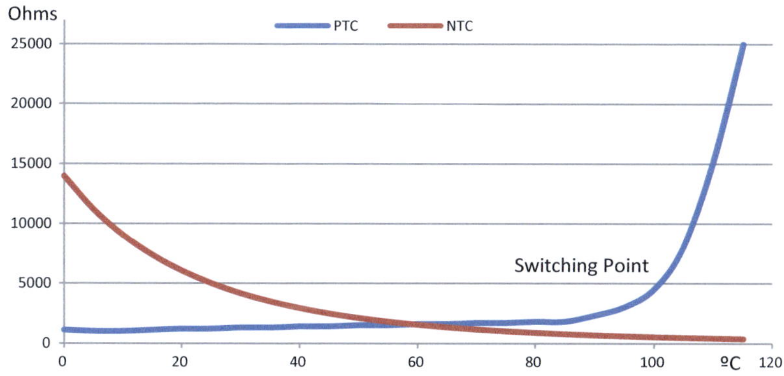

Fig. 6.10 Resistance variation by a temperature increase in an NTC and PTC

6.3.1.1 NTC as a Temperature Measurement

It is more common to use an NTC thermistor as a temperature sensor since no switching point is present. Although this is non-linear resistance variation, it is enough to have a certain accuracy for electric drive applications. If a high-performance microcontroller/DSP is used, it is possible to take a third or fourth-order polynomial function of the NTC curve with relatively low error (see Fig. 6.11). Another common way to achieve high precision results is through a look-up table. For a better optimization of the space used in the microcontroller/DSP, each position of the table should represent the temperature and its content its resistive value. For example, in the table position ten there is the resistance value which corresponds to 10 °C. By reading the NTC drop voltage with an A/D converter and knowing the transfer function, the microcontroller/DSP can calculate the resistance and look in the table at what

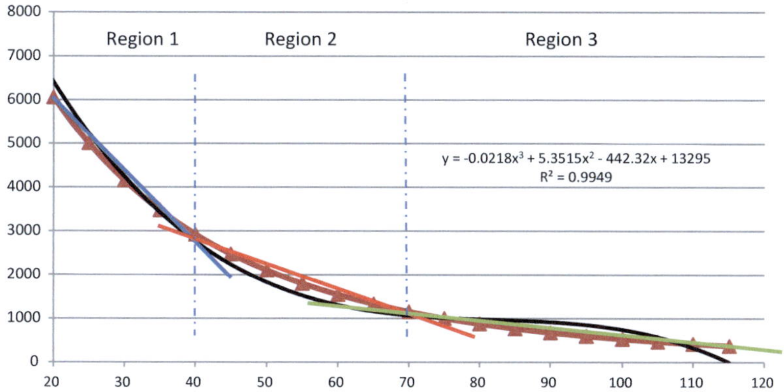

Fig. 6.11 Linear regions and polynomial function approximation for an NTC curve

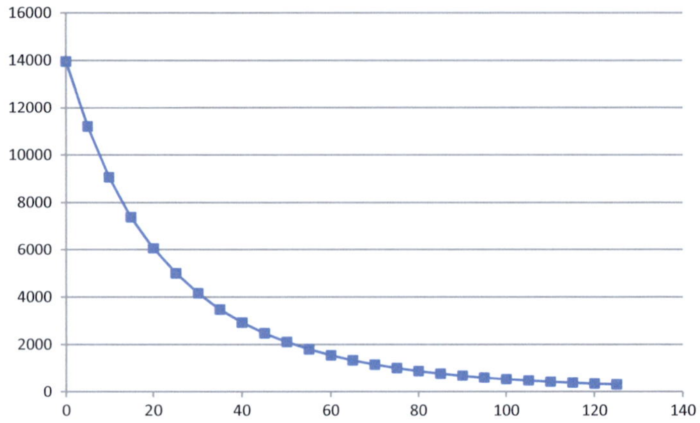

Fig. 6.12 Resistance variation in accordance with temperature increase in linear scale

temperature corresponds (position of the table). Through linear interpolation, intermediate values can be established to those of the table. On the other hand, if the high precision is not a requirement, it is also possible to take up to three different regions where the resistance varies in an approximately linear way—as shown in Fig. 6.11. Every region has a straight line, defined by the equation $y = m \cdot x + n$.

Fig. 6.12 shows resistance variation in accordance with temperature, where the resistance value at 25 °C is 5 kΩ.

There is a constant which depends on the NTC thermistor material. It is a measure of the material's resistance at one temperature compared with resistance at a different temperature. This is an exponential approximation between the two temperatures.

It is possible to consider two methods to determine the look-up table. By means of Beta equation of Eq. (6.8), or by means of the Steinhart–Hart equation of Eq. (6.9), where, T is the temperature in Kelvin, R is the resistance at temperature T in Ω, and A, B, and C are the Steinhart–Hart coefficients which vary depending on the type and model of thermistor and the temperature range of interest. Here, it is considered only the Beta equation to perform the look-up table.

Equation (6.8) expresses the constant, and its value may be expressed in degrees Kelvin (°K).

$$\beta = \left(\frac{T_1 \cdot T_2}{T_2 - T_1} \right) \ln \frac{R_1}{R_2} \tag{6.8}$$

In Fig. 6.12, material constant β is 3375 K for a range of between 25 and 50 °C. For a range of between 25 and 100 °C, that value is 3433 K.

$$\frac{1}{T} = A + B \ln(R) + C[\ln(R)]^3 \tag{6.9}$$

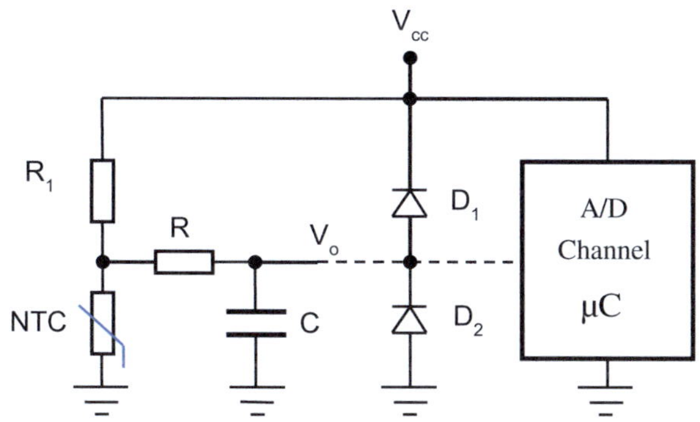

Fig. 6.13 Circuit diagram to connect an NTC to measure the temperature with a low-pass filter

The resistance value as a function of temperature can be expressed as follows:

$$R(T) = R_{25} \cdot e^{\beta \cdot \left(\frac{1}{T} - \frac{1}{298.15}\right)} \tag{6.10}$$

In Fig. 6.13, a circuit is shown in which V_0 varies according to the resistance variation of the NTC thermistor. The NTC values considered are in accordance with Fig. 6.12, that is to say, 5 kΩ at 25 °C, with β equal to 3433 K. It is also recommendable to use a low-pass filter as shown below. Cut-off frequency f_c can be set at a very low frequency since filter delay is not relevant. The cut-off frequency of 200 Hz is a good choice, for example.

As voltage measurement, V_0 is connected to a microcontroller/DSP analog input. The A/D in this example has a 12-bit resolution, and the maximum voltage is 3.3 V. With these parameters, the pull-up resistance R_1—which forms a voltage divider in accordance with the range 0 to 3.3 V—is 4.7 kΩ. As before, the leakage current of the A/D converter is negligible. Figure 6.14a shows the voltage and A/D counts for every temperature/resistance. According to above explanation, the look-up table to be stored in the microcontroller/DSP memory corresponds to the column R[Ω]. By using the linear interpolation is possible to get intermediates values, as discussed before. For higher precision, instead of temperature increments of 5 °C is possible to use 1 °C increments but with the disadvantages of the increment of five times the memory space needed in the microcontroller/DSP.

As commented, instead of using a look-up table, it is possible to use the polynomial characteristic of the NTC curve, as illustrated in Fig. 6.14b. The memory needed is lower compared with the look-up table. However, the microcontroller/DSP has to process some complex calculations as squares and cubic calculations.

(a)

Temperature [°C]	R [Ω]	V_o [V]	A/D Value
0	14359	2.49	3086
5	11453	2.34	2904
10	9208	2.18	2712
15	7459	2.02	2513
20	6086	1.86	2311
25	5000	1.70	2111
30	4134	1.54	1917
35	3440	1.39	1731
40	2879	1.25	1556
45	2423	1.12	1393
50	2050	1.00	1244
55	1743	0.89	1108
60	1490	0.79	986
65	1279	0.71	876
70	1103	0.63	779
75	955	0.56	692
80	831	0.50	615
85	725	0.44	548
90	635	0.39	488
95	559	0.35	435
100	493	0.31	389
105	437	0.28	348
110	388	0.25	312
115	346	0.23	280
120	309	0.20	252
125	277	0.18	228

(b)

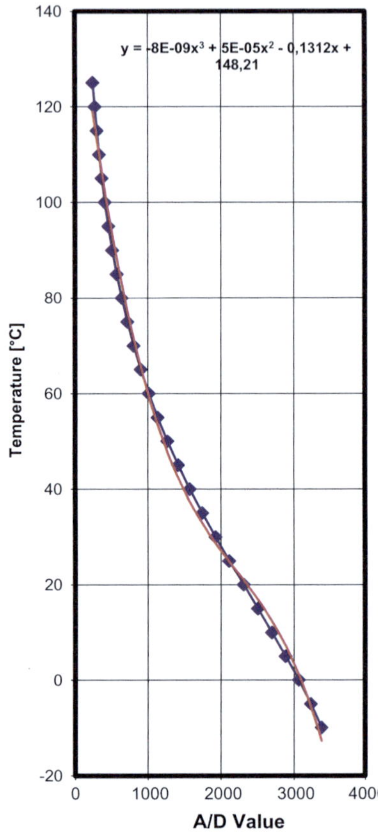

Fig. 6.14 **a** Voltage and A/D counts for every temperature/resistance. The look-up table to be stored in the microcontroller is the column of the resistance value. **b** Representation of the look-up table with polynomial approximation

6.4 Current Measurement

Current measurement is an essential task since it is a function of the machine's electromagnetic energy conversion (torque, for example), and it protects the electric drive itself and the machine when overload arises. In high-performance machine controls, an accurate and instantaneous measurement of the current is necessary in order to control the torque efficiently and to get a wide bandwidth. In this case, the current regulation loop is the inner loop of the high-performance control, which should be executed faster than the outer control loop. The bandwidth of the other outer control loop (speed or position, for example) depends on the bandwidth of the current regulation loop, and it is executed slower than the inner loop as discussed in Chap. 2.

6.4.1 Non-isolated Current Measurement

6.4.1.1 Shunt Resistor

By measuring the voltage drop in a known resistance, it is possible to determine the current through the application of Ohm's law. Resistance is in series to the circuit where the current is measured. The resistance of the resistor must be minimized in order to reduce the power losses, but on the contrary, the signal-to-noise ratio is reduced. Furthermore, measurement is not galvanically isolated, as the power converter is connected electrically to the control system. As a result, there exists common-mode noise and an unsafe system. Because of its simplicity, the most popular method has been to measure the current for home appliances and also for the low-cost variable speed drive system.

In Fig. 6.15, current measurement is shown using a shunt resistor where low voltage produced by low resistance must be amplified and filtered before connecting to the analog microcontroller/DSP pin, as illustrated in Fig. 6.16.

If positive and negative current must be measured (positive or negative V_R voltage)—for example, for sinusoidal currents in a machine—signal conditioning must apply an offset voltage to the output due to the fact that the voltage range for the A/D converter is only positive, and it is between 0 and V_{cc}.

The voltage in the microcontroller pin V_0 can be calculated as a function of the A/D converter counts AD_{counts}, number of the bits n, the supply voltage V_{cc}, the offset voltage V_{offset}, the gain G of the signal conditioning, and the shunt resistance R as

$$V_0 = AD_{counts} \cdot \frac{V_{cc}}{2^{n-1}} \cdot G \cdot R \cdot I_R + V_{offset} \tag{6.11}$$

Selection of the value for resistance depends on the maximum current through the resistor and the maximum allowed power dissipation. For very low resistance

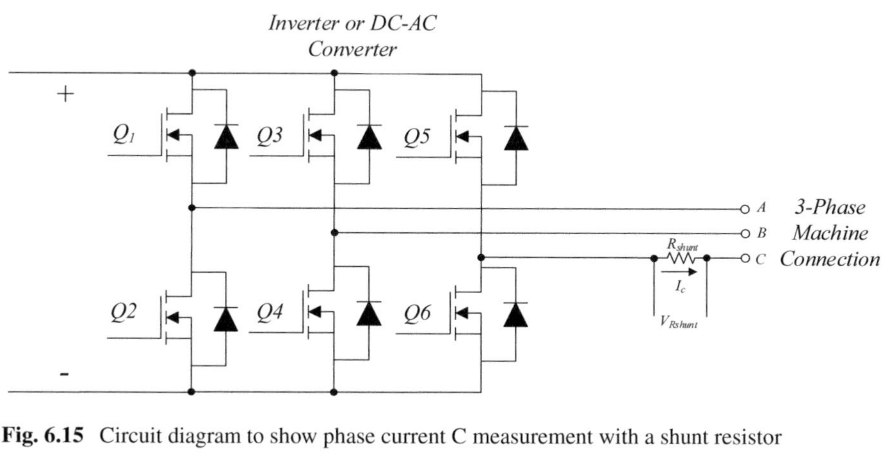

Fig. 6.15 Circuit diagram to show phase current C measurement with a shunt resistor

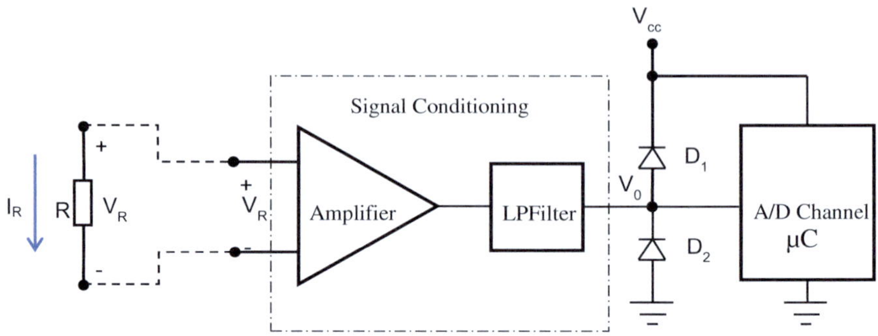

Fig. 6.16 Measurement of current using a shunt resistor with signal conditioning, diode protection, and microcontroller

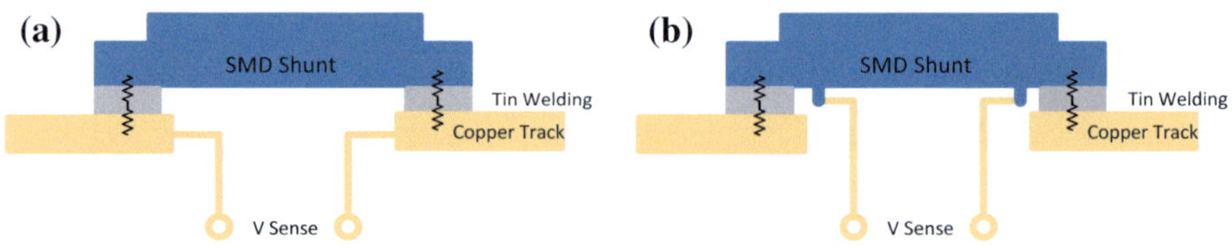

Fig. 6.17 SMD shunt resistor welded with tin on a PCB. **a** Two-terminal shunt resistor. **b** Four-terminal shunt resistor

shunt, e.g., less than 500 μΩ, tin welding, and copper tracks must be considered in the total resistance value for proper current measurement as illustrated in Fig. 6.17a. For example, experimental results for 200 μΩ shunt, the total resistance measured is 282 μΩ.

Moreover, there are four-terminal shunt resistors where the sense terminals are directly connected to the shunt avoiding the tin welding and its resistance as depicted in Fig. 6.17b. The four-terminal shunt resistors have a higher cost, and the parasitic inductance gets larger due to its construction, which is not recommended for sensing current with high d*i*/d*t*. However, it enables to detect current more precisely, with a smaller amount of error.

6.4.2 *Isolated Current Measurement*

6.4.2.1 Using a Current Transformer (CT)

One improvement of the aforementioned current measurement is to use a current transformer, which has the advantage that it is galvanically isolated with regard to the high voltage of the power converter. When AC current flows through the primary of the transformer, a flux variation is created in the core, which is induced in the

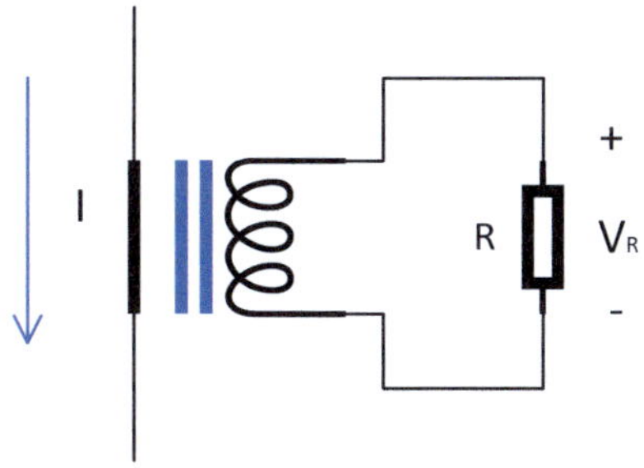

Fig. 6.18 Measurement of current using a current transformer

secondary winding. The current in the secondary can be measured as mentioned previously—with a resistor and with current being proportional to the current in the primary winding as illustrated in Fig. 6.18. One of the problems of the current transformers used is that they can only measure AC current because they operate with flux variation over time. Also, they are more expensive than a shunt resistor, and applications must be close to the power converters, with the necessity of having galvanic isolation. It is possible to use the same signal conditioning that was seen previously.

6.4.2.2 Current Measurement Using a Hall Effect Sensor

A current in a conductor generates a magnetic field. The magnitude of said field is proportional to the current, and no time delay is present between the current and the magnetic field. A Hall effect sensor varies its output voltage in response to the magnetic field. When a current is flowing through a semiconductor and a magnetic field is applied perpendicularly to the direction of the current flow, a voltage difference arises between the two sides of the semiconductor. Edwin Hall discovered the Hall effect in 1879.

The magnitude of the voltage is proportional to the magnitude of the flux under constant current. A Hall sensor measures the magnetic flux passing through the sensor, and the current is proportional to this flux. Thus, it is first necessary to generate flux by using a Ferrite ring core. Said flux would be proportional to the current, and the flux is measured by the Hall sensor. In Fig. 6.19, how current generates a magnetic flux in a Ferrite ring core and Hall effect sensor is shown.

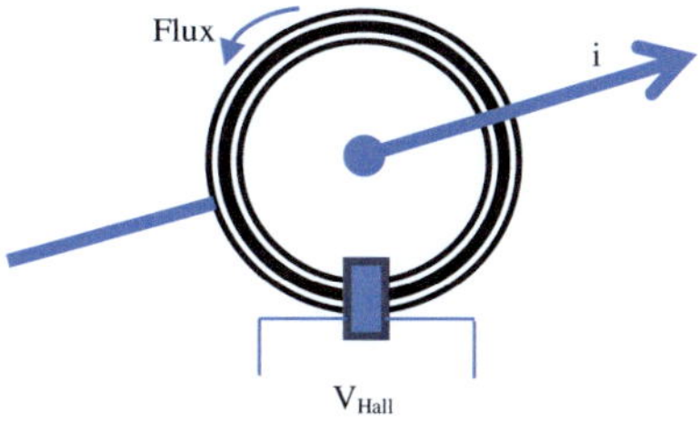

Fig. 6.19 Magnetic flux created in a Ferrite ring core by a conductor when current is flowing

One of the main benefits of Hall effect sensors is that they can be isolated with regard to the control system. They can measure DC or AC current and are widely used in current probes since ring cores can be opened and closed in order to insert the conductor and measure the current. Despite the temperature dependency and non-linearity of a simple Hall effect sensor, it is possible to improve the measurement reasonably—as well as the bandwidth—by using a closed-loop technique. For example, it is possible to find current sensors based on the Hall effect, with a bandwidth of 250 kHz, with $a \pm 1.5\%$ total output error at $T_a = 25$ °C. Furthermore, it usually used in the traction inverters for EV/HEV thanks to the high accuracy, the isolation level, and to the planar SOIC-8 package, where the current can be measured directly above a conductor. The robustness can be improved if U-shaped ferromagnetic shield is wrapped around the current conductor to protect the sensor from external fields as illustrated in Fig. 6.20. The ferromagnetic shield guides the magnetic field lines to generate a homogeneous magnetic field in the U-shape.

In Fig. 6.21, a typical linear curve which relates the output voltage versus the peak current is shown. There is more voltage resolution for the Hall solution when

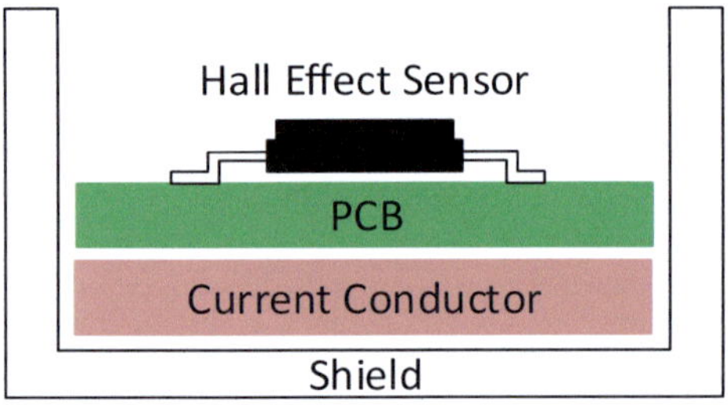

Fig. 6.20 U-shaped ferromagnetic shield wrapped around the current conductor to protect the sensor from external fields

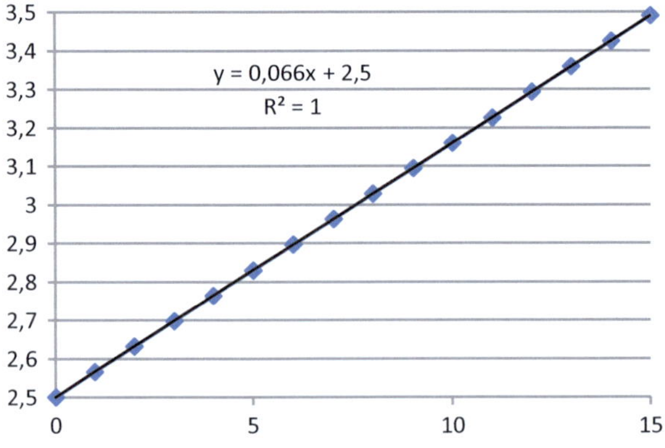

Fig. 6.21 Output voltage versus peak current in the Hall effect sensor

compared with the shunt solution. As can be observed, the sensitivity is 66 mV/A, with an offset of 2.5 V to measure negative currents.

As in the case of shunt and CT solutions, the performance of the Hall current sensing depends on the PCB design and layout, where the high voltage switching in electric machine control, increase the parasitic coupling capacitance generated by dv/dt transients.

6.5 Speed Measurement

6.5.1 *Tachometer Sensor*

Instantaneous rotor speed can be measured using a tachometer which is installed in the machine shaft. This is essential in a speed regulator when speed is not estimated. Rotor speed can also be measured using a rotor position sensor, such as an encoder in which the rotor speed will be the first derivative of the angle with regard to the time. As it will be seen in the next section, encoders are more expensive than tachometers. In some low-cost applications where high-resolution speed is not relevant, they are not preferable.

A tachometer works like a generator in which the output voltage and frequency are proportional to speed. A higher number of poles allow higher accuracy measurement speed. For instance, eight pair poles tachometer allows having a pulse every 45° of rotor movement. The output voltage of a tachometer is an analog signal that is sensitive to noise. Also, since permanent magnet characteristics vary depending upon temperature, voltage magnitude in the tachometer also varies. It is possible to use frequency as a speed measurement in order to avoid the variation of voltage with temperature and noise. Thus, the speed will be measured accurately with no temperature dependent.

A tachometer is still used in many simple speed regulation systems because it provides a direct measurement without the need for complex calculations.

In Fig. 6.22, the sinusoidal tachometer signal waveform and the output pulses of the acquisition circuit are shown for an electrical machine running at 750 RPM. The frequency of the signals is 100 Hz (eight pair poles), with 10 ms between rising edges.

The voltage/frequency law measured by a tachometer with eight pairs poles for high and low speeds is shown in Fig. 6.23. For example, for a mechanical rotor speed of 15,000 RPM, the tachometer shows a sinusoidal voltage of 48 V RMS at an electrical frequency of 2000 Hz. Equation (6.12) shows the relation between the mechanical and electrical frequency of the tachometer.

$$\begin{aligned} &Mechanical\_Freq = \frac{15{,}000\ rev}{1\ \text{min}} \cdot \frac{1\ \text{min}}{60\ \text{s}} = 250\ \text{rev/s} \rightarrow 250\ \text{Hz} \\ &Electrical\_Freq = 250\ \text{Hz} * 8\ pair\_poles = 2000\ \text{Hz} \end{aligned} \tag{6.12}$$

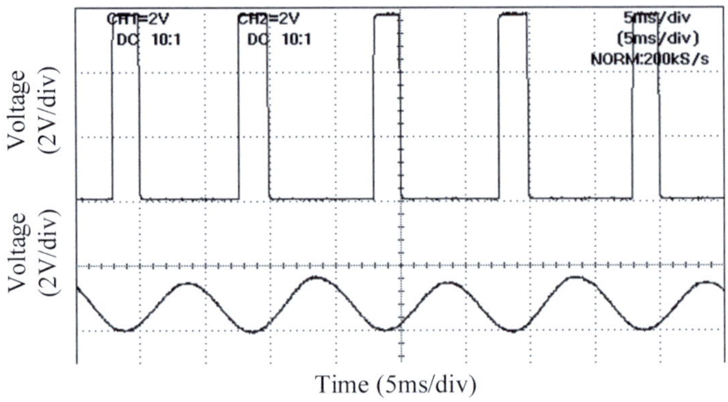

Fig. 6.22 Sinusoidal tachometer signal waveform and output of the acquisition circuit to obtain a square signal

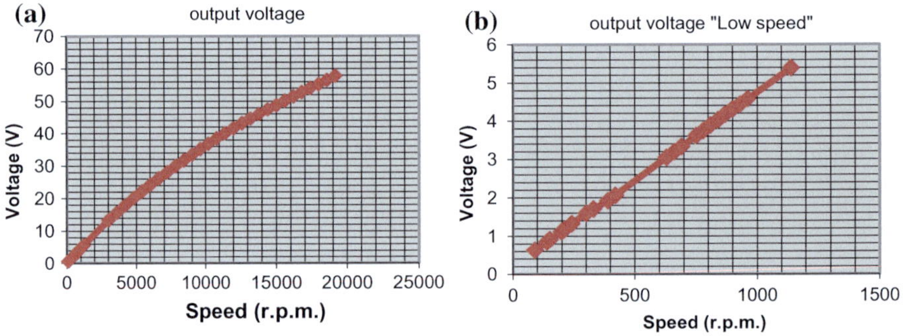

Fig. 6.23 Tachometer sensor voltage with regard to RPM speed for high and low speeds

The output of the tachometer acquisition circuit should be connected to a digital microcontroller/DSP input where an internal timer can be used in capture mode to measure the frequency. For example, the capture timer event can be configured to detect every rising edge to perform the speed calculation.

6.5.2 Speed/Position Measurement

6.5.2.1 Resolver

In specific applications, like an electric vehicle where demands are very rugged and reliable motor shaft position sensing is necessary, a resolver sensor is a right choice. This is like a rotating transformer in which the reluctance varies with the position. It is an expensive solution that the applications should merit using such as certain industrial, automotive, and military applications. By using resolvers, absolute shaft angle measurements can be obtained. The rotor of the resolver is attached to the rotor of the machine that requires position sensing.

The resolver rotor is a coil which is driven at a high carrier frequency, usually somewhere between 8 and 16 kHz. The stator of the resolver has two coils, which are wound in an orthogonal relationship to each other, as shown in Fig. 6.24a. A graphical representation of the excitation signal and the sine and cosine output signals is shown

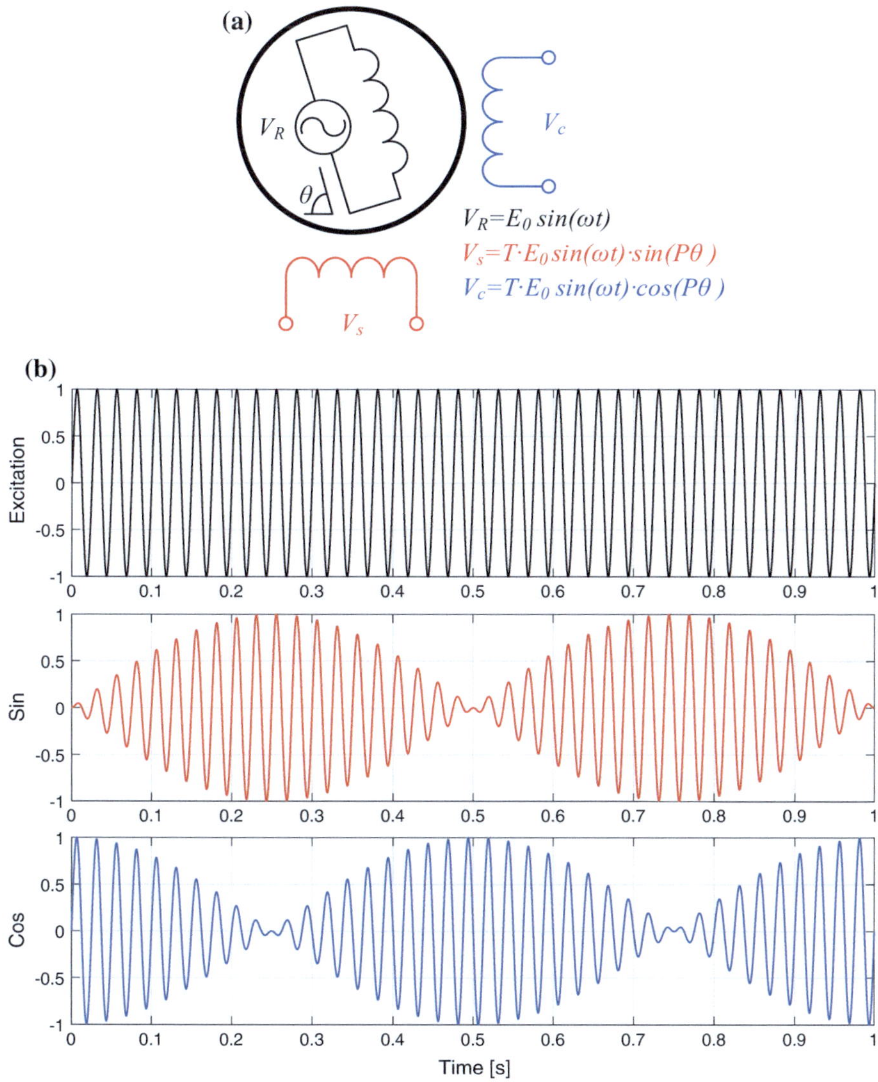

Fig. 6.24 **a** Two-pole resolver. **b** Excitation voltage, output voltages V_s, and V_c in accordance with the rotor position

in Fig. 6.24b. The sine signal has a maximum amplitude at 90° (at instant time $t =$ 0.25 s) and 270° (at instant time $t = 0.75$ s), and the cosine signal has a maximum amplitude at 0° (at instant time $t = 0$ s) and 180° (at instant time $t = 0.5$ s).

The equations that describe the resolver are shown below. As explained before, excitation is in the rotor via a sinusoidal signal with an amplitude E_0 and frequency ωt. The output is a modulated signal with the same frequency, in which amplitude depends on the turn ratio T. P is the number of pair poles the resolver has.

$$\begin{aligned} V_{R_1-R_2} &= E_0 \sin \omega t \\ V_{S_1-S_2} &= T \cdot E_0 \sin(\omega t) \cdot \cos(P\theta) \\ V_{S_2-S_4} &= T \cdot E_0 \sin(\omega t) \cdot \sin(P\theta) \end{aligned} \tag{6.13}$$

The carrier signal can be generated by a free timer output channel on the microcontroller/DSP using a PWM. The PWM follows a sinusoidal shape, and the first sinusoidal harmonic is taken through the use of a low-pass filter. The sinusoidal signal waveform can be amplified to be connected to the excitation coil of the resolver. On the other hand, by using an A/D converter with two simultaneously sampled inputs, the rotor shaft angle can be reconstructed by using a software angle demodulator. If the A/D converter does not have simultaneous sampling capability, there will be an angle error that is proportional to the conversion time of the A/D converter. However, the optimum solution is to use the two Delta-Sigma modulators, one for each channel, due to the very high sampling time, as it will be discussed in Sect. 7.3.4. Then, the Delta-Sigma modulation outputs can be averaged and filtered to get an acceptable resolution.

The resolver has many merits, like its robustness when faced with mechanical stress. However, the resolution of the position is poor compared with the optical encoder, which will be shown later. Thus, for high-performance speed regulations systems, the resolver's applications are limited.

6.5.2.2 Encoder Position Sensor

In advanced position control in electric drives, it is essential to use position detectors/sensor for proper control; for example, in permanent magnet synchronous motors (PMSM) for robotics arms.

As commented in previous sections, an encoder is a position sensor which is possible to obtain the speed and position of the rotor of the machine. It is made with a disc which contains opaque and transparent segments. The disc is attached to the rotor, and when the disc rotates with the rotor, there are devices like a phototransistor and a photodiode which detect the pulse light formed by the segments. There are two types of encoders: the incremental type, which contains uniform segments spaced equally in a radial way in order to produce pulses that can be counted. Pulse corresponds to the relative position of the rotor. The second type of encoder is the absolute type, in which pulses correspond to the absolute position of the rotor, providing rotor position in the form of digital bits.

Incremental encoders are getting more popular for the machine control, including synchronous machines since they are less expensive and signal processing with them is easier. Two optical sensors are usually used: *A* and *B*, yielding train pulses that are shifted 90°. *B* is 90° out of the phase of *A*, and it is used to determine the direction of rotation of the rotor. Usually, there is another output called *Z*, which is an index pulse. This *Z* pulse occurs once per revolution and provides a way to calculate the absolute position via *Z* pulse detection.

There are some microcontrollers dedicated for machine control applications with dedicated hardware designed explicitly for processing incremental encoder pulses. It has individual timers which can operate in phase counting mode, operating as an up/down counter. The phase difference between two external input clocks is detected, and the counter is incremented or decremented accordingly. Thus, position and speed can be measured by reading the counts of the internal microcontroller counter. Also, the direction of the machine is detected since the counter can increase or decrease. Figure 6.25 shows signal conditioning with low-pass filter and diode protection for proper connection between an incremental encoder and microcontroller. If the maximum speed for the rotor is 20,000 RPM and the incremental encoder has 1000 pulses per revolution, the maximum frequency for the clock signal will be 333 kHz. Cut-off frequencies for low-pass filter must be chosen above this frequency in the worst-case scenario. The filter will eliminate the noise produced mainly through PWM switching. With *R* equal to 2.2 kΩ, and *C* equal to 100 pF, the cut-off frequency is close to 723 kHz—more than two times the maximum frequency. Pull-up resistance R_1 is also 2.2 kΩ for a V_{cc} of 5 V, but this must be chosen accordingly with the datasheet of the encoder. Decoupling capacitor C_1 can be 100 nF and decouple the noise from one part of the circuit to another. It is essential to put the capacitor close to the encoder for the best decoupling results.

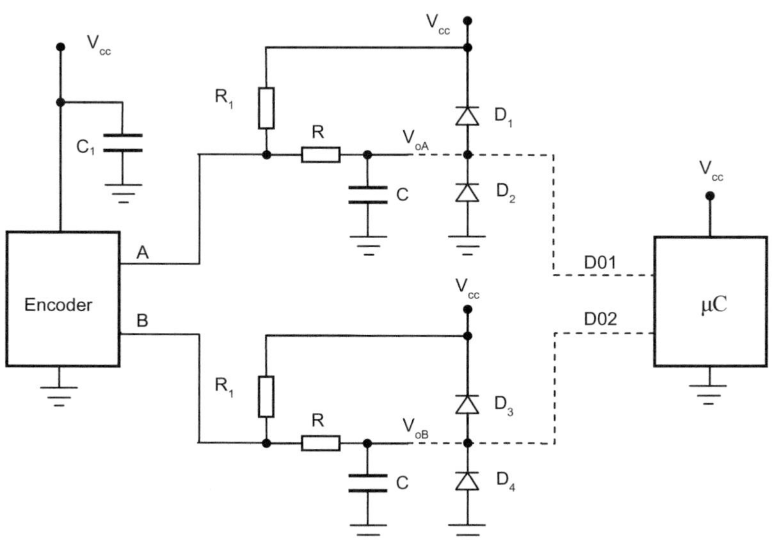

Fig. 6.25 Signal conditioning and filter circuit to connect an incremental encoder to a microcontroller

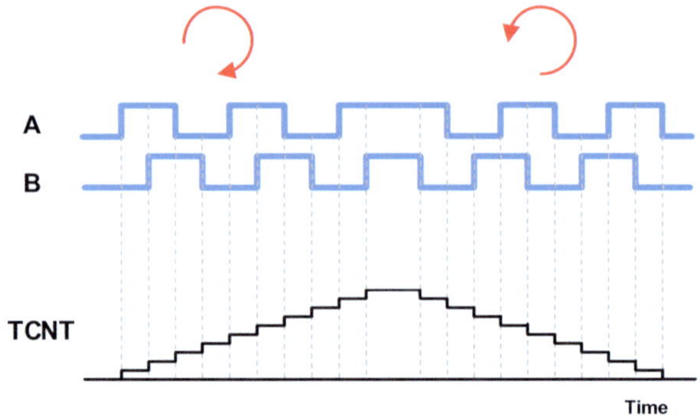

Fig. 6.26 Incremental encoder signal and internal microcontroller counter

Encoder signals can be observed in Fig. 6.26 for channels A and B for both rotation directions. The internal counter of the microcontroller is also shown, which increments or decrements depending on the rotation direction.

The microcontroller is configured to count every rising and falling edge for both channels to increment the resolution four times. For 1000 incremental pulses, the encoder has up to 4000 pulses per revolution. Angle resolution in degrees will be 360/4000; that is to say, 0.09° per pulse. Each change in the internal counter is 0.09° apart.

The delta angle is

$$\Delta\theta = N(m+1) - N(m) \tag{6.14}$$

It is mandatory to use a rapid loop, such as a sample time loop, to read if the counter has been changed to update the position (e.g., inner current loop).

Speed is computed by dividing the delta angle by the delta time ΔT:

$$\text{Speed} = \frac{N(m+1) - N(m)}{\Delta T} \tag{6.15}$$

where delta time ΔT is the time between pulses.

In motor control applications, it is enough to take at least one new speed every speed loop (outer loop). That is to say, for every speed loop, speed can be computed with the aforementioned equation, which is the derivative of the angle position.

Fig. 6.27 shows a routine flowchart for encoder acquisition and management in a closed-loop machine control application. The routine should be called every sample time (current loop), and *TCNT* is a microcontroller counter which is loaded in a variable called *EncValue*. As commented before, the microcontroller sets a bit to indicate the direction of the machine, depending on whether the counter increases or decreases. In this case, there are 4000 pulses per revolution, and the microcontroller is configured to reset the counter automatically when it reaches a value of 4000. The

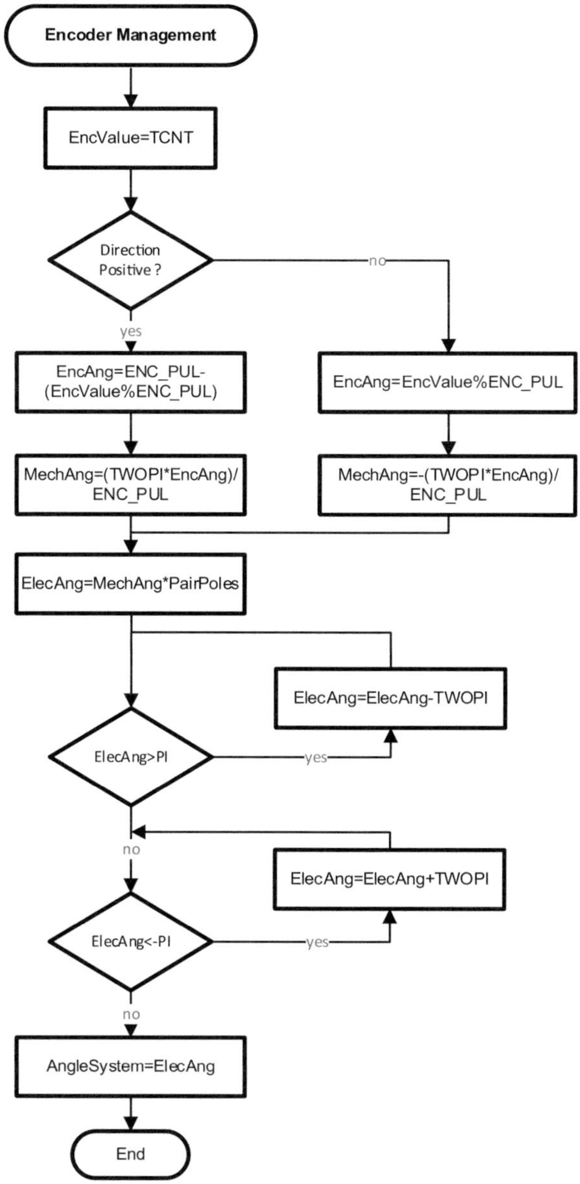

Fig. 6.27 Flowchart for encoder acquisition and management

ENC_PUL constant is 4000, and using module division, a proper value is calculated and loaded into the *EncAng* variable. This variable varies from 0 to 4000. Then, it is scaled to two times the number *pi* in the *MechAng* variable, which is the mechanical angle. Finally, the electrical angle is calculated since this is needed for loop control.

Chapter 7
Microcontroller Peripherals for Electric Drives

7.1 General Timer Module (GTM)

The Bosch general timer module (GTM) development started in 2009 and was derived from the Bosch microcontroller strategy, benefiting the OEMs, Tier1, and semiconductor suppliers. The first start of production (SOP) with GTM 1.5 was in 2014 with Infineon, ST, and NXP semiconductors manufacturers. The motivation was to cover the common requirements in the automotive market, such as:

- Multiple capture/compare of external signals and combination with time stamps
- Generation of complex PWM output signal waveforms
- Provide a common time base for the system
- Minimal CPU interaction
- Minimal interrupt requests to reduce CPU load
- Complex angle clock mechanism for powertrain with combustion engines
- Brushless DC machine (BLDC) support.

The GTM is not only a timer but also has a programmable logic with RISC instructions and an integrated RAM that allow direct interaction between all channels in order to reduce the CPU load of the microcontroller. In other words, between the digital inputs and the digital outputs, there is the possibility of adding lines of code to perform a task without the CPU having intervened. For example, in a BLDC machine control with Hall effect sensors, the GTM itself can be responsible for obtaining the rotor position by capturing the signals of the Hall effect sensors and processing the information to update the corresponding PWM outputs without the CPU does not have to interact. Only in the case that the PWM duty cycle had to be updated, that is, to vary the speed, the CPU would have to interact. GTM has a FIFO memory in which it is possible to fill different duty cycles through the CPU at a slow frequency and the GTM could pick up the FIFO memory data at a faster frequency to change the machine speed for example.

© Springer Nature Switzerland AG 2020

R. Molina Llorente, *Practical Control of Electric Machines*,
Advances in Industrial Control, https://doi.org/10.1007/978-3-030-34758-1_7

One of the main advantages of including the GTM IP (Intellectual Property) peripheral in the different microcontroller manufacturers is that the software development for its configuration is precisely the same, making it portable to different microcontroller suppliers. For example, for the generation of complementary PWM for a three-phase inverter, in the case of using the AURIX™ Tricore family from Infineon, it is possible to choose for either the CCU6 timer or the GTM peripherals. For the first case, the configuration and programming efforts will be exclusive for said microcontroller while, for the second case, it will be portable and exactly the same for another microcontroller with the GTM peripheral. The second option provides higher flexibility, even if initially it implies a higher development time due to its complexity.

The GTM has programmable FIFOs, multi-channel sequencer (MCS) memory, DMA integration, and hardware ready to capture complex time signals and also generate complex PWM outputs. The block diagram of the GTM is represented in Fig. 7.1. As can be observed, there are different sub-modules that are interconnected forming the GTM. At the center of the block diagram is the advance routing unit (ARU). It is a central routing unit, which manages all internal data movement between sub-modules. These sub-modules can be combined to form a complex timer module using the programmable multi-channel sequencer (MCS). In the ARU, it is possible to connect the sub-modules in a flexible way. The connectivity is programmable by software and can be configured during runtime.

The most relevant sub-modules for machine control are basically the deadtime module (DTM), the timer output module (TOM), ARU connected TOM (ATOM), the timer input module (TIM), the sensor pattern evaluation (SPE), and the advance routing unit (ARU). The SPE sub-module is a dedicated sub-module and is designed

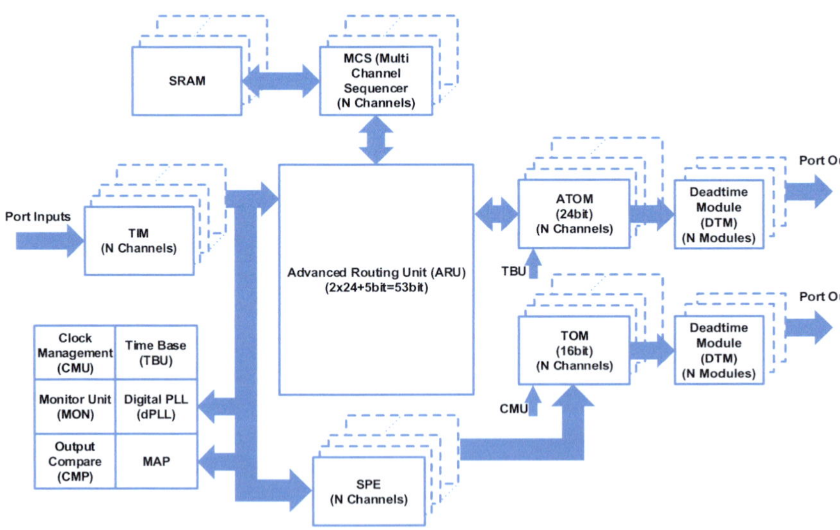

Fig. 7.1 GTM simplified architecture

only for reading Hall effect sensors for the BLDC machines control. The ARU is the heart of the GTM, where access to all sub-modules is possible as commented early. The TIM sub-module can be used in machine controls with position or speed sensor, either employing an encoder or a tachometer while the TOM/ATOM can be used to generate the PWM outputs patterns. For the previous example with the BLDC machines with Hall effect sensors, the PWM outputs can be generated with TOM/ATOM sub-modules, which are set by the position detection given by the SPE sub-module through the ARU without almost CPU intervention.

However, in machine controls without digital inputs to determine the position or the speed, the strictly necessary sub-module will be the TOM/ATOM. The DTM sub-module is only responsible for generating the complementary signals, but they can also be implemented with the TOM/ATOM sub-modules with a little more CPU calculation. The ARU sub-module will then be redundant for this machine control because there will be no routing between sub-modules. It is important to comment that, for this case, the rest of the GTM sub-modules can be used for other applications since they are independent sub-modules governed by the CPU. Without going further, the GTM allows to govern another three-phase BLDC machine without any restrictions and depending on the number of TIM/TOM channels, up to four three-phase machines can be governed.

7.1.1 GTM Sub-modules

7.1.1.1 Advanced Routing Unit (ARU)

As discussed, the ARU is responsible for transferring data from a source sub-module to destination sub-modules. The destination sub-module controls the data flow configuration within the ARU. That is, the destination sub-module defines where the data come from; the ARU only enables that particular data flow, since sequentially polls the data destinations of the connected sub-modules in a round-robin order. If a data destination requests new data from its configured data source and the data source has data available, the ARU delivers the data to the destination, and it informs both the data source and destination that the data is transferred. The data source marks the delivered ARU data as invalid, which means that the destination consumed the data. In other words, ARU access is a destructive read. This also means that if the source does not supply new data, the destination will not receive any additional information. The routing data width is 53 bits, where two times 24 bit is the data and five bits can be used as ARU control bits (ACB).

7.1.1.2 Timer Input Module (TIM)

The timer input module (TIM) is responsible for filtering and capturing the digital input signals of the GTM. Various characteristics of the input signals can be measured,

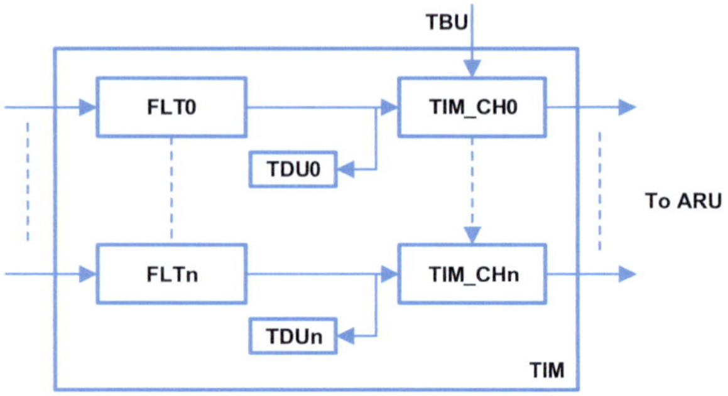

Fig. 7.2 *n*-channels TIM simplified block diagram

such as the frequency and duty cycle for fixed period signals, but also signals with varying periods and duty cycles. Also, it allows counting the rising or falling edges of a signal with the duration of each of them. TIM has a configurable input filter for each input that allows rejecting the signals with noise, and only when the signal is valid, is routed to the measurement unit. Also, it has shadow registers to store the measured data while a new signal is being processed. The result of the measurements can be read through the CPU and/or can be routed through the ARU to other sub-modules. The block diagram of the simplified TIM sub-module is shown in Fig. 7.2. In this case, up to *n* channels can be observed, each of which has a filtering unit (FLTn), a timeout detector (TDUn), and the TIM_CHn capture/measurement unit.

Transient pulses commonly called "glitches" are very rapid level changes that can generate false readings, for example, in an encoder position sensor. The FLTn filtering unit is capable of monitoring the input signal through internal counters to ensure that the validation of said signal is performed when there has been a change of edge or level. The FLTn filtering unit has several filtering modes, although here it only cites the individual deglitch time mode. In this case, the frequency of the counter should be higher than the frequency of level changes of the input signal. That is, transient pulses with duration less than the frequency of the selected CMU clock will not be detected. The individual deglitch time mode performs transient pulse detection for proper validation in a pre-configured time window by reaching a threshold, in this case, filter rise edge (FLT_RE) with the FLT_CNT counter as depicted in Fig. 7.3. The input signal to the TIM sub-module is the signal F_IN_SYNC. This has a series of glitches

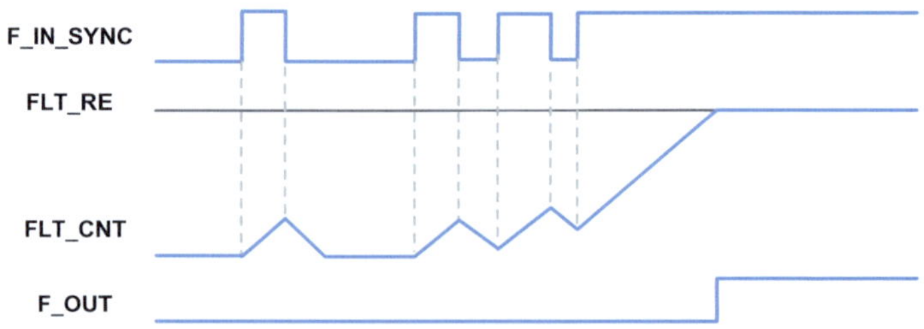

Fig. 7.3 Individual deglitch time mode (up/down counter) for rising edge detection

which are detected, and until it reaches the threshold, it is not validated, resulting in the output F_OUT. The FTL_CNT counter is incremented only when detecting rising edge while decreasing when detecting falling edge as can be observed.

Each TIM_CHn capture/measure channel consists of an ECNT event counter, a signal measure unit (SMU) with a CNT counter, CNTS shadow counter for the SMU, and two general-purpose registers GPR0 and GPR1 for storing values as described in (GTM-Cookbook, 2014).

The TIM sub-module has up to five operating modes, but the most commonly used for the electric machine control are the PWM TPWM (TIM PWM measurement mode), and the TIEM (TIM input event mode).

In TPWM mode, the TIM channel measures the period and duty cycle for a PWM-type input signal. It is possible to select the polarity of the signal to be measured, either the rising or falling edge with the Digital Signal Level (DSL) bit. In the case that DSL is 1, the TIM channel starts measuring after the first rising edge detected by the filter. An internal counter has a resolution preset by the clock frequency CMU_CLKx until the falling edge is detected where the content of the counter is transferred to a shadow register CNTS. The counter continues increasing to the next rising edge where it restarts to zero. Also, at this point, the content of the CNTS shadow register is transferred to the GPR0 register (duty cycle), and the captured value of the CNT counter is transferred to the GPR1 register (period). In the opposite case, that is to say for DSL equal to 0, the operation explained above is the same, but changing the rising edge by the falling edge. The TPWM mode is very interesting to monitor, for example, the PWM outputs that go to the three-phase inverter to ensure proper operation of the electric machine and perhaps complying with a degree of safety imposed by the application. Another example would be to read the direct PWM voltage signals in the three-phase machine terminals for deadtime compensation (Sect. 7.4.1.2) or to reconstruct the machine voltage as it will be discussed in Chap. 10.

The TIEM event counter mode permits to count the rising, falling, or both in a signal on a TIM channel. The CNT counter contains, in this case, the number of edges. The TIEM mode can be applied for the measurement of position and speed of an electric machine using, for example, digital encoders, where the number of counted edges determines the new position and the time between the edges and the speed.

For advanced data processing, the result of the captures of the previously explained modes of the TIM module can be through the ARU. That is, the measurement values can be read directly by the CPU through the AEI bus or they can be read through the ARU to other GTM sub-modules.

7.1.1.3 Timer Output Module (TOM) and ARU-TOM (ATOM)

The main difference between the ATOM and TOM is that the ATOM sub-module is connected to the ARU so that it can use the advantage of interconnection between, for example, the TIM sub-module and the ATOM. Moreover, the resolution of its counters is 24 bits compared to the 16 bits of the TOM so that very low-frequency values can be generated more accurately in its digital output. The following explanation will be valid for both sub-modules so that from now it will be designed as ATOM only.

The flexibility offered by the ATOM sub-module allows from generating serial peripheral interface (SPI) communication signals to complex PWM signals. Focusing on the control of a three-phase machine, the interesting signals are simple PWM signals with a sinusoidal duty cycle.

ATOM also allows selecting the activation time of the PWM in center-alignment to the period or the edge-alignment, where the turn-on time can be left or right of the period. For example, in left edge-aligned PWM, the outputs change to high level at the beginning of the period when the timer value is higher than zero, remaining at high level for a time determined by the duty cycle. However, the center-aligned PWM waveforms align the PWM signals to the half of the period, as illustrated in Fig. 7.4. Since the central alignment is the most common in the application of multiphase machine control, this chapter will be focused on center-aligned PWM, but the concepts presented here can be easily extended to the generation of edge-aligned PWM signals. Figure 7.4 illustrates the center-aligned PWM signals waveforms.

As can be observed in the figure, the PWM signal is generated with a periodic reference signal (also called carrier or master signal), which sets the PWM period. Moreover, the reference signal is used to set the points where the rise and falling edges should occur through a pair of registers per channel (e.g., RegA1 and RegA2). The ATOM sub-module can generate the reference signal by using one of its channels and programming a register with the desired PWM period value. For instance, for a GTM base clock of 100 MHz, to perform a PWM of 20 kHz of frequency, the value of said register should be set to 5000. That is, the internal counter of the ATOM will count to 5000, and then it will be reset. The timer resolution is 10 ns; that is, the duration of each increment is 10 ns forming the 50 μs period (20 kHz). Once the

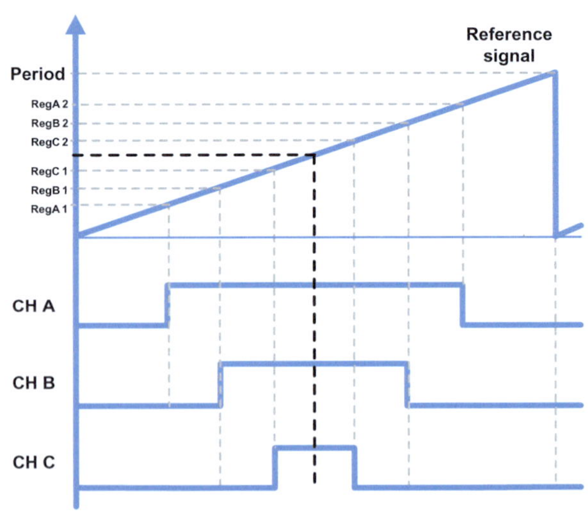

Fig. 7.4 Center-aligned PWM signals waveforms

period with an ATOM channel is set, another channel can be used to generate the PWM output. The complementary PWM output with deadtime required for the half-bridge structure can be generated with another channel in case the GTM does not have the DTM sub-module. In this way, with simple CPU calculations and thanks to the internal register update mechanism, the complementary signal can be generated without false activations or glitches. ATOM has a status configuration for the output level for each channel called SL (signal level); that is, a bit that indicates whether the level is high or low.

The ATOM mechanism which has for updating the internal registers that control the activation and deactivation of the PWM is based on shadow registers SR0 and SR1. The containing data of this registers (the value of the duty cycle) is transferred to the comparison registers CM0 and CM1 which performs the level change.

The update mechanism is synchronous when the CPU modifies the shadow registers SR0 and SR1 to vary the duty cycle in the next PWM period. On the other hand, if the CPU directly modifies comparison registers CM0 and CM1, the level change can be made during the same period. It is the asynchronous update mechanism.

An example of the asynchronous update operation is depicted in Fig. 7.5. In the first graph, the CPU does not perform any update, so the next period remains identical to the previous one. The period is set by CM0, which is equal to 120, while the duty cycle is just half set by CM1. The second graph is updated when the CN0 counter has a value of 70, but the falling edge has already occurred before so that the update is carried out in the following period. The last graph shows an interesting case where CM1 is updated to 20 while the CN0 counter was at 30. In this case, the level is immediately changed, but it is not until the next period when the duty cycle takes on the value assigned by CM1 (16.66%).

In the case of synchronous updating, all the previous cases would not make any level change until the next period, because the CM0 and CM1 registers are updated by the shadow registers SR0 and SR1 just in the reset of the period counter as it will be seen below.

When the pair of transistors of the half-bridge is activated by logic level "1," the high-side transistor PWM signal should be programmed with the SL bit equal to zero. Then the SR0 or CM0 registers SR0 or CM0 are used to set the falling edge (deactivation "1"→"0") while SR1 or CM1 registers are used to set the rising edge (activation "0"→"1"). On the contrary, for the low-side transistor PWM signal, the

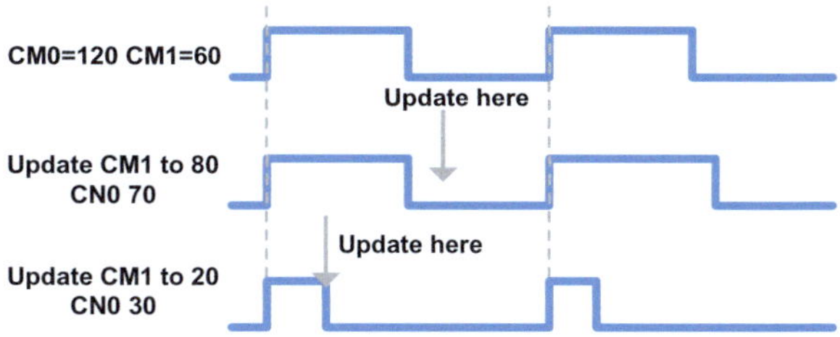

Fig. 7.5 Duty cycle asynchronous update mode

SL bit is set to one to reverse the operation, i.e., SR1 or CM1 registers are used to set a falling edge and SR0 or CM0 to set the rising edge.

Since both CM0 and CM1 registers are independent, the ATOM sub-module allows versatility, for example, for generating PWM signals shifted with respect to the beginning of the period, even transcending the period with the duty cycle. With a single reference channel (master), the rest of the channels (slaves) are generated with the corresponding assignment of the CM0 and CM1 registers of each of the channels. This particularity of the ATOM will not be necessary for three-phase machine inverters since the generation of the PWM outputs are not shifted, but they are typically center-aligned PWM waveforms as commented early. On the contrary, for some power converters with interleaving phase stages, it is essential. In these converters, the power are distributed in different phases of lower power optimized to work in an interleaved way. Figure 7.6 illustrates the experimental result of PWM signals generated with GTM for a four-phase power converter with 90° phase shift. The shift is given by the 360° division by the number of phases available, in this case, 360/4 = 90°. The figure shows the example with 50% of duty cycle. Respect to phase 1, phase 3 is at 90°, then phase 2 at 180°, and finally, phase 4 at 270°. Also, the deadtime between the normal and the complementary phase (e.g., D_0 high-side PWM signal and D_1 low-side PWM signal) can be appreciated.

The ATOM block diagram detail for one channel is also shown in GTM-Cookbook (2014). It consists of the following four units: CCU0, CCU1 (Counter Compare Unit), ACI (ARU Communication Interface), and SOU (Signal Output Generation). The CCU0 and CCU1 units compare the value loaded to the CM0 or CM1 operation register with the internal counter CN0 to make the corresponding level changes configured in the SOU unit. This is where the configuration of the SL performs its function since, through multiplexers, it exchanges the CCU0 outputs with the CCU1

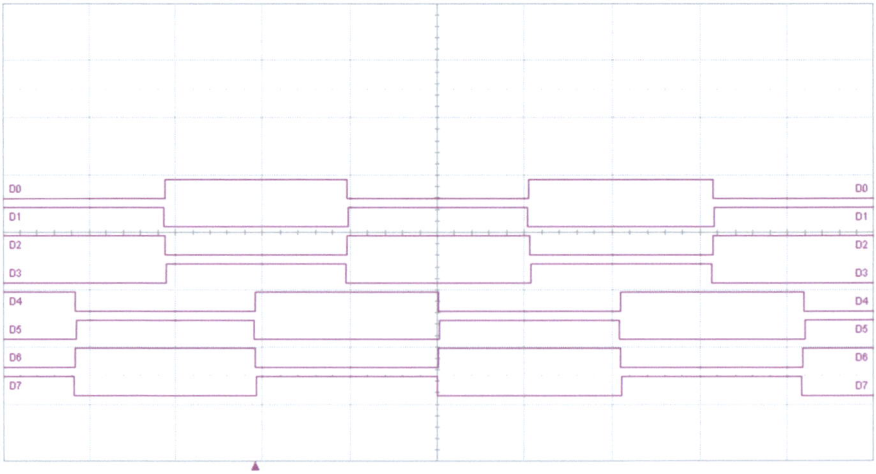

Fig. 7.6 Normal and complementary PWM signals shifted respect channels D_0 and D_1, 180°, 90°, and 270°

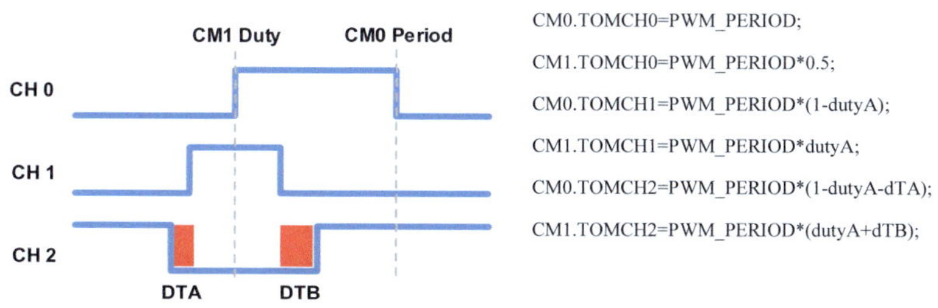

Fig. 7.7 Time base and two PWM channels. Complementary mode with deadtime

to assign the corresponding high or low level as the case may be, using an RS-type latch. Then, in its output, there is another multiplexer that enables the output to the output port. Finally, the ACI unit is responsible for taking the ARU data and then transferring it to the shadow registers SR0 and SR1. In the case of not using the ARU, the CPU can write directly to these registers as discussed early.

Figure 7.7 shows the operation of the three channels necessary to generate two PWM signals for a half-bridge with a preset deadtime DTA and DTB. As can be observed, the reference signal with channel 0 sets the period only configuring the CM0 register to the period value, that is, to PWM_PERIOD. When channel 0 is enabled, the CN0 of this channel starts to count, and when it reaches PWM_PERIOD, it will automatically restart to zero.

It is possible to observe how the CM0 and CM1 registers should be set to generate the desired PWM signals with deadtime. It is important to comment that the shadow registers updating of each of the channels should be done in the ISR interruption of the reference channel (CH0) to avoid "glitches" in the normal and complementary output. The ISR interruption is configured so that it is generated just in the middle of the period (rise edge of CM1 of CH0) being within the interruption where the shadow registers of all slave channels should be updated. The data transfer of the shadow registers to the comparison registers CM1 and CM0 is done automatically when the period ends as commented early, which is reset to zero to start another period once again.

7.1.1.4 SPE (Sensor Pattern Evaluation)

One of the most economical and simple ways of detecting the rotor position of an electric machine is to install a three Hall effect sensors 120° apart. As discussed in Chap. 4, BLDC-type machines can be controlled by these Hall effect sensors. The idea of controlling a BLDC machine is to know the position of the rotor to know what activation sequence of the outputs must be activated in the next step. This detection/actuation is also known as an electronic commutator since it has similar behavior to the mechanical commutator of the brushes DC machines.

The signals provided by the sensors can be evaluated directly by the SPE sub-module. The TIM sub-module can be used together with SPE sub-module, for example, to filter the digital inputs of the Hall effect sensors. As commented, the SPE sub-module, together with the ATOM/TOM sub-module, can control a BLDC machine with almost no CPU intervention.

When a new and valid input pattern is detected in the SPE sub-module, the SPE signals determine the outputs of the ATOM/TOM channels. In the most straightforward configuration, the connected ATOM/TOM channels load a new PWM pattern from their shadow registers and generate the PWM with this pattern. The CPU can then reload the following PWM pattern in the shadow registers.

It is also possible to define different output patterns. For example, for an input pattern, the change can define for the ATOM/TOM that some of the output channels change their output to a defined state while other channels generate a PWM signal at the output to vary, for example, the speed of the machine BLDC as will be seen in Chap. 8.

Figure 7.8 represents the different sub-modules involved in the BLDC machine control.

The three inputs S_1, S_2, and S_3 of the TIM sub-module are digital inputs of the Hall effect sensors. The most optimal pattern of these sensors is when the machine rotates, only one-bit changes. For example, the sequence during a complete mechanical turn clockwise would be $(S_1, S_2, S_3) = (1, 0, 0), (1, 1, 0), (0, 1, 0), (0, 1, 1), (0, 0, 1), (1, 0, 1)$, and again $(1, 0, 0)$. The arrangement of the above patterns can be defined in a programmable register by the software. It is also possible to mask an input signal path if, for example, one of the sensors stops working properly, making the system more robust. In the beginning, the software has to define where the rotor is to know where to start (which pattern is the current one). From that moment on, the SPE sub-module detects the direction of rotation and the change of the input signal by itself. When a new and valid input pattern is detected in the SPE sub-module, the

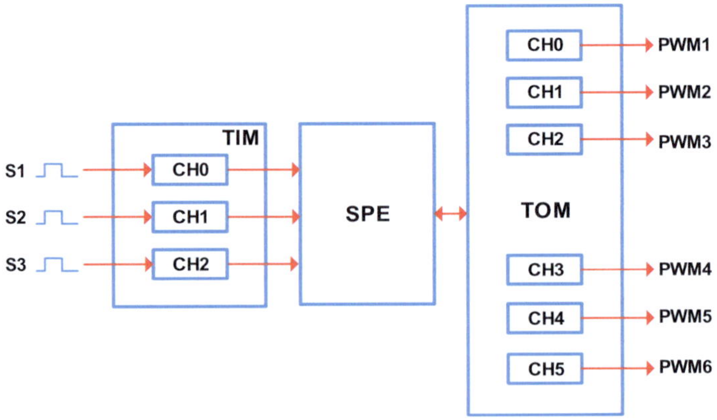

Fig. 7.8 BLDC engine control chain with TIM, SPE, and TOM sub-modules

SPE signals go to the TOM channels. In the simplest configuration, the connected TOM channels reload a new PWM pattern from their shadow registers to generate the output PWM in the next cycle. Then the CPU can then load the following PWM pattern in the shadow registers as commented before.

7.2 Analog-to-Digital Converter

7.2.1 Successive Approximation A/D Converter

The analog-to-digital successive approximations type converters are constant conversion time and are one of the most used at present for data acquisition systems because it allows a considerable conversion speed and a typical resolution of up to 12 bits at a low cost. The architecture is based on three blocks: a digital–analog converter (DAC), a comparator, and a successive approximation register (SAR). In order to keep the analog measurement stable during the approach process, it is necessary to use a sample-and-hold (S/H). Figure 7.9 represents the successive approximation A/D converter structure.

The successive approximation name is because there is a register (SAR) that changes until it approaches the analog input value. In the first clock cycle, the bit MSB of the approximation register is set to 1, and the rest remains zero. This register is the DAC input. The output of the DAC is equivalent to the reference voltage divided by two ($V_{ref}/2$) because only the MSB bit remains at 1. It is then compared with the analog signal that is previously maintained by the S/H. If the analog voltage is higher than $V_{ref}/2$, the MSB bit is definitely set to 1. If, on the contrary, the analog voltage is lower, then the MSB bit is reset. In the second cycle, the SAR sends the previous result to the DAC but adding a 1 in the second most significant bit to make comparisons again in a similar way to the first cycle. This process is repeated until the least significant bit LSB is reached. As can be seen, the conversion time is proportional to the number of bits of the converter, so that the more bits, the

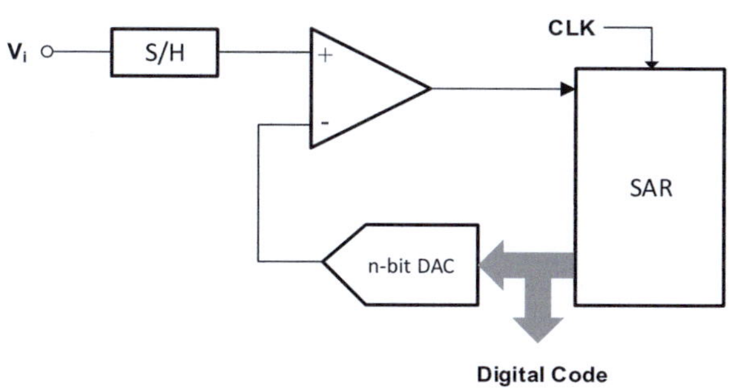

Fig. 7.9 Block diagram of A/D converter based on successive approximation

more conversion time will be necessary. As will be seen in Sect. 7.4.2, the general-purpose microcontroller RX63T of Renesas has a 12-bit successive approximation A/D converter with a conversion time of 1 μs.

7.2.2 Delta-Sigma Converter

As seen in the previous section, the successive approximation A/D converter has very efficient characteristics in terms of resolution and conversion speed. However, in the case of measuring the current of an electric machine, due to its limited resolution, it is not suitable for detecting overcurrents in very short times between 4 and 1 μs. The need for S/H, the anti-aliasing filter, and its resolution mean that, in case of overcurrent, it is not detected in time.

Today, the Delta-Sigma A/D converter (DSADC from now on) is increasingly available in microcontrollers due to its simplicity and low price. In the case of electrical machines control, the DSADC offers a very fast detection of overcurrent and allows to support the position-type position sensors, such as resolver, with a very high resolution almost without the need for complex circuits external to the microcontroller. The DSADC is also ideal for converting analog signals of a wide frequency range, from DC to several megahertz.

Basically, this converter consists of an oversampling modulator followed by a digital filter and downsampling that together produce a high-resolution result of the order of 24 bits with a low quantization noise. In conventional converters such as successive approximation A/D converter, oversampling translates into greater effective resolution, but to increase the resolution slightly, the oversampling frequency must be increased a lot. The A/D converter resolution is commonly evaluated based on the signal-to-noise ratio (SNR) of the signal output. For conventional A/D converters, each additional bit (b) of resolution increases the SNR by $6.02b + 1.76$ [dB]. The effective number of bits (ENOB) consists of clearing b from the previous equation to obtain (7.1).

$$\text{ENOB} = \frac{\text{SNR} - 1.76}{6.02} \tag{7.1}$$

For a one-bit converter, the ENOB is 0.71. If the signal to be converted is oversampled with an oversampling rate (OSR) of twice the Nyquist frequency, the SNR increases by 3.01 dB for each OSR duplication, which corresponds to a new ENOB of 1.79. The signal-to-noise ratio (SNR) as a function of the number of bits and the oversampling rate (OSR) can be observed in (7.2).

$$\text{SNR}_{ADC} = 6.02 \cdot b + 1.76 + 10\log_{10}(\text{OSR}) \tag{7.2}$$

However, for a Delta-Sigma converter with a first-order modulator, the SNR increase is approximately 9 dB (Harris 2004) for each duplication of the OSR and

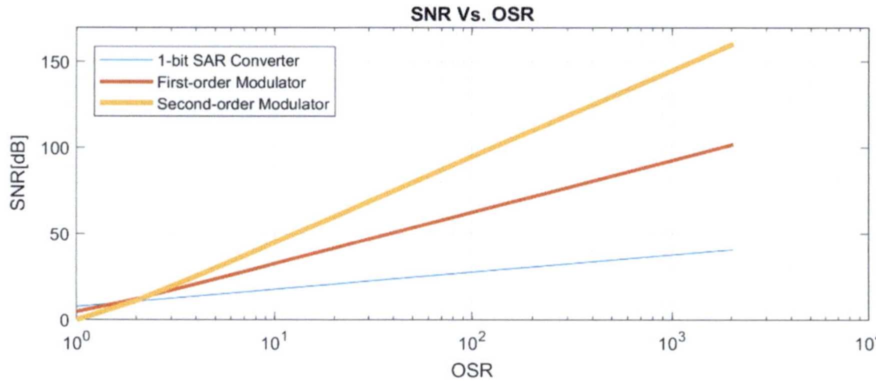

Fig. 7.10 SNR for one-bit SAR converter and DSADC with first- and second-order modulator

15 dB for a second-order modulator according to Eqs. (7.3) and (7.4), respectively.

$$\mathrm{SNR} = 6.02 \cdot b + 1.76 - 10\log_{10}\left(\frac{2}{\mathrm{OSR}} - \frac{2\sin\left(\frac{\pi}{\mathrm{OSR}}\right)}{\pi}\right) \tag{7.3}$$

$$\mathrm{SNR} = 6.02 \cdot b + 1.76 - 10\log_{10}\left(\frac{6}{\mathrm{OSR}} - \frac{8\sin\left(\frac{\pi}{\mathrm{OSR}}\right) - 2\cos\left(\frac{\pi}{\mathrm{OSR}}\right)\cdot\sin\left(\frac{\pi}{\mathrm{OSR}}\right)}{\pi}\right) \tag{7.4}$$

Equations (7.2), (7.3), and (7.4) can be plotted on a logarithmic scale on the x-axis to see how the SNR increases with the increase of the OSR. Figure 7.10 shows the signal-to-noise ratio (SNR) for a conventional one-bit converter join a Delta-Sigma converter with a first- and second-order modulator. Increasing the OSR improves resolution as seen previously, but also affects the duration of data conversion. It is an inherent feature of the DSADC. Between the conversion speed and the conversion accuracy, it is possible to choose the one that best suits the application.

As mentioned earlier, the DSADC modulator plays a significant role in increasing the signal-to-noise ratio (SNR). The Delta-Sigma modulator is a one-bit converter that allows an increase in resolution without oversampling.

The Delta-Sigma modulator has a negative feedback loop of the output signal to reduce the quantization noise produced by the one-bit quantizer. The error between the analog input and the modulator output previously converted to analog is integrated and quantified with the one-bit quantizer as can be seen in Fig. 7.11. The output of the modulator is a sequence of bits (bitstream). With this feedback structure and oversampling, the quantization noise is transferred to frequencies higher than the input signal's bandwidth. By means of a demodulator that basically consists of a digital low base filter, the input signal can be reconstructed, being the signal that will be digitally processed (Fig. 7.12).

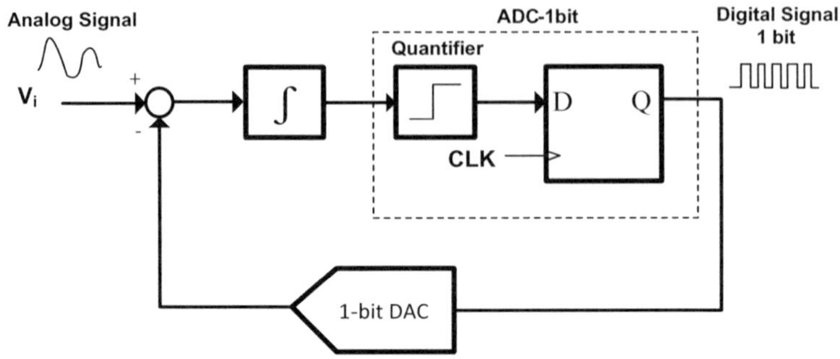

Fig. 7.11 Block diagram of Delta-Sigma modulator

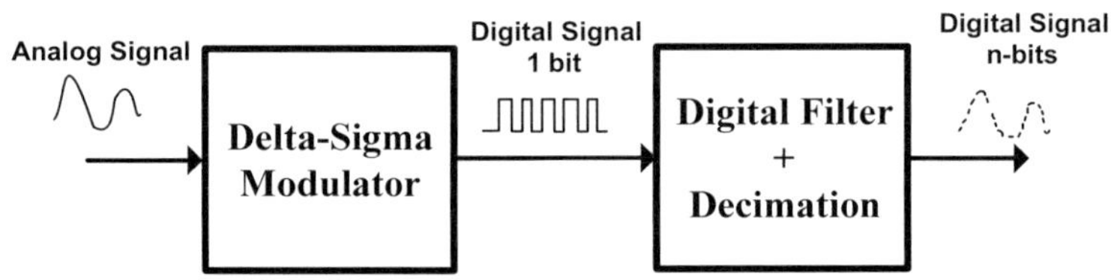

Fig. 7.12 Block diagram Delta-Sigma stages: modulator and digital filter with decimation (demodulator)

As commented, oversampling consists of sampling the signal at a higher rate than double its bandwidth and is represented by the acronym oversampling ratio (OSR). Oversampling in the control of electrical machines is a widespread practice since today's microprocessors allow it and because the bandwidth of the machine current is not usually higher than 2000 Hz. In addition, oversampling avoids anti-aliasing filters and can improve the signal-to-noise ratio (SNR).

Figure 7.13 shows a first-order digital Delta-Sigma modulator based on a discrete integrator and a one-bit quantizer. Delta-Sigma modulators of a higher order than one are implemented by first-order modulators in cascade connection, which are called multi-stage noise shaping (MASH). The first modulator converts the analog input signal, and each subsequent modulator converts the quantization error of the

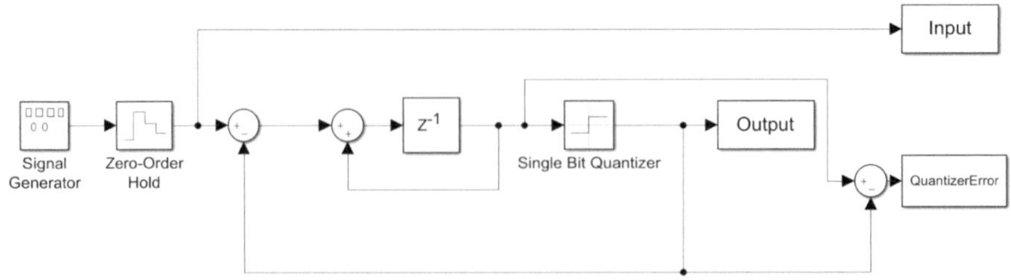

Fig. 7.13 Delta-Sigma first-order modulator implemented with Simulink blocks

previous modular by increasing the SNR for the same OSR as it will be observed in Fig. 7.35.

As previously mentioned, the digital Delta-Sigma modulator output is passed through a digital filter to eliminate the noise caused by the modulator quantizer that is in the upper frequency band of the processed signal. In order to optimize the digital processing data flow, the number of samples is usually reduced by decimating (downsampling). For example, if the output signal of the digital filter has 128 samples, and a decimation factor 4 is applied, the result is a signal of 32 samples. A digital filter that is usually used in microcontrollers is the Sinc-K-Filter since it is a filter that requires few hardware resources when compared to a digital FIR or IIR filter (Meyer-Baese 2014). The name Sinc is because its frequency response reminds of a cardinal sinus. The K represents the order of the filter. It has a single-stage that provides filtering of moving averages of the last D digital samples as represented in (7.5). If the left part of the equation is developed, it is possible to obtain two separate filters in cascade connection, as shown in the right part of Eq. (7.5). The left part is the integration part, while the right part is the differentiation part.

$$H(z) = \left(\frac{1}{D}\frac{1-z^{-D}}{1-z^{-1}}\right)^K = \frac{1}{D^K}\left(\frac{1}{1-z^{-1}}\right)^K \left(1-z^{-D}\right)^K \tag{7.5}$$

The implementation of the sinc filter of order K is very simple, as can be seen in Fig. 7.14. The reduction of the decimation rate D can be implemented by passing only a number of integrated samples D to the differentiation stages. This type of filter architecture is also known as the Cascade Integrator-Comb (CIC) (Hogenauer 1981) filter.

The frequency response is shown in Fig. 7.15, while the expression is in (7.6). It is possible to observe the shape of the frequency response, which is similar to a comb.

$$|H(\omega)| = \left(\frac{1}{\text{OSR}}\frac{\sin\left(\omega\frac{\text{OSR}}{4}\right)}{\sin\left(\frac{\omega}{4}\right)}\right)^K \tag{7.6}$$

After the previous CIC digital filter, it is possible to add an additional FIR-type digital filtering stage to filter further the high-frequency components that the CIC filter could not. The combination of both filters usually gives high-performance results when it comes to high-frequency noise reduction (higher SNR), but the conversion time can be considerable so that each case should be evaluated. As will be seen in

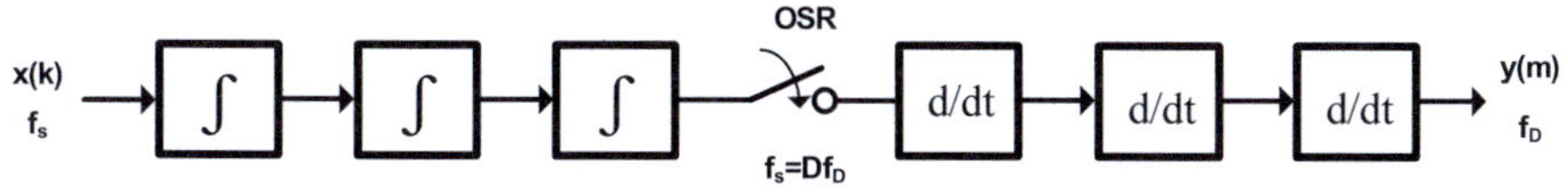

Fig. 7.14 Block diagram of CIC filter with two stages: integrator and differentiator with decimation in the middle

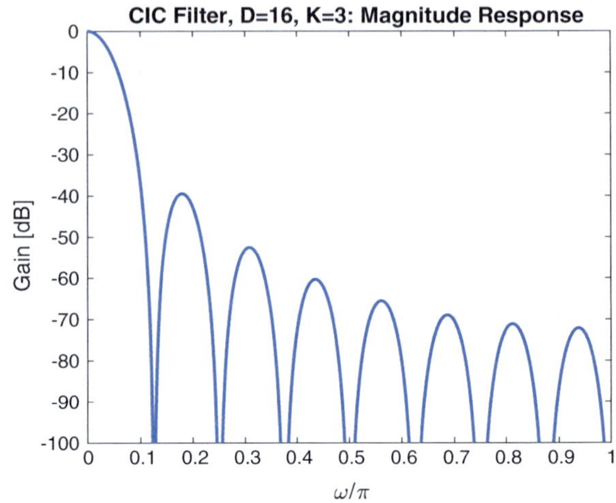

Fig. 7.15 Frequency response of third-order CIC filter

Sect. 7.3.4, the DSADC of the TriCore™ AURIX™ family from Infineon allows different configurable filter combinations according to the design needs.

7.3 Infineon AURIX™ Automotive Microcontroller

7.3.1 Introduction

Electronic systems based on microcontrollers/DSPs or FPGAs are most frequently used in the automotive industry in order to meet the new market requirements for safety, comfort, and connectivity. In addition, compliance with the new CO_2 emission reduction standards leads to the necessity to use new electrical and mechanical systems for the more efficient use of energy from combustion engines. To support the new electrical systems, the 48 V lithium batteries up to 15 kW of power have been fundamental since the power of conventional 12 V lead–acid batteries was very limited. The range of possibilities of 48 V technology has made it possible to design more efficient vehicles thanks to the reduction of mechanical loads to the combustion engine that have become electric-powered loads. For example, the air conditioning compressor is a typical load moved by the combustion engine, but some vehicles incorporate a compressor moved by an electric machine with an improvement of the whole efficiency. The electric machine for the compressor usually is a three-phase machine, where the efficiency should be higher for 48 V than for 12 V. The main reason is that the Joule effect losses are lower since the current is four times lower.

The 48 V technology is based on the energy recovery of vehicles in braking and/or downhill by an electric generator in solidarity with the combustion engine. Different vehicle manufacturers have been able to demonstrate that 48 V technology allows reducing emissions to meet the new CO_2 emission directives. This is the mild hybrid electric vehicles (MHEV) where the CO_2 is reduced between 13 and 22%.

The new electrical/electronic systems to meet this new technology along with the requirements of security, comfort, and connectivity makes microcontroller manufacturers have to develop microcontrollers with better computing power, more energetically efficient, and with a wide range of peripherals for security, comfort, and connectivity management.

The International Council on Clean Transportation (ICCT) has published the current and future standards for fleet CO_2 emissions, as illustrated in Fig. 7.16 in different countries. As can be observed from 2021 onward, the average fleet CO_2 emissions in the European Union will be limited to 95 g/km. The OEMs have to ensure that the new vehicle accomplishes with the CO_2 emission average. If the vehicle manufacturers exceed the average limits, they will have to pay fines.

The tendency is to reduce emissions gradually to allow time for the development of new technologies that make it possible.

On the other hand, security is increasingly dependent on electronic systems so that these systems must meet a series of safety requirements imposed by ISO 26262. ISO 26262 is the regulation of electrical and electronic systems for the automotive industry where they can find the standards to be met to ensure security in each of the different electronic systems. The heart of the electronic systems is basically the microcontrollers/DSPs, so they have been designed to comply with ISO 26262. The main manufacturers of automotive microcontrollers design their microcontrollers with an architecture, scalability, and with a large number of peripherals to allow application in the main divisions of the electronic systems in the automotive industry:

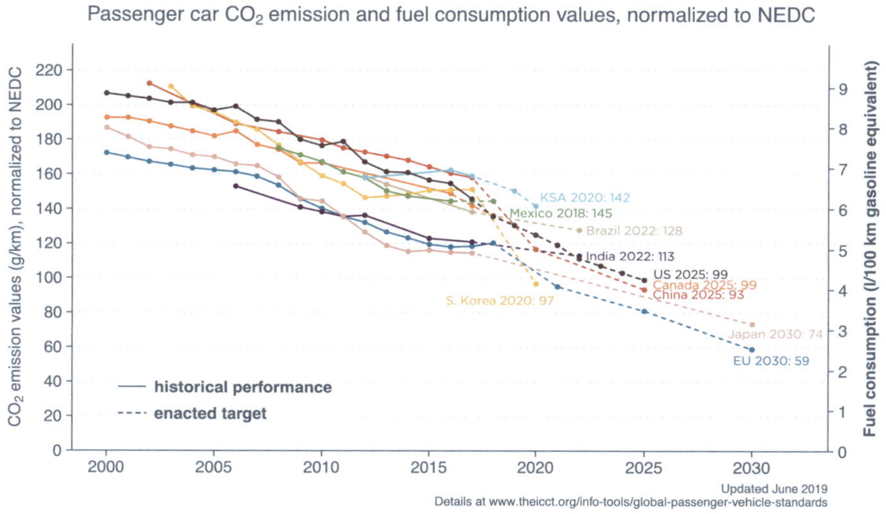

Fig. 7.16 Passenger car CO_2 emissions and fuel consumption normalized to NEDC. International Council on Clean Transportation (ICCT)

Powertrain, MHEV, (H)EV, Safety (Active and Passive), Body, and Drive Assistance Systems (DAS).

For example, within the powertrain division, it is possible to find applications where electric machine control is required, such as electric traction machines for hybrid electric vehicles and fully electric vehicles. Moreover, in the safety division, the electrical powered steering (EPS) is an example which can be found. In EPS systems, the application of the control of an electric machine may not be as clear as historically it was carried out with hydraulic energy, while, for some years, the conversion to electric power has been an improvement in terms of higher efficiency and control. The basic operation consists of a torque amplifier; that is, with a torque sensor placed on the steering wheel shaft, the amount of assistance needed to apply more or less torque to the electric machine is determined. Some of today's most sophisticated vehicles install electrical powered steering based on three-phase BLDC of PMSM instead of DC machines that further improve efficiency and control. With a position sensor, the position of the rotor is determined, which closes the position control loop and allows the improvement of efficiency and control with respect to DC machines. The electronic/mechanical system that governs the EPS must comply with the appropriate safety requirements since any failure of the system must not allow the address to be blocked in any of the cases. As can be observed, electric machines have a wide range of applications in automotive systems. For this reason, in this section, the most relevant peripherals for electric machine control of the AURIX™ microcontroller family of Infineon will be introduced.

7.3.2 Infineon AURIX™ Family

The 32-bit-Microcontroller TriCore™ family with its embedded safety and security features is the ideal platform for a wide range of automotive and industrial applications. AURIX™ TriCore™ unites the elements of a RISC processor core, a microcontroller, and a DSP in one single unit.

The AURIX™ microcontroller TC2xx family (Fig. 7.17a) has up to three 32-bit cores that can operate up to 300 MHz with up to two extra check cores (lockstep). The extra lockstep cores run the same set of instructions in parallel, but with a certain delay with the main core allowing redundancy for error detection and correction. The lockstep architecture reduces development efforts as it guarantees the safety standard required by ISO 26262.

The second-generation TC3xx family (Fig. 7.17b) with its up to hexa-core high-performance architecture and its advanced features for connectivity, security, and functional safety is ideally suited for a wide field of automotive and industrial applications. Four of the six cores feature additional lockstep cores. In addition to engine management and transmission control, targeted powertrain applications include new systems in electrical and hybrid drives.

The main cores share the same bus, each with its local RAM, but sharing a common FLASH ROM. Peripherals (timers, I/O ports, Ethernet, serial ports, etc.) are also

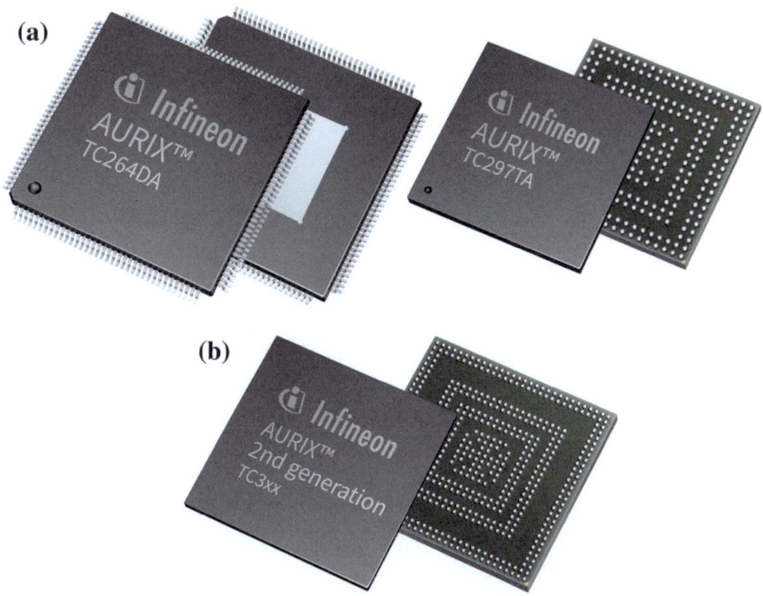

Fig. 7.17 **a** LQFP 144-pin TC264DA and LFBGA 292-pin TC297TA AURIX™ microcontrollers. **b** LFBGA-516-pin TC3xx second-generation microcontroller. Courtesy of Infineon

shared, with each core having full access to any peripheral. Also, there is a floating-point unit in each core, so the use of floating-point variables does not significantly decrease tasks.

In addition to standard peripherals, the AURIX™ family features the GTM peripheral and a Delta-Sigma analog-to-digital converter for fast and accurate measurements. Figure 7.18 represents the AURIX™ TC2xx family system architecture with three cores and two checker cores. The microcontroller cores are the so-called P (Performance Core) designed for high calculation performance having up to 30% more computing power than the E core. The efficiency E core is used in TC2xx microcontroller family to improve power consumption efficiency while maintaining excellent calculation performance. The microcontrollers TC2xx family with more than one core has inside the E core together with the P core as represented in Fig. 7.18.

Figure 7.19 represents the AURIX™ TC3xx family system architecture with six cores (all performance cores) and four checker cores.

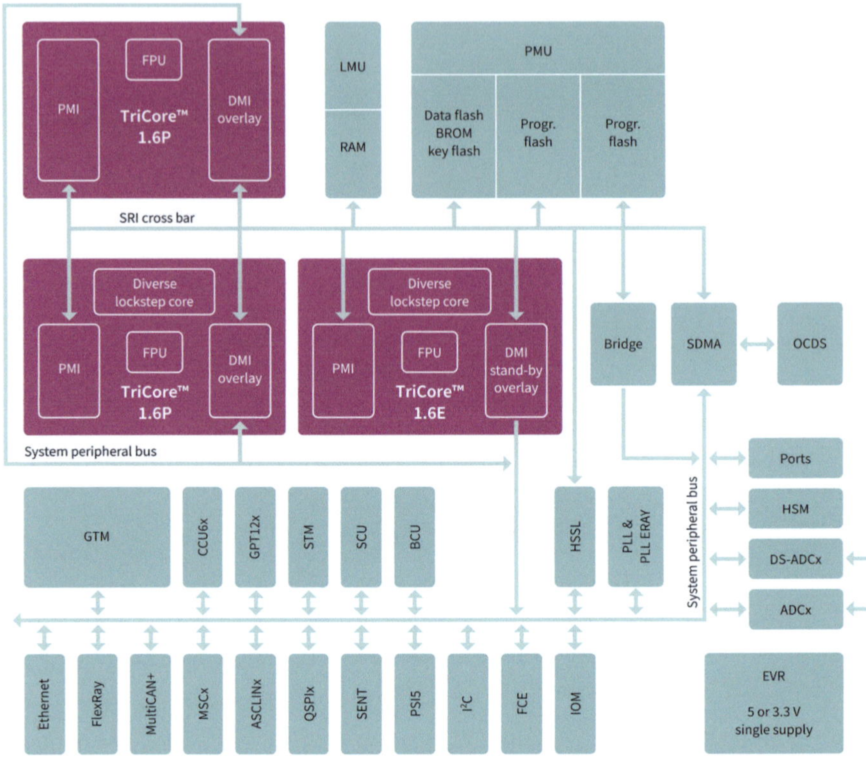

Fig. 7.18 First-generation AURIX™ TC2xx family architecture. Courtesy of Infineon

7.3.3 GTM in AURIX™ Family

Like many other microcontrollers manufacturers in the automotive industry, the AURIX™ family integrates the specific GTM peripheral as discussed in Sect. 7.1. As discussed, it was designed by Bosch for this industry, motivated mainly to offer a flexible solution for the different domains of application with minimal CPU interaction. Due to the market requirements of the automotive industry, different manufacturers integrate it into their microcontrollers. The GTM IP peripheral is identified by two data information, the IP configuration and the specification version. The IP configuration refers to the generation and its configuration while the version of the specification indicates the updates that are usually made of improvements and correction implicit errors of the module. It is important to comment that each IP configuration can have the same or different specification version since the IP only refers to the generation and configuration as commented before.

The same microcontrollers family could arrange different GTM versions. For example, it is possible to find the low-end microcontrollers of the same family with the newer GTM version while high-end microcontroller with an old GTM version.

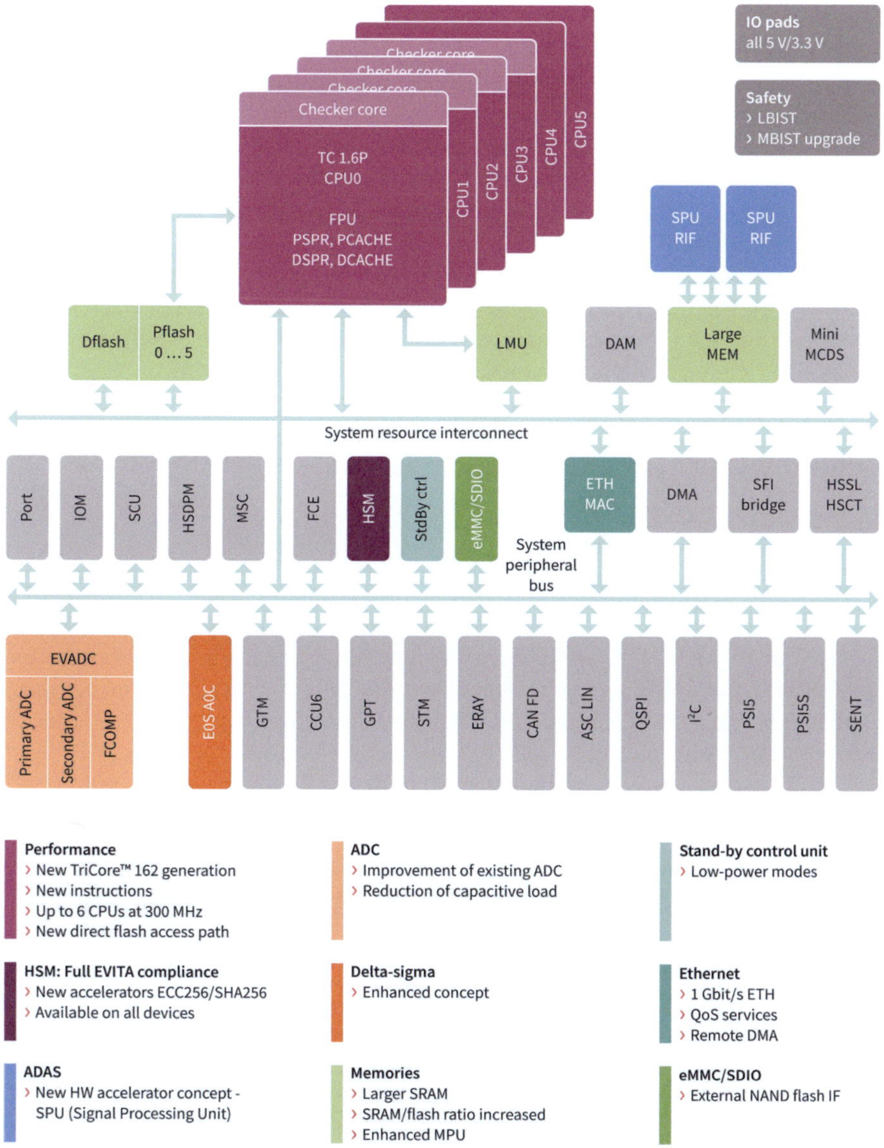

Fig. 7.19 Second-generation AURIX™ TC3xx family architecture. Courtesy of Infineon

Usually, this is because when a manufacturer launches a new generation of their family, they usually start with the high-end microcontroller and from there, they start to design the low-end microcontrollers with less performance and lower price to cover others applications that require less performance. Between the high-end microcontroller is launched to the market until the low-end microcontroller could be a period between two and three years. Thus, in this period, a new GTM version and IP can come out so that the microcontroller manufacturer can include it in the most recent microcontrollers. For example, the GTM version 1.5.5.5 does not have a sub-module that generates the complementary PWM with a programmable deadtime while version 2.02.1 does have that sub-module. With the deadtime sub-module, the CPU is unloaded a bit since fewer operations should be performed to calculate the new values of the complementary PWM registers. Also, for this example, it is only necessary to use three PWM channels instead of six since the deadtime sub-module internally generates the complementary channel with the pre-configured deadtime.

If, for a project, it is crucial to use the new GTM revision either because it has solved a configuration error or because they have simply added new functionality that makes the project simpler and safer, the selection of one of the microcontroller manufacturers can be determined. If, on the contrary, it is indifferent to use the old or new revision, the range of possibilities is even higher in terms of the number of manufacturers that it is possible to take into account.

7.3.4 *DSADC in AURIX™ Family*

The TriCore™ AURIX™ family from Infineon has a Delta-Sigma A/D converter. The Delta-Sigma ADC module provides several channels with a main and auxiliary demodulator that include configurable filters for decimation. The general description of the DSADC structure can be seen in Fig. 7.20.

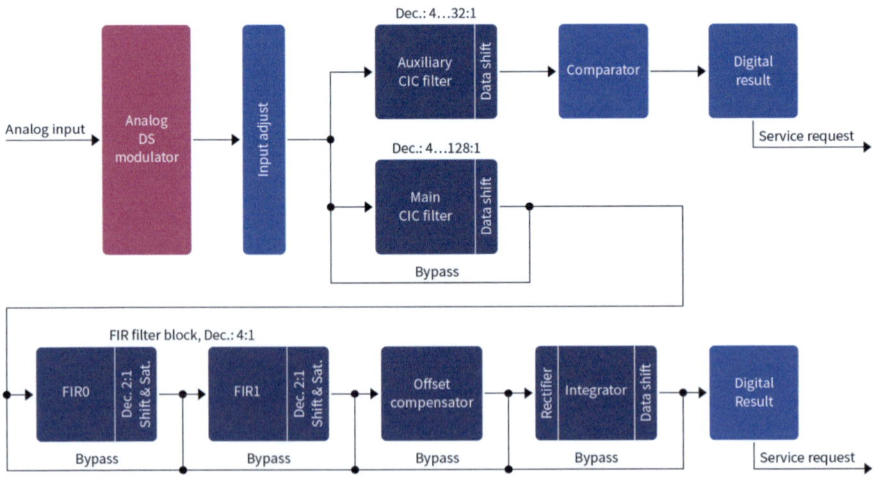

Fig. 7.20 DSADC structure of AURIX™ family. Usually, it corresponds to analog DS modulator, main CIC, and FIR filters. Courtesy of Infineon

DSADC has an integrated analog Delta-Sigma modulator that converts an analog input signal into a data stream. The main chain of digital filters builds the demodulator that produces result values at a configurable output rate. The elements of the filter chain can be selected according to the requirements of the application. The filter string configuration determines the attenuation and delay properties of the filter. Configurable decimation at different rates reduces the input sampling rate to a lower result data rate. The most outstanding elements are:

- A DS Analog Modulator consisting of a 1-1-1 MASH of order 3
- A configurable decimation filter of the Cycling Integrating Comb (CIC) type also called a three-stage synchronization filter that provides filtering and deletion of basic data with a selectable D decimation rate
- Two FIR filters, each with a decimated rate of 2, allow for effective signal configuration by attenuating the top of the signal spectrum frequencies
- A high-pass filter that provides offset compensation by eliminating the DC component from the input signal
- An integrator that accumulates a configurable amount of result values. The quantity of samples is programmable or can be controlled by a hardware signal. This function allows solving to obtain the baseband signal for the calculation of the machine position, and the integrator can also be used for shunt current measurement.

In order to understand the DSADC , it is possible to model the Delta-Sigma A/D converter as it will be shown in Sect. 7.5.3.

7.4 General-Purpose Renesas RX600 Microcontroller

As in the case of the Infineon AURIX™ family, the RX600 family features a series of advanced peripherals for electrical machine control applications. It has a high processing performance thanks to the speed of 100 MHz of its CPU and the inclusion of a floating-point unit by IEEE-754 hardware as part of the CPU core. The group of microcontrollers recommended for control applications of electric machines is the RX62T, RX63T, RX72T, RX66T, RX24x, and RX23T groups. In Fig. 7.21 is represented the RX63T microcontrollers with different packages.

The RX63T is the few-pin version designed to support appliances and industrial applications with a single interface for controlling electrical machines along with a 1 μs 12-bit conversion time A/D converter including three S/H circuits. The RX63T is capable of controlling two control applications of independent three-phase electric machines with two MTU3-4 and MRU6-7 timers blocks. For the 100-pin version or more, it has an improved independent 12-bit dual A/D converter unit, three proportional gain amplifiers (PGA) for signal conditioning, and three comparators with analog windows for each A/D converter unit. A third 10-bit A/D converter unit provides support for other analog-to-digital conversion needs.

Fig. 7.21 LQFP package of 48-, and 64-pin RX63T microcontrollers. Courtesy of Renesas

This section will be treated in detail the A/D and MTU3 peripherals, which are essential to perform a high-performance three-phase electric machine control with deadtime compensation.

7.4.1 Multi-function Timer Pulse Unit 3 (MTU3)

The MCU microcontroller includes a 16-bit 8-channel timer called MTU3. The functionalities are diverse from the capture of input pulses to generation of complementary PWM signals necessary for power structures in half-bridge. From the eight channels, the MTU3, MTU4, or MTU6, and MTU7 channels will be the ones that will allow complementary PWM signals generation, while the MTU5 channel is designed for deadtime compensation.

7.4.1.1 MTU3, MTU4 as Complementary PWM Mode

The basic characteristics of these channels are:

- Buffer operation
- Transfer of intermediate memory registers to temporary registers at the crest or valley of the carrier or reference signal
- Possibility of omitting interruptions in the crest and valley of the reference signal, as well as the A/D converter trigger. It is very useful in case it is not necessary to execute the control every PWM cycle
- A/D conversion starts delay function of the request allowing the A/D converter to start at any desired time and synchronized with PWM output
- POE0# input on the microcontroller pin that forces all PWM outputs in high impedance in case it is activated by an error detected such as an overcurrent.

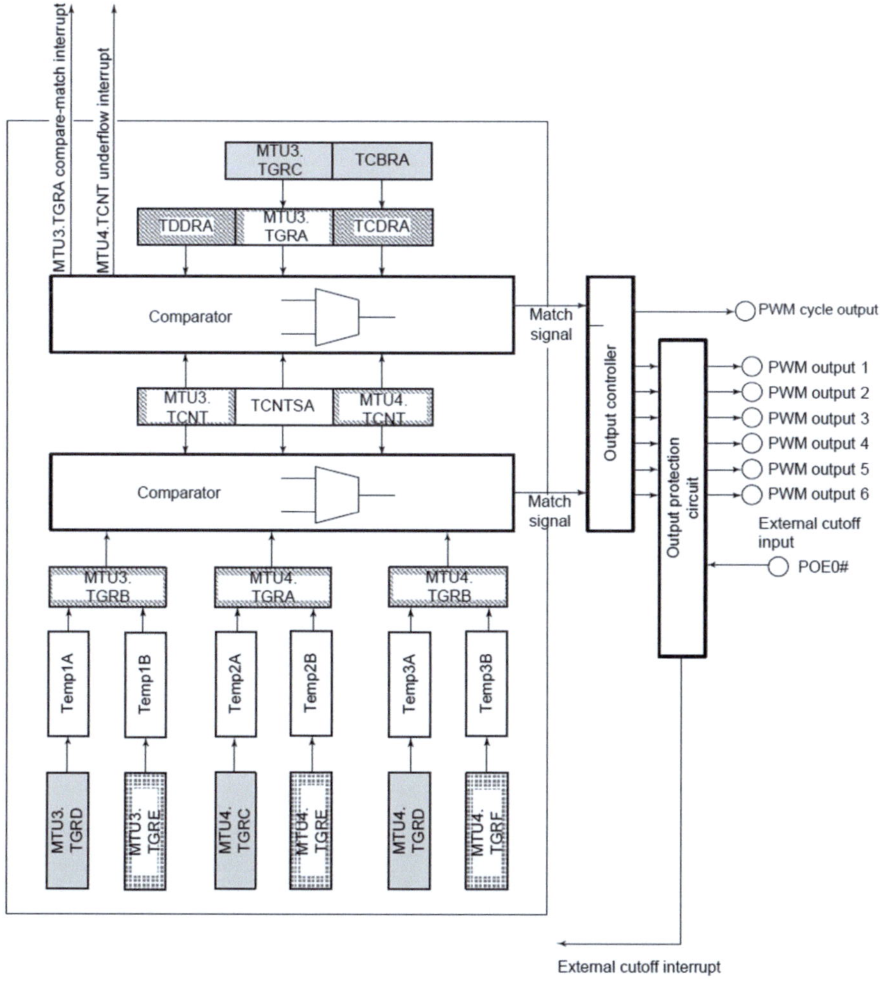

Fig. 7.22 Block diagram of MTU3-4 for complementary PWM mode. Courtesy of Renesas

Figure 7.22 shows the block diagram of the operation of the MTU3 and MTU4 channels in complementary PWM mode.

On the right side of the figure, the six physical PWM outputs can be seen next to a reference signal that can be used to trigger, for example, an external A/D converter. As a physical input, there is the POE0 signal that, if activated, forces all PWM outputs in high impedance. At the internal level of the microcontroller, there are two interrupt outputs, the first by comparison value reached and the second because of the MTU4.TCNT counter has reached zero (it has had an underflow). This last interrupt signal will be the one used to trigger the internal A/D converter of the microcontroller. The deadtime generation in the MTU unit is performed thanks to two counters that operate at the same time, but the first MTU3.TCNT initialized to the desired deadtime value while the second MTU4.TCNT counter is initialized to

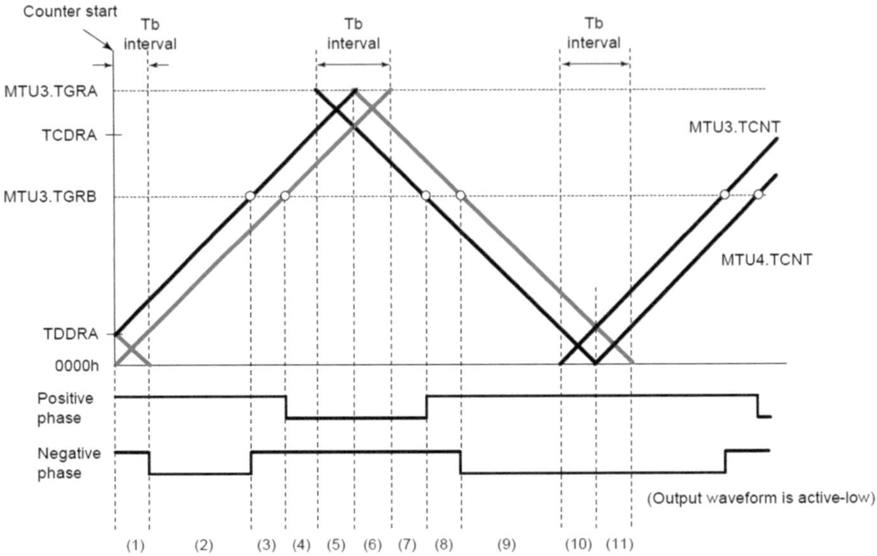

Fig. 7.23 PWM generation with deadtime according to two triangular signals delayed. Courtesy of Renesas

zero. With this, it is possible to have two triangular signals, one inscribed within the other, as can be seen in Fig. 7.23. When either of the two counters reaches their maximum set value, according to the desired PWM frequency and deadtime, the counter begins to decrease. The MTU4.TCNT counter will be incremented when it reaches zero, while the MTU3.TCNT counter will be incremented when it reaches the preset deadtime value. The registers that the CPU can modify to manage the duty cycle of each PWM signal are the MTU3.TGRD, MTU4.TGRC, and MTU4.TGRD registers, according to Fig. 7.22.

To generate the normal and its complementary pair PWM signals, it is sufficient to set a data value, for example, in the MTU3.TGRD register which will be transferred to the temporary register *Temp1A* (Fig. 7.22) and then to the MTU3.TGRB register. Figure 7.23 shows the operation of the normal and complementary PWM signal generation with a preset deadtime according to the value of MTU3.TGRB.

In this example, the zero output level indicates activated while the output level to one indicates deactivated. The output level is configurable using the TOCR1A.OLSP and TOCR1A.OLSN bits. If both are zero, a zero level means activated for both the normal and complementary signals. Otherwise, if they take the value of one, a high level means activated for both outputs. It is possible to observe that these bits allow the activation level to be configured separately for each of the outputs, being very useful when the high-side power switch is activated at the opposite level of the low-side power switch. In either case, these bits also indicate the level at which the output remains when the duty cycle is zero. Figure 7.24 shows the case where the TOCR1A.OLSP and TOCR1A.OLSN registers take the value of zero; that is, the

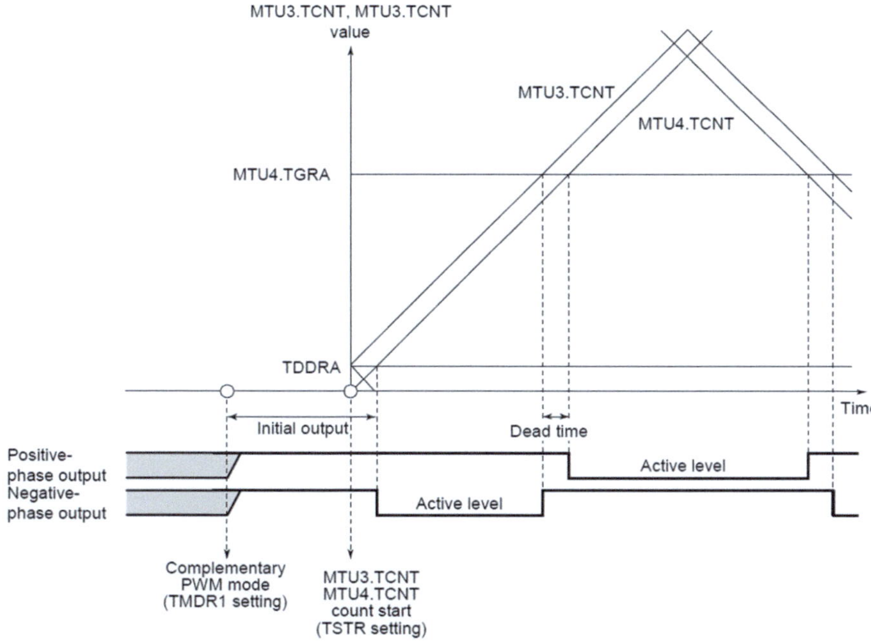

Fig. 7.24 Initial output of PWM after and before of its configuration OLSP = OLSN = 0. Courtesy of Renesas

activation is low. Also, it is possible to observe that just at the moment that these bits are configured, the PWM outputs take the configured value (initial output), and when the counters begin, the output level is already controlled by the MTU4.TGRA register for this channel.

7.4.1.2 MTU5 as Deadtime Compensation

The MTU5 channel is designed to compensate for the deadtime effect on half-bridges structures. It has up to three input channels that are dedicated to measuring real-time delays of the output waveform. Then the delay measurement can be reflected in the current duty cycle for its compensation. Figure 7.25 shows the PWM signals for a half-bridge structure.

For proper implementation, the measurement must be carried out at the midpoint of the half-bridge as can be seen in Fig. 7.26. In this case, a three-phase bridge is represented for the control of a three-phase machine.

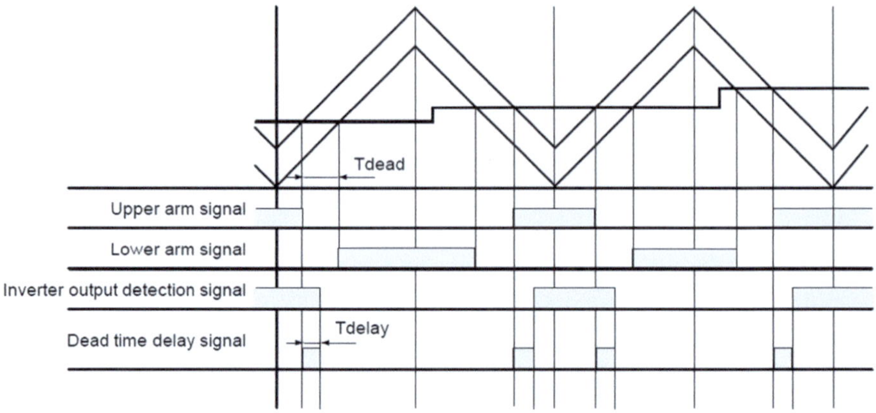

Fig. 7.25 Delay in the deadtime in complementary PWM. Courtesy of Renesas

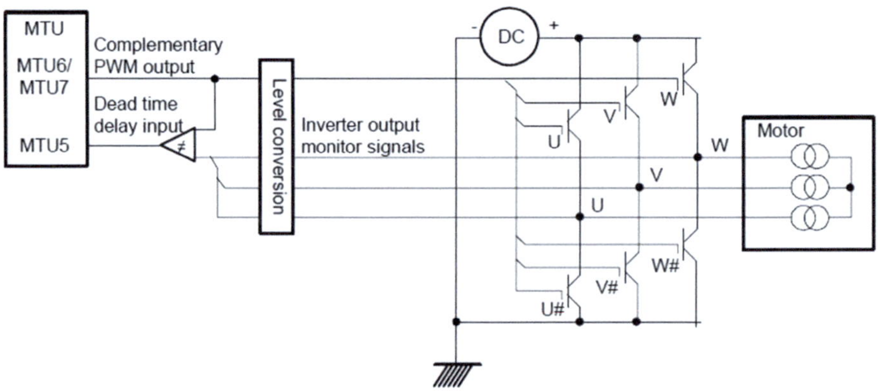

Fig. 7.26 Example of the three-phase machine control circuit configuration. Courtesy of Renesas

7.4.2 A/D Converter

The A/D converter consists of an eight-channel 12-bit successive approximation (SAR) converter and sample-and-hold. The A/D converter has three operating modes, single cycle scan mode where the conversion is only performed once, continuous scan mode, where the conversion is performed continuously, and group scan mode, where the conversion is performed at the same time as four channels for group A and four other channels for group B. The SAR converter allows triggering the conversion by different events, including software and hardware events. The hardware trigger can be performed by the MTU3 timer so that it is possible to synchronize the A/D conversion with the PWM timer. In addition, it is possible to add a delay time respect the PWM reference to avoid sampling in the switching, which is always noisily. The block diagram of the 12-bit converter in the 64- and 48-pin microcontroller version

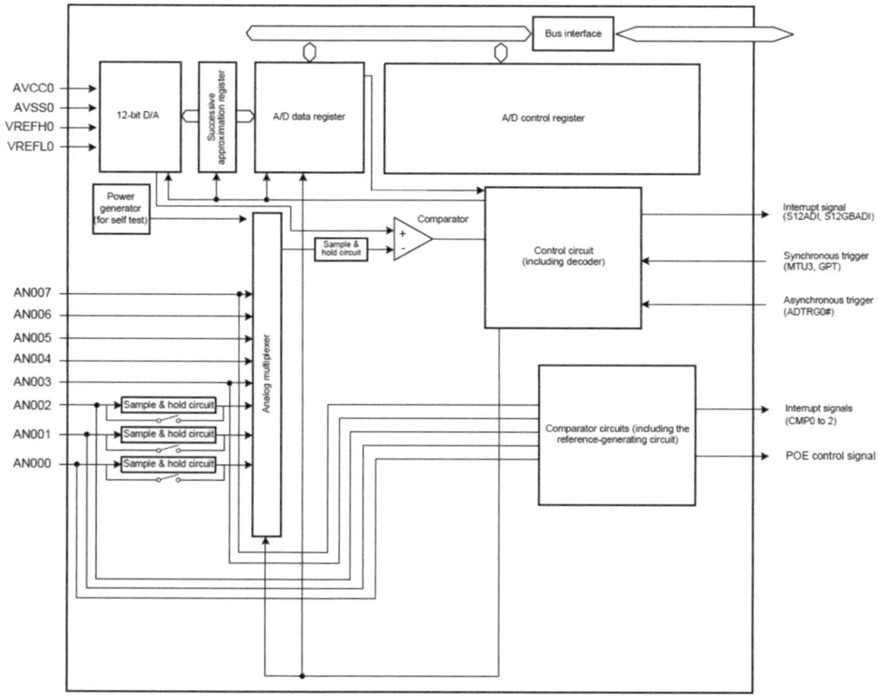

Fig. 7.27 Block diagram of 12-bit A/D converter for 64-, and 48-pin microcontroller version. Courtesy of Renesas

is shown in Fig. 7.27. In the figure, it is possible to observe that only the channels 0, 1, and 2 have sample-and-hold. This is a peripheral limitation, so that it must be taken into account in the design.

The A/D converter also has internal comparators that can be used to launch an event if the preset comparison level is exceeded. This is very useful, for example, for detecting overvoltages or overcurrents in an electric machine to quickly stop the control PWM through an ISR interruption or by the POE control signal as represented.

7.5 Modeling and Simulation

7.5.1 Modeling and Simulation of ATOM

The accurate PWM generation and the power plant models are essential when simulating the control of an electric machine to obtain the most realistic results. Electronic power devices are not ideal and therefore require correct PWM synchronization and maximum respect for switching times, especially deadtime. As will be seen in

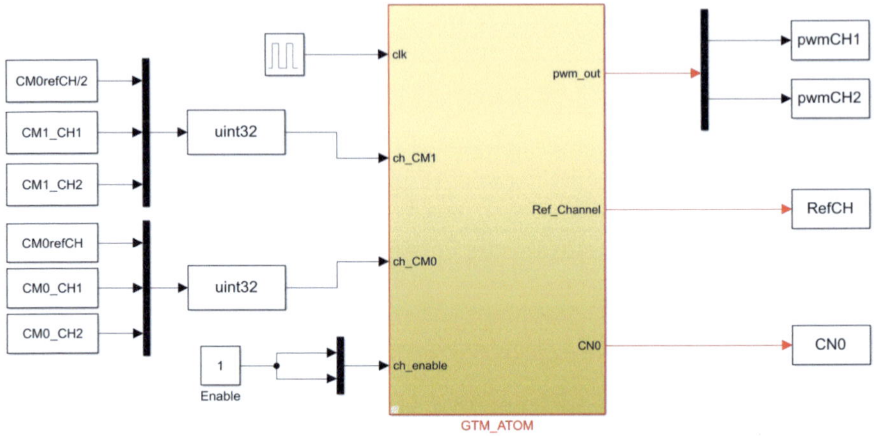

Fig. 7.28 Simulink block diagram for GTM ATOM sub-module to implement a complementary PWM with deadtime

Chaps. 8 and 9, deadtime is strictly necessary for power structures such as half-bridge or full bridge. The deadtime effect has an effect on the behavior of the power plant that is very important to consider since it reduces the performance to a degree not totally disposable. In this section, a model of the Simulink-based ATOM sub-module will be developed for use in the simulation of a three-phase DC or AC electric machine control system.

Based on the operation of the ATOM sub-module, it is possible to make the Simulink model using the Stateflow tool. Figure 7.28 represents the block of the ATOM model for a reference channel and two complementary PWM channels with the possibility of adding the deadtime. The reference channel or master channel sets the periodicity of the PWM and synchronizes the other channels as seen in Sect. 7.1.1.3. The clock signal is 100 MHz, and the PWM frequency considered in this example is 20 kHz. Therefore, the reference channel should be preset to 5000. The minimum realizable time is 10 ns, which is conditioned by the clock frequency of the GTM. The inputs CM1 and CM0 control on rising and falling edges respectively for SL = 0 and vice versa for SL = 1.

The Stateflow diagram of the reference channel is shown in Fig. 7.29. The initial state is the off state. Here, the output is initialized to zero. If the channel is enabled through the enable signal, it moves to the on state. Then, it enters in the sub-state LOW_LEVEL, where the counter CN0 is initialized to 1 and the output to 0. While it remains in this state CN0, it increases, and if it reaches the value of CM1, it goes to the sub-state HIGH_LEVEL. In this sub-state, the output is set to 1, and the counter CN0 is still increased. If the counter reaches the value of CM0, it goes back to the sub-state LOW_LEVEL, repeating the same process. As long as it remains in the on state if the enable input signal is zero, it will go to the off state by deactivating the output.

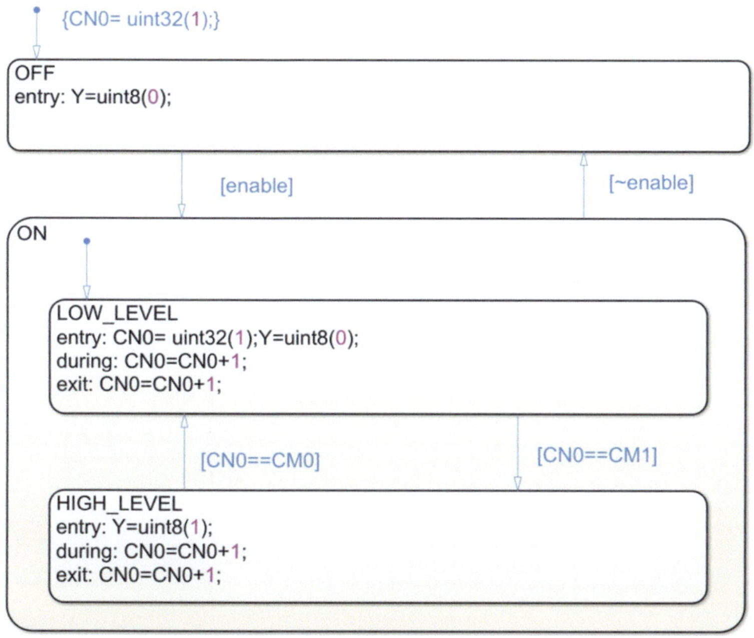

Fig. 7.29 Stateflow state machine for reference channel 0

The content of channel CH1 equivalent to CH2 is shown in Fig. 7.30. As in the previous case, if not enabled, the channel will remain in the off state. Here, the CM0 and CM1 registers are initialized to the desired value according to the duty cycle. In addition, the output is set to zero. If enabled, the channel will go to the on state, executing the TOGGLE_PWM sub-state and the SYNC_UPDATE sub-state. In the SYNC_UPDATE state, the internal variables CM0_CH and CM1_CH are updated with the values CM0 and CM1 defined for this channel according to the duty cycle. In the TOGGLE_PWM sub-state, the output is set to 1 or 0 depending on the status of the reference channel counter starting with high or low levels depending on the value of CTRL_SL. For example, if the counter reaches a value of CM1_CH and CM1_CH is higher than one, it will go to high sub-state by inverting the output.

The simulation result of the ATOM sub-module for two complementary PWM channels in central alignment is shown in Fig. 7.31. The CN0 counter of the reference channel is incremented every 10 ns, and when it reaches the default value of 5000, it is automatically reset. The reference signal corresponds to a signal with a 50% duty cycle that is set to one when the CN0 counter of the reference channel reaches 2500 and is reset to zero when it reaches 5000. This reference signal can be used to trigger, for example, an A/D converter so that the measurement is synchronized with the PWM. In addition, the reference channel counter is synchronized with the CN0 counters of the CH1 and CH2 channels. Channels CH1 and CH2 are complementary and with a deadtime of 1 μs so that it can be used in a half-bridge or full-bridge

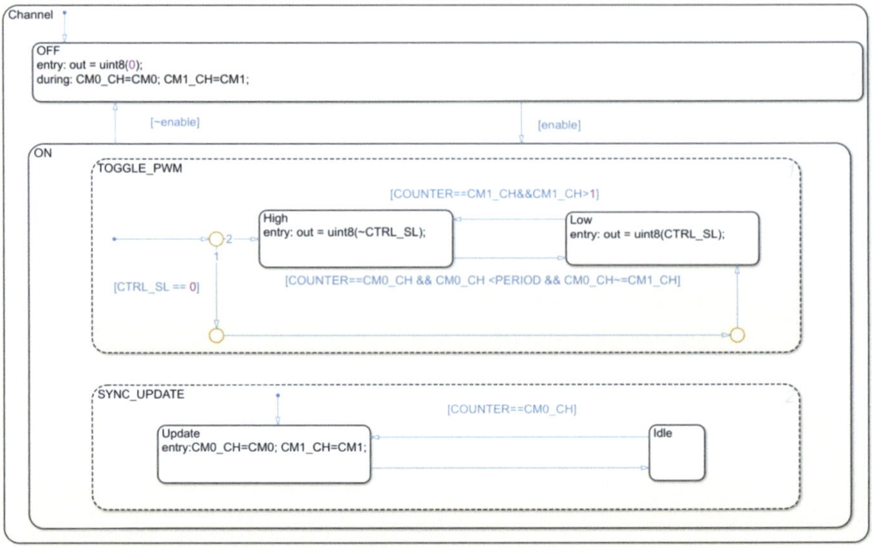

Fig. 7.30 Stateflow state machine for channel 1 or 2

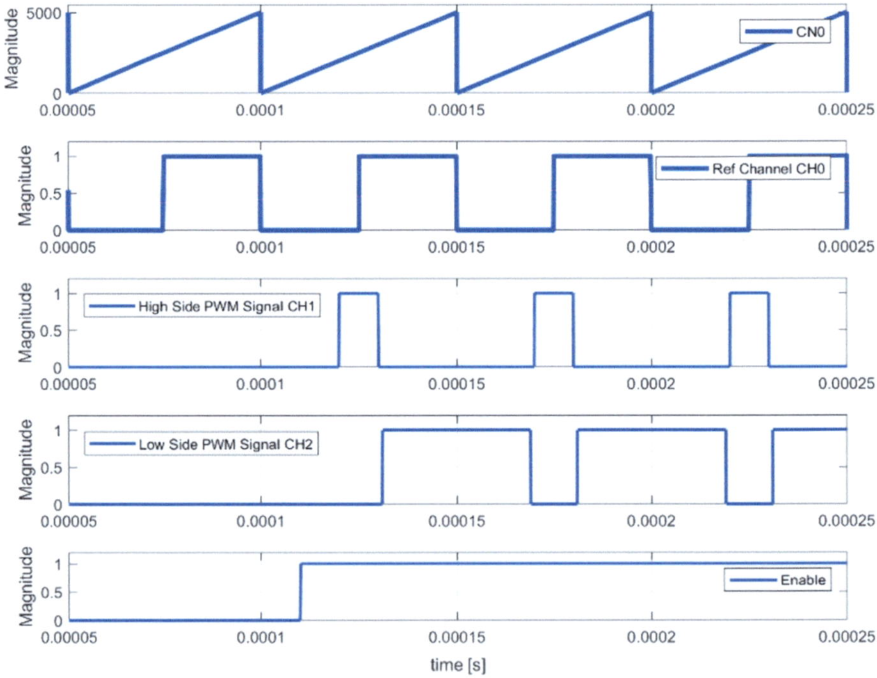

Fig. 7.31 High and low PWM signals at 20 kHz (5000) of switching frequency when enable signal is activated. CM0_CH1 = 3000 and CM1_CH1 = 2000 for SL = 0. CM0_CH2 = 3100, CM1_CH2 = 1900 for SL = 1

power structure. It is possible to appreciate that the duty cycle for the PWM high signal is 20%. Finally, the enable input signals are used to enable the corresponding channel. In this case, they are enabled at the same time at $t = 0.00011$ s.

7.5.2 ATOM Configuration

The basic configuration consists of first selecting the channels to be used according to the output pins of the microcontroller since the channels are associated with certain pins. In this case, channels 0, 1, and 2 of the ATOM0 are used. In particular, channel 0 is selected to be used as a reference channel, while channels 1 and 2 are selected to generate the high- and low-sides PWM signal. These last channels are associated with an output microcontroller pin. The ATOM sub-module configuration consists of eight steps, as illustrated in the flowchart of Fig. 7.32.

As can be observed in the figure, the first step is to define which ATOM0 channels receive a forced update event if there has been a trigger. In this case, channel 0 is the one that carries out the trigger because it is the master channel, and those that receive this trigger signal are channels 1 and 2. The second step is to select the operating mode of the channel (PWM mode) and select the source clock, as well as polarity SL bit. In the third step, reference values are imposed on the shadow registers of each channel. As commented, channels 1 and 2 are slaves of the reference channel 0. Thus, at the time instant which channel 0 is reset to zero to start counting again, channels 1 and 2 must also be reset in the same way. This is achieved in the fourth step where a trigger is configured at the time instant of reset of channel 0 that goes to channels 1 and 2.

In the fifth step, it is enabled if the PWM signal should be output or not in the microcontroller pin. In this case, all three channels are enabled. In the sixth step, the UPEN registers define for which ATOM channel the update of the CM0 and CM1 registers by means of the SRx shadow registers are enabled. That is, if enabled, the CM0 and CM1 registers will be updated on the reset of the CN0 counter with the values of the SRx registers (synchronous update). In the penultimate step, a CCU1TC interrupt is enabled so that it is launched just when CN0 >= CM1 is met. Also, the pulse interrupt notification mode is chosen. Finally, the last step enables triggering to update the ENDIS_STAT and OUTEN_STAT registers.

The SET_DUTY function of Fig. 7.33 is responsible for modifying the SRx shadow registers based on the desired duty cycle. The duty cycle has a range from 0 to 1. The value of the period, in the case of a switching frequency of 20 kHz, is 5000, as seen in the simulation of the ATOM sub-module in the previous section. If the duty cycle is higher than a preset minimum value and is enabled, the SRx registers will be modified according to the duty cycle value and deadtime DT. On the contrary, if neither is fulfilled, the registers are configured with the recommended values to impose a zero on both channels as can be seen in the flowchart of Fig. 7.33.

The channel 0 interrupt routine simply consists in deleting the interrupt notification bit by writing a “1” in the notification. As previously mentioned, the ISR

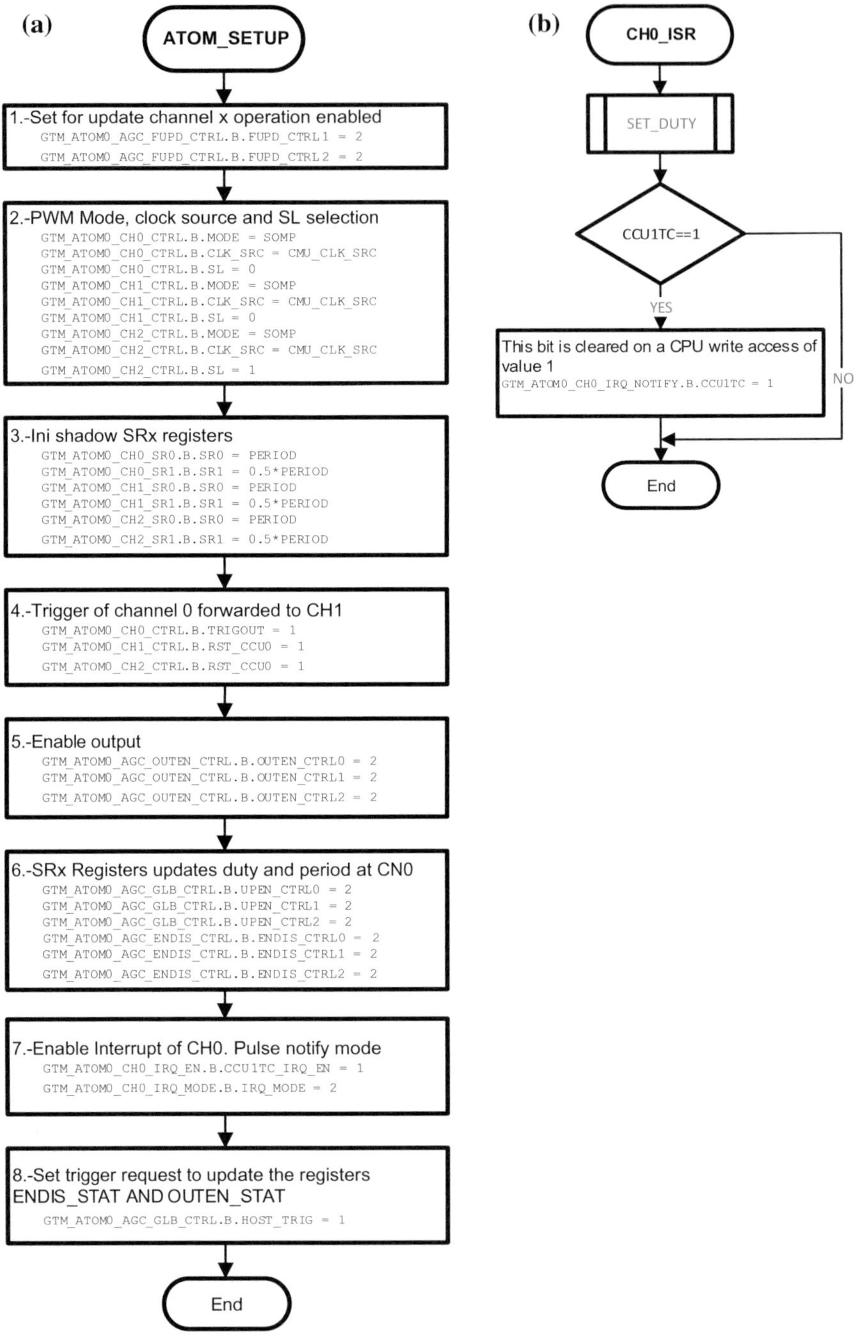

Fig. 7.32 **a** ATOM setup for PWM generation. It is called one time only. **b** Interrupt of CH0 which is called in the middle of the triangular signal

Fig. 7.33 Function which sets the shadow registers with the values calculated according to the duty cycle using center-aligned

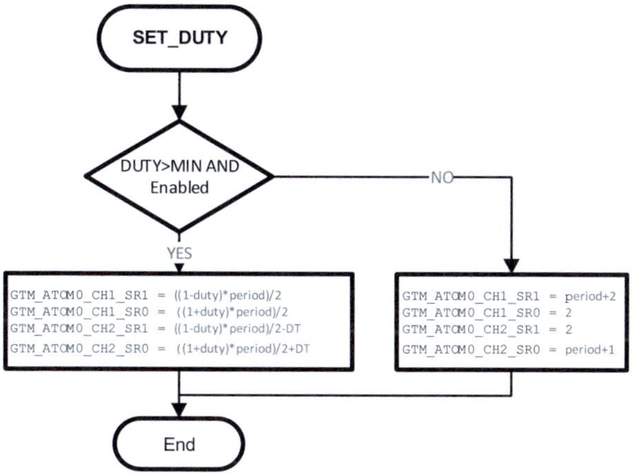

interruption will be launched in the middle of the triangular, that is, due to the duty (CN0 >= CM1). In order to avoid "glitches" in the PWM output, it is convenient to modify the SRx shadow registers within the interruption. In this case, the SET_DUTY function of Fig. 7.33 must be called within the interrupt, as shown in the flowchart of Fig. 7.32b. If this is not done in this way, it may happen that the interruption is executed when the SRx registers were being updated, resulting in a "glitch" during a PWM period.

Figure 7.34 represents an experimental result of the ATOM configuration for a three-phase inverter where PWM waveforms can be seen. The master signal named D_0 can be seen next to the six slave signals centered in the middle of the period

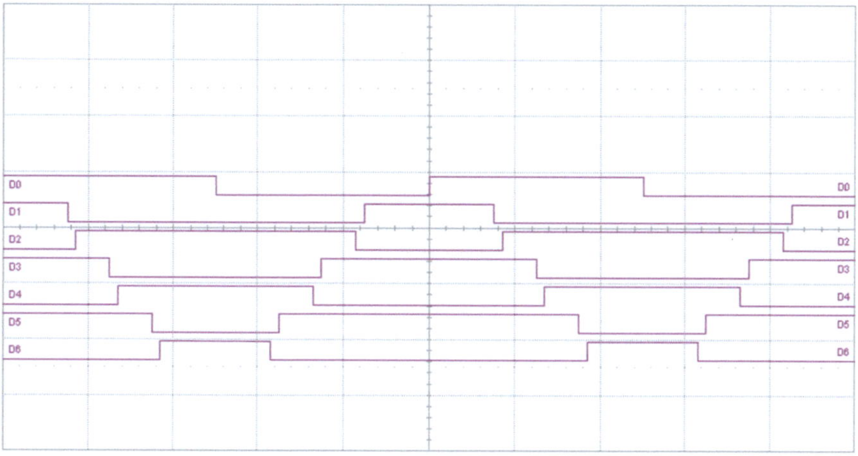

Fig. 7.34 Complementary PWM signals of six channels at 20 kHz with the toggle output signal in D_0 channel

named from D_1 to D_6. The frequency of the PWM is 20 kHz, the deadtime is 1 μs, and the duty cycles are 30, 50, and 70%.

7.5.3 Simulation of SDADC

The modulator used by the TriCore™ family consists of a third-order MASH 1-1-1 modulator made by cascaded first-order stages. The error of each stage is fed to the next stage as can be seen in Fig. 7.35. The outputs are then combined in a noise shaping block that cancels the noise of the first stages producing multi-bit outputs that exhibit the *n*th order characteristic noise formation. Then the total noise is molded and pushed at higher frequencies. The output levels that the MASH 1-1-1 modulator will produce are 2^3 levels, that is, eight levels.

The MAIN CIC filter in Fig. 7.36 corresponds to a three-stage sinc filter, as seen in Sect. 7.3.4. As can be seen in Fig. 7.20, it is possible to bypass the filter as is the case with FIR filters.

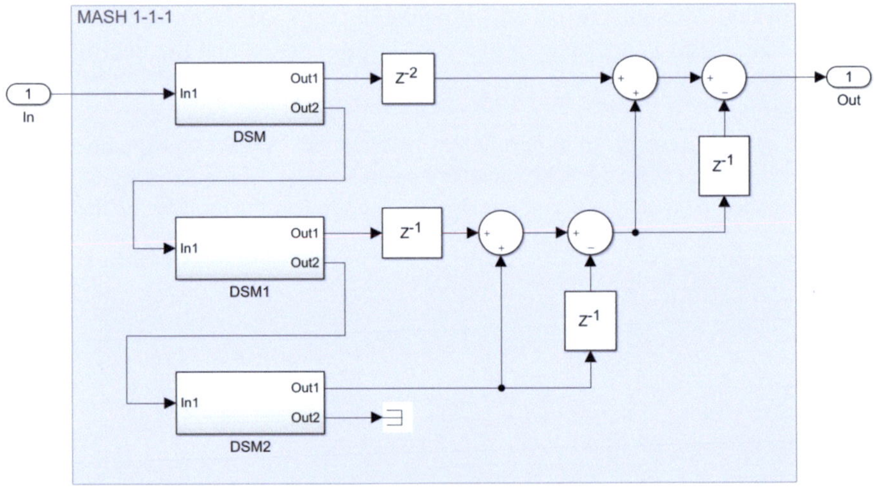

Fig. 7.35 Simulink MASH 1-1-1 model based on first-order stages in cascade connection

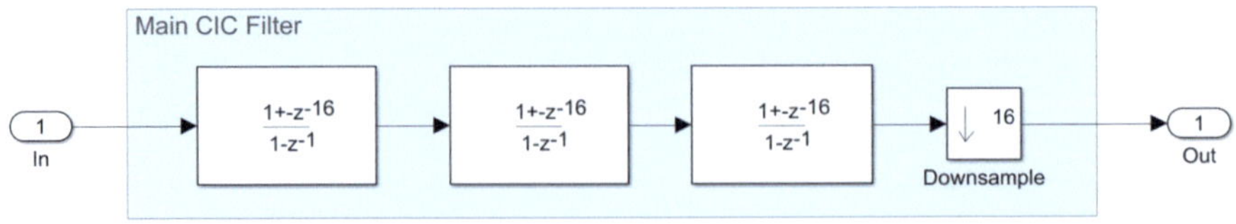

Fig. 7.36 Simulink MAIN CIC filter stage

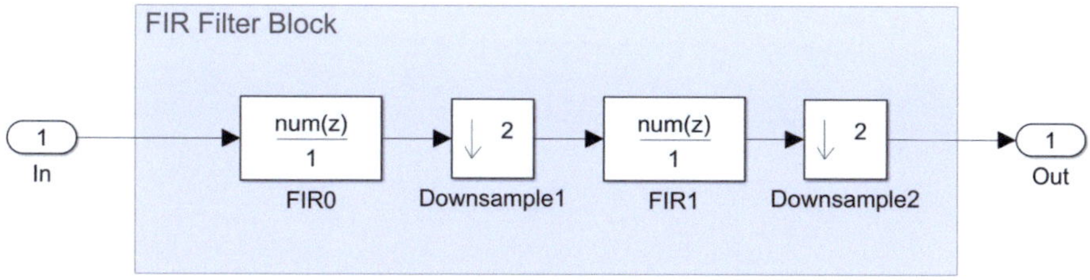

Fig. 7.37 Simulink FIR filters stage with decimation factor 2 for every filter

Finally, FIR filters provide additional filtering of those frequencies that the CIC filter could not filter. It consists of two FIR filters: the FIR0 consists of 8 coefficients and the FIR1 of 28 coefficients. Both have a decimation factor of 2. The block diagram of the FIR filters can be seen in Fig. 7.37.

The coefficients of both filters are fixed and correspond to those described in (7.7) and (7.8). The frequency response corresponds to a low-pass filter with a gain of 56 dB, as can be seen in Fig. 7.38.

$$H(z)_{FIR0} = 17 - 25z^{-1} + 83z^{-2} + 256z^{-3} + 256z^{-4} + 83z^{-5} - 25z^{-6} - 17z^{-7} \tag{7.7}$$

$$\begin{aligned} H(z)_{FIR1} = & -3 - 2z^{-1} + 2z^{-2} + 9z^{-3} - z^{-4} - 14z^{-5} - 8z^{-6} + 19z^{-7} \\ & + 25z^{-8} - 13z^{-9} - 55z^{-10} - 12z^{-11} + 126z^{-12} + 256z^{-13} + 256z^{-14} \\ & + 126z^{-15} - 12z^{-16} - 55z^{-17} - 13z^{-18} + 25z^{-19} + 19z^{-20} - 8z^{-21} \\ & - 14z^{-22} - z^{-23} + 9z^{-24} + 2z^{-25} - 2z^{-26} - 3z^{-27} \end{aligned} \tag{7.8}$$

If all the blocks are joined, it is possible to obtain the structure of the DSADC of the AURIX™ family of Infineon as can be seen in Fig. 7.39. A periodic signal generator with offset and white noise can be connected to the input as can be seen in

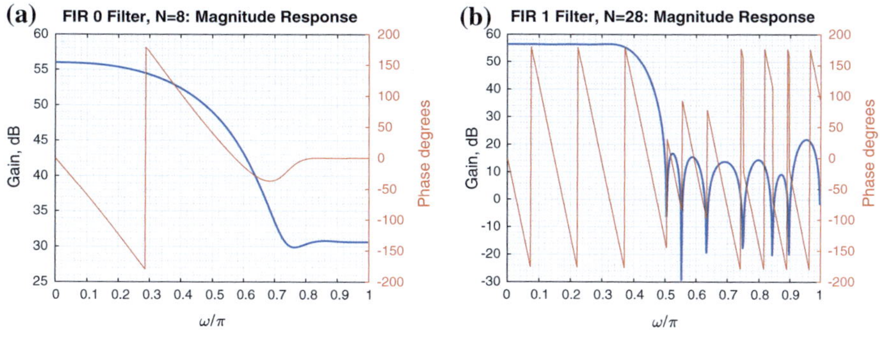

Fig. 7.38 Frequency response for FIR filters at normalized frequency. **a** 8 coefficients FIR0 filter. **b** 28 coefficients FIR1 filter

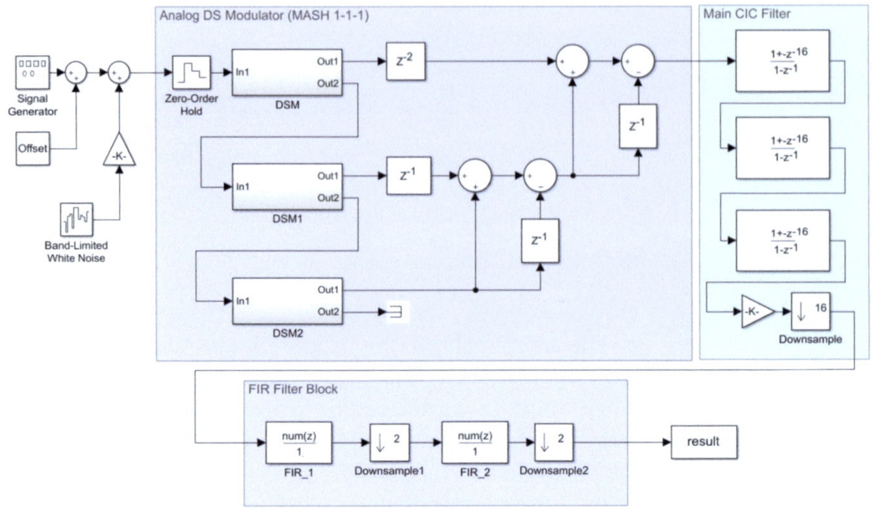

Fig. 7.39 Model-based block diagram of DSADC of AURIX™ family. Decimation factor is 64

order to test the model. It is also possible to observe that the total decimation factor corresponds to 64, 16 for the CIC filter, and 4 in total for FIR filters.

In an interesting case, it is possible to use the DSADC to measure the current of an RL circuit, which is the fundamental part of the electric machines. The RL circuit is connected to a full-bridge structure composed by MOSFET power switches, as shown in Fig. 7.40. In the full-bridge structure, it is possible to see positive and negative currents for a sinusoidal PWM modulation. To facilitate the interpretation of the results, the frequencies are very low, that is, 20 Hz for the main voltage

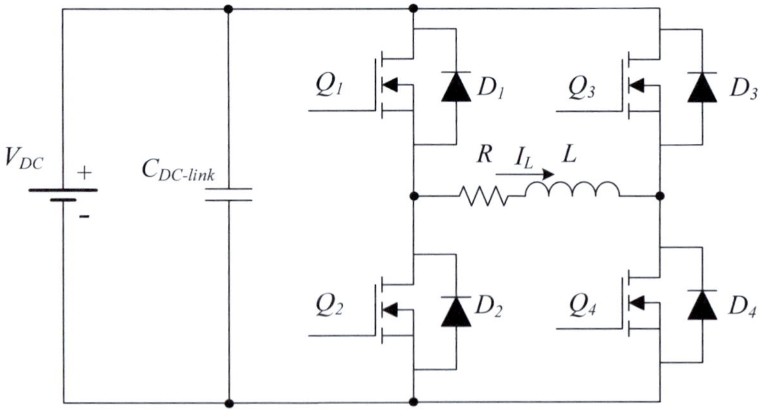

Fig. 7.40 Full-bridge MOSFET with RL circuit connected as load. The RL circuit is the typical load of a basic electric machine

harmonic and 300 Hz for the PWM. The full-bridge voltage supply is 1 V_{DC} voltage. The inductor value is 50 mH, and the resistance is 10 Ω. The sampling frequency of the selected DSADC is 20 MHz.

Figure 7.41 represents the PWM voltage waveform applied to the RL circuit with an amplitude of 1 V next to the inductor current. The current ripple, as can be seen, varies depending on the duty cycle of the PWM. As previously mentioned, the duty cycle varies with a sinusoidal form which increases from 0° to 90° and from 180° to 270°. As expected in an RL circuit, the inductor current is lag behind the voltage, as shown in Fig. 7.41. In this case, it is approximately 36°.

The result of the inductor current measurement is represented in Fig. 7.42. Due to the high sampling frequency of the DSADC, it is possible to reproduce precisely the shape of the inductor current with the included ripple, although the PWM frequency is much higher, for example, 20 kHz. The output variable consists of a 16-bit signed

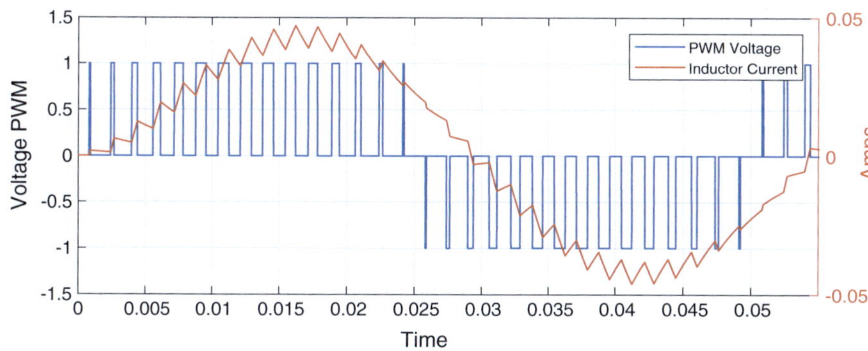

Fig. 7.41 Voltage drop in the RL load and inductor current. The voltage applied to the RL circuit is a sinusoidal PWM voltage. The fundamental frequency of the main harmonic of the inductor current is 20 Hz

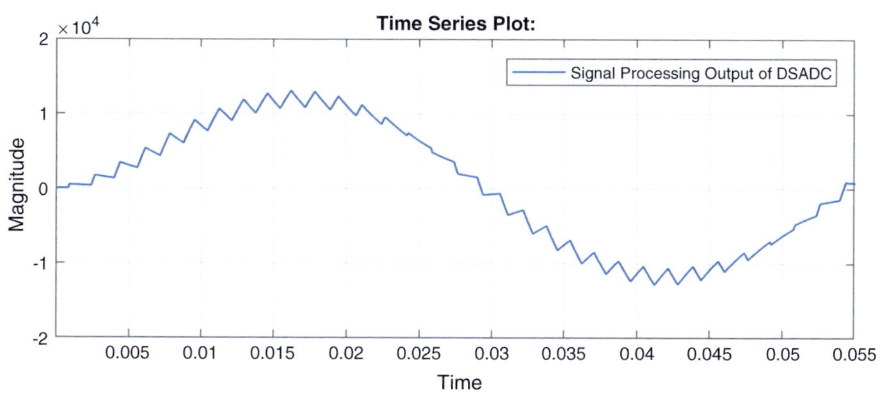

Fig. 7.42 Output simulation of DSADC with the inductor current acquisition at a sampling time of 20 MHz. The sampling ratio is 312.5 kHz

integer variable that can already be processed by the microcontroller. The resulting sampling rate corresponds to 312.5 kHz since the total decimated factor applied is 64. According to the DSADC hardware architecture, the bandwidth is one-third of the resulting sampling rate, that is, 104.17 kHz. The sampling rate can be lowered further by simply changing the decimation factor of the CIC filter without affecting the result as long as the bandwidth is higher than the signal to be measured. This reduces CPU processing since it will process the samples with a lower rate. For example, if it is desired to eliminate the current ripple in order to take average value, the digital filter processing will require less CPU performance for lower sampling rate. In the case of a decimating factor of 256, the sampling rate is 78.13 kHz, and the corresponding bandwidth is 26.04 kHz.

7.5.4 Simulation of MTU for Three-Phase Machines

As in the case of the ATOM sub-module seen previously, in the case of the Timer MTU3–4, it is also possible to perform the Simulink model. In this case, the objective is to model how PWM signals can be reproduced for a three-phase inverter with deadtime to be used in the Simulink environment.

Figure 7.43 shows the block diagram of the Simulink model of the MTU3–4 unit for the six required output channels. The inputs correspond to the CPU register where it can write the values of the corresponding duty cycles for each of the phases as previously seen.

Figure 7.44 represents the sub-blocks inside of Fig. 7.43 with six independent channels and MTU3 and MTU4 counters. MTU3–4 generates a periodic signal (triangular signal or reference signal) where the PWM periodicity is set for the two delayed reference signals separated with the desired deadtime. Finally, with a few simple comparators, the PWM outputs can be generated according to the set duty cycles. The comparison is made at each clock cycle so that, if the MTU4 counter value is larger than the desired duty cycle, the comparator output changes the PWM level. In the case of the MTU3 counter, the comparison is in a reverse way, since the PWM output of the low-side power switches devices must be inverted.

The result of the simulation can be seen in Fig. 7.45. The timer clock cycle is 100 MHz. In this example, the values of the MTU3.TGRD, MTU4.TGRC, and MTU4.TGRD registers are 2000, 1300, and 600, respectively. The PWM frequency is 20 kHz so that the peak of the triangular signal for MTU4 is 2500, that is, $100 \times 10e^6/20{,}000 = 5000/2$. The deadtime has been inflated a little to be able to see it in the graph and corresponds to 250, that is, 2.5 μs. The MTU3 counter has a peak of 2750, while its minimum is 250 as can be seen in Fig. 7.45.

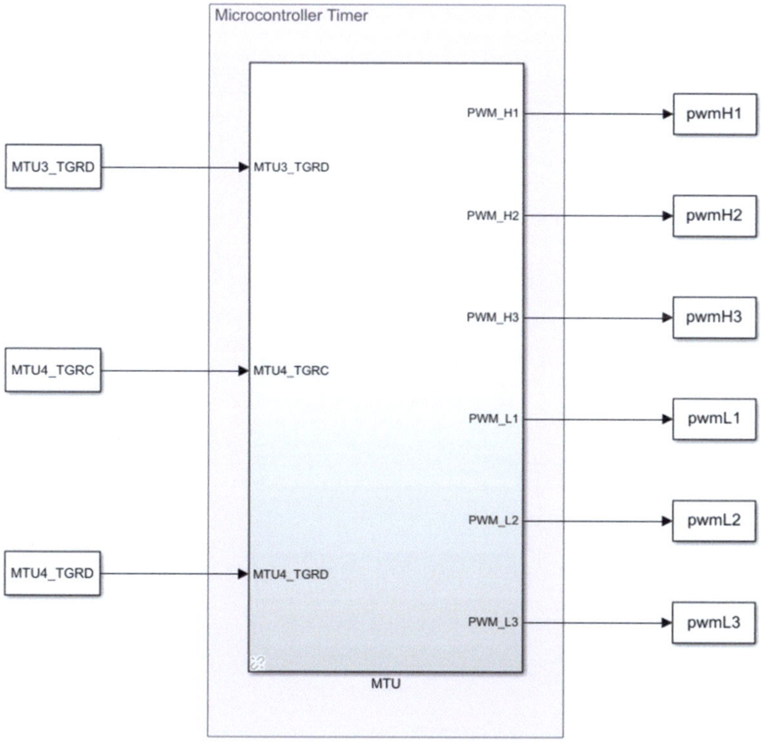

Fig. 7.43 Simulink block diagram of MTU3-4 for PWM generation with deadtime to be used in the Simulink environment

7.5.5 *MTU3-4 PWM Configuration*

The requirement specifications for setting the timer are as follows (Table 7.1):

The MTU3, and MTU4, or MTU6, and MTU7 configuration channels can be performed easily with a sequential order manner as recommended in the user manual of the RX63T microcontroller. The flowchart in Fig. 7.46 shows the configuration of the MTU3 and MTU4 channels for a complementary PWM configuration for six output channels. It is possible to observe that the configuration is done in eleven sequential steps. The first three steps correspond to the counters stop, clock selection, and initialization of the MTU3 and MTU4 counters. In steps four and five, the duty cycles are initialized, and the reference counters are configured with the deadtime. In steps six and seven, the activation level is set to one (one means power device activation), and the complementary PWM mode and the use of internal buffers are configured. In steps eight and nine, the output of the channels is enabled to be associated with the output port, and no interruption is skipped. That is to say; the interruption is caused to skip each PWM period, being able to specify if necessary how many PWM periods the interruption should be skipped. This is very useful when

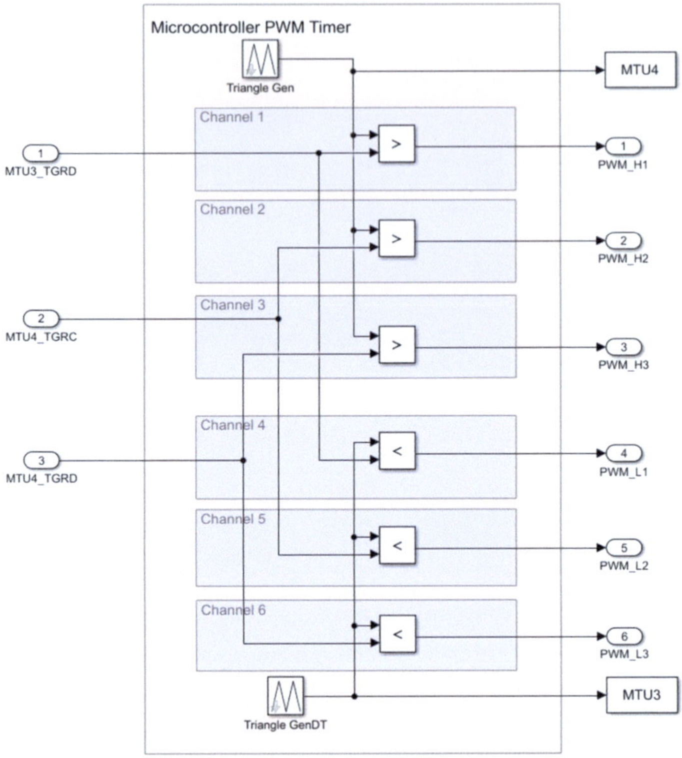

Fig. 7.44 Inside of Simulink block diagram of MTU3-4. Six channels are used, and two triangular signals are compared with the MTU3.TGRD, MTU4.TGRC, and MTU4.TGRD inputs

the control is to be carried out at a lower speed and multiple of the PWM period as discussed early. Finally, in steps ten and eleven, the A/D converter is configured to trigger in the valley of the triangular signal of the MTU4 counter. When the conversion of the configured A/D channels ends, the A/D ISR will be launched to process the data. On the other hand, the SET_DUTY function is responsible for updating the MTU3.TGRD, MTU4.TGRC, and MTU4.TGRD registers with the desired duty cycles. Only if the duty cycle exceeds, the minimum allowed is the duty cycle value copied to those registers. Otherwise, when the PWM outputs must be inactive, it is necessary to initialize the registers with a higher value than or equal to the crest of the MTU3 counter, i.e., 2500 + 100. This will mean that there is no change of level in the PWM output, and the outputs of the low-side devices remain at zero.

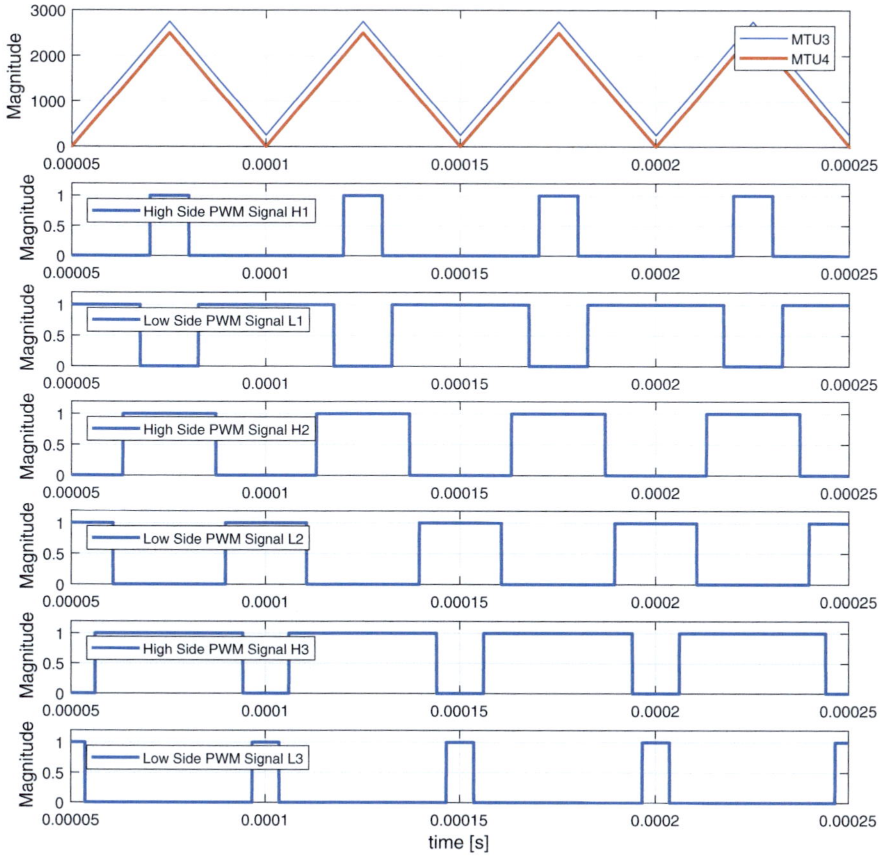

Fig. 7.45 PWM signals generated by Simulink simulation when MTU3.TGRD, MTU4.TGRC, and MTU4.TGRD values are 2000, 1300, and 600

Table 7.1 Specification of the complementary PWM

Item	Set value
PWM frequency	20,000 Hz
Deadtime	1 μs
A/D trigger	Valley (underflow of MTU4)
Active level	High
Registers transfer	At the crest and buffering
Skipped interrupt	No

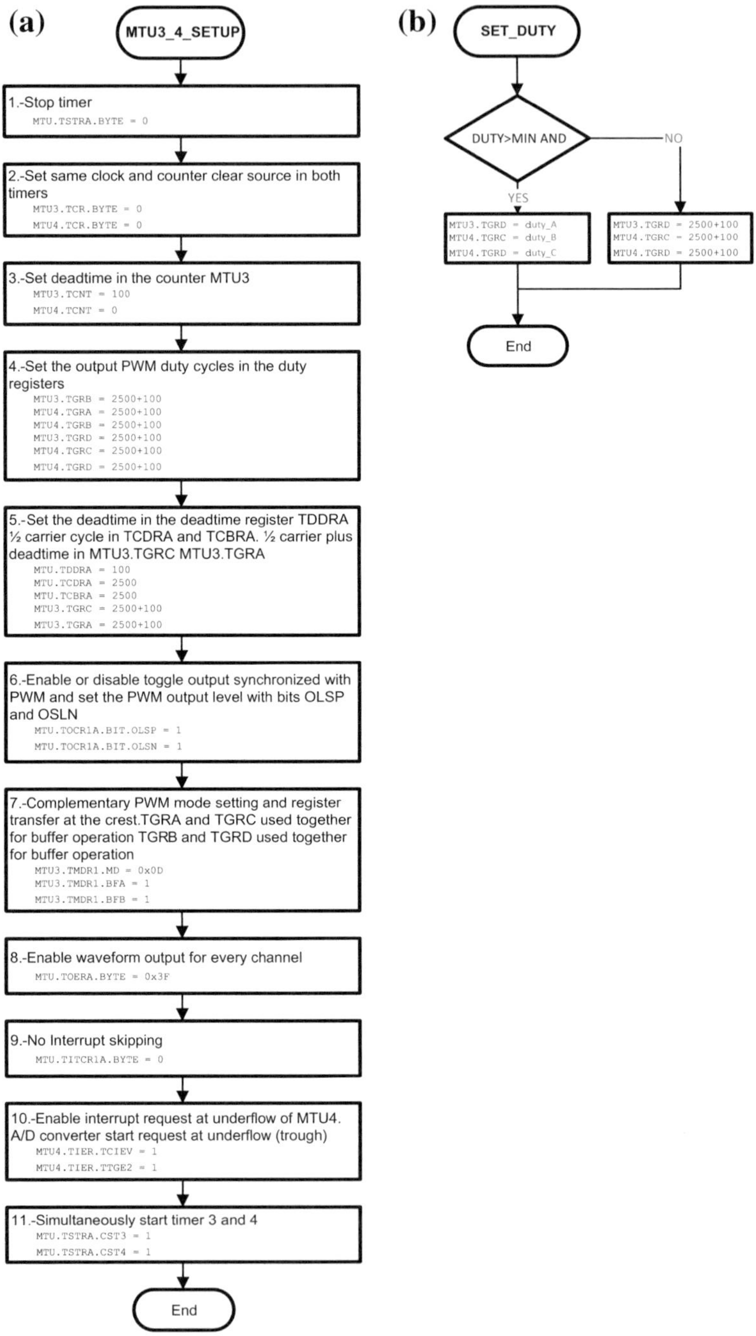

Fig. 7.46 **a** Flowchart of the configuration of MTU3 and MTU4 in complementary PWM. **b** Flowchart of duty cycle setting

7.5.6 MTU5 Configuration

The MTU5 channel configuration is much simpler than the previously seen configuration of the MTU3 and MTU4 channels. As the most essential point would be the second and third steps of the flowchart of Fig. 7.47a. Here, it is configured to measure the time while the pulse is low both in the crest and in the valley and the three counters are reset. It is in the ISR where the processing of the captured times is carried out. In this case, the A/D ISR is configured so that, when the A/D measurement is finished, the A/D ISR is launched, which has previously been triggered by the MTU4 counter underflow as seen in the previous section. In the A/D ISR, the data captured by each of the MTU5 channels corresponding to the three phases of the electric machine is read first, and the previous duty cycle values are subtracted as can be seen in Fig. 7.47b in step two. In step three, the new duty cycles corresponding to the control

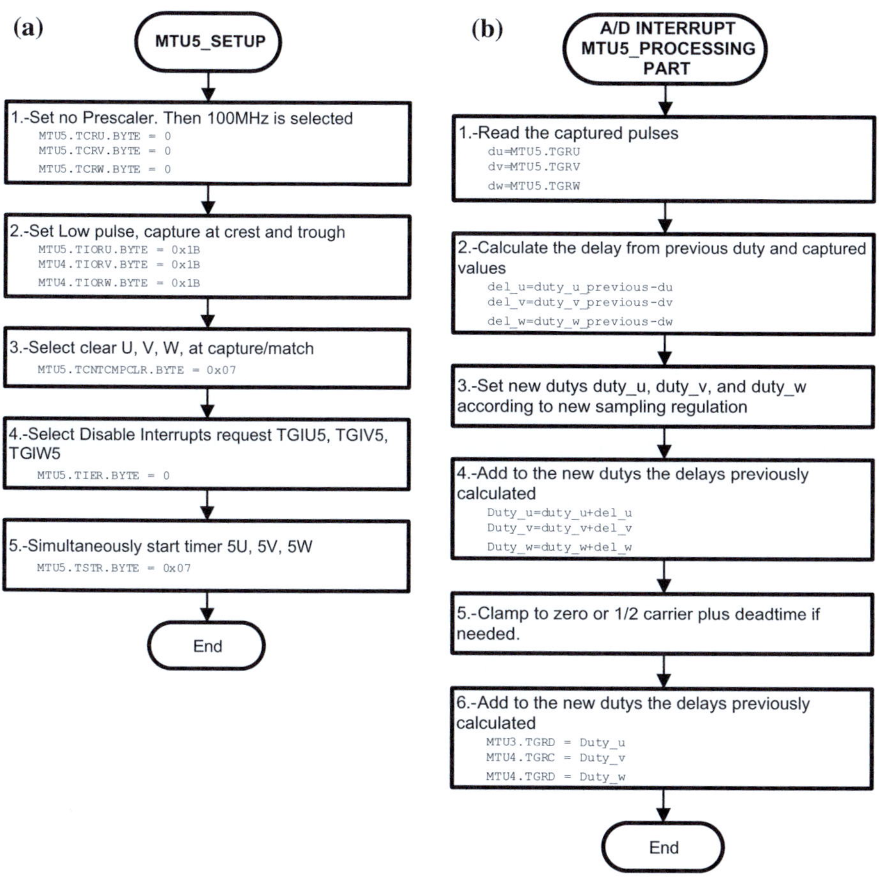

Fig. 7.47 **a** MTRU5 configuration flowchart for deadtime compensation. **b** MTU5 processing flowchart part in the A/D converter ISR interrupt

algorithm would be calculated, and in step four, the previously calculated values are added, thus compensating the deadtime. If necessary, saturation is performed if the maximum value is exceeded or if it is less than zero, it is set to zero. Finally, duty cycles are assigned to MTU3.TGRD, MTU4.TGRC, and MTU4.TGRD registers.

As will be seen in Chap. 8, the voltage applied to a three-phase electric machine can be reconstructed from the PWM duty cycle. In some cases, it is usually reconstructed using the same applied duty cycle. Then knowing the DC-link voltage of the three-phase bridge, it is possible to reconstruct the phase and line voltages of the electric machine. A particular case with better precision is to measure the duty cycles applied to the electric machine using the MTU5 timer. The advantage over the previous case is that the deadtime is already taken into account so that the measure will be more precise. Also, since the measurement is digital, there are no delays caused by the analog filters in the case of performing a conventional voltage measurement based on voltage dividers, low-pass filters, and an A/D converter. The delay in the voltage measurement in the machine terminals is extremely significant for high-performance control so that it must be minimized. It is important to keep in mind that deadtime compensation is only possible if MTU6 and MTU7 channels are used to generate PWM signals. In the previous examples, the configuration for the MTU3 and MTU4 channels has been used, but it is similar in the case of using the MTU6 and MTU7 channels.

7.5.7 Simulation of A/D Converter

Figure 7.48 shows the previously simulated MTU unit next to a half-bridge connected to an RL circuit. The objective is to measure the average current that passes through the inductor through the 12-bit A/D converter on a simple circuit. One of the most economical and effective ways to measure the current passing through the inductor is to place a shunt resistor between the source and ground of the low-side MOSFET as it can be seen. As will be seen in Chap. 8, the current sensing for this case can also be done by placing the shunt resistor in series with the RL circuit. It is possible to anticipate that the main advantage of sensing the current at this point is that the power that the shunt resistor must withstand can be lower because current only circulates when the low-side MOSFET is turned on. This forces the measurement of the A/D converter to be synchronized with the PWM, especially with the low-side PWM output. To do this, the A/D converter should be triggered in the underflow of the MTU4 signal as previously mentioned. The voltage drop in the shunt resistor is amplified by an inverter amplifier, and a voltage offset is added typically to measure negative currents. The offset is adjusted to half the full scale of the analog input to have the same measuring range for both positive and negative currents. In the case of the RX63T microcontroller, the half-scale voltage corresponds to 1.65 V and its RAW value will be 2048 ($2^{12}/2$). The value of shunt resistance is 1 mΩ.

The MTU4 signal and the trigger signal waveforms can be seen in Fig. 7.49. The A/D converter trigger can be observed at each underflow of the MTU4 signal.

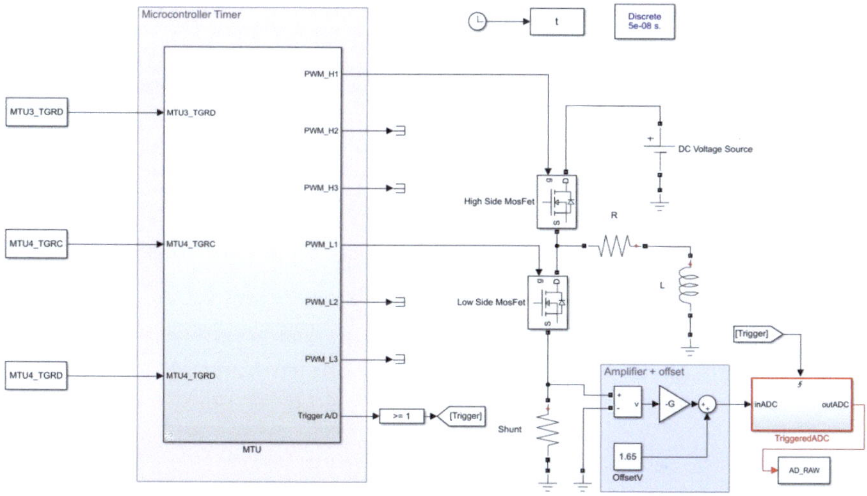

Fig. 7.48 A/D current measurement in a simple RL circuit with half-bridge controlled by the MTU3 timer

Between time instant $t = 0$ and $t = 0.0001$ s, the PWM signals remains inactive so that the A/D measurement is just half the full scale, that is, 2048 counts. From time instant $t = 0.0001$ s, the PWM is activated with a constant 20% duty cycle in which the inductor current begins to increase until it stabilizes at time instant $t = 0.0006$ s at about 2.5 A on average. As previously mentioned, the current through the shunt resistor is the inductor current when the low-side MOSFET is on. It is possible to observe that the current measurement is carried out just in the middle of the slope of the inductor current and the A/D RAW value increases until it stabilizes in approximately 2250 counts. It is important to comment that the A/D results could now be precisely the inductor current average value, but it is an approximation. To obtain exactly the inductor current average value, it should be necessary to oversampling the current, for example, with a Delta-Sigma converter (not available in the RX63T) as discussed in Sect. 7.5.3.

7.5.8 *A/D Configuration for Three-Phase Machines*

In this section, the 12-bit converter will be configured in group scan mode synchronized with the PWM signal to read the three currents of a three-phase machine. The A/D converter trigger must be performed by the MTU3 timer, and when the conversions of the three currents are finished, an ISR will be launched to process the control with the new measurements. The specifications to configure the A/D converter is in Table 7.2.

Figure 7.50a, shows the necessary configuration to acquire the three currents of

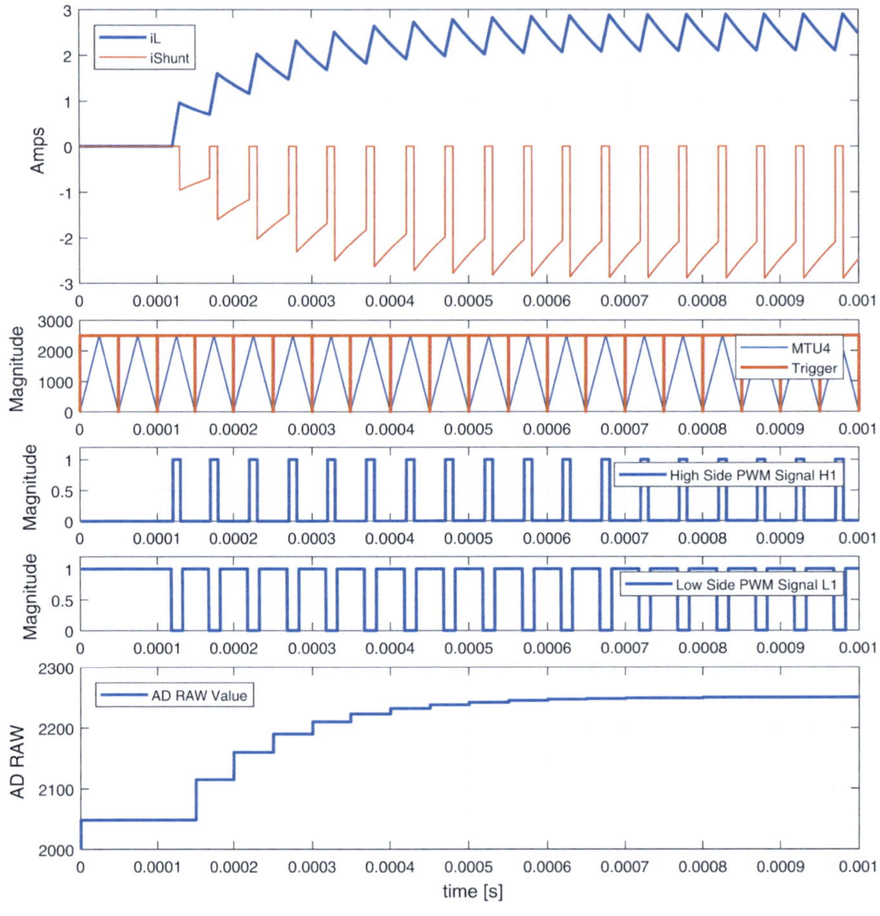

Fig. 7.49 Simulink simulation of the RL circuit with half-bridge controlled by MTU3 timer

Table 7.2 Specification of the A/D converter

Item	Set value
Trigger	MTU3 timer unit. Underflow of MTU4
A/D clock	50 MHz
S/H	Yes
Bits	12
Mode	Scan
A/D time conversion	<1 μs

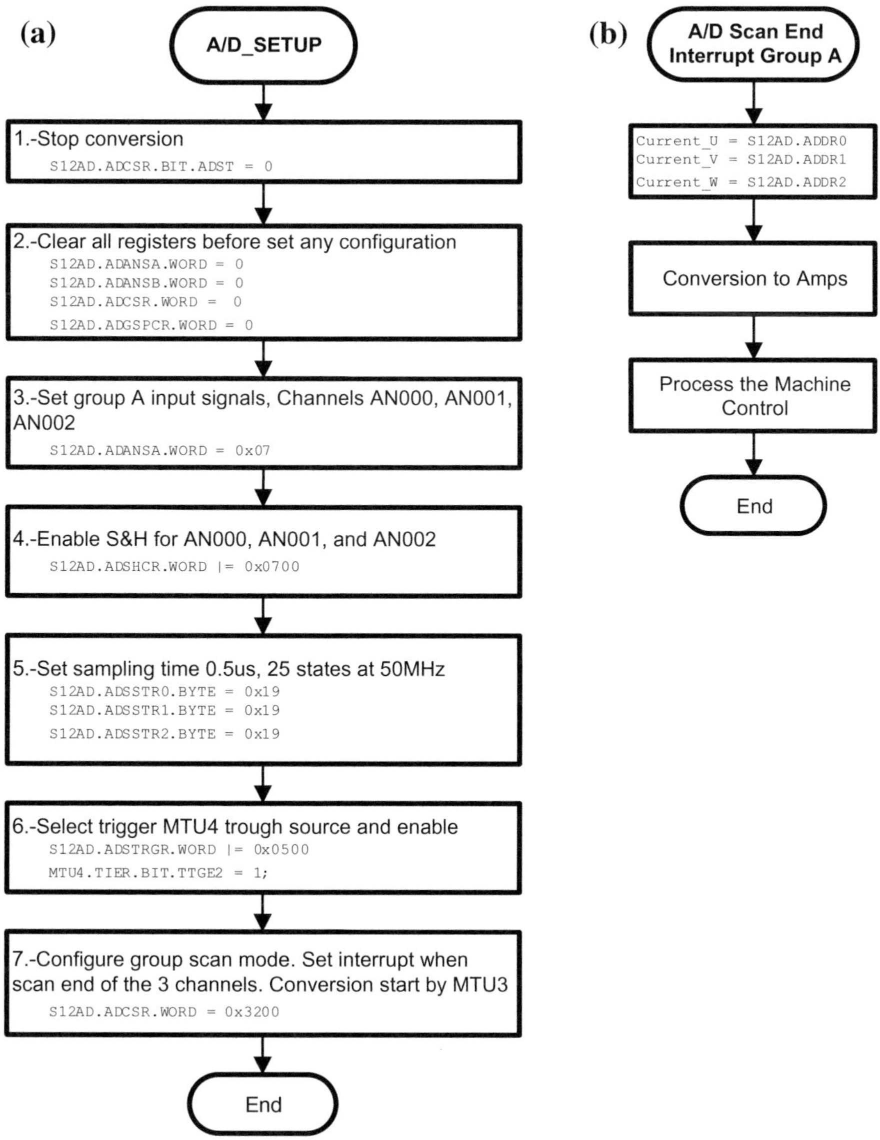

Fig. 7.50 **a** Flowchart of the configuration of S12ADB A/D converter unit. **b** Flowchart of scan end interrupt and sampling processing

the three-phase electric machine according to the specifications in Table 7.2. In steps one and two, the converter simply stops the A/D converter, and some configuration registers are reset before configuring them. In step three, the needed channels of group A are configured. In steps four and five, it is defined that the S/H is used for the three channels, the clock frequency of the A/D converter and the sampling time of 0.5 μs (25 states at 50 MHz). In step six, it is indicated that the converter is triggered by MTU4 channel in the valley or underflow. Finally, in step seven, it is configured that the A/D measurement begins according to the event of the MTU3 unit and it is configured to automatically launch the A/D ISR when the three conversions are finished. Figure 7.50b shows part of the A/D interruption where the A/D measurements are stored for processing by the control as well as converting them to current in amps.

References

GTM-Cookbook (2014) GTM-Cookbook: overview and application examples, Revision 0.5. Robert Bosch GmbH. http://www.bosch-semiconductors.com/media/ip_modules/pdf_2/gtm/gtm_cookbook_v05.pdf

Harris FJ (2004) Multirate signal processing for communication systems. Prentice-Hall PTR, USA, p 343

Hogenauer EB (1981) An economical class of digital filters for decimation and interpolation. IEEE® Trans Acoust Speech Signal Process ASSP 29(2):155–162

Meyer-Baese U (2014) Digital signal processing with field programmable gate arrays. Springer, Berlin

Chapter 8
Analysis of Three-Phase Voltage-Source Inverters

8.1 Introduction

The voltage-source inverter (VSI) topology is a DC-AC converter that transforms a DC voltage into an AC voltage at its output. Analogously, the current-source inverter (CSI) topology transforms a DC current into an AC current at its output. The DC source of voltage or current is obtained either through the electrical grid, single-phase, or three-phase through a bridge rectifier, or by an AC generator also with a bridge rectifier, or by a DC battery optionally, through a DC-DC boost converter. To improve the inverter system performance, the DC voltage or current source should be as ideal as possible. For instance, in the case of the VSI , the DC voltage source should have an ideally zero Thevenin impedance. In some cases, it is necessary to connect a high DC-link capacitance in parallel at the inverter input. On the other hand, for CSI , the DC current source should have an infinite Thevenin impedance, which can be approximated by placing a series inductance between the source and the inverter. The inductance should have a high inductance value in order to obtain an approximately an ideal current source. The power flow is reversible in the DC side; the voltage source in the VSI is unidirectional voltage bidirectional current, while the current source in the CSI is unidirectional current bidirectional voltage. The most common inverter structure used is single-phase and three-phase. The power switch semiconductors mostly used are metal-oxide semiconductor field-effect transistor (MOSFET) or the insulated gate bipolar transistor (IGBT), but gate turn-off thyristor (GTO) and silicon controlled rectifier (SCR) can be found. MOSFET has replaced the traditional bipolar junction transistor (BJT) in many inverters due to some relevant factors such as their high switching speed, low on-resistance, and ease of control.

Since frequency, and voltage (VSI) or current (CSI) can be controlled independently in their three-phase configuration, they are an excellent candidate to accomplish the three-phase AC machine requirements. For example, in VSI , the capability to adjust the frequency and the voltage allows to control the speed and to maintain a constant air-gap flux in the constant torque region. In the literature, it is possible to find different nomenclatures to specify a variable speed AC drive or inverter VSI such

© Springer Nature Switzerland AG 2020

R. Molina Llorente, *Practical Control of Electric Machines*,
Advances in Industrial Control, https://doi.org/10.1007/978-3-030-34758-1_8

as variable-frequency drive (VFD) and adjustable-frequency drive (AFD) where both provide a variable-voltage/variable-frequency (VVVF). The common part of these drives is the speed control in electromechanical systems where the speed is adapted according to the plant system requirements.

The CSI is commonly used in high-power applications because it has a simpler and more reliable topology. It does not require extra power circuits, and the shoot-through and short circuit of short duration in the AC machine terminals are supported without difficulties by the power switch devices. The shoot-through occurs when both switches of the same inverter leg are turned on simultaneously as it will be seen in this chapter. In addition, it is capable of operating inherently in all four quadrants, offers a current independent of the load at its output, and has a low voltage slew rate ($\mathrm{d}v/\mathrm{d}t$) (Beig and Ranganathan 2006). The power semiconductor devices used are self-commutated devices capable of withstanding reverse voltage such as the GTO or the SCR. The main drawback is the slow dynamics of the current, which makes a slow control due to the large input inductor in the inverter, but in high-power AC machines with higher inertias are not usually a concern. In these high-power applications, three-level or five-level multi-level topology inverters are preferred over two-level topology inverters, since they reduce switching losses and the phase current ripple in the AC machine terminals (Bhagwat and Stefanovic 1983). CSI are increasingly used in medium voltage and high-power applications thanks to the advance of semiconductors, where a rapid dynamic response is not required (Beig and Ranganathan 2006).

On the other hand, the VSI offers a voltage independent of the load and better performance for medium and low power being the most used today. In this chapter, it will focus on three-phase two-level VSI of PWM output for low voltage and medium power.

Figure 8.1 shows an AC power-sourced drives and battery-sourced drives for AC machine control where, in both cases, VSI is used. The first one is fed through the electrical grid or a generator, and the second one is fed by a battery. In Fig. 8.1a, the voltage source of the inverter is formed by the electrical grid, input filter, rectifier, and the DC-link. The input filter eliminates the harmonics generated by the rectifier and inverter to avoid its injection to the electrical grid. The rectifier can be single or three-phase uncontrolled diode bridge or controlled PWM boost rectifier, where both rectify the AC sinusoidal voltage to obtain a DC voltage. As commented before, the DC-link stage allows a quasi-ideally zero Thevenin impedance and filters the rectified DC signal to soften its shape and reduce its harmonics. The inverter is responsible for reproducing a variable three-phase voltage and frequency from a DC voltage source. Finally, the output filter has the function of filtering harmonics and reducing the slew rate or $\mathrm{d}v/\mathrm{d}t$ generated by the inverter that could damage the AC machine isolation windings, especially if the length of the connection power cable between the inverter and the machine is considerable. On the other hand, in Fig. 8.1b, the elements that change in the voltage source are the battery and the boost converter. The boost converter is optional, and its primary function is to maintain a constant voltage at its output, regardless of the state of charge (SOC) of the battery. In the same way as before, the input filter filters the harmonics generated by the inverter

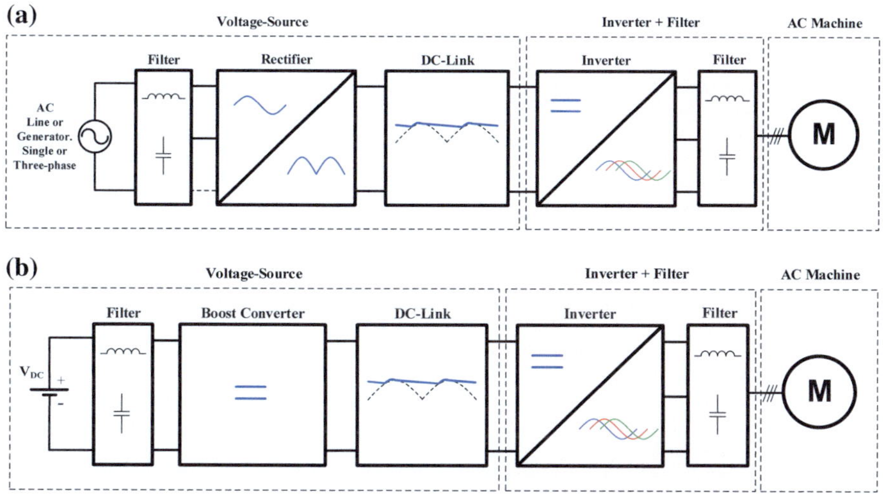

Fig. 8.1 AC machine control system. **a** AC power-sourced drives. **b** DC or battery-sourced drives

and boost converter, preventing them from affecting other elements connected to the battery as well as other electronic control units (ECU). It is possible to affirm that the MCU is integrated by the input filter, rectifier or boost converter, DC-link, inverter, and output filter, but sometimes can be found made-up by only the inverter and the output filter.

In the industry, the electrical grid is still an excellent source of voltage for fixed and variable speed in electric machines with three-phase winding such as the induction machine (IM). Although the frequency of the electrical grid is fixed, the variation of speed is performed by other methods such as the variation of the stator voltage, or the variation of the rotor resistance for winding rotor induction machines as seen in Chap. 4. However, to extract the maximum performance in terms of control and efficiency, the induction machine is increasingly connected indirectly to the power grid through an AC power-sourced drives.

Unlike the induction machine, the synchronous machines such as BLDC are designed to be connected indirectly to the power grid or a DC source through an inverter. In these machines, the synchronization between the stator and the rotor is fundamental for its correct operation, so that rotor position feedback is mandatory. Employing a rotor position control algorithm, it is possible to regulate the synchrony between the stator and the rotor. A correct and unique switching pattern is generated in the stator thanks to the VSI , where the voltage variation and frequency variation (VVVF) can be controlled independently. The VVVF is also essential for the control of other synchronous three-phase AC machines types, such as the brushless permanent magnet synchronous machine (PMSM) and the synchronous reluctance machine (SynRM), being optional for the induction machine (IM) as discussed above.

On the other hand, the evolution of semiconductors, in terms of switching frequency and conduction losses, has contributed significantly to the popularization of inverters and their evolution. In recent years, the field of digital electronics has undergone a significant change that has allowed microcontrollers to be available to designers at a very low cost and with powerful debugging and development tools. In this way, complex digital control strategies can be proposed without an appreciable increase in the final costs of the equipment. High number of applications in the different markets such as industrial, domestic, military, aerospace, and automotive have promoted the remarkable increase of VSI thanks to the capability of the control of torque, speed, and position in three-phase AC machines. The advantages of using VSI with digital control have been numerous. For example, the capability of wide range speed control and the efficiency increment of the system where the power delivered to the machine are adjusted instantaneously according to the load demand. Another interesting advantage is the possibility of four-quadrant operation of the torque versus speed plane. Moreover, in some cases, thanks to the digital control and the AC machine digital model, it is possible to remove the speed sensor or rotor position (sensorless), facilitating the installation task and reducing its cost at the same time.

8.2 VSI

Due to the versatile topology, the VSI can be applied in numerous applications such as AC and DC machine drives, uninterruptible power supplies (UPS), and AC power supply from the battery. According to the switching patterns, the most basic output is square wave used in some low-performance systems, while sine wave and PWM wave outputs are the most used for AC machine and DC machine drive, respectively. As mentioned before, the output can be single or polyphase and supports multi-level structures: In this section, single-phase and three-phase VSI are introduced in its two-level structure.

8.2.1 Single-Phase VSI

Previously to study the three-phase inverter, the single-phase inverter structure is introduced which is widely used not only in DC machines control but in DC-AC resonant converters.

The simplest possible inverter configuration is the two-level single-phase half-bridge inverter which consists of a pair of power switch devices as depicted in Fig. 8.2. The two-level term only applies two possible voltage levels in its output, positive or negative DC voltage thanks to the pair of MOSFETs (power switch devices). In Fig. 8.2a, the load is connected between point a and ground to have an AC output voltage which varies from $+V_{DC}$ and zero. In contrast, In Fig. 8.2b, the load is connected between point a and the center point 0 of a split-capacitor power supply to have an AC output voltage which varies from $+V_{DC}/2$ and $-V_{DC}/2$.

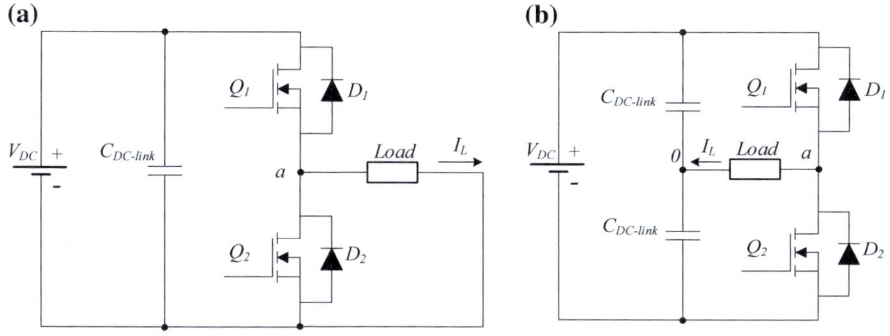

Fig. 8.2 Two-level single-phase half-bridge VSI . **a** Load connected between point a and ground. **b** Load connected between point a and 0

The half-bridge inverter can be used in DC machine drives where a single direction of rotation is a requirement such as a water pump. It supports two-quadrant control operation, where the current in motoring mode always flows in the same direction (first quadrant). In the second quadrant, the machine operates in direct regeneration or forward regenerative braking. This last case is when in the machine external torque forces to turn the rotor, in the same direction of the motoring mode, but the current changes its direction charging the battery.

On the other hand, the single-phase full-bridge structure, also known as H-bridge, consists of four power devices and can also be used in DC machine drives where two rotation directions is a requirement. In this case, the DC machine operates in the four quadrants of the torque versus speed plane. The rotation direction is reversed by changing the activation sequence of the power switches devices, which is essential in numerous servo applications controllers. The additional quadrants are the third for reverse motoring mode, and the fourth for the reverse regenerative braking, or inverse regeneration.

The DC machine terminals voltage can be regulated to perform a speed and torque regulation by using a PWM switching pattern applied to the power devices. Figure 8.3 represents the simplified single-phase full-bridge structure made with four MOSFET devices. As can be observed, in the output of the single-phase inverter, there is an inductive load formed by an RL circuit such as DC machine where back-EMF has been omitted. The DC voltage source consists of a battery together with a DC-link backup capacitance as can be seen in the input of the inverter. If device Q_1 is turned on, the positive supply source is connected to the corresponding inverter output terminal a. Now, if device Q_4 is turned on, the load terminal b is connected to ground. Thus, the current flows through the RL load from the positive supply source to ground. To reverse the current direction through the load, the devices Q_3 and Q_2 should be turned on. When Q_1 and Q_3 are on, or Q_2 and Q_4 are on, the output voltage in the load terminals is zero. Thereby, it is possible to observe that there are $2^2 = 4$ possible states since simultaneous activation of the device of the same leg is prohibited.

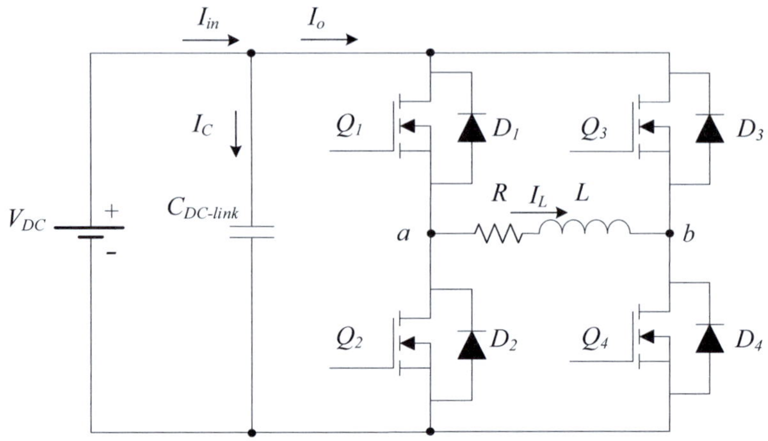

Fig. 8.3 Two-level single-phase full-bridge VSI integrated by four power semiconductor devices (MOSFETs) and battery

In order to generate a square wave voltage at output load terminals, devices Q_1Q_4 and Q_2Q_3 must be turned on alternately where the amplitude of the voltage is $+V_{DC}$. The current through the inductive load is the integration of the square wave of the voltage, that is, a triangular signal depending on the cut-off frequency of the RL circuit. Assuming perfect filtering, in Fig. 8.4, the positive half period when Q_1Q_4 remain active, the fundamental current through the inductor (main harmonic) is lag by an angle φ behind the voltage. The inductive load sees an active power when the current flows through Q_1Q_4 or Q_2Q_3, while a reactive power when the current flows through the diodes D_2D_3 or D_1D_4. The free-wheeling diodes parallel to the device ensure the continuity of the current, as shown in Fig. 8.4. When MOSFET is used, in some cases the freewheeling diode is not mounted externally since the physical structure is composed by a PN junction from source to drain. This diode is known as body diode. However, when IGBT semiconductor is used, the diode can be included inside the device packaging during its manufacture or can be included externally in

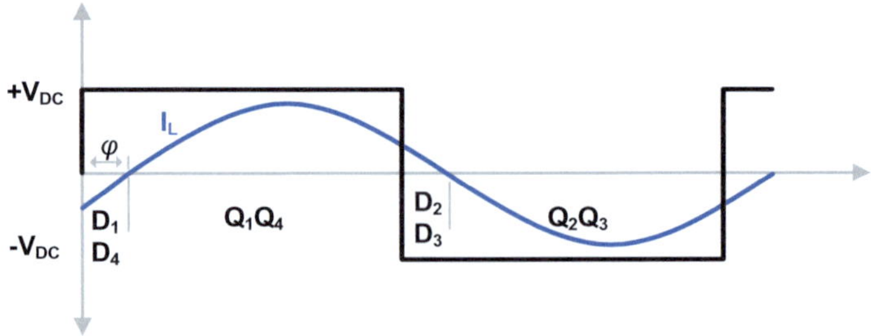

Fig. 8.4 Square output voltage and current waveforms, assuming harmonic-free load, in a single-phase full-bridge VSI

the circuit. If body diode of MOSFET does not meet the requirements for machine control applications, for example, because it has high turn-on voltage or it has a high reverse recovery time, an external Schottky diode can be mounted.

Therefore, the diode facilitates an alternative path of the current in the opposite direction when inductive loads are used when the semiconductor is turned off. In a purely resistive load, the free-wheeling or body diodes would not be needed.

In the negative half period something similar occurs, but in this case, the negative current flows through the diodes D_1D_4 when Q_2Q_3 is turned off. As previously mentioned, the current and voltage can be positive by activating Q_1Q_4 and negative by activating Q_2Q_3 so that if the load is a DC machine, the rotation can be controlled in both directions. The other two quadrants correspond to those of direct and inverse regeneration which can also be controlled with Q_2, together with D_1D_4, and Q_4, together with D_2D_3, respectively.

If a sinusoidal PWM pattern, as it will show in Chap. 9, is now applied to the Q_1Q_4 and Q_2Q_3 devices, the current through the inductive load has the shape of a sine wave. Each PWM cycle is integrated as before by the load resulting in the sine wave. The quality of the sine wave can be expressed in terms of total harmonic distortion (THD). THD is the percentage value between the total harmonic components with its fundamental component. High-quality values are when THD does not exceed 5%.

As commented before, the devices Q_1Q_2, and Q_3Q_4 cannot be turned on at the same time as it would cause a short circuit of the DC source that could damage the devices. The name that receives this phenomenon is shot-through. The way to avoid shot-through is by inverting the control signal in the devices of the same leg. Due to the power semiconductor stray capacitors, the activation from on to off state and vice versa is not instantaneous, so there is a very short time where the two devices of the same leg are conducting at the same time. Simultaneous driving, even in a very short time, would cause an increment of the junction temperature in the device unnecessarily, degrading the efficiency and the lifetime. To overcome the temperature increment, just when high- or low-side device must be turned on, a time delay t_{DT} should be imposed to assure that the complementary device is completely in off state. In other words, during t_{DT} time, both devices remain in off state, where is assured that both devices have reached their new stable off state. At this time, where both devices of the same leg are in off state, it is known as deadtime or blank time. The deadtime setting depends on the semiconductor device type and its specific characteristics of the manufacturer. The deadtime creates a non-linearity that can be significant since this time is added to the activation and deactivation times of the devices. In addition, it introduces a magnitude dependent on the load and a phase error in the output voltage as will be seen in Sect. 8.4.3. The deadtime range values can be from 5 ns for a high-performance HEMT technology MOSFET device such as low-voltage GaN (gallium nitride), up to 2–25 μs for a high-voltage silicon SCR.

Similarly, power devices need a minimum activation pulse width in order to activate the device. If the pulse width is too small, the device has not enough time to be turned on. Both constraints, the deadtime and the minimum width activation pulse, are some limitations that have to be taken into account during the VSI and AC machine control design.

8.2.2 Three-Phase VSI

Similar to the single-phase VSI, the three-phase VSI converts a DC voltage to three-phase AC voltage and current in the output. The three-phase output is synthesized by three half-bridge structures that are governed with a certain logic to achieve VVVF and a phase shift of 120° between the phases. Each of the half-bridge produces an output displaced by 120° with respect to each other. As mentioned, three-phase VSI is ideal for three-phase AC machine drive systems.

The inverter's power source is also a DC source, as discussed previously. The DC power source can be obtained from different configurations such as a DC battery, or the electrical grid through an AC-DC converter also known as a rectifier. The rectifier can be controlled, and bidirectional or simply an uncontrolled rectifier made with diodes. The AC source of the electrical grid can be single-phase, or three-phase, and variations of its voltage and/or nominal frequency can affect the AC machine regulation. The same happens when the power source is through a DC battery where its voltage decreases as the battery is discharged. In both cases, optionally a boost DC-DC converter type can be used to maintain a stable voltage DC in the VSI up to a minimum level of input voltage as mentioned before.

The connection of the AC machine is, in most cases, usually a direct connection of its terminals to the outputs of the inverter or through a filter as will be seen in Sect. 8.6. The semiconductor types to use mostly depend on the AC machine power requirement and the voltage. MOSFETs are preferred for high switching frequencies, moderate power level, and till 650 V VSI . However, the IGBT is preferred in VSI when high-power, high-voltage (400–1200 V), and switching frequencies less than 50 kHz are the requirements.

From the point of view of the control of the inverter, the semiconductor type is indifferent, and in many cases for its analysis of operation, switches are used instead of semiconductors. Figure 8.5 shows the simplified three-phase VSI made

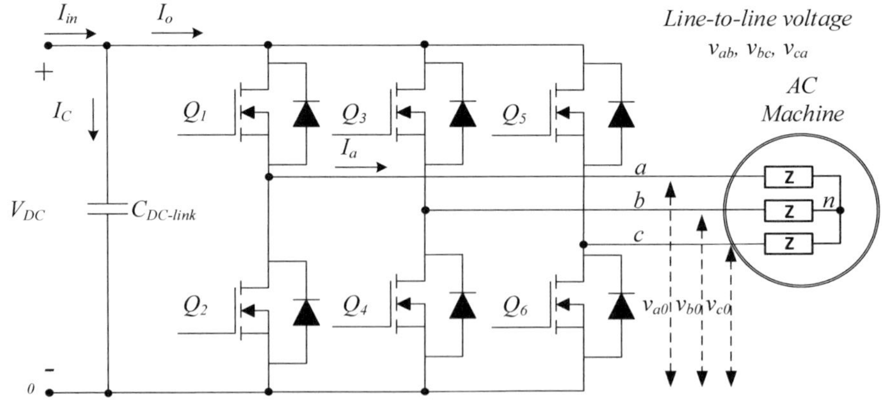

Fig. 8.5 Simplified three-phase VSI composed of six MOSFETs devices

with MOSFETs devices. The activation logic of power switches in three-phase VSI is also limited. Removing the cases that are not possible, such as the activations of the semiconductors of the same leg, only eight combinations remain. That is, the total possible states are $2^3 = 8$. These possible states set a known voltage in the output terminals, to zero or V_{DC} voltage, since there will always be at least one activated device in each of the inverter legs except in BLDC machine control as it will be shown below.

The DC voltage V_{DC} is known as the BUS or DC-link voltage. The name of BUS or DC-link comes from the passive element that connects the DC power source with the VSI corresponding to the capacitor also called the BUS or DC-link indicated in the figure as $C_{DC\text{-}link}$. The three half-bridges are formed by the MOSFETs Q_1Q_2, Q_3Q_4, and Q_5Q_6. For AC machine with the stator in star connection, as illustrated in Fig. 8.5, the phase voltages correspond to v_{an}, v_{bn}, and v_{cn}. The line voltages are v_{ab}, v_{bc}, v_{ca}, and the voltages with respect to ground or point 0 are v_{a0}, v_{b0}, and v_{c0}. The current coming from the DC source is the input current represented by I_{in} which is split into the capacitor current I_c and the current absorbed by the VSI I_o.

Many semiconductor manufacturers offer power modules (PM) in different configurations such as two-level half-bridge, full-bridge, and three-phase configuration. Some of them can include three-phase full-bridge diode rectifier and protection elements such as NTC to be able to measure the PM average temperature (see Fig. 6.9). The PM increases the power density since they usually reduce the space compared if, for example, the inverter is made with discrete devices. Sometimes, PM integrates also internal components such as logic control protection, and gate drive, given intelligent to the PM. They are known as intelligent PM (IPM). The IPM further optimizes the size of the printed circuit board (PCB) and the cost of the system. The IPM is widely used in small inverters of up to 2–3 kW, while the PM can reach up to hundreds of kW.

As a convenience, the control of self-synchronous brushless DC (BLDC) machine can be used to introduce the three-phase inverter operation. As discussed in Chap. 4, the BLDC machine has three-phase winding in the stator while the rotor is usually formed by permanent magnets. In this type of machines, the control is very simple and emulates the operation of brushed DC machine seen in the same Chap. 4. In the brushed DC machine, the mechanical switch (commutator and brushes) changed the N-S polarity to get the rotor to the next position, while in the BLDC machine the commutator becomes an electronic switch formed by the inverter. In BLDC machine, the knowledge of the instantaneous rotor position is mandatory to set the correct sequence in the stator winding. In Fig. 8.6 is shown the sequence of the switching variables for three-phase VSI . It is possible to observe that the sequence is repeated every six sectors (60°). The signals shown are the PWM switching control signals of each MOSFETs of Fig. 8.5 where the output average voltage values applied to the machine terminals are controlled.

The result of this sequence allows to change the direction of the current of each of the phases to cause the change of N-S polarity of the stator, and thus force the movement of the rotor.

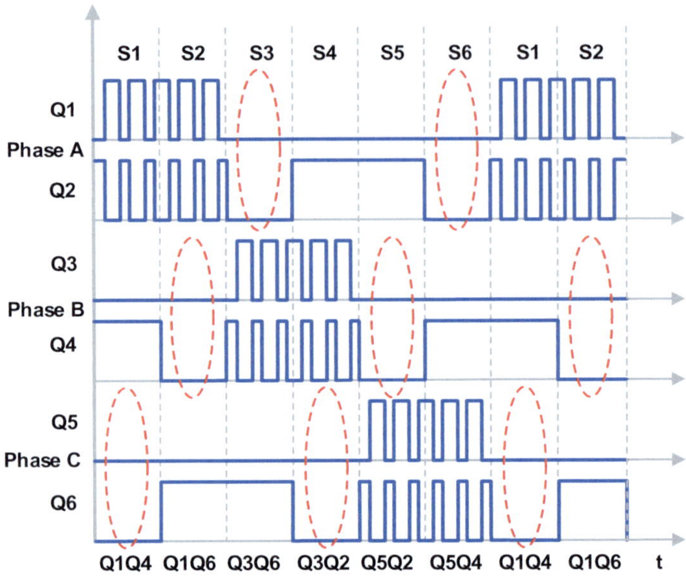

Fig. 8.6 PWM switching signals sequence for BLDC machine control according to the position of the rotor. The red circle indicates phase off during a sector (60°) which can be used to sense the back-EMF voltage to determine the rotor position (sensorless)

In the first sector, Q_1 and Q_4 are turned on so that the phase current I_a through the machine will have a positive direction. In the next sector, after the interval $\pi/3$, the current I_a remains in the same way as before since Q_4 has been turned off and Q_6 has been turned on. In sector 3, the current I_a is null since neither Q_1 nor Q_2 are active so that this phase remains off. Current I_a becomes negative in sectors 4 and 5. In sector 4, Q_2 and Q_3 are turned on, while in sector 5, Q_2 and Q_5 are turned on. In both cases, it is possible to observe how the current I_a has changed its direction. Finally, in sector 6, phase A is once again off. The same analysis can be carried out with the current I_b and I_c.

The decision to change the sector is based on the knowledge of the instantaneous rotor position by using sensors such as three Hall-effect sensors, or optical encoder, or the position can be estimated (sensorless), measuring the induced back-EMF voltage in the periods which the phases are off. Once the instantaneous rotor position is obtained, the next sector can be applied. The inverse rotation is achieved through the downward application of the sectors.

The frequency with which it changes from one sector to another depends on the machine speed, while the variation of the voltage and current can be made by regulating the duty cycle of the PWM. It is possible to observe that for this particular case, a VVVF control is being carried out, which is ideal for machine drive applications as commented before.

8.3 Power Semiconductor

8.3.1 Introduction

Nowadays, thanks to the evolution of semiconductors, it is possible to find a wide range of selection possibilities suitable for different applications. The silicon (Si) power semiconductors such an IGBT and MOSFET are still used in a wide variety of applications since their performance is further improved, the maturity of their well-known and proven technology, and because of their lower price. However, new wide bandgap (WBG) material technologies such as silicon carbide (SiC) and gallium nitride (GaN) are shown as the best candidates for high-power and high-voltage applications because of its higher performance.

As it will be shown in this section, the GaN, and SiC, and some Si power semiconductors allow to operate at higher switching frequencies with high-voltage slew rates (dv/dt). The VSI built with high-speed semiconductors and PWM modulation control could reach a high slew rate more than 12 kV/μs in the outputs. Some AC machines support maximum slew rates around 1 kV/μs. Above these limits, the loss of insulation winding in the machine can be produced. Thus, it is necessary to add elements to avoid high slew rates in the AC machine terminals as it will be shown in Sect. 8.6. Other issues of the high dv/dt are the bearing currents, and a noticeable harmonic content increment, increasing the electromagnetic interference (EMI) which can be conducted or radiated.

The bearing currents are currents which flow to ground through the machine bearing stray capacitances. This current circulates between the machine ground and the electrical grid ground. Higher dv/dt and higher PWM switching frequency increase the circulating currents. The effects of this current can damage the machine due to bearing wear.

The semiconductor physics structure has stray capacities which reduce the switching speed transitions since they have to be charged or discharged at every transition. On the other hand, the semiconductor terminal connection and the PCB layout (copper tracks) used to interconnect other components create stray inductances. The effects of stray inductance are more relevant for higher switching frequencies. They are transient overcurrent and overvoltage (high dv/dt and di/dt) with damped oscillations, which can produce severe EMI problems.

On the other hand, the semiconductors have different switching conditions that characterize the transition of the switching. During the switching transition (turn-on and turn-off transitions), if the voltage and current of the device are not zero, a hard-switching is produced. However, when the voltage or current is zero or near zero in the switching transition, a soft-switching is produced. The main advantage of the soft-switching is that the power loss due to the transition is null avoiding to stress the power semiconductor. However, sometimes it is challenging to get soft-switching without adding active components which increase the total cost of the power plant. For this reason, the hard-switching is a cost-effective solution for most of the drives where passive elements such as resistance and capacitors can be added to form snubber circuits which reduce the power loss during the transitions also protecting the device from switching stress. It is important to comment that the semiconductor power loss during the transitions on-off and vice versa are mostly related to the stray capacities and stray inductance.

In this section, the new WBG power semiconductors are introduced, and the power switch semiconductor issues for MOSFET and IGBT are deeply presented such as power losses, parasitic capacitance and inductances, gate charge, and dynamic characteristics.

8.3.2 Semiconductor Technology Overview

The silicon (Si) semiconductor devices have several limitations, such as high losses, low switching frequency, and poor high-temperature performance. The relatively low bandgap energy of 1.1 eV and the low critical electric field of 30 V/μm involve large thickness for high-voltage applications, which means higher on-resistance and associated conduction losses. To keep lower conduction losses in Si MOSFET, large die areas is required, which results in high gate capacitance producing losses at high frequencies. However, the Si IGBT has smaller die than MOSFETs because the minority carriers are injected into the drift region to reduce the on-resistance, but when the transistor is turned off, these minority carriers should be recombined reducing the useful switching frequency range. The Si MOSFET contains a parasitic (body) diode formed in the P-N junction which conducts the current in a reverse direction very useful for inductance loads, but with high reverse recovery losses.

On the other hand, the WBG power semiconductors devices such as SiC and GaN have approximately ten times higher critical electric fields. It allows ten times thinner die, which reduces the on-resistance and the parasitic capacitances. Moreover, their higher electron mobility and electron saturation velocity allow higher-frequency operation compared with a Si device. The high-energy bandgap of approximately three times compared with Si allows a more stable and low leakage current over a temperature range, that is, higher energy is required to have free electrons. In terms of temperature, the WBG devices have three times higher thermal conductivity, which permits faster heat transfer.

The SiC devices are fabricated analogous to silicon devices since SiC epitaxy layer is formed on a SiC substrate as illustrated in Fig. 8.7. SiC exists in a variety of polymorphic crystalline structures called polytypes such as 3C-SiC, 6H-SiC, and 4H-SiC. In power devices, the 4H-SiC structure is generally preferred for manufacturing. As can be observed in Fig. 8.7 is a vertical device where current flows vertically. It has a good electrically and thermally conductive path from the top to the bottom of the wafer and can handle large voltages. It is also possible to observe the body diode formed by the P-N junction between source and drain.

SiC MOSFET has superior performance compared to Si MOSFET, especially in the increase of the switching frequency. In high-power application where Si IGBTs are widely used, the SiC MOSFET is a promising alternative device with higher

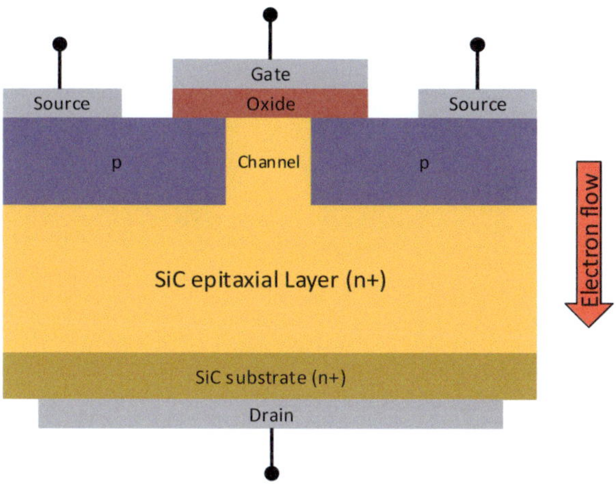

Fig. 8.7 Vertical SiC MOSFET cross-sectional structure

performance. Compared to silicon, SiC offers a combination of higher voltage-blocking capacity, higher switching frequency due to smaller parasitic capacitances, and reduced on-resistance, further adding its potential for high-temperature operations. As mentioned, the SiC MOSFET also contains a parasitic (body) diode formed in the P-N junction. However, the body diode has a higher threshold voltage, and a relatively large forward voltage drop since the bandgap of SiC is three times larger than that of Si. The high forward voltage drop of the diode can be reduced by turning on the SiC device, which is very common in the synchronous rectification technics. Fortunately, in the normal operation of the inverter, the current only flow through the body diode that is during deadtime, so it will no cause complications.

Many EV/HEV traction VSI are made with SiC MOSFET semiconductors due to their high-power density (ideal for voltages above 650 V and very high currents) and low switching losses making the inverter more efficient and therefore increasing the autonomy of the EV/HEV.

On the other hand, the HEMT technology of GaN FET semiconductors can operate at a higher switching frequency than SiC, mainly due to the fact that it needs a lower gate voltage and because it has reduced parasitic capacitances.

GaN FET is made a horizontal form with a smaller die, leading fewer parasitic capacitance and inductances. In Fig. 8.8 is illustrated a cross section of the GaN FET structure where current flows in a horizontal way. The lateral or horizontal construction limits the maximum voltage of the device. A buffer layer made with aluminum nitride (AlN) is used to accommodate the crystallographic differences. The AlN also provides electrical isolation but implies a complex process and as a consequence, an increment of the cost. A "2DEG" channel is formed between the undoped GaN layer and the n-doped AlGaN layer. The "2DEG" means a two-dimensional electron gas where electrons are free to move in two directions creating a channel. The high mobility of the 2DEG results in a very-low on-resistance.

When a gate-source positive voltage is applied, it causes a field effect which attracts the free electrons. Then, the electrons are pooled, and this accumulation

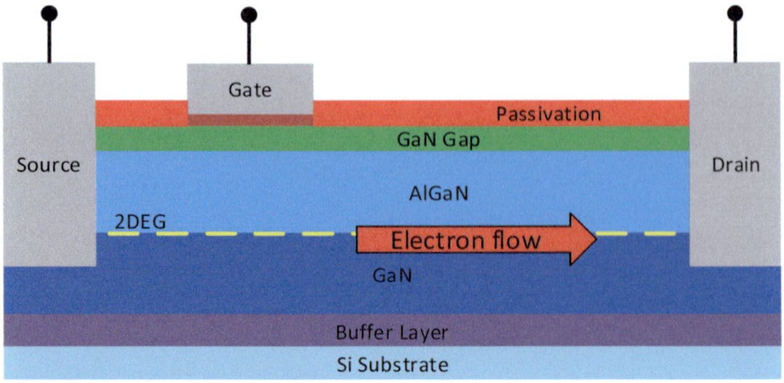

Fig. 8.8 Lateral GaN FET cross-sectional structure

leads to reduce the resistance channel where high current can flow (turn-on). When the voltage applied is removed, the electrons under the gate are dispersed recreating the depletion region blocking the current (turn-off). Unlike the Si power MOSFET, in an enhancement mode GaN FET, the reverse conduction occurs when there is a positive voltage between gate and drain. It is because the electron channel turns on at this positive voltage and turns-off instantly when the voltage is removed without any stored charge to dissipate. Thus, the reverse recovery charge and the reverse recovery time are zero, meaning zero reverse recovery losses.

As in the case of SiC, GaN improves conduction losses with respect to silicon and also offers low on-resistance. On the contrary, the blocking voltage is lower, up to 650 V compared to the SiC.

Table 8.1 represents the comparison of the main parameters between Si and WBG power semiconductor. These parameters will be explained in this section.

Table 8.1 Main characteristics comparison between similar power Si and GaN devices

Parameter	Si MOSFET	SiC MOSFET	GaN FET
R_{ON}	Low	Low	Very Low
Q_{rr}	High	Medium	0
Q_G	High	Medium	Low
C_{iss}	High	Low	Low
C_{oss}	Low	Low	Low

8.3.3 *Parasitic Effect in Semiconductor Switches*

8.3.3.1 Parasitic Capacitance

The power semiconductors such as MOSFET and IGBT have internal parasitic capacitances, which affect the dynamic characteristics. The internal structure of the power semiconductors establishes these inherent parasitic capacitances as discussed early.

At every turn-on or turn-off switching, this capacitance should be charged or discharged as fast as possible to reduce the switching losses. This requires a driver circuit that can provide or absorb a high current to charge or discharge the parasitic capacitances quickly. Figure 8.9a shows the MOSFET semiconductor with three parasitic capacitances. C_{GS} is the dielectric capacitance between the gate electrode and the source. C_{GD} is the feedback capacitance between the gate electrode and drain. C_{DS} is the capacitance of the body diode between the drain and source.

The C_{DS} capacitance is due to the PN junction that forms the body diode, while the capacitances C_{GS} and C_{GD} are due to the geometry of the device. The capacitance C_{GS} is formed by the superposition of the source and the region of the channel by the gate electrode. Its value is defined by the actual geometry of the regions and remains constant under different conditions. The C_{GD} capacitance has a non-linear behavior that depends on the voltage drain–source (V_{DS}). The capacitance C_{DS} also has a non-linear behavior and again depends on the V_{DS} voltage.

The capacitors that are often provided in the MOSFET data sheet are the input capacitance C_{iss} (small-signal input capacitance), the output capacitance C_{oss} small-signal, and the reverse transfer capacitance C_{rss} (small-signal reverse transfer capacitance). These are easily measurable experimentally. The correspondence between the capacitance discussed earlier and those provided by the data sheet is explained in Fig. 8.9. The C_{iss} capacitance is measured between the gate and source when the

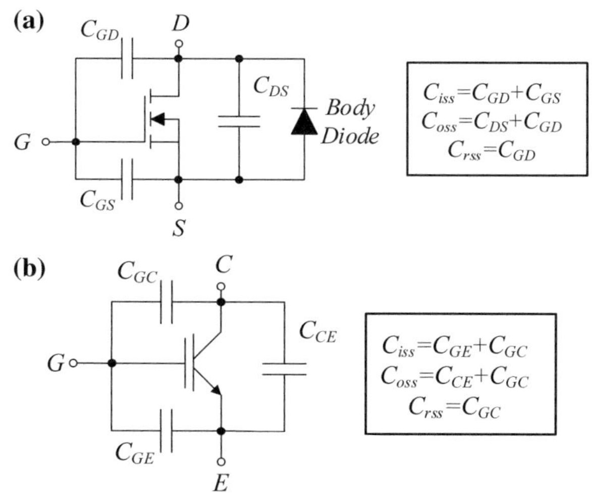

Fig. 8.9 **a** MOSFET semiconductor parasitic capacitance C_{GD}, C_{GS}, and C_{DS}. **b** IGBT semiconductor parasitic capacitance C_{GE}, C_{GC}, and C_{CE}

drain–source terminals are shorted. C_{iss} has a direct influence on the switching speed since it must be charged and uncharged on the MOSFET activation/deactivation. The C_{oss} capacitance is measured between the drain and source when the gate-source terminals are short circuited. C_{oss} can affect the resonance of the circuit since together with the parasitic inductance form a resonant LC circuit. C_{rss} capacitance is measured between the drain and gate with the drain connected to ground. Frequently, it is known as Miller capacitance. It affects the rise and fall time of the voltage during switching.

Similarly, the IGBT (Fig. 8.9b) has the input capacitance C_{iss}, the reverse transfer capacitance C_{rss}, and the output capacitance C_{oss}. The capacitances C_{iss} and C_{rss} are used to dimensioning the gate driver circuit, while C_{oss} limits the dv/dt at switching transitions.

8.3.3.2 Parasitic Inductance

The power semiconductors have internal inherent parasitic inductances in series with each of its terminals due to the connection cable between the pin of the device and the exact point where the semiconductor begins. The parasitic inductances are also present externally to the power semiconductor due to the PCB layout design. The copper tracks dimensions (length and width), as well as the path of the loops, forms stray circuit inductance which should be minimized. The minimization of stray circuit inductance is achieved by minimizing the length of the connection pin, by using smaller semiconductor packages, and by designing the most optimal layout where the paths are not too long. At higher switching frequencies above 100 kHz in VSI , the stray circuit inductance effects are more relevant. The main effects are surge voltage and transient's overvoltage with damped oscillations. With the help of the simulation packages where the PCB layout and power semiconductor can be modeled accurately, the effects can be evaluated in order to optimize some paths and components to reduce the transient's overvoltage. Different switching frequency can be evaluated to see how affects the intensity of the damped oscillation produced by the stray circuit inductance.

The parasitic inductance that has the most noticeable effect on switching performance is the gate and source inductance. In a MOSFET where the gate connects to a resistor R_G (Fig. 8.10a) and its source is connected to ground, the parasitic inductance external to the device is present at the connection between the resistor R_G and the gate pin of the device, and between the ground and the source. In addition, if a shunt resistance is added between source and ground, for example, to measure the current, it also adds a parasitic inductance. These parasitic inductances present in the driving MOSFET control produces a slow current pulse, and consequently, a slower charging/discharging of the C_{iss} input capacity.

In Fig. 8.10b, the RLC circuit in simplified form is composed by the total gate resistance R_{GT}, the sum of all the parasitic inductances L_T, external and internal to the device, and the input parasitic capacitance C_{iss} of the MOSFET. The gate resistance

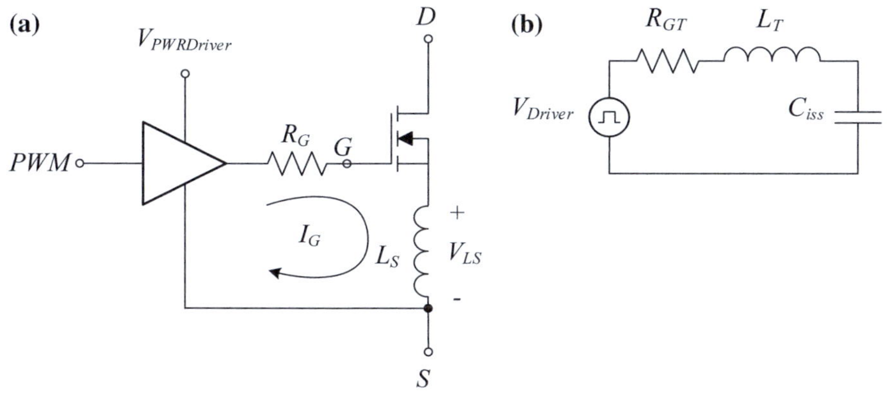

Fig. 8.10 **a** Equivalent circuit formed by the MOSFET, parasitic inductance, R_G and driver. **b** Equivalent RLC circuit formed by the total parasitic inductance, C_{iss}, and R_{GT}

R_{GT} is equal to the sum of the external resistance R_G plus the internal resistance of the MOSFET R_{GI}.

The resonant circuit RLC is responsible for the oscillations and surge voltage during the activation and deactivation of the MOSFET. To reduce oscillations and overshoots, the resistance R_G can be fine-tuned. A small value implies a voltage overshoot in the gate, but also a quick activation of the MOSFET. Higher values attenuate the oscillation and dampen the overshoot, but increases the switching time and therefore the switching losses. If the damping coefficient is chosen to be the unit, that is, $\zeta = 1$ to have zero oscillations, the resistance R_G can be calculated as in Eq. (8.1).

$$R_G = 2\sqrt{\frac{L_T}{C_{iss}}} \tag{8.1}$$

The MOSFET/IGBT semiconductor packages with 4-pin configuration offer a Kelvin connection that significantly reduces the inductance of the source in the case of the MOSFET or the emitter in the case of the IGBT. The Kelvin connection is used as a reference potential for the gate driving voltage where the voltage drop in the source inductance is eliminated. Thus, the oscillation in the gate-source or gate-emitter loops is reduced. Figure 8.11a shows a 4-pin MOSFET device, while Fig. 8.11b shows the gate-source loop with a gate driver where the stray inductance of spout S_2 is much smaller than S_1.

In addition, the TOLL (TO LeadLess) SMD packages of Fig. 8.11c still further reduce the internal parasitic inductances, since they offer a much shorter connection to the semiconductor as can be seen.

To conclude, in Fig. 8.12 is illustrated the MOSFET with the most relevant parasitic components such as capacitance, inductance, and resistance.

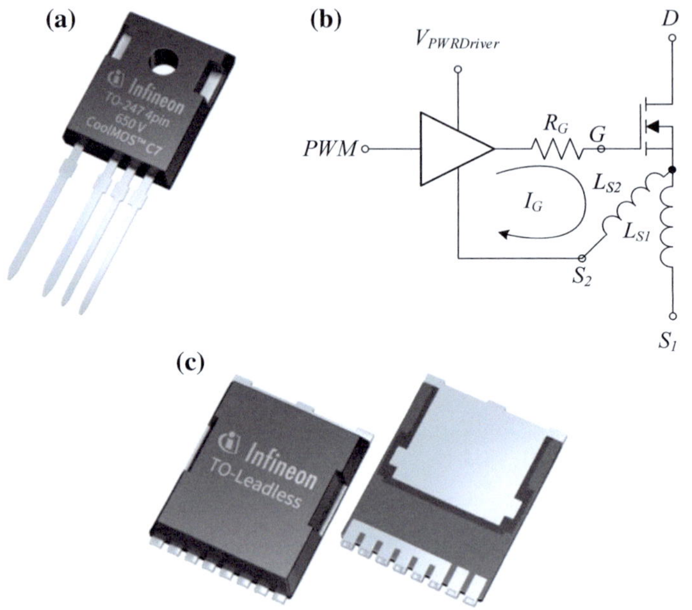

Fig. 8.11 **a** 4-pin MOSFET TO-247 package. Courtesy of Infineon. **b** Equivalent circuit formed by the MOSFET, parasitic inductance, R_G and driver. **c** TOLL package with reduced internal parasitic inductor. Courtesy of Infineon

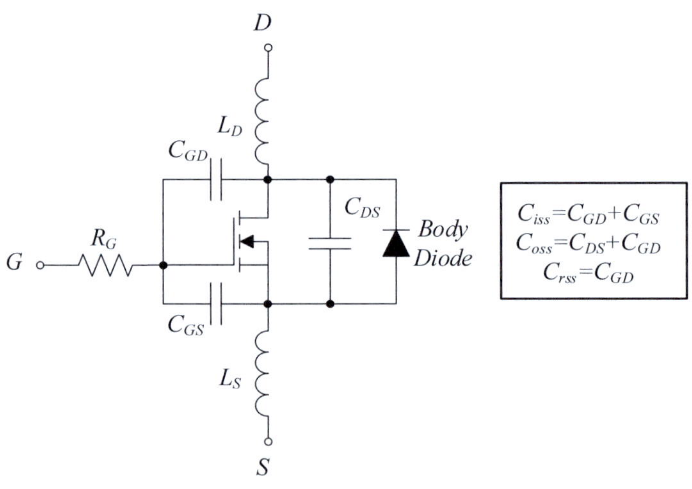

Fig. 8.12 MOSFET semiconductor parasitic capacitance C_{GD}, C_{GS}, and C_{DS}, parasitic resistance R_G, and parasitic inductance L_D and L_S

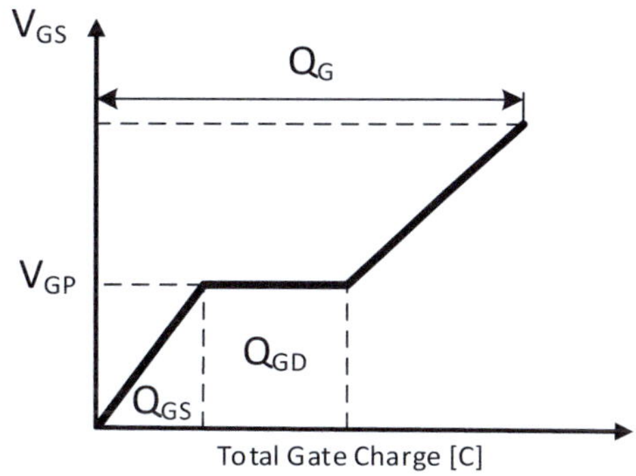

Fig. 8.13 Gate-to-source voltage versus total charge waveform

8.3.4 Gate Charge

The total gate charge Q_G is a parameter that usually is specified in the manufacturer's data sheet of the MOSFET and IGBT. The parameter Q_G is directly related to the switching speed and provides a better indication of the switching capability. Low values allow faster switching and consequently, lower switching losses. Q_G depends only on the parasitic capacitances of the device. This parameter is weakly dependent on the drain current, the supply voltage, and the temperature. The data sheet also specifies the gate-source charge Q_{GS} and the gate-drain charge Q_{GD}.

Figure 8.13 shows the gate charge waveform when the MOSFET is turned on. Q_{GS} is defined as the charge from the origin to the Miller Plateau V_{GP} level. In other words, Q_{GS} is the charge required to charge the capacitance C_{iss} to a gate voltage where a drain current starts to conduct. Q_{GD} is defined as the charge from V_{GP} to the end of the plateau level. It is also defined as the gate charge required to charge C_{rss} to the same voltage of C_{iss}. Finally, the charge required to charge the MOSFET input capacitance C_{iss} to the applied gate voltage is the total charge Q_G.

8.3.5 Dynamic Characteristic

8.3.5.1 Turn-On, Turn-Off Time Definition

The charging or discharging process of the internal capacitors discussed early is directly related to the turn-on and turn-off switching times of a power semiconductor. With a resistive load switching circuit is possible to characterize the MOSFET switching times such as turn-on delay time $t_{d(on)}$, rise time t_r, turn-off delay time $t_{d(off)}$, and fall time t_f. In Fig. 8.14 is represented as a waveform with the above switching times, which are usually specified in the MOSFET data sheet.

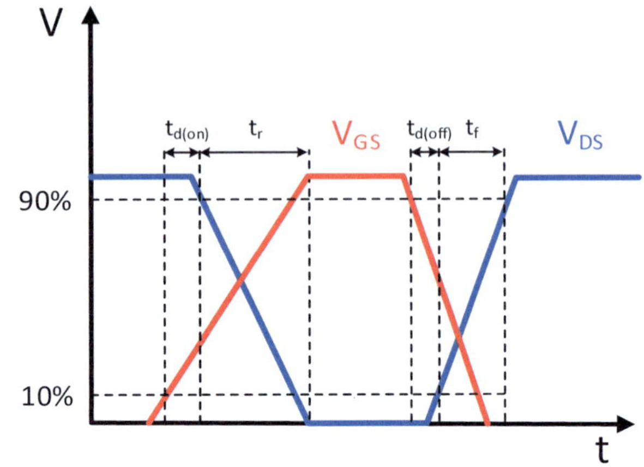

Fig. 8.14 Turn-on and turn-off characteristics in a MOSFET

These times are dependent on the resistance, and parasitic inductances, as well as the parasitic capacitances, explained early. This information allows selecting an appropriate deadtime between the turn-on and turn-off in the half-bridge topology.

The delay time $t_{d(on)}$ corresponds to the turn-on delay time defined as the time since the gate-source voltage reaches 10% of V_{GS} until the drain–source voltage reaches 90% of V_{DS}. t_r is the rise time of the drain current when it rises from 10 to 90%. The total turn-on time would be the sum of the two previous times. However, the total turn-off time is the sum of the delay time t_f plus the delay time $t_{d(off)}$. Similarly, the time $t_{d(off)}$ corresponds to the delay time defined as the time since the voltage V_{GS} reaches 90% until the voltage V_{DS} reaches 10%. Finally, the time t_f is the time required to pass from 10 to 90% of the V_{DS}.

8.3.5.2 Turn-On, Turn-Off Dynamic Characteristic

The analysis of the turn-on and turn-off device dynamic characteristic can be carried out with the half-bridge and an inductive load, as represented in Fig. 8.15.

Starting from the deadtime, where the two MOSFETs Q_1 and Q_2 are deactivated, during this time, the inductor current flows through the diode of the low-side MOSFET, increasing the voltage V_{DS2} a little. When Q_1 is turned on at time t_1, the voltage V_{GS1} begins to increase up to the *plateau* voltage $V_{GS(p)}$ at time t_2, as can be seen in Fig. 8.16a. During the interval I_1, the majority of the gate current is used to charge the C_{GS} capacitance and in smaller quantity the C_{GD} which produces a delay time equal to $t_2 - t_1$. When the voltage reaches the $V_{GS(th)}$ level, the MOSFET is ready to enable to conduct the I_D current through the drain, being at this point when the current begins to increase.

The interval I_2 is the linear zone where the current is proportional to the gate voltage. When the gate-source voltage V_{GS1} reaches the Miller plateau voltage at time t_3, the voltage V_{DS1} begins to fall, deactivating the body diode, allowing that

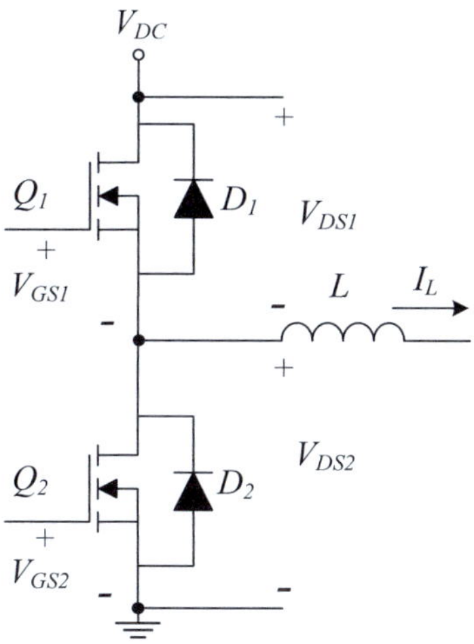

Fig. 8.15 Half-bridge circuit with inductive load

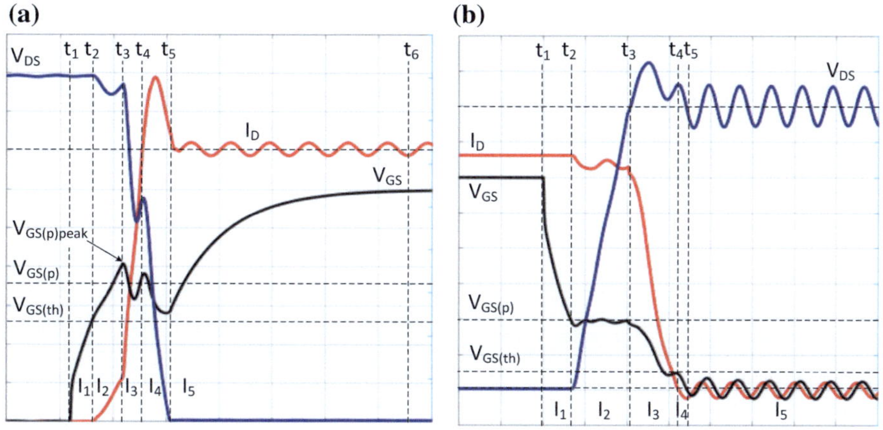

Fig. 8.16 Low-side MOSFET switching. **a** Turn-on. **b** Turn-off

voltage to fall further. The current reaches its maximum and constant level at time t_4 limited by the load. The voltage V_{GS1} in the interval $I_3 - I_4$ remains approximately constant while the voltage V_{DS1} is falling to zero in time instant t_5. The constant V_{GS1} region is referred to as the Miller plateau region. At time t_5, the voltage V_{GS1} starts rising again, entering the MOSFET in the ohmic zone defining the on-resistance of the MOSFET. During this interval (I_5), both C_{GS} and C_{GD} capacitances are charged through the I_G gate current, which will decrease while these capacitances are being charged. Also, during this charging process of the C_{GS} and C_{GD} capacitances, the drain current remains constant, and the V_{DS1} voltage decreases slightly as the on-resistance of the MOSFET is reduced. Table 8.2 shows a summary of the intervals

Table 8.2 Turn-on intervals resume

Interval	Region	Description
I_1 ($t_1 \rightarrow t_2$)	Delay period	Low C_{iss} $\rightarrow$ low gate drive loss and shorter time
I_2 ($t_2 \rightarrow t_3$)	di/dt	Low C_{iss} and parasitic inductance $\rightarrow$ faster di/dt and lower switching loss
I_3 ($t_3 \rightarrow t_4$)	dv/dt (Miller Plateau)	Low C_{rss}, C_{oss}, and Q_{rr} $\rightarrow$ faster dv/dt and lower switching loss
I_4 ($t_4 \rightarrow t_5$)	Stabilization	Low parasitic inductance $\rightarrow$ less overshoot
I_5 ($t_5 \rightarrow t_6$)	Conduction	Effective power $\rightarrow$ $R_{DS(on)}I_D^2$

described above.

The description of the deactivation of the MOSFET is similar, but in the opposite direction, and can be seen in Fig. 8.16b. The first time interval I_1 is the shutdown delay that is required to discharge the C_{iss} capacitor from its initial value to the Miller plateau level.

During this interval I_1, the gate current is supplied by the capacitor C_{iss} and flows through the C_{GS} and C_{GD} MOSFET capacitances. The drain current has not changed during this interval. In the Miller region, interval I_2, the drain–source voltage V_{DS1} grows linearly due to the resistance R_{DON} that increases its value. The final value is saturated to the DC-link voltage plus the voltage of the body diode V_{BD}. Upon reaching this limit, the drain current starts to decrease to near zero, just as the gate voltage reaches the $V_{GS(th)}$ level. During this interval I_3, the capacitor C_{GD} is charged because the voltage in the drain is higher than the gate voltage, which is constant. Part of the drain current charges this capacity. From the gate voltage $V_{GS(th)}$ to zero, the C_{GS} capacity is discharged being responsible for the weak gate current. In this interval I_4, the drain current reaches its zero value, where it is possible to have an overshoot due to the parasitic inductances. Finally, in the I_5 interval, the MOSFET is completely OFF. Table 8.3 shows a summary of the intervals described above.

In the deadtime interval, the current circulates through the body diode causing a voltage drop due to forward biased of the diode. This voltage drop along with the diode current causes losses due to deadtime. The larger the deadtime, the longer the duration and therefore, the more significant losses that will increase the temperature of the MOSFET.

Table 8.3 Turn-off intervals resume

Interval	Region	Description
I_1 ($t_1 \rightarrow t_2$)	Delay period	Low C_{iss} $\rightarrow$ low gate drive loss and shorter time
I_2 ($t_2 \rightarrow t_3$)	Miller Plateau	Low C_{iss} and parasitic inductance $\rightarrow$ faster di/dt and lower switching loss
I_3 ($t_3 \rightarrow t_4$)	dv/dt	C_{oss} charging
I_4 ($t_4 \rightarrow t_5$)	di/dt	Low parasitic inductance $\rightarrow$ less overshoot and ringing
I_5 ($>t_5$)	OFF	MOSFET in off state

The parasitic capacitances of the MOSFET play a decisive role in the time duration each one of the intervals commented previously. The speed with which these capacitances are charged and discharged will also depend on the capacity of the driver and the value of the gate resistance. Sometimes, to increase the charge or discharge capacity if the driver does not provide a high current, a totem-pole topology is usually used that acts as a non-inverting current amplifier that helps to charge or discharge the C_{GD} and C_{GS} capacitances in a fast way, reducing switching losses.

8.3.6 Snubber Circuits

The snubber circuits protect the power semiconductors against transient voltages that occur during turn-on and turn-off. Also, it can help to reduce the EMIs. As commented before, in a hard-switching, the snubber circuits reduce the switching stress and the power loss at every turn-on/turn-off transition. The most probably widely used snubber circuit consists of resistance with a capacitor in series. Thus, this circuit is connected in parallel to the power MOSFET/IGBT to cut the current in the circuit that causes its voltage to increase considerably due to stray inductances. That is the snubber damps or absorbs the surge voltage protecting the power semiconductor and the nearby components. In Fig. 8.17 is represented a half-bridge with an RC snubber in parallel with the two MOSFETs.

The snubber circuit analytical design sometimes is laborious since the stray capacitance and inductance are no set till the VSI circuit is not mounted in the PCB. For this reason, the RC components are fine-tuned experimentally where different combinations of RC values can be tested and mounted in the PCB spaces dedicated for the snubber. The first criteria of the snubber capacitance selection are that it should be larger than the parasitic capacitance of the circuit.

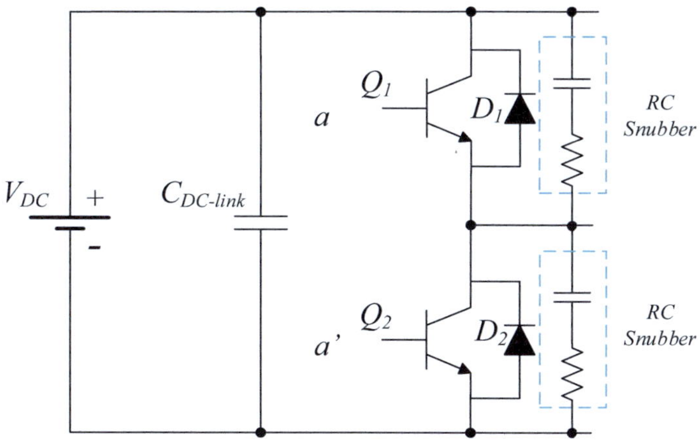

Fig. 8.17 Half-bridge circuit with RC snubber in parallel to both MOSFET

It is important to comment that the RC snubber circuit produces losses which should be added to the total power losses of the VSI . The power losses are mostly due to the power dissipation of the resistor. At every switching interval, the snubber capacitor is charged from 0 V to V_{DC} or discharge from V_{DC} to 0 V through the resistor. The power losses are proportional to the switching frequency, the snubber capacitance, and the square of V_{DC} voltage.

It should be noted that depending on the power to dissipate in the resistor, sometimes is interesting to split the power in two or four resistances in a series/parallel connection.

8.3.7 Semiconductor Power Losses

The IGBT is a semiconductor that provides the advantages of a MOSFET in terms of its high input impedance and the advantage of the BJT, capable of handling a very high current. The low impedance and the high current features have allowed developing VSI with relatively small losses. Unlike the MOSFET, it does not include an intrinsic PN junction between the emitter and collector so that an external diode must be added between emitter and collector to allow the continuity of the inductive load current in the half-bridge topology. The advantage over the MOSFET is that the external diode can be optimized to improve the switching of the device. Silicon MOSFET devices are more suitable than the silicon IGBT for a high switching frequency of more than 20 kHz since it offers less switching losses. They are suitable for powers up to 150 kW and maximum voltage up to 900 V,while the IGBT can reach up to 7000 V and a maximum power of up to 3 MW. On the other hand, the IGBT is preferable in VSI when it is necessary to handle high current density at lower switching frequency (<50 kHz). As mentioned, the SiC and GaN semiconductors have higher performance than silicon-based devices, where higher switching frequency than 50 kHz is possible without excessive power losses.

The selection of the switching frequency is an essential factor in the design since it determines the power losses and therefore, its efficiency. At the same time, the frequencies below 20 kHz are audible so that depending on the application, it is not advisable, for example, in EV/EHV. In industrial environments and very high powers, the switching frequency can reach up to 2 kHz in order to reduce the power losses. However, higher switching frequency is used in a high-performance AC machine drives to achieve faster dynamic response since the control bandwidth is also higher as discussed in Chap. 2. It is possible for VSI implemented with SiC MOSFETs or GaN FETs, with high processing units such as DSP or SoC.

It is important to mention that the improvement of conduction and switching losses is a priority of the different power semiconductor manufacturers due to market demand, especially in the automotive sector.

8.3.7.1 Static Losses

Static losses are also known as semiconductor conduction losses which it is due to non-ideality of the device. Semiconductors offer a non-zero resistance when it is turned on, which creates a voltage drop when current flows through the resistance. Then, a non-zero power is given during their conduction. Power is basically calculated by knowing the voltage and current flowing through the semiconductor when it remains active.

In case of an IGBT, in on state, the power losses can be calculated according to the expression (8.2). In the case of a MOSFET, the expression (8.3) should be used. It can be seen that the IGBT has two differentiated terms. The first terms refer to the threshold voltage collector–emitter V_{CE0} that depends on the junction temperature T_j. Higher currents become this term more significant. On the other hand, the second term refers to the on-resistance that also depends on the junction temperature T_j. This last term is common for both IGBT/MOSFET semiconductors.

$$P_{IGBT\_Cond} = V_{CE0}(T_j) \cdot I_C(t) + r_{CE}(T_j) \cdot I_C^2(t) \tag{8.2}$$

$$P_{MOS\_Cond} = r_{DS\_ON}(T_j) \cdot I_D^2(t) \tag{8.3}$$

In the IGBT data sheet, it is possible to find the device characteristic curve that defines the current as a function of the V_{CE} voltage for different temperatures. The conduction losses can be calculated by using this curve when the IGBT operates at a constant current. On the other hand, it is also necessary to consider the free-wheeling diode conduction losses. These losses only exist when the IGBT/MOSFET is in off state where current flows through the diode generating voltage drop in the PN junction. Knowing the voltage drop and current, it is possible to calculate the diode conduction losses similarly.

In a three-phase VSI consisting of IGBT transistors, the phase output current flows in a complementary manner between the high- and low-side transistors. In the positive half period of the phase current, the high-side transistor conducts while the low-side transistor the current only passes through the diode. In the negative half period, it is just the opposite; the low-side transistor conducts while the high-side transistor the current flows through the freewheel diode. For sinusoidal PWM (SPWM) modulation, the conduction losses depend on the modulation index m and the phase angle φ (or power factor angle) between the AC machine phase current and the voltage.

In a pure sinusoidal PWM, according to (Kolar et al. 1991) the instantaneous duty cycle can be written as in Eq. (8.4).

$$d(t) = \frac{1}{2}(1 + m \cdot \sin(\omega t + \varphi)) \tag{8.4}$$

This is derived from an extrapolation of the intersections of the sawtooth waveform signal with the sinusoidal reference. The current shape through the high-side

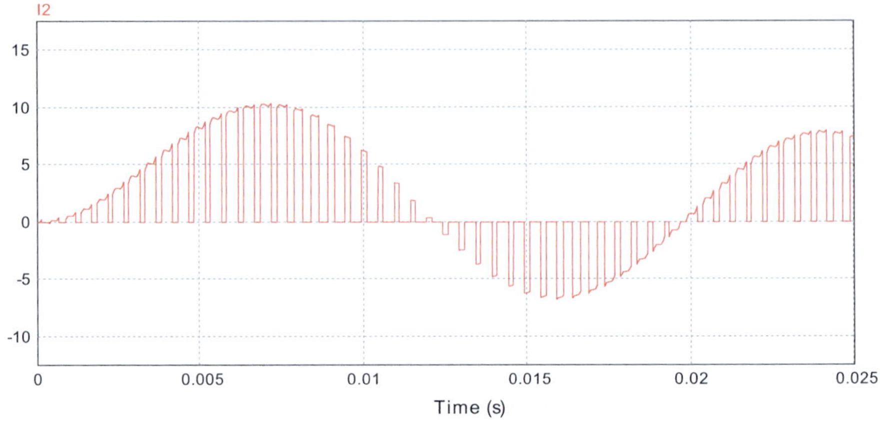

Fig. 8.18 Current waveform through high-side transistor when SPWM modulation is used in the three-phase VSI. During positive half period, the current flows through the high-side transistor (on state). During negative half period, the current flows through the body diode of the high-side transistor (off state)

semiconductor in a VSI is a sinusoidal waveform as expected, but following a kind of sawtooth waveform pulses as depicted in Fig. 8.18. The positive half period corresponds when the high-side semiconductor is in on state.

The conduction losses can be obtained performing the average of the sinusoidal waveform current. Since one of the devices of the same leg only conducts in a half period of the phase current, the conduction losses can be calculated integrating the losses in direct according to (Kolar et al. 1991; Bierhoff and Fuchs 2004) as

$$P_{IGBT\_Cond} = \frac{1}{T_0}\int_0^{T_0/2}\left(V_{CE0}\cdot \hat{I}_{Load}\sin(\omega t) + r_{CE\_ON}\hat{I}^2_{Load}\sin^2(\omega t)\right)\cdot \mathrm{d}(t) \quad (8.5)$$

Substituting Eqs. (8.4) in (8.5) and performing the integral calculation, it is possible to obtain (8.6) of the conduction losses averaged for the IGBT device.

$$P_{IGBT\_Cond} = V_{CE0}\frac{\hat{I}_{Load}}{2\pi} + r_{CE\_ON}\frac{\hat{I}^2_{Load}}{8} + m\cdot\cos\varphi\left(V_{CE0}\frac{\hat{I}_{Load}}{8} + r_{CE\_ON}\frac{\hat{I}^2_{Load}}{3\pi}\right) \quad (8.6)$$

The above equation is valid for the linear modulation zone, i.e., for $m \leq 1$.

On the other hand, in the remaining half period, the current flow through the diode which generates a forward voltage drop V_{F0} plus a voltage drop in its resistance r_F. These cause power losses in the free-wheeling diode. This can be calculated as in the previous case, but taking into account that the diode only is on when the IGBT remains in off state. In this case, the term that depends on the modulation index m must be subtracted as represented in Eq. (8.7).

$$P_{D\_Cond} = V_{F0}\frac{\hat{I}_{Load}}{2\pi} + r_F\frac{\hat{I}_{Load}^2}{8} - m\cdot\cos\varphi\cdot\left(V_{F0}\frac{\hat{I}_{Load}}{8} + r_F\frac{\hat{I}_{Load}^2}{3\pi}\right) \tag{8.7}$$

In the same way as before, the above expressions can be derived for a MOSFET. This is shown in Eq. (8.8). Equation (8.8) can also be used to calculate the losses in the diode.

$$P_{MOS\_Cond} = r_{DS\_ON}\frac{\hat{I}_{Load}^2}{8} + m\cdot\cos\varphi\cdot r_{DS\_ON}\frac{\hat{I}_{Load}^2}{3\pi} \tag{8.8}$$

In Fig. 8.18 is illustrated the current through the body diode, which is during the negative half period.

It is important to comment that the conduction losses increase with the temperature because of the on-resistance increase also with the temperature. Moreover, as mentioned, the voltage threshold V_{CE0} varies with the temperature varying also the conduction losses.

8.3.7.2 Dynamic Losses

8.3.7.2.1 Gate Charge Losses

The IGBT/MOSFET gate losses are related to the charging and discharging input capacitance C_{iss} (Balogh 2013). If the switching frequency is not very high, they are usually not relevant compared to the switching or conduction losses. The IGBT and MOSFET devices need only a few short current pulses, positive to charge the capacitance to turn-on the device, or otherwise, a negative current pulse to discharge the capacitance to turn-off. The current magnitude of the pulses is usually high for fast switching but must be limited by a gate resistance R_G between the gate driver circuit and the gate. The gate resistance has the dual function of limiting the current flow and fine-tuning the device switching. At higher resistance, the charge capacity is slower, incrementing the switching time, which increases the switching losses. At lower resistance, the charge capacity is faster, decreasing the switching time that reduces the switching losses. As will be seen in Sect. 8.4.1, in some cases it is crucial to be able to control the charge/discharge capacity separately, dividing the charging and discharging paths each with their respective optimum R_{GON} and R_{GOFF} resistance.

The average gate current can be calculated as $Q_G\cdot f_{SW}$, which is the average polarization current required by the power switch semiconductor gate. This current at the same time generates losses in the gate resistance and in the gate driver that should be taken into account when designing the driver circuit. Losses at the gate can be calculated as described in Eq. (8.9) where V_{driver} is the gate driver circuit voltage supply.

$$P_{GATE} = V_{driver} \cdot Q_G \cdot f_{SW} \tag{8.9}$$

8.3.7.2.2 Switching Losses

As commented before, the power semiconductors transition switching can be hard- or soft-switched. The most common switching transition is known as hard-switching. This is characterized because, in turn-on and turn-off transitions, the voltage and current of the device are not zero. For example, on the turn-off, the voltage rises before the current decays, while on the turn-on, the current rises before the voltage decays. This overlap results in power ($V \cdot I$) not null in each switching known as switching losses. These are due to the non-idealities of the device as parasitic capacitance discussed in Sect. 8.3.3 and are proportional to the frequency. That is, the higher the switching frequency, the greater the number of transitions from on to off and vice versa, which translates into more power losses. In Fig. 8.19 is illustrated the transitions from off to on and on to off for IGBT semiconductor.

It is possible to observe how, in the transitions, the power switch semiconductor current and voltage is not zero. The longer the transition lasts, the more power losses

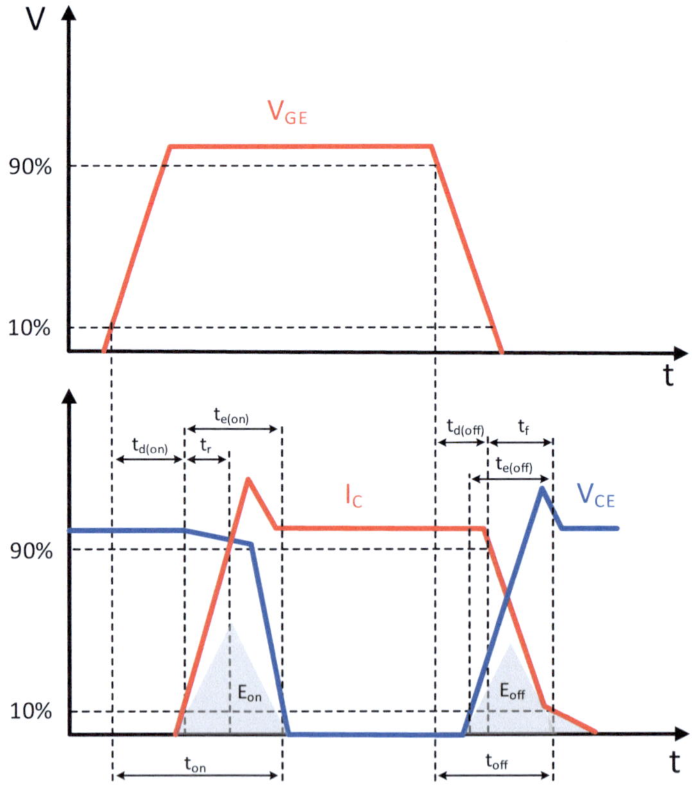

Fig. 8.19 Turn-on, turn-off hard-switching transitions of V_{CE} and I_C signals of IGBT

the device will have. The shaded area indicates the switching losses, where the energies E_{ON} and E_{OFF} are not zero.

The switching energy ESW is defined as the area below the power curve ($V_{CE} * I_C$) and is usually used to estimate the switching losses. For IGBT, ESW is a data sheet parameter which is provided as a function of the gate resistance R_G for different temperature conditions. Higher resistance in the gate denotes a more significant duration of time in the transition on to off or vice versa and therefore more lost energy. Therefore, attention should be paid to the design of the gate resistance, as discussed early. If the ESW is not described in the manufacturer's data sheet, simulations can be performed, as will be seen in Sect. 8.3.7.4.

Unlike the conduction losses, the switching power depends on the switching frequency and not on the modulation index. It is possible to estimate the average switching power losses in a VSI as:

$$P_{IGBT\_SW} = \frac{1}{\pi} \cdot f_{SW}(E_{ON} + E_{OFF})K \tag{8.10}$$

The value of the E_{ON} and E_{OFF} energies is extracted from the IGBT device data sheet, which usually corresponds to a fixed operation point. The K factor (Bierhoff and Fuchs 2004) can help to perform a better approximation since, as shown in Eq. (8.11), it takes into account the nominal values I_{NOM} and V_{NOM} for which E_{ON} and E_{OFF} were measured, together with the voltage and current at the operation point of interest of the design.

$$K = \frac{I_{peak}}{I_{NOM}} \frac{V_{DC}}{V_{NOM}} \tag{8.11}$$

The soft-switching transition attempts to cancel the switching losses, decreasing the stress of the device, i.e., cancel the E_{ON} and E_{OFF} energies. Soft-switching can be achieved by making the voltage zero or near zero in the transition, Zero Voltage Switching (ZVS), or by making the current zero or near zero, Zero Current Switching (ZCS). Both are achieved by adding external elements that achieve a commutation either with zero voltage or with zero current during the transition, increasing the efficiency of the device. In a VSI , the transitions are hard-switching, and most VSI designs do not use methods to get ZVS or ZCS.

On the other hand, it is important to consider the free-wheeling diode reverse recovery losses. Both power switch semiconductors, MOSFET and IGBT, when the free-wheeling diode switches from its on state to the off state, have reverse recovery spikes. In this transition, the charge of the minority carriers must be removed, either by a negative current or passively by recombination within the device. Fast and soft recovery diodes reduce the reverse recovery losses. In the MOSFET/IGBT data sheet, usually, there are three characteristics which define the diode reverse recovery. These are:

- t_{rr}: Reverse recovery time
- I_{rr}: Peak inverse current
- Q_{rr}: Reverse recovery charge.

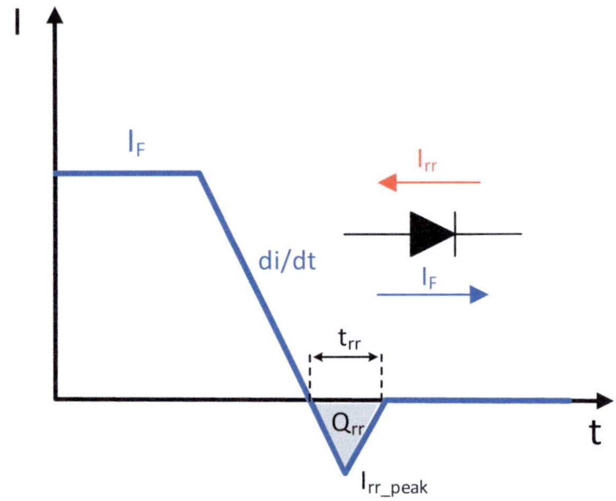

Fig. 8.20 Diode reverse recovery time t_{rr}, charge Q_{rr} and peak current I_{rr}

The Q_{rr} is defined as the area within the negative portion of the diode current as can be seen in Fig. 8.20.

The reverse recovery losses can be estimated by knowing the reverse recovery energy E_{rr} and the switching frequency. The reverse recovery energy depends on the emitter current I_E in an IGBT, or the source current I_S in a MOSFET. The higher the emitter current, the greater the energy of the losses due to reverse recovery. The reverse recovery power losses depend on the switching frequency and are estimated, as shown in (8.12). As before, the K factor fine-tunes the reverse recovery power losses out the nominal conditions.

$$P_{D\_SW} = \frac{1}{\pi} \cdot f_{SW} E_{RR} K \tag{8.12}$$

8.3.7.3 Total Power Losses

Equations (8.6), (8.7), (8.9), (8.10), and (8.12) correspond to the conduction and switching losses for a three-phase VSI for each of the IGBTs. The overall power losses are calculated according to Eq. (8.13) taking into account the number of devices that in the case of the two-level VSI is 6 devices ($n = 6$).

$$P_{T\_loss} = n \cdot \left(P_{D\_Cond} + P_{IGBT\_Cond} + P_{D\_SW} + P_{IGBT\_SW} + P_{GATE}\right) \tag{8.13}$$

Therefore, the VSI efficiency can be calculated as the quotient of the output and input power as shown in (8.14)

$$\eta = \frac{P_{out}}{P_{out} + P_{T\_loss}} \tag{8.14}$$

8.3.7.4 Power Losses Simulation

As discussed before, some manufacturer's data sheet parameters are needed for the power loss analytical method calculation. They are not always available doing complicate and laborious the power losses calculation in a three-phase VSI . For this reason, it is possible to use accurate circuit and a software package simulation as PSIM, to evaluate the power losses in a VSI instead of to use the previous analytical equations. The advantage of this method is that once the circuit simulation is validated, it can be used for power losses estimation and in other phase design, as machine control evaluations, for example.

To overcome the simulation, it is used a three-phase RL load in star connection instead of an AC machine to fix easily different working points. In Fig. 8.21 is shown a typical EV/EHV traction VSI . As can be seen, the inverter is powered by a DC 400 V battery with its parasitic elements. Moreover, the battery is supported by a DC-link capacitance of 1000 μF, also with some relevant parasitic elements. The precise modeling of the VSI is a very meticulous task, and in the example, that is presented, the parasitic elements such as the inductance and resistance in

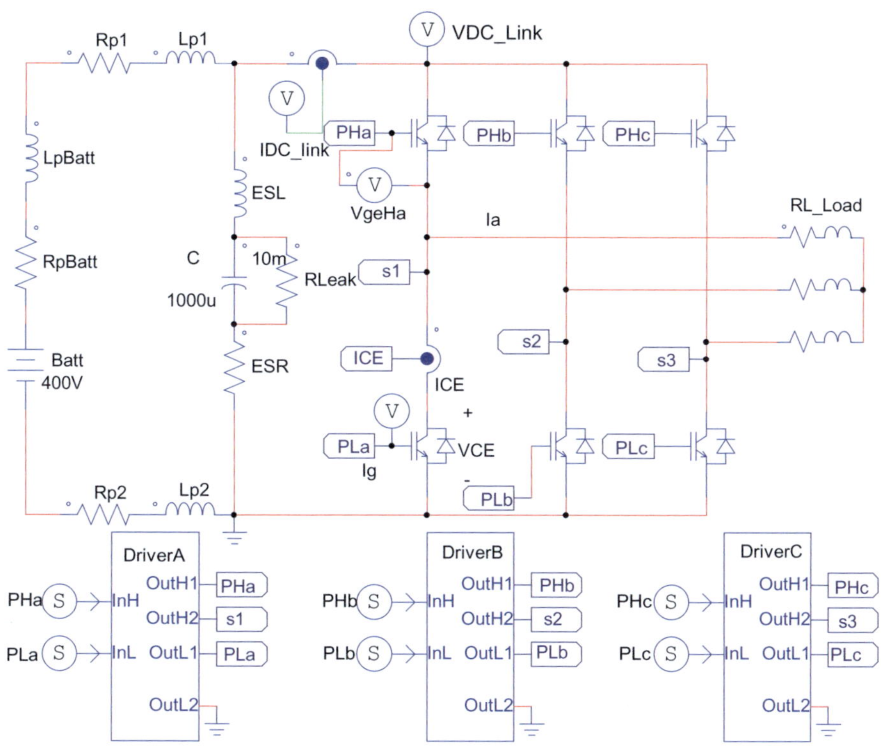

Fig. 8.21 Typical EV/EHV traction VSI circuit implemented in PSIM environment for simulation purpose. Gate driver circuit for IGBT's transistors are also considered in the simulation

the semiconductors, as well as in the layout of the PCB itself, have been ignored. However, both the battery and the DC-link capacitor have parasitic resistances and inductances since they are easier to obtain from the manufacturer's data sheet. In Fig. 8.21, the resistance R_{pBatt} and the inductance L_{pBatt} are the parasitic battery elements provided by the manufacturer. The resistors R_{p1}, R_{p2} and the inductances L_{p1} and L_{p2} are the parasitic elements of the connection between the battery and the inverter (connection power cable between battery and inverter). The DC-link capacitor has its parasitic elements such as the effective series inductance (ESL), effective series resistor (ESR), and its leakage resistance R_{Leak}. These elements are also provided by the capacitor manufacturer. The inverter consists of 600 V IGBT's capable of delivering a current of 145 A_{RMS} at an encapsulation temperature T_C of 100 °C. The maximum active power of the inverter is 84.4 kW, that is 28.1 kW per phase.

The gate-emitter voltage control of the IGBT transistors is achieved with a gate driver circuit designed for the IGBT model. This provides the current and voltage necessary to turn-on/turn-off the IGBT as recommended by the manufacturer for optimum switching losses. The gate driver input is a logical PWM signal coming from the microcontroller/DSP, and this case is connected to the Simulink models through the PSIM SimCoupler module. Figure 8.22 shows the Simulink model of the space vector PWM (SVPWM) modulation, together with the SimCoupler module. Inside of the SimCoupler module, there is the VSI PSIM model of Fig. 8.21, where its inputs are the PWM signals of each IGBT transistor. In the same way, the outputs of the PSIM model can be passed to the Simulink environment, for example, to feedback some measurements to perform a closed-loop control. In this case, the power of the IGBT, the line voltage V_{LLa}, the collector–emitter current I_{CE}, the phase current I_a, and the DC-link voltage can be observed.

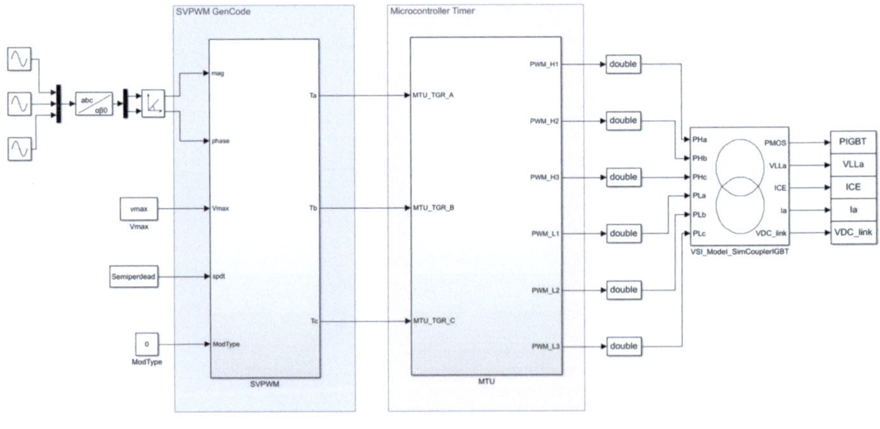

Fig. 8.22 Simulink models connected to VSI PSIM model. Microcontroller Timer Simulink model was seen in Chap. 7, and SVPWM model will be seen in Chap. 9

The dynamic and static losses are obtained for the worst case, that is at maximum power delivered to the RL load. It is considered a sinusoidal voltage waveform with a frequency of interest of 50 Hz. Thus, the power factor which sets the chosen RL load at this voltage frequency is 0.86. The switching and conduction losses are evaluated in a single device, and the total result of the losses must be multiplied by six, as shown in Eq. (8.13). Conduction losses are evaluated at two operation points, at maximum power, and at half power, while switching losses are evaluated at two different frequencies of 5 and 10 kHz.

Figure 8.23 shows the result of the simulation obtained, according to the VSI of Fig. 8.21, for its maximum output power of 84.4 kW. The line-to-line voltage of phase A is represented together with the three-phase currents that flow through the RL load in Fig. 8.23a. It is important to comment that the line-to-line voltage is a sinusoidal PWM waveform which should be filtered by using a low-pass filter as it will be discussed in Chap. 9. However, the phase current is a sinusoidal waveform because of the RL load acts as a low pass filter removing the high-frequency switching components of the PWM. As can be seen in Fig. 8.23b, the low-side IGBT collector–emitter current I_{CE} of phase A is inverted with respect to the output phase current A as expected. In the negative half period of the I_{CE} current, the diode conducts, while in the positive half period, the IGBT conducts as discussed early. The conduction and switching power losses of the low-side IGBT of leg A are shown in Fig. 8.23c. The power losses are more noticeable in the positive half period of the I_{CE} current because it is when the IGBT conducts for a longer time. In the negative half period, the losses are lower because the IGBT conducts for less time and the current flow through the freewheeling diode. High-power switching peaks of up to 70 kW mask the IGBT conduction losses.

The deadtime t_{DT} is set low as possible (500 ns) since the IGBT has a low Q_G along with times $t_{d(ON)}$, $t_{d(OFF)}$, t_r, and t_f of 33, 300, 43, and 33 ns, respectively. It is satisfied that the deadtime t_{DT} is approximately 20% of the sum of the times $t_{d(ON)}$, $t_{d(OFF)}$, t_r, and t_f.

In Fig. 8.24 is represented the turn-on and turn-off IGBT transistor simulation result. The current I_G and the voltage V_{GE} during turn-on/turn-off are shown in Fig. 8.24a, b. In Fig. 8.24a, the first pulse of current I_G is very abrupt and high as expected because it is the initial charging of the IGBT input capacitance. Once the input capacitance has been charged, the V_{GE} voltage reaches the plateau voltage level, and the current decreases considerably at time instant $t = 2.8039$ ms. Then the evolution of the V_{GE} voltage is much slower, reaching 15 V as can be observed at time instant $t = 2.8042$ ms. In the off time (Fig. 8.24b), something similar happens, but in the opposite direction, that is, the current pulse is negative to be able to quickly discharge the input capacitance reaching almost -3 A of peak. As soon as the V_{GE} voltage falls, the current is already relaxed until it reaches zero along with the V_{GE} voltage. There is a small oscillation due to the parasitic inductances of the gate driver and the IGBT transistor loop.

In Fig. 8.24c, d are also shown the V_{GE} voltage multiplied by 20 to facilitate visualization, together with the I_{CE} current, and the V_{CE} voltage in the turn-on/turn-off. It is important to mention that there is a current overshoot in turn-on and a voltage

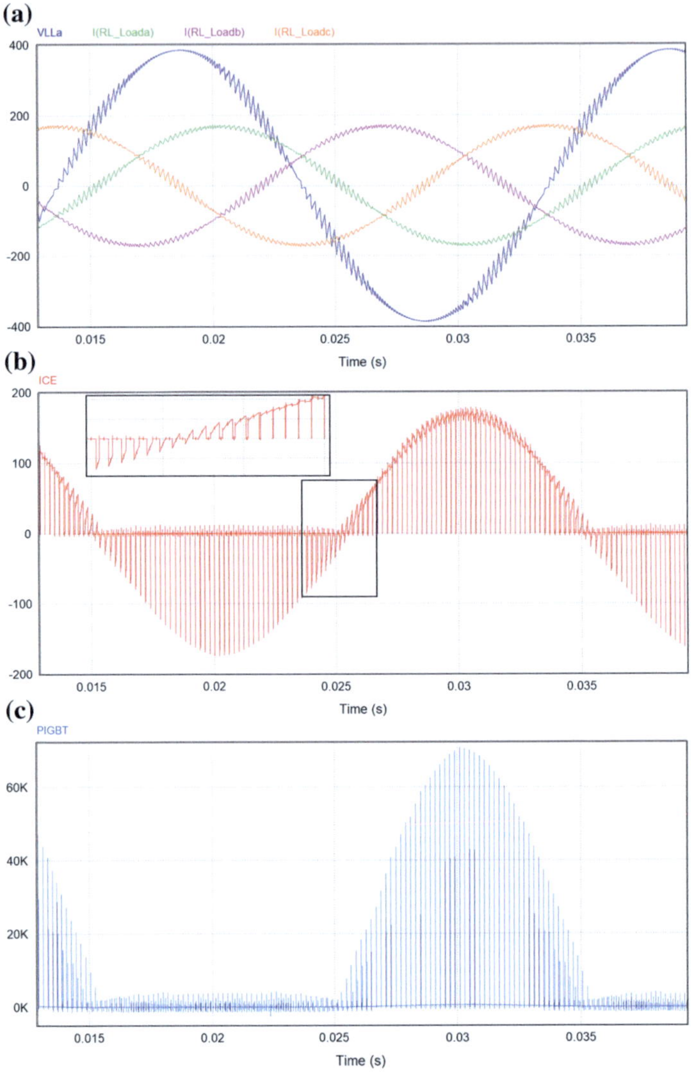

Fig. 8.23 Simulation result of VSI at maximum power with a three-phase RL load. **a** Line-to-line phase A voltage and phase current I_a, I_b, I_c. **b** Collector–emitter current at low-side IGBT of A leg. **c** Low-side IGBT power losses. Modulation index: $m = 0.9$, output frequency: $\omega_c = 50$ Hz. Deadtime: $T_{DT} = 500$ ns. Switching time: $T_{PWM} = 200$ µs. Simulation step PSIM: $T_{spP} = 1$ ns. Simulation step Simulink: $T_{spS} = 10$ ns

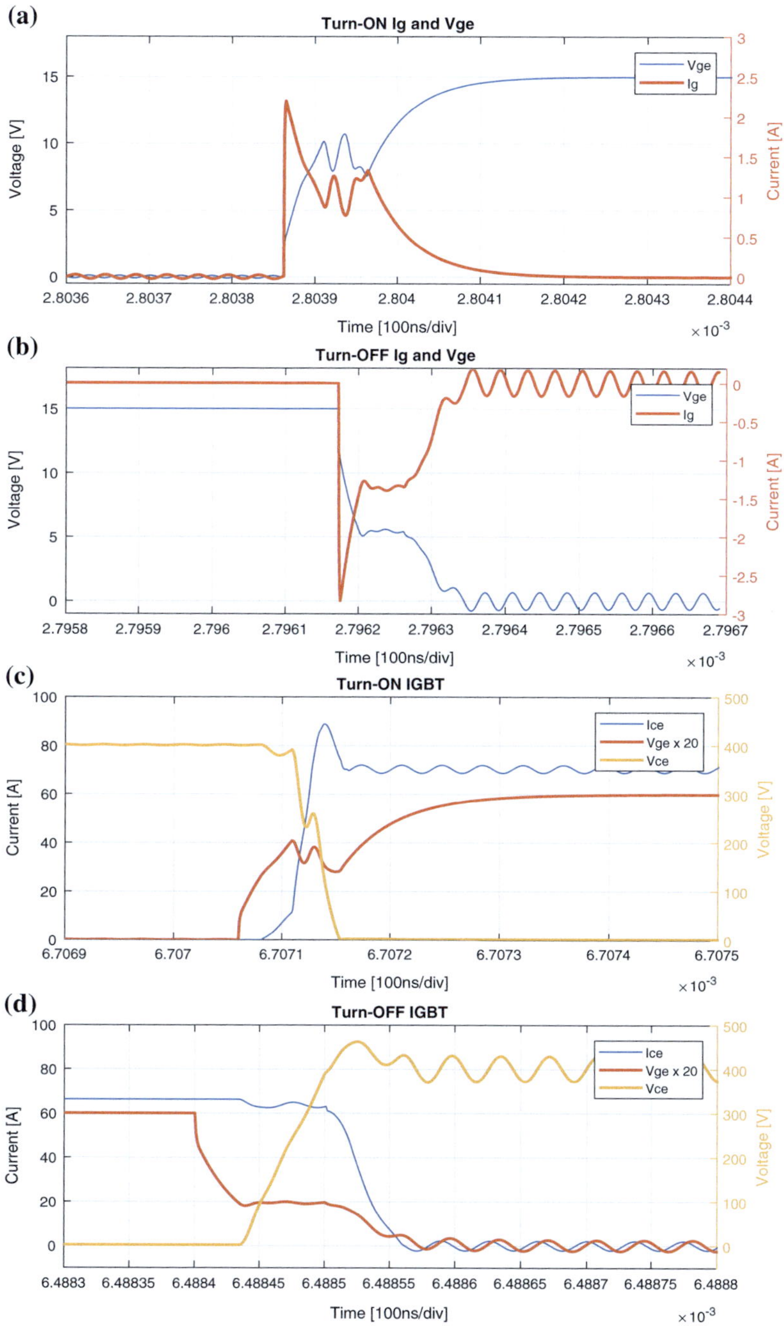

Fig. 8.24 Simulation result of VSI at maximum power. **a** Turn-on I_G and V_{GE}. **b** Turn-off I_G and V_{GE}. **c** Turn-on I_{CE}, V_{GE}, and V_{CE}. **d** Turn-on I_{CE}, V_{GE}, and V_{CE}. Modulation index: $m = 0.9$, output frequency: $\omega_c = 50$ Hz. Deadtime: $T_{DT} = 500$ ns. Switching time: $T_{PWM} = 200$ µs. Simulation step PSIM: $T_{spP} = 1$ ns. Simulation step Simulink: $T_{spS} = 10$ ns

overshoot in the turn-off. The first is due to the reverse current of the freewheeling diode while the second is due to the parasitic inductance of the loop, as it will be seen below.

It is possible to observe that there is an area where the I_{CE} current and V_{CE} voltage are not null. For example, in Fig. 8.25 is shown a detail of how the power peak reaches 70 kW in a very short period of time during turn-off. This is the switching power losses, and the energy is approximately 3.5 mJ.

In the previous figures was shown the switching detail of the IGBT transistor. The following figure shows a longer simulation to have the turn-on and turn-off in the same capture. Figure 8.26 shows the IGBT turn-on/turn-off of the same inverter leg where the deadtime can be seen. The gate-emitter voltages of both IGBTs are represented together with the collector–emitter voltage V_{CE} of the low-side IGBT transistor.

The V_{CE} voltage surge from on to off state in Fig. 8.27 is approximately the di/dt by the total loop parasitic inductance according to Eq. (8.15). As discussed, this is due to the IGBT parasitic inductance join the cabling and/or layout between the DC-link voltage and the IGBT. Minimizing this parasitic inductance minimizes the voltage surge and the damped oscillations from collector–emitter and therefore reduces the losses by switching. In addition, these peaks and the frequency of the damped oscillations at each switching generate electromagnetic interference, so it is always advisable to reduce them to the maximum. Figure 8.27 shows the overshoot voltage in V_{CE} as well as the instantaneous power in the switching.

The parasitic inductances lead to transient voltages in the switching transients (surge voltage) and are an important source of EMI noise (electromagnetic interference) as previously mentioned.

$$\Delta V = L_s \frac{\mathrm{d}i}{\mathrm{d}t} \tag{8.15}$$

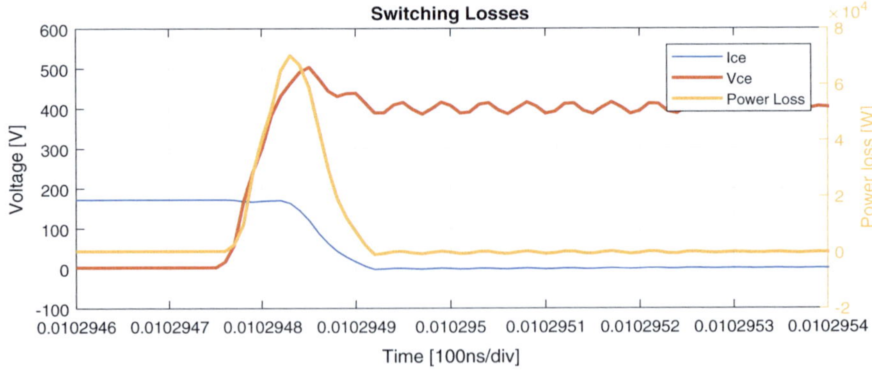

Fig. 8.25 Power loss in Turn-off, I_{CE}, and V_{CE}. The energy is approx. 3.5 mJ. Modulation index: $m = 0.9$, output frequency: $\omega_c = 50$ Hz. Deadtime: $T_{DT} = 500$ ns. Switching time: $T_{PWM} = 200$ µs. Simulation step PSIM: $T_{spP} = 1$ ns. Simulation step Simulink: $T_{spS} = 10$ ns

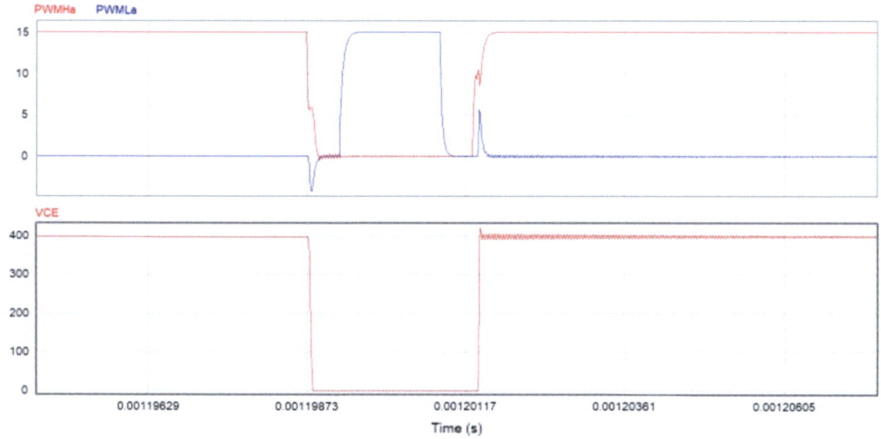

Fig. 8.26 Gate-emitter voltage of high (P_{WMHa})- and low (P_{WMLa})-side IGBT of leg A with 500 ns of deadtime. Low-side IGBT collector–emitter voltage (V_{CE}) of leg A. Modulation index: $m = 0.9$, output frequency: $\omega_c = 50$ Hz. Deadtime: $T_{DT} = 500$ ns. Switching time: $T_{PWM} = 200$ µs. Simulation step PSIM: $T_{spP} = 1$ ns. Simulation step Simulink: $T_{spS} = 10$ ns

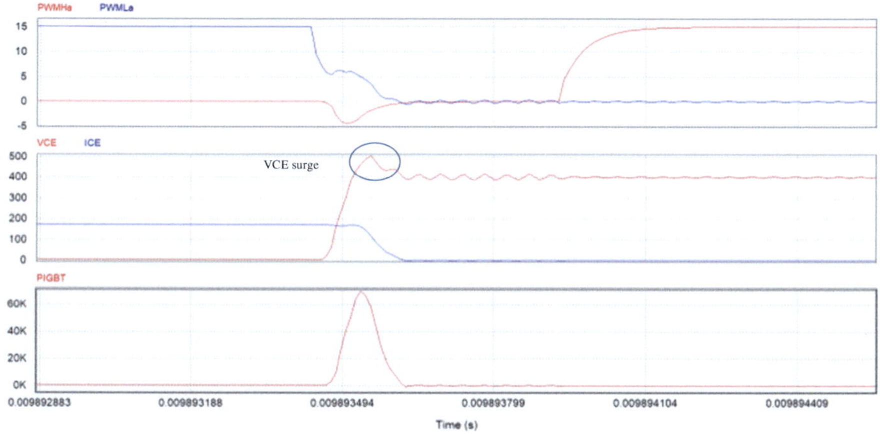

Fig. 8.27 Turn-off process showing the surge voltage, the damped oscillations, and the instantaneous power switching. Modulation index: $m = 0.9$, output frequency: $\omega_c = 50$ Hz. Deadtime: $T_{DT} = 500$ ns. Switching time: $T_{PWM} = 200$ µs. Simulation step PSIM: $T_{spP} = 1$ ns. Simulation step Simulink: $T_{spS} = 10$ ns

In the previous simulation, it has a di/dt of 1.6 kA/µs and a ΔV of 100 V. Then, the total parasitic loop inductance can be calculated according to the (8.15), resulting 62.5 nH.

Figure 8.28 shows the turn-on/turn-on transitions where gate-emitter voltages of both IGBTs are represented together with the collector–emitter voltage V_{CE} of the low-side IGBT transistor, the collector–emitter current I_{CE}, and the power switching

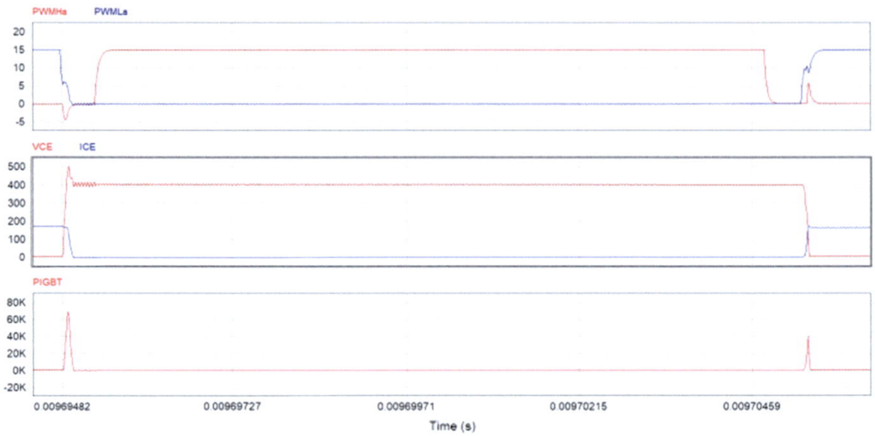

Fig. 8.28 Turn-on/turn-off transitions. Gate-emitter voltages, V_{CE}, and instantaneous power switching is shown. Modulation index: $m = 0.9$, output frequency: $\omega_c = 50$ Hz. Deadtime: $T_{DT} = 500$ ns. Switching time: $T_{PWM} = 200$ μs. Simulation step PSIM: $T_{spP} = 1$ ns. Simulation step Simulink: $T_{spS} = 10$ ns

losses. In Fig. 8.29 is observed the current overshoot in collector–emitter current I_{CE} due to the inverse recovery current of the freewheeling diode as previously mentioned. A diode with better characteristics minimizes this overshoot and therefore the diode power losses.

Finally, it is important to mention that the IGBT/MOSFET conduction losses are an important part of the total losses in AC machine control applications. Table 8.4

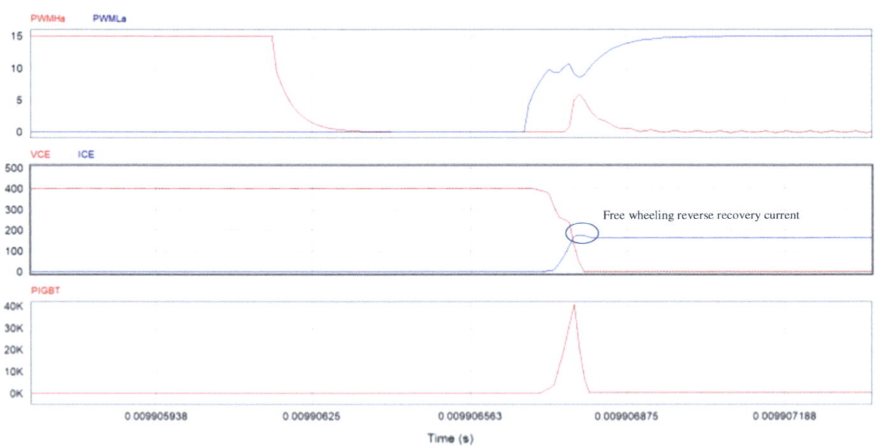

Fig. 8.29 Current overshoot due to freewheeling reverse recovery current during turn-on. Gate-emitter voltages, V_{CE}, and instantaneous power switching are shown. Modulation index: $m = 0.9$, output frequency: $\omega_c = 50$ Hz. Deadtime: $T_{DT} = 500$ ns. Switching time: $T_{PWM} = 200$ μs. Simulation step PSIM: $T_{spP} = 1$ ns. Simulation step Simulink: $T_{spS} = 10$ ns

Table 8.4 Calculated and simulated conduction losses with two modulation index

SVPWM	Calculated		Simulated	
M	P_{C_IGBT} (W)	P_{C_DIODE} (W)	P_{C_IGBT} (W)	P_{C_DIODE} (W)
1	109.92	13.32	97.62	11.28
0.6	83.2	27.5	74.52	24.35

Table 8.5 Calculated and simulated switching losses with two switching frequencies

SVPWM	Calculated		Simulated	
F_{SW} (kHz)	P_{SW_IGBT} (W)	$P_{SW\,DIODE}$ (W)	P_{SW_IGBT} (W)	P_{SW_DIODE} (W)
5	54	3.27	45.93	3.19
10	108	6.54	90.94	5.29

shows a comparison of the conduction power losses for two modulation index with RMS current of 120 A in a 10 kW VSI . It is shown the analytic and simulations results.

Table 8.5 shows a comparison for the same current as in the previous case, but for two different switching frequencies. As can be seen, the switching frequency has a stronger influence than that produced by the variation of the modulation index in conduction losses.

8.4 VSI Design Considerations

8.4.1 Gate Driver

The PCB layout must be designed accurately to minimize the parasitic inductances between the output of the gate driver and the semiconductor. As mentioned, a stray inductance is formed by the loop between the gate and the return ground path of the source in the low-side semiconductor. However, in the high-side semiconductor, the return path of the loop is not made by the ground, but by a floating point as it will be seen in this section. The stray inductances of both cases cause a reduction in switching time and high overvoltage spikes with oscillations in the loop. With a wide and flat copper tracks ground, it is reduced significantly. As seen in Sect. 8.3.3.2, the 4-pin encapsulation also helps to minimize stray inductances. Short and direct connection between gate driver and the semiconductor benefit said reduction. On the contrary, if the gate driver is far away and flying leads must be used, it is convenient to use twisted pair or shielded cables.

Additional measures should be necessary for higher switching frequencies than 100 kHz since the effects of stray inductance are more noticeable. For these cases, the design of the layout (Reusch and Strydom 2014) should be as optimal as possible to reduce the inductances of the gate-source loop, and the common source inductances present in the half-bridges between the high- and low-side semiconductors.

The gate driver circuit has the main function of providing an appropriate current in the activation and deactivation of the power switch semiconductor, to obtain the maximum performance of the device. The appropriate current must cause that the power switch is saturated or desaturated in a short period of time.

The input signals of the gate driver come from the control circuitry such as microcontroller or DSP, which can be with or without isolation. Gate drives can be with or without isolated supplies and include protection circuits such as short-circuit protection and under voltage protection in the supplied voltage. The supply voltage must be very stable so that typically a bulk capacitor higher than 10 μF, close to the driver, is connected to avoid voltage drops during the transients produced in the activation/deactivation of the power switch semiconductor.

Often there are applications where isolation between the references of high and low voltages are strictly necessary as in EV or industrial sectors. In addition, isolation reduces EMI transfer. Safety standards such as VDE 0884, IEC 60950, and similar requirements of other agencies require the isolation of potentially lethal voltages and current s from possible human contact. Galvanically isolated topologies use the isolation between the primary and secondary coils of a transformer to carry out the isolation of the power supplies. In the market, there are also gate driver's packages where internally they have an isolation barrier that can be either opto- or transformer-isolated, which isolate the control signals and supply voltage. In some applications such as high-voltage systems, the fiber optic cables are often used to provide isolation and to have more flexibility in locating the control circuity. The gate driver should be close to the power switch as commented before, but the control circuity can be a few meters if fiber optic is used.

In this section, it will see the driver for half-bridges with its necessary elements for better performance and the technique of parallel devices currently used to increase efficiency in VSI .

8.4.1.1 Half-Bridge Driver

Today, there are a wide variety of gate driver's packages prepared to control half-bridges that significantly simplify the design task. Some of them have protection elements such as the inhibition of identical activation pulses on the high and low sides of the half-bridge (shoot-through prevention). Other protection mechanisms such as overcurrent and undervoltage detectors in their power supply can be found. Figure 8.30 shows the basic schematic of the elements necessary for a standard gate driver without isolation to control a half-bridge composed of MOSFET semiconductors.

The gate driver, in this case, is activated by a high level in the V_{iH} and V_{iL} inputs. These inputs are the control circuity PWM signals that should be filtered through a simple RC low-pass filter, as observed in Fig. 8.30. The filter should be placed very close to the driver and is composed by R_{FL} and C_{FL}. The low-pass filter avoids the

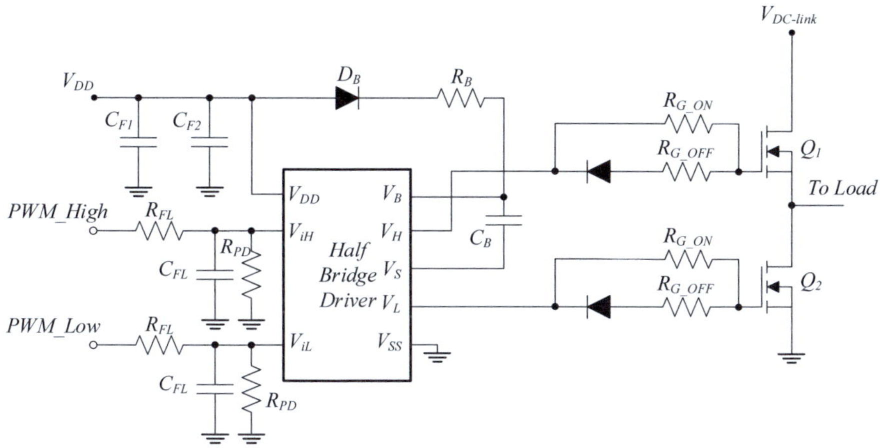

Fig. 8.30 Non-isolated half-bridge gate driver

false activation when high-frequency short transient noise is present in the circuit, mainly due to high stray inductances. It should be noted that the low-pass filter could cause a slow down in turn-on or turn-off switching where switching losses can be incremented. However, it can help to reduce the voltage slew rate, i.e., some switching problems as overshoot voltages due to stray inductance, which caused a stress situation in the machine insulation winding.

The pull-down resistance R_{PD} sets zero voltage level in the input of the driver when the control circuity sets high impedance level (no control signals). This resistance is externally placed if the gate driver does not have a pull-down resistance. The high impedance level is caused during a microcontroller/DSP reset where all its outputs tend to remain in high impedance during reset time.

The high-side power switch (MOSFET) requires a floating power supply because its source is not referenced to ground. This floating power supply is generated with the circuit consisting of D_B, R_B, and C_B (bootstrap circuit). When the low-side MOSFET turn-on, the C_B capacitor is charged to V_{DD} minus the voltage drop in the diode via D_B and R_B. The resistor R_B limits the charge current of the C_B capacitance. When the low-side MOSFET is turned off, the diode stops conducting the current, and the charge is stored in the C_B capacitor. The charge of the capacitor C_B is used as a floating power supply for the internal buffer of the high-side MOSFET gate driver. The capacitors C_{F1} and C_{F2} are high-frequency decoupling capacitors to absorb current transients reducing the noise in the driver supply voltage.

The gate resistance must be dimensioned according to Sect. 8.3.7 but also following the recommendations of the semiconductor manufacturer. To accelerate the deactivation of the MOSFET/IGBT, a totem-pole circuit can be added that improves the discharge time of the C_{iss} capacitance. The totem-pole circuit is composed of a BJT transistor for each MOSFET/IGBT that provides a higher current capacity than the driver to discharge the C_{iss} capacitance. It can be placed near the MOSFET/IGBT

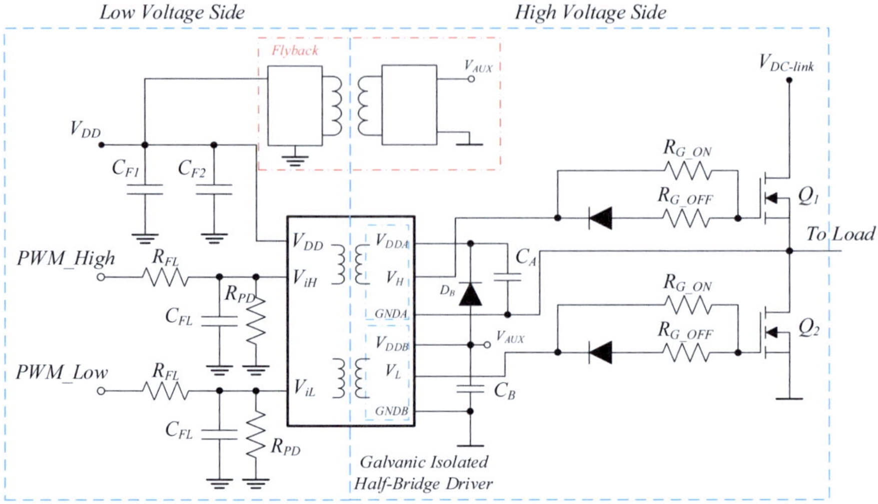

Fig. 8.31 Galvanic isolated half-bridge gate driver

to reduce the loop current of the turn-off and suppresses the Miller effect that will be seen next. The totem-pole circuit is usually used in the activation/deactivation of parallels MOSFET/IGBT since they need a higher current for turn-on and turn-off.

On the other hand, in Fig. 8.31 is shown the galvanically isolated version of the half-bridge driver. As can be seen, on the high voltage side, the part that feeds the driver internal circuit needs an auxiliary power supply source for its operation. That is, the driver needs two power supplies, one on each side of the low voltage and high voltage. Both PWM signals high and low, which come from control circuity, are internally isolated from the high-voltage side. The auxiliary voltage source is usually taken from the low voltage side through a galvanically isolated flyback switched source using a transformer as shown in the figure. It can also be observed that the grounds of both sides are separated.

8.4.1.2 Miller Effect

The Miller effect is due to the parasitic capacitance, also called Miller capacitance, located between gate-drain (MOSFET), or gate-collector (IGBT). Through Miller capacitance, a current can flow when there is a very high dv/dt. If it is considered that initially, the high-side power switch of Fig. 8.32 is off and the low-side power switch Q_2 is in on state. If Q_2 is turned off, and Q_1 is given an activation signal, the drain voltage to Q_2 has a transient that causes a very high dv/dt. Due to this high dv/dt, the current called Miller current can flow through the Miller capacitance C_{GS}. This current passing through the gate driver can cause the voltage V_{GS2} to increase to a certain level, which could cause Q_2 to turned on in a short period next while

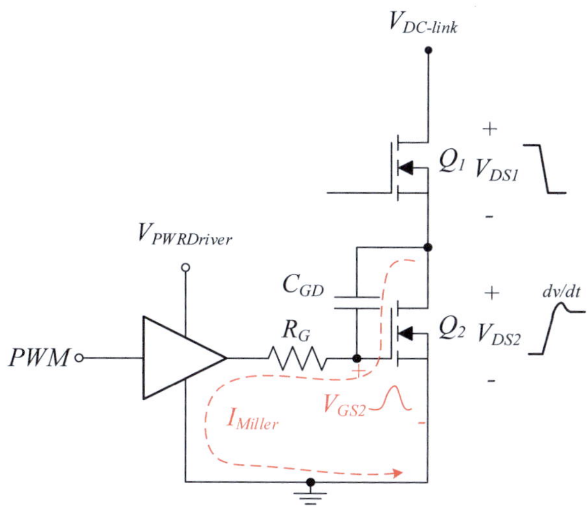

Fig. 8.32 Miller Effect during high d*v*/d*t* which creates a current through Miller capacitance

power switch Q_1 is on. This simultaneous conduction of both transistors in a short period causes a short circuit of the DC-link voltage that can damage both devices.

The most common way to avoid the Miller effect is to divide the on and off paths into two by means of two resistors, R_{GON}, R_{GOFF}, and two diodes that force the direction of the current as shown in Fig. 8.33. The resistance R_{GOFF} must be smaller than the original resistance R_G so that the voltage V_{GS2}, during the off state of Q_2, does not grow as much as in the previous case. Another way to avoid the Miller effect is to use the totem-pole topology as discussed early.

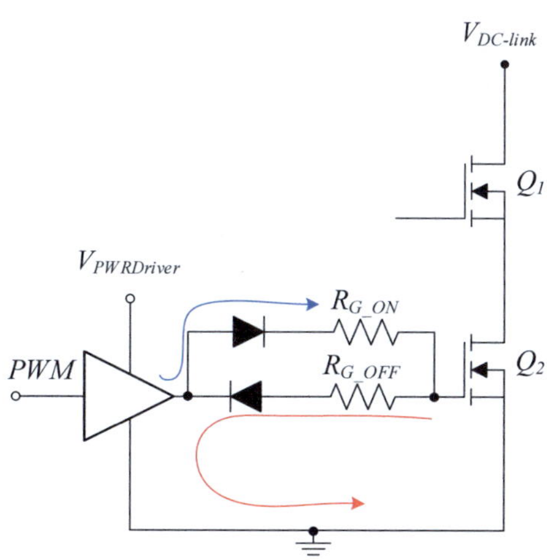

Fig. 8.33 Gate current split to avoid the Miller effect during high d*v*/d*t*

8.4.1.3 Paralleling Power Switch Semiconductor

The technique of parallel power switch semiconductors offers a series of advantages such an increment of the VSI power and a reduction of the total conduction losses. However, the current for each power switch should be well distributed, and the layout design complexity is incremented.

For m devices in parallel, Eq. (8.13) is rewritten, as shown in Eq. (8.16):

$$P_{T\_loss} = n \cdot \left(\frac{P_{D\_Cond}}{m} + \frac{P_{IGBT\_Cond}}{m} + P_{D\_SW} + P_{IGBT\_SW} + m \cdot P_{GATE} \right) \quad (8.16)$$

The conduction losses are reduced while the gate losses increase according to the number of devices in parallel. As discussed early, the gate losses were much lower than the conduction and switching losses, so it can be said that the total losses are considerably reduced. Then, the VSI efficiency is incremented at the expense of increasing the number of devices and therefore, its cost.

On the other hand, the parasitic resistances and inductances are also reduced, although the design of the driver requires more effort to achieve an identical parallel activation of the devices connected in parallel. For example, there are two effects by which currents called circulation current flows, as shown in Fig. 8.34 for two

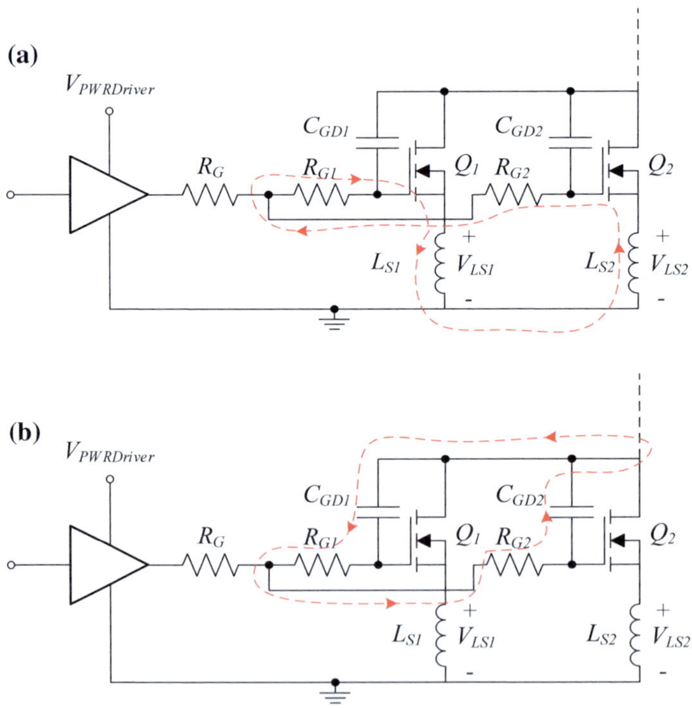

Fig. 8.34 Gate circulating current. **a** Due to different parasitic inductance in the source. **b** Due to skew and Miller effect

MOSFETs in parallel. The first is due to the natural difference between the common parasitic inductance of both devices. When the MOSFET is turned on, an identical di/dt current flows through both MOSFETs, which generate different voltages V_{LS1} and V_{LS2}. It simultaneously generates a current that circulates as shown in Fig. 8.34a. The second is due to the Miller capacitance previously seen. Both can be reduced by separating the gates of both MOSFETs through the resistors R_{G1} and R_{G2}, also reducing the oscillations in the gate-source loop. The resistance R_G is common for both devices, which allows fixing the ignition current of the two power switch at the same time.

In order that all power switch semiconductors in parallel are turned on at the same time, the gate driver circuit must be very similar. That is, the gate current I_G and voltage V_{GS} for each of them must be identical. One way that offers excellent results is to use only one gate driver for all the devices in parallel. The R_G resistance tolerance of each of the gates must be very low (0.1%) to allow identical gate current I_G for all devices. Stray inductances such as source loop affect the activation and deactivation times, as seen in Sect. 8.3.3. If a MOSFET device has a greater source inductance than the others, a higher voltage is induced at the inductance terminals, reducing the V_{GS} voltage and slow down its activation. The way to solve this problem is basically to design the layout as symmetrically as possible so that each of the parasitic inductances has a very close value. The V_{TH} voltage in a MOSFET can vary between 2 and 4 V so that the MOSFET with a V_{TH} of 2 V will be activated sooner. To avoid this, MOSFET devices must be chosen with a V_{TH} voltage with a very low tolerance. The gates' connections without any resistance must be avoided since in addition to the effect of the circulating currents commented previously, the MOSFET that reaches its V_{TH} first enters the Miller region faster blocking the gate of the other MOSFETs to the Miller voltage. Then, it slows down, even more, the gate voltages of the rest of the MOSFETs. The temperature affects the V_{TH} voltage, so it is vital to keep all the MOSFET at the same temperature.

Although identical symmetry in the driver circuit, if a MOSFET has a lower C_{iss} input capacity, it reaches its V_{TH} voltage before, driving most of the current on activation. On deactivation, the MOSFET with a higher C_{iss} capacitor will reach the voltage V_{TH} the latter conducting the current for a longer period of time. The individual gate resistances of each of the small value MOSFETs help to reduce this effect. The selection of gate resistances results in a compromise between the suppression of circulating currents and the effect of variations in C_{iss} and V_{TH}.

As mentioned before, the paralleling technique allows increasing the efficiency of the inverter, but it is also used to design higher-power VSI with discrete semiconductors.

8.4.2 Current Measurement

As discussed in Chap. 6, current measurement can be performed using different types of sensors such as shunt resistor or Hall effect sensors. The current measurement is fundamental in power stage topologies such as DC-DC converters and inverters for protection and control purposes. In electrical machines, the current measurement is also used to protect the power stage and the machine in case of failures, and of course, for control purposes. When the DC or AC machine is controlled by a PWM, the current can be seen as two components: an average component and a switching component due to ripple. The current ripple is not more than the natural filtering of the PWM voltage in the machine terminals which present higher or lower ripple depending among other factors of the equivalent machine inductance *L*. In machine control, the current component of interest is the average component. In Fig. 8.35 is shown an example of a sinusoidal phase current portion for three-phase AC machine with a fundamental average current of 110.5 and 16 A of ripple approximately. As can be observed, depending on the sampling instant, the obtained result can be different from the average value of the current. For this reason, the sampling instant should be synchronized with the PWM.

In a high-performance torque control for DC and three-phase AC machines, the current measurement should be the finest as possible where the measurement effects should be minimized. In the digital control based on microcontrollers, DSP's, or SoCs, the analog-to-digital converters have good linearity throughout the measurement range with 12-bit resolutions, and a negligible delay compared to the processing of the measurement. As discussed in Chap. 6, the processing of the measurement consists of different stages such as current sensor, an amplifier, and a low-pass filter, whose output is connected to the analog input of the microcontroller. In the processing of these stages, errors, delays, and random noise are introduced in the current measurement without the analog-to-digital converter stage having participated yet. The total bandwidth of the different stages should be higher to avoid losing the current ripple shape since lower bandwidth reduces the ripple introducing an excess of delay that also reduces the control performance.

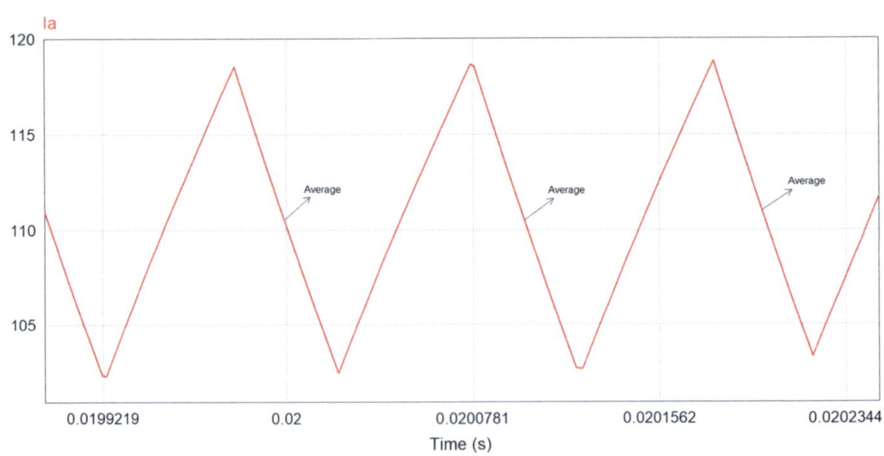

Fig. 8.35 Current ripple envelope about fundamental average current sinusoidal portion

It is important to note that the low-pass filter stage is designed to eliminate the high-frequency noise of the PWM harmonics and not to eliminate the current ripple. The low-pass filter cut-off frequency can be designed with the criterion that it is at least ten times higher than the PWM frequency avoiding bandwidth limitation. The set of errors and delays caused by current processing and by the analog-to-digital converter itself limit the bandwidth of current control, and in extreme cases, can generate instability in the current loops (Sepe et al. 1994).

In the case of a high-performance control such as vector control in AC machines, the cycle-to-cycle phase current measurement is essential to obtain the highest dynamic response where the torque can be controlled instantaneously. The cycle-to-cycle control involves executing the torque and flux loops each PWM period with the actual *dq* current components measurements. According to this, depending on the PWM frequency, the processor requirement to process the control loops can be higher or lower. It is essential to mention that there is some application where the cycle-to-cycle current control is not required, being able to run n times slower than the frequency of the PWM. Thus, low-performance microcontrollers can be used without reducing the PWM frequency. The opposite case, that is, executing the control n times faster ($n < 1$) than the PWM frequency implies an unnecessary oversampling without any advantage. In general, it is possible to assume

$$\begin{aligned} T_s &= n \cdot T_{PWM} \\ n &\geq 1 \end{aligned} \tag{8.17}$$

where T_s is the sampling period of the current acquisition and the current loop. When n takes the value of 2, it means that the current sampling runs at half the speed of the PWM frequency, and therefore the control loops of the currents *dq* also.

The main factors that influence when choosing the PWM frequency are:

- Power devices switching losses
- Torque and flux periodicity control (bandwidth of current control)
- Electric machine equivalent inductance L
- Current and torque ripple desired
- Acoustic noise produced by the switching frequency.

Low PWM frequency mainly means higher current ripple, which causes a torque ripple in the electric machine. The torque ripple can be critical in high-precision positioning systems such as servo motors used in robotic arms where a very low-inductance PMSM or BLDC machines can be usually found. In these cases, preferable are higher PWM frequencies to improve the precision positioning by reducing the torque ripple. Moreover, higher precision sinusoidal currents (low THD) above 1 kHz of frequency is possible with higher PWM frequencies, which are an ideal solution for high-speed drone machines.

The acoustic noise produced in the electric machine due to the audible switching frequency of less than 20 kHz can be a problem in non-industrial environments such as domestic and automotive. In this case, frequencies greater than 20 kHz are more comfortable, especially in household appliances such as refrigerators with three-phase inverters.

The high PWM frequency to avoid acoustic noise and to improve the precision positioning in servo motors are contradicted by the power devices switching losses that are proportional to the frequency as discussed. However, the improvement of the power devices means that the switching losses are increasingly reduced as discussed previously in WBG devices that are alternatives to silicon, such as GaN and SiC.

As discussed early, higher-frequency and higher performance processing are needed to execute the current loops each PWM period. Then, high-performance microcontrollers or DSP's and in the worst case, programmable logic such as SoC or FPGA will be necessary. FPGAs and SoCs have more and more use in the automotive market, especially in (H)EV vehicles because it allows increasing the power density thanks to the PWM frequency increment in DC-DC and AC-DC converters. The direct consequences of PWM frequency increment are the inductors' size reduction used in the DC-DC and AC-DC converters. PWM frequencies and cycle-to-cycle control of more than 100 kHz and up to 500 kHz can be supported by most current FPGAs. However, in the case of VSI , the increase of the PWM frequency does not mean to increase the power density winding, but few advantages like lower motor torque ripple (current ripple reduction), close to ideal sinusoidal motor current waveform (lower THD), smaller filter size, and lower cost filter. In the most of applications this advantages are not enough to strengthen the idea of increasing the PWM frequency since the switching losses for inverters over 10 HP will be considerable, not increasing its power density. The inverter power density should be increased with other techniques for example by using GaN and SiC devices.

On the other hand, it is intuitive to think that the higher the power of the machine, the lower the PWM frequency can be to reduce the switching losses, but also because they require a less dynamic control (lower bandwidth) due to its inertia and that of the system to which it is connected. In other words, the response to the step speed command with higher inertia will always be slower than the response of a machine with lower inertia. For example, in electric robotic arms, the size AC machines are low, with very-low equivalent inductance, where high PWM frequency is typically used to have an extensive bandwidth control.

8.4.2.1 Phase Current Measurement in a Three-Phase Machine

To reconstruct the *dq* currents, it is only necessary to measure the current of phase A and B, as was demonstrated in Chap. 3, although for higher safety level, especially in the automotive sector, is usually measured in each phase. The measurement of any of these phases can be carried out by means of a shunt-type current sensor or alternatively a Hall effect sensor. In case of a shunt, the sensor can be placed either between the low-side power switch and ground (low-side $R_{A\text{-}Shunt2}$) or directly between the inverter and the AC machine (in-line $R_{A\text{-}Shunt1}$) as shown in Fig. 8.36.

In the case of low-side $R_{A\text{-}Shunt2}$ option is used, the low-side MOSFET should be turned on so that the current of that phase of the machine passes through of the

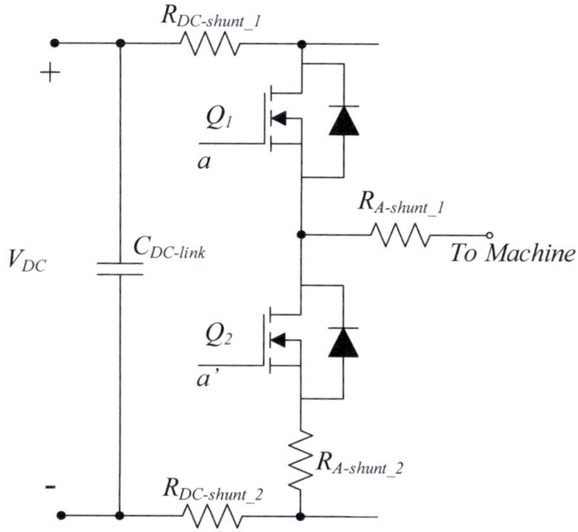

Fig. 8.36 Different possibilities of current measurement for a half-bridge configuration. The phase current A can be measured directly by using two possibilities $R_{A\text{-}Shunt1}$ and $R_{A\text{-}Shunt2}$, while with a single shunt, the phase current of phases A, B, and C can be reconstructed from the $I_{DC\text{-}link}$ current by using either $R_{DC\text{-}Shunt1}$ or $R_{DC\text{-}Shunt2}$

current sensor. This is one of the standard options due to its fast response time, high bandwidths, and low common-mode input voltages for the signal conditioner. However, the main drawback is the shunt resistor parasitic inductance which in series with the gate-source loop as discussed in Sect. 8.3.3, where possible voltage overshoots and ringing have to be filtered out, especially for higher PWM switching frequencies. This extra filtering signal could limit the current regulation bandwidth.

The in-line shunt sensor option is exactly the current to measure independent of the state of the power switches. It provides higher precision, higher response, and less processing in the microcontroller. However, the shunt resistor sees high voltage as it is connected to DC-link voltage when high-side power switch is turned on. Thus, higher common-mode and isolated signal conditioner should be used, which usually both represent a higher-cost solution.

Another technique with a lower cost which requires high processing performance consists of using only one current sensor ($R_{DC\text{-}Shunt1}$) placed between the ground of the three inverter legs junction and the DC-link ground, or between DC-link positive voltage and the three inverter legs junction ($R_{DC\text{-}Shunt2}$). As can be observer, one shunt sensor and one signal conditioner are only needed. In this case, the reconstruction of the *dq* currents is not trivial and can affect the regulation and bandwidth of the current loop. This technique is commonly used in low-cost vector controls that do not require maximum regulation.

On the other hand, the Hall effector sensors are usually placed in-line between the inverter and the AC machine since this is where the current is continuous, not dependent on the state of the devices, and because its bandwidth is limited. As discussed in Chap. 6, these sensors offer a proportional and linear output voltage response to the current flowing through the phase so that the signal conditioning is more straightforward than in the case of using a shunt-type sensor. In addition, they offer inherent insulation between the high- and low-voltage sides. If the insulation is necessary for the design, but for different reasons a shunt-type sensor must be used, the signal conditioning must be isolated, either galvanically or optically.

Considering the low-side and in-line options with a cycle-to-cycle control as discussed early, it is necessary to execute the current loops each PWM period with the actual measurements $ia(k)$, $ib(k)$, $ic(k)$, acquired in the previous duty cycle $da(k-1)$, $db(k-1)$, $y\ dc(k-1)$. The usual way to execute the control loops is inside of A/D converter interruption itself, which is configured to generate it once the acquisition of at least the phase currents is completed. In Fig. 8.37 is illustrated the synchronization between the PWM and the A/D converter for each phase for low-side current shunt option. As can be observed, the samples are taken when the low-side power switch is turned on (complementary of shown PWM signals). As it will be seen in Chap. 9, all low-side power switches are on when vector V_0 is applied in a space vector modulation. After the three conversions are completed in a T_c period, as commented before, the A/D converter interrupt is launched to execute the current control loop and other tasks. That is, the current control loop is executed and everything that needs to be executed at this speed as failure detections, flux observer, etc. All calculations must not exceed the PWM period to avoid losing control cycles and not saturating the CPU. As a criterion, it should not be higher than 60% of the PWM period, or what is the same; the CPU load should not higher than 60% in applications where the microcontroller has to perform other tasks (background tasks) such as safety tasks and serial communications.

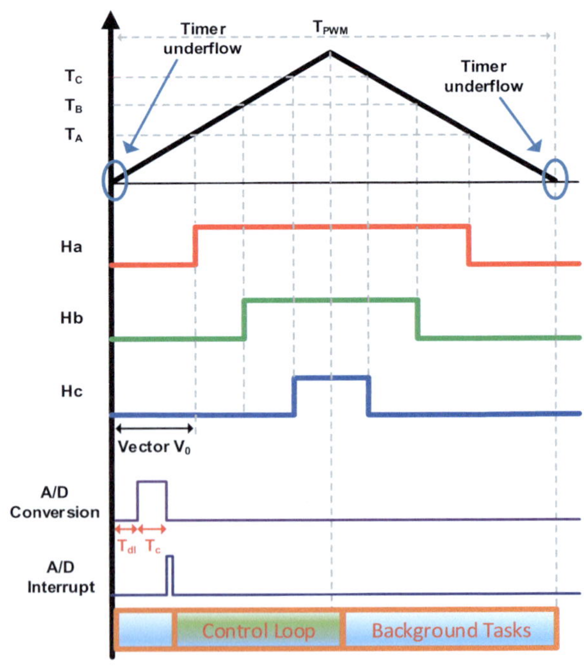

Fig. 8.37 Scheduling of control algorithm in a three-phase machine. It is shown the high-side PWM signals, A/D measurements for every phase, the A/D converter trigger interruption, and the processing tasks

If the A/D converter remains synchronized with the PWM period as seen above, the maximum regulation is achieved according to (Yamamoto et al. 1992; Blasko et al. 1998). Most of microcontroller allows a programmable delay time after PWM synchronization to perform the A/D conversion with a stable current after the PWM switching. This programmable delay time is represented in Fig. 8.37 as T_{dl}. For low-side sensors, it is important to consider the situation when vector V_0 time is too short for one of the phases, that is when one of the low-side power switches is active for a very short time. In this case it is possible that the A/D converter has no enough time to perform the conversion during this interval. This situation can be detected by an algorithm which evaluates the vector V_0. Then, if a predefined short time is detected, the current of this phase is calculated according to the others phase measurement with Kirchhoff law. As can be observed, this algorithm requires three low-side sensors.

In the following figures are represented a VSI simulation with three-phase RL load for the best understanding of the above explanation. In Fig. 8.38, the phase current A, the voltage in low-side shunt $R_{A\text{-}Shunt2}$, and the high-side PWM pulses are illustrated. It is possible to observe the high ripple (10 App) in the phase current due to the PWM frequency is set to a lower value of 5 kHz for easy view. The deadtime is 2 μs, and the shunt value is 1 mΩ. Note that the acquisition should be sampled when low-side power switch is turned on, that is when complementary of the illustrated PWM signals takes high level (vector V_0). For an easy view, it is shown the voltage drop in the shunt multiplied by 800 and inverted. The PWM signals are shown with an offset of −16 V and −32 V for phases B and C, respectively.

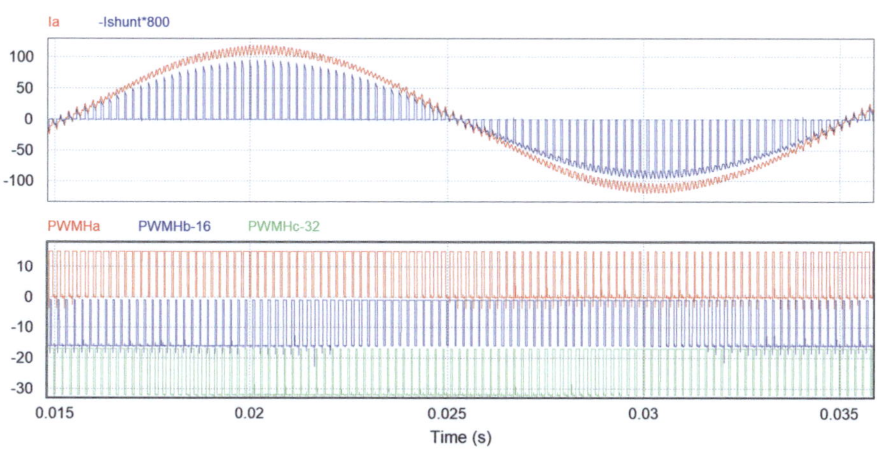

Fig. 8.38 Phase current A with the current through low-side shunt resistor, and high-side PWM signals during a complete sinusoidal cycle for an inductive load. Output frequency: $\omega_c = 50$ Hz. Deadtime: $T_{DT} = 2$ μs. Switching time: $T_{PWM} = 200$ μs. Simulation step: $T_{sp} = 100$ ns

It is possible to get a zoomed view in order to see the phase current A and the current through the low-side shunt resistor, as illustrated in Fig. 8.39. As can be observed, the current flows through the shunt resistor of phase A when low-side power switch is turned on. Also, the current acquisition magnitude can vary depending on where the A/D converter trigger is performed due to the current ripple.

In Fig. 8.40 shows another portion of Fig. 8.38 to see exactly where the samples are taken to get approximately the average current.

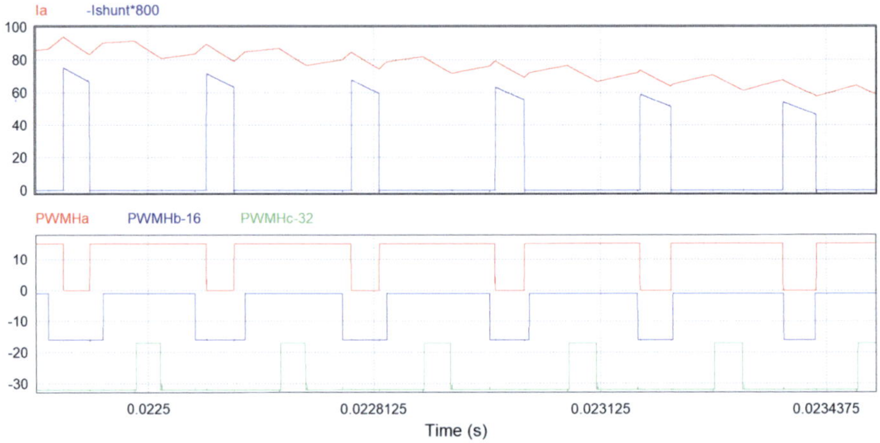

Fig. 8.39 Detail of phase current A with the current through low-side shunt resistor, and high-side PWM signals for one portion of the sinusoidal signal. Output frequency: $\omega_c = 50$ Hz. Deadtime: $T_{DT} = 2$ µs. Switching time: $T_{PWM} = 200$ µs. Simulation step: $T_{sp} = 100$ ns

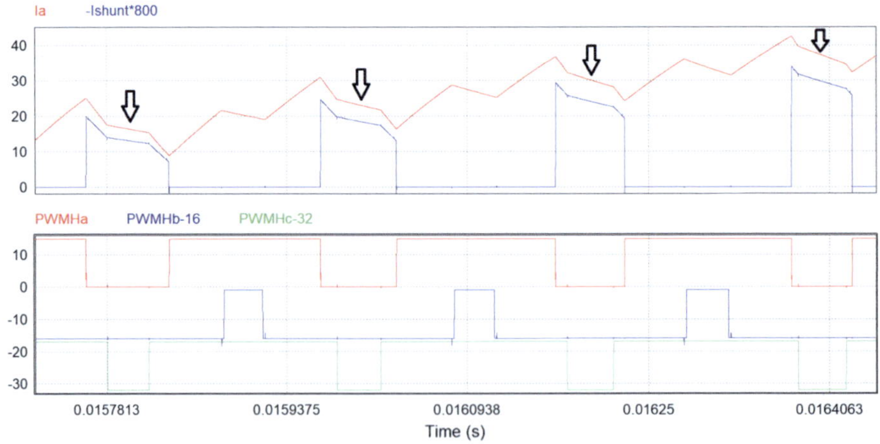

Fig. 8.40 Detail of phase current A with the current through low-side shunt resistor, and high-side PWM signals. It is shown the instant moment to acquire the current. Output frequency: $\omega_c = 50$ Hz. Deadtime: $T_{DT} = 2$ µs. Switching time: $T_{PWM} = 200$ µs. Simulation step: $T_{sp} = 100$ ns

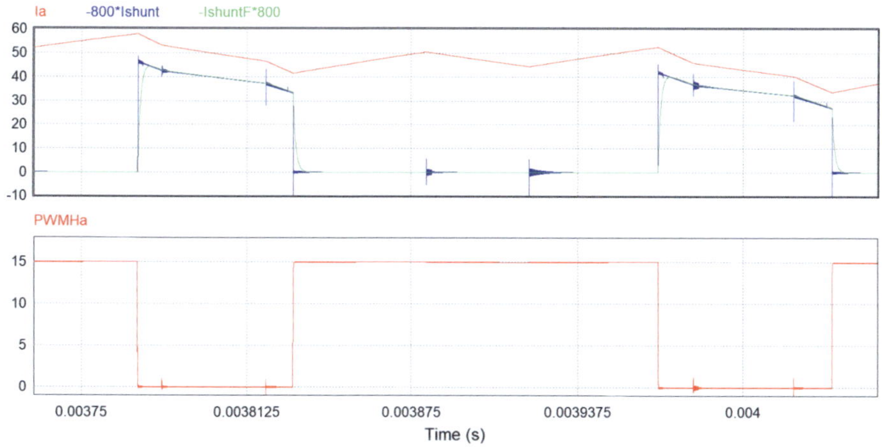

Fig. 8.41 Detail of phase current A with the current through low-side shunt resistor when parasitic inductance is set in series with shunt resistor. There is noise at every switching of the other power switches. It is also shown the filter result signal and the high-side PWM signal. Output frequency: $\omega_c = 50$ Hz. Deadtime: $T_{DT} = 2$ μs. Switching time: $T_{PWM} = 200$ μs. Simulation step: $T_{sp} = 10$ ns

Now, if a parasitic inductance of 10 pH is added in the low-side shunt of phase A, the voltage drop in the shunt has a low-level noise as represented in Fig. 8.41. It is important to say that noise is present at every power switching and by means a low-pass filter the noise can be suppressed as observed in green signal. The cut-off frequency of the low-pass filter is set to 150 kHz.

Figure 8.42 shows the experimentally current acquisition for each of the three phases at the output of signal conditioner formed by an inverter amplifier and low-pass filter. It is also shown the experimentally phase current A measured with a current probe (green color), where its RMS value is 3.63 A. The amplifier is supplied with a voltage source of 3.3 V and sets an offset of 1.65 V in order to measure positive and negative currents.

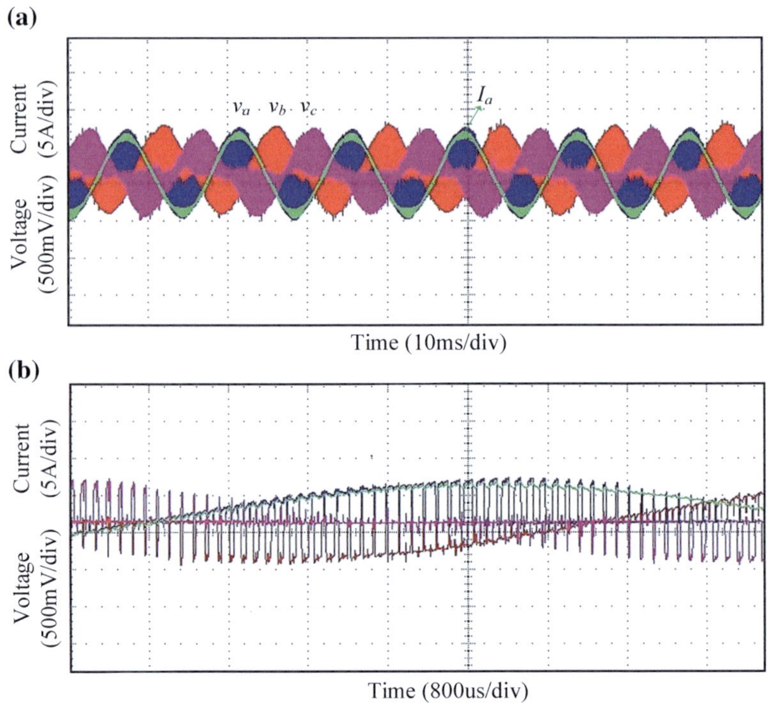

Fig. 8.42 **a** Experimentally phase current in a three-phase AC machine, which is shown the output of the signal conditioning and phase A measured with a current probe (green color). **b** Zoom detail where current probe signal matches the signal conditioning of phase A (green and blue colors)

8.4.3 *Output Voltage Distortion*

8.4.3.1 Deadtime, Turn-On, and Turn-Off Effect

As previously mentioned, the deadtime supposes a non-linearity and, therefore, a half-bridge architecture limitation, which affects the three-phase VSI . The limitation becomes significant if the deadtime represents more than 8% of the switching period. Thus, the error between the actual voltage in the machine terminals and the theoretical voltage may be higher than 20%. The distortion in the voltage increases with the switching frequency, introducing harmonic components that can cause instabilities and additional power losses in the AC machine. In the VSI output, the effect is more significant for small voltages (Munoz-Garcia and Lipo 1998; Chen and Peng 2008; Sukegawa et al. 1991; Leggate and Kerkman 1997). It is essential to add that the deadtime also adds a distortion in the current of the electric machine degrading the control performance.

During the deadtime period in a half-bridge (Fig. 8.43a), both power switches of the same inverter leg remain deactivated so that the voltage output of the inverter at

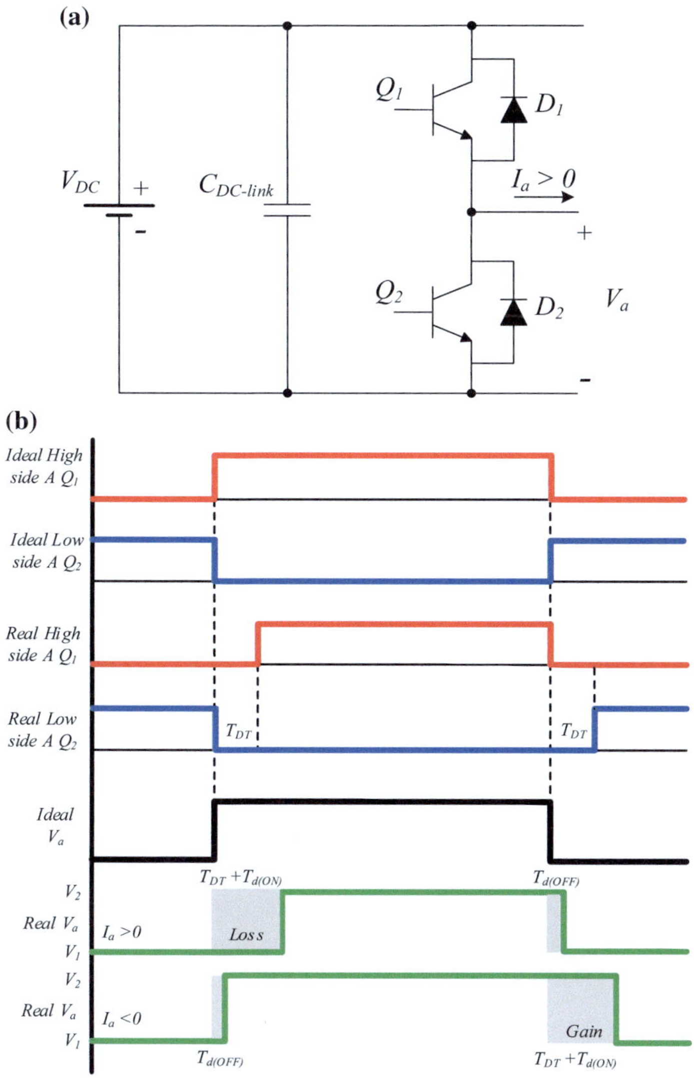

Fig. 8.43 **a** Half-bridge circuit. **b** High- and low-side switching patterns with real output voltage V_a according to the direction of the phase current I_a

the midpoint depends on the sign of the current of that phase. In this interval, when the phase current is positive, the current flows through the diode of the low-side power switch. In this case, the voltage value at the midpoint is V_1, as can be seen in Fig. 8.43b. On the other hand, when the phase current is negative, the current circulates through the diode of the high-side power switch, causing the voltage of the midpoint to be the voltage V_2, as can also be seen in Fig. 8.43b. In the figure, it is also possible to observe the ideal high and low switching pattern signals, the high and low switching pattern with the deadtime added, and the ideal voltage V_a. The turn-on and turn-off delay that is $T_{d(ON)}$ and $T_{d(OFF)}$ times are also included in the real signal of Fig. 8.43b.

When the Q_2 PWM signal changes to low level and the current is positive, the actual voltage at the midpoint, V_a does not change its value from V_1 to V_2 until the deadtime T_{DT} plus the delay time $T_{d(ON)}$ elapse. That is, after the deadtime T_{DT}, Q_1 changes its level to a high level, but it is not until $T_{d(ON)}$ elapses when the device is actually turned on changing the V_a magnitude. In the same way, when Q_1 changes from high to low level, until $T_{d(OFF)}$ elapses, the voltage V_a does not change to V_1 again as can be seen in Fig. 8.43b. On the other hand, if the current is less than zero when Q_2 changes to low level, it is not until $T_{d(OFF)}$ has elapsed when it effects on V_a changing its level from V_1 to V_2. Finally, when Q_2 changes to a high level again, until the time $T_{d(ON)}$ plus deadtime T_{DT} elapses, it does not change the voltage V_a to the value V_1 again.

From Fig. 8.43, it is possible to deduce the error of the voltage V_a degraded by the effect of the deadtime as shown in Eq. (8.18).

$$V_a = V_a^* - \Delta V_a \tag{8.18}$$

where V_a^* is the ideal voltage.

A first approximation to calculate the error ΔV_a is considering V_1 equal zero and V_2 equal V_{DC}, as described in Eq. (8.19). In this equation, the error depends on the sign of the current, as explained previously.

$$\Delta V_a = \left(\frac{T_{DT} + T_{d(ON)} - T_{d(OFF)}}{T_{PWM}}\right) V_{DC}\,\mathrm{sign}(i_a) \tag{8.19}$$

Furthermore, as expected, the error in the output voltage is proportional to the deadtime and the switching frequency. If the deadtime increases, the output voltage error will be higher. In the same way, the higher the switching frequency for the same deadtime, the greater the error ΔV_a.

8.4.3.2 Voltage Drops in the Semiconductors

The power switch semiconductors such as MOSFET and IGBT have non-ideal behavior during on state of transistor and on state of freewheeling, causing a voltage drop. As a consequence, the inverter output voltage is degraded.

The voltage drop can be modeled with a constant forward voltage V_d in case of the diode, and voltage V_{SAT} in case of the MOSFET or IGBT, which does not depend on the current through the device. More precise modeling consists of adding a series resistance R_d in case of the diode, and R_{ON} in case of MOSFET or IGBT. The series resistance models a variation of the voltage drop that depends on the current through the device. Although it seems that the voltage drop in the power switch may be negligible with respect to the DC-link voltage V_{DC}, high-power switch devices can become very important due to their high current. In most cases, to obtain a good approximation, the effect of the resistance R_d and R_{ON} can be neglected in the calculation. In this case and according to (Munoz-Garcia and Lipo 1998) above Eq. (8.19) can be rewritten as in (8.20) as:

$$
\begin{aligned}
I_a > 0 \rightarrow \Delta V_a &= \left(\frac{T_{DT} + T_{d(ON)} - T_{d(OFF)}}{T_{PWM}}\right)(V_{DC} - V_{SAT} + V_d) + \frac{V_{SAT} + V_d}{2} \\
I_a < 0 \rightarrow \Delta V_a &= \left(\frac{-T_{DT} - T_{d(ON)} + T_{d(OFF)}}{T_{PWM}}\right)(V_{DC} - V_{SAT} + V_d) + \frac{V_{SAT} + V_d}{2}
\end{aligned}
\tag{8.20}
$$

It is important to mention that, although the series resistance is neglected, the delay times $T_{d(ON)}$ and $T_{d(OFF)}$, as well as the saturation voltage of the MOSFET/IGBT V_{SAT} and forward voltage V_d of the diode, depend on the temperature and current through the device so that the voltage error ΔV is dependent on the load current. These parameters are usually provided in the manufacturer's data sheet for different currents and temperature conditions so that the most unfavorable case for the application must be considered.

In order to get an idea of the deadtime effect and the voltage drop effect in the power switch devices, it is possible to calculate the voltage error ΔV. In this example, the voltage does not vary with load or temperature, but different deadtimes are considered. Table 8.7 shows the result of the analytic calculation, according to Eqs. (8.19) and (8.20), for a 400 V DC-link voltage and a PWM switching period of 50 μs. The parameters used are for silicon IGBT of 600 V and 120 A shown in Table 8.6.

Table 8.6 Parameters extracted from silicon IGBT device data sheet

	Si IGBT Device
$T_{d(ON)}$	33 ns
$T_{d(OFF)}$	310 ns
V_{SAT}	1.5 V
V_d	1.65 V

Table 8.7 Voltage error for different deadtimes with and without considering the IGTB device voltage drop

Deadtime	Without voltage drop		With voltage drop	
T_{DT} (μs)	ΔV ($i > 0$) (V)	ΔV ($i < 0$) (V)	ΔV ($i > 0$) (V)	ΔV ($i < 0$) (V)
0.5	1.784	−1.784	3.36	−0.21
1	5.784	−5.784	7.36	−4.21
1.5	9.784	−9.784	11.36	−8.21
2	13.784	−13.784	15.36	−12.21

As can be observed, the voltage drop is kept constant as expected, and the deadtime has a higher effect than the voltage drop in the power switch.

8.4.3.3 Simulation Results

Figure 8.44a shows the reference voltage V^*_{a0}, the real V_{a0}, the current when the modulation index m is 0.5, and a fundamental frequency of 100 Hz for a high-power IGBT VSI . The switching frequency is 2 kHz with a deadtime of 40 μs for easy view. The forward diode voltage V_d is 10 V, and the on-resistance R_{ON} of the IGBT is 150 mΩ. The delay times $T_{d(ON)}$ and $T_{d(OFF)}$ have been neglected. As it is possible to observe, the voltage drop in the IGBT and the diode cannot be neglected since it represents an important percentage error.

In both cases of positive and negative load current, it is possible to appreciate the ideal voltage V^*_{a0} and the actual voltage V_{a0}. In this simulation, the variation of the forward voltage V_d and voltage V_{SAT} with the load current is taken into account as can be seen. At the point of highest current amplitude in the positive or negative half period, the difference is much more noticeable because the higher current magnitude causes a higher voltage drop in the diode and IGBT. As commented before, this is due to its internal resistance. Figure 8.44a also shows the error ΔV_{a0}, where the absolute mean value of the positive half period of the current roughly matches Eq. (8.20). The error ΔV_{a0} is positive for positive load current, whereas it is negative for a negative load current as expected.

In Fig. 8.44b, c, a part of Fig. 8.44a of the reference voltage V^*_{a0} and the actual V_{a0} for a positive and a negative current, respectively, are shown in detail.

In this section, it has been observed how the output voltage of the VSI is degraded by the deadtime and the voltage drop effects. If the compensation of these non-idealities is not made, the control performance will be affected since, on the one hand, the available voltage will not be delivered to the AC machine, and on the other hand, the voltage and current will have significant distortion. With a deadtime compensation, the deadtime effect in the VSI output can be suppressed. For high-performance machine control applications, it is advisable to measure the machine terminal voltage. However, thanks to the continuous improvement of Si, SiC, and GaN devices technology, better switching and conduction characteristics significantly reduce the deadtime and voltage drop. In some cases, the deadtime can be reduced to less than 0.5% of the switching period. Thus, the voltage and current distortion depend more on the power switch voltage drop and its parasitic capacitances, where the maximum reachable voltage will be close to the maximum possible.

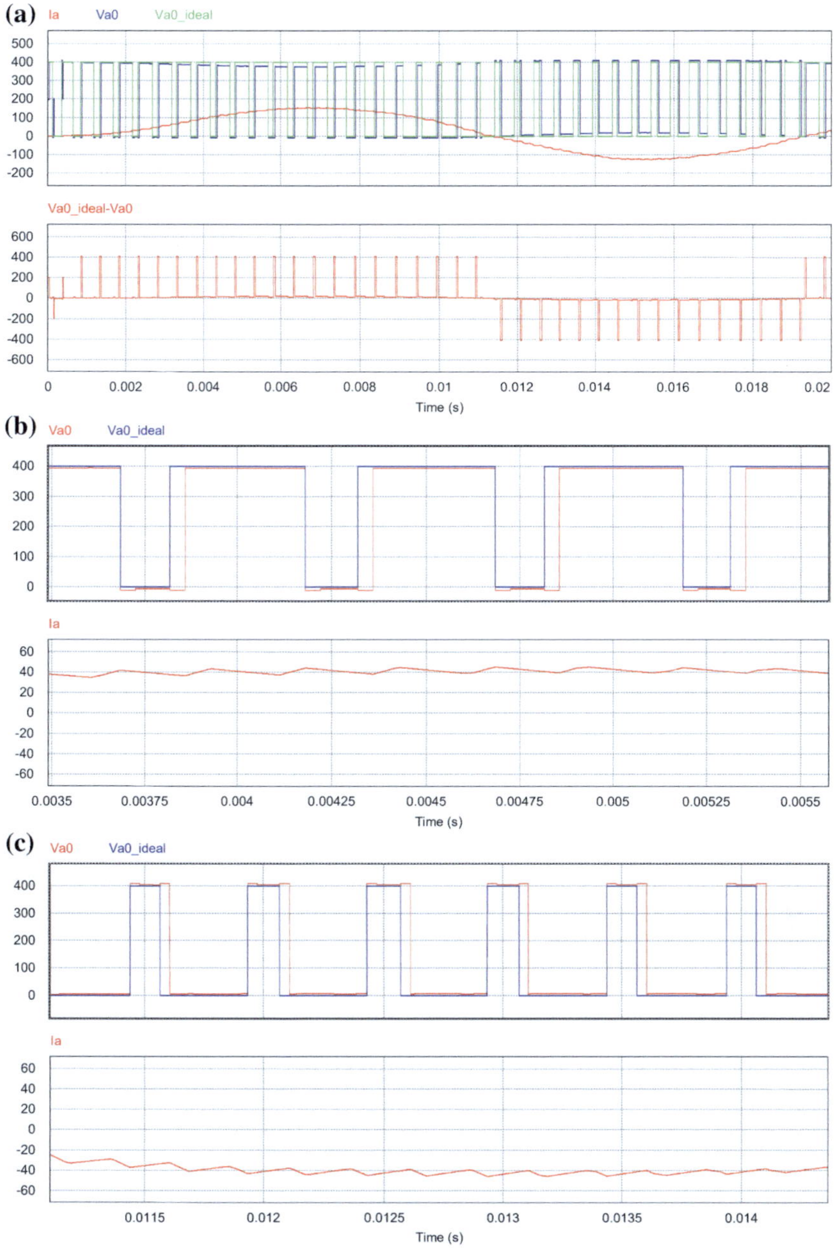

Fig. 8.44 Simulation result. **a** Inverter leg output voltage $V_{a0}*$, V_{a0} and voltage error. **b** Inverter leg output voltage $V_{a0}*$, V_{a0} detail with positive current. **c** Inverter leg output voltage $V_{a0}*$, V_{a0} detail with negative current. Modulation index: $m = 0.5$, output frequency: $\omega_c = 100$ Hz. RL load used $R = 0.1\ \Omega$, $L = 5$ mH. Deadtime: $T_{DT} = 40\ \mu$s. Switching time: $T_{PWM} = 500\ \mu$s. Simulation step: $T_{sp} = 10$ ns

8.4.4 DC Voltage Source

As mentioned, the DC source stabilization is fundamental to be able to generate a stable AC voltage in the outputs of the VSI . The PWM switching frequency plays an essential role in the machine phase current ripple, which is smoother for higher frequencies. The phase current ripple depends on the voltage V_{DC}, the modulation type, the modulation index, the equivalent inductance of the AC machine, the back-EMF, and the switching frequency. In each switching period, the terminal of the machine winding takes the voltage V_{DC}, or 0 V. When a V_{DC} is applied, the machine current increases linearly since the voltage step is integrated by the equivalent inductance of the AC machine. However, when zero voltage is applied, the machine current also drops linearly in the same way as before. This rise and fall of current constitute the high-frequency ripple. The high-frequency ripple in each of the phases disturbs the DC voltage source, so it is not advisable to have a phase current ripple higher than 20%. The worst case is when the AC machine rotates at low speed, in the constant torque region, where its back-EMF is also low. If the DC source is provided by a battery, high-frequency ripple will not be welcome, as it may reduce the battery lifetime. For this reason, although the DC source is provided by a battery, it is important to add a DC-link capacitor to absorb the ripple and part of the AC machine load fluctuations, extending the lifetime of the battery.

On the other hand, if the DC source is obtained from the electrical grid through a rectifier bridge, the DC-link capacitor sees the same behavior as previously explained. Moreover, it must also filter the low-frequency ripple voltage present in the output of the rectifier. The output voltage frequency is two times the input frequency for single-phase full-bridge rectifier, while six times for full-wave three-phase bridge rectifier. In the case of 50 Hz grid frequency, the frequencies are 100 Hz and 300 Hz, respectively. This low-frequency ripple must be filtered with a DC-link capacitor typically with considerable capacity value. However, if a PWM boost rectifier bridge is used, the reduction of the DC-link capacitor is possible, reducing its cost, but increasing the complexity and cost of the controlled rectifier. In this section, the DC-link capacitor needed for VSI is estimated for both cases: for DC source battery and for AC input source with uncontrolled rectifier bridge.

8.4.4.1 DC-Link Capacitor Selection

The selection of DC-link capacitor type depends on the requirements of the application and circuit parameters. If the bus voltage ripple should be minimized, the electrolytic capacitors are ideal due to the high capacity per volume. Their voltage rating is generally low (500 V or less), so for higher voltages, they must be placed in series which implies an increase in their equivalent series resistance (ESR).

On the other hand, if the inverter plus machine set suppose a high ripple current, the film capacitors are more suitable due to its low ESR. Generally, the electrolytic capacitors are of lower cost compared to the film capacitors. However, film capacitors offer a higher lifetime compared to electrolytic, so in many applications, EV/EHV inverters are preferable.

It is important to differentiate between current ripple and voltage ripple. The current ripple depends on the inductive load, the bus voltage, modulation index, and the inverter PWM frequency as commented before. The current ripple intervenes, among other factors, in the capacitor lifetime. However, the voltage ripple affects the stability of the DC-link voltage due to its low frequency. This plays a very important role in the performance of the AC machine at high speed when the DC voltage comes from an uncontrolled full-wave rectifier. For example, excess voltage ripple can lead to problems when the machine operates at high speeds at constant power, where the maximum speed could not be reached, even if the load torque is lower than the machine's rated value. The set formed by VSI and AC machine ideally must be designed to meet the requirements of maximum speed. For it, the DC-link capacitor must be designed for the worse case ensuring that the voltage provided to the AC machine is the rated voltage which it was designed in a wide speed range. Some AC machine manufacturer's design the machine to achieve the rated torque at a lower voltage than the AC source because they are exclusive to be controlled by an inverter. In this way, it compensates for the reduction of the average output voltage by different factors such as the DC-link voltage ripple, voltage drop in the devices, deadtime, and type of modulation.

An optimal compromise between the voltage ripple, current ripple, along with the performance requirements of the AC machine, the cost and size can be minimized.

8.4.4.1.1 Capacitance Estimation in AC Power-Sourced Inverter

For a single-phase (Fig. 8.45) or three-phase AC-DC converter based on an uncontrolled full-wave rectifier with semiconductor diodes, the capacitor value required can be estimated by equalizing the energy absorbed by the load to the energy that the capacitor yields during the hold time as shown in Eq. (8.21).

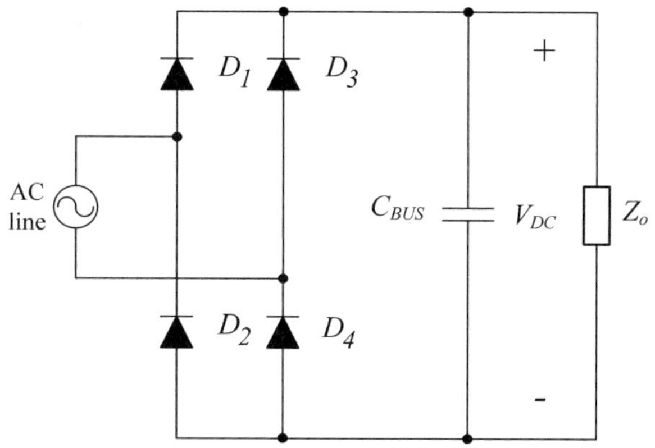

Fig. 8.45 Single-phase AC voltage source, uncontrolled full-bridge rectifier, DC-link capacitor, and load

$$t_h \frac{P_{load}}{\eta} = \frac{1}{2} C \cdot \left(V_{peak}^2 - V_{\min}^2 \right) \tag{8.21}$$

where the efficiency η represents the VSI losses in the case of an AC load machine, while the voltage ripple is the difference between V_{peak} and $V_{\min}$.

The hold time t_h is the time in which the stored energy is transferred to the load without the AC input injects current, as shown in Fig. 8.46. The hold time can be calculated according to (8.22). The first term of the equation (K) depends on whether the AC source is single-phase or three-phase, while the second term depends on the ripple and the frequency of the grid. This second term corresponds to the time elapsed from the AC voltage zero-crossings until the voltage reaches $V_{\min}$. As expected, the hold time t_h is lower for a three-phase AC source than for a single-phase AC source. As a consequence, the capacitor value needed for the same output power and same voltage ripple is also lower.

$$t_h = K + \frac{\sin^{-1}\left(\frac{V_{\min}}{V_{peak}}\right)}{2\pi f_G} \rightarrow K = \begin{cases} \frac{1}{4f_G} \rightarrow 1\text{phase} \\ \frac{1}{12f_G} \rightarrow 3\text{phase} \end{cases} \tag{8.22}$$

By using (8.21), the capacitor value C for a given voltage ripple can be estimated. It is possible to calculate different capacitors values for different voltage ripples, as shown in Table 8.8. The conditions are:

- Single-phase and three-phase AC voltage
- 2 kW load power
- VSI efficiency of 90%
- Grid frequency f_G of 50 Hz.

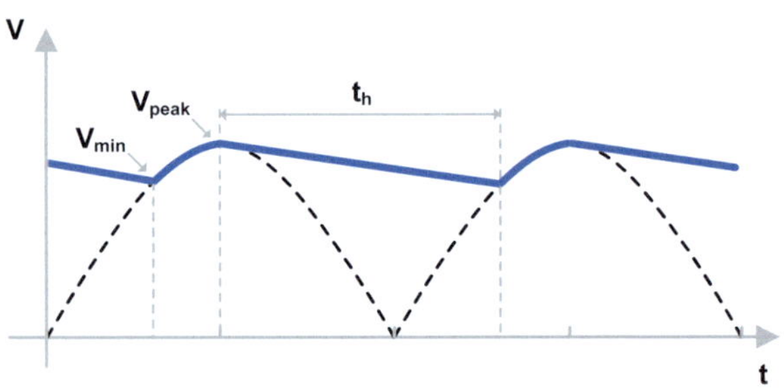

Fig. 8.46 Single-phase full-bridge rectifier output voltage. Hold time when AC input does no power the DC-link capacitor

Table 8.8 Different capacitors values at different voltage ripples

Option	V_{peak} (V)	V_{min} (V)	V_{ripple} (V)	$t_{h\ 1phase}$ (ms)	$C_{\ 1Phase}$ (mF)	$t_{h\ 3phase}$ (ms)	$C_{\ 3Phase}$ (mF)
1	340	325	15	9.05	4.0	5.72	2.55
2	340	315	25	8.77	2.38	5.44	1.48
3	340	300	40	8.44	1.46	5.11	0.886

In the first option, the ripple is approximately 4% of the peak voltage. To achieve this, a capacitor of 4 mF is required for a single-phase AC voltage. However, 2.55 mF capacitor is required for a three-phase AC voltage. The difference is notable between options 1 and 3, where the ripple is 4% and 12%, respectively, and the required capacitor values vary from 4 mF to 1.46 mF for single-phase and from 2.55 mF to 886 μF for three-phase AC voltage. As can be observed, the capacitor values have a higher value and should withstand a minimum voltage of 400 V. In this case, electrolytic capacitors are more suitable than film capacitors since their volume/capacity ratio is more optimal, as discussed previously. In the first option, it would be necessary to use four capacitors in parallel to reach that capacity increasing the cost, but the performance of the machine control will be higher. However, in option 3, the cost is lower, but the benefits compared with the previous case will not be the same at high speed. It is important to mention that the previous calculations are estimations since DC current in the load has been assumed without PWM modulation. In AC machine control, the input current in the inverter is not a constant DC current, but a pulsating current that depends on the type of modulation, modulation index, and power factor among other factors. That is, the input current only exists and is positive when any of the high-side power switches are turned on.

The previous estimation can be used as a starting point, which can be corroborated and fine-tune with simulation, as shown in Fig. 8.47. The simulation performed is for option 2 of Table 8.8 with a three-phase VSI . The AC input source is a single-phase 240 V_{RMS} and 50 Hz source. The capacitor value is 2250 μF corresponding to three standard 750 μF and 400 V capacitors each connected in parallel. The DC-link voltage ripple is approximately 25 Vpp, as can be seen in Fig. 8.47a. The active power absorbed by the inverter is 2 kW. The capacitor current I_c can also be seen in Fig. 8.47a, which is positive when the capacitor is charged through the rectifier bridge and negative when it supplies power to the inverter. The most significant of this result is the peak-to-peak I_c current that is almost 105 App that the DC-link capacitor must support. As there are three capacitors in parallel, the peak-to-peak current for each of them will be 35 App, and its RMS value is 7 A_{RMS} according to the simulation. The RMS value, together with other parameters, is used to calculate the capacitor lifetime. In Fig. 8.47b is shown the current absorbed by the inverter and the three-phase currents through the high-side power switches. As can be observed, the current only exists when the high-side power switch is turned on. In Fig. 8.47c is shown a zoomed detail of Fig. 8.47b.

If the modulation index becomes the maximum, the maximum power will be delivered to the load so that the current absorbed by the inverter will be higher. Figure 8.48 shows the result of the simulation for the maximum output power, representing the DC-link voltage ripple, the phase current I_a, and the line-to-line voltage V_{ab}. It is possible to appreciate that the voltage ripple has increased considerably to 80 Vpp. This causes a fluctuation in the line-to-line voltage of the load, as can be seen in Fig. 8.48. The higher the DC-link voltage, the greater the peak value of voltage V_{ab} and vice versa. This fluctuation is also generated in the load current, and in the case of using an AC machine as load, this fluctuation is unwelcome since it

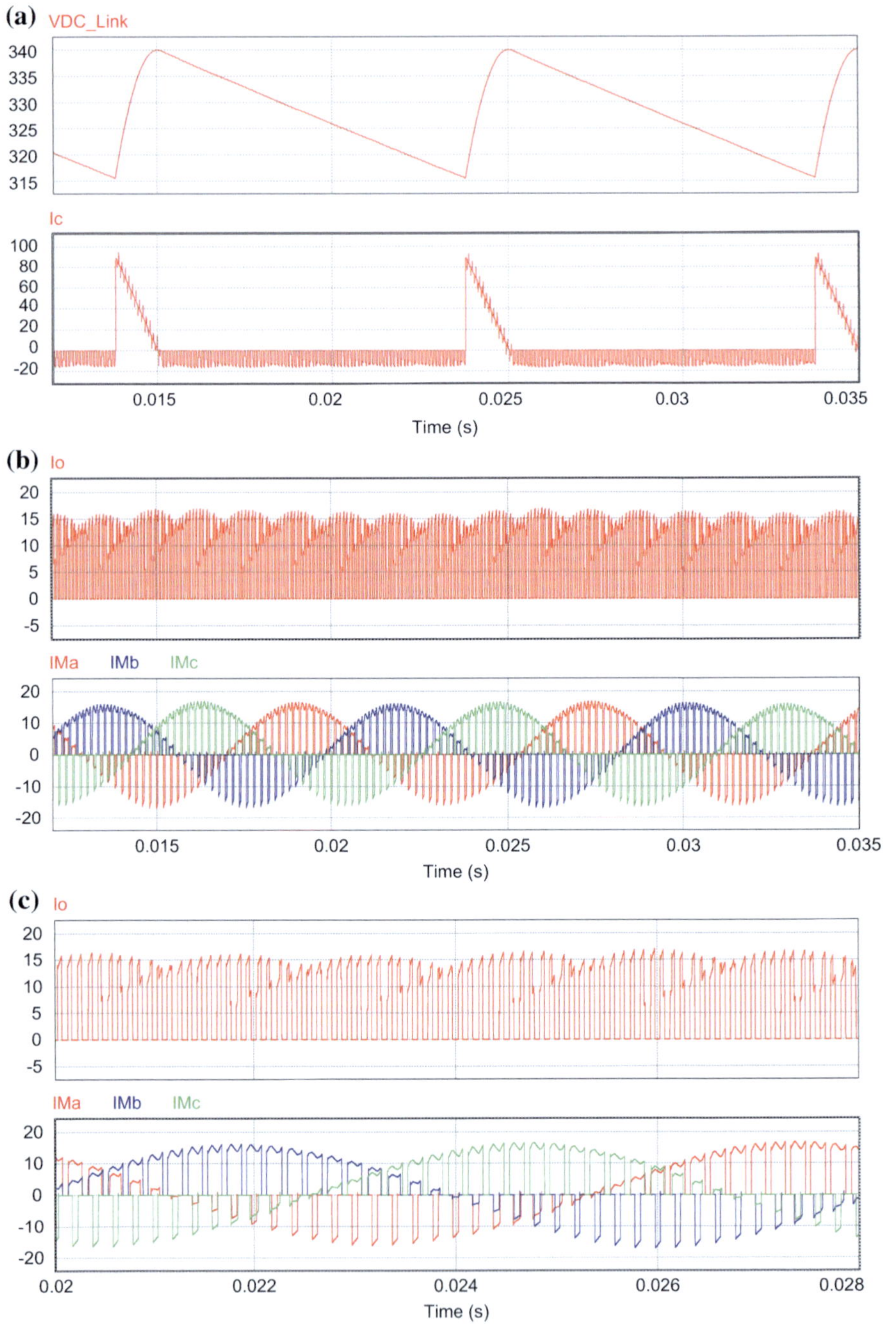

Fig. 8.47 Simulation. **a** DC-link voltage ripple, and capacitor current I_c. **b** Output current I_o, and phase currents I_{Ma}, I_{Mb}, I_{Mc}. **c** Zoom detail of (**b**). Modulation index: $m = 0.5$. Output frequency: $\omega_c = 120$ Hz. RL load used $R = 5.7\ \Omega$, $L = 2.1$ mH. PF $= 0.85$. Deadtime: $T_{DT} = 1\ \mu$s. Switching time: $T_{PWM} = 200\ \mu$s. Simulation step: $T_{sp} = 500$ ns

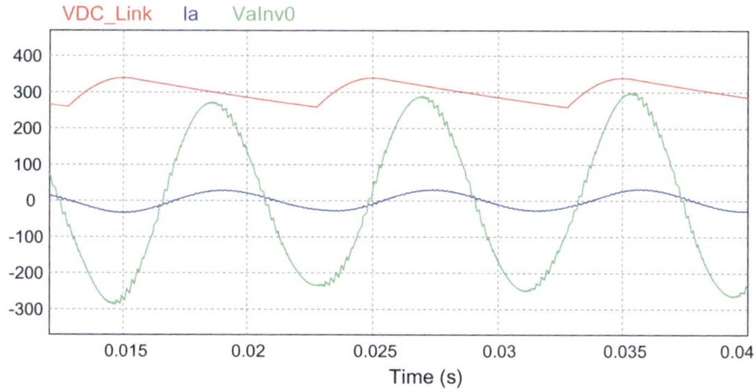

Fig. 8.48 DC-link voltage ripple, phase current A, and line-to-line voltage V_{ab}. Modulation index: $m = 0.95$. Output frequency: $\omega_c = 120$ Hz. RL load used $R = 5.7\ \Omega$, $L = 2.1$ mH. PF = 0.91. Deadtime: $T_{DT} = 1\ \mu$s. Switching time: $T_{PWM} = 200\ \mu$s. Simulation step: $T_{sp} = 500$ ns

affects the torque and therefore to the machine performance. The previous simulations have been performed for a maximum power of 2 kW, but now the power has become 7.3 kW. If the same calculations are made for the new power, the resulting DC-link capacitor is 7.8 mF. The new capacitor value improves the voltage ripple, and the capacitor sees lower RMS current, decreasing the voltage fluctuations in the load and increasing the capacitor lifetime.

Figure 8.49 shows an experimental result with two different capacitors values and maximum modulation index for a PMSM. It is shown the DC-link voltage and phase current. Figure 8.49a shows a ripple of 43.34 Vpp for a capacitor value of 240 μF when speed command is the nominal. As can be observed, the phase current has an electrical frequency of 163.7 Hz, which means approximately 4911 RPM. The instability of the voltage and current in the machine terminals reduces the machine performance where nominal speed is not achieved. However, if the capacitor value is changed to 470 μF, the voltage ripple is reduced to 21.33 Vpp, and machine performance is improved reaching the 5214 RPM nominal speed with a more stable voltage and current, as shown in Fig. 8.49b.

8.4.4.1.2 Capacitance Estimation in Battery-Sourced Inverter

When the VSI is powered through a battery, as shown in Fig. 8.50, the DC-link voltage ripple usually is lower and of a higher frequency. Therefore, the previous analysis of the energy balance between the capacitor and the VSI is not applicable.

The capacitor does not yield energy as in the previous case, but in AC machine control it has the function of absorbing the phase current ripple and part of the fluctuations of the load extending the battery lifetime. This complicates its analysis a bit more since it depends on many factors such as type of modulation, load power factor,

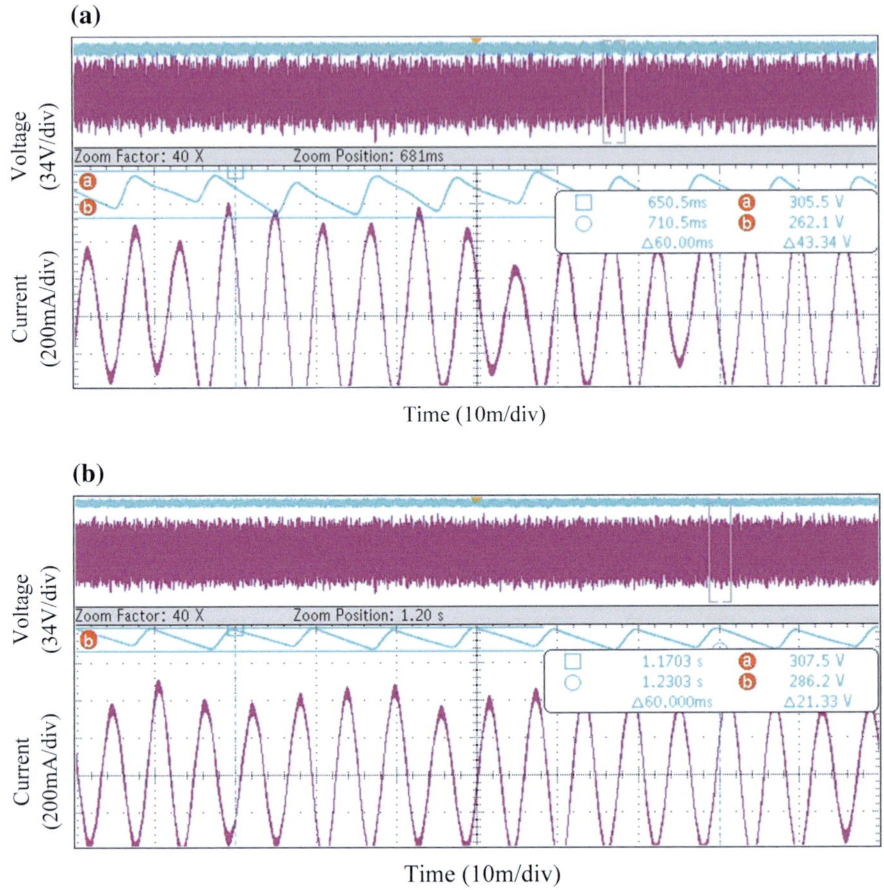

Fig. 8.49 PMSM phase current fluctuation due to high ripple voltage in the DC-link capacitor. **a** Ripple voltage of 43.34 Vpp with 240 μF. **b** Ripple voltage of 21.33 Vpp with 470 μF

Fig. 8.50 Battery and DC-link capacitor

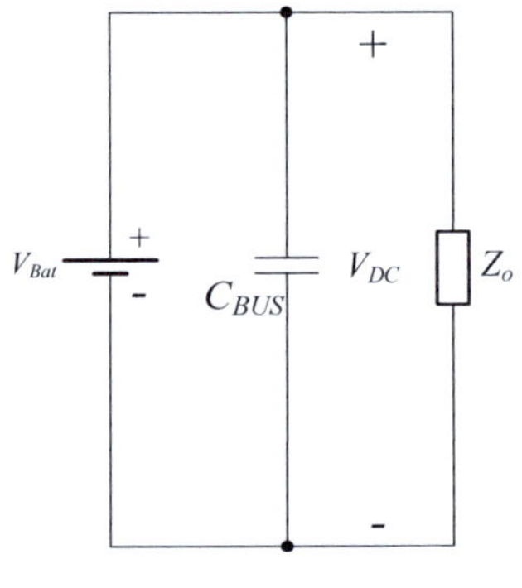

modulation index, equivalent inductance of the AC machine, and the magnitude of the current absorbed by the inverter.

As Vujacic et al. (2018) suggests, the voltage ripple of the capacitor can be estimated by integrating the current flowing through the capacitor, in a specific time interval. This is true as long as the impedance of the capacitor, at the switching frequency, is much lower than the impedance of the battery's RL circuit. In this case, the voltage ripple in the capacitor depends on its capacity. Equation (8.23) allows establishing the minimum capacitor that complies with the maximum ripple required, and that depends on the phase current of the machine and the switching frequency. The one-fourth factor comes from the analysis suggested by Vujacic et al. (2018) that depends on the power factor and the modulation index, which in this case are approximately 0.86 and 0.5, respectively.

$$C \geq \frac{1}{4} \frac{I_m}{f_{sw} \Delta v_{DC}^{\max}} \tag{8.23}$$

With below design parameters, the minimum capacitor can be calculated by using Eq. (8.23).

- 2 kW output power
- Phase current of 11.3 A_{RMS}
- Maximum voltage ripple of 3 Vpp
- PWM switching frequency of 5 kHz.

The result is a capacitor of 188 μF. Normalizing the capacitor to a standard value of 220 μF, the new ripple voltage is 2.56 Vpp. As it is possible to observe, the required capacity is much lower than in the case of the AC source as expected. Also, if the switching frequency is increased to twice, the capacitor value is reduced by half. This is one of the most significant advantages of inverters powered by batteries where the required capacity is inversely proportional to the switching frequency. Hence, in some AC machine control applications is interesting to use high switching frequency of more than 50 kHz in order to reduce the capacitance by using power devices with high performance as GaN semiconductors. Since lower capacitor values are required, it can be more interesting to use a film capacitor which increases the lifetime and reduces the losses due to its low ESR.

Figure 8.51 shows the result of the simulation with the previous parameters to verify the estimation. It can be observed in Fig. 8.51a that the voltage ripple is basically the integration of the current that flows through the capacitor as previously mentioned. The ripple is approximately 1.5 Vpp, less than that specified at this operating point. It is possible that the AC machine operates in another operating point where the power factor and modulation index can be different so that the ripple can increase, but it will not be in any case higher than previously calculated. In Fig. 8.51a, the DC-link voltage ripple has a double switching frequency. It is possible to observe the currents that circulate through the high-side power switches I_{Ma}, I_{Mb}, and I_{Mc}, where the sum of the three composes the current absorbed by the inverter I_o. Due to this instantaneous variation of I_o, this causes a variation of the magnitude in the

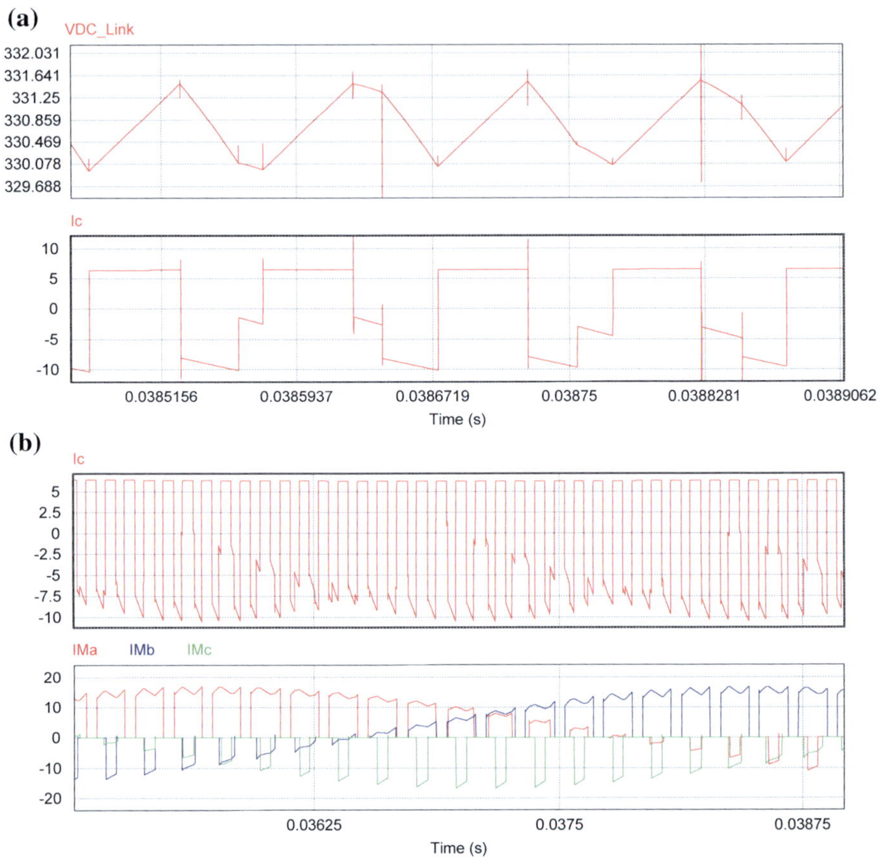

Fig. 8.51 Simulation. **a** DC-link voltage ripple, and capacitor current I_c. **b** Capacitor current and machine phase currents of high-side power switches. Modulation index: $m = 0.5$. Output frequency: $\omega_c = 120$ Hz. RL load used $R = 5.7\ \Omega$, $L = 2.1$ mH. PF = 0.85. Deadtime: $T_{DT} = 1\ \mu$s. Switching time: $T_{PWM} = 200\ \mu$s. Simulation step: $T_{sp} = 50$ ns

capacitor current I_c as can be observed. However, the input current I_{in} does not see these instantaneous variations but a DC current so that the battery will operate in more optimal conditions extending its lifetime as discussed previously.

8.4.5 DC-Link Pre-charge

Under initial conditions, the DC-link capacitor can remain discharged or with a very low voltage. When starting the AC machine, the VSI should be ready. That is, the DC-link voltage should match the DC voltage source. It means that the DC-link capacitor should be charged before machine start-up. The pre-charging process

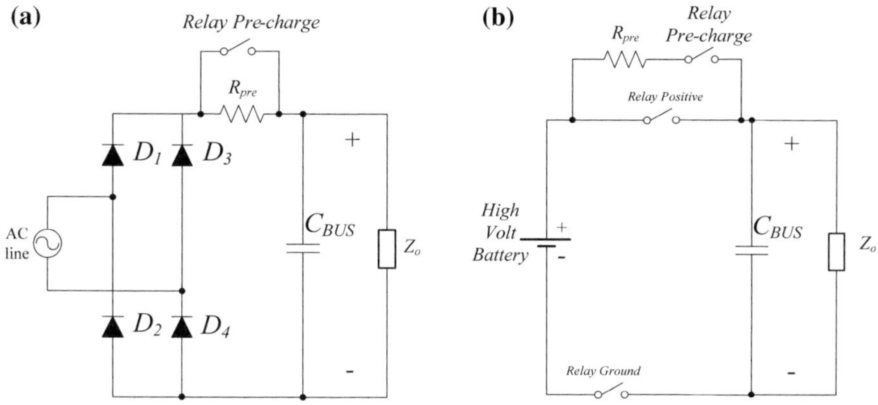

Fig. 8.52 Pre-charge circuit. **a** Using AC voltage grid. **b** Using DC voltage battery in EV/EHV

charges the DC-link capacitor in a controlled manner to avoid a high current transient that could damage the active elements such as diodes and PCB copper tracks. The high current transient is called in-rush current and is reduced by a slow charge of the capacitor. The inverter can be fed by a battery or through a rectifier as previously mentioned. If rectifier is used, the in-rush current can be limited by a resistor placed after the rectifier, as shown in Fig. 8.52a. Since there are active devices such as the diode, the pre-charge current should avoid exceeding the breaking current of the same. Similarly, if the inverter is fed by a battery, pre-charging is strictly necessary, even though there are no active devices between the battery and the capacitor since the in-rush current could damage the PCB copper tracks. In Fig. 8.52b is shown the in-rush current limitation made with a resistor in series with DC-link capacitor which is typically used in EV/HEV. Here it is also shown two additional safety relays that disconnect and isolate, in case of error, the high voltage battery (400 V) from the inverter mostly used in EV/HEV.

Considering the circuit of Fig. 8.52b, the sequences are first to close the ground contactor and then the pre-charge contactor. At this moment, the circuit is connected, and the energy starts to be stored in the capacitor. Once the DC-link capacitor is almost charged, the positive contactor should be closed before the machine start-up. It is important to comment that in some cases, the relay or contactors can be replaced by power switches, e.g., a MOSFET in parallel to the resistance. In this case, special attention should be paid with the body diode, since it does not block the reverse voltage in off mode. A solution for this, most used in battery switches, is to use a back-to-back configuration with two MOSFETs in common source which provides in off mode the bidirectional current blocking, i.e., it acts as a relay or contactor.

The capacitor energy can be expressed according to Eq. (8.24):

$$E = \frac{1}{2}C \cdot V_C^2 = V_{HVBAT} I_{in} \cdot t \tag{8.24}$$

Observing (8.24), if the capacitor charging time t is very small, and the applied voltage is fixed V_{HVBAT}, the input current I_{in} should be very high. The most commonly used methods that allow a slow charge of the capacitor (limitation of I_{in}) are through passive elements such as the resistance or the NTC thermistor. Depending on the conditions of maximum current and maximum voltage, either NTC or a power resistance can be used.

The capacitor voltage can be calculated according to Eq. (8.25):

$$V_c(t) = V_{HVBAT}\left(\frac{Z_o}{R_{pre}+Z_o}\right)\left(\left[V_{C_{t=0}}\left(R_{pre}+Z_o\right) - Z_o V_{HVBAT}\right]e^{\frac{-t(R_{pre}+Z_O)}{R_{pre}Z_O C}}\right) \tag{8.25}$$

Moreover, the current through resistance is:

$$i_{Rpre}(t) = \frac{V_{HVBAT} - V_c(t)}{R_{pre}} \tag{8.26}$$

The power and the energy that the resistance must support can be calculated with the expressions (8.27) and (8.28), respectively.

$$P_{Rpre}(t) = i_{Rpre}^2(t)\cdot R_{pre} \tag{8.27}$$

$$E_{Rpre}(t) = \int_0^t P_{Rpre}(t)\cdot \mathrm{d}t \tag{8.28}$$

As NTC thermistor has a negative temperature coefficient, the resistance decreases when the temperature increases. At ambient temperature or in a cold state, the resistance value of NTC is high, and if at that moment the power supply is turned on, it will limit the peak power current. The amount of energy that must be supported in the thermistor during the pre-charge period depends on the voltage waveform of the input source. However, a good approximation for energy calculation through NTC is according to Eq. (8.29).

$$E = \frac{1}{2}C\cdot V_C^2 \tag{8.29}$$

If the inverter is operating in stationary mode and if it is disconnected, the NTC will cool down to ambient temperature, increasing its resistance again so that when it is reconnected perform its protection function.

An alternative to the NTC is to use power resistance in series with the inverter, as was shown in Fig. 8.52. For higher-power applications or where the solution with NTC has a higher cost, the power resistance can be the right solution. It is important to comment that the power resistance should be fireproof to avoid possible flames in abnormal operation.

If the impedance Z_o is infinite and the initial voltage of the capacitor V_c is zero, Eq. (8.25) can be simplified as:

$$V_c(t) = V_{HVBAT}(1 - e^{\frac{-t}{R_{pre}C}}) \tag{8.30}$$

Thus, the current that flows through the resistance is:

$$i_{Rpre}(t) = \frac{V_{HVBAT} - V_c(t)}{R_{pre}} = \frac{V_{HVBAT}e^{\frac{-t}{R_{pre}C}}}{R_{pre}} \tag{8.31}$$

The power and energy that resistance must support can be calculated as expressed in (8.32) and (8.33) respectively.

$$P_{Rpre}(t) = i^2_{Rpre}(t) \cdot R_{pre} = \frac{V^2_{HVBAT}e^{\frac{-2t}{R_{pre}C}}}{R_{pre}} \tag{8.32}$$

$$E_{Rpre}(t) = \int_0^t P_{Rpre}(t) \cdot \mathrm{d}t \tag{8.33}$$

Figure 8.53 shows the capacitor voltage, current through the resistance, power of the resistor, and total energy accumulated during the pre-charge process for 2 mF capacitor, 10 Ω resistance, and 400 V DC voltage.

If the inverter is able to measure the current in the resistance, it can estimate the energy according to (8.33) integrating it in time. With this estimation, it can be evaluated, for example, whether the resistance has suffered an excessive overload in the pre-charge to detect abnormal situations. For example, the energy estimated during a pre-charge can be used to abort abnormal pre-charge. Then, the new pre-charge attempt can be performed after a waiting time in order to have time to cold down the resistance.

On the other hand, one of the advantages of using an NTC instead of a resistor is that for small VSI , the NTC is not strictly necessary to be short-circuited by a relay or contactor once the capacitor is charged. Basically, it is due to its negative temperature coefficient, the higher the temperature, the lower its resistance, and therefore offers less resistance to current. However, the resistance, having a fixed value, must be short-circuited by means of a relay so that there is no voltage drop in the resistance.

The opening and closing management of the relay, together with DC-link voltage monitoring, can be managed by means of a pre-charge management algorithm. The pre-charge management is performed one time before starting the electric machine to set the VSI in a ready state. It is only repeated after a power off and power on, that is, when the mains or battery has been disconnected and reconnected again. After the pre-charge mode, the DC-link voltage is monitored to detect some failures as undervoltage or overvoltage, and relay failures. Some relays have position feedback which can be used to detect stuck or welded failures. If the control algorithm is not

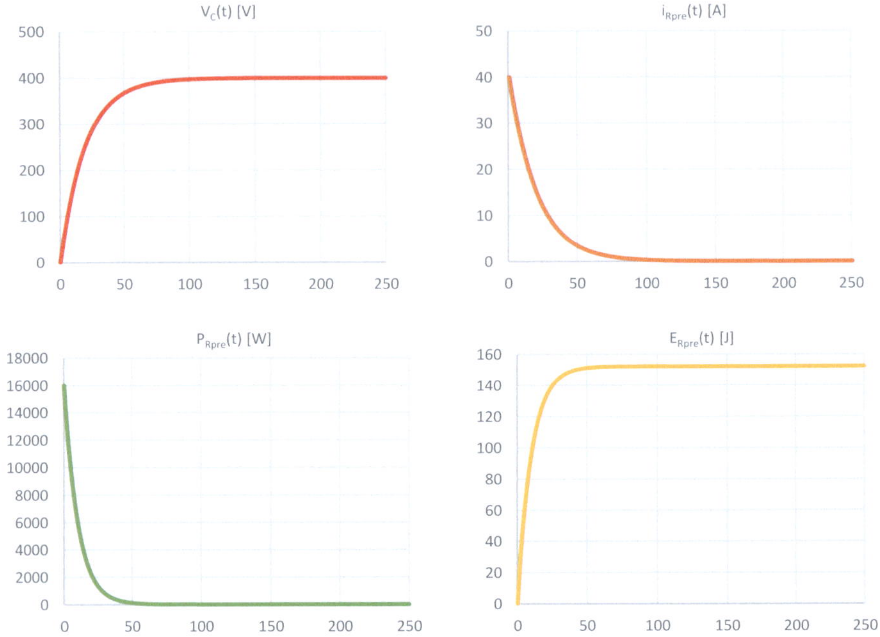

Fig. 8.53 DC-link pre-charge process. Capacitor voltage, pre-charge resistance current, power of pre-charge resistance in Watts, and pre-charge resistance accumulated energy in Joules

able to detect stuck or welded conditions, the PCB tracks or the active devices such as diodes can be damaged in abnormal situations.

In an emergency situation, where the AC machine should be stopped, the algorithm must first stop the inverter, that is, to put all the power switches in off state, and then if necessary open the relays. The reverse case is not recommended as it could generate electrical arcs in the relay contacts degrading its lifetime and causing a possible contactor welded. The pre-charge algorithm is mostly managed by a microcontroller/DSP and plays a decisive role in assuring the correct operation and extends the lifetime of the VSI and AC machine.

Figures 8.54, 8.55, and 8.56 show the flowchart diagrams corresponding to the pre-charge management algorithm of Fig. 8.52b with resistance energy monitoring. For a higher resolution in energy estimation, the algorithm should take current samples every millisecond ($T_s = 1$ ms). As it is possible to observe the pre-charge attempts are up to N times. The pre-charge can only be performed when the state is READY, and the accumulated energy is less than LIMIT_MIN. If this is not the case, the energy accumulated in a previous pre-charge will be reduced with the K ratio. The purpose is to give time to cool down the resistance, avoiding overheating the resistance in an abnormal situation where continuous pre-charges has been performed. The K factor

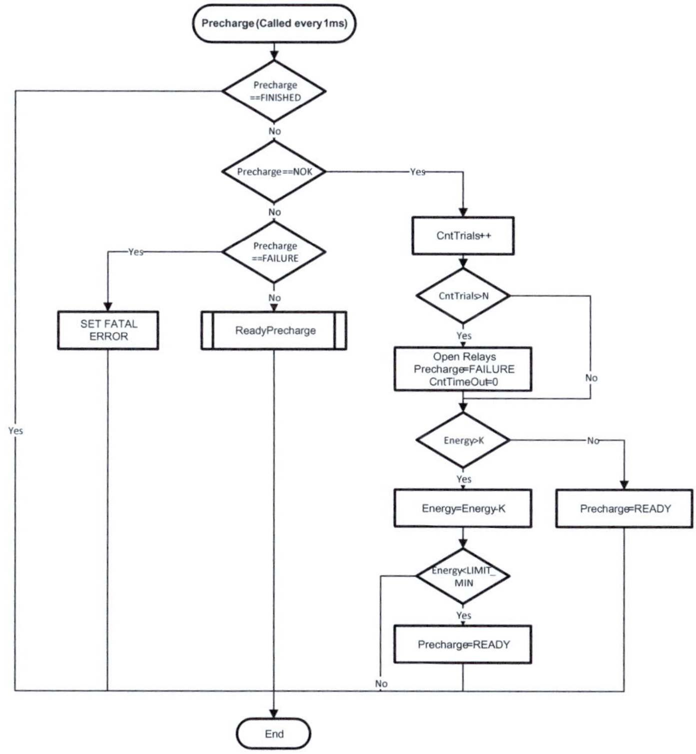

Fig. 8.54 Primary flowchart of pre-charge algorithm

must be correctly selected according to the resistance used and the worst ambient temperature condition. An experimental test can be performed to fine-tune the K factor.

The relay feedback position is monitored to know their position. If the relay position is not correct during TimeOutFeedBack period time, the subroutine of Fig. 8.55 sets irreversible fatal error. This can happen, for example, if the contacts have been welded as commented before.

When the pre-charge relay has been closed, the energy is accumulating since the current will pass through the resistor to charge the DC-link capacitor. If the capacitor voltage does not reach the VBUS_MIN value at a time controlled by TimeOut of Fig. 8.56, the pre-charge will be aborted. At the same time, if the accumulated energy exceeds LIMIT_MAX, the pre-charge will also be aborted. Otherwise, the pre-charge will be validated by passing its status to FINISHED.

On the other hand, once the pre-charge has finished, another algorithm can monitor the DC-link voltage to detect undervoltage and overvoltage failures. In case of any error event, the relay can be opened to proceed to repeat, if necessary, the pre-charge sequence. For example, if the DC-link voltage falls below a defined lower limit, the relay is opened setting the low voltage error. On the other hand, if the voltage

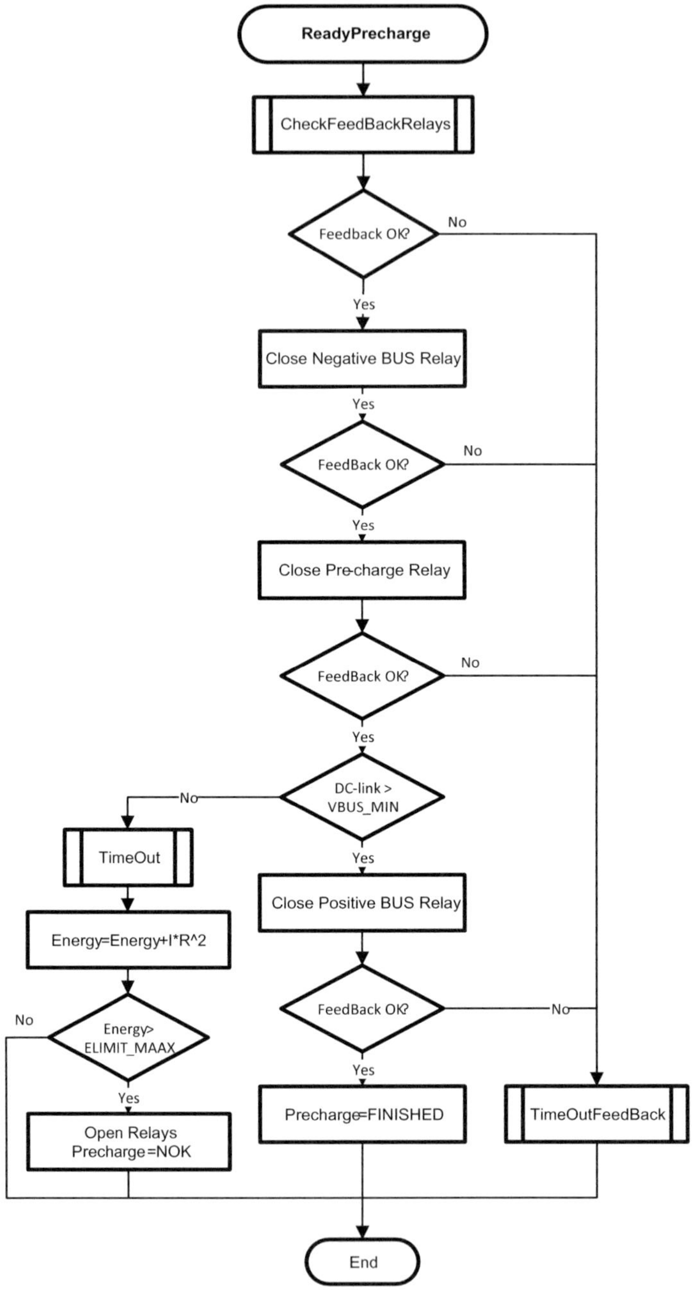

Fig. 8.55 Secondary flowchart pre-charge algorithm

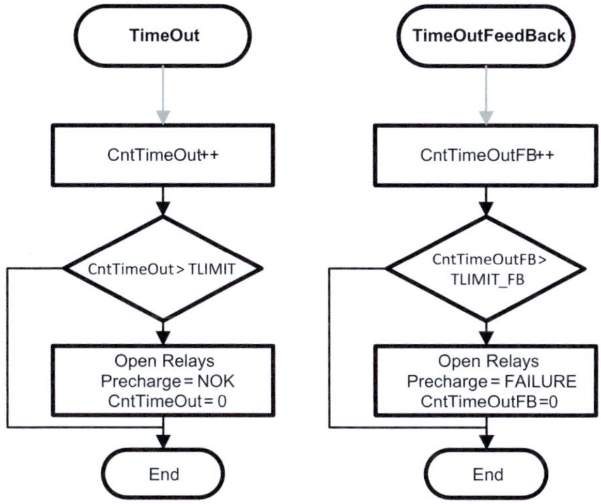

Fig. 8.56 Flowchart of time out management

rises above the defined upper limit, the overvoltage error can be activated. Both errors disconnect the inverter; that is, all the power switches are turned off, moving to a safe state. If the voltage source comes from the electrical grid, the electrical grid voltage detection is usually used to know if the pre-charge sequence should be repeated or not again, for example, during power off and power on of the inverter as commented before.

As an alternative, a controlled full-bridge rectifier can be used to control the flow of charge energy of the DC-link capacitor when the inverter is supplied by an electrical grid. It is mostly used for high-power machine control. The advantage is that it is not necessary any passive element like the NTC or the resistor, nor the relay, and is more feasible if the PCB space is limited. The disadvantage of the system is that it is necessary to add elements to perform the rectifier control, resulting in a higher cost solution.

Figure 8.57a shows the experimental result of the DC-link voltage pre-charge. It is controlled by an algorithm similar to that explained above and according to the circuit of Fig. 8.52a but with NTC instead of resistance. It also shows the current flowing through the NTC and the relay. The relay closes after approximately 2.9 s as can be seen. Figure 8.57b, c shows the detail of the capacitor charge along with the current flowing through the NTC, and the relay current, respectively. In Fig. 8.57b, the charge current decreases as the capacitor is charged. The pulses are every 100 Hz since the full-wave rectifier is connected to the 50 Hz single-phase power grid.

8.4.6 DC-Link Discharge

In some applications, it may be interesting, even mandatory, to discharge the DC-link capacitor when the inverter is disconnected from the mains or from the battery. In this

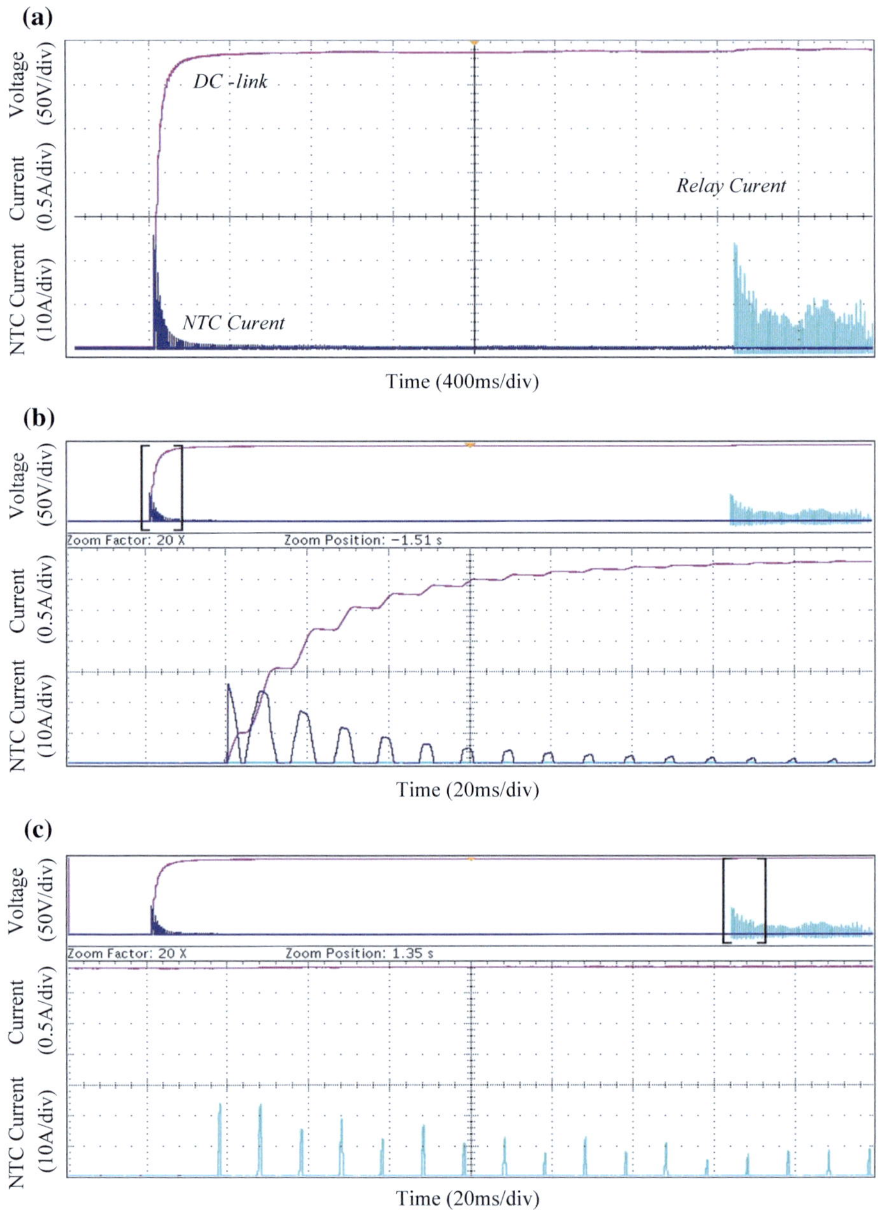

Fig. 8.57 DC-link pre-charge with NTC. **a** Capacitor voltage, NTC input current, and relay current. **b** Detail of charge voltage and input current. **c** Detail of relay current when the relay is closed

way, the electrical risk by contact is suppressed, and maintenance can be performed without electric shock risk. In EV/HEV, the DC-link discharge is mandatory since, in case of an accident, the electrical shock risk must be limited only in the battery and not in the vehicle's electronics. A signal called a "crash" signal detects the accident and is used to activate the DC-link discharge algorithm and to open all the high voltage relays. The slow passive discharge consists of a power resistor which is connected in parallel to DC-link voltage by using a relay or a power switch semiconductor. The discharge time should be less than 120 s and can be calculated with (8.30).

On the other hand, the fast active discharge consists of using the AC machine as a discharge circuit of the DC-link voltage. The inverter injects a DC current into the AC machine until the voltage level of the DC-link capacitor reaches a safe level. The injection of DC current is achieved in a controlled manner, where DC voltage is generated in the machine terminals with a PWM signal. The required discharge time will vary depending on the application and the capacitor value, but it should not be too small (>100 ms) since electrolytic capacitors do not tolerate fast discharges very well.

8.5 VSI in Dynamic and Regenerative Braking Mode

As seen in this chapter, the VSI allows that the AC machine can operate in all four quadrants of the torque versus speed plane, which is a requirement for many applications. As discussed in Chap. 4, they are forward motoring, reverse motoring, regenerative braking, and reverse regenerative braking. The AC machine in operation as a motor can undergo a series of speed changes due to the load torque variation, entering in generator mode. In this mode, the AC machine transfers energy to the DC-link voltage through the VSI which behaves like a rectifier. This excess energy can be used or not depending on the application. For example, in any of the configurations of electric vehicles, EV, MHEV, HEV, and electric train, the regenerative brake can be used to recover energy during braking.

The regenerative brake is the process by which the energy from the electric machine is stored in an electric storage system, such as battery, supercapacitor, and flywheel during the braking process. In (H)EV, the electric traction machine operates as a generator, producing negative torque on the wheels recovering electric energy. The total braking torque required in an EV is usually divided between the mechanical and electrical brake. However, sometimes the electric machine can produce sufficient torque without the help of the mechanical brake. For this reason, a controller is necessary to calculate the braking torque for both braking systems.

For example, in BLDC machine, the way to limit the regeneration energy is through the controlled activation, through a PWM, of the low-side power switch, as indicates in Fig. 8.58. Thus, the transfer energy flow to the battery is controlled.

In Fig. 8.58, only one of the six possible sectors of the BLDC machine controlled by a VSI is shown, as seen in Sect. 8.2.2. If an excess of regenerative braking energy is detected, it is possible to choose two of the low-side power switches where PWM

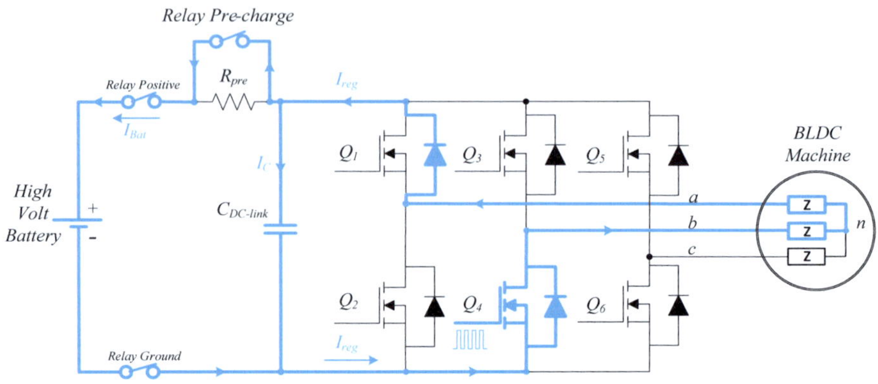

Fig. 8.58 EV/HEV regenerative braking by using the VSI

signal is applied so that the energy is derived to ground. The activation of these two low-side devices is alternate depending on the position of the rotor. The electric braking torque is calculated according to the duty cycle of the PWM, and the rest of braking torque is fulfilled with the mechanical brake.

In regenerative braking, the battery is viewed as a load by the machine so that it provides a braking force in the vehicle. Electric vehicles that use the regenerative brake can increase their driving autonomy up to 15% compared to an EV that uses only a mechanical brake. However, not all of the braking energy can be recovered to the battery. Some limitations are considered. For example, if the state of charge (SOC) of a Li-ion battery is more than 80%, the regenerative brake cannot be applied. This type of batteries offers a high discharge capacity and low charge capacity so that it is necessary to limit the regeneration energy. In Fig. 8.59 is illustrated an EV drive cycle with accelerations and deacceleration with braking where current, voltage, and power are represented. It is possible to observe a time instant $t = 32$ s a high acceleration where current reaches approximately 850 A, and the battery pack voltage drops from 400 to 358 V approximately. Till time instant $t = 37$ s, the power delivered by the battery is 308 kW. On the other hand, a time instant $t = 44$ s, regenerative braking is producing. The current is negative, which means that the current flows from the electric machine to the battery. It is limited to -212 A approximately due to the high SOC of the battery pack. The power delivered to the battery pack is -86 kW. During this regeneration, the battery pack maximum voltage reaches 409 V and stabilizes to 397.5 V approximately.

As explained above, the current was limited during braking to -212 A due to the high SOC of the battery pack. However, supercapacitors, also called ultra-capacitors, can be employed in the regenerative braking to recover the energy when the battery is almost charged. The supercapacitor has higher power density than a battery, supports high braking currents, has an excellent life cycle, and has low heating losses. However, the supercapacitor has a high leakage current, which means the energy cannot be held for a long time (Khodaparastan and Mohamed 2017).

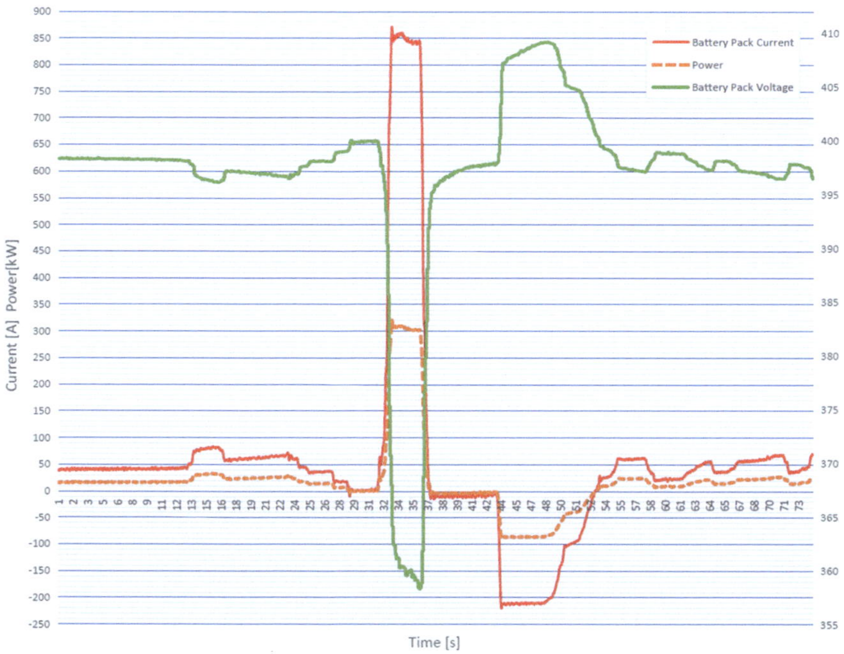

Fig. 8.59 Acceleration and braking conditions in an EV drive cycle

On the other hand, the flywheel is an electromechanical system that stores and delivers kinetic energy when it is needed. The flywheel is composed by an electrical machine with high mass connected in the rotor axle. The mass is driven at very high speed, and the energy stored depends on the mass and the speed. The bearing friction losses are reduced by using magnetic bearing, while the air friction losses are reduced, inserting the rotor in a vacuum chamber. In the EV application, the electrical machine charges the flywheel by accelerating the mass by using the regenerative braking during vehicle slowdown. Then, during acceleration or climbing hills, the energy stored can be supplied back by using the same electrical machine which acts as a generator. As a consequence, the rotor speed is decreased during discharge. The electrical machine used in the flywheels can be induction machine, BLDC, PMAC, and SynRM. Similar to supercapacitors, the flywheels have a longer lifetime, high power density, high efficiency, and frequent charge–discharge capability.

The use of flywheel or supercapacitors mainly prolongs the lifetime of the battery and allows more efficient energy storage. However, both solutions suppose a more complex system where additional power converter will be necessary such as bidirectional DC-DC converter for supercapacitors, and bidirectional VSI for the flywheel.

On the other hand, when VSI is connected to the electrical grid, and a unidirectional uncontrolled rectifier is used, the regenerated energy must be dissipated in a resistive load since the current cannot circulate in the opposite direction. In Fig. 8.60,

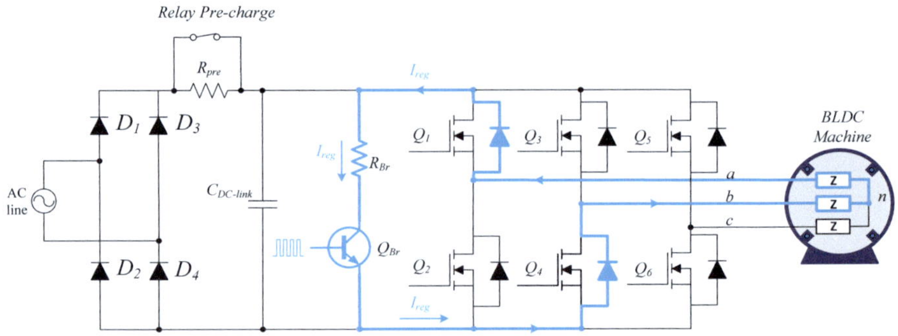

Fig. 8.60 Dynamic braking in a VSI . The energy of the machine is dissipated in a resistance load controlled by PWM

the transistor Q_{Br} has the function of preventing high DC-link voltage by means of its control in chopper mode according to a PWM signal. The excess energy is dissipated in the R_{Br} resistance until the regeneration is completed. This mode is also known as dynamic braking since it constitutes a dynamic brake of the AC machine. It is important to mention that the resistance R_{Br} limits the DC-link voltage and that the power of the dynamic brake V^2/R_{Br} must be higher than the regeneration power of the AC machine.

In low-cost applications or those that do not need a dynamic brake, the excess energy produced by regeneration, if it exceeds a predetermined limit, can be avoided by inhibin the VSI PWM signals. It is valid for induction machines since they produce no back-EMF voltage when it is running without control. In the case of machines with permanent magnets, the deactivation of the VSI would lead to generating a voltage proportional to its speed according to Faraday's law, that is, a possible high back-EMF voltage. If the back-EMF voltage is higher than the DC-link voltage, the voltage difference ensures that the diodes act like a rectifier incrementing the DC-link voltage uncontrollably. A low-cost solution to avoid damaging the VSI in overvoltage situations due to the back-EMF of permanent magnets machines consists of turn-on the three low-side devices. That is, the machine terminals are short-circuited, as shown in Fig. 8.61 imposing voltage to zero. The short circuit of the phases produces a dynamic brake of the machine in a controlled manner since the current which flows through the machine is not higher than its nominal current. The steady-state short-circuit current can be expressed as:

$$I_{SC} = \left| \frac{\Psi_m \omega_r}{R_s + j\omega_r L_s} \right| = \frac{\Psi_m}{\sqrt{\frac{R_s^2}{\omega_r^2} + L_s^2}} \tag{8.34}$$

The main drawback is that the dynamic brake energy is converted in heat in the machine and the braking capacity is poor.

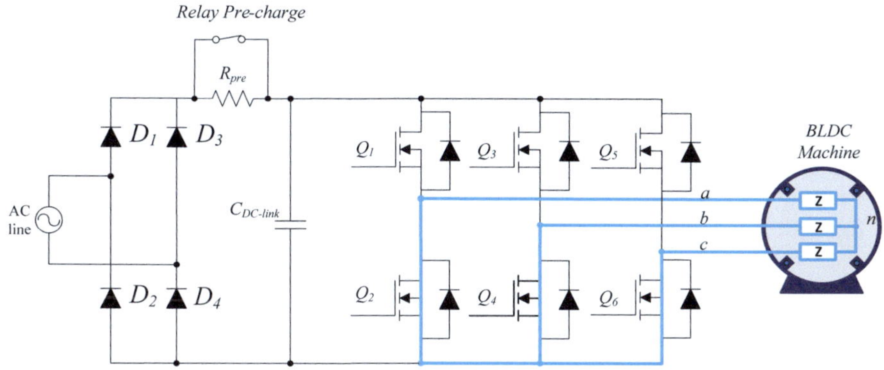

Fig. 8.61 Low-cost dynamic braking in a VSI for permanent magnet machines such as BLDC and PMSM

8.6 Machine Terminal Overvoltage

Sometimes in an industrial environment, the inverter and the AC machine are installed with a long power cable between them. The high dv/dt produced by the PWM inverter travels through the power cable to the machine, which can be reflected in the terminals of this, depending on impedance imbalance between the machine and the cable. As a result, it leads to overshoot voltage stress in the machine due to the reflected wave and produces shaft voltages, bearing currents, and severe problems with EMI.

As the impedance of the cable increases with the length of the wiring, the transient overvoltage is also increased, causing a stress situation in the insulation of the winding of the machine. The mismatch impedance between the cable and the machine terminal increase the transient overvoltage because the reflection coefficient is also increased. To overcome this transient overvoltage and to reduce the EMI, suitable filters can be used to clamp the machine terminal reflected voltage in safety level. The filters can also solve the bearing current effects. Figure 8.62 illustrates a transient voltage overshoot generated in a PWM pulse both on the rising and falling edges in the AC machine terminals when long power cable is used.

Fig. 8.62 Voltage step at the output of VSI produce an overvoltage in machine terminals

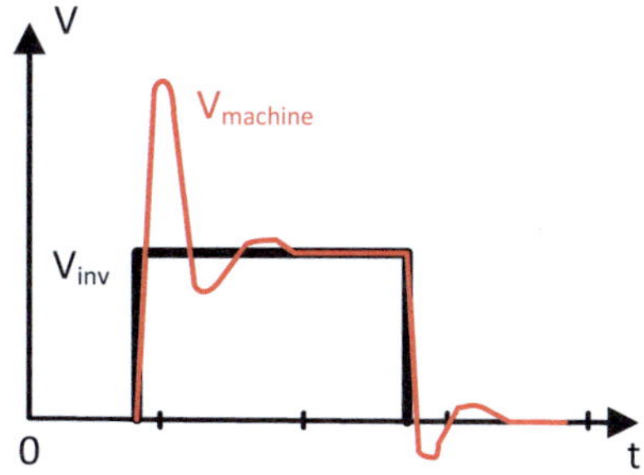

In general, the usage of filters has the following advantages:

- Smooth the PWM d*v*/d*t* to protect the machine winding
- Reduction of transient overvoltage when the cable length is large
- Reduction of AC machine losses
- Reduction of EMI
- Reduction of the acoustic noise produced by the switching frequency
- Reduction of machine bearing currents.

8.6.1 Involved Impedance

The voltage in the machine V_{in} can be calculated based on the reflection coefficient on the load side, as indicated in Eq. (8.35):

$$V_{in} = V_{Inv}(1 + \Gamma_L) \tag{8.35}$$

where Γ_L is calculated as follows:

$$\Gamma_L = \frac{Z_L - Z_o}{Z_L + Z_o} \tag{8.36}$$

The Z_o represents the cable characteristic impedance and Z_L the machine impedance.

The power cable impedance varies among other parameters with the length, type of insulation, number of conductors, and coupling between phases. The common-mode cable impedance depends on the insulation material inductance, the capacitance to ground and the stray capacitance between the different conductors. In a three-phase power cable, the impedance characteristic depends on the type of cable and the distance between the phase cables. For example, for tight cables between them, the capacitance is more significant than in separate cables.

The physical constitution of three-phase power cables is composed of a set of three inner power cables in trefoil formation, as shown in Fig. 8.63. There are different insulating materials and a conductive layer called armor which has the main function of protecting the set.

It is possible to find different impedances such mutual impedance between the sheath and the conductors, self-impedance of phase conductor, capacitance between phases and sheath, which describes the behavior of the cable.

The three-phase power cable can be modeled with parameters lumped in a PI section (Deng et al. 2019). The *R*, *L*, and *C* lumped elements can describe the long cable adequately where the inductive and capacitive coupling between the three-phase conductors and ground is represented.

The common-mode impedance of the AC machine is usually higher than the common-mode impedance of the cable (von Jouanne and Enjeti 1997). This is

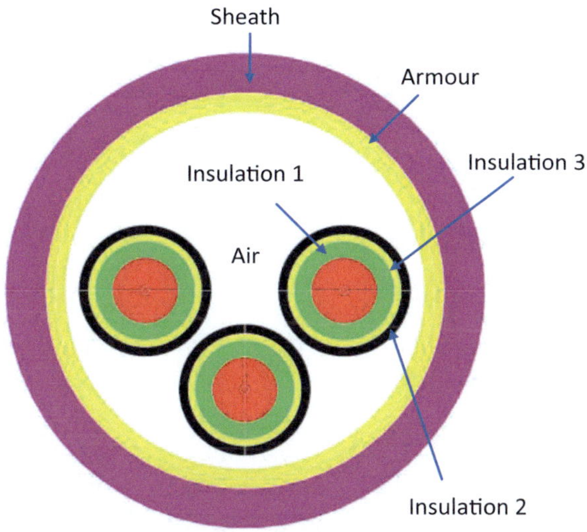

Fig. 8.63 Set of three core cables in trefoil formation of a three-phase power cable. Image developed using Flux2D provided courtesy of Altair Engineering, Inc.

complicated to calculate analytically. Experimental results show impedance values between 400 and 2000 Ω depending on the power. Therefore, the reflection coefficient on the load side Γ_L usually takes a maximum value of 0.95 for low-power machines. In this case, and according to (8.35), the reflected voltage is almost the double of the inverter output voltage.

On the other hand, the impedance of the inverter is formed by the DC-link capacitor and the freewheeling diodes. The first reflection of the wave toward the inverter passes through the diodes, driving the reflected voltage toward the DC-link capacitor, which is equivalent to a short circuit for the fast rise time pulses.

To obtain an accurate simulation of the transient overvoltage generated in the machine terminals, it is necessary to have a high-frequency machine model join inverter and cable models. For AC induction machine, the model is based in a low-frequency model as discussed in Chap. 5 with a high-frequency model to reproduce the variation of its impedance with the frequency (Grandi et al. 1997; Boglietti and Carpaneto 1999). The most relevant elements of the high-frequency model are the capacity C_g and resistance R_g between the windings and ground, and between the neutral and ground. The capacity C_g and resistance R_g represent the impedance "surge" of the machine. The resistance R_g models the losses of the resistance "frame" of the machine, while the capacity C_g, is the capacity between the windings/neutral and ground. The resistor R_e models the losses introduced by eddy currents inside the magnetic core. Figure 8.64 illustrates the most important impedances discussed above that contribute to the high-frequency response (high-frequency model) with the low-frequency model. The values of these impedances vary with the power of the machine and can be measured experimentally.

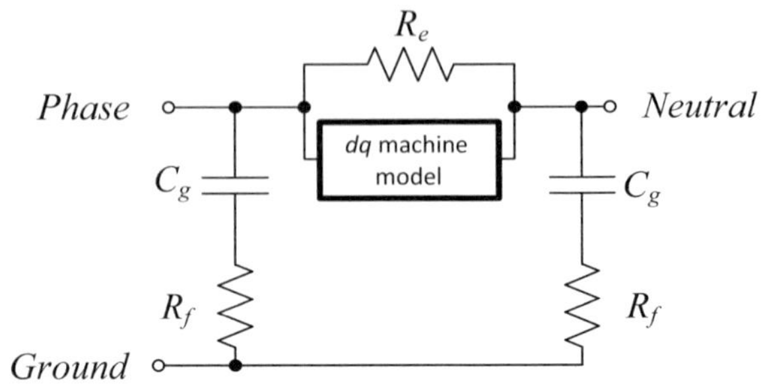

Fig. 8.64 High-frequency machine model including *dq* low-frequency and high-frequency models

8.6.2 Sine-Wave Low-Frequency Output Filter

The sinusoidal filter is based on an LC low-pass filter which is installed on the inverter output and before the power cable, in each of the phases as can be seen in Fig. 8.65. In series to each phase, there is an inductance L and a capacitor C with a common floating star connection. The PWM pulse signals in the VSI output voltage are converted into a smooth pure sinusoidal signal at the terminals of the AC machine. This type of filter due to its structure is also known as a differential mode sinusoidal filter. The primary filter function is the following:

- Smooth the PWM dv/dt
- Reduce transient overvoltage when the cable length is large
- Reduce losses in the AC machine: Reduction of harmonics and electromagnetic interference (EMI)
- Reduce the acoustic noise produced by the switching frequency
- Reduce the machine bearing currents.

For applications sensitive to acoustic noise, the selection of a switching frequency higher than 16 kHz is usually a solution since it is less sensitive to humans. The sources of acoustic noise generated by the machines are due to the noise of the magnetic core, the bearings, and the fan if it is incorporated. The first can be reduced according to the switching frequency as described above. However, this solution

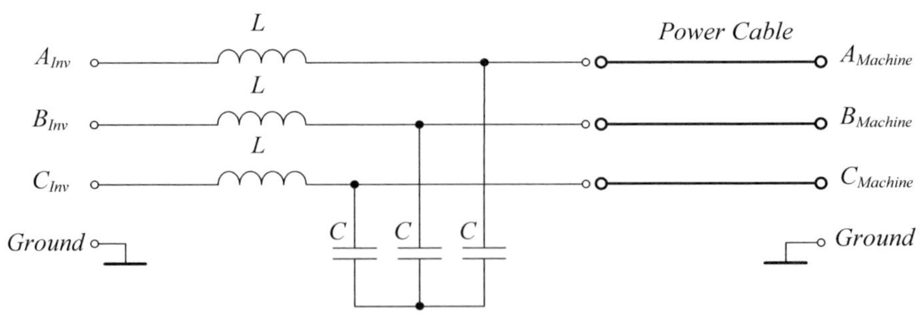

Fig. 8.65 Differential model LC sine-wave filter

creates more switching losses and therefore, an increase in temperature and reduction in efficiency, as discussed in Sect. 8.3.7.2. For this reason, the sinusoidal filter can be used to reduce the acoustic noise without increasing the switching losses.

Depending on the filter cut-off frequency and its power, the cost and size may vary. Higher cut-off frequency usually leads to a reduction in cost and size. However, the harmonic content will not be filtered efficiently if the PWM switching frequency is low. On the other hand, it is important to mention that the filter has associated losses, although they may be less than the losses produced in the AC machine without the sinusoidal filter.

The design criteria for the filter cut-off frequency must be at least five times smaller than the switching frequency. With this criterion and selecting an adequate inductance value that can withstand the nominal phase current, the capacitor can be calculated as shown in Eq. (8.37).

$$C_{swf} = \frac{1}{(2\pi \cdot f_c)^2 L_{swf}} \tag{8.37}$$

8.6.3 High-Frequency Output Filter

Unlike the low-frequency filter, the high-frequency filter can be placed in the inverter output before the power cable, but it is also possible to place it in the machine terminals, i.e., after the power cable. In both cases, the high-frequency filter has the objective of reducing voltage surges in the machine terminals and high-frequency EMI emissions with much lower losses in the filter compared with the sinusoidal filter. However, the acoustic noise or losses of the AC machine are no reduced. Since the filter has a high cut-off frequency, the size and cost are smaller than the sinusoidal filter. There are different filter networks, some more suitable to connect them to the machine terminals, and others for the inverter output terminals. The LRC filter like the one in Fig. 8.66, is a filter that is placed in the inverter output before the power cable.

If the inductance L_f is suppressed, the filter can be placed on the machine terminals. The filter has only six components, and therefore, the cost is lower. In addition, the filter installation in the machine terminals could be easier than in the inverter output. The design that follows is a machine terminal RC filter, as shown in Fig. 8.67.

The combination of the series resistance with the capacitor, as can be seen in Fig. 8.67, reduces losses due to the Joule effect since it allows a fine-tune of the damping coefficient ζ. The capacitance C_f behaves like a short circuit at high frequencies, and the resistance R_f is designed to match the impedance arises Z_o of the cable to absorb the reflected energy according to Eq. (8.38).

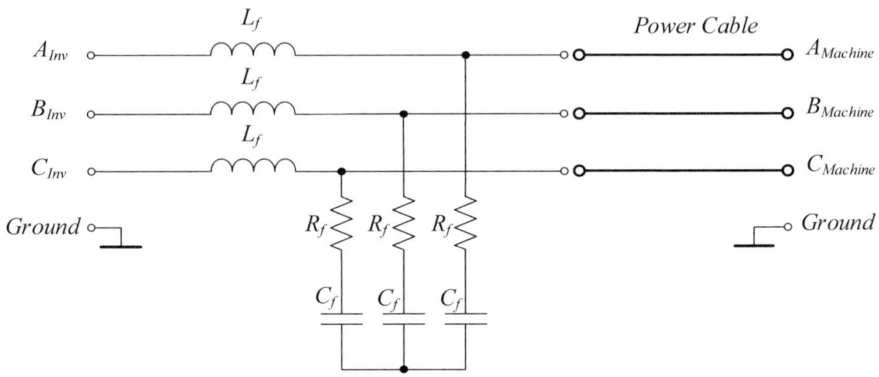

Fig. 8.66 Second-order LRC filter to be connected at the inverter output terminals

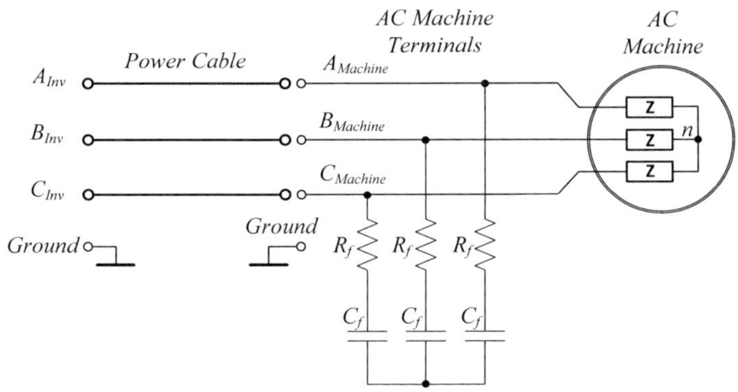

Fig. 8.67 First-order RC filter connected at the machine terminals

$$R_f = Z_o = \sqrt{\frac{L_c}{C_c}} \tag{8.38}$$

In this equation, it has been assumed that the cable does not have losses since the resistive components of the cable have been ignored.

When the cable impedance Z_o matches the filter impedance Z_f, the reflection coefficient is zero, as can be seen in (8.39).

$$\Gamma = \frac{Z_f - Z_o}{Z_f + Z_o} = 0 \tag{8.39}$$

As a consequence, the reflected wave is suppressed. To calculate the capacitance C_f, it is necessary to know the rise time t_r of the inverter output pulse. At the first instant at $t = 0$, the RC filter acts as a pure resistance since the capacitor is discharged. After the pulse rise time, the capacitor is fully charged, acting as an open circuit and avoiding power dissipation. The RC filter response to an amplitude voltage step $V_{DC\text{-}link}$ at the capacitor terminals is described in (8.40) as:

$$V_C = V_{DC-link}\left(1 - e^{\frac{-t_r}{R_f C_f}}\right) \tag{8.40}$$

If a maximum voltage peak of 15% is considered, the capacitor C_f can be calculated according to (8.41) as:

$$C_f = \frac{-t_r}{\ln(0.85)R_f} \tag{8.41}$$

8.6.4 d*v*/dt Simulation

The VSI circuit of Fig. 8.68 can be used for voltage overshoots and voltage peaks simulation due to high d*v*/d*t*. This is basically composed of the following models:

- Voltage source
- VSI
- Shielded three-phase power cable
- Induction machine.

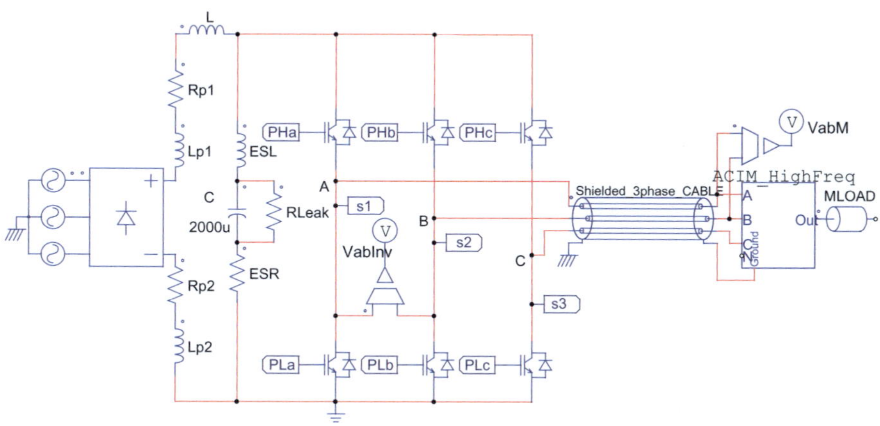

Fig. 8.68 VSI circuit with a three-phase long cable and high-frequency induction machine models. IGBT's transistors and driver with an accuracy model are considered for the most realistic simulation

The voltage source is provided by the electrical grid through a three-phase rectifier bridge made with diodes and a DC-link capacitance. The VSI is made with IGBTs with a very precise model that can provide a very realistic voltage pulse at the output of the inverter. As can be observed, the inverter ground is independent of the main ground (or earth). In the main ground, is connected the electrical grid ground, the three-phase cable shield, the machine frame, and the ground inverter chassis. The cable shield is connected at both ends to the inverter ground terminal and the other one to the machine grounding terminal. The electric machine consists of a 10 Hp induction machine. The machine model is a standard PSIM library model but with high-frequency elements as described in Sect. 8.6.1. The machine neutral remains unconnected. The power cable model used corresponds to three conductors # 6 AWG (13.3 mm^2) with shielding also provided by the PSIM library, which takes into account the inductive and capacitive coupling between the phases. The shielding is commonly used in inverter-fed AC machine drives to reduce the conducted and radiated EMI emissions to the neighbor components located near the power cable.

The first simulation uses a 25-m power cable between the inverter and the machine in order to see the voltage overshoot and voltage peaks at the induction machine terminals. The PWM switching frequency is 10 kHz, and the inverter voltage output frequency is 120 Hz. Figure 8.69a shows the voltage at the induction machine terminals for a complete cycle where it is possible to observe the voltage peaks in each of the PWM pulses. Figure 8.69b shows a zoomed view where high-frequency ringing occurs at the machine terminals in each PWM pulse. The excessive overvoltage at every PWM pulse could break the machine insulation as discussed. Finally, Fig. 8.69c shows the voltage output spectrum where at approximately 1.25 MHz, there are a large number of harmonics with a significant magnitude. It corresponds to the damping frequency of Fig. 8.69b.

In the inverter output, before the power cable, the harmonics of the voltage are reduced, as shown in Fig. 8.70a. It is possible to observe that the main harmonic at frequency 1.25 MHz do not exist. Figure 8.70b shows a section where it is possible to see the harmonics resulting from the switching frequency and its replicas.

The filters studied previously reduce the voltage overshoots and voltage peaks significantly, improving the spectrum as they will be seen in the following simulations.

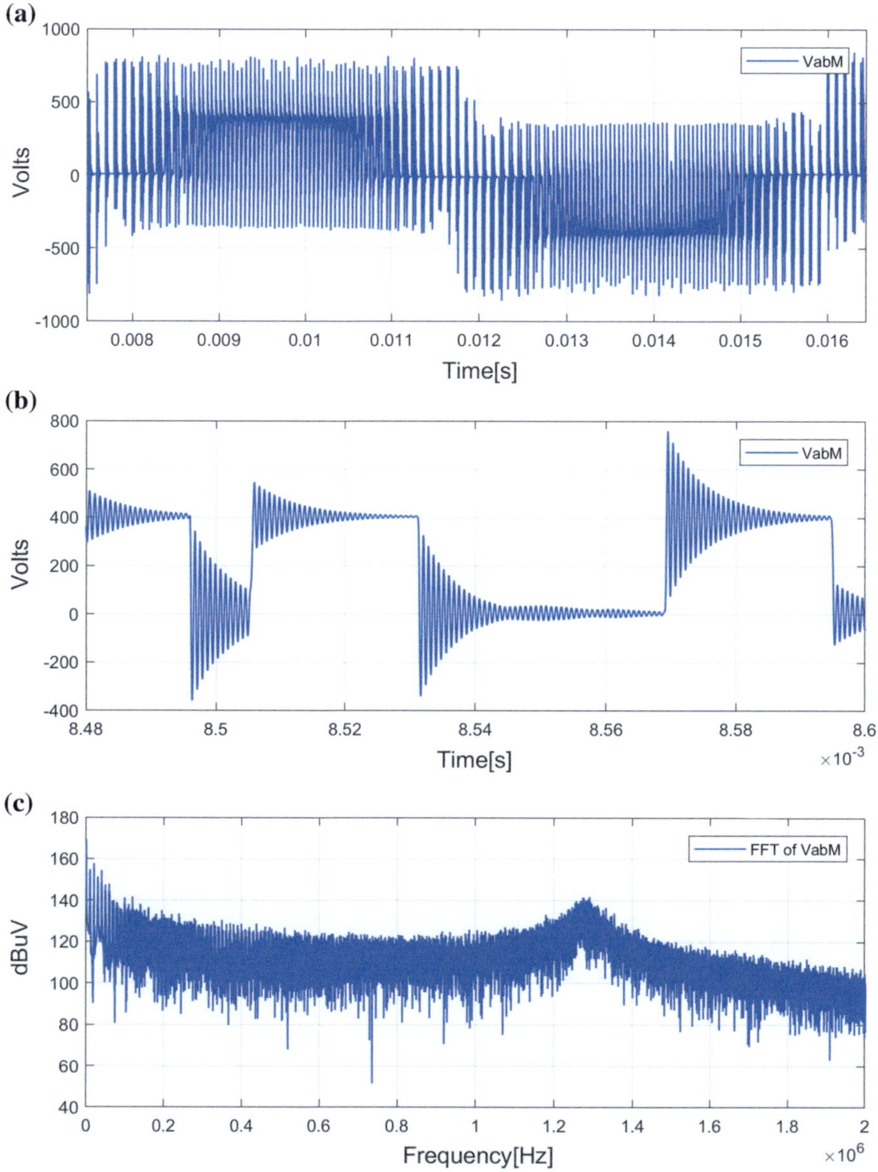

Fig. 8.69 **a** Line-to-Line voltage at machine terminals. **b** Detail section. **c** Frequency components. Modulation index: $m = 0.8$, output frequency: $\omega_c = 120$ Hz. Deadtime: $T_{DT} = 1$ µs. Switching time: $T_{PWM} = 100$ µs. Simulation step: $T_{sp} = 10$ ns

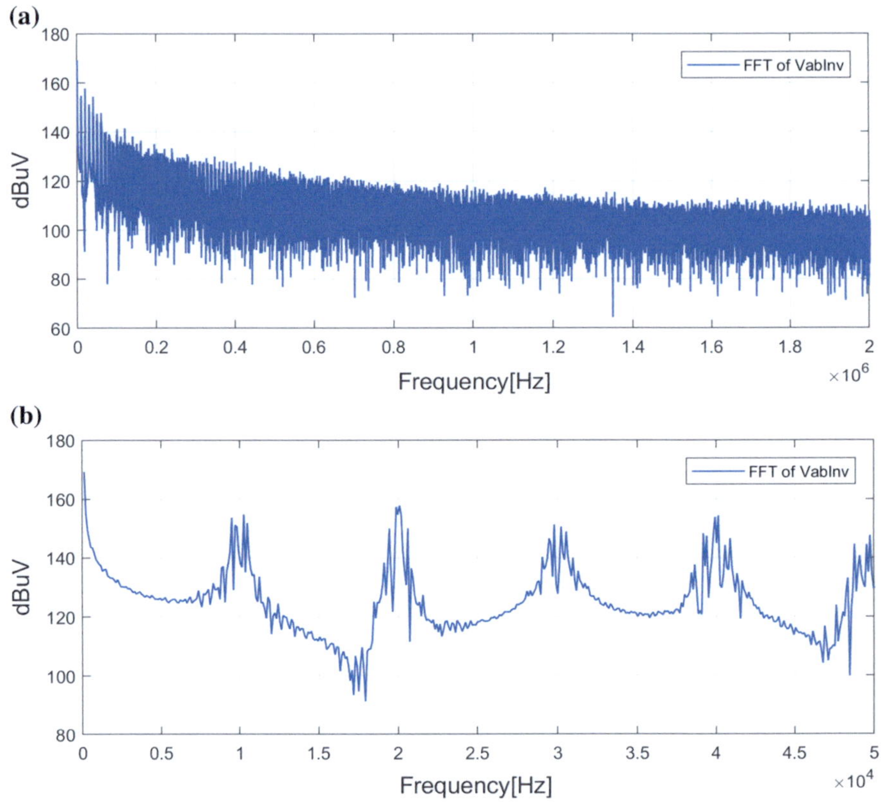

Fig. 8.70 **a** Line-to-line inverter output voltage frequency components. **b** Detail section to show the harmonics of PWM frequency. $m = 0.8$, output frequency: $\omega_c = 120$ Hz. Deadtime: $T_{DT} = 1$ μs. Switching time: $T_{PWM} = 100$ μs. Simulation step: $T_{sp} = 10$ ns

8.6.4.1 Sine-Wave Filter at Inverter Output Terminals

As mentioned early, the sinusoidal filter has advantages over the high-frequency RC filter such as the reduction of electromagnetic emissions, reduction of acoustic noise, smoothing the dv/dt, but, on the contrary, the filter size is larger and more expensive. In Fig. 8.71 is illustrated the same schematic shown above, but with the sinusoidal filter connected to the inverter output. It is possible to observe that the inductance L_{SWF} has to support all the current flowing to the induction machine, which can give an idea of the inductance size.

The switching frequency for this simulation is also 10 kHz. The filter cut-off frequency is chosen to be at least ten times smaller, i.e., approximately 1 kHz. According to (8.37), it is possible to calculate the capacitance required for said cut-off frequency by choosing an inductance of 1 mH.

Figure 8.72 shows the result of the simulation where it is illustrated the PWM inverter output line-to-line voltage modulated according to a sinusoidal PWM and

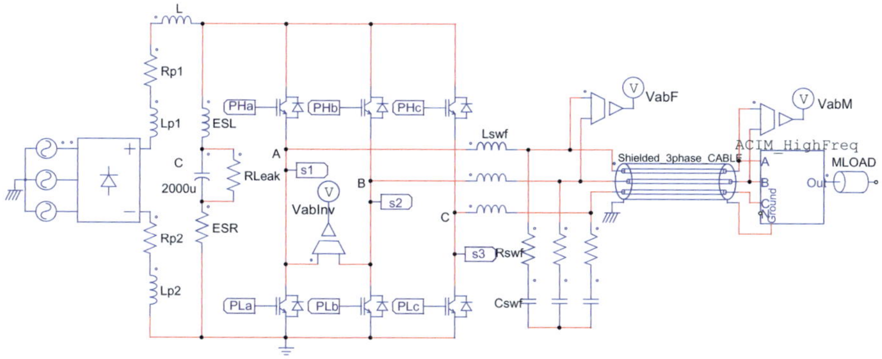

Fig. 8.71 VSI circuit with sine-wave filter, three-phase cable, and induction machine models

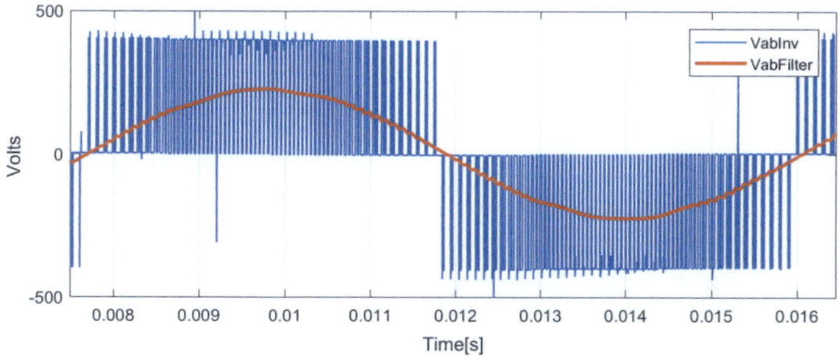

Fig. 8.72 Line-to-Line voltage before and after sine-wave filter with a cut-off frequency of 1 kHz. Modulation index: $m = 0.8$, output frequency: $\omega_c = 120$ Hz. $L_{SWF} = 1$ mH, $R_{SWF} = 2\ \Omega$, $C_{SWF} = 22$ nF. Deadtime: $T_{DT} = 1\ \mu$s. Switching time: $T_{PWM} = 100\ \mu$s. Simulation step: $T_{sp} = 100$ ns

the resultant sinusoidal filter. As can be seen, the filter output voltage is very smooth, which corresponds to the fundamental harmonic at 120 Hz of the PWM modulation signal. Moreover, it has a very-low dv/dt so that the voltage overshoot and voltage peaks in the induction machine terminals are not present. The length of the cable affects this type of signal in a lower proportion. However, the filter adds some delay, and the amplitude in the machine terminals is reduced due to the voltage drop in the filter inductance.

The power cable losses may be higher because a low-frequency signal travels through the cable instead of a higher-frequency sinusoidal PWM signal. In addition, the losses, size, and cost are higher in the sinusoidal filter than in the high-frequency filter.

8.6.4.2 High-Frequency RC Filter at the Machine Terminals

As depicted in Fig. 8.73, the high-frequency RC filter calculated according to (8.41) is connected in the induction machine. In the same way as before, the RC filter is connected in star with its common floating point.

The simulation has been performed with four different lengths of power cables such as 5, 10, 25, and 50 m to see the differences. The rise time is 121 ns, where the d*v*/d*t* for 400 V is approximately 2670 V/μs. This high d*v*/d*t* produces a voltage peak of 37% on the DC-link voltage of 400 V, as can be seen in Fig. 8.74a for 5-m power cable. As the length of the power cable increases, the voltage peak increase but not more than twice the DC-link voltage, but on the contrary decreases its frequency as can be seen in Fig. 8.74b, c, and d. For example, for 50-m power cable of Fig. 8.74d, the voltage overshoot reaches 766 V, that is 91.5% over the DC-link voltage. It is important to mention that the three-phase power cable model provided by PSIM is not frequency-dependent and, therefore, the experimental result could have some discrepancy with the simulation performed.

On the other hand, if the RC filter is installed for each of the phases as explained above, the simulation results are very different compared to the previous one. In this case, as can be seen in Fig. 8.75, the voltage overshoot in the induction machine terminals is much lower. For example, when the power cable length is 25 m (Fig. 8.75a), the voltage overshoot is approximately 476 V with a d*v*/d*t* of 914 V/μs.

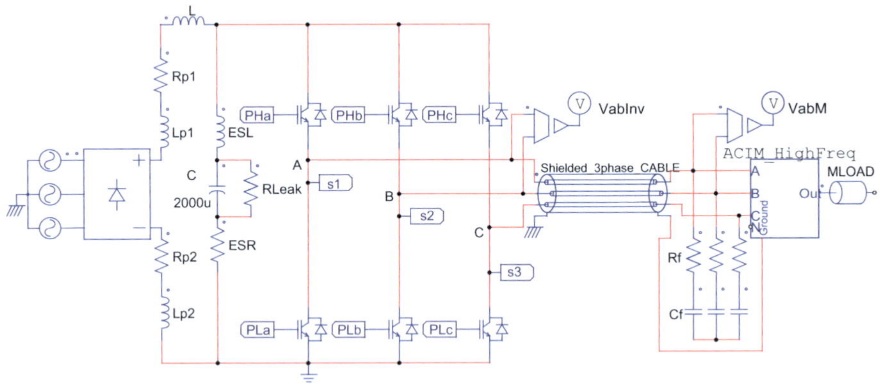

Fig. 8.73 VSI circuit with three-phase cable, high-frequency RC filter, and induction machine models

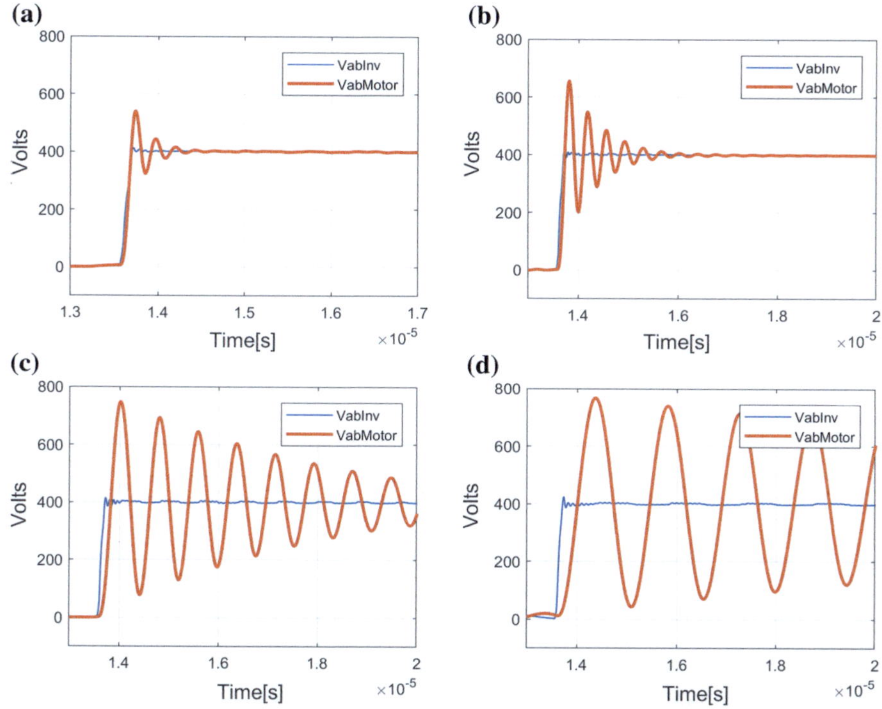

Fig. 8.74 Line-to-Line voltage at machine terminals when step voltage dv/dt at the inverter output terminals is 2.67 kV/μs without filter. **a** Cable length is 5 m. **b** Cable length is 10 m. **c** Cable length is 25 m. **d** Cable length is 50 m. Modulation index: $m = 0.5$, output frequency: $\omega_c = 60$ Hz. $R_g = 23\ \Omega$, $C_g = 700$ pF. Deadtime: $T_{DT} = 1\ \mu$s. Switching time: $T_{PWM} = 200\ \mu$s. Simulation step: $T_{sp} = 10$ ns

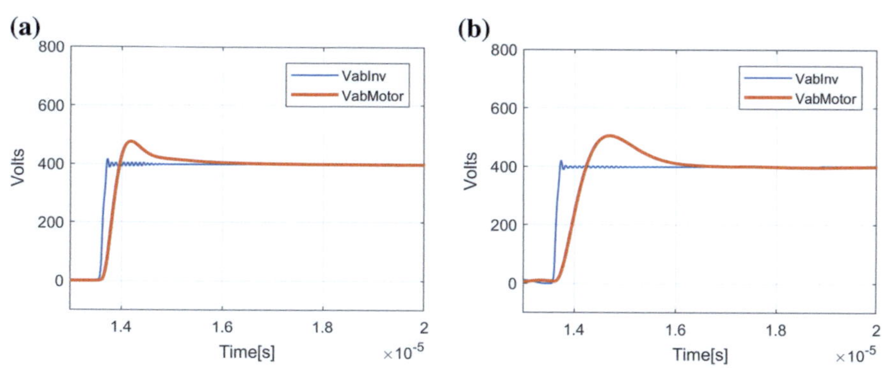

Fig. 8.75 Line-to-Line voltage at machine terminals when step voltage dv/dt at the inverter output terminals is 2.67 kV/μs with RC filter. **a** Cable length is 25 m. **b** Cable length is 50 m. Modulation index: $m = 0.5$, output frequency: $\omega_c = 60$ Hz. $R_g = 23\ \Omega$, $C_g = 700$ pF. $R_f = 40\ \Omega$, $C_f = 28$ nF. Deadtime: $T_{DT} = 1\ \mu$s. Switching time: $T_{PWM} = 200\ \mu$s. Simulation step: $T_{sp} = 10$ ns

8.7 VSI Self-protection

The VSI must be protected against an abnormal operating environment, such as a high ambient temperature or incorrect operation of the AC machine. The most common incorrect operation is the overload (which cause overheating), energy regeneration, and excessive current and voltage transients.

When the VSI is operating in an overload condition, the temperature of the power switches is increased. The overload is typically caused when the AC machine is operating above its nominal conditions. This will be discussed in more detail in the next Sect. 8.8.2. Energy regeneration can produce a high-overvoltage situation damaging the inverter if an adequate protection mechanism is not available. Finally, current and voltage transients can damage the most sensitive devices such as MOSFET/IGBTs and its freewheeling diode if a specific limit is exceeded. The protection mechanism must protect these more sensitive devices so as not to damage the inverter. In this section, it is discussed different protections that protect the inverter against the abnormal conditions previously discussed.

8.7.1 Short-Circuit Protection (Surge Current Detection)

The short-circuit or high current transient's protection is mandatory for a robust VSI design and to comply with safety standards. However, to implement this protection feature, sometimes it is necessary to add some extra board components. In the market, it is possible to find half-bridge gate driver circuits with fully integrated protection elements which simplify the PCB design. In this section, two types of short-circuit protections will be discussed, such as desaturation detection and the short circuit detection by using a shunt sensor.

The simplest example of a high current transient is when phase-to-phase short circuit in the AC machine is produced, for example, due to the loss of insulation in the stator windings, or due to a fault in the wiring between the inverter and the machine. In this case, the DC-link voltage will short-circuit to ground through the windings of the machine being less dangerous than direct short circuit (shoot-through). The shoot-through is when in the same inverter leg (half-bridge topology) the high- and low-side devices are turned on simultaneously by error or by a false activation, producing a direct path between the DC-link voltage and ground through the devices. Under short-circuit conditions, extremely high current flows through the power switch devices, where an excessive junction temperature can destroy it completely. The time to destroy the device depends on many factors such as the current flowing, device temperature just before the short circuit, the device, etc. It is usually a very short time such less than 50 μs. The protection circuit must detect the high current and act in time to prevent device destruction. These protection circuits are usually designed with discrete hardware circuits because implementation based on a microcontroller/DSP and an A/D converter with a control algorithm would not be fast enough to detect

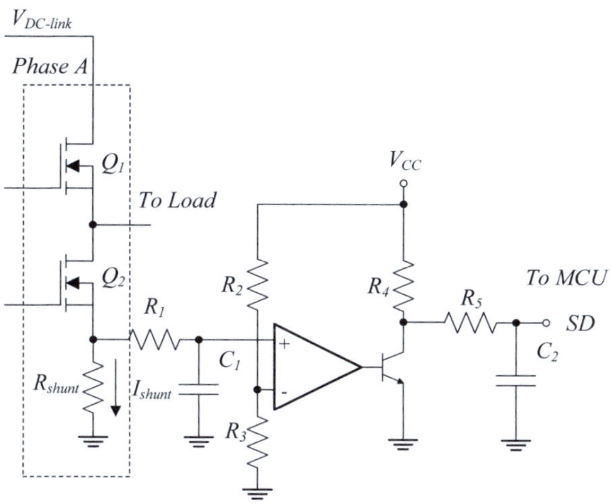

Fig. 8.76 Overcurrent detection circuit to protect the IGBTs or MOSFETs devices by measuring the current in every half-bridge

the short circuit in time. Some microcontroller/DPS has dedicated digital inputs as shutdown (SD) signal with internal hardware which can be programmed to inhibit the PWM signals in less than 1 μs protecting the inverter. In any case, it is usually used as a control algorithm to count the number of short circuits or trials to prevent consecutive short circuits. Although the protection circuit acts in time, preventing the device destruction if, for any reason, there is consecutive short circuit, the device has no sufficient cooling time where the destruction is also possible. For this reason, it is necessary to set a minimum pause time of some seconds by the control algorithm.

The power switch short-circuit protection typically uses a comparator which compares a reference with a voltage drop in a shunt resistance as depicted in Fig. 8.76. It is capable of detecting either high current transient or shoot-through. The shoot-through, because its high current, can be detected easily since the voltage drop in the shunt resistance gets the reference voltage easily inhibit the PWM signals. However, the high current due to a phase-to-phase short circuit is usually lesser magnitude and should be possible that the voltage drop on the resistance is not enough to inhibit the PWM signals. Moreover, in both cases, it is necessary that the power switch is on during a minimum time longer than the detection time of the circuit. If this is not the case, the high current transient could not be detected overheating the MOSFET until it is completely destroyed. This can occur at the machine start-up where the voltage and frequency are very low. The first PWM pulses, as they follow a sinusoidal profile, have a very short duration which may be below the detection time of the circuit. Then, if there is phase-to-phase short circuit, a certain number of short pulses can irreversibly destroy the MOSFET device due to the gradual increase in temperature. A technique used to detect the possible phase-to-phase short circuits consists of applying sufficiently wide activation pulses before the start-up so that the circuit can detect it. The verification process has a very short duration of a few milliseconds, so it will not affect normal operation.

In Fig. 8.76, the current transient passing through the shunt resistor generates a voltage drop. The R_1 and C_1 components perform a high-frequency low-pass filter to avoid false activations due to ground noise. The voltage drop is compared with the reference value set by the resistors R_2 and R_3. If it exceeds the reference value, the BJT transistor is turned on, and the SD output signal falls to zero.

The SD output can be connected to microcontroller/DSP SD input in a way that inhibits the PWM outputs in the event of short-circuit or high current transient avoiding the device's breakdown. If the microcontroller/DSP does not have SD input or similar, the SD signal can be used to inhibit the PWM outputs of the microcontroller to protect the inverter with additional external hardware. In this case, the SD signal can also be connected to a digital standard microcontroller input for the sole purpose of failure notification.

In Fig. 8.77a has represented an experimental test with a high current transient when high-side IGBT is short-circuited. The three-phase currents are shown, although the high current transient only flows through one of them as expected. The maximum value reached is 123 A, and the pulse width is 985 ns. On the other hand, the falling edge of the SD signal that is used to turn-off all the IGBTs is also shown. When this signal is close to zero volts, all IGBTs are open, and the current level falls to zero.

Figure 8.77b shows the same signals, but for a phase-to-phase short circuit where the maximum value reached in this case is lower, 79 A in a pulse width of 1.73 μs. Since di/dt is smaller, the pulse time in the phase short-circuited is longer. The detection of the current transient has a delay time of 0.4 μs, and the threshold is set to 65 A. In both cases, the pulse time is not dangerous to damage any IGBT according to the manufacturer specification.

On the other hand, the desaturation detection uses the device as a measure of current. Detection is achieved with a simple circuit, one for each device, as can be seen in Fig. 8.78. As commented before, there are gate drivers in the market that offer this feature where internally it has a fast comparator that after a prudent time internally inhibits the device in question. These usually have an output where the microcontroller can monitor to know if there has been any type of failure. In this case, externally, only three components should be placed, the capacitance, the resistor, and the diode for each device. In case of using a conventional gate driver, the part framed in red would be necessary to include it in the design for each device in addition to the three components discussed above.

In Fig. 8.78, the diode ensures that the IGBT collector–emitter voltage is only monitored by the detection circuit during the on time. That is when the collector–emitter voltage is very low. However, if a short-circuit event occurs, the IGBT collector current increases to a level that causes the IGBT to leave the saturation region and move to the linear region. This causes a rapid increase in the collector–emitter voltage. In this case, the capacitor is quickly charged up to a voltage threshold V_{DESAT}, which causes the device to turn-off via the SD signal and the microcontroller. Care should be taken when implementing desaturation detection to prevent false triggering. This can occur particularly during the transition from the IGBT off state to the on state

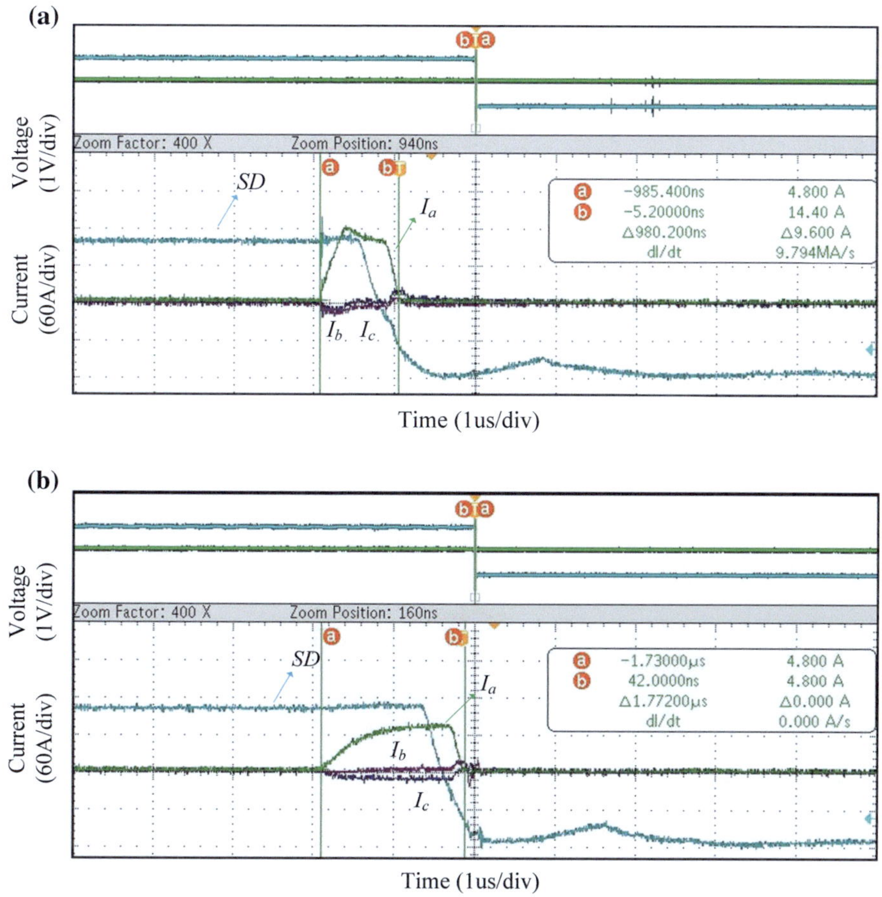

Fig. 8.77 **a** Overcurrent detection when high-side transistor is short-circuited (shoot-through). **b** Phase-to-phase short-circuit overcurrent detection

when the IGBT is not completely in saturation. Generally, a blinding time is inserted between the start of the ignition and the point at which the detection of desaturation is activated in order to avoid false detection. The RC filter introduces a short time constant in the detection to filter spurious introduced by the pickup of noise and to insert the appropriate time of blinding.

The high current transient detection by means of a shunt is an optimal and simple solution for those applications where the cost must be reduced since the current sensing through the shunt can also be used to read the machine phase currents. However, as an alternative to current sensing by means of a shunt, some inverters are designed with Hall effect current sensors. This offers increasingly fast responses comparable to those of a shunt, and also for high current inverters is usually more attractive as the efficiency of the VSI is improved. That is to say, the dissipation of

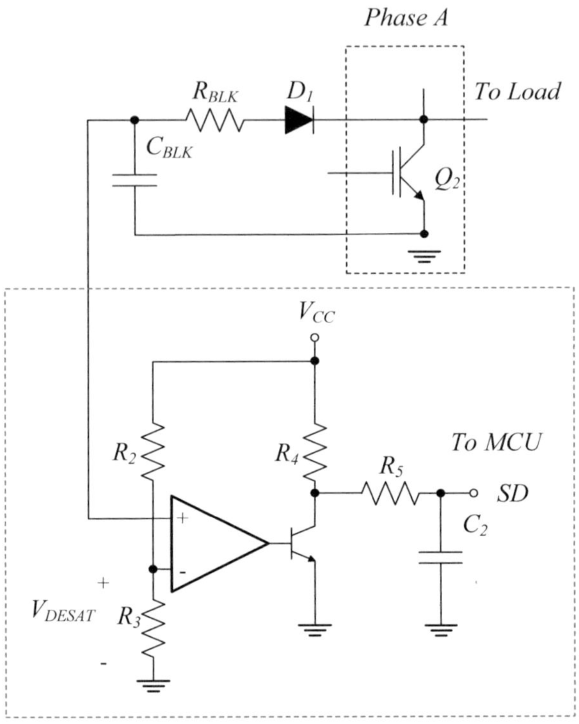

Fig. 8.78 Overcurrent desaturation detection circuit to protect the IGBTs or MOSFETs devices by monitoring the saturation voltage

power of the shunts implies a reduction, although small of the total efficiency. That is why some inverters such as in EV/HEV use Hall effect current sensors where the short-circuit detection is through the desaturation circuit.

8.7.2 *Overcurrent Detection*

The overcurrent situation is not usually detected by the previous circuit detectors. Overcurrent is usually detected by reading the phase current because they are slow current transients (few seconds). Overcurrent protection is essential both for the lifetime of the inverter and for the electrical machine. Moreover, if the inverter is powered through a battery as in EV/EHV, the battery life can be degraded without overcurrent protection.

The overcurrent can be detected as a control algorithm implemented in the microcontroller/DPS. That is to say, by reading the phase currents by an A/D converter, it is possible to determine if a current limit has been exceeded, which could damage the inverter if it is maintained for a long time. In addition to the aforementioned protection, fuse protection method is used to protect the battery, inverter, and electrical machine from extreme conditions. Compared to intelligent protection, fuse protection is more reliable and direct because fuse protection does not require the detection circuit. In general, the fuses are in series with the power supply elements either by means of a battery or by an AC source of the electrical grid.

8.7.3 Overvoltage and Undervoltage Detection

The inverter must also present protection against overvoltage and undervoltage, to avoid the destruction of its components, especially in the case of overvoltage. If an unusual voltage is detected that could damage the inverter, the protection system deactivates the outputs to stop the machine.

DC-link voltage monitoring protects the inverter from surges due to the regeneration of the AC machine itself, or from a voltage surge in the power grid. A high voltage rise with a duration of a few seconds can cause irreversible damage in the inverter. In the induction machines, as there are no permanent magnets, a rapid reaction to the overvoltage is usually not required and can be detected by an algorithm in the microcontroller by constant monitoring of the DC-link voltage. If the overvoltage limit is detected, the protection mechanism consists of opening all the power switches devices. However, in AC machines with permanent magnets, the back-EMF overvoltage protection is crucial. In this machines, if for any reason the synchronization control is lost, the back-EMF voltage can generate a high-voltage which could be much higher than the DC-link voltage in the case that all VSI switched devices are opened. In this case, the protection mechanism must have a faster reaction time. To overcome the dangerous high voltage, the machine terminals can be short-circuited, reducing the voltage to zero, as discussed in Sect. 8.5. The phase-to-phase short circuit is usually done by activating the three low-side power switches and is usually implemented by a hardware circuit which quickly detects the overvoltage as shown in Fig. 8.79. If the voltage in any of the phases exceeds the limit set by the resistors R_2 and R_3, the comparator changes its state to force the falling edge of the OV output. This can be used to activate the low-side power switches devices, and furthermore, if it is connected to a microcontroller input, an interruption can be generated to inhibit the inverter devices and to generate the overvoltage alarm by regeneration or synchronization lost.

On the other hand, undervoltage protection is less dangerous and is usually detected by the VSI by sensing the DC-link voltage to prevent that the AC machine operates under poor performance conditions.

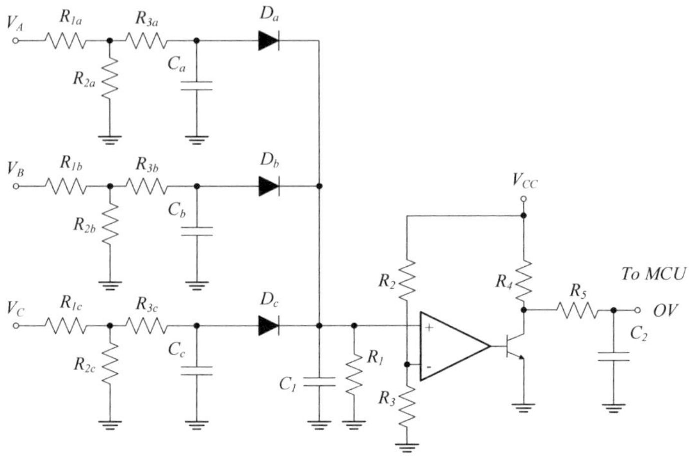

Fig. 8.79 Overvoltage detection circuit in PMSM

8.7.4 *Overheating Detection*

The inverter overheating is usually due to an inverter overload or when the ambient design temperature has been exceeded. For the first case, the inverter may be operating in overload because the AC machine is also operated in overload. In this case, it is possible to detect it as it will be seen in the next Sect. 8.8.2. For the second case, if the ambient temperature exceeds the design temperature, the inverter will overheat. Over temperature, detection can be performed by measuring the temperature of the power switches devices through an NTC (hottest point). Different temperature limits can be set to relax the temperature increase. That is, if a non-dangerous alarm threshold T_1, e.g., 110 °C is exceeded, the performance of the inverter/machine can be reduced (degrade state) to prevent the temperature from rising without stopping the machine. In the applications that it is possible to stop the machine, it can be stopped to cool down a bit, and then restart. Otherwise, it is possible to set a degradation operation mode which basically consists either of a reduction of power or a reduction of the PWM switching frequency. The reduction of the PWM switching frequency is more interesting because it reduces switching losses. Some microcontrollers such as the Infineon Aurix^{TM} allow for varying the PWM switching frequency in runtime. The reduction of the PWM switching frequency does not affect much the machine performance and will relax the increase of temperature, even in some cases it will be possible to return to the original switching frequency if a limit T_2, e.g., 95 °C lower than T_1 is exceeded. If this is not the case, and the temperature continues to increase, a degradation of the operation can be made, reducing the power delivered to the AC machine.

8.8 Machine Fault Detection

Before AC machine start-up, a test procedure can be performed in order to check if there is, for example, an overheating, a phase-to-phase short circuit, open circuit or open phase, and a locked rotor detection. By measuring the stator resistance in the three windings and averaging is possible to estimate the machine temperature. For this, it is necessary to know the winding material (copper or aluminum) and the nominal resistance at a fixed temperature, typically 25 °C. The resistance can be measured just before starting the machine by injecting a DC current through the windings using the VSI . The phase-to-phase short circuit can be detected, as discussed in Sect. 8.7.1. The open circuit can be detected by injecting a DC current as used to measure the winding resistance. In fact, with the measuring resistance test if possible to decide if some phase is in open circuit. The short-circuit and open-circuit faults can be caused, for example, by mechanical connections failures, and insulation failures due to machine stress in overvoltage, overcurrent, and overheating conditions. The locked rotor detection can be detected during the start-up, also with the VSI . The rotor can be locked, for example, by excessive loading, by mechanical faults such as air-gap irregularities, broken rotor bars, cracked rotor end-rings, bearing faults. In this situation, it is common to have an excessive current with a very-low back-EMF voltage.

On the other hand, when the AC machine is running, it is possible to detect overload and open-phase fault. The machine overload will lead to an excess current provided by the VSI , raising its temperature by Joule effect considerably. In the induction machines, the manufacturers supply in a nameplate (NP) the service factor (SF) where the machine can operate within the specifications but for a short period of time. For example, commonly it has an SF of 1.15 which in case of the nominal current means that the motor can operate up to 15% above the current specified in the NP for a short period of time. The same can be applied to other NP parameters, such as voltage and torque. Above the nominal specifications, the machine could operate in overload, increasing its temperature and reducing its lifetime.

The open phase can occur during the machine vibration, which could cause the break or disconnection of some of the cable phases. In an induction machine rotating at high speed (high inertia) if a phase is opened the machine continues to rotate continuously if the torque that offers the load is still below the torque of the machine in that new condition. If there is no detection mechanism, the machine and the inverter could be damaged.

As will be seen in this section, locked rotor, overload, overheating, and open-phase detection are easily implemented in a VSI microcontroller/DSP if the voltage and phase current measurements are available.

8.8.1 Locked Rotor Detection

The locked rotor detection is very similar to the overcurrent detection of Sect. 8.7.2. This consists of measuring the current of each one of the phases during the AC

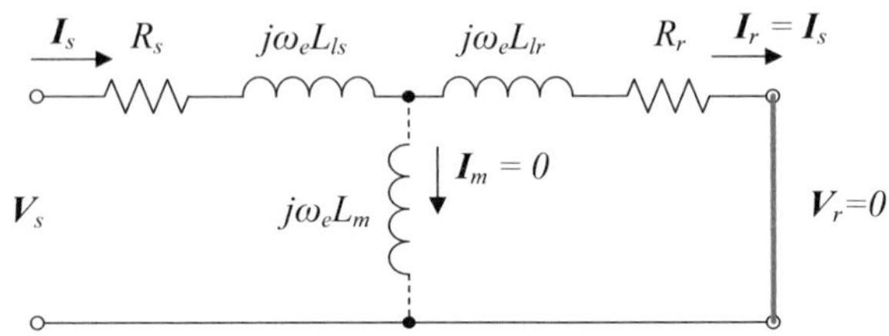

Fig. 8.80 Equivalent circuit of induction machine when the rotor is locked. Slip is the unity, and magnetizing current is zero

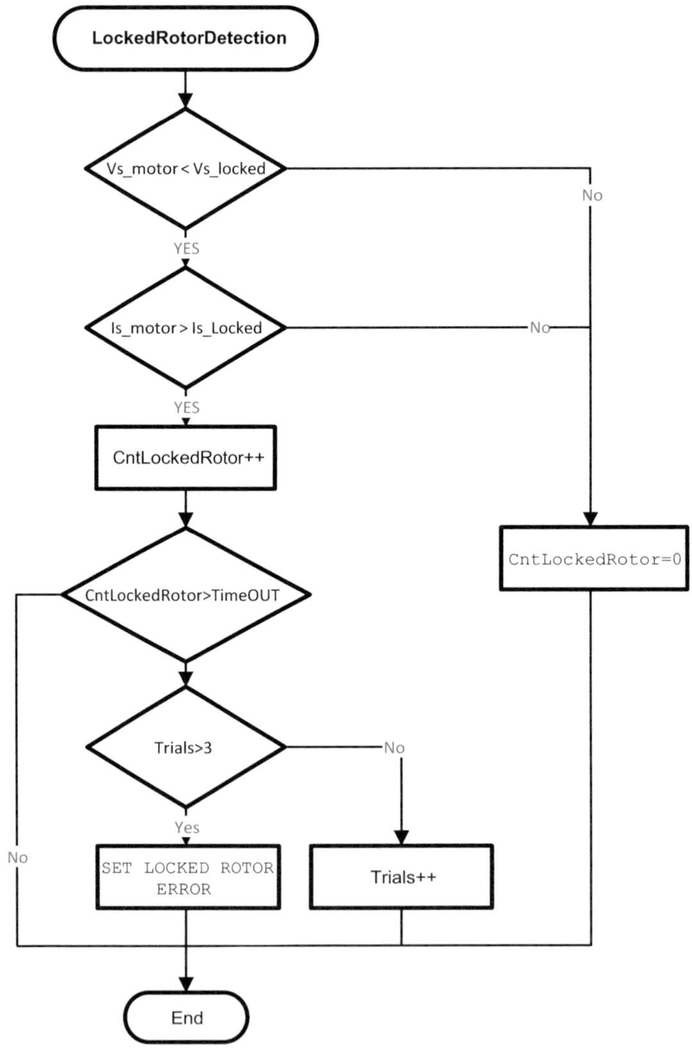

Fig. 8.81 Flowchart of rotor locked detection algorithm

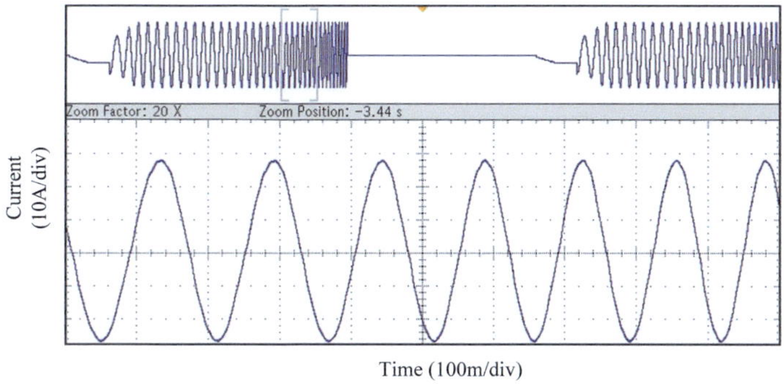

Fig. 8.82 Machine phase current in locked rotor condition

machine start-up to evaluate if the rotor is locked or not. If the rotor is locked, the back-EMF voltage is close to zero, but the phase current is very high, although a low voltage is applied to the stator. In the case of the induction machine, it is said that the machine is short-circuited since the slip is the unit and the magnetizing current is zero, as shown in Fig. 8.80. In practice, the current which flows per phase is equivalent to the phase voltage divided by the sum of the impedances.

The algorithm that detects the locked rotor during start-up basically consists of monitoring the current and voltage per phase of the AC machine. To avoid false alarms, it is recommended that the evaluation of the locked rotor is performed for a few seconds (5–10 s), and if possible, perform at least three attempts in different machine rotation directions as shown in the flowchart diagram of Fig. 8.81. Here only the voltage V_s and the current I_s are represented to simplify. At machine start-up, if a voltage limit set by $V_{sLOCKED}$ is not exceeded and the current exceeds the $I_{sLOCKED}$ limit, the locked rotor counter is increased. If the counter reaches a defined value, the trials counter is increased, and machine start-up is aborted. After three attempts, a fatal locker rotor failure error is set.

Figure 8.82 shows the experimental starting current of a 5 HP induction machine with the rotor locked where two trials can be seen. It is possible to observe how the current achieve 28 A of peak quickly during the start-up.

8.8.2 *Overload Detection*

A microcontroller/DSP algorithm can monitor the power flow delivered to the AC machine in order to prevent overload without the need to measure the temperature of the machine directly. There are some sophisticated algorithms which can provide an estimation of the AC machine temperature according to the power flow delivered to the machine. Here it is only considered the protection to avoid the overheating using a simple algorithm for overload detection.

Most of the energy dissipated in a machine in the form of heat is due to the losses in the windings of the stator. This energy loss, which is not converted into mechanical energy, can be calculated according to (8.42), where R_L is the stator winding effective resistance, and I is the RMS current flowing through the winding.

$$E = I^2 R_L \cdot \text{Time} \tag{8.42}$$

The AC machine equilibrium temperature is reached when the energy remains balanced between the energy delivered to the machine and the energy converted to mechanics plus the energy losses as heat. The rated continuous machine current I_{cont} is defined as the current where the machine can be running without its temperature exceeding its maximum nominal temperature. That is, the machine is capable of dissipating the energy of the previous Eq. (8.42). In transients, the machine can tolerate some excess energy from the previous continuous limit. Equation (8.43) represents the energy during the transient, which is nothing more than the injected energy minus the nominal continuous energy I_{cont}.

$$E_{trans} = I^2 R_L \cdot \text{Time} - I_{cont}^2 R_L \cdot \text{Time} \tag{8.43}$$

If the injected current exceeds the continuous current I_{cont}, there will be an excess of energy that can be monitored to evaluate whether the machine is in overload or not. In an induction machine, it is possible to find the following parameters provided by the manufacturer:

- Nominal current that can be applied continuously (A_{RMS})
- SF (Service Factor)
- Current transient peak (A_{RMS})
- Maximum duration of the current transient peak.

Since the resistive term R_L is on both sides of (8.43), it is possible to avoid this term since it is not necessary to calculate the energy. In this way, both terms will have units of A^{2}s. For example, it is possible to calculate the limit value for the following conditions which, if exceeded, the machine should be stopped:

- Continuous Current Limit: 8.0 A_{RMS}
- Overload Current Limit: 9 A_{RMS}
- Max. Time Duration of this overload current: 300 s
- Sample Time: 0.04 s

$$\text{OL\_Limit} = \frac{\left(9^2 - 8^2\right) \cdot 300}{0.04} = 127{,}500\ \text{A}^2\text{s} \tag{8.44}$$

If the phase current is higher than 9 A_{RMS}, the above limit will be reached in less than 300 s. On the contrary, if the current is less than 8 A_{RMS}, there will be no accumulation, and therefore, it will not reach the limit. For example, the time needed to reach the previous limit with 10 A_{RMS} can be calculated according to (8.45).

$$T = \left(\frac{127,500}{10^2 - 8^2}\right) \cdot 0.04 = 141.66\,\text{s} \tag{8.45}$$

Figure 8.83 shows the algorithm that allows detecting overload in the AC machine according to the previous description. As can be observed, there is a variable that multiplies the current called "Duty" that varies between 0 and 1 and allows to extend the operation of the algorithm in cases where the machine operates with ON/OFF

Fig. 8.83 Flowchart of the overload detection algorithm

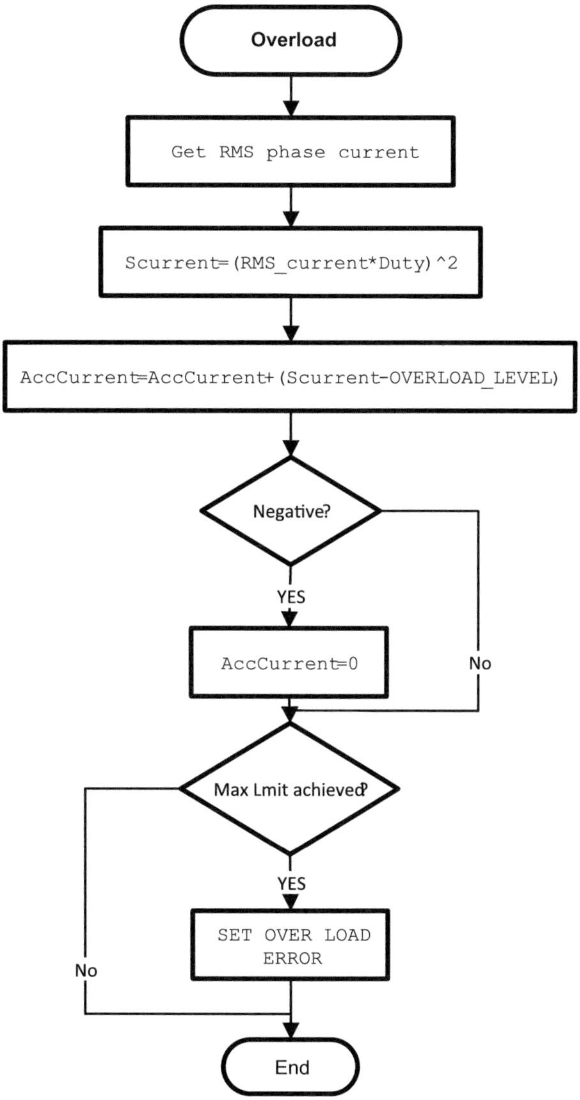

cycles continuously as in the case of a washing machine. A value of one means that the machine has an infinity ON time. If the square of the current multiplied by the duty exceeds the OVERLOAD_LEVEL limit, it will begin to accumulate if it is positive so that if it exceeds a predetermined limit, the overload fault error will be set.

8.8.3 Overheating Detection

As commented before, by knowing the stator winding material and the nominal resistance at a known temperature, the temperature of the machine can be estimated according to (8.46). This equation is a linear approximation of the resistance increment with the temperature.

$$R_1 = R_0\left(\frac{|K_m| + T_1}{|K_m| + T_0}\right) \tag{8.46}$$

where $|K_m|$ is the absolute zero value of the winding material, i.e., at what temperature the material resistance is zero. For copper, $K_m = -234.5$ °C, and for aluminum, it is −236 °C.

The resistance in a three-phase AC machine can be measured by injecting a DC current through the PWM and setting a suitable modulation index for the desired command current. Since the applied voltage is known and assuming that the inverter has a current measurement, the resistance can be calculated according to Ohm's law. To obtain a more precise measurement without the effect of the voltage drop in the semiconductors or the deadtime, two measurements can be made at different currents. Then, the resistance can be calculated by making the voltage difference divided by the difference in current, according to the (8.47). However, if a voltage measurement is available with a single injection of current will be sufficient. For the more precise case, the error between estimated temperature and actual temperature measured with different thermocouples can reach a maximum ±10 °C error.

$$R_s = \frac{V_1 - V_2}{I_1 - I_2} \tag{8.47}$$

By using (8.46) is possible to deduce the temperature T_1 as (8.48) to calculate the temperature of the winding with the measurement of the resistance.

$$T_1 = R_s\left(\frac{|K_m| + T_0}{R_s}\right) - |K_m| \tag{8.48}$$

Figure 8.84 shows the experimental resistance measurement using two current pulses of 4 and 2 A of 1.6 s duration each. As previously mentioned, the two pulses counteract the voltage drop in the devices and the deadtime. Only the current of one

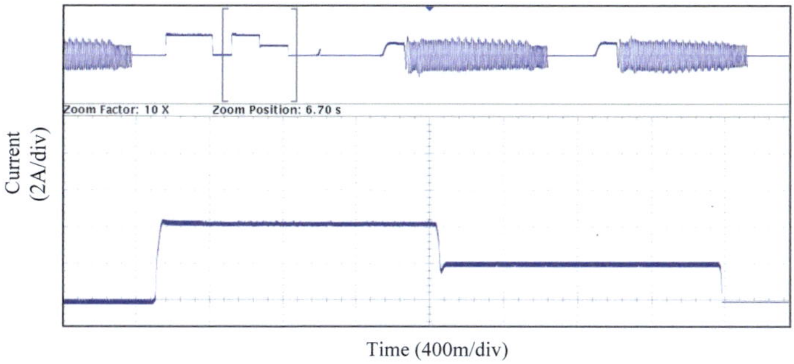

Fig. 8.84 Resistance measurement before AC machine start-up with two current pulses

phase is shown, while the method where the voltage is measured with a single current pulse is presented Fig. 8.85. For best accuracy, it is possible to measure the three resistances of the three phases where the average value can be used to estimate the temperature. In this way, Eq. (8.47) is derived in (8.49):

$$R_s = \frac{1}{3}\left(\frac{V_{A1} - V_{A2}}{I_{A1} - I_{A2}} + \frac{V_{B1} - V_{B2}}{I_{B1} - I_{B2}} + \frac{V_{C1} - V_{C2}}{I_{C1} - I_{C2}}\right) \tag{8.49}$$

Figure 8.85 shows the experimental AC machine start-up where a resistance measurement was previously made for 600 ms by injecting a DC current of −10 A in phase A while 4 and 6 A for phase B and C, respectively. To avoid unnecessary current surges, the DC injection current increased as a ramp until it reaches the limit of 10 A. For high-power machines, the stator resistance is usually less than 1 Ω. Therefore, care must be taken with the injected DC voltage since high currents are

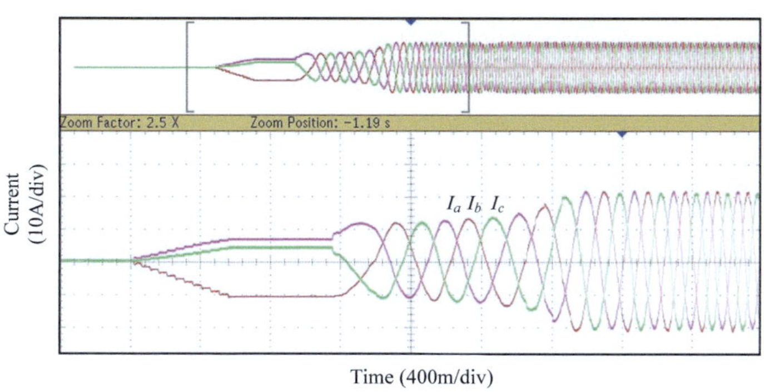

Fig. 8.85 Resistance measurement before AC machine start-up with a single current pulse

reached easily with low voltage. In any case, the injected current must not exceed the rated current of the machine. Once the measurement of the resistance is finished, the temperature can be calculated according to the (8.47), and if it is inside the limits, the normal start-up is performed.

It is important to mention that the measurement of the resistance can only be performed with machine stopped, limiting its application. The time it takes to perform the measurement depends on the stabilization time necessary to perform a correct measurement and the number of samples. On the other hand, there are more sophisticated algorithms to estimate the temperature based on the estimation of the resistance while the machine is running. In any case, having the stator resistance and current temperature roughly can serve to compensate the control to improve its performance, especially in high-performance controls such as the vector control.

8.8.4 Open-Phase Detection

The open-phase detection can be carried out in a simple way if a measurement of current is available in each of the phases. If the sampling period is higher than at least ten times of the current frequency, detection will be faster and safer. To avoid negative values of the negative half periods, the square of the current per phase can be calculated in the microcontroller/DSP algorithm. If any of the currents has a small value and its value is different compared to the average of the three currents during a prudent time, an alarm can be set as shown in the algorithm of Fig. 8.86. The algorithm checks each of the phases named I_X by calculating the error with respect to the arithmetic mean of the three currents. If this exceeds I_LIMIT1, it means that there is a deviation. Also, it is checked if current I_X has a value below the limit I_LIMIT2. If both conditions are met, the counter is incremented, and if it exceeds the limit set by TimeOut, the error is set.

An experimental test performed on an induction machine running at high speed where phases A and B are measured, and one of them is opened as represented in Fig. 8.87. As can be observed, when the open phase is detected, the PWM control is stopped after 40 ms approximately.

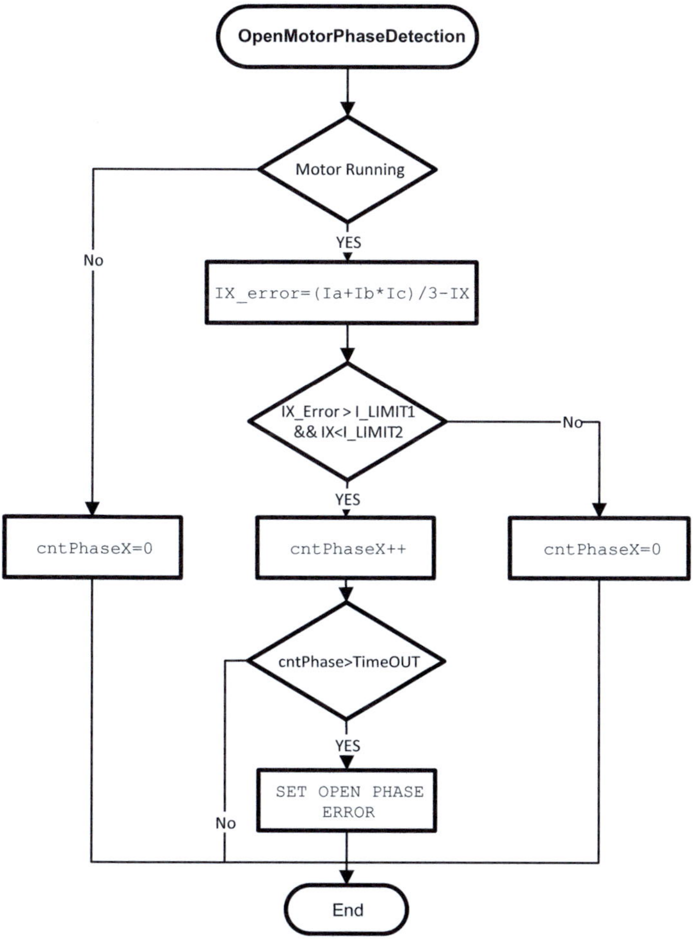

Fig. 8.86 Flowchart of the open-phase detection algorithm

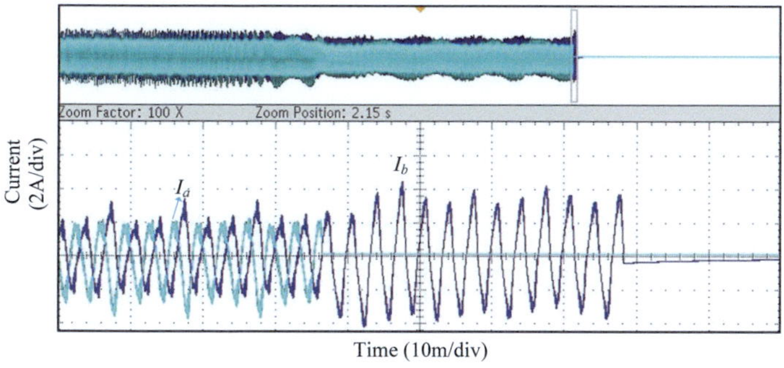

Fig. 8.87 Phase current A and B, where phase A is opened. The algorithm detects the failure in 40 ms

8.9 VSI Power Plant Model

In the previous sections was shown a series of simulations to treat different important topics of VSI . For AC machine control purposes is possible to complete the power plant model seen with new necessary features such as current and voltage phase signal conditioning measurement, DC-link voltage measurement, and the pre-charge relay circuit control. The VSI power plant model implemented in PSIM and packed in the SimCoupler module is depicted in Fig. 8.88. As discussed in Chap. 1, the SimCoupler module allows an interface between PSIM and MATLAB/Simulink® environments. As can be seen, the inputs of the SimCoupler module are the PWM signals for each power switches devices, and the control signals of the pre-charge, positive and negative DC-link relays. These signals take the logical "0" and "1" states so that within the model, the signal is conditioning to drive the relays and the IGBT's.

On the other hand, the outputs are the three-phase AC machine connection, and the voltage and current sensing. The model includes the signal conditioning of the phase current I_a and I_b sensing, the measurement of the DC-link voltage, and the measurement of the voltages of each of the phases V_a, V_b, and V_c.

In Fig. 8.89 is shown the PSIM model detail of previous Fig. 8.88. Here, it is possible to appreciate two shunts to measure the phase current and the DC-link voltage

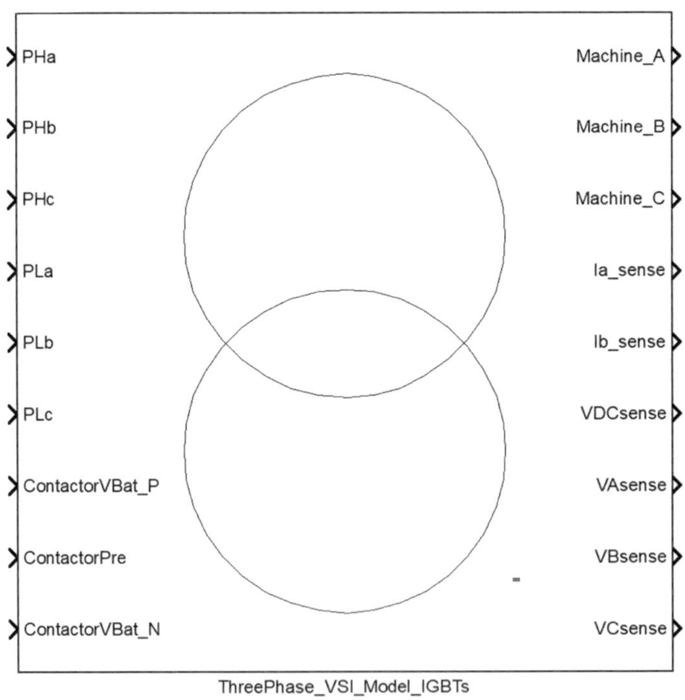

Fig. 8.88 VSI model inside of a SimCoupler block

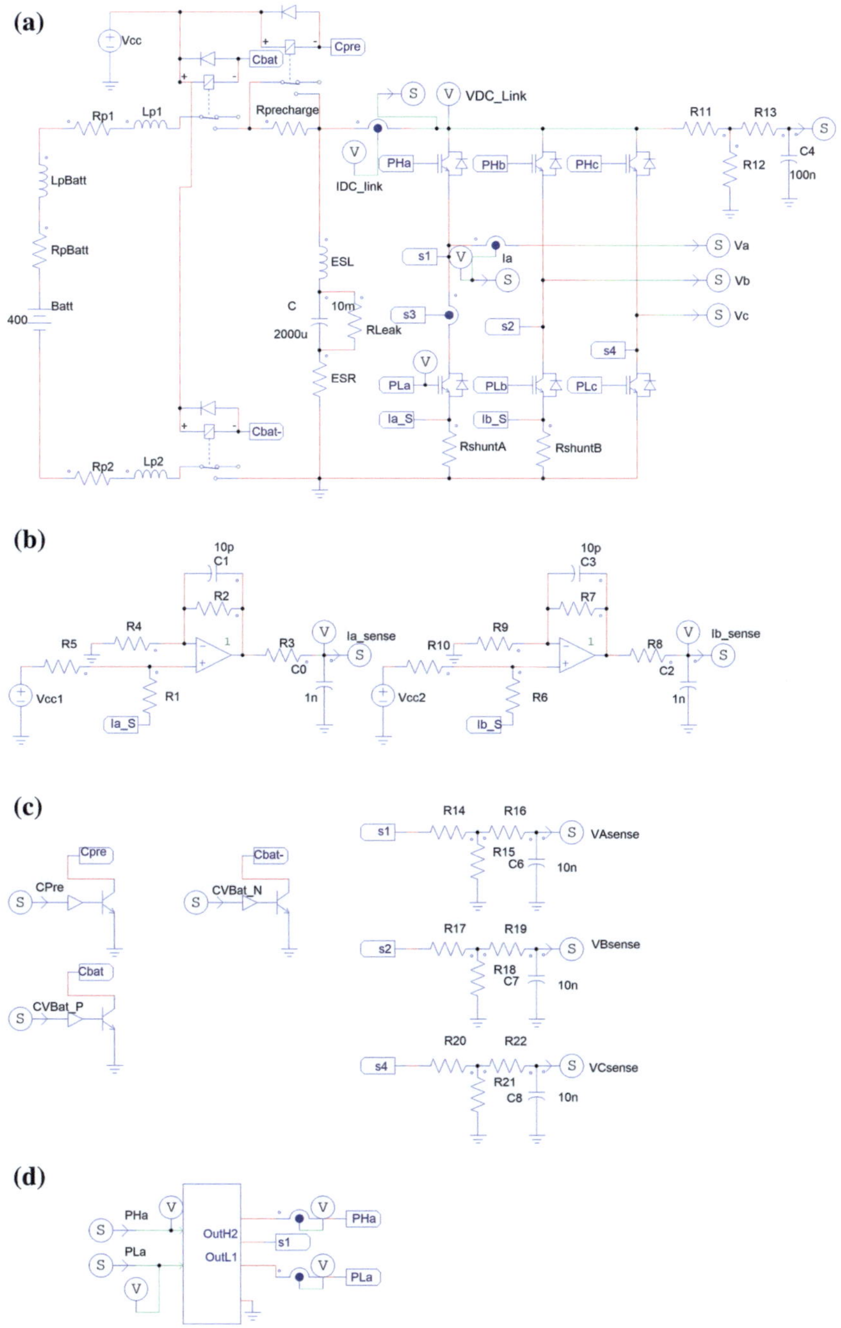

Fig. 8.89 Detail of VSI with signal conditioning for current and voltage phases, DC-link voltage, pre-charge circuit, and DC-link relays. **a** VSI power plant and DC-link signal conditioning. **b** Signal conditioning for phase current. **c** Signal conditioning for voltage phases with relay drivers. **d** Phase A IGBT gate driver

measurement. Moreover, the battery and the different relays are also observed. The current measurement signal conditioning of Fig. 8.88b is made by a non-inverting operational amplifier that can measure positive and negative current thanks to the preset offset. Its power supply is V_{cc1} and V_{cc2} which can be 3.3 or 5 volts depending on the microcontroller/DSP A/D converter.

By performing the analysis of the signal conditioning circuit of Fig. 8.89b, Eqs. (8.50), (8.51), and (8.52) can be obtained:

$$G = 1 + \frac{R_2}{R_4} \tag{8.50}$$

where G is the gain of the non-inverting amplifier.

$$\begin{aligned} V_{BIAS} &= \frac{V_{CC}}{\left(\frac{1}{R_1} + \frac{1}{R_5}\right) R_5} \cdot G \\ V_{SIGNAL} &= \frac{I_a R_{shuntA}}{\left(\frac{1}{R_1} + \frac{1}{R_5}\right) R_1} \cdot G \end{aligned} \tag{8.51}$$

V_{BIAS} is the imposed offset voltage to measure negative currents, and V_{SIGNAL} is the voltage that varies with the current flowing through the shunt resistance R_{shuntA}, this case, phase A. Finally, V_{OUT} is the sum of both voltages:

$$V_{OUT} = V_{BIAS} + V_{SIGNAL} \tag{8.52}$$

The DC-link voltage conditioning and the phase voltages are simply made by a voltage divider and a low-pass filter as depicted in Fig. 8.89c. It is also possible to observe the drivers for relay control such as pre-charge relay, positive and negative DC-link relays. In Fig. 8.89d is illustrated the gate driver of phase A. The accurate model of the gate driver allows to have realistic simulations since the dynamic behaviors of the power switches can be performed in a precise way. The non-ideal voltage supply and the stray inductances and resistances are taken into account in the model.

References

Balogh L (2013) Design and application guide for high speed MOSFET gate drive circuits

Barkhordarian V (1997) Power MOSFET basics. International Rectifier, Technical Paper

Beig AR, Ranganathan VT (2006) A novel CSI fed induction motor drive. IEEE Trans Power Electron 21(4):1073–1082

Bhagwat PM, Stefanovic VR (1983) Generalized structure of a multilevel PWM inverter. IEEE Trans Ind Appl IA-19:1057–1069

Bierhoff MH, Fuchs FW (2004) Semiconductor losses in voltage source and current source IGBT converters based on analytical derivation 4:2836–2842

Blasko V et al (1998) Sampling of discontinuous voltage and current signals in electric drives-a system approach. IEEE Trans Ind Appl 34(5):1123–1130

Boglietti A, Carpaneto E (1999) Induction motor high frequency model. In: Proceedings of 34th IEEE industry applications society conference (IAS'99), October 3–7, Phoenix, AR
Chen L, Peng FZ (2008) Dead-time elimination for voltage source inverters. IEEE Trans Power Electron 23(2):574–580
Deng Y, Liang Z, Xia P et al (2019) J Electr Eng Technol 14:219. https://doi.org/10.1007/s42835-018-00023-7
Grandi G, Casadei D, Massarini A (1997) High frequency lumped-parameter model for AC motor windings. In: Proceedings of 7th European conference on power electronics and applications (EPE'97), vol 2, pp 578–583, Trondheim, Belgium
Khodaparastan M, Mohamed A (2017) Supercapacitors for electric rail transit system. In: 6th International conference on renewable energy research and application, vol 5, pp 1–6
Kolar JW, Ertl H, Zach FC (1991) Influence of the modulation method on the conduction and switching losses of a PWM converter system. IEEE Trans Ind Appl 27:1063–1075
Leggate D, Kerkman RJ (1997) Pulse-based dead-time compensator for PWM voltage inverters. IEEE Trans Ind Electron 44(2):191–197
Munoz-Garcia A, Lipo TA (1998) On-line dead-time compensation technique for open-loop PWM-VSI drives. In: Proceeding of IEEE Applications Power Electronics Conference, Feb. 1998, pp 95–100
Reusch D, Strydom J (2013) Understanding the effect of PCB layout on circuit performance in a high frequency gallium nitride based point of load converter. In: APEC 2013, IEEE transactions on power electronics 2014
Sepe RB et al (1994) Implementation of discrete-time field-oriented current control. IEEE Ind Appl 30(3):723–727
Sukegawa T, Kamiyama K, Mizuno K, Matsui T, Okuyama T (1991) Fully digital, vector-controlled PWM VSI-Fed ac drives with an inverter dead-time compensation strategy. IEEE Trans Ind Appl 21(3):552–559
von Jouanne A, Enjeti P (1997) Design considerations for an inverter output filter to mitigate the effects of long motor leads on ASD applications. IEEE Trans Ind Appl 33(5):1138–1145
Vujacic M, Hammami M, Srndovic M, Grandi G (2018) Analysis of dc-link voltage switching ripple in three-phase PWM inverters. Energies 11(2):471
Yamamoto Y et al (1992) Digital current control method of induction motor using synchronous current detection with PWM signals. Trans IEE-Japan 112-D(7):613–622

Chapter 9
Space Vector Modulation

9.1 Space Vector Modulation

9.1.1 Introduction

As was seen in Chap. 8, the VSI can generate a three-phase variable amplitude voltage, frequency, and phase angle thanks to the correct pattern of the activations of the power switches' devices. The most basic three-phase voltage generation with a variable-frequency can be performed by six-step modulation. The six-step uses a sequence of six switching patterns that generate a complete three-phase voltage cycle at the output of the VSI . The eight possible states of the VSI , six are active states that connect the midpoint to zero or DC-link voltage for each of the inverter legs. The two remaining states are states that connect either all the high side devices to DC-link or all low side devices to zero. The frequency can be controlled, while the voltage amplitude is fixed in the six-step modulation since not PWM signals are used. In a 360° cycle, for every 60°, there is a change in the states to be able to generate a three-phase voltage output. In Fig. 9.1, it is possible to observe the VSI schematic with MOSFET's devices which are fed by the DC-link voltage V_{DC}. The output of the inverter is connected to a three-phase AC machine. The on state is when the MOSFET is turned on and designated with the logical value 1, i.e., $a = 1$ or $a' = 1$. On the contrary, the off state is when the MOSFET remains open (turn-off), being designated with the logical value 0 ($a = 0$ or $a' = 0$).

The six active states can be assigned spatially as shown in Fig. 9.2 where they are at a distance of 60°. The complete period of the output wave is generated with an imaginary voltage vector which should go through all the vertices until a complete 360° turn is made. The speed with which a full turn is executed defines the frequency, while the voltage cannot be varied as discussed above.

Assuming that a three-phase AC machine in star connection is connected to the VSI as in Fig. 9.1, and taking into account what is explained in Chap. 8, it is possible to carry out Table 9.1 where the combinations of the active states with the line-to-line and phase voltage are shown.

© Springer Nature Switzerland AG 2020

R. Molina Llorente, *Practical Control of Electric Machines*,
Advances in Industrial Control, https://doi.org/10.1007/978-3-030-34758-1_9

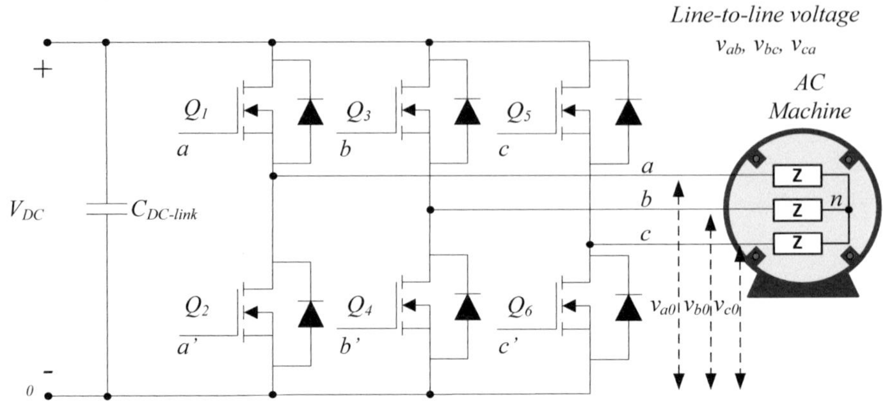

Fig. 9.1 Three-phase VSI implemented with six MOSFETs. Line-to-line voltages and voltages reference to ground are shown

Fig. 9.2 Spatial vector position according to the switching combination in six-step modulation

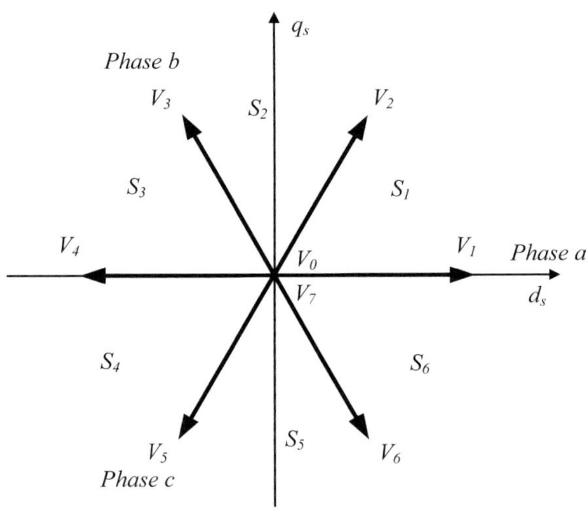

If Table 9.1 is represented graphically, Fig. 9.3 is obtained. For the sake of simplicity, only the control signals of the high side devices and a single voltage relative to the neutral point v_{an} are shown. The control signals of the low side devices are their corresponding complementary signal with deadtime delay added.

The maximum utilization of the DC-link voltage, as it is a square waveform with a higher RMS value, is possible with the six-step modulation, as can be seen in Fig. 9.3 in the last graph. That is why it is usually used as a reference for other types of modulation such as PWM sinusoidal modulation, giving a quantitative value, sometimes called the modulation index of the percentage of the six-step modulation. It is essential to avoid the confusion with the modulation index m which is defined as the amplitude of the modulating signal V_p, normalized concerning the maximum

Table 9.1 Switches active states with output voltages, line-to-line, and phase voltage

c	b	a	v_{an}	v_{bn}	v_{cn}	v_{ab}	v_{bc}	v_{ca}	Vector
0	0	0	0	0	0	0	0	0	V_0
0	0	1	$2V_{DC}/3$	$-V_{DC}/3$	$-V_{DC}/3$	V_{DC}	0	$-V_{DC}$	V_1
0	1	1	$V_{DC}/3$	$V_{DC}/3$	$-2V_{DC}/3$	0	V_{DC}	$-V_{DC}$	V_2
0	1	0	$-V_{DC}/3$	$2V_{DC}/3$	$-V_{DC}/3$	$-V_{DC}$	V_{DC}	0	V_3
1	1	0	$-2V_{DC}/3$	$V_{DC}/3$	$V_{DC}/3$	$-V_{DC}$	0	V_{DC}	V_4
1	0	0	$-V_{DC}/3$	$-V_{DC}/3$	$2V_{DC}/3$	0	$-V_{DC}$	V_{DC}	V_5
1	0	1	$V_{DC}/3$	$-2V_{DC}/3$	$V_{DC}/3$	V_{DC}	$-V_{DC}$	0	V_6
1	1	1	0	0	0	0	0	0	V_7

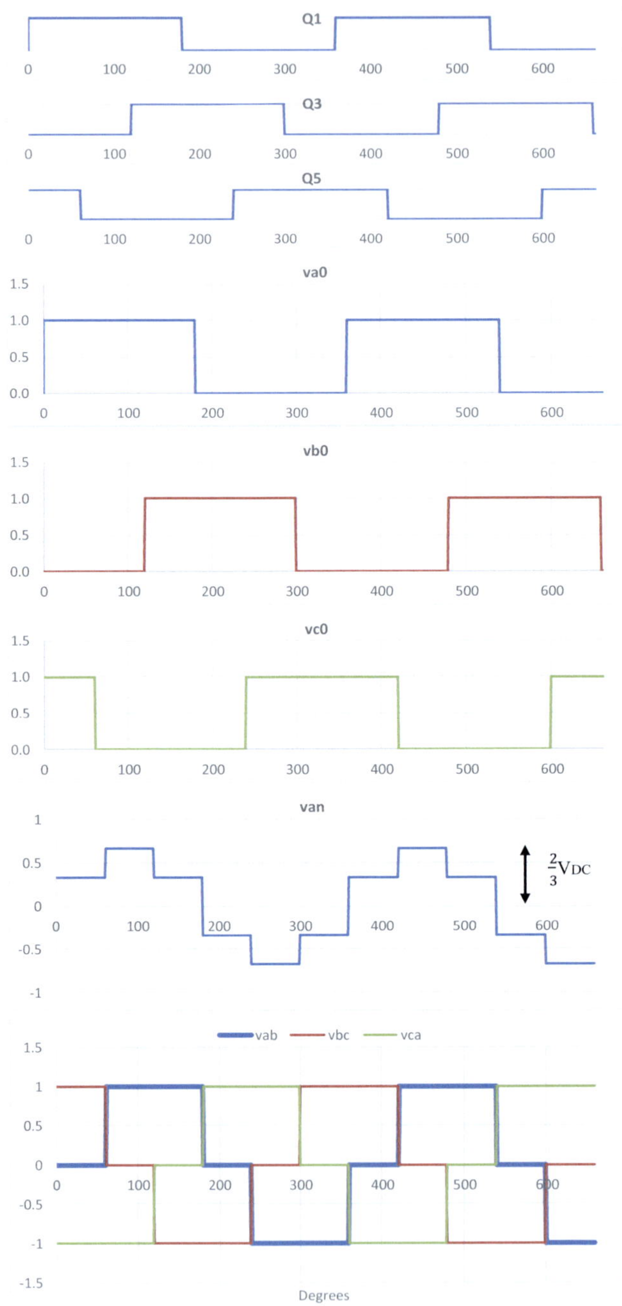

Fig. 9.3 Switching signals Q_1, Q_3, and Q_5 of the high side device, phase voltage referenced to ground v_{a0}, v_{b0}, and v_{c0}, phase voltage v_{an}, and line-to-line voltage v_{ab}, v_{bc}, and v_{ca} in six-step modulation

amplitude of the carrier V_{tri} according to Eq. (9.1). Therefore, the modulation index m' is defined as the magnitude of the reference voltage vector V_{ref} of the modulating signal divided by the maximum amplitude of the six-step modulation as indicated in Eq. (9.2).

$$m = \frac{V_p}{V_{tri}} \tag{9.1}$$

$$m' = \frac{\hat{V}_{ref}}{\hat{V}_{sixstep}} \tag{9.2}$$

The maximum amplitude of the output voltage (peak value) per phase in the six-step modulation is given by the first harmonic (9.3):

$$\hat{V}_{sixstep} = \frac{2V_{DC}}{\pi} \tag{9.3}$$

Moreover, the line-to-line voltage is

$$\hat{V}_{sixstep\_ab} = \sqrt{3}\frac{2V_{DC}}{\pi} \tag{9.4}$$

Based on (9.4), it is possible to deduce the line-to-line RMS voltage of the VSI output, as a function of the input voltage V_i, assuming that the supply source comes from a full-bridge rectifier such as

$$V_{sixstep\_abRMS} = \frac{\sqrt{3}}{\sqrt{2}}\frac{2V_{DC}}{\pi} = 0.78V_{DC} = 0.78\sqrt{2}V_{iRMS} = 1.1V_{iRMS} \tag{9.5}$$

It can be seen from (9.3) that the RMS output voltage of the inverter is 10% higher than the input RMS voltage if six-step modulation is used. However, it is not the case with the space vector modulation (SVM) as will be shown in the next section. On the other hand, the main disadvantages of the six-step modulation are that it is not possible to vary the amplitude of the voltage, and it generates an important amount of harmonics. These aspects have a negative influence on the control of the electric machine causing less performance and an electromagnetic torque ripple, affecting the life of the machine. Even so, due to its simplicity, it can be a valid strategy for economic control systems of AC machines with modest control features.

9.1.2 Space Vector Modulation

Sinusoidal PWM modulation (SPWM) improves the main disadvantages of the six-step modulation, that is, allows amplitude modulation, and reduces the amount of

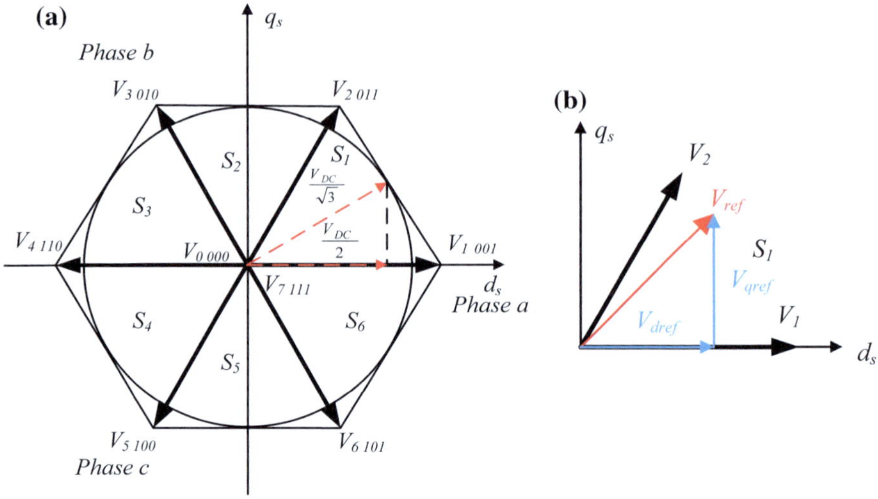

Fig. 9.4 **a** Voltage space vector formed by the three-phase VSI showing the switching combination of the high side devices and sectors. It is also represented the inscribed circle and its radius value. **b** Representation of the reference voltage V_{ref} with *dqs* components

harmonics. However, the maximum output voltage is 78.55% of the fundamental wave of the six-step modulation. The SVM is the substitute for classical SPWM in three-phase inverters since it uses the DC-link voltage more efficiently. The maximum output voltage is 90.7% of the fundamental wave of the six-step modulation as it will be seen below.

SVM is a modulation method based on the reconstruction of the reference voltage vector V_{ref} from its two adjacent vector components with a V_0 and/or a V_7 vector in a balanced three-phase VSI . From the eight possible states, the six states that activate a high side device with a low side device of another inverter leg are taken, and a hexagon with six sectors is formed, as illustrated in Fig. 9.4a. The representation of the reference voltage vector V_{ref} can be with the orthogonal components such as α, β, or also *dqs* axis transformation as illustrated in Fig. 9.4b. Figure 9.4a represents the digital combination for each of the states where the on state is when the MOSFET is closed (activated) and is designated as in the case of six-step modulation, i.e., with $a = 1$ or $a' = 1$. The states that are not possible which are not considered are when the high side device and low side device of the same leg are activated at the same time. It is possible to observe that for the voltage vector V_{ref} rotates on the hexagon to generate a sinusoidal voltage in the output of the VSI , the sectors have to advance in a correlative way where the advance speed will fix the frequency, while the amplitude of the voltage will be fixed by the amplitude of the reference voltage vector V_{ref}, that is, for its components. The combination of states only varies one device between sectors as can be seen. The two remaining states, vector V_7 $(a, b, c) = (1, 1, 1)$ where all the high side devices are activated, and vector V_0 $(a', b', c') = (1, 1, 1)$ where all the low side are activated, serve to complete the PWM cycle.

Moreover, different combinations for vector V_7 and V_0 can be employed to perform different SVM techniques (Chung et al. 1998; Zhou and Wang 2002; Varma and Narayanan 2015) as it will see in this section.

The maximum voltage amplitude V_{ref} that can be obtained with the SVM in the linear zone corresponds to the circle inscribed in the hexagon. The radius of the inscribed circle is $V_{DC}/\sqrt{3}$. Therefore, the modulation index concerning the modulation six-step m' can be derived easily according to Eq. (9.6) as

$$m' = \frac{\hat{V}_{ref}}{\hat{V}_{sixstep}} = \frac{V_{DC}\big/\sqrt{3}}{2V_{DC}\big/\pi} = 0.907 \tag{9.6}$$

That is, the SVM in the linear zone can reach 90.7% of the fundamental wave of the six-step modulation. However, by performing a similar analysis, the classical sinusoidal modulation is only able to reach 78.55% as commented before. Therefore, with the SVM, efficient use of the DC-link voltage is demonstrated, even in the linear zone, as mentioned above.

In the same way as Eq. (9.5), it is possible to deduce the line-to-line RMS voltage in the inverter output, according to the SVM, as a function of the input voltage V_i:

$$V_{SV\_abRMS} = \frac{\sqrt{3}}{\sqrt{2}}\frac{V_{DC}}{\sqrt{3}} = \frac{1}{\sqrt{2}}V_{DC} = \frac{1}{\sqrt{2}}\sqrt{2}V_{iRMS} = V_{iRMS} \tag{9.7}$$

In this case, the RMS output voltage of the inverter is identical to the RMS input voltage.

On the other hand, according to Fig. 9.1, the output voltages referred to ground are v_{a0}, v_{b0}, and v_{c0} and depend on the state of the devices, that is, if $a = 1$ and $a' = 0$, then $v_{a0} = V_{DC}$. The instantaneous mean value can be written as $(v_{a0} + v_{b0} + v_{c0})/3$ or what is the same as $V_{DC}(a + b + c)/3$. If instead of referring the voltage to ground, it is referred to the instantaneous mean value, the Eq. (9.8) is obtained:

$$\begin{bmatrix} v_a \\ v_b \\ v_c \end{bmatrix} = \frac{V_{DC}}{3}\begin{bmatrix} 2 & -1 & -1 \\ -1 & 2 & -1 \\ -1 & -1 & 2 \end{bmatrix}\begin{bmatrix} a \\ b \\ c \end{bmatrix} \tag{9.8}$$

As described above, the reference voltage vector V_{ref} can be represented using the *dqs* components (stationary reference frame). Thus, the Clarke transformation can be performed in (9.8) to obtain the V_{ds}, V_{qs} vectors referred to a fixed system in phase with v_a. Table 9.2 shows a summary of the values obtained according to the possible states of the devices. These are the instantaneous values of the phase output voltages, the line-to-line voltages, and the instantaneous values referred to the system dq_s, V_{ds}, and V_{qs}. The corresponding voltage vector name is shown in the last column.

For any reference voltage vector V_{ref} in the *dqs* axes, it is convenient to obtain it employing a linear combination of its nearest vectors to reduce the switching of the

Table 9.2 Switches states, output voltages, and Clarke transformation

c	b	a	v_{an}	v_{bn}	v_{cn}	v_{ab}	v_{bc}	v_{ca}	V_{ds}	V_{qs}	Vector
0	0	0	0	0	0	0	0	0	0	0	V_0
0	0	1	$2V_{DC}/3$	$-V_{DC}/3$	$-V_{DC}/3$	V_{DC}	0	$-V_{DC}$	$2V_{DC}/3$	0	V_1
0	1	1	$V_{DC}/3$	$V_{DC}/3$	$-2V_{DC}/3$	0	V_{DC}	$-V_{DC}$	$V_{DC}/3$	$V_{DC}/\sqrt{3}$	V_2
0	1	0	$-V_{DC}/3$	$2\ V_{DC}/3$	$-V_{DC}/3$	$-V_{DC}$	V_{DC}	0	$-V_{DC}/3$	$V_{DC}/\sqrt{3}$	V_3
1	1	0	$-2V_{DC}/3$	$V_{DC}/3$	$V_{DC}/3$	$-V_{DC}$	0	V_{DC}	$-2V_{DC}/3$	0	V_4
1	0	0	$-V_{DC}/3$	$-V_{DC}/3$	$2V_{DC}/3$	0	$-V_{DC}$	V_{DC}	$-V_{DC}/3$	$-V_{DC}/\sqrt{3}$	V_5
1	0	1	$V_{DC}/3$	$-2V_{DC}/3$	$V_{DC}/3$	V_{DC}	$-V_{DC}$	0	$V_{DC}/3$	$-V_{DC}/\sqrt{3}$	V_6
1	1	1	0	0	0	0	0	0	0	0	V_7

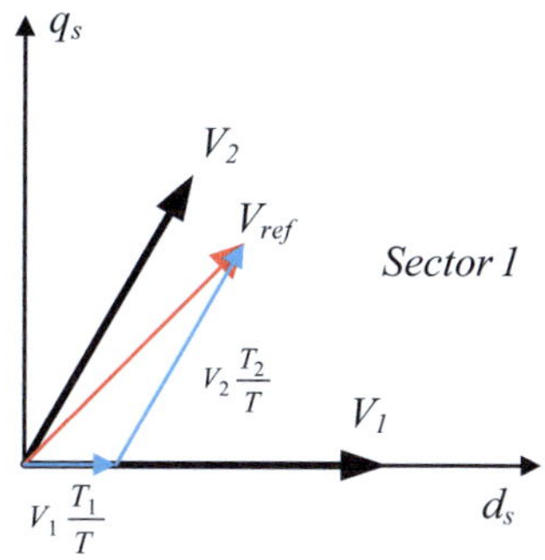

Fig. 9.5 Construction of V_{ref} vector in sector 1 composed by the nearest *dqs* components

power devices (Zhou and Wang 2002). For example, if the reference voltage vector V_{ref} in Fig. 9.5 is in sector 1, in a period *T*, the vector V_{ref} can be written as a function of its adjacent vectors as expressed in Eq. (9.9):

$$\begin{aligned} \overline{V}_{ref} &= \frac{T_1}{T}\overline{V}_1 + \frac{T_2}{T}\overline{V}_2 + \frac{T_0}{T}\overline{V}_0 + \frac{T_7}{T}\overline{V}_7 \\ T_1 + T_2 + T_0 + T_7 &= T \end{aligned} \tag{9.9}$$

That is, the reference voltage vector V_{ref} of Fig. 9.5 is instantaneously achieved by combining the state of the devices in sector 1, with the duration of times T_1, T_2, T_0, and T_7 where their respective vectors are applied. The sum of all duration times T_1, T_2, T_0, and T_7 is equal to period *T*. Period *T* is the PWM period as expected and usually coincides with the sampling period. According to Zhou and Wang (2002), the duration time T_1 and T_2 can be calculated according to the Eq. (9.10).

$$\left.\begin{aligned} T_1 &= \frac{\sqrt{3}}{2}m'T\cos\left(\omega t + \frac{\pi}{6}\right) \\ T_2 &= \frac{\sqrt{3}}{2}m'T\cos\left(\omega t + \frac{3\pi}{2}\right) \end{aligned}\right\rangle \rightarrow 0 \le \omega t \le \frac{\pi}{3} \tag{9.10}$$
$$T_0 + T_7 = T - (T_1 + T_2)$$

It is possible to observe that the above equation is only valid when the reference voltage vector V_{ref} is between 0 and 60° and in the linear zone. Table 9.3 shows a summary of the calculation of the corresponding duration times T_1 and T_2 for each of the sectors.

In previous Table 9.3, it has been seen that for T_1 and T_2 calculation, it is necessary to know in which sector the vector V_{ref} is located. This can be determined with simple geometric considerations, that is, a vector is in a particular sector if its components satisfy the following conditions:

Table 9.3 Space vector modulation times calculation for every sector

Sector I $0 \leq \omega \cdot t \leq \pi/3$	Sector II $\pi/3 \leq \omega \cdot t \leq 2\pi/3$	Sector III $2\pi/3 \leq \omega \cdot t \leq \pi$
$T_1 = \frac{\sqrt{3}}{2} m' T \cos\left(\omega \cdot t + \frac{\pi}{6}\right)$	$T_2 = \frac{\sqrt{3}}{2} m' T \cos\left(\omega \cdot t + \frac{11\pi}{6}\right)$	$T_3 = \frac{\sqrt{3}}{2} m' T \cos\left(\omega \cdot t + \frac{3\pi}{2}\right)$
$T_2 = \frac{\sqrt{3}}{2} m' T \cos\left(\omega \cdot t + \frac{3\pi}{2}\right)$	$T_3 = \frac{\sqrt{3}}{2} m' T \cos\left(\omega \cdot t + \frac{7\pi}{6}\right)$	$T_4 = \frac{\sqrt{3}}{2} m' T \cos\left(\omega \cdot t + \frac{5\pi}{6}\right)$
$T_0 + T_7 = T - (T_1 + T_2)$	$T_0 + T_7 = T - (T_2 + T_3)$	$T_0 + T_7 = T - (T_3 + T_4)$
Sector IV $\pi \leq \omega \cdot t \leq 4\pi/3$	Sector V $4\pi/3 \leq \omega \cdot t \leq 5\pi/3$	Sector VI $5\pi/3 \leq \omega \cdot t \leq 2\pi$
$T_4 = \frac{\sqrt{3}}{2} m' T \cos\left(\omega \cdot t + \frac{7\pi}{6}\right)$	$T_5 = \frac{\sqrt{3}}{2} m' T \cos\left(\omega \cdot t + \frac{5\pi}{6}\right)$	$T_6 = \frac{\sqrt{3}}{2} m' T \cos\left(\omega \cdot t + \frac{\pi}{2}\right)$
$T_5 = \frac{\sqrt{3}}{2} m' T \cos\left(\omega \cdot t + \frac{\pi}{2}\right)$	$T_6 = \frac{\sqrt{3}}{2} m' T \cos\left(\omega \cdot t + \frac{\pi}{6}\right)$	$T_1 = \frac{\sqrt{3}}{2} m' T \cos\left(\omega \cdot t + \frac{11\pi}{6}\right)$
$T_0 + T_7 = T - (T_4 + T_5)$	$T_0 + T_7 = T - (T_5 + T_6)$	$T_0 + T_7 = T - (T_1 + T_6)$

- S_0: $\sqrt{3}V_{ds} \geq V_{dq} \geq 0$
- S_1: $V_{dq} > \sqrt{3}V_{ds} \geq 0$, $V_{dq} \geq -\sqrt{3}V_{ds} > 0$
- S_2: $-\sqrt{3}V_{ds} > V_{dq} \geq 0$
- S_3: $0 > V_{dq} \geq \sqrt{3}V_{ds}$
- S_4: $0 \geq \sqrt{3}V_{ds} > V_{dq}$, $\sqrt{3}V_{ds} > 0 > V_{dq}$
- S_5: $0 > V_{dq} > -\sqrt{3}V_{ds}$.

Then, given the desired output vector in the *dqs* reference frame, to determine T_0, T_1, and T_2 are necessary to follow the following steps:

1. Obtain the *dqs* components of the vector V_{ref}.
2. Determine the sector in which the vector is located.
3. Calculate T_1 and T_2 according to Table 9.3.
4. Calculate $T_0 = T_7 = (T - T_1 - T_2)/2$ for symmetric switching for example.
5. Depending on the sector, assign the duty cycles to T_a, T_b, and T_c according to Table 9.4 corresponding to symmetric switching.

The symmetric switching is characterized in that both vectors V_0 and V_7 are applied during a single switching sequence. The vector V_0 is applied at the beginning and end of the switching period *T*, while the vector V_7 is applied in the middle of the switching cycle as shown in Fig. 9.6. In Fig. 9.6, it is possible to observe the "on" times in case of the high side devices for each of the *a*, *b*, and *c* phases. In total, in a switching period, there are seven differentiated portions.

The harmonic content of the output voltage generated for the symmetric switching is lower than asymmetric switching, as will be seen in Sect. 9.1.2.2. On the other hand, the non-linear zone of the SVM is during overmodulation, i.e., when the peak value of the modulation signal exceeds the carrier. In other words, in overmodulation, the reference voltage vector exceeds the inscribed circle where the limit of the magnitude

Table 9.4 T_a, T_b, and T_c patterns according to T_0, T_1, T_2, and sector for symmetrical switching

Sector	T_a, T_b, T_c
1	$T_a = T_0$ $T_b = T_0 + T_1$ $T_c = T_0 + T_1 + T_2$
2	$T_a = T_0 + T_2$ $T_b = T_0$ $T_c = T_0 + T_1 + T_2$
3	$T_a = T_0 + T_1 + T_2$ $T_b = T_0$ $T_c = T_0 + T_1$
4	$T_a = T_0 + T_1 + T_2$ $T_b = T_0 + T_2$ $T_c = T_0$
5	$T_a = T_0 + T_1$ $T_b = T_0 + T_1 + T_2$ $T_c = T_0$
6	$T_a = T_0$ $T_b = T_0 + T_1 + T_2$ $T_c = T_0 + T_2$

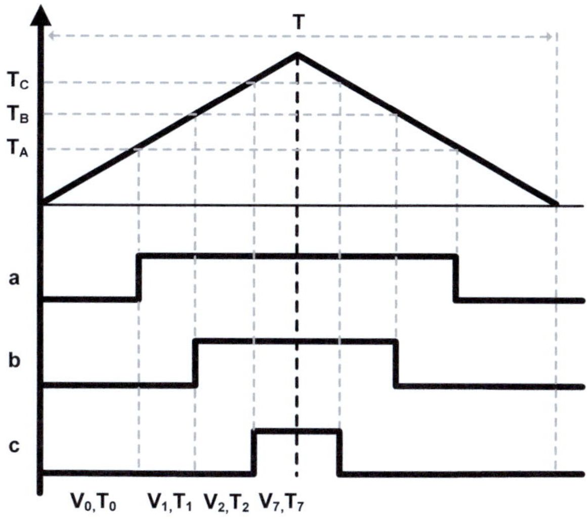

Fig. 9.6 Symmetrical PWM switching signals for every phase when reference vector V_{ref} is in sector 1

will be the hexagon. Thus, the three-phase voltage is not sinusoidal increasing the amount of harmonics, but it is capable of delivering a higher voltage at the output. Different overmodulation strategies exist in the literature as in Lee and Lee (1998).

9.1.2.1 Continuous SVM ($T_0 = T_7$)

The SVM continuous mode ensures a reduction of the harmonic content compared to the discontinuous mode. Also, it is characterized in that at each period cycle T at least one switching is generated. In the continuous mode, the times T_0 and T_7 of the vectors V_0 and V_7 must be calculated according to Eq. (9.11), which correspond the time remaining between the commutation period and the times T_1 and T_2.

$$T_0 = T_7 = \frac{(T - T_1 - T_2)}{2} \tag{9.11}$$

In other words, symmetric times are applied to T_0 and T_7 to generate the vector. According to Fig. 9.6, the switching pattern of a sector is represented, where it is possible to see that the vector is generated according to V_1, V_2, V_0, and V_7.

9.1.2.2 Discontinuous SVPWM ($T_0 \neq T_7$)

The SVPWM discontinuous mode reduces switching losses in the power devices. For an identical carrier frequency, the switching of the devices is reduced by a factor 2/3. The disadvantage of this method is the increase of the harmonic content in the output waveforms as commented before. The main characteristics are the following:

1. Only two phases are changed. No change during a 120° period T cycle.
2. There is only one null vector for the period of change.
3. There is a reduction in the number of commutations.

It is possible to consider two variants within discontinuous modulation. The first one is when T_0 takes the value zero, while T_7 takes the value of the Eq. (9.12). For this case, the modulation is also known as discontinuous PWM MAX (DPWMMAX) (Taniguchi et al. 1988) where the high side device of the half-bridge remains activated during two sectors, i.e., 120° (1/3 of the cycle period of the output voltage of the VSI). In Fig. 9.7a, the PWM signals for the high side devices are shown for a period T where the PWM signal of phase C does not change and remains at a high level. This high-level state is repeated for 120°, while the outputs of phases A and B switch with the duty cycle applied. If the high level means the activation of the device, it

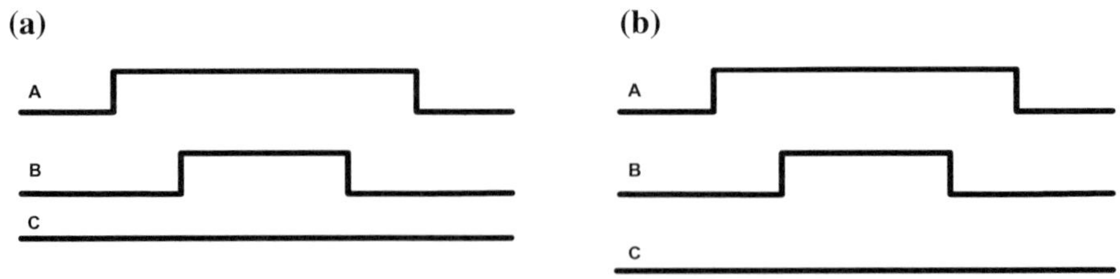

Fig. 9.7 Discontinuous SVM pulse pattern of voltage vectors in sector 1 for high side devices. **a** DPWMMAX. **b** DPWMMIN

is possible to observe that the high side device has higher conduction losses when being turned on during 120°, but, on the contrary, the low side device does not switch in that period, reducing its commutation losses.

The second one is known as DPWMMIN modulation (Kolar et al. 1990), where the device that remains activated for 1/3 of the cycle period of the output voltage is the low side device. The time T_7 takes the value zero, while T_0 takes the value of the Eq. (9.13). Figure 9.7b shows the PWM signals in the case of high side device where the PWM signal of phase C does not switch being its low-level state. In this case, the high side device is the one that remains disabled, without switching for 120° of the T period.

$$T_7 = T - T_1 - T_2 \tag{9.12}$$

$$T_0 = T - T_1 - T_2 \tag{9.13}$$

The discontinuous modulation DPWMMIN is more convenient to use since it guarantees the bootstrap capacitors charge, while one of the phases does not switch. The phase that does not switch for DPWMMIN keeps the low side device in the half-bridge activated, guaranteeing the charge of the bootstrap capacitor as seen in Sect. 8.4.1.1.

On the other hand, the disadvantage presented by both DPWMMIN and DPWM-MAX modulations is that phase currents in the electric machine result in higher AC component current ripple. The increase of the ripple is double compared to the continuous modulation SVPWM as seen in Eqs. (9.14) and (9.15). The current ripple in the electric machine depends on its inductance, the modulation index m', the DC-link voltage, and the switching period. It is possible to graph both equations as illustrated in Fig. 9.8.

It is possible to observe how the ripple increases when the modulation index increases, having a maximum when the modulation index reaches the value of 0.5. Also, usually the current ripple increment in the discontinuous modulation supposes

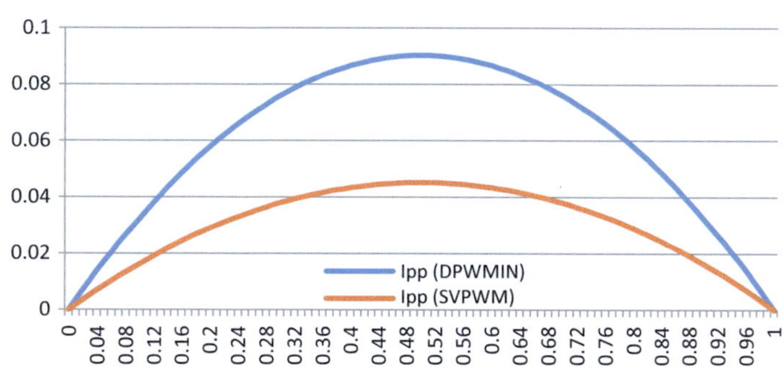

Fig. 9.8 Peak-to-peak variation as a function of the modulation index for continuous and discontinuous modulation

a higher acoustic noise compared with continuous modulation, limiting some applications where the acoustic noise should be minimized. However, the discontinuous modulation reduces the switching losses as previously mentioned in a significant factor, reducing the heat generated by the power switch device.

$$I_{ppDPWMIN} = \frac{2V_{DC}}{3L}(1 - m')m' \cdot T_s \tag{9.14}$$

$$I_{ppSVPWM} = \frac{V_{DC}}{3L}(1 - m')m' \cdot T_s \tag{9.15}$$

In spite of the improvements described concerning the SPWM sinusoidal modulation, the SVM presents the same issues where the maximum performance of the three-phase VSI is not obtained. On the one hand, it has the limitation imposed by the power devices in terms of the necessary deadtime between switching the high side device with the low side device. On the other hand, there is a minimum turn-on time T_{min} and a voltage drop in the power switch device, seen in Chap. 8, which imposes additional limitations. These overall limitations lead to the fact that Eq. (9.7) is not satisfied; that is, the line-to-line RMS output voltage of the VSI will be lower than the input RMS voltage V_i. However, in the literature (Bergas 2000), different SVM optimization techniques are described to get the full performance of the VSI which compensates the issues discussed except the voltage drop.

9.1.2.3 Voltage Resolution and Restriction Time

As shown in Chap. 7, the resolution of the timers of both the Infineon microcontroller and that of Renesas was 10 ns. That means that the minimum pulse that can be generated is 10 ns (pulse width of 10 ns). Taking into account that the internal circle of the hexagon is used, the resolution can be calculated according to below Eq. (9.16):

$$\Delta_u = \frac{2}{\sqrt{3}} \frac{\Delta t}{T_p} V_{DC}[\mathrm{V}] \tag{9.16}$$

If the switching period is 50 μs, the resolution is 78.52 mV when DC-link voltage V_{DC} is 340 V. This resolution is real as long as the switching times of the power devices can be switched from off to on, and then, from on to off again, in 10 ns. The zero vectors V_0 and V_7 become almost zero when the voltage reaches its maximum amplitude. It is equivalent to an immediate on or off of the pair of transistors in question after it has been turned off or on. In most cases, the power devices will not be able to switch in this switching time, and then, it should be limited. For example, if the minimum time T_{min} of activation is limited to 1 μs, the new resolution will then be 100 times smaller, that is to say 7.852 V. In the case of the SVM, if the algorithm fulfills the condition of $T_0 + T_7 = 2 \cdot T_7 > T_{min}$, the width of the pulse applied to the power device will be as minimum T_{min}.

On the other hand, if the output voltage of the VSI is tiny, that is to say, that the time vectors T_a and T_b are minimal, and a similar limitation should be implemented. Hence, times T_a and T_b should be limited in the same way as before.

9.1.2.4 Deadtime Compensation

As discussed in Chap. 8, the deadtime is a limitation of the half-bridge topology and therefore in the three-phase VSI . The limitation can be relevant if the deadtime represents more than 8% of the switching period. The error between the actual voltage applied to the electric machine and the theoretical voltage may be higher than 20%. Moreover, the deadtime also adds a distortion in the current of the electric machine affecting the control performance as discussed in this chapter. Therefore, for those applications where the deadtime represents more than 8% and high-performance control is desired, the deadtime compensation will be mandatory.

There are several deadtime compensation techniques. One of them was seen in Chap. 7 where PWM voltage pulses applied to the electric machine were captured with the microcontroller. This technique uses an additional timer to measure the applied voltage pulses, but the additional processing by the microcontroller to compensate the deadtime was almost negligible. However, the processing required for the technique presented in this section is higher, although it does not require an additional timer. The technique basically consists of a software algorithm that optimizes the SVM to compensate the deadtime by knowing in advance the reference voltage vector position and capturing the sign of the phase currents of the electrical machine.

It is possible to represent the deadtime effect by using Fig. 9.9. In Fig. 9.9, the desired reference voltage pulse v_a^* with the actual pulse applied in the high and low side devices of half-bridge is illustrated. In the pulse of the left side of the figure, the distortion is not so significant, but in the right side, it can be observed that the pulse for the high side device a is suppressed. It happens when the duration of the pulse width approaches twice the value of the deadtime T_{DT}. Therefore, it is convenient to set a minimum pulse time below which it does not make sense to apply a voltage pulse since it will be suppressed.

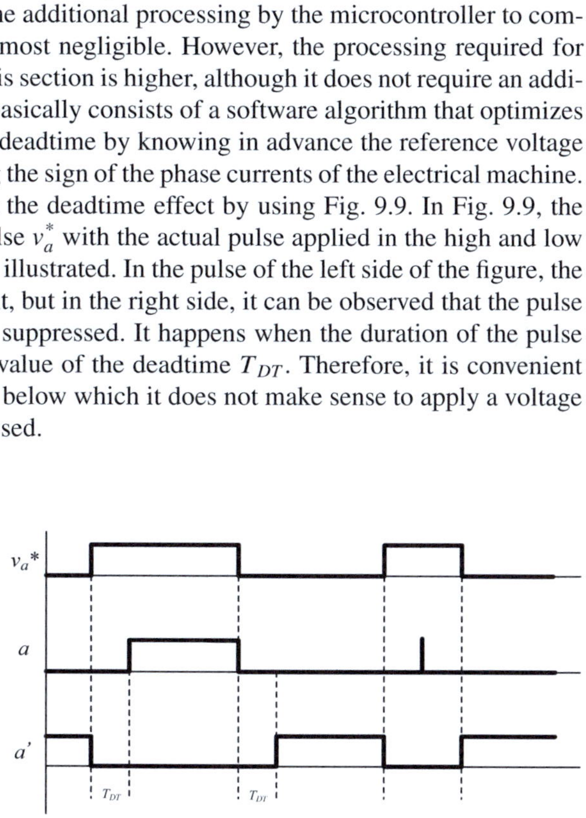

Fig. 9.9 Voltage reference pulse and high and low side switching pulses in a half-bridge of VSI . In the right pulse, high side pulse could be suppressed since the width of low side signal is two times the deadtime T_{DT}

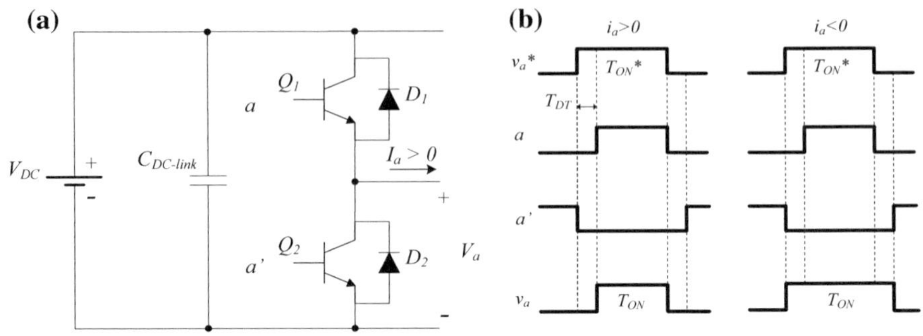

Fig. 9.10 **a** Half-bridge phase A of VSI . **b** Voltage reference pulse, high and low side switching pulses, and the actual pulse applied in the half-bridge output for positive and negative current

Assuming that the current flow in the designated positive direction of Fig. 9.10a and that the low side transistor Q_2 at that moment turn-on, the current does not flow through this transistor but through the anti-parallel diode D_2 (the transistor only drives in one direction). When the low side transistor turns off, the current remains flowing through the diode, and the voltage applied to the load remains as before. As soon as the high side transistor Q_1 turns on, the current can flow through it so that the voltage becomes positive. On the contrary, when the current is negative, when the low side transistor Q_2 is deactivated, the current, in this case, does not flow through the diode but through the transistor Q_2. At the moment when it has the command to open the low side transistor Q_2, the only path through which the current can flow is through the diode D_1 of the high side transistor Q_1, while the voltage becomes positive. Finally, when the high side transistor turns on, the current remains in the same direction as it was, that is, it flows through the diode of the high side transistor, while the voltage remains positive.

As mentioned, and with the help of Fig. 9.10b, the duration of the actual voltage pulse depends on the sign of the current flowing through the half-bridge. Based on this sign, the duration of the actual voltage pulse T_{ON} is the command pulse T^*_{ON} plus the deadtime T_{DT} (as a function of the sign of the current) as can be seen in Eq. (9.17).

$$
\begin{aligned}
T_{ON} &= T^*_{ON} - T_{DT} \rightarrow i_a > 0 \\
T_{ON} &= T^*_{ON} + T_{DT} \rightarrow i_a < 0
\end{aligned}
\tag{9.17}
$$

It is possible to observe that the method of Chap. 7, Sect. 7.4.1.2, the capture pulses applied to the VSI is based precisely on the previous Eq. (9.17). In order to compensate the deadtime employing a software algorithm, Eq. (9.17) can be developed for a three-phase inverter taking into account the already explained operation of the SVM.

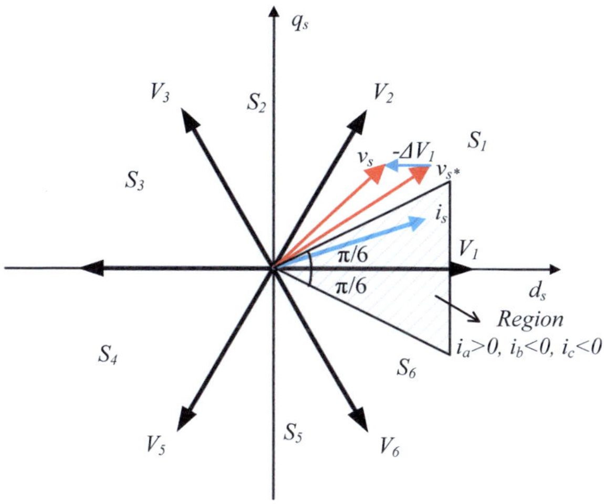

Fig. 9.11 Space vector command v_s^* and actual v_s due to the deadtime effect. The vector of the stator current i_s is in the region between angles $\pi/6$ and $-\pi/6$, where the current of phase a is positive, and the current for phases b and c is negative

With the help of the vector diagram represented in Fig. 9.11, it is possible to deduce the Eq. (9.18) that relates the actual voltage applied v_k to the three-phase inverter with the command voltage v_k^* and the deadtime.

$$v_k = v_k^* - \Delta v_k \tag{9.18}$$

where k takes the a, b, c values.

It is easy to deduce that the variation Δv_k depends on the deadtime proportion with respect to the switching period T, the DC-link voltage V_{DC}, and the sign of the current as can be observed in Eq. (9.19).

$$\Delta v_k = \frac{T_{DT}}{T} V_{DC} \text{sign}(i_k) \tag{9.19}$$

Now, if the i_k current is transformed with the Clarke transform, it is possible to obtain two component vectors in the stationary reference frame dqs:

$$\begin{aligned} \text{sign}(i_{ds\_s}) &= \frac{2}{3}\left(\text{sign}(i_a) - \frac{1}{2}\text{sign}(i_b) - \frac{1}{2}\text{sign}(i_c)\right) \\ \text{sign}(i_{qs\_s}) &= \frac{2}{3}\left(\frac{\sqrt{3}}{2}\text{sign}(i_b) - \frac{\sqrt{3}}{2}\text{sign}(i_c)\right) \end{aligned} \tag{9.20}$$

If the expression of the previous vectors inside of the parentheses is divided by two, a unit module vector is obtained. Thus, Eq. (9.19) can be rewritten as

$$\begin{aligned} \Delta V_{ds\_s} &= \frac{T_{DT}}{T}\frac{2}{3} V_{DC}\left[\text{sign}\left(i_{ds\_s}\right)\right] \\ \Delta V_{qs\_s} &= \frac{T_{DT}}{T}\frac{2}{3} V_{DC}\left[\text{sign}\left(i_{qs\_s}\right)\right] \end{aligned} \tag{9.21}$$

Table 9.5 Values of dqs components according to the position of the i_s current vector

i_a	i_b	i_c	sign(i_{ds_s})	sign(i_{qs_s})
>0	<0	<0	1	0
>0	>0	<0	1/2	$\sqrt{3}/2$
<0	>0	<0	−1/2	$\sqrt{3}/2$
<0	>0	>0	−1	0
<0	<0	>0	−1/2	$-\sqrt{3}/2$
>0	<0	>0	1/2	$-\sqrt{3}/2$

Table 9.5 shows the value of the components of the dqs unit vector. These are the values that depend on the sector where the current i_s is located and that vary the magnitude of the voltage vectors ΔV_{ds_s} and ΔV_{qs_s}.

For the deadtime compensation, it will be enough to apply the above voltage vectors according to Eq. (9.21) to the desired command voltage $V^*_{ds\_s}$ and $V^*_{qs\_s}$ as shown in Eq. (9.22).

$$\begin{aligned} V_{ds\_scomp} &= V^*_{ds\_s} + \Delta V_{ds\_s} \\ V_{qs\_scomp} &= V^*_{qs\_s} + \Delta V_{qs\_s} \end{aligned} \tag{9.22}$$

The most critical point of this algorithm resides in the capture of the sign of the actual phase currents a, b, and c. The measuring noise at zero crossings of the phase currents can deteriorate the detection when an analog-to-digital converter is used. As an alternative option, an external circuit that exclusively uses hardware to detect the zero crossing can be used. The same problems of noise will exist but can be reduced employing hysteresis comparators, for example, for each phase. The rise edge of the comparators outputs can be used to launch different interruption routines (ISR) in the microcontroller for each zero-crossing detection, to process the deadtime compensation algorithm.

9.2 Model Design

9.2.1 Introduction

In Chap. 7, a Simulink block was designed which generated six PWM channels with configurable deadtime according to the RX and Aurix™ family microcontrollers timers of Renesas and Infineon, respectively. As it was seen, with the switching times duration (duty cycle) for each of the phases, the corresponding PWM with the programmed deadtime was performed. In this section, a model for the SVM that can operate in continuous and discontinuous mode will be designed and evaluated with overmodulation with hexagonal saturation and deadtime compensation.

9.2.2 SVPWM Model

Nowadays, thanks to the new generations of microcontrollers/DSPs, the implementation of the SVM is a more straightforward task. Most of the microcontrollers dedicated to electric machine control incorporate the sine, cosine, and tangent trigonometric functions, together with a floating-point calculation unit. The processing time of these mathematical functions has been reduced in such a way that it can be used within the SVM algorithm, avoiding cumbersome calculations and look-up tables. The processing time of the sine function for this microcontrollers usually is less than 1 μs. As can be observed, the processing time is minimal compared to the strictest sampling time. Furthermore, by using the sine and cosine functions with a floating-point unit, the accuracy obtained is higher. In some low cost applications with low-performance microcontrollers, where the trigonometry functions and a floating-point unit are not available, it is possible to perform an optimization algorithm for the SVM implementation. The optimization algorithm does not need trigonometric calculations, but a look-up table and basic functions such as multiplication and division are necessary for the implementation. In this section, it is only discussed the implementation with high-performance microcontroller with a floating-point unit and trigonometric functions such as the RX63T and TC264.

The requirements for SVM implementation are

1. Adjustable switching frequency
2. Adjustable deadtime
3. Type of modulation selection.

Figure 9.12 shows the SVM model and the model already studied in Chap. 7 based on the Renesas Rx63T microcontroller timer. To meet the above requirements, the SVM model needs threeputs such as V_{max}, *Semiperdead*, and *ModType* as shown in Fig. 9.12. The first input defines the maximum allowable voltage based on whether or not overmodulation is used. That is, V_{max} takes the value of $V_{DC}/\sqrt{3}$ in the linear zone and $2V_{DC}/3$ in overmodulation with hexagonal saturation. The *Semiperdead* input defines the switching frequency and corresponds to the value of the half of the switching period plus the deadtime according to the pre-scaler of the microcontroller timer. The *ModType* input takes the value 0 for continuous modulation while 1 and 2 for the modulation DPWMMIN and DPWMAX, respectively. Finally, the two remaining inputs correspond to the magnitude and phase of the voltage vector v_s. As can be seen, the magnitude and phase are achieved through the axis transformation (Clarke transform). The output of the said transformation is the components *dqs* or $\alpha\beta$ of the vector v_s, i.e., V_{ds} and V_{qs}. Additionally, if the deadtime compensation seen in Sect. 9.1.2.4 is used, the compensation value V_{comp_d} and V_{comp_q} is added

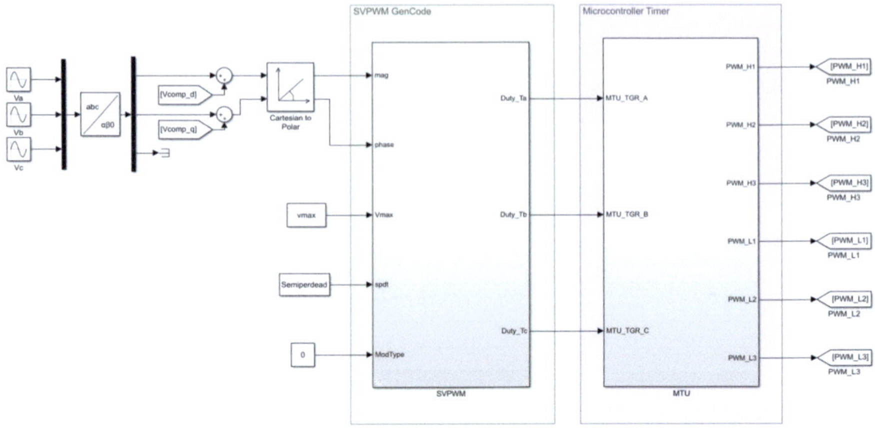

Fig. 9.12 Simulink space vector modulation model (called SVPWM) able to generate C-code and microcontroller timer model

to each component. Lastly, the outputs of the SVM model correspond directly to the switching times T_a, T_b, T_c that define the duty cycle for each of the phases.

The content of the SVM model is divided into two submodules made with MATLAB M-functions, as illustrated in Fig. 9.13. The first submodule (left side) is the most complex since it is the one that generates the overall time periods according to the theory of SVM. As can see, three additional outputs are only used for debugging purposes and correspond to the outputs T_1, T_2, and T_0. The second submodule (right

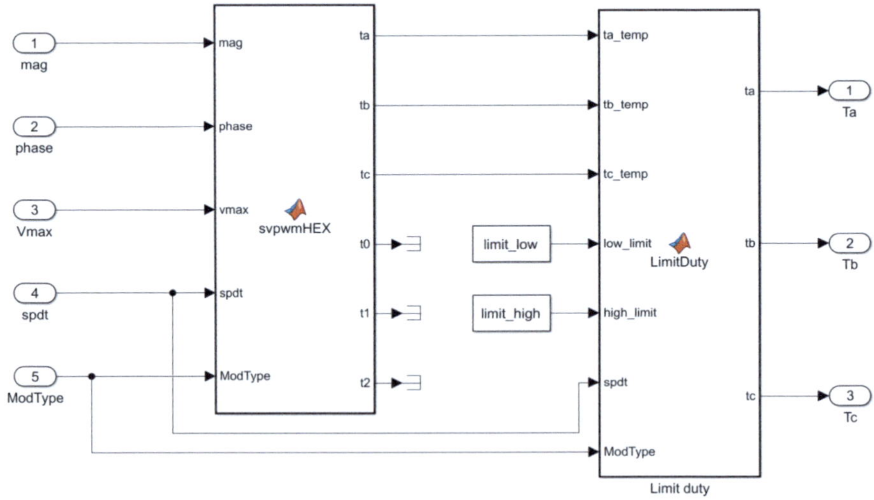

Fig. 9.13 Simulink space vector model which consists of two M-function blocks

side) can be optional and consists of applying a limitation of the duty cycles according to the inputs *limit_low* and *limit_high*, that is, minimum and maximum width, respectively. For example, the minimum width of the pulse should be such that the power device can be activated and deactivated, depending on the driver's delay and the parasitic capacitances that cause the device to switch more slowly as discussed in Sect. 9.1.2.3.

The SVM flowchart algorithm model is shown in Fig. 9.14. The first part of the algorithm is to determine whether or not to apply saturation, i.e., the voltage vector is limited. Next, it is determined in which sector it is and calculates the times T_1 and

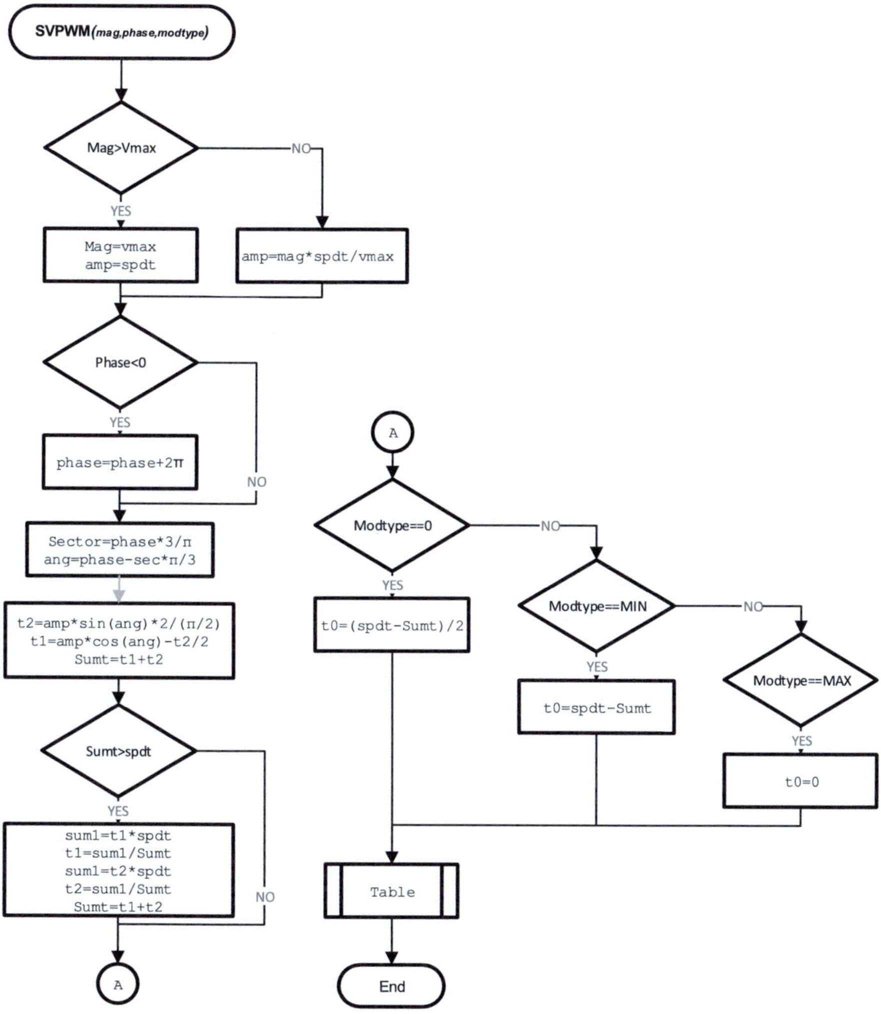

Fig. 9.14 Flowchart algorithm used in the M-function of space vector model

T_2. Then, the hexagonal saturation is applied if necessary, and T_0 and T_7 are assigned according to the modulation type. Finally, depending on the sector, the times T_a, T_b, and T_c corresponding to each of the phases according to Table 9.3 are applied.

9.2.3 Deadtime Compensation Model

As was seen in Sect. 9.1.2.4, the deadtime compensation discussed uses the current sign in order to define the sector of the sinusoidal voltage. As discussed, the detection of current sign is not trivial because the possible noise in the current zero crossing may disturb the detection. There are several zero-crossing detection techniques, including prediction techniques that improve the detection essentially. However, in the model presented, the zero-crossing conventional technique detection is used. It is based on analog-to-digital converter of the phase currents where the comparison of the last filtered sample $i_a[n-1]$ and the actual current filtered sample $i_a[n]$ is used to define whether or not it has opposite signs. This technique is valid for most AC machine control applications. Figure 9.15 shows the deadtime compensation model where its inputs are the filtered currents of each of the phases, and its outputs are the two *dqs* components vectors V_{comp} transformed to volts. Within the *SignCurrentCalculation* model, Table 9.5 is implemented.

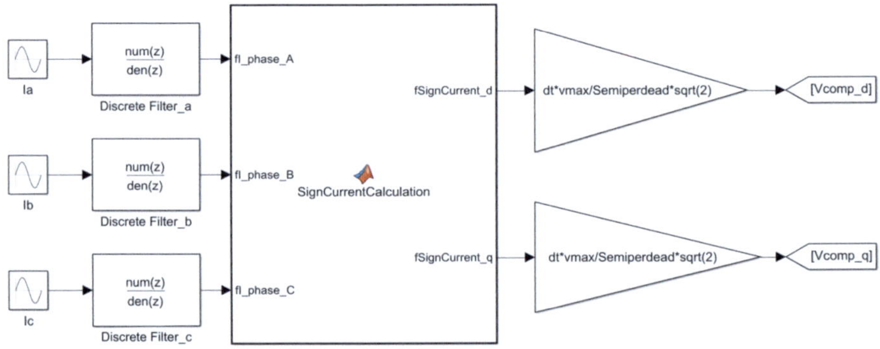

Fig. 9.15 Deadtime compensation model-based in the conventional zero-crossing detection

9.2.4 Simulation Results

The simulation of the previous modules allows validating the model in multiple situations in the MATLAB/Simulink environment before its use in real hardware. Here, only one of these situations will be presented.

The simulation results that are presented below are based on the following configuration parameters:

```
vmax=vbus/sqrt(3);
f1=150;                        %Freq of sinusoidal output voltage in Hz
Fpwm=5000;                     %Freq of PWM in Hz
dt=0;                          %Dead time of 1us
ts=1/Fpwm;                     %Sampling time
MIN_ON_TIME=0;                 %Min ON time of 1us
SEMIPER=100e6/(2*Fpwm);        %CLK of microcontroller is 100MHz. No prescaler is used
limit_low=MIN_ON_TIME+(dt/2);
limit_high=SEMIPER-(MIN_ON_TIME/2);
Semiperdead=SEMIPER+dt;        %Half period or half of the triangular plus dead time
```

As it is possible to observe, the selected switching frequency is low for easy visualization. The resolution used in the microcontroller's timer is the maximum being 100 MHz (clock frequency) so that the number of timer counts in a half-period is 10,000 for the PWM frequency of 5 kHz. The peak of the triangular signal is 10,000, and the total width is 20,000, corresponding to a PWM period of 200 μs.

9.2.4.1 SVPWM Simulation Results

The following Figs. 9.16, 9.17, 9.18 show the SVM module simulation results when the deadtime and minimum "on" time are zero.

In Fig. 9.16, the SVM module outputs T_a, T_b, and T_c in counts of each one of the timers from 0 to its maximum number 10,000 are illustrated. The fundamental frequency of these signals is 150 Hz. In the right side of the figure, the high side switching PWM signal of phase a is represented that is, T_a. Also, the outputs T_1, T_2, and T_0 are represented, which are the ones that form each of the times T_a, T_b, and T_c, depending on the sector, as shown in Table 9.3. In Fig. 9.16a, the maximum modulation index $m = 1$ is shown, while Fig. 9.16b shows just half that before ($m = 0.5$).

If the modulation is changed from continuous to discontinuous modulation, it is obtained the results as shown in Fig. 9.17a, b for a modulation DPWNMIN and DPWMMAX, respectively. It is possible to observe that during 1/3 of the period, there is no switching of the high side device remaining off for the DPWMMIN modulation and on for the DPWMMAX modulation. The modulation index m is equal to one.

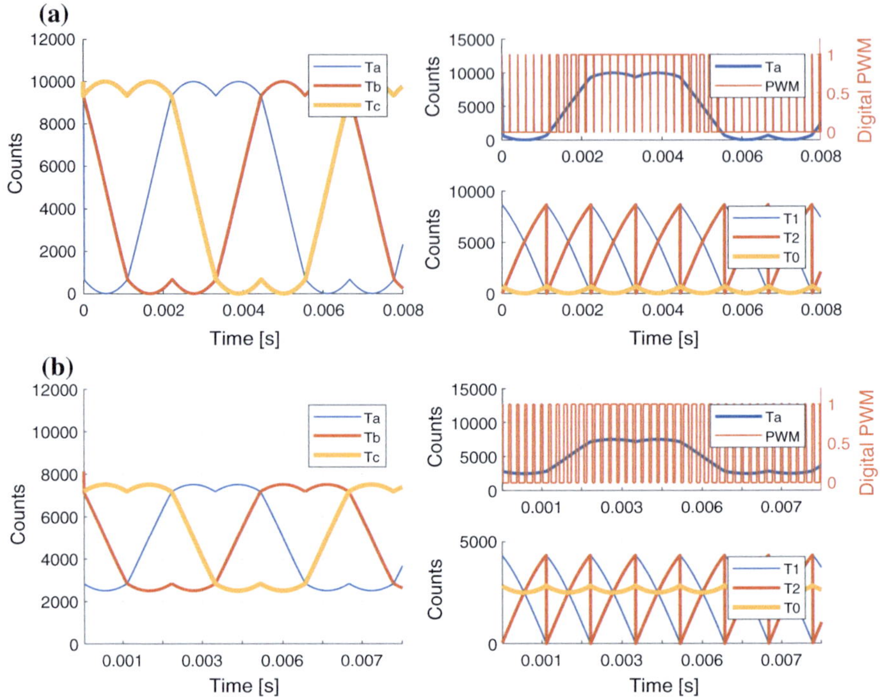

Fig. 9.16 Continuous SV modulation interval times T_a, T_b, and T_c of every phase, and interval times T_1, T_2, and T_0, both in counts. **a** $m = 1$. **b** $m = 0.5$. Output frequency: $\omega_c = 150$ Hz. Deadtime: $T_{DT} = 0$ µs. Switching time: $T_{PWM} = 200$ µs. Simulation step: $T_{sp} = 100$ ns

If the modulation index is greater than unity, the non-linear zone is obtained, that is, the overmodulation as shown in Fig. 9.18. It is possible to appreciate that the waveform shape of the signal tends to saturate since the voltage vector is limited according to the hexagon. This phenomenon causes non-sinusoidal waveform with higher harmonic content as discussed.

In the results of the previous simulations, the forbidden switching zones of the hexagon described in Quang and Dittrich (2015) have not been taken into account. These are determined by the deadtime and the minimum "on" time (on pulse) so that the device can be activated and deactivated properly. If a deadtime and an "on" time of 5 µs are defined for easy visualization, the result of Fig. 9.19 is obtained. Here, it is possible to see that unlike the previous cases, the T_a, T_b, and T_c timer counts do not reach the range of before, but it is reduced by the deadtime and the minimum "on" pulse. The minimum value of the T_0 does not reach zero, but a certain

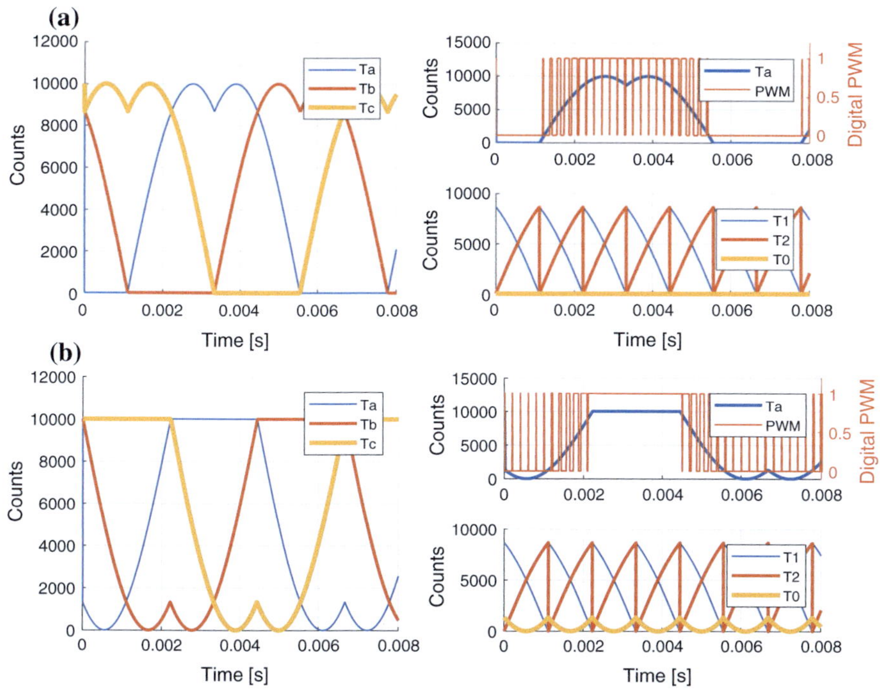

Fig. 9.17 Discontinuous SV modulation interval times T_a, T_b, and T_c of every phase, PWM digital signal, and interval times T_1, T_2, and T_0, both in counts. **a** DPWMMIN. **b** DPWMMAX. $m = 1$, Output frequency: $\omega_c = 150$ Hz. Deadtime: $T_{DT} = 0$ µs. Switching time: $T_{PWM} = 200$ µs. Simulation step: $T_{sp} = 100$ ns

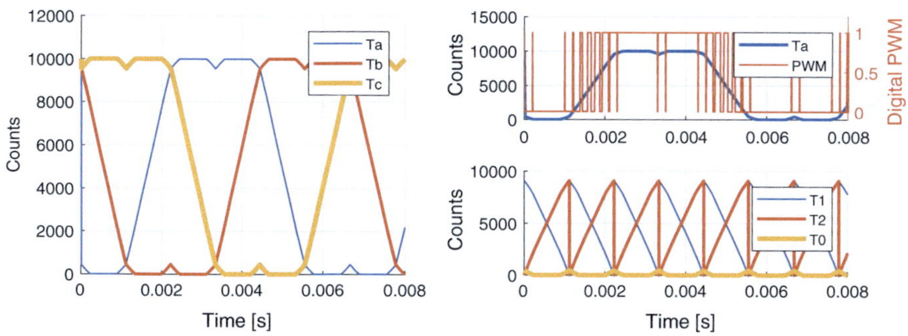

Fig. 9.18 Overmodulation region1 showing interval times T_a, T_b, and T_c of every phase, PWM digital signal, and interval times T_1, T_2, and T_0, both in counts. Output frequency: $\omega_c = 150$ Hz. Deadtime: $T_{DT} = 0$ µs. Switching time: $T_{PWM} = 200$ µs. Simulation step: $T_{sp} = 100$ ns

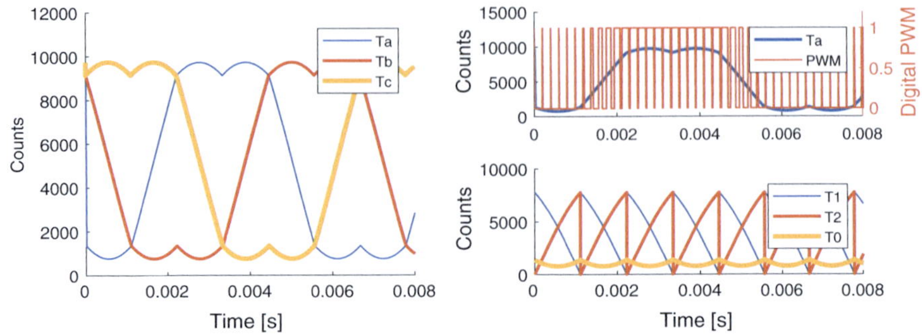

Fig. 9.19 Continuous SVM interval times T_a, T_b, and T_c of every phase, PWM digital signal, and interval times T_1, T_2, and T_0, both in counts. Output frequency: $\omega_c = 150$ Hz. Deadtime: $T_{ONmin} = 5$ µs. $T_{DT} = 5$ µs. Switching time: $T_{PWM} = 200$ µs. Simulation step: $T_{sp} = 100$ ns

minimum value is maintained to guarantee a minimum "on" time. This limit affects the modulation performance, affecting the inverter output voltage as discussed in Chap. 8.

9.2.4.2 Deadtime Compensation Simulation Results

Before activating the deadtime compensation block, some simulations are shown with the previous SVM model and a VSI to see the effect it produces on the line-to-line voltage and the phase current. The line-to-line voltage v_{ab} in case of different deadtimes is shown in Fig. 9.20a. As can be seen, as the deadtime increases, the magnitude of the voltage and current decreases, further increasing the distortion. It is also possible to appreciate that, in the zero crossing of the phase current i_a of Fig. 9.20b is where the effect of distortion in the voltage v_{ab} is observed.

If the deadtime compensation is enabled in the SVM model, the result of Fig. 9.21a is obtained for the case of deadtime of 1.5 µs. The line-to-line voltage v_{ab} and the current i_a have improved their harmonic content exceptionally as expected, although there is still a THD of 1.5% and 1.02%, respectively. In Fig. 9.21b, the inputs of the SVM model V_{ds_s} and V_{qs_s} that compensates the deadtime are represented. Also, it is possible to observe how in each sector (1/6 of the sinusoidal), there is a level change since it corresponds to the zero crossing of each of the phase currents i_a, i_b, and i_c according to Table 9.5.

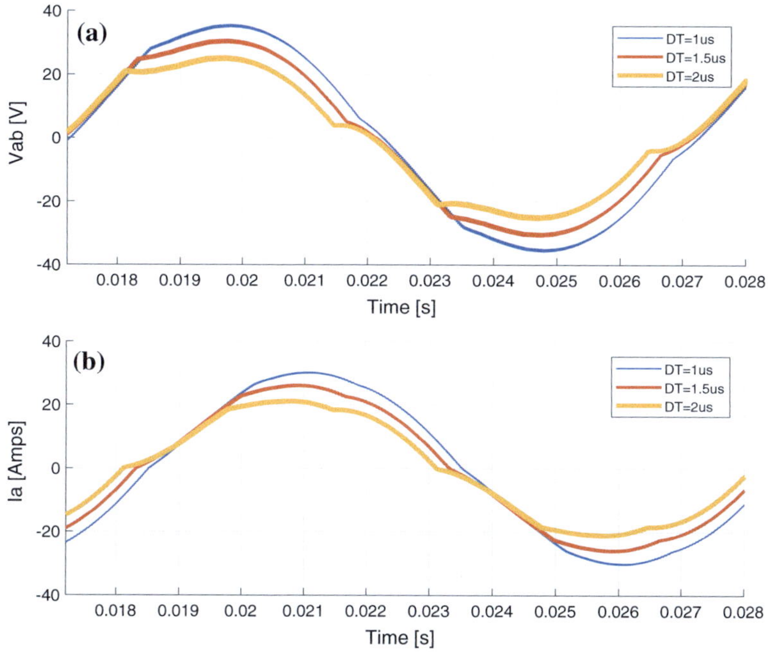

Fig. 9.20 Distortion effect with different deadtimes. **a** Line-to-line voltage v_{ab}. **b** Phase current i_a. $m = 0.15$, Output frequency: $\omega_c = 100$ Hz. Switching time: $T_{PWM} = 50$ µs. Simulation step: $T_{sp} = 100$ ns

9.3 Experimental Results

9.3.1 Continuous SVPWM

The continuous SVM is characterized by its symmetry as discussed where the vector times 0 and 7 were identical. The output PWM voltage v_{a0} and v_{b0} referenced to the ground can be filtered by an *RC* low-pass filter for debugging purposes to see the sinusoidal waveform. The line-to-line voltage is then the difference of the phase voltages v_{a0} and v_{b0}. The filtered v_{a0}, v_{b0}, and the line-to-line voltages are represented in Fig. 9.22a. The sinusoidal frequency is 65 Hz, and the modulation index is the maximum of the linear zone, that is, 0.907. The maximum line-to-line RMS voltage is 223 V with a DC-link voltage of 315 V, but as can be observed, the output in the VSI is 205 V RMS approximately. The difference of 18 V is mainly due to the deadtime, the minimum "on" time seen in Fig. 9.19, semiconductor voltage drop, and the little attenuation of the low-pass filter at the 65 Hz frequency. The unfiltered line-to-line voltage for a switching frequency of 8 kHz is shown in Fig. 9.22b. As can be observed, the maximum voltage is approximately 315 V.

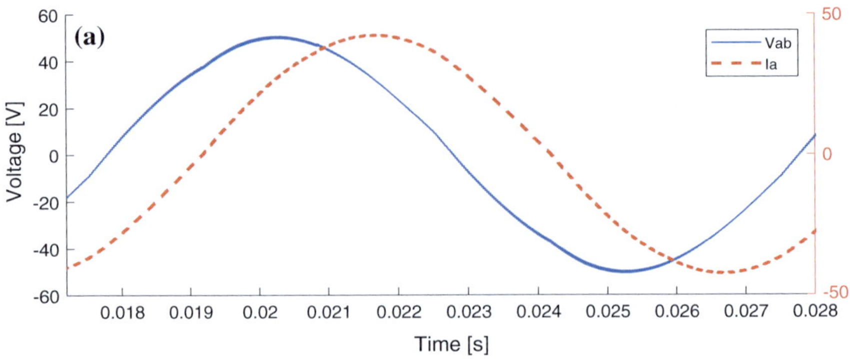

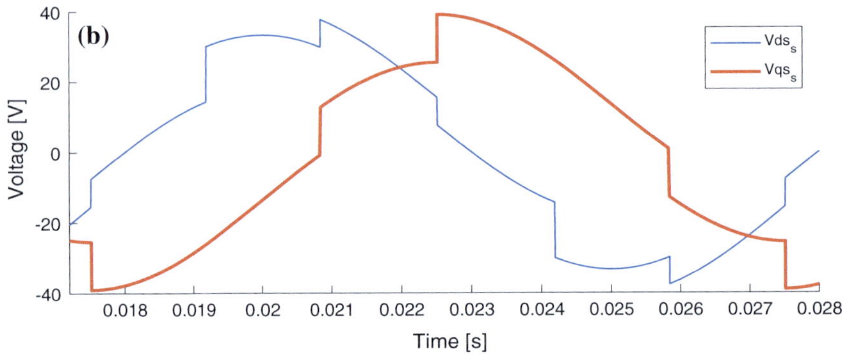

Fig. 9.21 Deadtime compensation result when deadtime is set to 1.5 μs. **a** Line-to-line voltage v_{ab} and phase current i_a. **b** Voltage vector V_{ds_s} and V_{qs_s} with the compensation. $m = 0.15$, Output frequency: $\omega_c = 100$ Hz. Switching time: $T_{PWM} = 50$ μs. Simulation step: $T_{sp} = 100$ ns

Thereby if the frequency of the reference vector is increased to 430 Hz, it is possible to appreciate in Fig. 9.23 with greater detail the effect of the *RC* low-pass filter. Since the frequency is closer to its cut-off frequency, the voltage magnitude is attenuated. For this higher frequency test, the cut-off frequency of the low-pass filter should be higher, but the obtained signal will have more harmonics. As can be observed, the RMS voltage in the output VSI is now of 198 V approximately.

Finally, if the modulation index is increased to leave the linear zone, the overmodulation is obtained as shown in Fig. 9.24. In Fig. 9.24, it is possible to appreciate that the voltage waveform v_{a0} is flatter due to hexagonal saturation. As can be observed, there are regions without PWM pulses switching, i.e., the high side device is permanently activated during 1/3 of the sinusoidal frequency and no activated during 1/3 as expected in the hexagonal saturation. The filtered signal resembles a trapezoidal signal due to hexagonal saturation. The line-to-line RMS voltage when the overmodulation is used is 215 V which is 10 V higher than without overmodulation. Note that the frequency of the reference vector is 125 Hz.

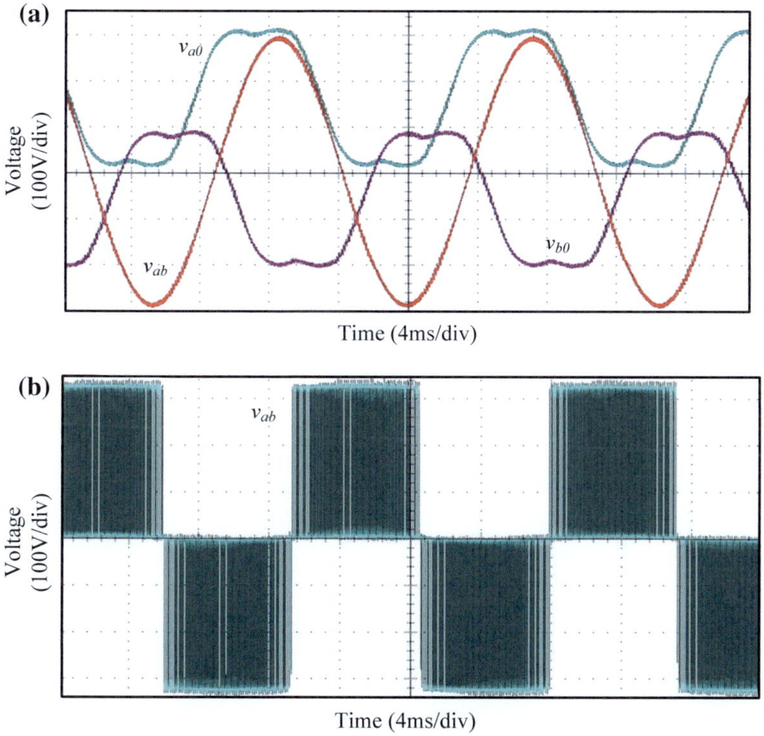

Fig. 9.22 Example of space vector PWM voltage generation close to voltage limit of the driver. **a** Voltages v_{a0} and v_{b0} referenced to ground of three-phase VSI . Line-to-line voltage v_{ab}. **b** Non-filtered line-to-line voltage. Switching frequency is 8 kHz

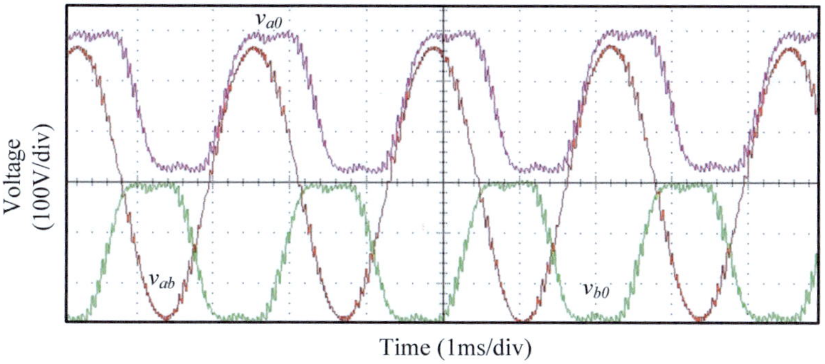

Fig. 9.23 Example of space vector PWM voltage generation at 430 Hz and maximum voltage available with a switching frequency of 8 kHz

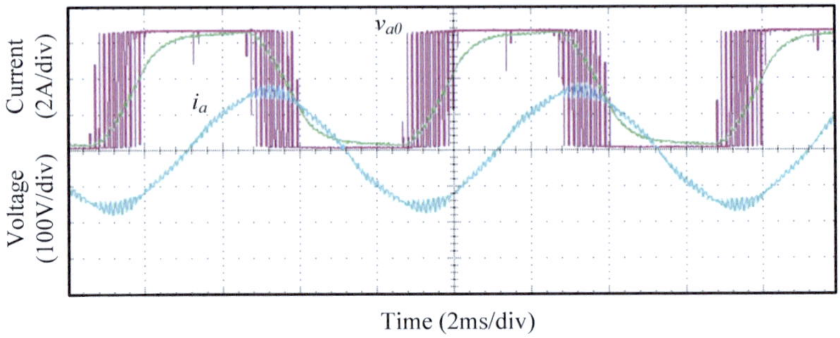

Fig. 9.24 Voltage and current in the machine terminals with hexagonal saturation (overmodulation). Phase current a, unfiltered PWM signal, and filtered v_{an}

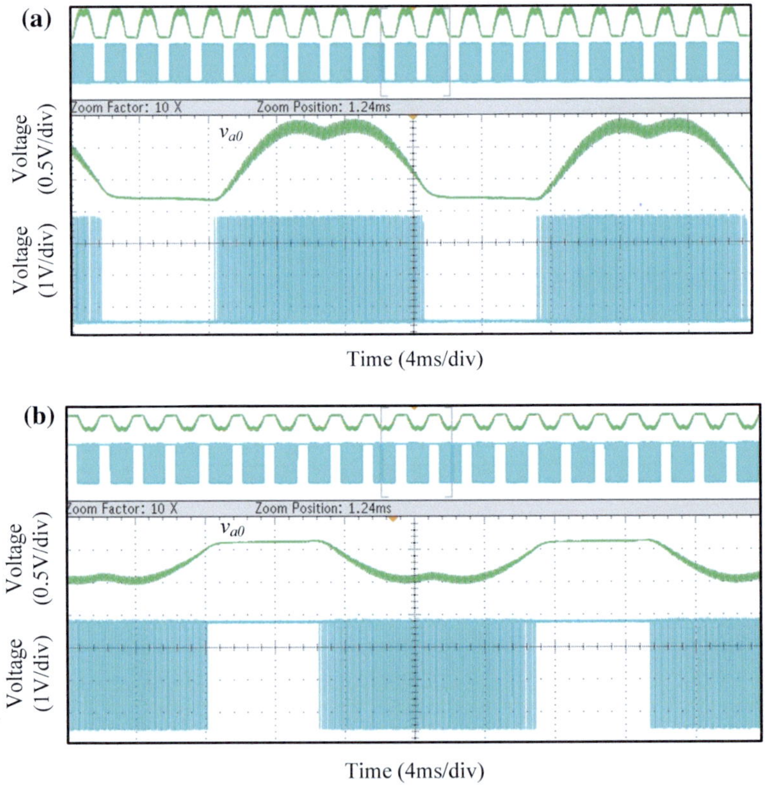

Fig. 9.25 High side PWM signal for phase a with its filtered result. **a** DPWMMIN modulation. **b** DPWMMAX modulation

9.3.2 Discontinuous SVPWM

Figure 9.25 shows the DPWMMIN and DPWMMAX modulations. It is illustrated the high side PWM signal for phase *a* with the filtered voltage referenced to the neutral. As mentioned, during 120°, there is no switching in the inverter leg of phase *a*. For example, if the high PWM level means device activation, for the DPWMMIN modulation, the high side device is "off", and the low side device is "on". On the contrary, the high side device is "on," while the low side device is "off" in case of DPWMMAX modulation.

The reconstruction of the line-to-line voltage is illustrated in Fig. 9.26 for the DPWMMAX modulation.

Finally, in Fig. 9.27, the PWM signals *a* and *a*′ of the high and low side devices are represented in the case of DPWMMIN modulation.

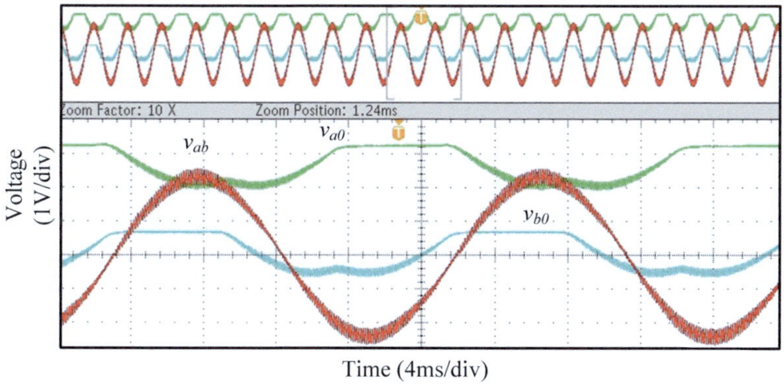

Fig. 9.26 Phase voltage v_{a0}, v_{b0} referenced to ground and line-to-line voltage

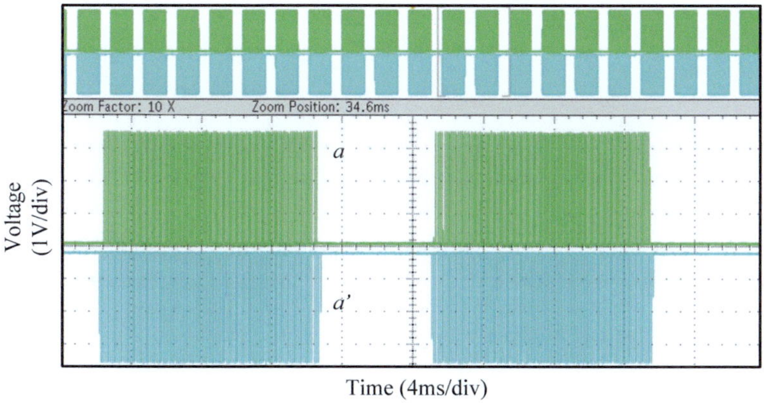

Fig. 9.27 High and low side signals waveforms of the same inverter leg for DPWMMIN discontinuous modulation

9.3.3 Distortion Effect in the AC Current

As discussed in Sect. 9.1.2.2, in an AC machine, the phase currents result in a higher AC component current ripple when discontinuous modulation is used. The experimental test is illustrated in Fig. 9.28 for a PMSM electric machine with continuous SVM and discontinuous modulation DPWMMIN. The modulation index is 0.6, the

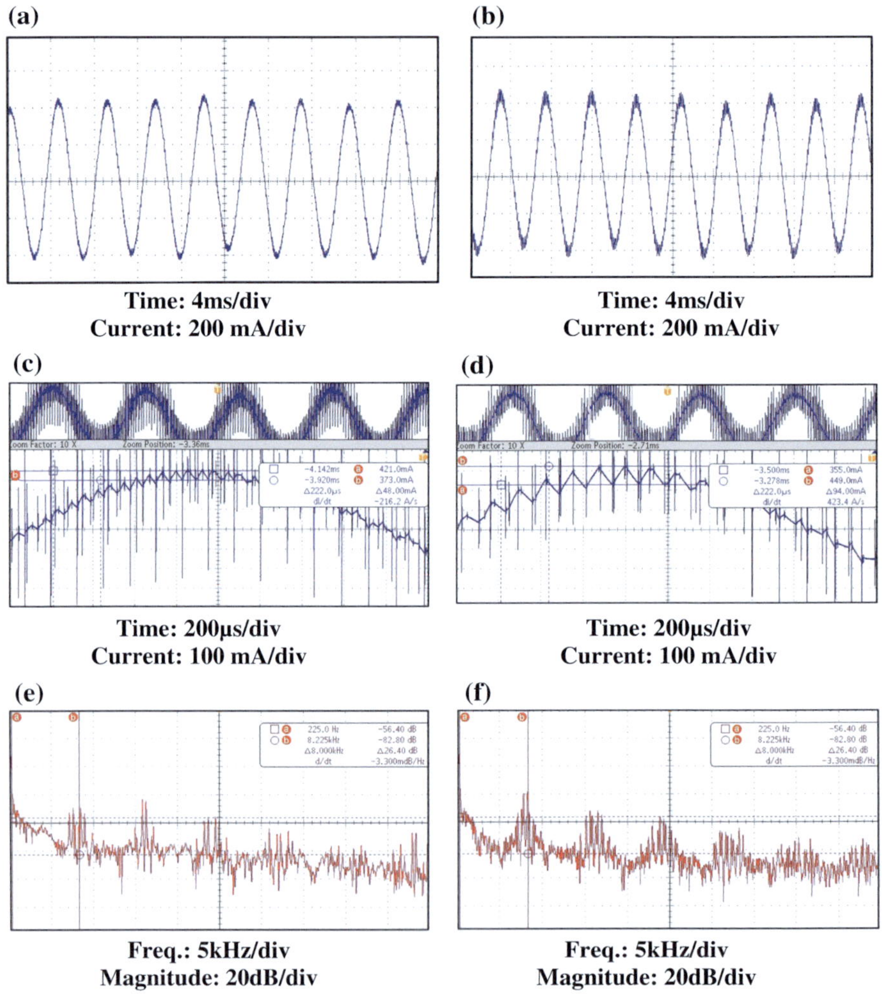

Fig. 9.28 Effect of the discontinuous modulation in the phase current a of a three-phase PMSM. Left side is continuous SVM, and the right side is discontinuous modulation. **a** Phase current a for SVM. **b** Phase current a for DPWMMIN modulation. **c** Zoom in phase current a for SVM. **d** Zoom in phase current a for DPWMMIN modulation. **e** Spectral analysis of phase current a for SVM. **f** Spectra of inverter phase current a for DPWMMIN modulation

switching frequency is 8 kHz, and the phase current frequency is 225 Hz. In Fig. 9.28a, b, the distortion in the current is noticeable. In Fig. 9.28c, d, the ripple difference is approximately the double as can be observed. Finally, the spectra of the inverter phase current i_a for continuous SVM and discontinuous DPWMMIN modulation can be observed in Fig. 9.28e, f. Both spectra illustrate a pronounced magnitude at the 225 Hz and PWM frequency and its multiples. Because of the natural low-pass filter of the machine winding, harmonics beyond the 4th are reduced. It is possible to observe that discontinuous DPWMMIN modulation has a higher number of harmonics with a higher magnitude.

9.3.4 Semiconductor Temperature Effect

As mentioned, the discontinuous modulation reduces the commutations of the power devices reducing the switching losses. In order to verify, the switching losses reduction is possible to measure the temperature in the power devices. In this case, it uses a power module (PM) which integrates a three-phase bridge in a single package where the maximum temperature operation is 120 °C. The load consists of a PMSM operating at rated power.

In Fig. 9.29, it is shown infrared thermographic camera pictures of the power module. The continuous modulation SVM results are illustrated in Fig. 9.29a, and Fig. 9.29b illustrates the discontinuous modulation DPWMMIN. It is possible to observe that the difference in temperature between the two modulation types is 16.5 °C when load and ambient temperature are identical. The ambient temperature is 25 °C where the temperature increment is 68.3 °C for the continuous modulation and 51.8 °C for the discontinuous modulation. In both cases, there is no heatsink in the power module, but in case of ambient temperature of 60 °C, the maximum temperature in the continuous modulation would be 128.3 °C, while for discontinuous modulation, the maximum temperature would be 111.8 °C. As can be observed, if the continuous modulation is used, it is necessary an element like a fan or heatsink

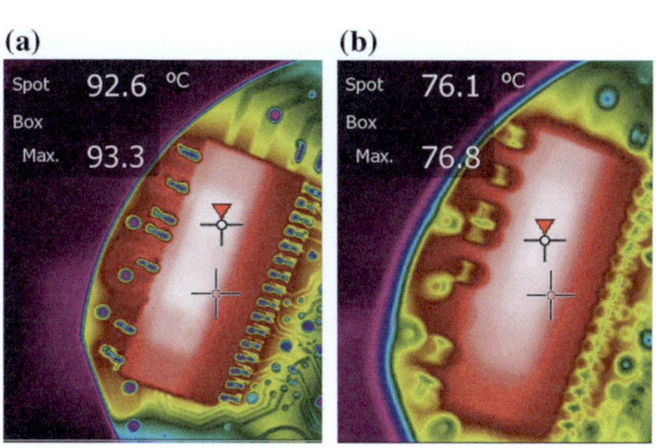

Fig. 9.29 Thermographic pictures of a three-phase bridge integrated into a single power module package when a PMSM is operating at rated power. **a** Continuous SVPWM modulation. **b** Discontinuous DPWMMIN modulation

to reduce the temperature below the 120 °C. Since most of the power module has internally an NTC, it is possible to add an algorithm which monitors the temperature in order to change the modulation type, while the machine is running as discussed in Sect. 8.7.4. For example, if the temperature reached is higher than a threshold of 110 °C, the modulation type can be changed from continuous to discontinuous to reduce the temperature avoiding to place any heatsink or fan. Then, when the temperature is below 100 °C, it can back to the continuous modulation. However, as commented before, the current ripple in the machine will be higher for discontinuous modulation but only when the temperature is inside of the hysteresis window.

9.3.5 Deadtime Compensation

As mentioned, the deadtime compensation enhances the performance of the control since the voltage and current sinusoidal waveform in the machine terminals are improved essentially.

Figure 9.30 shows the reference voltage v_{an}^* and the actual voltage v_{an} in the machine terminals when the deadtime is set to 1 μs with 8 kHz of switching frequency. As seen from Fig. 9.30, there is some distortion zones at zero crossing of the phase currents i_a, i_b, and i_c.

The actual voltage v_{an} difference respects the reference voltage is considerable so that it will also affect the current and the electromagnetic torque on the electrical machine. The phase current i_a and the voltage v_{an} in an electrical machine without deadtime compensation are shown in Fig. 9.31. It is possible to observe in the zero crossing of the phase current the distortion in the voltage $v_{an.}$

If the deadtime is extended to 3 μs, it represents 4.8% of the half PWM period. In this case, there is a considerable distortion according to Fig. 9.32a in the voltage and current of the machine. If the deadtime compensation algorithm is applied, the distortion of both signals decreases considerably, as can be seen in Fig. 9.32b.

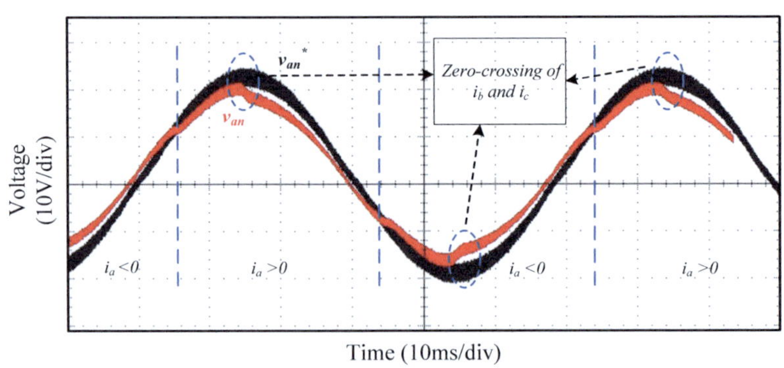

Fig. 9.30 Reference voltage and actual voltage distortion respect neutral due to the deadtime effect

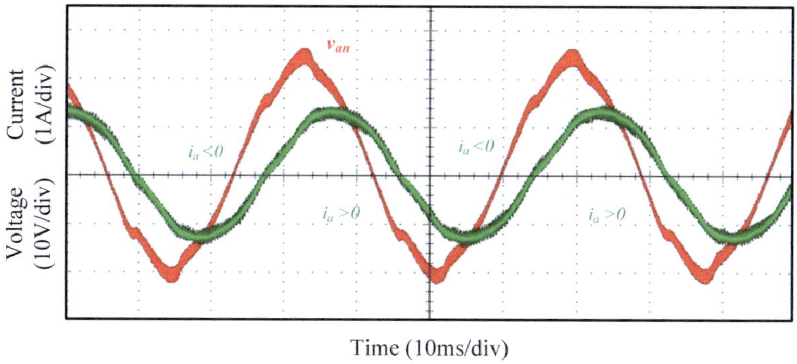

Fig. 9.31 Reference voltage respects neutral v_{an} and phase current i_a where the sign of the current is shown

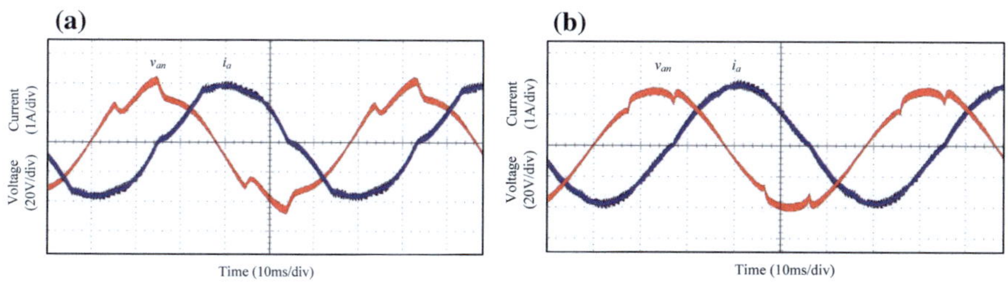

Fig. 9.32 Voltage and phase current when deadtime is 3 μs with a PWM switching frequency of 8 kHz. **a** Without deadtime compensation. **b** With deadtime compensation

In Fig. 9.33, it is possible to observe the compensation voltage values, extracted experimentally, that compensate a deadtime of 3 μs. As can be seen, the compensation values are added to the command voltage variables V_{ds_s} and V_{qs_s}.

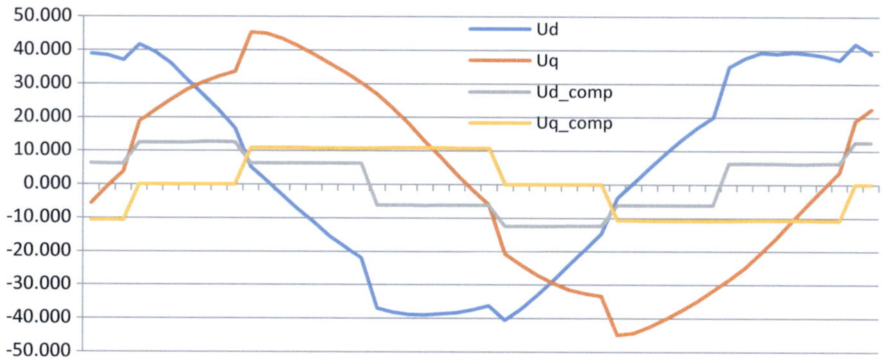

Fig. 9.33 Voltage to compensate V_{ds_comp} and V_{qs_comp}, which is added to the V_{ds_s} and V_{qs_s} voltages in order to reproduce the reference voltage desired

References

Bergas J (2000) Control del motor d inducció considerant els limits del convertidor i del motor. Universidad Politecnica de Catalunya, Barcelona

Chung D-W, Kim J-S, Sul S-K (1998) Unified voltage modulation technique for real-time three-phase power conversion. IEEE Trans Ind Appl 34(2):374–380

Kolar J, Ertl W, Zach F (1990) Minimization of the harmonic RMS content of the mains current of a PWM converter system based on the solution of an extreme value problem. Budapest, Hungary

Lee D-C, Lee G-M (1998) A novel overmodulation technique for space-vector PWM inverters. IEEE Trans Power Electron 13(60):1144–1151

Quang NP, Dittrich J-A (2015) Vector control of three-phase AC Machines

Taniguchi K, Ogino Y, Irie H (1988) PWM technique for power MOSFET inverter

Varma PS, Narayanan G (2015) Space vector PWM as a modified form of sine-triangle PWM for simple analog or digital implementation. IETE Journal of Research 52(6):435–449

Zhou K, Wang D (2002) Relationship between space-vector modulation and three-phase carrier based PWM: a comprehensive analysis. IEEE Trans Industr Electron 49(1):186–196

Chapter 10
Practical Control of AC Machine

10.1 Introduction

The knowledge of the magnetic and electrical model of the machine is of crucial importance for its control, especially in AC machines. The PMSM synchronous AC machines and SynRM/PMASynRM reluctance machines have non-linear relationships between the flux linkages and the dq current components. Moreover, it has their corresponding cross-coupling and the temperature sensitivity of the permanent magnets discussed in Chap. 5. Similarly, it occurs with the induction machine where the leakage and magnetization inductance saturation effects impact on torque and speed dynamics (Otýpka et al. 2016). Without the exact knowledge of the machine model, the machine control can lead to power losses, torque efficiency reduction, and deteriorate the dynamic response. Most of the PMSM and SynRM/PMASynRM and some IM are custom-designed for the application and are increasingly designed using FEA software that facilitates their design. The dq flux linkages parameters as L_d, L_q, L_{ls}, L_{lr}, and L_m are the state variables of the electrical model as shown in Sect. 4.5.1.2. It can be described with a single value or using look-up tables, with the dependency of the current component magnitude. In both cases, it can be obtained experimentally or using FEA simulation. In this chapter, the DC machine control is analyzed as an introduction to vector control for AC machines, where the precision of the machine parameters is the key to obtain maximum control performance.

© Springer Nature Switzerland AG 2020

R. Molina Llorente, *Practical Control of Electric Machines*,
Advances in Industrial Control, https://doi.org/10.1007/978-3-030-34758-1_10

10.2 Control Overview in an Electrical Machines

10.2.1 Rotating Load Speed Control Design

10.2.1.1 Open-Loop Speed Control

In the case of a DC machine, it is assumed that torque is controlled through the armature current I_a neglecting the armature reaction. However, in AC machines, by using vector control, torque is also controlled similarly through the current torque component i_{qse}, which will be described in Sect. 10.3 (Seung-Ki Sul 2011).

The K_m constant of Fig. 10.1 depends on the machine in the study, and it is called the torque constant, which models the conversion of current in torque, as shown in Sect. 4.3.2. In the open-loop control for rotating load systems, there is no feedback from the speed sensor or speed estimator as represented in Fig. 10.1. To analyze the open-loop transfer function of the rotating load system, the PI speed controller is considered.

The block diagram of Fig. 10.1 is valid for AC and DC machines, where torque current I_T can be the armature current for DC machine or the torque component i_{qse} for AC machines. After the PI controller, there is a non-ideal current controller modeled as a first-order low-pass filter with a bandwidth defined by ω_{cr}. Thus, the open-loop transference function $G_{oc}(s)$, taking into account the current controller, is as follows:

$$G_{oc}(s) = K_p\left(1 + \frac{1}{\tau_i s}\right) \cdot \left(\frac{\omega_{cr}}{s + \omega_{cr}}\right)\left(\frac{K_m}{J_T s}\right) \tag{10.2}$$

For simplicity's sake, the non-ideal current controller can be modeled as the ideal current controller, with unitary gain, taking into account the fact that the cut-off frequency of the PI controller is below the cut-off frequency of the current controller ω_{cr}:

$$G_{cr}(s) = \frac{\omega_{cr}}{s + \omega_{cr}} \cong 1 \tag{10.3}$$

Then, the open-loop transference function $G_{oc}(s)$, taking into account unity gain in the current controller can be simplified as:

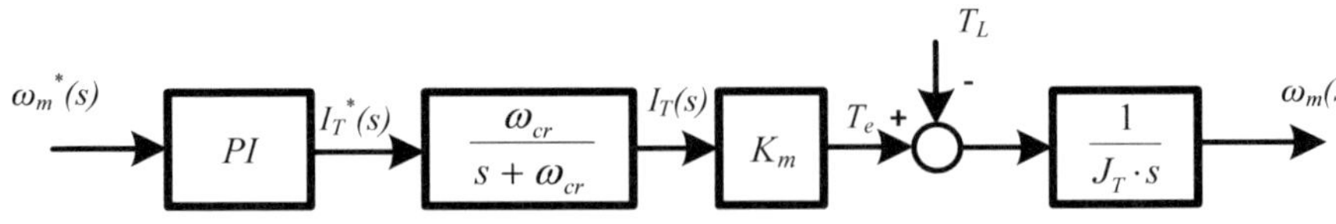

Fig. 10.1 Block diagram of speed open-loop control in a rotating load system

$$G_{oc}(s) = K_p\left(1 + \frac{1}{\tau_i s}\right) \cdot \left(\frac{K_m}{Js}\right) \tag{10.4}$$

The PI block is a proportional-integral controller as discussed in Chap. 2, where the transfer function can be written in Laplace form as follows:

$$G_{PI}(s) = K_p\left(1 + \frac{1}{\tau_i s}\right) \tag{10.1}$$

Both PI gain parameters can be set independently. It should be noted that, in the steady-state sinusoidal analysis, the Laplace operator s can be considered to be a complex variable; that is to say, $s = j\omega$.

The cut-off frequency of the PI regulator ω_{pi} is defined as:

$$\omega_{pi} = \frac{1}{\tau_i} \tag{10.5}$$

It is possible to use a numerical example to represent a Bode diagram. The open-loop transfer function $G_{oc}(s)$ Bode diagram for PI gain parameters $Kp = 5$, $\tau_i = 0.04$, inertia $J = 0.05$ kg $\cdot$ m^2 and with a bandwidth of $\omega_{cr} = 200$ rad/s for the current controller is shown in Fig. 10.2. In Fig. 10.2, the cut-off frequency for PI is likewise shown, as well as the current controller, with different slopes by the poles and zeros of the system. In Eq. 10.6, the numerical $G_{oc}(s)$ is shown based upon prior

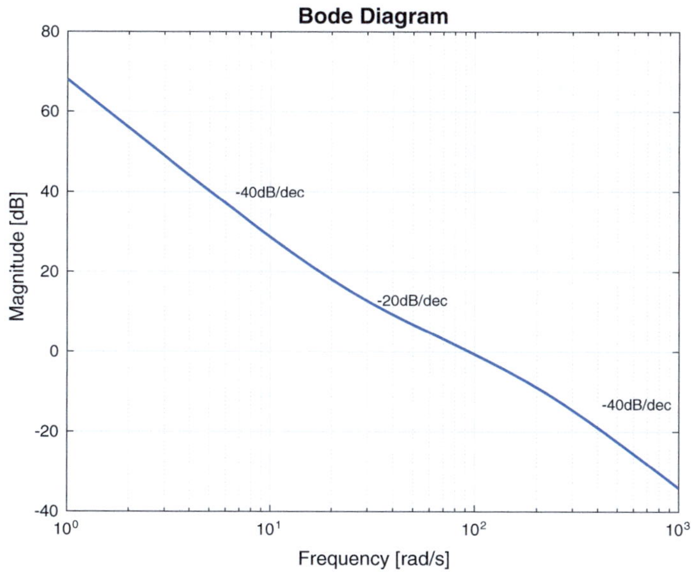

Fig. 10.2 Open-loop Bode diagram with $K_P = 5$, $\tau_i = 0.04$, $J = 0.05$ kg $\cdot$ m^2, and current regulator, with $\omega_{cr} = 200$ rad/s of bandwidth

parameters. The pole is set at the real axis: −25 rad/s; one zero is at real axis: −200 rad/s; and two poles are at the origin.

$$G_{oc}(s) = \frac{40s + 1000}{0.002s^3 + 0.4s^2} \tag{10.6}$$

In the left-hand side of the figure, the slope with −40 dB/dec comes from the inertia of the mechanical load and integral gain of the speed controller. In the middle of the figure, the slope with −20 dB/dec also comes from the inertia but with the proportional gain of the speed controller. This region starts at ω_{pi} which set a pole at −25 rad/s changing the slope. The last slope of −40 dB/dec comes from the current controller and the load, which starts in $\omega_{cr} = 200$ rad/s. From ω_{cr} frequency, the speed controller acts as a proportional controller while the current controller acts as an integral controller.

10.2.1.2 Closed-Loop Speed Control Design

The rotating load system described above can be controlled in a closed-loop manner by using the above PI controller but by using the feedback which comes from a speed sensor. In Fig. 10.3 is depicted the previous system but with the feedback. As can be seen, the instantaneous torque is controlled by using the speed as input. Said in another way, by controlling the torque, speed is controlled.

The transfer function of the complete system described in Fig. 10.3, taking into account unity gain in the current controller, is expressed in (10.7).

$$\frac{\omega_m(s)}{\omega_m^*(s)} = \frac{\frac{K_m K_p}{J_T} s + \frac{K_m K_p}{J_T} \frac{1}{\tau_i}}{s^2 + \frac{K_m K_p}{J_T} s + \frac{K_m K_p}{\tau_i} \frac{1}{J_T}} \tag{10.7}$$

alternatively:

$$\frac{\omega_m(s)}{\omega_m^*(s)} = \frac{K_m K_p(\tau_i s + 1)}{J_T \tau_i s^2 + K_m K_p \tau_i s + K_m K_p} \tag{10.8}$$

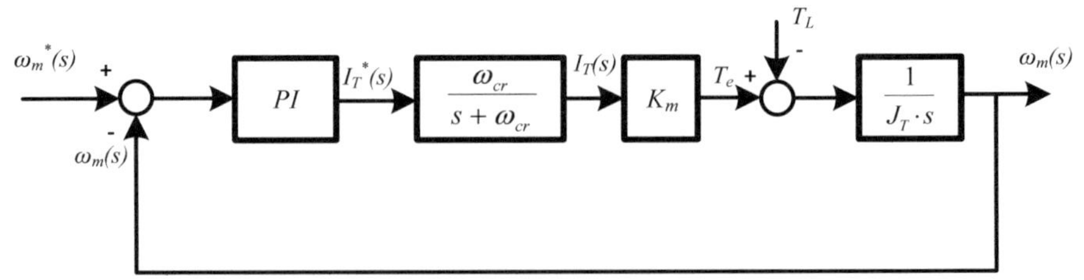

Fig. 10.3 Closed-loop speed control in a rotating load system

This system has one zero and two poles. By observing the transference function of Eq. 10.8 for high frequency, it is possible to find a simplified equation as:

$$\left|\frac{\omega_m}{\omega_m^*}\right| \cong \frac{K_m K_p}{\omega \cdot J_T} \tag{10.9}$$

The frequency ω defines the bandwidth of the speed control loop; that is to say, when the aforementioned transference function reaches a gain equal to unity, the gain is $\omega = \omega_{Bsc}$. Subsequently, the proportional term of the speed controller is calculated as follows:

$$K_p = \frac{J_T \cdot \omega_{Bsc}}{K_m} \tag{10.10}$$

where ω_{Bsc} is the bandwidth of the speed control loop.

The speed control loop bandwidth can be set to 100 rad/s for common rotating load control, and if total inertia is known, the proportional gain constant of PI controller can be calculated by Eq. 10.10.

Now, integrator time constant τ_i can be designed by the expected response to the unitary step. From Eq. 10.8, a denominator can be written as $s^2 + 2\zeta\omega_o + \omega_o^2$, given the damping coefficient and natural frequency, respectively:

$$\zeta = \frac{1}{2}\sqrt{\frac{K_m K_P}{J_T}\tau_i} \tag{10.11}$$

$$\omega_o = \sqrt{\frac{K_m K_P}{\tau_i J_T}} \tag{10.12}$$

By substituting (10.10) into (10.11) and (10.12) is possible to obtain the damping coefficient and natural frequency as a function of the desired speed control bandwidth ω_{Bsc} and the integrator time constant τ_i:

$$\zeta = \frac{1}{2}\sqrt{\omega_{Bsc}\tau_i} \tag{10.13}$$

$$\omega_o = \sqrt{\frac{\omega_{bsc}}{\tau_i}} \tag{10.14}$$

According to control theory, the damping coefficient can be selected as, for example, $\zeta < 1$, if an oscillatory response decaying exponentially (underdamped) is desired. Taking $\zeta = 1$ causes critical damping since this value is on the boundary between overdamped and underdamped. Figure 10.4 shows a step response simulation of the system represented in Fig. 10.3 with three damping coefficients: $\zeta = 1$, $1/\sqrt{2}$, and $\sqrt{2}/4$.

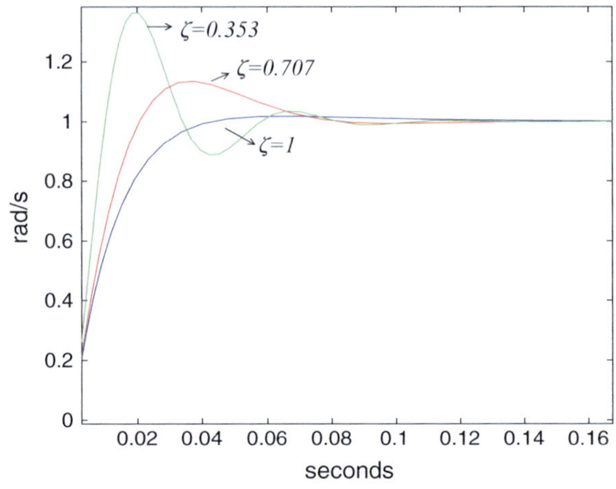

Fig. 10.4 Step response with $\zeta = 1, 1/\sqrt{2}$, and $\sqrt{2}/4$

In four-quadrant electric drives, it is mandatory to select a proper damping coefficient to avoid over speed in the step response, sometimes causing regeneration. In over reference speed, with damping coefficient close to $\sqrt{2}/4$, the speed controller in a VSI for AC machines tries to brake the load a bit, causing regeneration to the DC-link bus capacitor and thus incrementing the voltage as seen in Chap. 8. By observing Fig. 10.4, it is possible to appreciate that a damping coefficient equal to one is an optimal choice according to the previous explanation. On the contrary, as can be seen, the time response is a little higher than the value it takes in cases of overdamping.

In resume, the PI gain parameters are calculated as illustrated in Eqs. (10.15) and (10.16), with $\omega_{Bsr} = 100$ rad/s of bandwidth for the speed loop control, and selecting the damping coefficient. For simplicity's sake, the selected torque constant K_m is one as commented before. The total inertia of the system J_T is 0.05 kg $\cdot$ m^2.

$$K_p = \frac{J_T \cdot \omega_{Bsc}}{K_m} \tag{10.15}$$

$$\tau_i = \frac{4\zeta^2}{K_m \omega_{Bsc}} \tag{10.16}$$

In Table 10.1 is represented different damping coefficients for different combinations of PI gain parameters when total inertia is 0.05 kg $\cdot$ m^2.

Table 10.1 Values used in Simulink simulation

ζ	K_p	τ_i
$\sqrt{2}/4$	5	0.005
$1/\sqrt{2}$	5	0.02
1	5	0.04

In Fig. 10.5 is shown the Simulink® closed-loop dynamic model of rotating load system with a speed PI controller used in the simulations

In Fig. 10.6 is shown the root locus and Bode diagram closed-loop simulation results for two damping coefficients. In case of ζ =1, the system has two poles on the complex s plane, exactly at $-\omega_{Bsr/2}$. On the contrary, for $\zeta = 1/\sqrt{2}$ there are two complex conjugate poles located on the complex s plane at $-\omega_{Bsr/2} \pm j\omega_{Bsr/2}$. Here, zero is located at $-\omega_{Bsr/2}$. According to the Bode diagrams, the cut-off frequency is

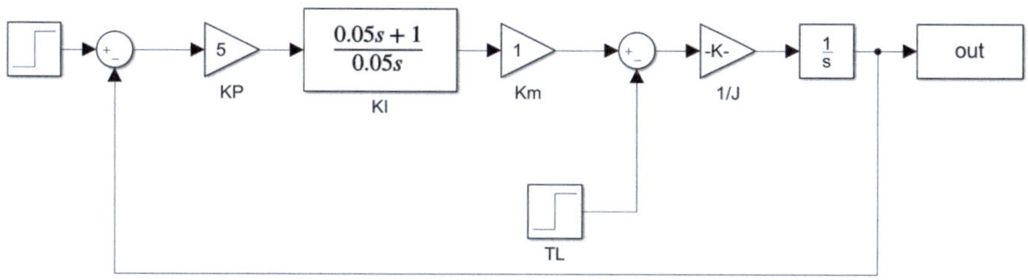

Fig. 10.5 Closed-loop dynamic model of rotating load system with a speed PI controller

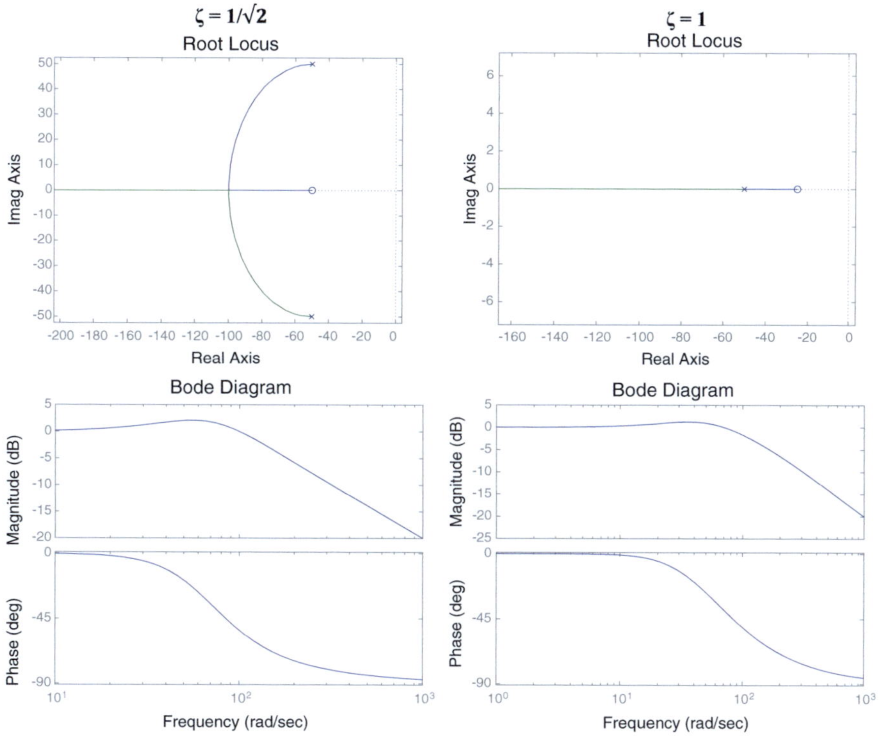

Fig. 10.6 Bode diagram and root locus simulation results for two damping coefficients, $\zeta = 1$ and $1/\sqrt{2}$

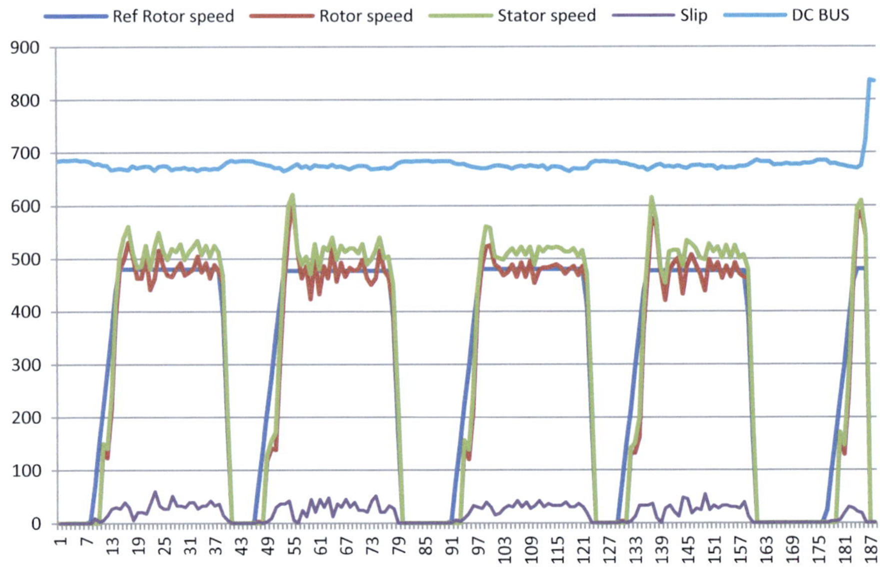

Fig. 10.7 Different start-up processes for induction machine with belt load where over speed is present in some cases

100 rad/s, as expected, which corresponds to a unitary gain. From this frequency, the transference function is calculated by Eq. 10.9. That is to say, the gain is inversely proportional to the frequency. On the other hand, likewise shown is how the zero increases the gain before the cut-off frequency.

As mentioned before, if the PI gain parameters are not chosen correctly, response to the speed step can cause over speed and sometimes overvoltage in the DC-link line, as shown in Fig. 10.7. Figure 10.7 shows a real test in an induction machine and VSI showing the DC-link voltage and machine speed in Hz. The DC-link voltage average is around 670 V_{DC} for a 480 VAC input voltage grid. The machine speed is multiplied by a factor equal to 30 to be able to be compared with the DC-link voltage. The last start-up shows how the rotor speed is higher (underdamped) than the reference rotor speed (speed reference of 16.1 Hz), causing—in this case—an overvoltage to 837 V_{DC}, when control brakes a bit to rapidly return to the speed reference. It should be noted that in this real example, the plant has a timing belt system, with the load having randomly low or high inertia. Thus, selecting the PI gain parameters is a difficult task, as total inertia J_T is not fixed. In this case, it is necessary to study the plant with the previous analysis and fine-tune the PI gain parameters using simulation or experimental tests.

10.2.2 PI Current Control Design

10.2.2.1 Current Control Design for a DC Machine

In Fig. 10.8, the current regulator made with a PI controller for a DC machine is shown. A new constant K_c is added by the power stage, which it will discuss in Sect. 10.2.3, used to represent how the voltage command of the DC machine varies with the output of the current PI controller.

Note that the friction coefficient B is assumed to be zero for simplification.

In order to focus on the current controller loop, it is necessary to avoid the dynamic parts of the DC machine model such as inertia and load torque. In accordance with the aforementioned, it only has the back-EMF voltage V_a affecting the current controller loop, but this can be neglected if the inertia of the machine is high enough to consider the back-EMF as a constant. Then, the DC machine is modeled such as simple *RL* circuit as shown in Fig. 10.9.

The open-loop transference function $G_{oc}(s)$ considering the current controller is as follows:

$$G_{oc}(s) = K_p\left(1 + \frac{1}{\tau_i s}\right) \cdot K_c\left(\frac{1}{R_a + sL_a}\right) \tag{10.17}$$

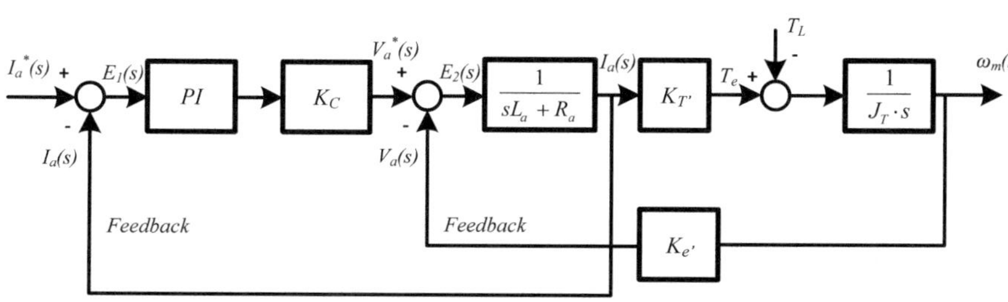

Fig. 10.8 DC machine with a PI current regulator

Fig. 10.9 DC machine modeled with an RL circuit with a PI current controller

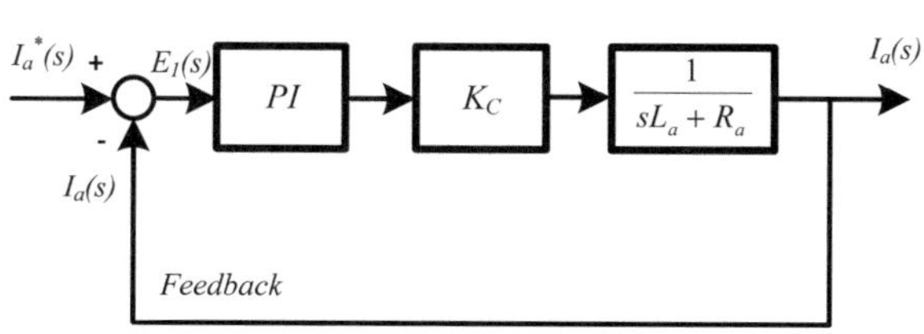

The pole of the system caused by the *RL* circuit of DC machine model can be canceled with a zero as the PI controller:

$$G_{oc}(s) = K_p\left(1 + \frac{1}{\tau_i s}\right) \cdot K_c\left(\frac{\frac{1}{R_a}}{1 + \frac{L_a}{R_a}s}\right) \tag{10.18}$$

$$\tau_i = \frac{L_a}{R_a} \tag{10.19}$$

This method, which cancels the pole by using the zero of the controller, is called the magnitude optimum (MO) method. The aim is to obtain a control system with a frequency characteristic close to unity and the flattest maximum bandwidth possible. An alternative to the MO tuning method with better performance, when compared with load disturbance, is the symmetric optimum (SO) tuning method. This method maximizes the phase margin of the control system's leading symmetrical phase and amplitude characteristics. Both tuning methods are widely used in cascade control systems, like in machine control (Perić 1979; Deur 1999).

By introducing (10.19) into (10.18), it is possible to get the open-loop transference function (10.20) as specified below:

$$G_{oc}(s) = K_p K_c\left(\frac{1}{sL_a}\right) \tag{10.20}$$

Finally, the closed-loop transference functions can be written as follows:

$$\frac{i_a(s)}{i_a^*(s)} = \frac{\frac{K_p K_c}{L_a}}{s + \frac{K_p K_c}{L_a}} \tag{10.21}$$

It should be noted that the simple closed-loop transference function corresponds to the first-order low-pass filter, where the bandwidth of the regulator is the cut-off frequency defined as (10.22):

$$\omega_c = \frac{K_p K_c}{L_a} \tag{10.22}$$

By selecting the bandwidth of the current controller properly, in the end, the PI gains (Table 10.2) can be calculated by (10.19) and (10.23):

$$\begin{aligned} K_p &= \frac{\omega_c L_a}{K_c} \\ \tau_i &= \frac{K_p K_c}{R_a \omega_c} \end{aligned} \tag{10.23}$$

Table 10.2 Values used in the Simulink simulation

L_a	R_a	ω_c	K_p	τ_i
25 mH	1.5 Ohms	100 rad/s	2.5	0.016

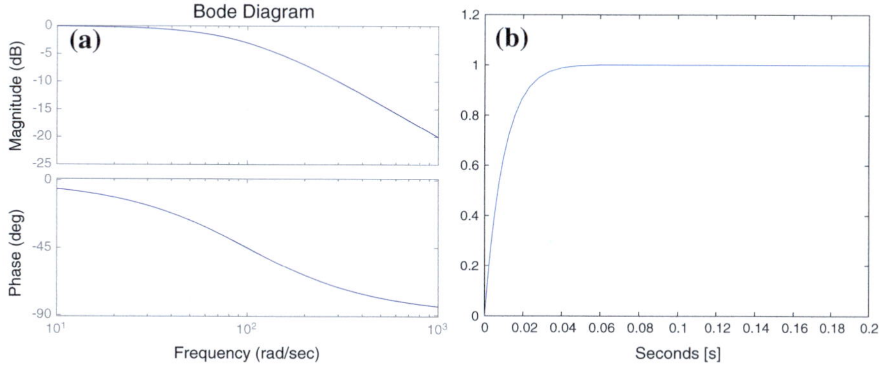

Fig. 10.10 **a** Closed-loop control bode diagram. **b** Unitary current step response

A Bode diagram simulation result is shown in Fig. 10.10, which corresponds to a first-order low-pass filter with a cut-off frequency of 100 rad/s. Note that the margin phase is exactly $-45°$, as expected.

In accordance with the PI parameter selection, the response obtained from the current regulator when faced with a unitary current step does not have overshoot, and its error in the steady-state is zero, as can be seen in Fig. 10.10b.

10.2.3 DC Servo Motor Drive Model-Based Simulation

This section intends to put into practice those discussed above for the DC machine control with two models: the model of state-space equations discussed in Sect. 5.3.1 and the Simscape™ model of Simulink® of a DC machine together with its power stage. The model in state-space equations is a model based on differential equations as it was seen that does not take into account the voltage drop in the power devices, and the control is not done with a PWM but a percentage of the voltage V_a according to the duty cycle. This model is used to perform simulations and check stability, also, to extract the transient response, the frequency response, as well as the disturbances in a faster way. The most accurate model is achieved with the power stage and a more accurate model of the DC machine, but as will be seen below, both models present very similar results. The power stage consists of a single-phase half-bridge, as illustrated in Fig. 10.11, which is composed of two MOSFET devices. The switching frequency of the MOSFET devices is usually much larger than the frequency of the bandwidth of the current regulator so that the power converter can be modeled as a voltage amplifier with a constant gain K_c. The half-bridge restricts the operation of the DC machine to a single direction of rotation in both motor mode and generator mode (2-quadrants). In other words, in motor mode, the current cannot be inverted as it would be with a full-bridge inverter, as discussed in Chap. 8.

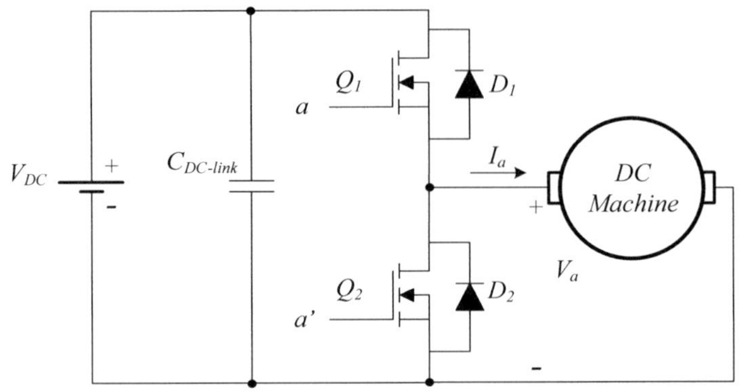

Fig. 10.11 DC machine control in two quadrants by using a single-phase half-bridge inverter

The DC machine corresponds to a low power machine of 50 W and 12 V of base voltage. The parameters are shown in Table 10.3. By using these parameters and setting a current controller bandwidth of 100 rad/s and a damping coefficient of $\zeta = 1$, it is possible to calculate the PI gain parameters of the current loop and the speed loop according to Eqs. (10.15), (10.16), (10.19), and (10.23) described above. The sampling time of the current loop is equal to the PWM frequency, that is 100 μs, while the sampling time for the speed loop is 1 ms. This PWM frequency helps to have a lower ripple in the armature current I_a as will be seen below.

Table 10.4 shows the parameters of the PI speed and current controllers calculated according to Eqs. (10.15), (10.16), (10.19), and (10.23).

Table 10.3 DC machine parameters

Parameters for DC machine	Symbol	Value	Unit
Armature inductance	L_a	0.276	mH
Armature resistance	R_a	0.57	Ohms
Back-EMF constant	K_e	0.0175	V/(Wb turn rad/s)
Torque constant	K_T	0.0175	Nm/(Wb turn A)
Friction torque	T_f	0.005	Nm
Rotor inertia	J	5.5e−6	$kg \cdot m^2$
Damping coefficient	B	0	$kg \cdot m^2/s$

Table 10.4 Controller PI parameters

PI controllers	Symbol	Value
Speed proportional part	K_{Ps}	0.0886
Speed integral part	K_{Is}	2.2857
Current proportional part	K_{Pc}	0.0276
Current integral part	τ_{Ic}	0.00048421

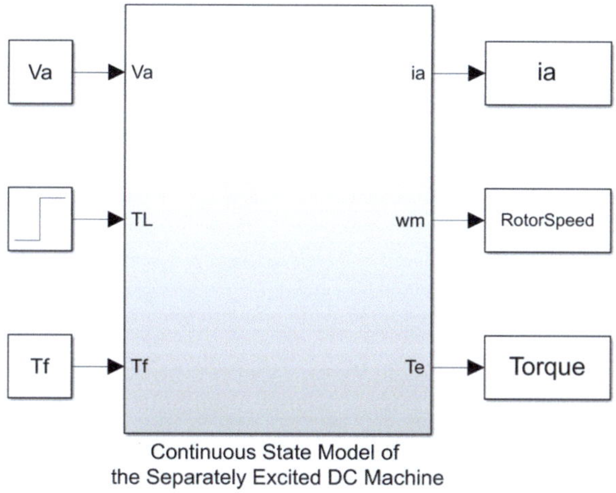

Fig. 10.12 State-space DC machine model in open loop

10.2.3.1 Open-Loop Simulation

In Fig. 10.12 is shown the state-space model discussed in Sect. 5.3.1, while Fig. 10.14 illustrates the model based on a half-bridge and the model of DC machine corresponding Simscape™ toolbox from Simulink®. Both models are used in open-loop simulations with the load torque disturbance input.

The open-loop response when the duty cycle is fixed to 0.2 is illustrated in Fig. 10.13 for both models. The armature current I_a, the speed in RPM, and a detail of the current I_a are represented. The results of the simulation are very similar for both models, as can be seen.

In the state-space model, the speed achieved is slightly higher since the model does not take into account the voltage drop in the MOSFET devices, diodes, and in the internal resistance R_i of the voltage supplied source. On the other hand, the ripple of armature current can be observed when used the more realistic model with Simscape components.

10.2.3.2 Closed-Loop Simulation

In Fig. 10.15 is shown the model for the closed-loop simulation of the two loops, the current and speed loop through the model of the PID controller seen in Chap. 2. The current and speed measurement is filtered through a low-pass filter. Also, in the model, a delay is applied to take into account the delay in signal processing. The current is sampled every 100 μs for the current loop and every 1 ms for the speed loop. Both have a zero-order hold model as can be observed. The PI output of the current loop corresponds to the duty cycle, which is used to generate the PWM signal for the high- and low-side MOSFET devices.

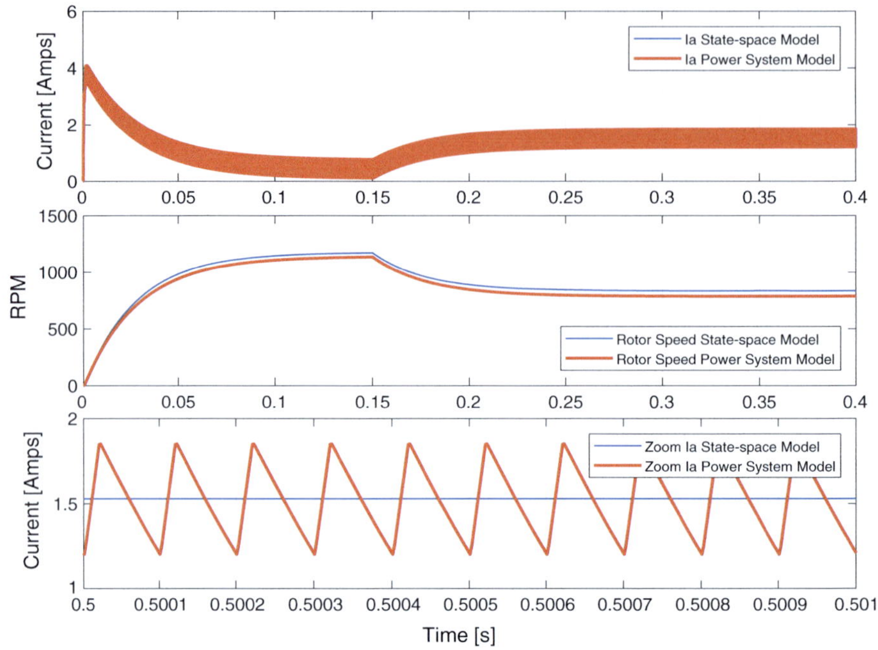

Fig. 10.13 Open-loop response with no load (only $T_f = 0.005$ Nm) for both models when the duty cycle is set to 0.2. At $t = 0.15$, disturbance torque of 0.02 Nm is applied to reduce the machine speed

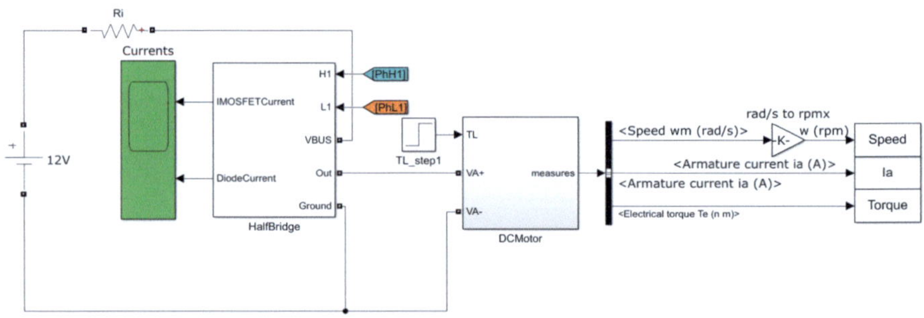

Fig. 10.14 Power system and DC machine Simscape models

The closed-loop response is illustrated in Fig. 10.16 for both models. The armature machine current I_a, the speed in RPM, the duty cycle, and the electromagnetic torque are represented. The results of the simulation are very similar for both models, as was the case in the previous case. The response to the reference speed step of 1000 RPM is very fast and without overshoot as expected according to the analysis discussed in this section. The duty cycle reaches the value of 0.9 during the first instants until the speed remains close to the reference speed, at which point it rapidly drops to

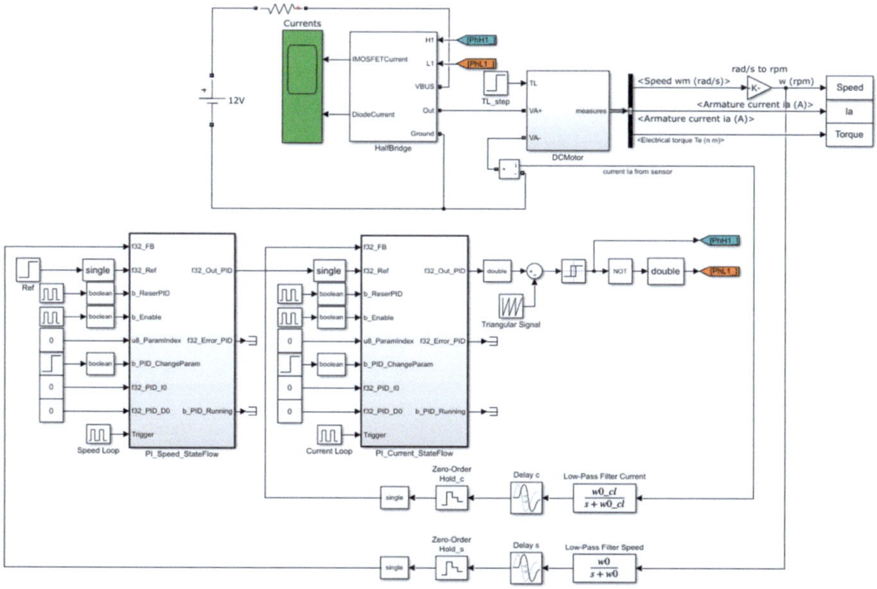

Fig. 10.15 Closed-loop control system using the Stateflow® PID of Chap. 2

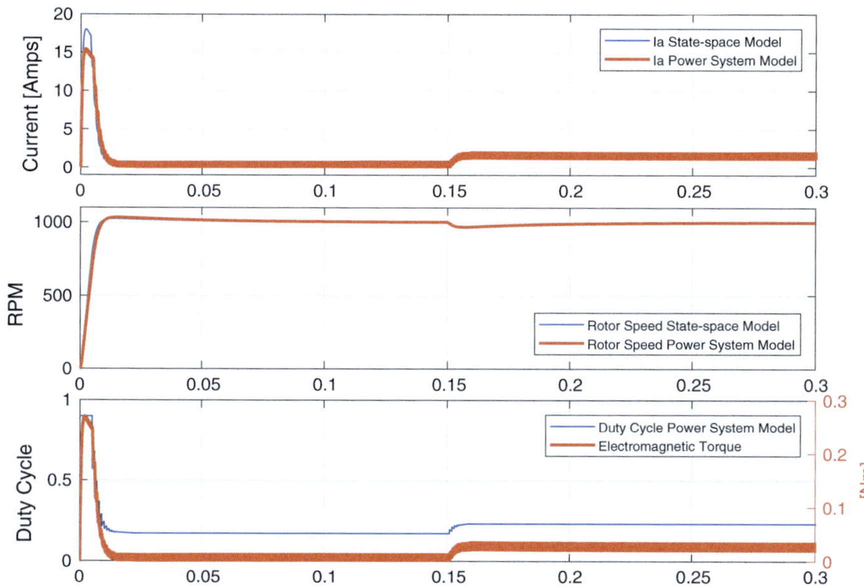

Fig. 10.16 Closed-loop response with no load (only $T_f = 0.005$ Nm) for both models. At $t = 0.15$, disturbance torque of 0.02 Nm is applied where the machine speed is controlled

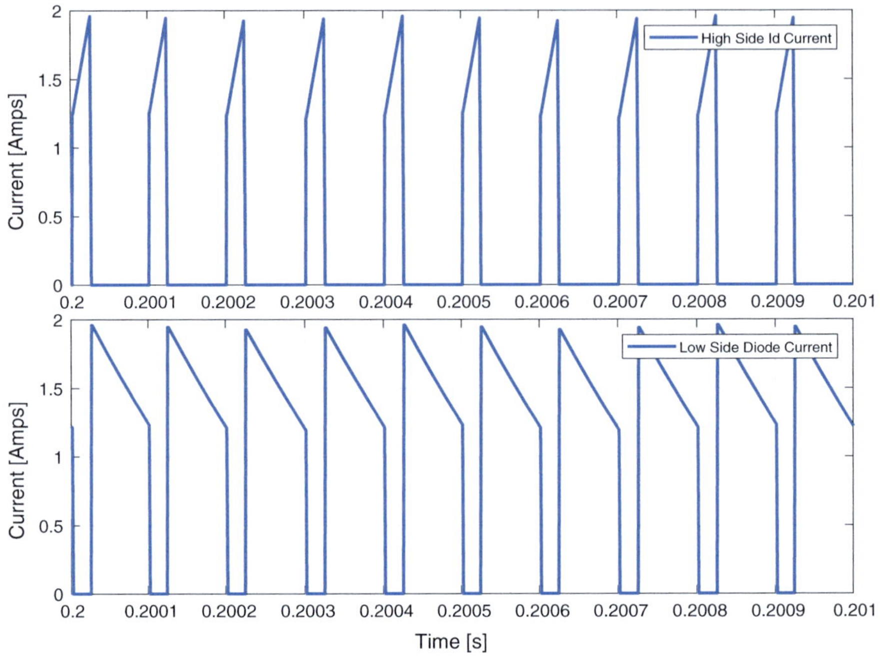

Fig. 10.17 Detail of drain current of high-side MOSFET and the diode of the low-side MOSFET

approximately 0.2. When a load torque disturbance is applied at time instant $t = 0.15$ s, the machine speed is not almost affected.

Lastly, Fig. 10.17 shows the detail of the drain current flowing through high-side MOSFET and the current flowing through the diode of the low-side MOSFET.

10.3 Principle of Vector Control

The independent control between the excitation current and the armature current is possible in separately excited DC machines as seen in the present chapter and Chap. 4. In these machines, the stator current i_f sets the magnetic flux, and the rotor armature current i_a is used as direct torque control. The physical position of the brushes concerning the field of the stator ensures the optimum torque condition under all possibilities. The most basic operation of the DC machine is shown in Fig. 10.18 in a simplified way. The rotor armature is formed by a winding with three conductors which rotates counterclockwise. The pair of forces F rotates the rotor in the counterclockwise direction switching the commutator, to change the direction of the current in the loop and therefore its orientation of the field. As a consequence, the direction of the pair of forces changes.

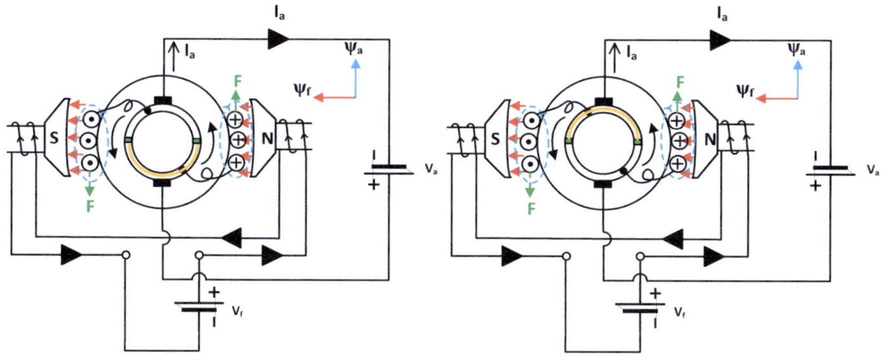

Fig. 10.18 Mechanical field orientation using the brushes

This process continues and is repeated every turn of the rotor. In this way, the field is always perpendicular to the armature current and ensures an optimum torque and an always oriented field.

In brushless AC machines, no mechanical device fixes the relative positions between the stator and rotor fluxes, being no longer linear the expression of the torque, unlike the DC brush machine. In induction machines, the armature windings, as discussed, are in the rotor, while the field is generated by the currents of the stator windings. The rotor current was generated from the electromotive force that was induced in the rotor winding as a consequence of the relative movement of the rotor conductors concerning the rotor field. The rotor was normally inaccessible because it was a fully closed or short-circuited, either by short-circuited bus bars (squirrel cage) or by windings also short-circuited. In the case of the synchronous machines as PMAC and SynRM, the field is generated by the currents of the windings of the stator as in the case of the IM, while the rotor is based on permanent magnets, or with a determined structure section that it formed a variable reluctance respectively. However, in the BLDC machine, as was seen in Sect. 8.2.2, the correct orientation of the stator magnetic field with respect to the rotor was guaranteed thanks to the rotor position sensor. It emulated the DC machine operation with an electronic commutation of the stator winding phases according to the rotor position.

The principle of field-oriented control (FOC) defines the conditions for separating field control from torque control. That is, both the magnetic field and the torque developed by the machine can be controlled independently. It allows the optimum torque production conditions (maximum torque per unit of current) when the machine operated in steady-state and when the machine is in transient operating conditions.

One of the objectives of the FOC control is to achieve a torque expression for AC machines linear like DC machines. The linear expression of the torque developed from a DC machine as discussed can be expressed as a function of the rotor armature currents i_a, and the current of the stator i_f, or using the magnetic fluxes λ_a of the rotor, and the stator λ_f

$$T_{em} = Ki_a i_f = K'\lambda_a \lambda_f \tag{10.24}$$

where K and K' are constants. The flux is controlled by adjusting the stator field current, i_f, and the torque is controlled independently by adjusting the current i_a. Usually, i_a is selected as a torque control variable, while i_f it is kept constant, to have a constant and linear flux.

In general, torque control in an IM and a SynRM is not as straightforward as in a separated excited DC machine or a PMAC machine, because the field current and torque current must be generated from the stator windings. However, in the IPMSM and SPMSM, the current for torque control is controlled using the stator voltages, while the field current is imposed by the permanent magnets.

The high performance of DC machine control can be extended to AC machines if the control uses the *dqe* synchronous reference frame, where the variables appeared as DC quantities. Therefore, the electromagnetic torque of the induction machine can be expressed as (10.25):

$$T_{em} = Ki_{ds}^e i_{qs}^e = K'\lambda_r i_{qs}^e \tag{10.25}$$

The stator current components of the quadrature axis q and the direct axis d rotate with the synchronously rotating reference frame. In the vector control theory, i_{dse} is analogous to the field current i_f and i_{qse} is analogous to the armature current i_a of the DC machine.

The emulation of the DC machine on the AC machines is only possible if i_{dse} is oriented or aligned with the direction of the rotor flux and if i_{qse} is always perpendicular to the flux vector. For a perfect decoupling between both currents, if i_{qse} varies, it affects only the current i_{qse} but does not affect the flux component. Likewise, when i_{dse} varies, only the flux is controlled without affecting the current component i_{qse}.

10.4 Sensored Vector Control

10.4.1 Induction Machine

There are two methods of vector control, indirect and direct for induction machines. In the indirect method, the flux and current pair components are indirectly adjusted through the slip, which is controlled. In other words, the position of the rotor flux θ_e of Eq. 10.26 is indirectly controlled by adjusting the slip. According to (10.26), it is essential to measure or estimate the instantaneous speed ω_r of the rotor.

$$\theta_e = \int_0^t \omega_e \mathrm{d}t = \int_0^t (\omega_r + \omega_{sl}) \mathrm{d}t \tag{10.26}$$

On the other hand, the direct method consists of identifying first the position and rotor flux magnitude to control the torque instantaneously directly, using the flux and torque current components. These two components are the decomposition of the stator current, where both are controlled by its magnitude. For the identification of the position and magnitude of the rotor flux, vector is not a simple task so that there are different methods. One of the first methods more intuitive without advanced digital processing is to measure the air gap flux through Hall effect sensors. These can measure the density of magnetic flux, but it is often difficult to install this type of sensors in the induction machine. However, thanks to the advance in digital processing in microcontrollers and DSPs, the position and magnitude of the rotor flux vector can be identified through the measurement of the phase current and the voltage at the machine terminals as it will be seen in Sect. 10.6.2.

The indirect vector control method is usually used in sensored vector control in IM since it requires few machine parameters for its implementation. It needs the instantaneous measurement of the rotor speed, which is measured by a speed sensor (sensored). The high-performance control is achieved with a speed sensor based on an incremental optical encoder or a resolver, both preferred with a high resolution. The sensored solution of vector control is a more expensive solution, but it requires less performance in the microprocessor, although it involves the placement of the speed sensor on the axis of the machine where it is sometimes inaccessible or very complicated. However, it is also possible to obtain the rotor speed employing a speed estimator based on the machine model, as will be seen in Sect. 10.6.2, where the performance requirements of the microprocessor will be magnified. The speed estimation supposes a lower precision in the regulation concerning the obtained one with an optical encoder, but for the majority of applications, the requirements of the load are fulfilled.

For the analysis of the indirect control of the induction machine is possible to use the expressions in the synchronously rotating reference frame deduced in Sect. 4.4.1 with $\omega = \omega_e$, where the reference axes *dqe* rotated at the synchronous speed ω_e of the magnetic field. Thus, the expression of the electromagnetic torque can be expressed as (10.27):

$$T_e = \frac{3}{2}P \cdot \frac{L_m}{L_r}\left(i_{qs}^e \cdot \lambda_{dr}^e - i_{ds}^e \cdot \lambda_{qr}^e\right) \tag{10.27}$$

Under the assumption that the rotor flux linkage exists only on *d*-axis, the electromagnetic expression can be represented in the synchronously rotating reference frame as represented in (10.28):

$$T_e = \frac{3}{2}P \cdot \frac{L_m}{L_r} i_{qs}^e \cdot \lambda_{dr}^e \tag{10.28}$$

where flux in the *q*-axis is

$$\lambda_{qr}^e = 0 \tag{10.29}$$

Equations (4.39) and (4.40) derived in Sect. 4.4.1 can be written in the synchronously rotating reference frame as:

$$v_{ds}^e = R_s i_{ds}^e + \frac{d\lambda_{ds}^e}{dt} - \omega\lambda_{qs}^e \tag{10.30}$$

$$v_{qs}^e = R_s i_{qs}^e + \frac{d\lambda_{qs}^e}{dt} + \omega\lambda_{ds}^e \tag{10.31}$$

$$0 = R_r i_{dr}^e + \frac{d\lambda_{dr}^e}{dt} - (\omega - \omega_r)\lambda_{qr}^e \tag{10.32}$$

$$0 = R_r i_{qr}^e + \frac{d\lambda_{qr}^e}{dt} + (\omega - \omega_r)\lambda_{dr}^e \tag{10.33}$$

From (10.33), the slip can be derived assuming that derivative terms are zero in steady-state:

$$\omega_{sl} = \omega - \omega_r = -\frac{R_r i_{qr}^e}{\lambda_{dr}^e} \tag{10.34}$$

Since rotor currents are not measurable, it is common to eliminate from the equation system. The rotor flux linkage was expressed as

$$\lambda_{dr}^e = L_{lr} i_{dr}^e + L_m (i_{ds}^e + i_{dr}^e) \tag{10.35}$$

$$\lambda_{qr}^e = L_{lr} i_{qr}^e + L_m (i_{qs}^e + i_{qr}^e) \tag{10.36}$$

By substituting (10.29) into (10.36), and using $L_r = L_{lr} + L_m$, the rotor current in the q-axis can be expressed as:

$$i_{qr}^e = -\frac{L_m i_{qs}^e}{L_r} \tag{10.37}$$

Then, Eq. 10.34 can be rewritten as

$$\omega_{sl} = \omega - \omega_r = \frac{R_r}{L_r}\frac{L_m i_{qs}^e}{\lambda_{dr}^e} \tag{10.38}$$

As can be observed, the slip can be controlled with q-axis component of the stator current and the d-axis of the rotor flux linkage. Now, if rotor flux linkage in the d-axis λ_{dr} does no change (constant flux), its derivative term is zero. By using this condition and the previous ($\lambda_{qr} = 0$), Eq. 10.32 becomes:

$$0 = R_r i_{dr}^e \tag{10.39}$$

It means that the rotor current d-axis component is forced to be zero when previous conditions are assumed. Then, the rotor flux linkage only exists in the d-axis and is expressed as

$$\lambda_{dr}^{e} = L_m i_{ds}^{e} \tag{10.40}$$

The slip expression (10.38) can be described by the two dq stator current components, by substituting (10.40) into (10.38):

$$\omega_{sl} = \omega - \omega_r = \frac{R_r}{L_r}\frac{i_{qs}^{e}}{i_{ds}^{e}} \tag{10.41}$$

Then, the electromagnetic torque can be rewritten as:

$$T_e = \frac{3}{2}P \cdot \frac{L_m^2}{L_r} i_{qs}^{e} \cdot i_{ds}^{e} \tag{10.42}$$

In practice, the magnitude of the rotor flux vector is adjusted by controlling i_{dse}, and the orientation of the d-axis of the rotor field is maintained by either keeping the slip speed or i_{qse} according to (10.41).

With the proper orientation of the field, the dynamics of λ_{dre} is limited to the axis d and is determined by the rotor time constant T_r of the rotor circuit L_r/R_r. It can be seen in (10.32) with i_{dre} replaced by

$$i_{dr}^{e} = \frac{1}{L_r}(\lambda_{dr}^{e} - L_m i_{ds}^{e}), \quad \lambda_{qr}^{e} = 0 \tag{10.43}$$

That is,

$$\frac{\mathrm{d}\lambda_{dr}^{e}}{\mathrm{d}t} + \frac{R_r}{L_r}\lambda_{dr}^{e} = \frac{R_r}{L_r}L_m i_{ds}^{e} \tag{10.44}$$

The rotor time constant has a significant role since the machine dynamic depends on this constant. It can give an idea of the current control loops bandwidth. As a consequence, the sampling frequency for the digital–analog converter, when abc phase currents and voltages in control are feed, can be chosen approximately. On the other hand, the rotor time constant is directly linked to the field orientation control. That is, the change of temperature of the rotor can affect directly to the precision of the control since the resistance R_r varies with the temperature. When the machine has been stopped for some time, the resistance of the rotor has a specific value according to ambient temperature. However, when the machine has been running for hours, the rotor has a higher temperature, and as a consequence, its resistance is also higher than the initial one. The rotor resistance may increase by 50% with the temperature when the machine is running and can be increased a few times more with the skin effect if the rotor current frequency is high.

In addition, the rotor inductance changes with magnetic saturation, which is a nonlinear relationship between this change value and the magnetic saturation degree. Thus, if the rotor time constant changes with the machine temperature and magnetic saturation, the decoupling conditions of the flux and torque components are not ideal, degrading the IM dynamic performance. There are different methods to compensate this deviation (Zai et al. 1992; Foulon et al. 2007) which consist in the online parameter identification (when the machine is running). For example, it is possible to cite the observer-based technology, based on extended Kalman filter (EKF), and the model reference adaptive system (MRAS) identification technology. There are sophisticated algorithms which require the model machine and some measurements to be processed in a high-performance CPU processor.

Using the previous analysis, the rotor flux λ_{dre} and the torque current component i_{qse} are octogonal, achieving the optimal torque production, maintaining the analogy with the DC machine with separate excitation. In Fig. 10.19 is depicted the schematic of the indirect vector control of the IM with the feedforward compensation.

It is possible to observe that the indirect vector control uses the rotor speed and the slip frequency. The rotor flux position θ_e is to be processed according to (10.26). The angle θ_e is used to perform the axis transformation. For example, the synchronously rotating reference frame is transformed to the *abc* components by using the inverse of the rotation matrix and the inverse of the Clarke transformation. There are two current controllers one for each *dq* component, which regulates the direct rotor flux current and the quadrature torque current, respectively. Hence, the commands are the rotor and the torque current references. The direct rotor flux current reference is

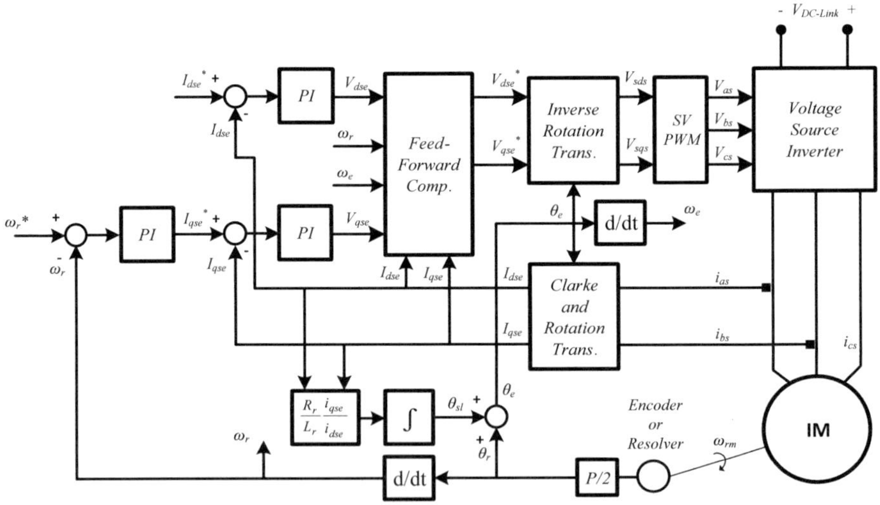

Fig. 10.19 Block diagram of the indirect vector control of the IM with feedforward compensation

constant during constant torque region till induction machine rated speed. Thus, the rotor flux linkage is also a constant according to:

$$i_{ds}^{e} = \frac{\lambda_{dr}^{e}}{L_m} \quad (10.45)$$

At higher speeds, the flux weakening should be applied as it will be shown in Sect. 10.5.1. It is not trivial to set the correct value of the rated rotor flux. With the help of FEA simulations it is possible to find an optimal value of rated rotor flux. On the contrary, as a reference value can be approximated by knowing the rated power factor and the rated current of the machine. Thus, the stator direct component i_{dse} can be approximated with the imaginary part:

$$i_{ds}^{e} = I_n \sin \varphi \quad (10.46)$$

where $cos\ \varphi$ is the power factor and I_n is the rated current of the IM.

It should be noted that when the load torque is small, or machine is operating without load, the efficiency of the machine is reduced. In this situation, the flux λ_{dre} is the rated (and also the flux current component i_{dse}), but the current torque component is close to zero to regulate the no-load torque. To avoid this situation, it is possible to set the flux current i_{dse} with a dependence of the torque current $i_{qse.}$ That is, the more torque demand there is, the more torque current is required, and therefore, a higher flux current component is needed. For example, within limits, the flux current component i_{dse} can be K times the torque current component $i_{qse.}$

The feedback of the current controllers is the measurement of the phase current of the machine. For balanced IM where 0-component is null, it is only necessary to measure two currents as observed in Fig. 10.19. The machine phase current a and b are converted in the synchronously rotating reference frame since the current controllers operate in DC quantities. For this reason, there is a block which applies the Clarke and rotation transformation based in the angle θ_e. The torque reference comes from another controller which regulates the speed based on the speed reference ω_{ref} and the speed measured as can be observed. The controllers can be, for example, a common PIs controllers with anti-windup as shown in Fig. 10.19. Lastly, the feedforward compensation is added to cancel the cross-coupling of the stator current components, inherent to the machine model, which improves the control performance as seen in Chap. 5 for the SynRM. The parameters needed to implement the sensored indirect vector control are R_r, L_r, and L_m to set the desired rotor flux. The accuracy of these parameters joins to the signal processing which determines the accuracy of the indirect vector control performance.

10.4.2 SynRM/PMASynRM

The SynRM and their PMASynRM variant use the principle that electromagnetic torque is produced to minimize the resistance of magnetic paths. The three-phase stator currents produce a flux that rotates at the synchronous speed in the air gap. When the rotor rotates in synchrony with this flux, a torque is produced which attempts to align the minimum reluctance path of the rotor with the rotational flux of the air gap. When a load torque is applied, the rotor begins to delay the rotating flux in the air gap, producing misalignment of the minimum path between the reluctance and the rotating flux of the air gap. In this way, an electromagnetic torque is produced to minimize the resistance that attempts to maintain the alignment. When this torque is equal and opposite to the load torque in the rotor, the rotor will rotate again with synchronous speed. As discussed in Chap. 4, the expression of electromagnetic torque developed by the PMASynRM was:

$$T_e = \frac{3}{2}P\Big[(L_d - L_q)i_{ds}^e i_{qs}^e + \Psi_m i_{ds}^e\Big] \tag{10.47}$$

where the flux of the permanent magnet Ψ_m is zero for the SynRM. It is possible to observe, as in the case of the induction machine, that the torque can be controlled by the two stator current components in the synchronously rotating reference frame. Also, the same strategy of keeping the flux component i_{dse} up to the flux weakening region can be applied as will be seen below. On the other hand, according to the equation, the optimum angle between the stator current *dqe* components to obtain the maximum torque per ampere (MTPA) is 45° since in this case, both components have identical magnitudes.

The vector control with position sensor is simple to implement because of once the position of the rotor is known, the stator current components can be regulated to maintain the desired torque and speed. Unlike the induction machine, the rotor rotates at synchronous speed so that the slip concept disappears from the control. The schematic of the vector control of the SynRM/PMASynRM is depicted in Fig. 10.20.

The phase current *a* and *b* must be measured to be regulated through a current controller, one per component as in the indirect vector control of the induction machine. Also, the same axis transformations are necessary since both controllers operate in the synchronously rotating reference frame. The speed regulation sets the necessary torque component through the speed obtained by the derivative of the position. Finally, the feedforward compensation is added to cancel the cross-coupling of the stator currents' components of the machine model.

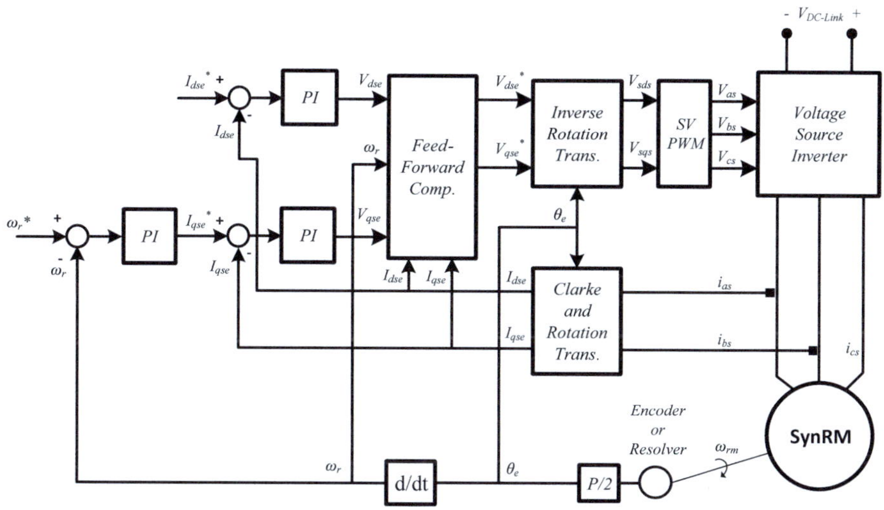

Fig. 10.20 Block diagram of the vector control of SynRM and PMASynRM with feedforward compensation

10.4.3 PMSM

In the PMSM , the vector control with position sensor is also easy to understand as shown for the SynRM. In the same way, once the position of the rotor is known, the components of the stator current can be regulated to maintain the desired torque and speed. The schematic of the vector control of the PMSM is shown in Fig. 10.21, specifically for the SMPMSM.

In order to get MTPA in the SMPMSM, the magnitude of the current reference i_{dse} must be zero up to the speed where the flux weakening begins. From this speed, the current reference i_{dse} is reduced to negative values to weak the magnetic field of the rotor. In the case of an IMPMSM machine, to obtain MTPA, the current reference i_{dse} is usually varied depending on a look-up table. This table is optimized for that machine using a magnetic model extracted either experimentally or through an FEA simulation.

10.5 Flux Weakening Control

In specific applications such as HEV/EV, the AC machine should operate at speeds higher than base speed one by a factor of up to 5 times higher. In order to operate above speeds, the field (flux) weakening strategy is required. In vector control, the conventional flux weakening technique consists in decreasing the stator flux linkage reference in inverse proportion to the rotor speed. However, this technique does not

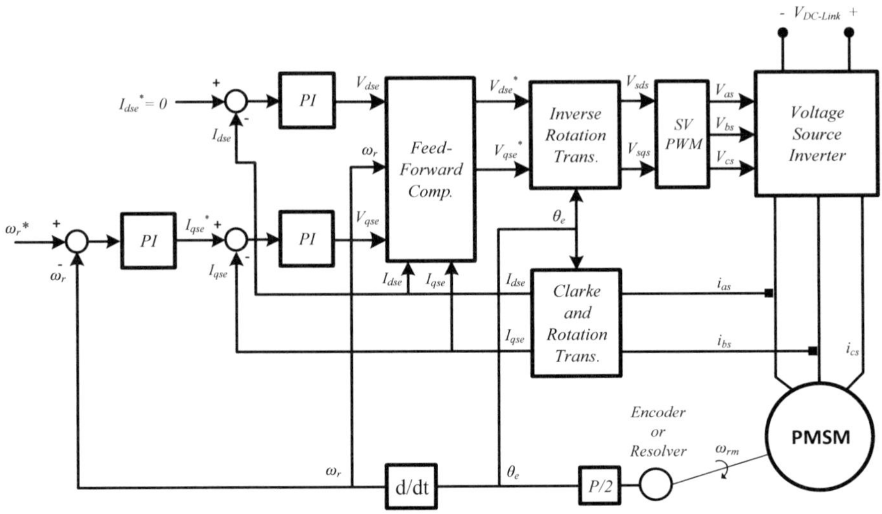

Fig. 10.21 Block diagram of the vector control of SMPMSM with feedforward compensation

allow reaching the maximum torque capacity over the entire speed region, since the curve that provides the optimal flux level as a function of the rotor speed may differ from the inverse of the rotor speed. To optimize the torque produced at speeds above the rated speed, it is essential to adjust the magnetic flux while maintaining the maximum voltage and current. In this way, maximum torque can be guaranteed over a wide speed range (Kim et al. 1995).

In Chap. 8 was shown how the three-phase VSI allowed to vary the voltage and frequency independently, which was ideal for three-phase AC machine control. As discussed, the VSI had a series of non-idealities that made their efficiency degraded by the nature of electronic devices and passive devices. On the one hand, the output voltage was degraded by a factor η that could vary between 0.9 and 0.95, due to the voltage drop of the devices and the deadtime. On the other hand, energy losses were generated that were converted into heat, which had to be evacuated, either by natural and forced convection (fan) through aluminum heat sinks or by water cooling systems. For this reason, the maximum power that the VSI must deliver to the machine must be limited. This limit must be chosen so that the inverter operates in the safe operation area, without breaking, in all the conditions for which it was designed including the worst cases, such as its maximum operating ambient temperature. Due to the above explanation, there are two fundamental parameters of the VSI that must be known for the control of AC machines, such as the maximum current $I_{\max}$ and the maximum voltage $V_{\max}$.

On the other hand, the AC machine also offers a maximum limitation of operation based on its rated current and voltage; although due to its typical thermal constant higher than that of the inverter, it can withstand voltages and currents higher than the rated ones in a short time. Some of the AC machines are designed to be controlled by a VSI , especially those that usually operate above their rated speed so that the rated voltage of the design machine is chosen a little lower than the grid voltage to take into account the inverter $V_{\max}$ limitation. Under these conditions, the machine has its rated voltage, although the factor η takes the value of 0.9.

Therefore, for a space vector modulation and operating in its linear zone, the maximum voltage of the VSI can be rewritten as:

$$V_{\max} = \frac{V_{DC}}{\sqrt{3}}\eta \quad (10.48)$$

In vector control, the voltages and currents in each of the dq axes, both in the synchronously rotating reference frame and in the stationary reference frame, have a constraint and must satisfy:

$$V_{ds}^{e2} + V_{qs}^{e2} = V_{ds}^{s2} + V_{qs}^{s2} \leq V_{\max}^2 \quad (10.49)$$

$$i_{ds}^{e2} + i_{qs}^{e2} = i_{ds}^{s2} + i_{qs}^{s2} \leq I_{\max}^2 \quad (10.50)$$

Unlike the voltage $V_{\max}$, the selection of the $I_{\max}$ value may be due either to the inverter as previously mentioned or to the rated current of the AC machine. In the latter case, the inverter can be designed to withstand a rated current higher than the rated current of the machine, so that the machine can absorb more current in the transients, improving its dynamic performance in acceleration and deceleration without the inverter being affected.

10.5.1 Flux Weakening Control of Induction Machine

In the induction machine, the stator voltage can be rewritten in the synchronously rotating reference frame as:

$$v_{ds}^e = R_s i_{ds}^e + L_s \frac{di_{ds}^e}{dt} + L_m \frac{di_{dr}^e}{dt} - \omega_e \left[L_s i_{qs}^e + L_m i_{qr}^e \right] \quad (10.51)$$

$$v_{qs}^e = R_s i_{qs}^e + L_s \frac{di_{qs}^e}{dt} + L_m \frac{di_{qr}^e}{dt} + \omega_e \left[L_s i_{ds}^e + L_m i_{dr}^e \right] \quad (10.52)$$

The above equations can be expressed in terms of rotor flux linkage and stator currents instead of rotor currents by using (10.53) and (10.54):

$$i_{dr}^e = \frac{1}{L_r}\left(\lambda_{dr}^e - L_m i_{ds}^e\right) \quad (10.53)$$

$$i_{qr}^{e} = \frac{1}{L_r}\left(\lambda_{qr}^{e} - L_m i_{qs}^{e}\right) \tag{10.54}$$

Thus, Eqs. (10.51) and (10.52) can be rewritten as:

$$\begin{aligned} v_{ds}^{e} &= R_s i_{ds}^{e} + \left(L_s - \frac{L_m^2}{L_r}\right)\frac{\mathrm{d}i_{ds}^{e}}{\mathrm{d}t} + \frac{L_m}{L_r}\frac{\mathrm{d}\lambda_{dr}^{e}}{\mathrm{d}t} - \omega_e\left[i_{qs}^{e}\left(L_s - \frac{L_m^2}{L_r}\right) + \frac{L_m}{L_r}\lambda_{qr}^{s}\right] \\ &= R_s i_{ds}^{e} + \sigma L_s \frac{\mathrm{d}i_{ds}^{e}}{\mathrm{d}t} + \frac{L_m}{L_r}\frac{\mathrm{d}\lambda_{dr}^{e}}{\mathrm{d}t} - \omega_e\left[i_{qs}^{e}\sigma L_s + \frac{L_m}{L_r}\lambda_{qr}^{s}\right] \end{aligned} \tag{10.55}$$

$$\begin{aligned} v_{qs}^{e} &= R_s i_{qs}^{e} + \left(L_s - \frac{L_m^2}{L_r}\right)\frac{\mathrm{d}i_{qs}^{e}}{\mathrm{d}t} + \frac{L_m}{L_r}\frac{\mathrm{d}\lambda_{qr}^{e}}{\mathrm{d}t} + \omega_e\left[i_{ds}^{e}\left(L_s - \frac{L_m^2}{L_r}\right) + \frac{L_m}{L_r}\lambda_{dr}^{s}\right] \\ &= R_s i_{ds}^{e} + \sigma L_s \frac{\mathrm{d}i_{qs}^{e}}{\mathrm{d}t} + \frac{L_m}{L_r}\frac{\mathrm{d}\lambda_{qr}^{e}}{\mathrm{d}t} - \omega_e\left[i_{ds}^{e}\sigma L_s + \frac{L_m}{L_r}\lambda_{dr}^{s}\right] \end{aligned} \tag{10.56}$$

where it is assumed that vector control is based on the flux linkage orientation rotor. If the derivative terms are canceled due to the steady-state operation, the term voltage drop in the stator resistance is neglected, and according to the vector control theory, the flux in the q-axis λ_{qre} is null, and then (10.55) and (10.56) can be rewritten as:

$$v_{ds}^{e} \cong -\omega_e \sigma L_s i_{qs}^{e} \tag{10.57}$$

$$v_{qs}^{e} \cong \omega_e \sigma L_s i_{ds}^{e} + \omega_e \frac{L_m}{L_r}\lambda_{dr}^{e} \tag{10.58}$$

If it is now conveniently approximated as in (10.59):

$$\lambda_{dr}^{e} = L_r i_{dr}^{e} + L_m i_{ds}^{e} \cong L_m i_{ds}^{e} \tag{10.59}$$

Equation (10.58) is simplified as:

$$v_{qs}^{e} \cong \omega_e L_s i_{ds}^{e} \tag{10.60}$$

Therefore, Eq. (10.49) can be rewritten by replacing (10.60) and (10.57) as:

$$(\omega_e L_s i_{ds}^{e})^2 + \left(\omega_e \sigma L_s i_{qs}^{e}\right)^2 \le V_{\max}^2 \tag{10.61}$$

Equation (10.61) is an ellipse equation where the size by the voltage constraint decreases as the operating machine speed ω_e increases. It is essential to mention that (10.61) has been obtained conveniently, where the stator voltage of each of the axes depends only on the stator current and the flux linkage rotor. Therefore, this can be represented in the current plane, as shown in Fig. 10.22. In Fig. 10.22, the obtained result is represented with four different speeds corresponding to 5 HP induction machine according to the nameplate constant of Table 10.5.

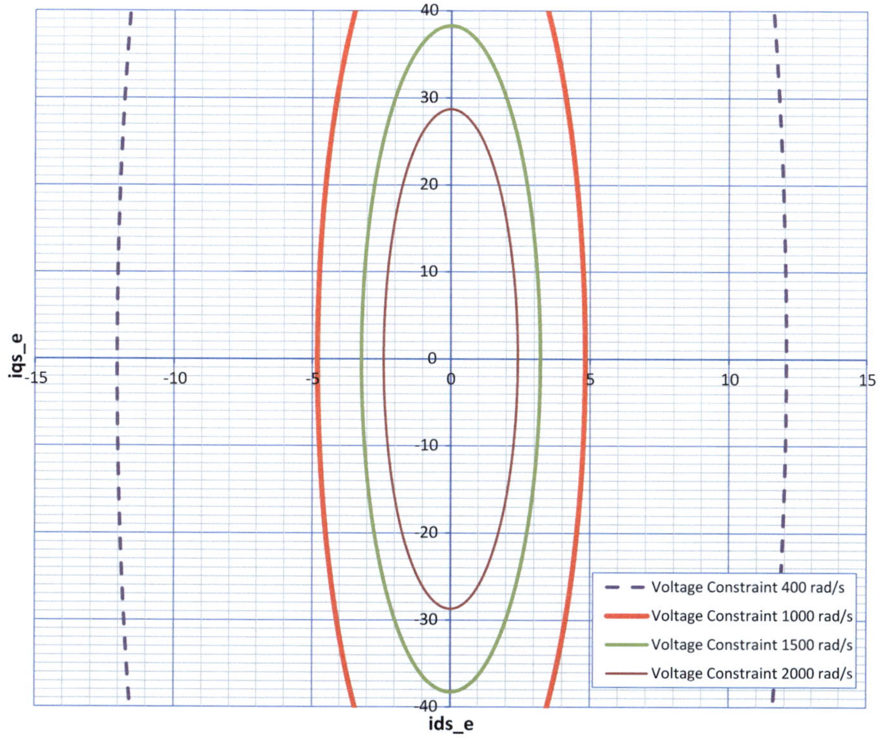

Fig. 10.22 Voltage constraint ellipse at four different operating speeds of 5 HP induction machine in the current plane

Table 10.5 Nameplate data of 5 HP induction machine

Nameplate	Symbol		Unit
Nominal power	P_N	5	HP
Line-to-line nominal voltage	V_N	200	V
Nominal current	I_N	14.2	A
Nominal frequency	f_N	60	Hz
Nominal speed	N_N	1762	RPM
Nominal power factor	$\cos\varphi$	0.8	
Service factor	SF	1.15	

The current constraint is a circle of $I_{\max}$ radius that limits the stator current vector space according to (10.50). The voltage and current constraints can be represented in the current plane to deduce different operating regions, such as the constant torque region, constant power region, and high-speed or constant slip region as it will be discussed in the following section.

10.5.1.1 Constant Torque Region

As described in Sect. 4.4.1.6, the constant torque region in an induction machine is below the rated speed. Up to this rated speed, the machine can provide constant torque in infinite speeds between zero and the rated speed with the proper regulation of the stator current components. The limit can be represented by using the current plane and the voltage and current constraints equations, as shown in Fig. 10.23. In Fig. 10.23, the ellipse represented in dashed line corresponds to the voltage constraint at rated speed, while the current constraint circle is represented in a continuous line. As can be observed, there is a crossing point between the I_{max} current circle and the voltage constraint where d-axis rated current $i_{dseRated}$ is set. Moreover, at this same crossing point, there is the maximum constant torque curve for the 5 HP induction machine, corresponding to 23.5 Nm.

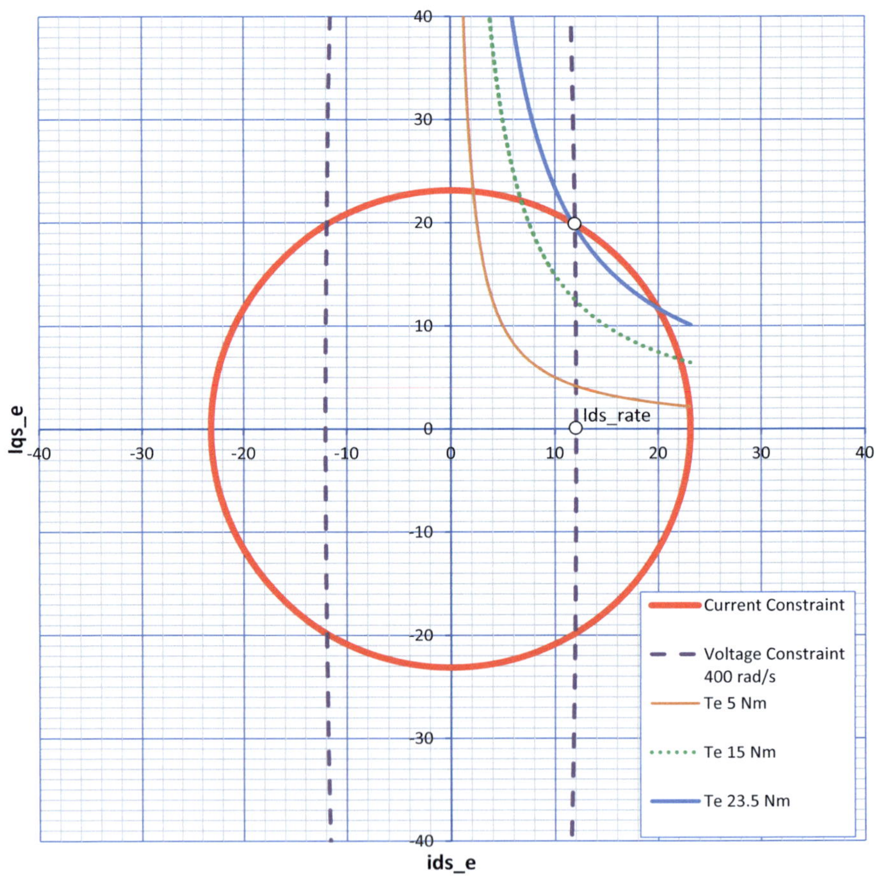

Fig. 10.23 Constant torque region: voltage constraint ellipse at rated speed, current constraint circle, and different constant torque locus

The current i_{dse} cannot be higher than $i_{dseRated}$ since it would suppose a magnetic saturation of the machine so that the maximum available torque is decided only by the maximum current i_{qse} as:

$$\begin{aligned} i_{ds}^{e} &= i_{ds\_rate}^{e} \\ i_{qs}^{e} &= \sqrt{I_{\max}^{2} - i_{ds\_rate}^{e2}} \end{aligned} \tag{10.62}$$

where rated rotor flux linkage can be expressed as

$$\lambda_{dr\_rate}^{e} \cong L_m i_{ds\_rate}^{e} \tag{10.63}$$

The rated speed can be deduced according to (10.61), (10.62) and the rotor flux linkage (10.63), where the angular slip frequency is not considered:

$$\omega_b = \sqrt{\frac{V_{\max}^2}{(\sigma L_s I_{\max})^2 + \frac{L_s^2(1-\sigma^2)\lambda_{dr\_rate}^{e}}{L_m^2}}} \tag{10.64}$$

In the current example, the rated speed is 400 rad/s, about 1910 RPM, the i_{dse} is 12 A, and the i_{qse} approximately 20 A.

10.5.1.2 Constant Power Region

The constant power region specifies that the maximum voltage $V_{\max}$ available from the VSI has reached its limit that coincides with the rated speed and ends at the critical top speed ω_c. It is also from the rated speed where the flux weakening starts; that is, the current in the d-axis should be below the $i_{dseRated}$ if the machine must operate at speeds above the rated speed. In this region, the induction machine can deliver the rated power. In Fig. 10.24, the voltage constrain ellipse is represented at 4.5 times the rated speed, that is, 1800 rad/s. With the crossing point B, with the current constraint circle of radius $I_{\max}$, it is possible to deduce the new current i_{dse} which much lower than the $i_{dseRated}$ as expected.

According to Fig. 10.24 and the (10.64), it is possible to obtain the current i_{dse} which maximizes the torque, while the current in the q-axis is derived from the current constraint:

$$\begin{aligned} i_{ds}^{e} &= \sqrt{\frac{\left(\frac{V_{\max}}{\omega_e}\right)^2 - \sigma^2 L_s^2 I_{\max}^2}{L_s^2 - \sigma^2 L_s^2}} \\ i_{qs}^{e} &= \sqrt{I_{\max}^2 - i_{ds}^{e2}} \end{aligned} \tag{10.65}$$

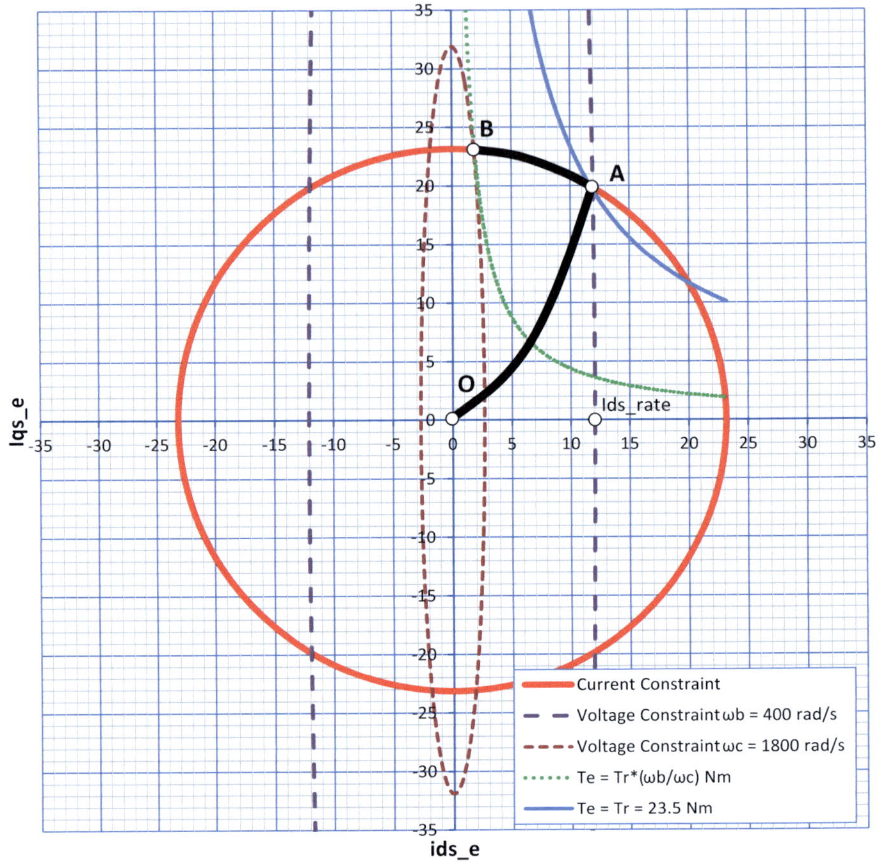

Fig. 10.24 Constant power region: Voltage constraint ellipse at base ω_b and ω_c speeds, current constraint circle, and different torque locus

In Fig. 10.24, to obtain maximum torque in the constant power region, the current vector must slide on line A–B (10.65), which is an intersection between the current circle limit, the voltage ellipse limit, and the constant torque curve. In this case, for a speed of 1800 rad/s, about 8594 RPM, the current components i_{dse} is 1.9 A and the i_{qse} approximately 23 A.

10.5.1.3 Constant Slip Region

From power region, if the machine speed continues increasing, the voltage constraint will be inside the current constraint circle as represented in Fig. 10.25. The constant slip region begins when the circle meets the ellipse at a single crossing point.

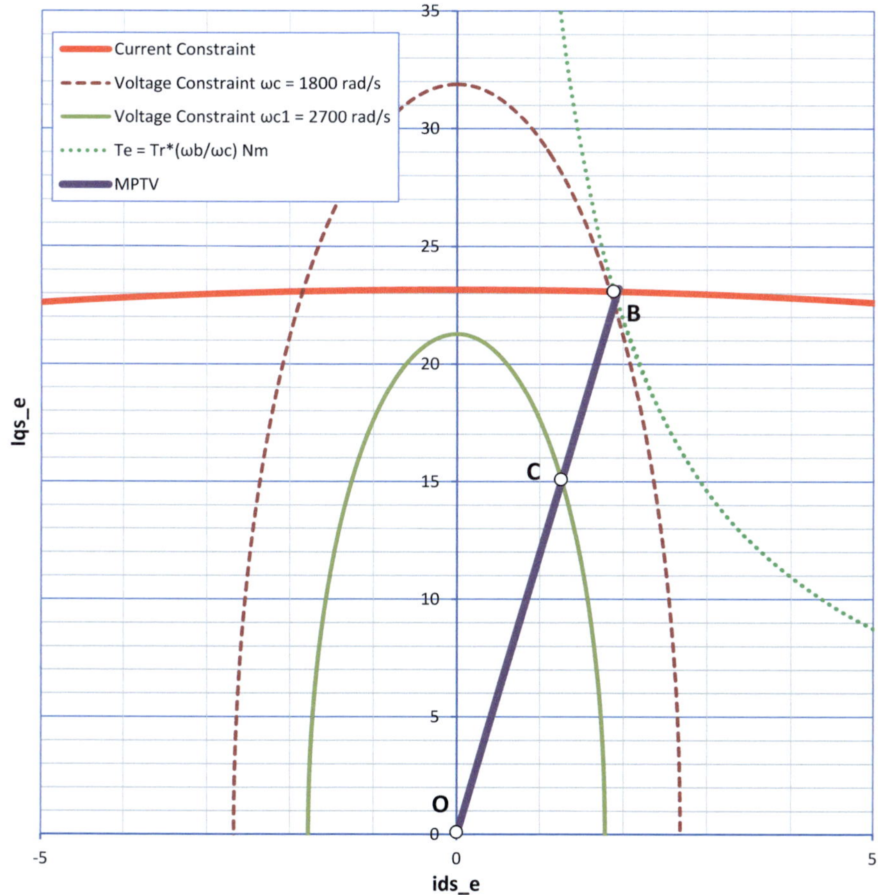

Fig. 10.25 Constant slip region: Voltage constraint ellipse at critical speed ω_c and higher speed with current constraint circle

The critical speed ω_c is where this region begins and can be calculated according to Eq. 10.66:

$$\omega_c = \frac{V_{\max}}{I_{\max}}\sqrt{\frac{1+\sigma^2}{2\sigma^2 L_s^2}} \tag{10.66}$$

From point B, the components of the optimal reference currents that maximize the torque are calculated from the maximum voltage and the parameters of the machine:

$$
\begin{aligned}
i_{ds}^{e} &= \frac{V_{\max}}{\sqrt{2}\omega_e L_s} \\
i_{qs}^{e} &= \frac{V_{\max}}{\sqrt{2}\omega_e \sigma L_s}
\end{aligned}
\tag{10.67}
$$

This optimal point subject to only the voltage-limit condition is called the maximum torque per voltage (MTPV) operating point.

For example, if the induction machine speed should increase from the critical velocity ω_c to a higher speed ω_{c1}, the path followed by the current components i_{dse} and i_{qse} is from point B to point C as shown in Fig. 10.25. In an ideal case, without friction and load, if the speed continues increasing, a point will arrive at which both components at the center of the ellipses are zero, that is the point O. At this point (center of the ellipses), the speed will be considering infinite due to both components which takes the zero value, and then, the speed must be infinite according to (10.67).

10.5.2 Flux Weakening Control of SynRM and PMASynRM

Like the case of the induction machine, if the SynRM/PMASynRM should operate at a higher speed than the rated speed, the flux must be weakened. There are also three operating regions.

By performing the same analysis as with the induction machine, it is possible to obtain Eq. 10.68 which describes the ellipses of the voltage constraint for the generic case of the PMASynRM:

$$
\left(\omega_r L_d i_{ds}^{e}\right)^2 + \left(\omega_r L_q\right)^2\left(i_{qs}^{e} + \frac{\Psi_m}{L_q}\right) \le V_{\max}^2 \tag{10.68}
$$

Equation 10.68 corresponds to an ellipse that gets smaller every time the speed of the machine ω_r increases. In this case, the ellipses are centered on $(0, \Psi_m/L_q)$. If there is no permanent magnet, i.e., the flux is zero, as in the case of the SynRM, the origin of the ellipses is (0,0) as in the case of the induction machine.

For SPMSM and IPMSM machines, Eq. (10.69) is slightly different compared with (10.68):

$$
\left(\omega_r L_d\right)^2\left(i_{ds}^{e} + \frac{\Psi_m}{L_d}\right) + \left(\omega_r L_q i_{qs}^{e}\right)^2 \le V_{\max}^2 \tag{10.69}
$$

In the SPMSM machine, as $L_d = L_q = L_s$, the shape is not an ellipse, but a circle with center in $(\Psi_m/L_s, 0)$.

However, the current constraint is the same for all machines:

$$i_{ds}^{e2} + i_{qs}^{e2} = i_{ds}^{s2} + i_{qs}^{s2} \leq I_{\max}^2 \tag{10.70}$$

Therefore, there are crossing points between both curves in the current plane where different operating regions can be defined. Hence, the torque can be maximized as it happened with the induction machine.

The ellipses and circles diagram for the PMASynRM is shown in Fig. 10.26. It is represented the case when the magnet flux Ψ_m is less than $L_q I_{\max}$ (infinite speed is reached at the origin of the axis). From the origin O to A, the MTPA path is followed, corresponding to the region I or constant torque region ending at the rated speed ω_b. Once the nominal speed ω_b has been reached, the torque optimization is achieved following the path A–B up to the critical speed ω_c (region II). From higher speeds than the critical speed (region III), the current vector follows the trajectory toward the center of the ellipse C (MTPV) as can be seen in Fig. 10.26.

In the region *I*, or constant torque region, the current components that optimize the torque according to MTPA for the SynRM are when the angle between the components is 45° as discussed. However, in the case of the PMASynRM, the MTPA path is described according to (10.71):

$$\sin\delta = \frac{\sqrt{\Psi_m^2 + 8(L_d - L_q)^2 I_{\max}^2} - \Psi_m}{4(L_d - L_q)I_{\max}} \tag{10.71}$$

It is possible to observe that if the flux Ψ_m is zero, Eq. 10.71 is reduced to:

$$\sin\delta = \frac{\sqrt{2}}{2} \tag{10.72}$$

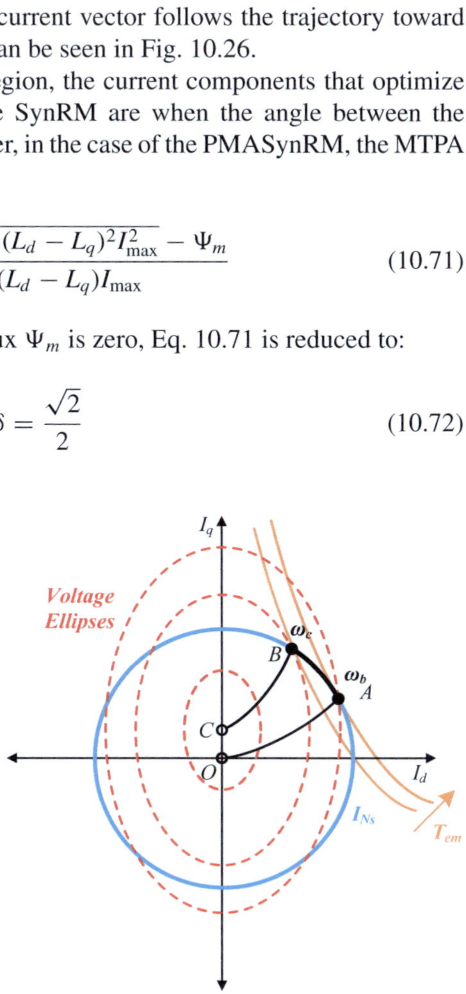

Fig. 10.26 Ellipse and circle diagram of PMASynRM when $\Psi_m < L_q I_{\max}$

That is, $\delta = 45°$.

Therefore, the components are calculated as:

$$i_{ds}^{e} = \frac{\sqrt{\Psi_m^2 + 8(L_d - L_q)^2 I_{\max}^2} - \Psi_m}{4(L_d - L_q)}$$
$$i_{qs}^{e} = \sqrt{I_{\max}^2 - i_{ds}^{e2}} \tag{10.73}$$

The rated speed can be calculated as:

$$\omega_b = \sqrt{\frac{V_{\max}^2}{(L_d I_{ds}^e)^2 + (L_q I_{qs}^e + \Psi_m)^2}} \tag{10.74}$$

The maximum speed can be calculated by imposing i_{dse} equal to zero and i_{qse} to the maximum:

$$\omega_{\max} = \frac{V_{\max}}{L_q I_{\max} - \Psi_m} \tag{10.75}$$

On the other hand, if $\Psi_m > L_q I_{\max}$, the center of the ellipses is outside the current limit, so point B does not exist.

10.5.3 Flux Weakening Control Strategy

The field weakening strategy is in general not sensitive to the AC machine parameters. Thus, it can be extended to all AC machines when the machine voltage feedback is used (Kim and Sul 1997). However, the higher dynamic response is achieved with the feedforward compensation strategy, which uses some AC machine parameters. In Fig. 10.27, the field weakening control strategy based on the AC machine voltage feedback is represented. As can be observed, the voltage feedback can be either the output voltage of the current regulators or the voltage measured at the machine terminals.

The strategy consists of monitoring the machine voltage to be compared with the maximum voltage available in the VSI . As it approaches the maximum voltage, the regulator, which can be a simple PI controller, causes its i_{dse} output to decrease, weakening the flux, while the torque component i_{qse} is limited as a function of the new i_{dse} value as it can be seen. As mentioned early, there is a dependency on the machine parameters if the feedforward compensation is implemented so that they affect the performance of the current regulator. In Fig. 10.27, the current regulators have this compensation. On the other hand, if the voltage measurement at the machine terminals is used, it must be as accurate as possible to increase the flux weakening dynamic control. In both cases, the voltage drops in the power switches devices,

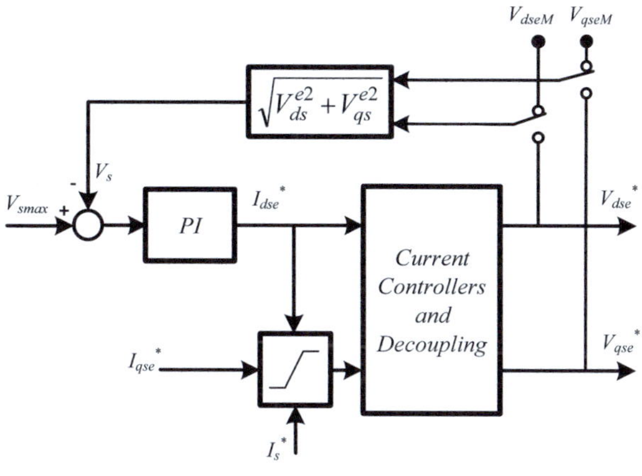

Fig. 10.27 Flux weakening control strategy using the machine voltage feedback. The voltage feedback can be either the output voltage of the regulators or the voltage measured at the machine terminals

and the deadtime should be compensated. Finally, to avoid that this new controller does not interfere with the current controllers, the bandwidth of this regulator should be lower. Choosing a bandwidth identical to the speed regulator bandwidth usually offers good results.

10.6 Sensorless Control

10.6.1 Introduction

As previously discussed, sensorless vector control avoids using a position or speed sensor, which is a simpler solution to implement, although it requires high-performance microcontrollers/DSPs. It prevents the placement of the sensor on the rotor shaft and its corresponding wiring where it is sometimes very complicated and expensive. In the sensorless concept, it is then necessary to estimate the speed or position of the machine. As will be seen in this section, it will be necessary to use the machine model together with the measurement of the voltages and currents in the machine terminals. As discussed in Chap. 8, the voltage and current measurements at the machine terminals are made from the VSI itself without the intervention of any special wiring. That is, the sensorless concept commonly called in the literature means that the machine is connected to the VSI only with the three-phase power wires. The microprocessor or DSP of the inverter should estimate the machine speed or position, based on the model and the corresponding measurements in the terminals of the machine. In some applications where the speed sensing implies a security level for people as in HEV/EV, the estimation is usually combined with the measurement of speed to corroborate that the estimation and measurement are coherent at overall times. In case one of the two mechanisms to obtain the machine speed fails, the vehicle can continue to operate, but with certain restrictions such as the limitation of its maximum speed. If these two mechanisms fail, the vehicle would be totally out of control, implying a critical safety problem for people. In this section, the flux estimation, position, and the speed of the machine valid for steady-state and transient for high-performance vector control will be described.

10.6.2 Rotor Flux Linkage Estimator in IM, PMSM, SynRM, and PMASynRM

In order to have a fast and precise control of the induction machine, it is essential to know the stator and rotor flux. As previously mentioned the direct vector control, the flux measurement by Hall effect sensors is not a simple task and increases the cost of the application. It adds an excess of complexity and a lack of robustness in the mechanics associated with inserting said sensors in the machine air gap. Instead of measuring the flux, the flux can be observed based on measurements of voltage, current, and angular speed by an observer (Jansen et al. 1994). In Fig. 10.28, the flux linkage rotor observer general block diagram is shown. The inputs are voltages and currents which are generally measured, and the outputs are the estimated values of the two vector flux components and the rotor flux linkage angle position. The variables shown are in the stationary reference frame axes. A good precision in the signal processing of the voltage and the currents in the machine terminal is fundamental for a flux and position angle correct estimation.

Technically, the word "observer" implies an estimator that uses, on the one hand, the integration of the model and, on the other, feedback to correct the error that improves the accuracy of the estimation. It is possible to define an observer in open loop as one that does not have feedback for its correction, while the closed-loop observer (Jansen and Lorenz 1994) does have the feedback mechanism for its correction, improving its accuracy.

10.6.2.1 Open-Loop Flux Observers

The open-loop flux observer does not have a correction using feedback but offers a capacity of almost instantaneous rotor flux linkage tracking in the induction machine. It is possible to distinguish two methods: The first is based on the voltage model, while the second is based on the current model of the induction machine. This will

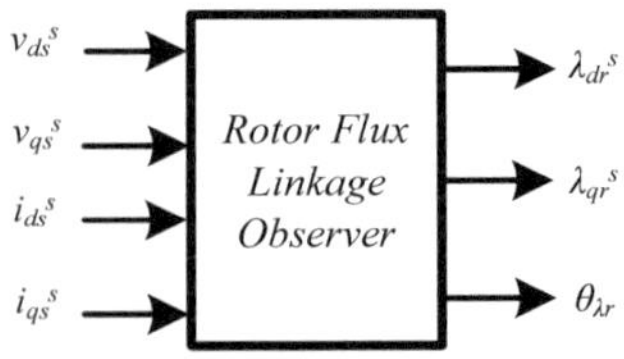

Fig. 10.28 Rotor flux linkage observer based in the measurement of the voltage and current *dq* components in the stationary reference frame

be discussed in the following sections with the help of Eqs. (10.76–10.83) in the stationary reference frame.

$$v_{ds}^{s} = R_s i_{ds}^{s} + \frac{\mathrm{d}\lambda_{ds}^{s}}{\mathrm{d}t} \tag{10.76}$$

$$v_{qs}^{s} = R_s i_{qs}^{s} + \frac{\mathrm{d}\lambda_{qs}^{s}}{\mathrm{d}t} \tag{10.77}$$

$$0 = R_r i_{dr}^{s} + \frac{\mathrm{d}\lambda_{dr}^{s}}{\mathrm{d}t} + \omega_r \lambda_{qr}^{s} \tag{10.78}$$

$$0 = R_r i_{qr}^{s} + \frac{\mathrm{d}\lambda_{qr}^{s}}{\mathrm{d}t} - \omega_r \lambda_{dr}^{s} \tag{10.79}$$

Moreover, flux linkages:

$$\lambda_{ds}^{s} = L_s i_{ds}^{s} + L_m i_{dr}^{s} \tag{10.80}$$

$$\lambda_{qs}^{s} = L_s i_{qs}^{s} + L_m i_{qr}^{s} \tag{10.81}$$

$$\lambda_{dr}^{s} = L_r i_{dr}^{s} + L_m i_{ds}^{s} \tag{10.82}$$

$$\lambda_{qr}^{s} = L_r i_{qr}^{s} + L_m i_{qs}^{s} \tag{10.83}$$

Voltage Model

In the voltage model, the stator flux linkage can be estimated merely by measuring the voltage and the currents in the machine terminals, according to (10.84):

$$\hat{\lambda}_s = \int \left(v_s - \hat{R}_s i_s\right) \mathrm{d}t \tag{10.84}$$

where ^ superscript denotes estimated values. In the stationary reference frame, the two stator flux linkages are:

$$\hat{\lambda}_{ds}^{s} = \int \left(v_{ds}^{s} - \hat{R}_s i_{ds}^{s}\right) \mathrm{d}t \tag{10.85}$$

$$\hat{\lambda}_{qs}^{s} = \int \left(v_{qs}^{s} - \hat{R}_s i_{qs}^{s}\right) \mathrm{d}t \tag{10.86}$$

and the module of the stator flux linkage is:

$$\hat{\lambda}_s = \sqrt{\hat{\lambda}_{ds}^{s2} + \hat{\lambda}_{qs}^{s2}} \tag{10.87}$$

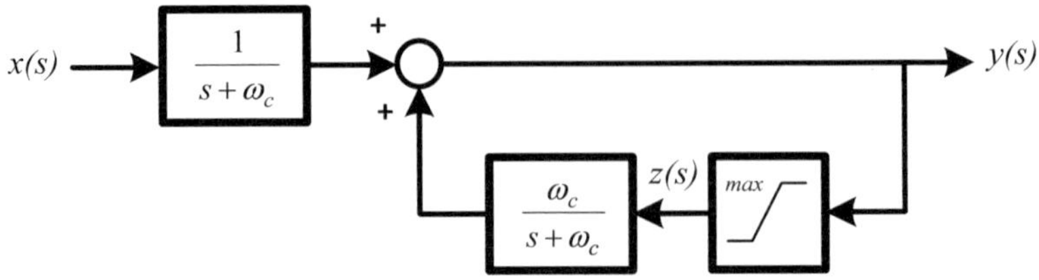

Fig. 10.29 Block diagram of a modified integrator

If the stator resistance $\hat{R}_s$ is known and identical to that of the machine ($\hat{R}_s = R_s$), and the voltage and current measurement are ideal, a perfect estimation of the flux linkage stator can be obtained. However, since the resistance varies with temperature and errors in the measurement are almost inevitable, the estimation will suffer a considerable error. For example, at low frequencies or low speeds, the resistance of the stator R_s plays a significant role, since the machine voltage and its back-EMF are usually low, comparable to the voltage drop in the resistance. That is, this estimator requires a good precision of the stator resistance, and its operation should be limited for a frequency not lower than 5 Hz. On the other hand, the ideal integrator $1/s$ of the estimate itself has an infinite bandwidth where the high-frequency components were amplified as seen in Chap. 2. However, the real integrator based on an operational amplifier overcomes this effect, because the bandwidth was limited, where it acted as an integrator up to the fundamental design frequency and as a low-pass filter from a specific frequency. The integration process also integrates the DC component, creating large values if a reset or anti-windup is not used. The integrator of Fig. 10.29 (Hinkkanen and Luomi 2003) is often used in flux estimation since it prevents the problems of the ideal integrator, such as high-frequency amplification and DC components.

The transfer function is:

$$y(s) = \frac{1}{s + \omega_c} x(s) + \frac{\omega_c}{s + \omega_c} z(s) \tag{10.88}$$

If $y(s)$ exceeds the MAX limit, then $z(s) = \text{MAX}$. Otherwise, $z(s) = y(s)$.

From (10.82) and (10.83) is possible to represent the rotor currents as:

$$i_{dr}^{s} = \frac{1}{L_r}\left(\lambda_{dr}^{s} - L_m i_{ds}^{s}\right) \tag{10.89}$$

$$i_{qr}^{s} = \frac{1}{L_r}\left(\lambda_{qr}^{s} - L_m i_{qs}^{s}\right) \tag{10.90}$$

By substituting (10.89) and (10.90) into (10.80) and (10.81), respectively, the stator flux in the stationary reference frame can be described as:

$$\lambda_{ds}^{s} = L_s i_{ds}^{s} + \frac{L_m}{L_r}\left(\lambda_{dr}^{s} - L_m i_{ds}^{s}\right) \tag{10.91}$$

$$\lambda_{qs}^{s} = L_s i_{qs}^{s} + \frac{L_m}{L_r}\left(\lambda_{qr}^{s} - L_m i_{qs}^{s}\right) \tag{10.92}$$

Now, if the rotor flux linkage is represented in terms of the stator flux, Eqs. (10.93) and (10.94) are obtained. It represents the estimation of the rotor flux linkage in the stationary reference frame:

$$\hat{\lambda}_{dr}^{s} = \frac{\hat{L}_r}{\hat{L}_m}\hat{\lambda}_{ds}^{s} - \frac{\hat{\sigma}}{1-\hat{\sigma}}\hat{L}_m i_{ds}^{s} \tag{10.93}$$

$$\hat{\lambda}_{qr}^{s} = \frac{\hat{L}_r}{\hat{L}_m}\hat{\lambda}_{qs}^{s} - \frac{\hat{\sigma}}{1-\hat{\sigma}}\hat{L}_m i_{qs}^{s} \tag{10.94}$$

In Fig. 10.30, the voltage model block diagram of the induction machine which used Eqs. (10.85), (10.86), (10.93), and (10.94) is represented. It can be observed that by measuring the voltage and current in the machine terminals, it is possible to estimate the stator and rotor flux linkage.

Current model
The estimator based on the current model can operate below 5 Hz, even in a standstill. It is based on Eq. 10.95, where it is assumed that the rotor current is zero:

$$\hat{\psi}_s = \hat{L}_m i_s \tag{10.95}$$

As can be observed, (10.95) does not use any integral terms for the flux estimation.

In the estimator based on the current model, a correct flux estimation requires a correct value of the magnetizing inductance L_m. Due to the magnetic saturation in the machine, the inductance varies, which makes it challenging to obtain a correct value of L_m. Also, this estimator offers an unfortunate result when the rotor current differs

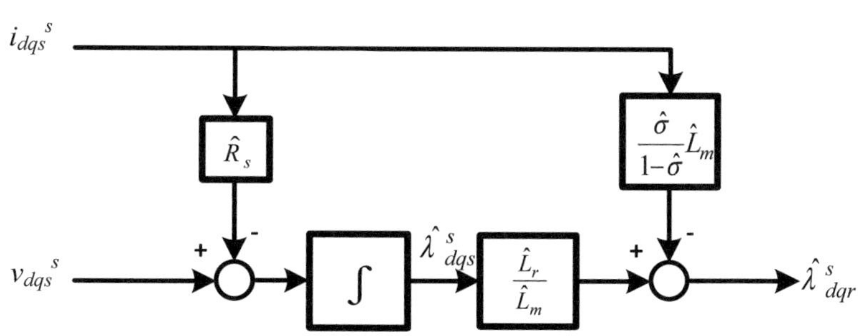

Fig. 10.30 Voltage model of IM where the stator and rotor flux linkage is estimated by means an open-loop observer with the stator voltage and stator current measured in the machine terminals

from zero (when the machine has a load), but can be compensated by measuring the angular speed of the rotor as explained below.

If in the rotor voltage Eqs. (10.78) and (10.79), the rotor current is replaced by (10.89) and (10.90), and it is possible to obtain:

$$\frac{d\lambda_{dr}^{s}}{dt} = R_r \frac{L_m}{L_r} i_{ds}^{s} - \omega_r \lambda_{qr}^{s} - \frac{R_r}{L_r} \lambda_{dr}^{s} \quad (10.96)$$

$$\frac{d\lambda_{qr}^{s}}{dt} = R_r \frac{L_m}{L_r} i_{qs}^{s} + \omega_r \lambda_{dr}^{s} - \frac{R_r}{L_r} \lambda_{qr}^{s} \quad (10.97)$$

Removing the derivative terms is possible to obtain the rotor flux linkages in the stationary reference frame as:

$$\hat{\lambda}_{dr}^{s} = \int \left(\hat{R}_r \frac{\hat{L}_m}{\hat{L}_r} i_{ds}^{s} - \omega_r \hat{\lambda}_{qr}^{s} - \frac{\hat{R}_r}{\hat{L}_r} \hat{\lambda}_{dr}^{s} \right) dt \quad (10.98)$$

$$\hat{\lambda}_{qr}^{s} = \int \left(\hat{R}_r \frac{\hat{L}_m}{\hat{L}_r} i_{qs}^{s} + \omega_r \hat{\lambda}_{dr}^{s} - \frac{\hat{R}_r}{\hat{L}_r} \hat{\lambda}_{qr}^{s} \right) dt \quad (10.99)$$

The above equations give the rotor flux linkages as a function of the stator current and the rotor speed ω_r, without depending on the stator voltage. As in the case of the voltage model, the integral term is needed for its calculation, and there is a cross-coupling of the flux as can be observed. With this current model, the rotor flux estimation can operate at very low speeds, even at zero speed, being ideal for starting the machine. However, it needs the measurement of the rotor speed, and its accuracy depends on some machine parameters, especially with the rotor resistance, which varies with the temperature and according to the skin effect.

In Fig. 10.31, the open-loop observer is represented based on the current model of the induction machine.

The estimator based on the voltage model does not specify the rotor speed, unlike the estimator based on the current model. It is possible to affirm that the estimator based on the voltage model offers excellent results at higher speeds, where the effect of the stator resistance R_s does not prevail, while the estimator based on the current model offers excellent results at low speeds even at zero speed. It is easy to intuit

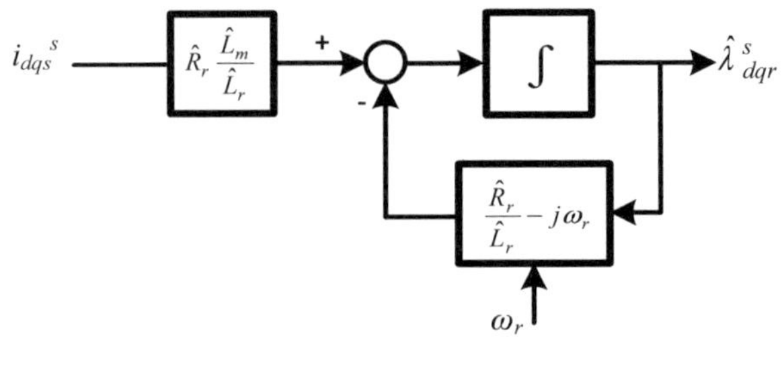

Fig. 10.31 Current model of IM where the rotor flux linkage is estimated by means an open-loop observer with the stator current and rotor speed measurement

that the two estimators can be used for estimating the rotor flux linkage where one or the other is used depending on the rotor speed. However, there are different studies where it is possible to use a hybrid model where the transitions between one and the other are smooth, as explained by Jansen et al. (1994).

10.6.2.2 Closed-Loop Flux Observer Model

As previously mentioned, closed-loop flux observers have a corrective feedback mechanism that acts to improve accuracy when changing parameters. The accuracy of the estimation is determined by the accuracy of the open-loop observer model used, what input variables are measured and what is estimated, and by the gain of the controller that applies the correction. In Lascu et al. (2000), a closed-loop observer is proposed that operates in a wide speed range, where the stator and rotor flux linkage is estimated. The closed loop is performed with a PI controller which adapts the voltage model using the associated errors in the pure integrator and the stator resistance variation with the temperature. The closed-loop observer contains the two previous models: the open-loop current model that produces a precise value represented by the subscript i, especially for low-speed operations, and the adaptive voltage model represented by the subscript v, for a wide range of operating speeds. Both models operate at all times applying the corrective actions as some machine parameters vary, such as the resistance of the stator $\hat{R}_s$ or rotor resistance $\hat{R}_r$, where supposedly they are known (or estimated) parameters and fixed in the observer, but that it may suffer changes due to temperature.

The effect of the cross-coupling of the previous open-loop rotor flux observer based on the current model can lead to instabilities at high speeds. If (10.96) and (10.97) are written in the synchronously rotating reference frame, the dependence of the rotor speed and cross-coupling is removed:

$$\frac{\mathrm{d}\lambda_{dr}^{e}}{\mathrm{d}t} = R_r \frac{L_m}{L_r} i_{ds}^{e} - \frac{R_r}{L_r} \lambda_{dr}^{e} \rightarrow \frac{\mathrm{d}\hat{\lambda}_{dr}^{e}}{\mathrm{d}t} = \hat{R}_r \frac{\hat{L}_m}{\hat{L}_r} i_{ds}^{e} - \frac{\hat{R}_r}{\hat{L}_r} \hat{\lambda}_{dr}^{e} \tag{10.100}$$

Moreover, if the rotor flux linkage is aligned on the d-axis, then:

$$\lambda_{qr}^{e} = 0 \tag{10.101}$$

In this case, the current model is implemented in the synchronously rotating reference frame in such a way that it requires transformations of axes where the rotor position must be known.

The rotor flux linkage can be converter to the stator flux linkages by using (10.93) and (10.94), where the stator flux linkage is represented as a function of the rotor flux linkages:

$$\hat{\lambda}_{ds}^{s,i} = \frac{\hat{L}_m}{\hat{L}_r}\hat{\lambda}_{dr}^{s,i} + \frac{\hat{\sigma}}{1-\hat{\sigma}}\frac{\hat{L}_m^2}{\hat{L}_r}i_{ds}^s = \frac{\hat{L}_m}{\hat{L}_r}\hat{\lambda}_{dr}^{s,i} + \frac{\hat{L}_s\hat{L}_r - \hat{L}_m^2}{\hat{L}_r}i_{ds}^s \tag{10.102}$$

$$\hat{\lambda}_{qs}^{s,i} = \frac{\hat{L}_m}{\hat{L}_r}\hat{\lambda}_{qr}^{s,i} + \frac{\hat{\sigma}}{1-\hat{\sigma}}\frac{\hat{L}_m^2}{\hat{L}_r}i_{qs}^s = \frac{\hat{L}_m}{\hat{L}_r}\hat{\lambda}_{qr}^{s,i} + \frac{\hat{L}_s\hat{L}_r - \hat{L}_m^2}{\hat{L}_r}i_{qs}^s \tag{10.103}$$

Finally, to determine the instantaneous position of the rotor flux linkage, the voltage model is used, which is converted to rotor flux linkage using (10.93) and (10.94), that is:

$$\hat{\lambda}_{dr}^{s,v} = \frac{\hat{L}_r}{\hat{L}_m}\hat{\lambda}_{ds}^{s,v} - \frac{\hat{\sigma}}{1-\hat{\sigma}}\hat{L}_m i_{ds}^s \tag{10.104}$$

$$\hat{\lambda}_{qr}^{s,v} = \frac{\hat{L}_r}{\hat{L}_m}\hat{\lambda}_{qs}^{s,v} - \frac{\hat{\sigma}}{1-\hat{\sigma}}\hat{L}_m i_{qs}^s \tag{10.105}$$

Then, the instantaneous angular position can be represented as:

$$\hat{\theta}_{\lambda r} = \tan^{-1}\left(\frac{\hat{\lambda}_{qr}^{s,v}}{\hat{\lambda}_{dr}^{s,v}}\right) \tag{10.106}$$

And the estimated rotor flux linkage is:

$$\hat{\lambda}_r^{s,v} = \sqrt{\left(\hat{\lambda}_{dr}^{s,v}\right)^2 + \left(\hat{\lambda}_{qr}^{s,v}\right)^2} \tag{10.107}$$

Note that the instantaneous angle $\hat{\theta}_{\lambda r}$ is used to the rotation transformation which is needed in the rotor flux linkage observer based on the current model.

Figure 10.32 shows the block diagram of the closed-loop flux observer, which uses the previous current model and the voltage model.

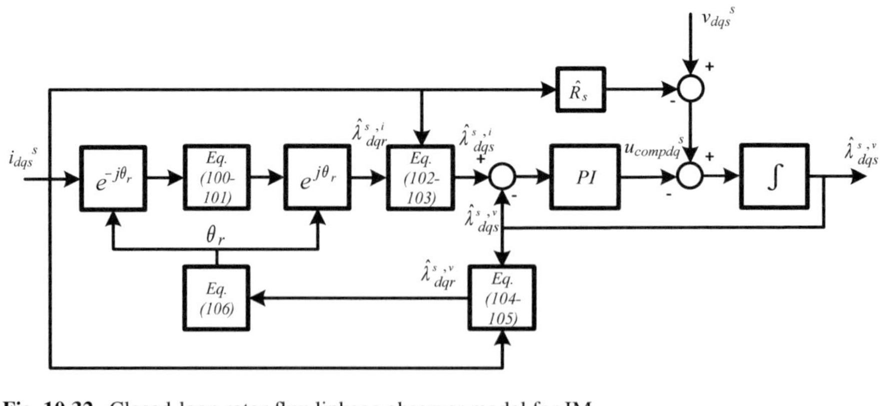

Fig. 10.32 Closed-loop rotor flux linkage observer model for IM

As it is possible to observe, the PI controller input is the error between the stator flux linkage of the voltage model and the current model. If the error is zero means that two models are estimating precisely the same, then voltage model is used. If the result is different to zero, for example, in the case of the stator resistance changes, compensated voltage u_{comp} is introduced by PI, which compensates the variation in this case of stator resistance parameter. The equations of the stator flux linkage based on the voltage model are modified as:

$$\hat{\lambda}_{ds}^{s,v} = \int \left(v_{ds}^{s} - \hat{R}_s i_{ds}^{s} - u_{comp\_d} \right) \mathrm{d}t \tag{10.108}$$

$$\hat{\lambda}_{qs}^{s,v} = \int \left(v_{qs}^{s} - \hat{R}_s i_{qs}^{s} - u_{comp\_q} \right) \mathrm{d}t \tag{10.109}$$

Moreover, voltage compensation can be written as:

$$u_{comp\_d} = K_P \left(\hat{\lambda}_{ds}^{s,v} - \hat{\lambda}_{ds}^{s,i} \right) + K_I \int \left(\hat{\lambda}_{ds}^{s,v} - \hat{\lambda}_{ds}^{s,i} \right) \mathrm{d}t \tag{10.110}$$

$$u_{comp\_q} = K_P \left(\hat{\lambda}_{qs}^{s,v} - \hat{\lambda}_{qs}^{s,i} \right) + K_I \int \left(\hat{\lambda}_{qs}^{s,v} - \hat{\lambda}_{qs}^{s,i} \right) \mathrm{d}t \tag{10.111}$$

The K_P and K_I gain parameters can be obtained as recommended by Lascu et al. (2000).

The flux observer presented in this section requires up to five parameters of the induction machine: stator resistance, rotor time constant, stator transient inductance, rotor transient inductance, and stator self-inductance. The cited parameters should be as accurate as possible for correct estimation of the rotor flux linkage magnitude and its position. As previously mentioned, these parameters may change due to temperature, magnetic saturation, and the aging of the machine. Small variations in the parameters are tolerable by the presented flux observer. There are online identification techniques that identify some of the previous parameters during a self-commissioning stage where the machine is at a standstill (Vas 1993). For example, in Sect. 8.8.3 was presented how to measure the stator resistance by injecting a DC current into the stator windings and measuring the voltage at the machine terminals in steady-state. The rotor time constant $\hat{L}_r/\hat{R}_r$ can be identified from the exponential voltage waveform when a step change in the stator current is applied, and the transient stator time constant $\hat{L}_s/\hat{R}_s$ is identified from the slope of the stator currents when a step voltage is applied. For correct identification, it is logical to think that the injection of voltage and current must be controlled and measurable by the VSI with the highest possible accuracy and to its corresponding calibration. On the other hand, as can be seen, the flux observer does not need to know the rotor resistance $\hat{R}_r$, but the rotor time constant which can be identified as previously mentioned. The closed-loop rotor flux linkage observed presented can operate in a wide speed range with 2–8 poles IMs, and the instantaneous angular position is generated at 0 RPM.

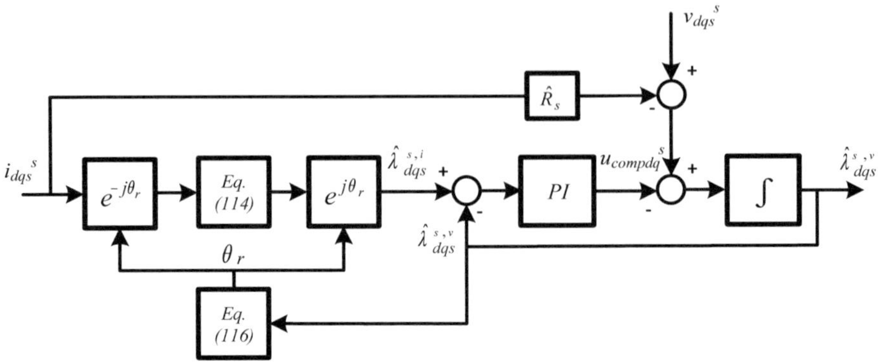

Fig. 10.33 Closed-loop rotor flux linkage observer model for SynRM/PMASynRM

The closed-loop rotor flux linkage observer can be derived for the PMSM and SynRM/PMASynRM as depicted in Fig. 10.33. As it is a synchronous machine, there is no slip, and the stator and rotor fluxes are identical simplifying the model (Yoo and Sul 2008).

The equations of the closed-loop flux observer of the voltage model are identical to Eqs. (10.108)–(10.109), that is:

$$\hat{\lambda}_{ds}^{s,v} = \int \left(v_{ds}^{s} - \hat{R}_s i_{ds}^{s} - u_{comp\_d} \right) \mathrm{d}t \tag{10.112}$$

$$\hat{\lambda}_{qs}^{s,v} = \int \left(v_{qs}^{s} - \hat{R}_s i_{qs}^{s} - u_{comp\_q} \right) \mathrm{d}t \tag{10.113}$$

Moreover, the current model for PMSM in the synchronously rotating reference frame is

$$\begin{aligned} \hat{\lambda}_{ds}^{e} &= \hat{L}_d i_{ds}^{e} + \hat{\Psi}_m \\ \hat{\lambda}_{qs}^{e} &= \hat{L}_q i_{qs}^{e} \end{aligned} \tag{10.114}$$

On the other hand, the current model for SynRM/PMASynRM is:

$$\begin{aligned} \hat{\lambda}_{ds}^{e} &= \hat{L}_d i_{ds}^{e} \\ \hat{\lambda}_{qs}^{e} &= \hat{L}_q i_{qs}^{e} - \hat{\Psi}_m \end{aligned} \tag{10.115}$$

The instantaneous rotor flux linkage angular position can be described by (10.116) for both machines:

$$\hat{\theta}_{\lambda s} = \tan^{-1} \left(\frac{\hat{\lambda}_{qs}^{s,v}}{\hat{\lambda}_{ds}^{s,v}} \right) \tag{10.116}$$

10.6.3 Rotor Flux Linkage Estimator PMSM

As an alternative to the previous closed-loop rotor flux linkage observer, the rotor flux linkage in a PMSM can be performed with the machine model equations in the stationary reference frame. It requires less computation so that it is easier to implement in a microcontroller with lower performance. It only uses one proportional controller to reduce the error between the nominal rotor flux of the permanent magnets Ψ_m, and the rotor flux observed. Especially in steady-state, the actual flux linkage vector is synchronized to the rotor given an exact rotor position. The known problems as the drift of the integral terms and DC amplification are compensated by this P controller. The output of the P controller is a compensated voltage u_{compdq} which acts on the PMSM voltage model. Although the estimator has a mechanism to correct the integration problems, it is not enough to compensate the measurements errors, and some experimental compensation in the estimated angle should be applied due to the difference between the real position and the estimated angle.

From (10.112) and (10.113), it is possible to derive the below voltage model equations for stator flux of PMSM:

$$\hat{\lambda}_{ds}^{s} = \int \left(v_{ds}^{s} - \hat{R}_s i_{ds}^{s} - u_{comp\_d} \right) \mathrm{d}t \tag{10.117}$$

$$\hat{\lambda}_{qs}^{s} = \int \left(v_{qs}^{s} - \hat{R}_s i_{qs}^{s} - u_{comp\_q} \right) \mathrm{d}t \tag{10.118}$$

So rotor flux is:

$$\hat{\Psi}_m^s \cos(\hat{\theta}) = \hat{\lambda}_{ds}^{s} - \hat{L} i_{ds}^{s} \tag{10.119}$$

$$\hat{\Psi}_m^s \sin(\hat{\theta}) = \hat{\lambda}_{qs}^{s} - \hat{L} i_{qs}^{s} \tag{10.120}$$

where L is known as synchronous inductance. If the synchronous inductance is small, the current terms can be neglected, if not they have to be considered. This model is ideal for SMPMSM where both synchronous inductance L_d and L_q are identical and the equal to L. In case of IPMSM machines, where L_d are different to L_q, is possible to use the same model with one inductance L, only under the condition that direct current is slowly varying in the time. Thus, the synchronous inductance L can be equal to L_q. The main difference with SMPMSM is that the maximum torque per ampere (MTPA) is obtained giving a rated value of negative direct current (reluctance torque) as seen in Sect. 4.4.2.4.

By substituting the expressions (10.117) and (10.118) into the rotor flux (10.119) and (10.120) is possible to determine the expression of the rotor flux, which uses the voltage and current components, and the machine parameters as synchronous inductance and stator resistance:

$$\hat{\Psi}_m^s \cos(\hat{\theta}) = \int \left(v_{ds}^s - \hat{R}_s i_{ds}^s - u_{comp\_d} \right) \mathrm{d}t - \hat{L} i_{ds}^s \tag{10.121}$$

$$\hat{\Psi}_m^s \sin(\hat{\theta}) = \int \left(v_{qs}^s - \hat{R}_s i_{qs}^s - u_{comp\_q} \right) \mathrm{d}t - \hat{L} i_{qs}^s \tag{10.122}$$

The rotor instantaneous angular position can be found by dividing both Eqs. (10.121) and (10.122) as:

$$\frac{\cos \hat{\theta}}{\sin \hat{\theta}} = \frac{1}{\tan \hat{\theta}} = \frac{\int \left(v_{ds}^s - \hat{R}_s i_{ds}^s - u_{comp\_d} \right) \mathrm{d}t - \hat{L} i_{ds}^s}{\int \left(v_{qs}^s - \hat{R}_s i_{qs}^s - u_{comp\_q} \right) \mathrm{d}t - \hat{L} i_{qs}^s} \tag{10.123}$$

Then, the instantaneous angular position of the rotor flux linkage is:

$$\hat{\theta} = \tan^{-1} \left(\frac{\int \left(v_{qs}^s - \hat{R}_s i_{qs}^s - u_{comp\_q} \right) \mathrm{d}t - \hat{L} i_{qs}^s}{\int \left(v_{ds}^s - \hat{R}_s i_{ds}^s - u_{comp\_d} \right) \mathrm{d}t - \hat{L} i_{ds}^s} \right) \tag{10.124}$$

With the angle position, it is possible to obtain the rotor speed by using the derivative of the angle.

Because the integral at time instant $t = 0$ requires an initial condition, the rotor must be aligned with a known position by applying DC current. However, in some applications such as EV is not possible since the rotor needs to move to the desired aligned position.

In Fig. 10.34 is shown the complete flux estimator for PMSM where $L = L_q$ (Large inductance). In case of SMPMSM, $L = L_q = L_d$, and for IPMSM machines where $L_q > L_d$, then L_q should be taken as L.

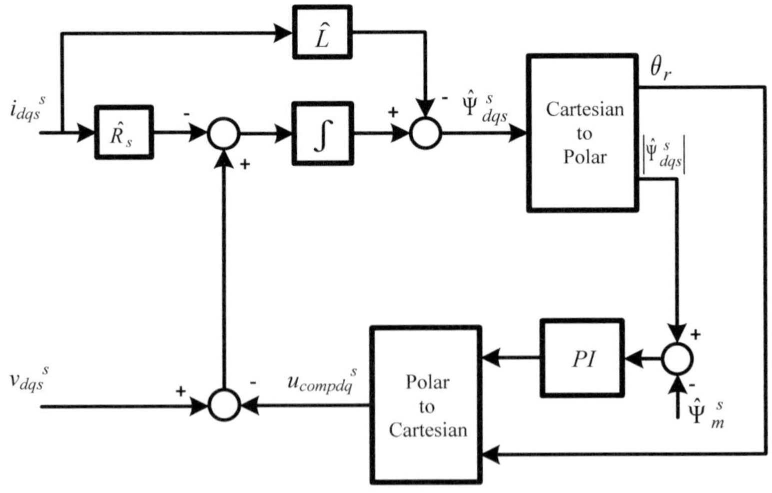

Fig. 10.34 Rotor flux linkage estimator model for PMSM

As can be observed, it uses Eq. (10.124) and includes a P controller which compensated the drift problem of ideal integrators with a voltage compensator. The total flux amplitude feedback is used to keep the integral limited and aligned in amplitude. Moreover, the rotor flux estimation is not possible at very low speeds since it uses the voltage model of the machine, which needs a minimum voltage to start to estimate the rotor flux and angle. Then, an open-loop start-up of the machine should be used to have enough back-EMF to estimate the rotor flux and the rotor position correctly. When the open-loop angle matches approximately the estimated angle, the transition to using the estimated rotor position can be performed. It should be noted that this transition should be taken as smoothly as possible, to avoid over currents peaks or to lose the synchronization. The rotor flux linkage estimator presented works in a wide speed range with 2–32 poles PMSM and electrical frequencies between 5 and 1500 Hz.

10.6.4 Instantaneous Slip and Speed Estimator for IM

The closed-loop flux observed presented in Sect. 10.6.2.2 provides the instantaneous position of the rotor flux linkage for IM, PMSM, and SynRM/PMASynRM. If the derivative is made at this angle, the speed of the rotor flux linkage is obtained which is the synchronous speed of the induction machine while the rotor speed for the SynRM/PMASynRM and PMSM:

$$\hat{\omega}_{\mathrm{SynRM/PMASynRM}} = \frac{\mathrm{d}\hat{\theta}_{\lambda s}}{\mathrm{d}t} \tag{10.125}$$

If the synchronous speed of the induction machine is subtracted from the slip, the rotor speed is obtained as:

$$\hat{\omega}_{IM} = \hat{\omega}_e - \hat{\omega}_{sl} \tag{10.126}$$

where

$$\hat{\omega}_e = \frac{\mathrm{d}\hat{\theta}_{\lambda r}}{\mathrm{d}t} \tag{10.127}$$

Therefore, unlike the SynRM/PMASynRM where the rotor rotated at the same speed as the stator, to know the speed of the induction machine, it is essential to know the slip. In this section, the angular instantaneous slip frequency and the angular rotor speed will be described under steady-state and transient conditions of the induction machine. The description is based on the induction machine equations and some of the parameters already discussed in the previous sections.

In the indirect vector control of Sect. 10.4.1, it was found that the slip can be estimated by knowing the rotor time constant and the components of the stator current i_{dse} and i_{qse} in the synchronously rotating reference frame as:

$$\hat{\omega}_{sl} = \frac{\hat{R}_r}{\hat{L}_r}\frac{i_{qs}^e}{i_{ds}^e} \tag{10.128}$$

It is also possible to obtain the slip from the stationary reference frame model derived in Sect. 10.6.2 with expression (10.129) as follows:

$$\begin{aligned} \frac{\mathrm{d}\lambda_{dr}^s}{\mathrm{d}t} &= \frac{L_m}{T_r} i_{ds}^s - \frac{1}{T_r}\lambda_{dr}^s - \omega_r \lambda_{qr}^s \\ \frac{\mathrm{d}\lambda_{qr}^s}{\mathrm{d}t} &= \frac{L_m}{T_r} i_{qs}^s - \frac{1}{T_r}\lambda_{qr}^s + \omega_r \lambda_{dr}^s \end{aligned} \tag{10.129}$$

The instantaneous angle position was represented as:

$$\theta_{\lambda r} = \tan^{-1}\left(\frac{\lambda_{qr}^s}{\lambda_{dr}^s}\right) \tag{10.130}$$

Thus, it is possible to develop the derivative of (10.130) as:

$$\omega_e = \frac{\mathrm{d}\theta_{\lambda r}}{\mathrm{d}t} = \frac{\mathrm{d}\left(\tan^{-1}\left(\frac{\lambda_{qr}^s}{\lambda_{dr}^s}\right)\right)}{\mathrm{d}t} = \frac{1}{1+\left(\frac{\lambda_{qr}^s}{\lambda_{dr}^s}\right)^2}\frac{\mathrm{d}\left(\frac{\lambda_{qr}^s}{\lambda_{dr}^s}\right)}{\mathrm{d}t} = \frac{\lambda_{dr}^s \frac{\mathrm{d}\lambda_{qr}^s}{\mathrm{d}t} - \lambda_{qr}^s \frac{\mathrm{d}\lambda_{dr}^s}{\mathrm{d}t}}{\left(\lambda_{dr}^s\right)^2 + \left(\lambda_{qr}^s\right)^2} \tag{10.131}$$

By replacing (10.129) into the synchronous speed expression (10.131):

$$\begin{aligned} \omega_e &= \frac{\lambda_{dr}^s\left(\frac{L_m}{T_r} i_{qs}^s - \frac{1}{T_r}\lambda_{qr}^s + \omega_r\lambda_{dr}^s\right) - \lambda_{qr}^s\left(\frac{L_m}{T_r} i_{ds}^s - \frac{1}{T_r}\lambda_{dr}^s - \omega_r\lambda_{qr}^s\right)}{\left(\lambda_{dr}^s\right)^2 + \left(\lambda_{qr}^s\right)^2} \\ &= \omega_r + \frac{L_m}{T_r}\left(\frac{\lambda_{dr}^s i_{qs}^s - \lambda_{qr}^s i_{ds}^s}{\left(\lambda_{dr}^s\right)^2 + \left(\lambda_{qr}^s\right)^2}\right) \end{aligned} \tag{10.132}$$

where slip is:

$$\omega_{sl} = \frac{L_m}{T_r}\left(\frac{\lambda_{dr}^s i_{qs}^s - \lambda_{qr}^s i_{ds}^s}{\left(\lambda_{dr}^s\right)^2 + \left(\lambda_{qr}^s\right)^2}\right) \tag{10.133}$$

Equation 10.133 uses the rotor flux linkage and the monitored stator current components i_{ds} and i_{qs} in the stationary reference frame. Note that from the previous closed-loop observer, the rotor flux linkage was estimated. Then, the expression of the slip estimated of (10.133) can be rewritten with the subscript ^ which meant variables and parameters estimated:

$$\hat{\omega}_{sl} = \frac{\hat{L}_m}{\hat{T}_r}\left(\frac{\hat{\lambda}_{dr}^{s,v} i_{qs}^{s} - \hat{\lambda}_{qr}^{s,v} i_{ds}^{s}}{\left(\hat{\lambda}_{dr}^{s,v}\right)^2 + \left(\hat{\lambda}_{qr}^{s,v}\right)^2}\right) \tag{10.134}$$

The disadvantage concerning the slip estimator of Eq. 10.128 is that it uses an additional machine parameter, the magnetizing inductance, but offers better performance when used with the flux observer of Sect. 10.6.2.2. In fact, the slip estimator of Eq. 10.128 is usually used when the speed of the machine is measured through, for example, an encoder for indirect vector control, while the slip estimator of Eq. 10.134 is usually used in direct vector control when the magnitude and position of the flux linkage rotor are known, either through a flux observer or through Hall effect sensors. In the case of using the closed-loop observer of Sect. 10.6.2.2, the magnitude, the position of rotor flux linkage, and the slip are obtained with five parameters join the utilization of the voltage and current machine terminals. The measurement of the current in the terminals can be carried out as was discussed in Sect. 8.4.2. However, the voltage measurement at the machine terminals can be performed by reconstructing from the duty cycle of the PWM output voltage of the VSI as introduced in Sect. 7.4.1.2. By using the instantaneous measurement of the DC-link voltage of the VSI, and knowing the theoretical duty cycle applied, it is possible to reconstruct the phase and line voltages of the electric machine as discussed. A particular case with higher accuracy was to measure the duty cycles applied to the electric machine just at its terminals using the MTU5 timer set in capture mode. The advantage concerning the previous case of the theoretical duty cycle is that the deadtime is already taken into account, and the measurement will be more precise. Furthermore, as the measurement is digital, there are no delays caused by the signal processing such as analog filters when used the conventional voltage measurement based on voltage dividers, a low-pass filter, and an A/D converter. If the measurement has a considerable delay, the vector control performance is degraded. Therefore, if the MTU5 timer is configured in the same way as was seen in Sect. 7.4.1.2 for deadtime compensation, it will be sufficient to apply Eqs. 10.135 and 10.136 to reconstruct the voltage at the terminals of the machine, where it is assumed that the machine is balanced where the 0-component is omitted.

$$v_{an} = \left(\frac{2}{3}D_a - \frac{1}{3}D_b - \frac{1}{3}D_c\right)\frac{V_{\text{DC}}}{\text{PeakTriangular}}$$

$$v_{bn} = \left(\frac{2}{3}D_b - \frac{1}{3}D_a - \frac{1}{3}D_c\right)\frac{V_{\text{DC}}}{\text{PeakTriangular}}$$

$$v_{cn} = \left(\frac{2}{3}D_c - \frac{1}{3}D_a - \frac{1}{3}D_b\right)\frac{V_{DC}}{\text{PeakTriangular}} \tag{10.135}$$

where

$$\begin{aligned} D_a &= \text{MTU5.TRGU} \\ D_b &= \text{MTU5.TRGV} \\ D_b &= \text{MTU5.TRGW} \\ \text{PeakTriangular} &= \text{MTU3.TGRC} \end{aligned} \tag{10.136}$$

The values captured in the registers of the MTU5 timer for each of the phases of the machine join the peak value of the triangular used to generate the PWM at its corresponding frequency which will be sufficient to reconstruct the machine voltage terminals.

It is important to mention that, in case of insulation requirement between the high voltage and the low voltage of the control, being digital signals, it will be a more straightforward task than in the case of using analog signals and A/D converter. The isolation can be performed with common optical isolation using an optocoupler without losing precision in the measurement.

10.7 Simulations Results

After describing the different estimation methods of the rotor position as well as the flux linkage rotor, in this section, different simulations are presented to evaluate their behavior for IMs and PMSMs. Open-loop control is used for the induction machine where the rotor flux linkage and the slip estimator are evaluated. However, for the PMSM, a sensored field orientation control is used, due to its unstable zones when controlled in open loop as seen in Sect. 4.5.2.1. The PMSM simulations in closed loop will allow seeing the rapid dynamics of the control when the field orientation control is used, being the prelude to what it will see in Chap. 11. The models of both machines are the ones studied in Chap. 5, corresponding to continuous models of state-space equations. Observers and estimators are represented in a discrete domain since the objective is to implement it in a microcontroller or DSP.

10.7.1 Flux Observer and Slip Estimator Simulations in IM

The flux observer and the slip estimator can be validated by connecting it to an induction machine model seen in Sect. 4.5.1.2. The model of the machine can be controlled in an open loop by merely applying a suitable voltage/frequency curve. Being an open-loop control, using a voltage–frequency curve, the start-up of the

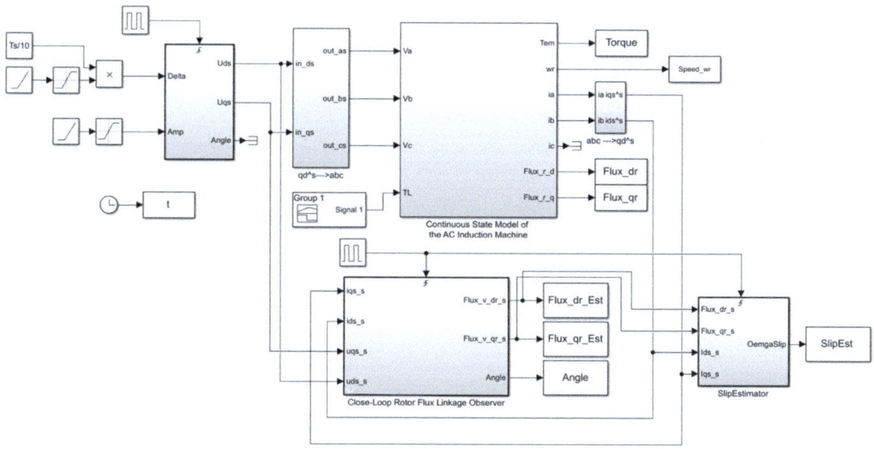

Fig. 10.35 System simulation of IM in open-loop control where closed-loop rotor flux linkage observer and slip estimator are connected to the machine

induction machine, as well as load changes in the stationary regime, will cause the torque and speed to have an oscillatory response. The flux observer and the slip estimator can be verified in both transients and steady-state. In Fig. 10.35, the model of the induction machine is shown together with the flux observer and the slip estimator. Both the flux observer and the slip estimator are implemented in discrete time. The sampling period T_s, in which the voltages and currents are sampled, affects the accuracy of the estimations, mainly when the machine operates at high speeds where the electrical frequency of the voltages and currents gets closer to the sampling frequency. The sampling delay of the signals is neglected, as well as the zero-order hold. The sampling period chosen for the discrete models is 100 μs.

The contents of the closed-loop flux observer block are shown in Fig. 10.36. As can be seen, the structure is similar to the one seen previously in Fig. 10.32. The inputs are the *dqs* currents and voltages in the stationary reference frame, and the outputs are the rotor flux linkage, also in the *dqs* axes, and the synchronous position angle. On the other hand, the slip estimator directly implements Eq. (10.134).

The results obtained with the flux observer are shown in Figs. 10.37, 10.38, and 10.39 when the unloaded induction machine is supplied with a gradual voltage of up to 140 V in the v_{ds} and v_{qs} voltage components. The reference speed consists of a ramp that reaches 6000 RPM in 2.5 s. At start-up, an oscillation can be observed in the developed torque and the *dqs* currents due to the open-loop control. When the oscillation ends, the torque remains constant until it reaches the speed of 6000 RPM where it decreases up to the friction torque.

The detail of the estimated instantaneous angle and the rotor flux components measured and estimated at the start of the machine are shown in Fig. 10.38. It is possible to observe that the magnitude of the measured and estimated rotor flux components has a minimal error during start-up.

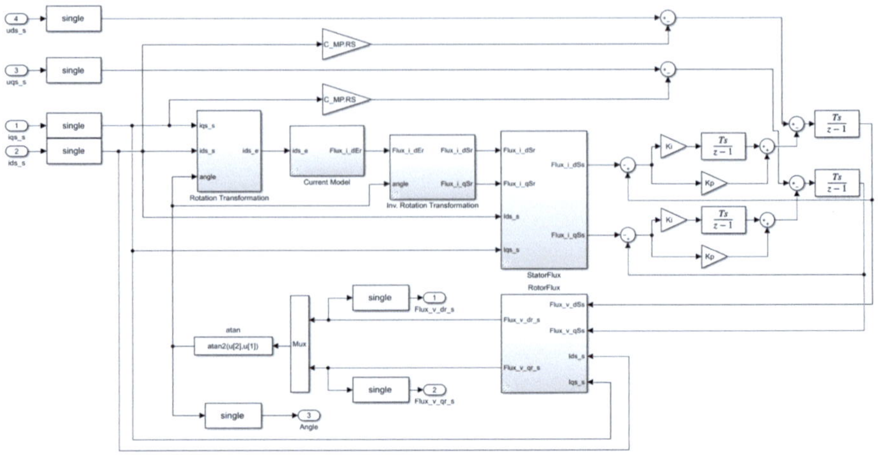

Fig. 10.36 Closed-loop rotor flux linkage observer for IM based on the voltage and current model of the machine equations

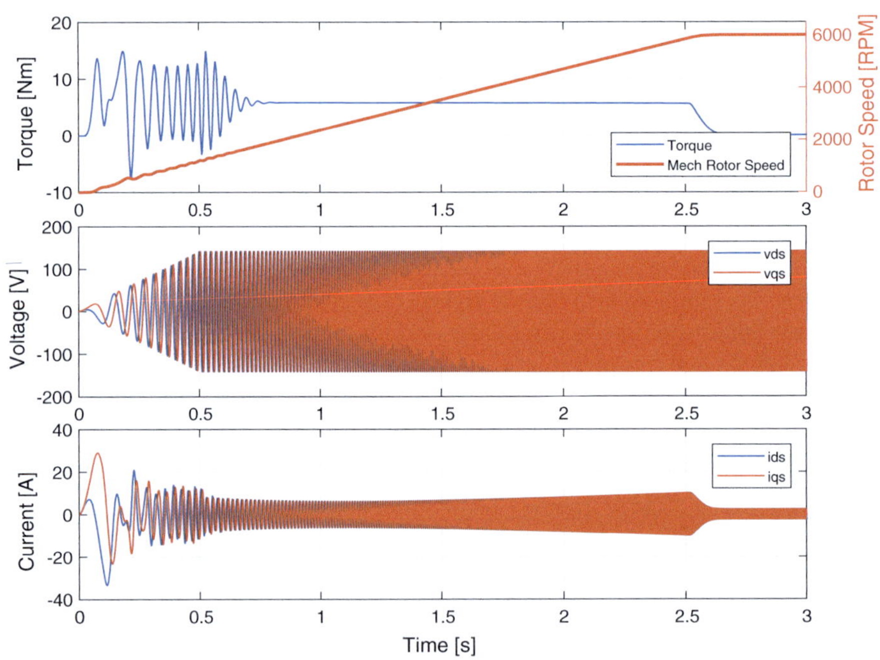

Fig. 10.37 Run-up machine speed, torque, voltage and current components in the stationary reference frame with no-load torque. Simulation variable-step size. Minimum step $T_{sp} = 1\ \mu s$, and automatic solver selection

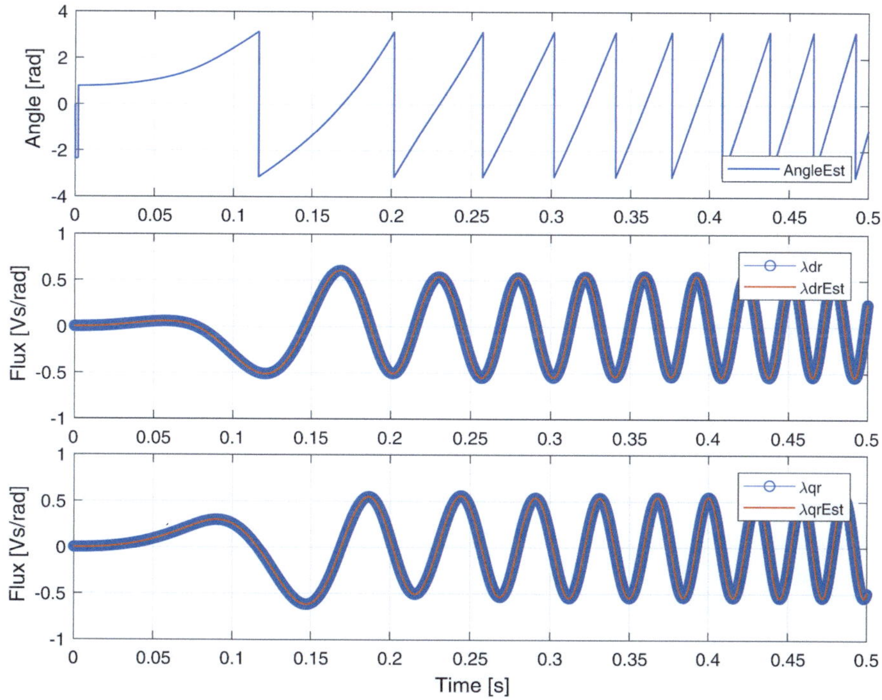

Fig. 10.38 Run-up machine speed, torque, voltage and current components in the stationary reference frame with no-load torque. Simulation variable-step size. Minimum step $T_{sp} = 1\ \mu s$, and automatic solver selection

The error is more significant at high speed, as shown in Fig. 10.39, where the speed is already 6000 RPM, and the discrete time of the estimated signals can be appreciated in greater detail.

A quality factor or error factor can be obtained with the division between the estimated rotor flux modulus and the measured rotor flux modulus. In Fig. 10.40 is represented the quality factor according to the speed of the machine. Note that although the x-axis of the graph is a time in seconds, it should be considered the speed profile of Fig. 10.37. The quality factor of both fluxes is observed in Fig. 10.40, where the ideal would be to obtain a constant and equal value to the unit. This figure also shows the quality factor when the flux observer uses an incorrect value of 0.5 and 1.5 times the stator resistance. The quality factor increases as the speed increases, while when the reference speed is reached at $t = 2.5$ s, the error decreases considerably for the three cases as can be seen in Fig. 10.40. The closed-loop rotor flux observer offers an almost negligible error in steady-state while the transient regime increases above all in high speeds and while the machine accelerates. In steady-state, at 6000 RPM, the error is very negligible and independent of the value of the stator resistance.

The behavior of the quality factor at low speeds is shown in Fig. 10.41 where a simulation with different stator resistances has also been carried out. The speed

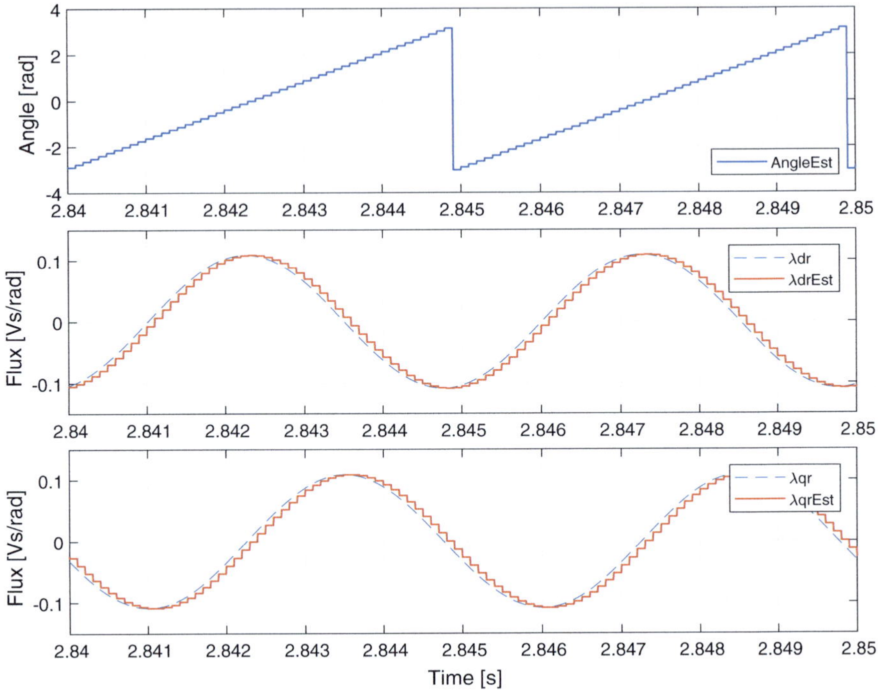

Fig. 10.39 Estimated angle, measured and estimated *dq* flux components when the machine is in steady-state at 6000 RPM. Simulation variable-step size. Minimum step $T_{sp} = 1$ μs, and automatic solver selection

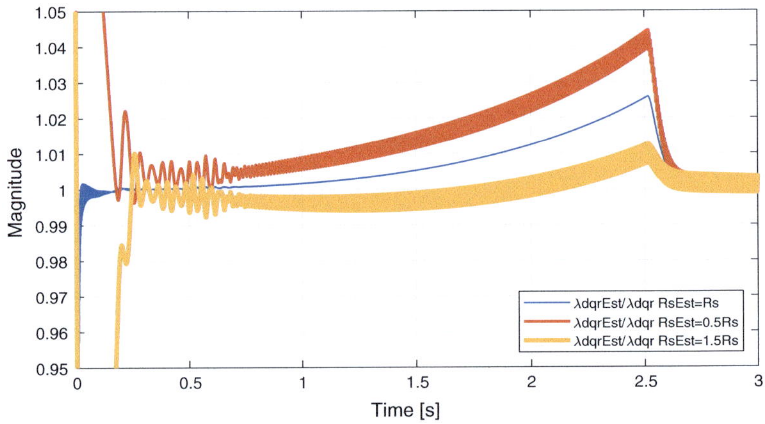

Fig. 10.40 Module of the estimated flux divided by the module of the measured flux when estimate resistance is different from the stator resistance of the machine. Minimum step $T_{sp} = 1$ μs, and automatic solver selection

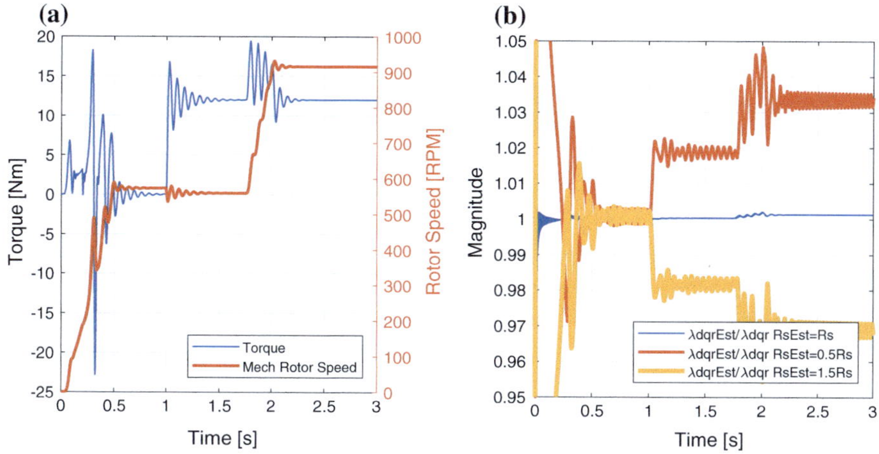

Fig. 10.41 **a** Torque and mechanical rotor speed at low speeds. **b** The module of the estimated flux divided by the module of the measured flux when estimate resistance is different from the stator resistance of the machine. Simulation variable-step size. Minimum step $T_{sp} = 1$ μs, and automatic solver selection

profile is shown in Fig. 10.41a. The load torque is zero until $t = 1$ s, and from here, it is changed to 12 Nm in a step manner. At $t = 1.75$ s, the speed is again increased to 915 RPM. The amplitude of the voltage components v_{ds} and v_{qs} remains constant once the speed of 550 RPM has been reached. For an estimated resistance with a correct value precisely to the real resistance of the stator, the error is a very negligible as can be seen in Fig. 10.41b, while for resistances different from the current resistance of the machine, the absolute error increases, especially when the machine rotates at 915 RPM.

The result of the slip estimator is shown in Fig. 10.42. In this case, the machine starts with no load up to a speed of 600 RPM. At $t = 1.5$ s, the torque is increased to 20 Nm in ramp manner, and at $t = 2.5$, the load torque is decreased from 20 to 5 Nm. The error in the slip estimation increases at the start of the machine as can be seen. Being an open-loop control, the torque is not controlled so that it oscillates during start-up. In this situation, the real and estimated slip oscillates in line with the oscillation of the developed torque so that its estimation is very complicated. However, in a vector control where the torque is controlled instantaneously, the oscillation in the slip and the developed torque disappear so that the estimate of the slip will be correct as well.

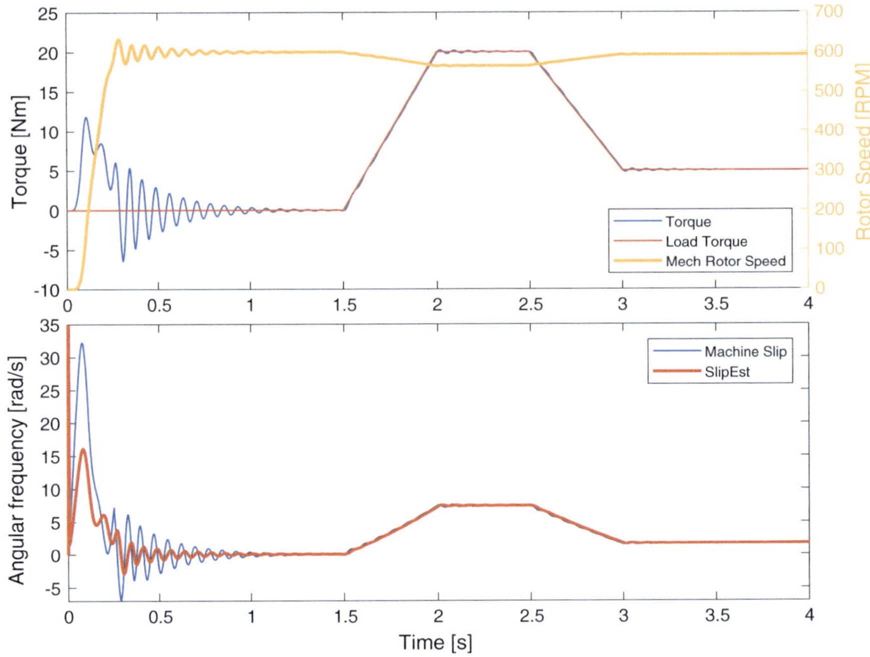

Fig. 10.42 Run-up machine speed, the torque developed, load torque, real slip, and estimated slip. Simulation variable-step size. Minimum step $T_{sp} = 1\ \mu s$, and automatic solver selection

10.7.2 Flux Observer in PMSM

10.7.2.1 Vector Control Simulation

The estimator presented in Sect. 10.6.3 and the closed-loop rotor flux linkage observed of Sect. 10.6.2 can be used to estimate the rotor flux linkage and the angle position as described. In Sect. 4.5.2.1, an open-loop simulation was carried out to demonstrate that the PMSM has zones of instability. The objective of the simulation presented in this section is to evaluate both estimators. To do this, a PMSM controlled by vector control is used to guarantee its stability and be able to extract enhanced results from the estimators. The machine is a 5 HP SMPMSM where its parameters were described in Chap. 5. The model of the machine of Sect. 4.5.2.1 is modified slightly to provide as inputs the components v_{dse}, v_{qse}, and load torque T_L, as can be seen in Fig. 10.43. It is also possible to observe the vector control structure, formed by two current control loops together with the feedforward compensation to eliminate the coupling effect, and a PI controller to control the speed of the machine. Besides, the possible delay in the measurements of the i_{dse} and i_{qse} current components, as well as the rotor speed, is modeled. The control is implemented in discrete time so that zero-order holds will be necessary.

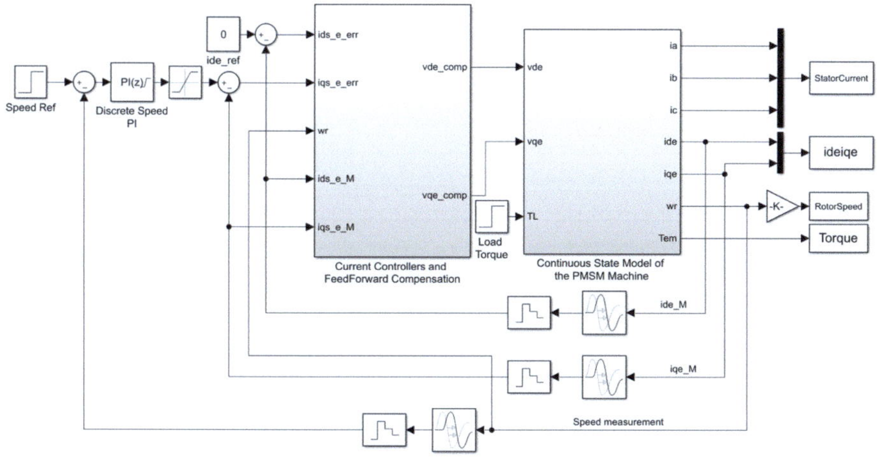

Fig. 10.43 System simulation of SMPMSM with sensored vector control

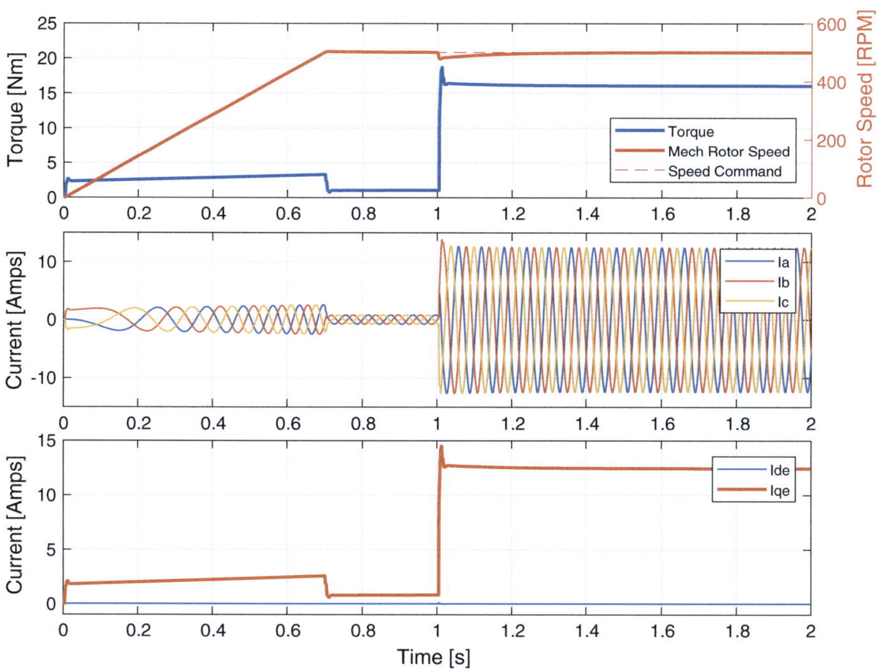

Fig. 10.44 Run-up speed reference, rotor speed, torque, phase currents, and i_{dse} and i_{qse} components of unloaded PMSM till $t = 1$ s with electrical set point speed of 16.66 Hz (only friction coefficient as load torque). At $t = 1$ s 15 Nm of load torque is applied. Simulation variable-step size. Minimum step $T_{sp} = 1$ μs, and automatic solver selection

The sampling time T_s is set at 400 μs where it matches the execution speed of the internal current loops, while the external speed control loop is executed every 4 ms.

The result of the simulation is shown in Fig. 10.44 when a reference speed of 500 RPM is applied without load (only load torque due to friction) up to $t = 1$ s. At time instant $t = 1$ s, a load step of 15 Nm is applied. As it can be observed the response to the load step is very fast, about 12 ms and its speed only falls about 20 RPM. The magnitude of the phase currents only with the load due to friction is 565 mA RMS between $t = 0.7$ s and $t = 1$ s, whereas when the load step is applied at $t = 1$ s, it rises to 8.8 RMS approximately. Thanks to the feedforward compensation, the regulation of the *dqe* current components improves considerably increasing the control dynamics. The i_{dse} current component is regulated at zero amperes in overall time as shown, while i_{qse} is regulated to the current reference i_{dseRef} according to the torque demand.

In order to see in more detail the vector control dynamics for the PMSM, it is possible to simulate fast speed changes even at negative speeds. As discussed, the positive speed is understood when the machine operates in the first quadrant, while the speed in negative or in the opposite direction is when the machine operates in the third quadrant. The negative speed supposes a change of direction of the machine so

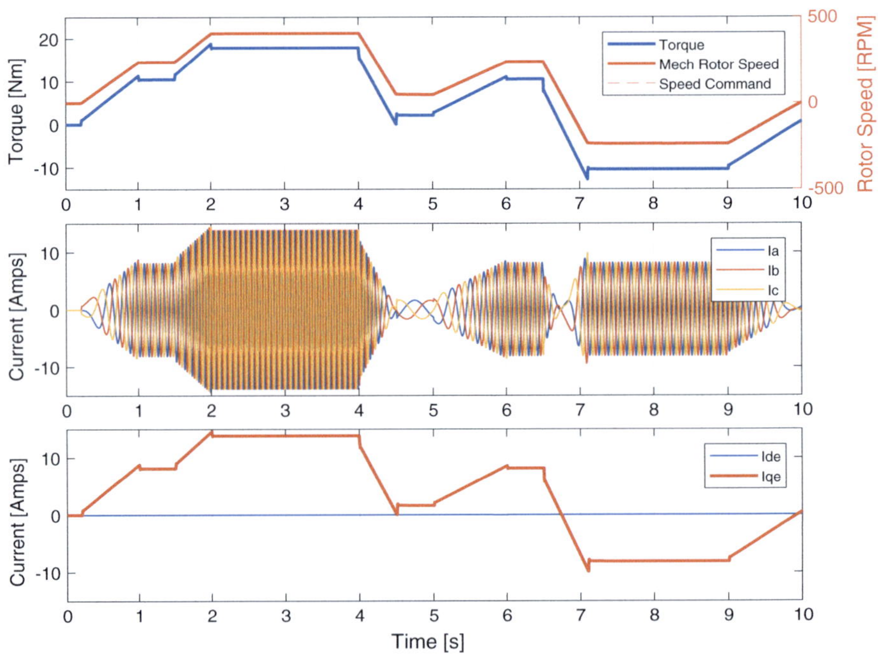

Fig. 10.45 Speed reference profile with positive and negative speeds with load torque proportional to the rotor speed. It shows the developed torque, speed reference, machine speed, phase currents, and i_{dse} and i_{qse} components. Simulation variable-step size. Minimum step $T_{sp} = 1$ μs, and automatic solver selection

that to obtain it, a negative i_{qse} torque current component must be applied as it will be seen below. The axis transformation automatically changes the sequence order of the three-phase *abc* phases, when the q component changes from positive to negative. In Fig. 10.45, a reference speed profile is shown together with the electromagnetic torque developed as a function of speed. The machine changes direction at $t =$ 7 s approximately, being the developed torque negative. In the phase currents, it is possible to observe in the time instant $t = 6.7$ s, the change of order passing from a sequence *abc* to *acb*. The i_{dse} component is adjusted to zero while the torque component is regulated to achieve the required torque.

The voltage and current components in the stationary reference frame can be seen in Fig. 10.46. As in the case of phase currents, it also changes its sequence when the machine changes direction. Finally, the voltage components in the synchronously rotating reference frame corresponding to the outputs of the current controllers with feedforward compensation are shown.

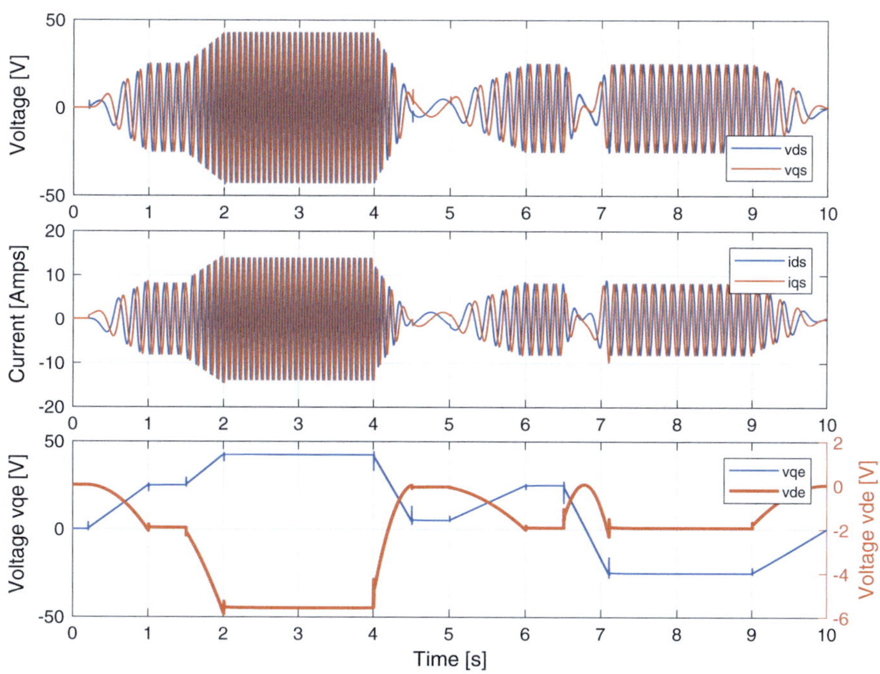

Fig. 10.46 Speed reference profile with positive and negative speeds with load torque proportional to the rotor speed. It shows the v_{ds}, v_{qs} voltage components, i_{ds}, i_{qs} current components, and v_{dse}, v_{qse} voltage components. Simulation variable-step size. Minimum step $T_{\mathrm{sp}} = 1$ μs, and automatic solver selection

10.7.2.2 Rotor Flux Estimator

If both estimators are connected to the previous model, it is possible to evaluate the behavior of the estimation of the angle and the rotor flux linkage. In Fig. 10.47 is shown both estimators, the flux linkage estimator, and the closed-loop flux linkage observer. In Figs. 10.48 and 10.49, the content of both estimators is represented, respectively.

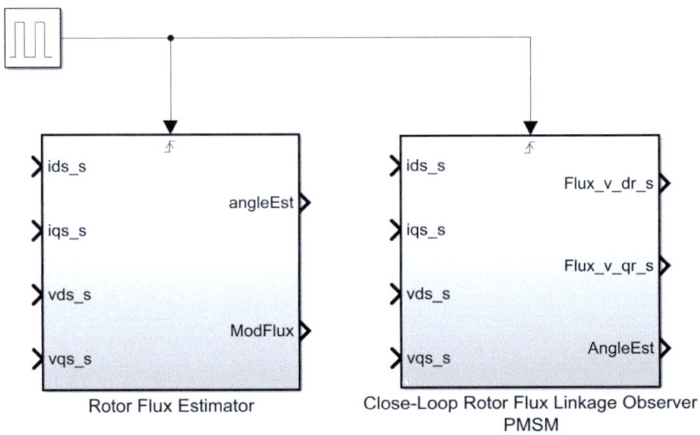

Fig. 10.47 Rotor flux linkage estimator and closed-loop rotor flux linkage observer for PMSM

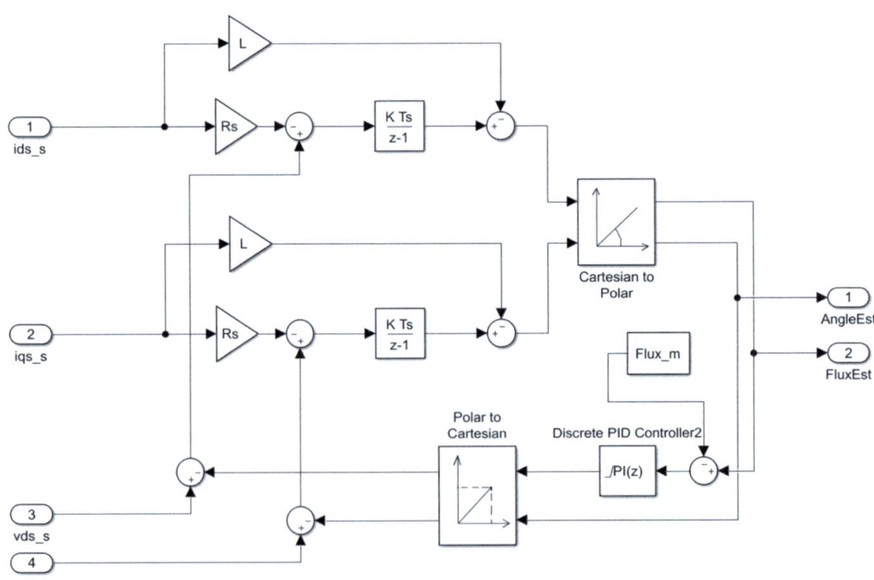

Fig. 10.48 Rotor flux linkage estimator for PMSM based on single *P* controller

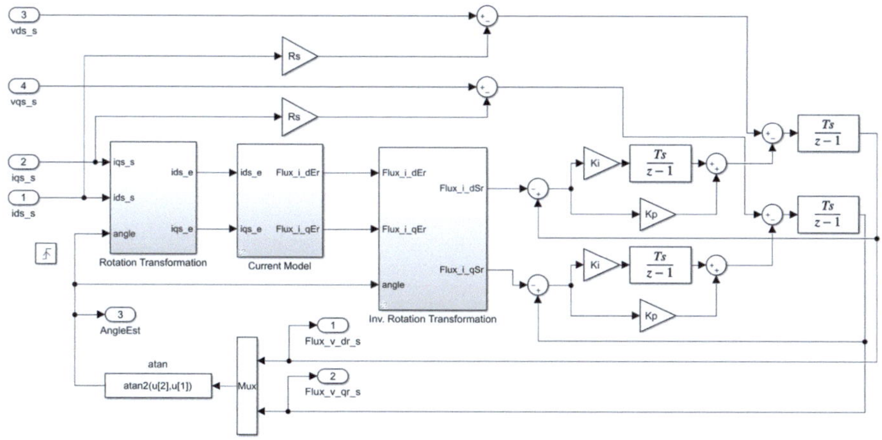

Fig. 10.49 Closed-loop rotor flux linkage observer for PMSM based on the voltage and current model of the machine equations

Figure 10.50 shows the start-up of the SMPMSM controlled by the vector control seen previously. The load torque increases linearly from time instant $t = 0$ to $t =$ 0.05 s as can be seen and up to its maximum value of 8 Nm. The reference speed ramp increases linearly to 239 RPM in 0.5 s. The torque developed by the machine has to overcome the load torque plus the torque due to friction. The measured angle of the rotor position together with the estimated angle and the angle observed by the closed-loop flux linkage observer are also shown. It is possible to appreciate the error between the angle of the position of the rotor and the angles of both estimators, where the angle of the closed-loop flux linkage observer converges until it matches with the measured angle. However, the angle of the estimator based on a single controller P maintains a constant error between the measured angle and the estimated angle. As discussed in Sect. 10.6.3, this can be compensated experimentally by adding a shift angle in the estimated angle.

In Fig. 10.51, the same previous signals are shown but with a load torque step from 0 to 18 Nm at $t = 2$ s to observe the behavior of both estimators. The estimator based on a single controller P up to $t = 2$ s has an error in degrees of 11.46° due mainly to the fact that the machine has no load, only the load by friction, whereas when the load step is applied in $t = 2$ s, the error in degrees with respect to the measured angle increases considerably up to 28.66°. On the other hand, the angle observed by the closed-loop rotor flux linkage observer offers a better result since the error concerning the measured angle increases from 1.43° to 5.73° when the load step is applied.

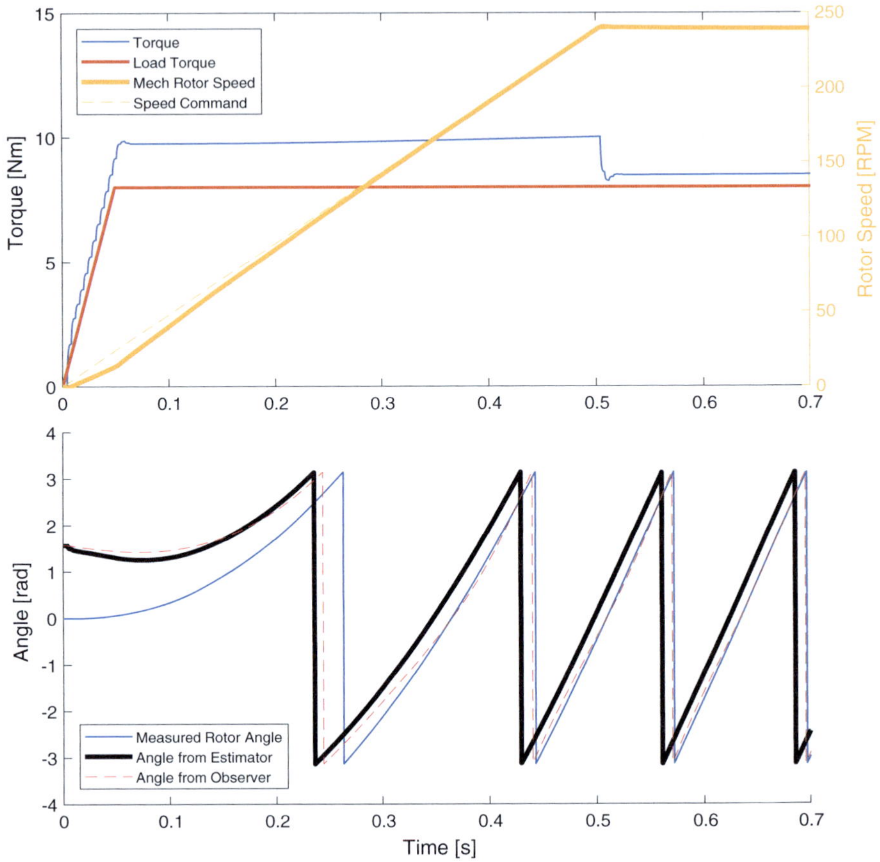

Fig. 10.50 Start-up of SMPMSM controlled with vector control. It shows the developed torque, load torque, speed reference, machine speed, measured rotor angle position, estimated angle, and observed angle from closed-loop rotor flux linkage observer. Simulation variable-step size. Minimum step $T_{sp} = 1\ \mu s$, and automatic solver selection

10.7.2.3 Permanent Magnet Synchronous Generator (PMSG)

One of the disadvantages of wind turbines that use permanent magnet machines is the effect of cogging torque, which complicates the start-up of the wind turbine through the aerodynamic torque. The cogging torque, as discussed in Chap. 4, is the torque due to the interaction between the permanent magnets of the rotor and the stator slots. When an axis of a PMSM/PMSG machine is rotated without power, it is possible to feel it, since it is perceived as a jump from the current point and settles in another. The

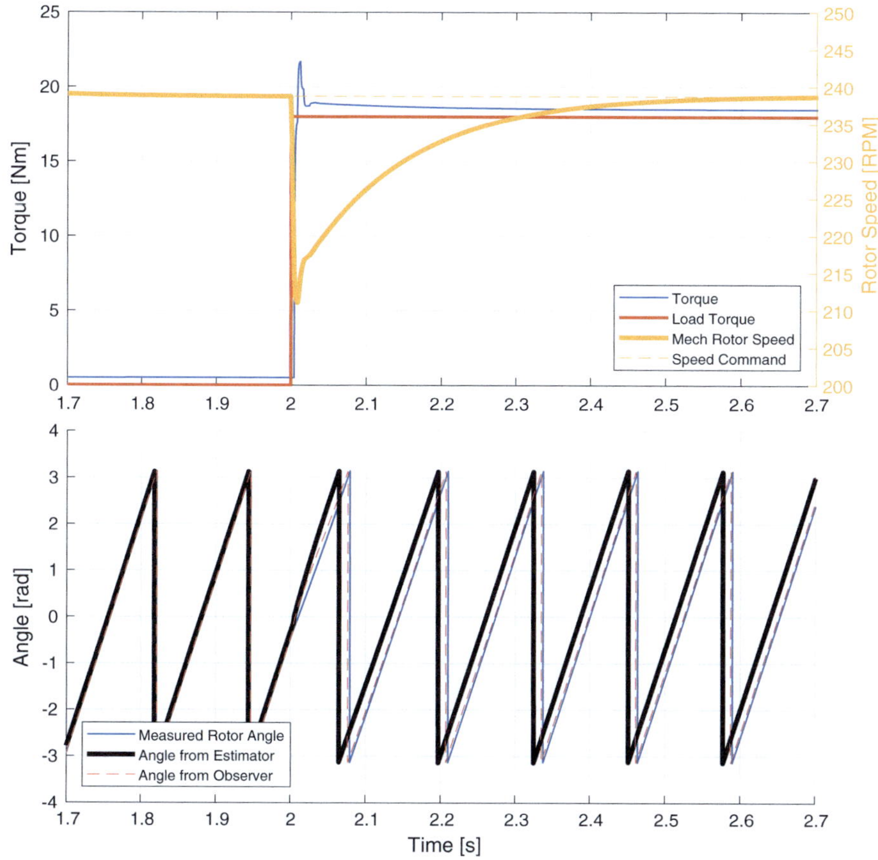

Fig. 10.51 Load step at $t = 2$ s detail. It shows the developed torque, load torque, speed reference, machine speed, measured rotor angle position, estimated angle, and observed angle from closed-loop rotor flux linkage observer. Simulation variable-step size. Minimum step $T_{sp} = 1\ \mu s$, and automatic solver selection

torque experienced in these jumps is the torque required to separate from the current rotor/stator alignment and move on to the next rotor/stator alignment. The cogging torque in small wind turbines based on PMSM can be large enough to prevent the start-up of the wind turbine when the wind speed is not sufficient. One technique that is usually employed is to operate the PMSG machine in motor mode, for a short time, to supplement the aerodynamic torque with additional torque, which also allows reducing the starting time of the turbine. The optimal reference speed in motor mode can be calculated based on the current wind speed, or it can just be a predetermined speed.

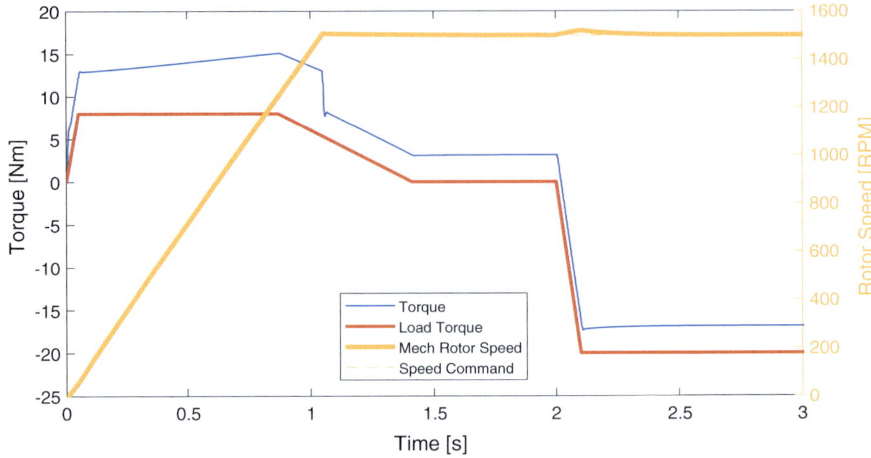

Fig. 10.52 PMSM is working in motor mode till time instant $t = 2$ s. From this time, the machine is working in generator mode. It shows the developed torque, load torque, speed reference, and machine speed. Simulation variable-step size. Minimum step $T_{sp} = 1$ μs, and automatic solver selection

During start-up in motor mode, when the optimum reference speed is reached, it switches to generator mode by merely reversing the torque developed by the machine. Due to the inertia of the propellers, the inertia of the rotor, the wind speed, and the control allow maintaining the constant reference speed generating energy toward the electrical grid.

In the previous model of the vector control of the PMSM, something similar can be reproduced by merely inverting the sign of the load torque when the machine operates in motor mode. The speed controller ensures that the speed is the same as the reference speed, even if the load torque has been reversed.

In Fig. 10.52, the result of the simulation is shown when the machine enters in generator mode, as previously mentioned. First, the machine starts in motor mode with a load torque of 8 Nm, reaching a speed of 1500 RPM approximately. This load torque is reduced to zero at $t = 1.4$ s. At $t = 2$ s, a negative torque of 20 Nm is applied through a ramp, and the vector control applies the necessary corrections to follow the reference speed which increases up to 1518 RPM, but decreases rapidly to the reference speed as can be observed.

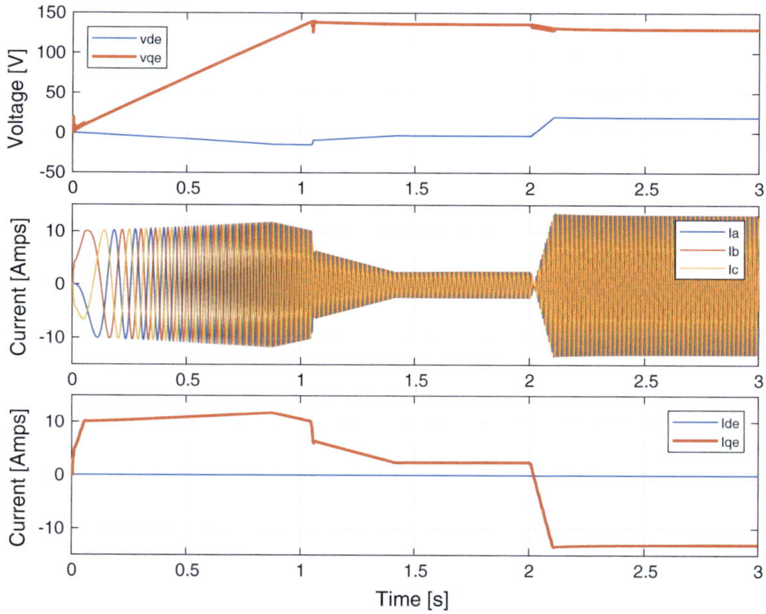

Fig. 10.53 PMSM working in motor mode till time instant $t = 2$ s. From this time, the machine is working in generator mode. It is shown v_{dse}, v_{qse} voltage components, phase currents, and i_{dse} and i_{qse} components currents. Simulation variable-step size. Minimum step $T_{\text{sp}} = 1$ μs, and automatic solver selection

On the other hand, Fig. 10.53 shows the v_{dse} and v_{qse} voltage components, the phase currents, and the i_{dse} and i_{qse} components. In generator mode, the v_{dse} component becomes positive while the v_{qse} component remains positive. However, the i_{qse} component becomes negative as expected, while the i_{dse} component remains at zero for the case of the SMPMSM.

According to Eqs. (3.66)–(3.67), and considering $i_{dse} = 0$, the active power in generator mode delivered to the electrical grid takes an approximate value of −2976 W, while the reactive power is 537 VAr.

References

Deur J (1999) Kompenzacija učinaka elastičnosti i trenja u prijenosnim mehanizmima slijednih sustava (in Croatian), Doctoral thesis, University of Zagreb, Zagreb, Croatia

Foulon E, Forgez C, Loron L (2007) Resistances estimation with an extended Kalman filter in the objective of real-time thermal monitoring of the induction machine. IET Electr Power Appl 1(4):549–556

Hinkkanen M, Luomi J (2003) Modified integrator for voltage model flux estimation of induction motors. IEEE Trans Ind Electron 50(4)

Jansen PL, Lorenz RD (1994) A physically insightful approach to the design and accuracy assessment of flux observers for field-oriented I.M. drives. IEEE Trans Ind Appl 30:101–110

Jansen PL, Lorenz RD, Novotny DW (1994) Observer-based direct field orientation: analysis and comparison of alternative methods. IEEE Trans Ind Appl 30:945–953

Kim S-H, Sul S-K (1997) Voltage control strategy for maximum torque operation of an induction machine in the field-weakening region. IEEE Trans Ind Electron 44(4):512–518

Kim S-H, Sul SK, Park MH (1995) Maximum torque control of an induction machine in the field weakening region. IEEE Trans Ind Appl 31(4):787–794

Lascu C, Boldea I, Blaabjerg F (2000) A modified direct torque control for induction motor sensorless drive. IEEE Trans Ind Appl 36:122–130

Otýpka J et al (2016) The enlarged d-q model of induction motor with the iron loss and saturation effect of magnetizing and leakage inductance. In: Department of electrical engineering FEECS, technical university of Ostrava. The Czech Republic. Springer International Publishing Switzerland

Perić N (1979) Optimiranje sistema za pozicioniranje s reguliranim istosmjernim elektromotornim pogonom (in Croatian), Master thesis, University of Zagreb, Zagreb, Croatia

Sul S-K (2011) Control of electric machine drive systems. Wiley-IEEE Press

Vas P (1993) Parameter Estimation, Condition Monitoring, and Diagnosis of Electrical Machines (Monographs in Electrical and Electronic Engineering)

Yoo A, Sul S-K (2008) Design of flux observer robust to parameter variation of interior permanent magnet synchronous motor. In: Proceedings of industry applications society annual meeting, vol 1, pp 1–7, 5–9

Zai L-C, DeMarco CL, Lipo TA (1992) An extended Kalman filter approach to rotor time constant measurement in PWM induction motor drives. IEEE Trans Ind Appl 28(1):96–104

Chapter 11
Model-in-the-Loop Development in a Vector Control of Induction Machine

11.1 Introduction

The model-based design (MBD) approach is adopted to test the inverter, and the induction machine control with the benefits explained in Chap. 1. As shown in this chapter, MiL consists of combining multi-physics simulation software with control software in an offline environment. In this scenario, physical elements such as the induction machine and the inverter are replaced by models performed, for example, in Simulink or PSIM environments. The MiL is where the development of the different software components necessary for the control implementation of the machine and the inverter is validated.

The VSI model can be implemented using PSIM plus SimCoupler, as seen in Chap. 8 or using the toolbox Simscape Electrical™ from Simulink®. If the PSIM plus SimCoupler solution is chosen, the model of the induction machine should also be the one offered by PSIM. Otherwise, it is possible to use the model of the induction machine offered by the toolbox Simscape Electrical™ of Simulink. During the VSI model implementation, in parallel, the software architecture can be designed in detail, where the necessary software components (SWCs) for the vector control application are defined. The MiL begins at this moment, where each SWCs is implemented and tested using a unit test.

The unit test verifies how the outputs of the model respond to the different inputs in a simulation environment such as Simulink. When the different SWCs operate under minimum requirements, the MiL scenario can be developed with the VSI and the induction machine models. Numerous testing loops should be necessary for this phase to fine-tune the control of the inverter and the machine under minimum expectations where it is safe to go through the following phases. In this case, the testing options are diverse. If the hardware prototype of the VSI is already available, the HiL scenario allows evaluating the generated code and the hardware on an emulated induction machine. In this case, it is not necessary the machine nor all the mechanical systems that it entails (dynamometer, belts, gears), as well as all its security elements since it is a rotating machine that can cause harm to the user. On the other hand, if the

© Springer Nature Switzerland AG 2020

R. Molina Llorente, *Practical Control of Electric Machines*, Advances in Industrial Control, https://doi.org/10.1007/978-3-030-34758-1_11

hardware is not available or a HiL scenario is not available, the generated code can be tested in an evaluation target microprocessor/DSP connected to a platform where the inverter and machine are simulated in an FPGA as it will be introduced in Appendix 12.1. The last step is to use the real environment, that is, the real VSI and the real machine where most of the problems encountered during development have been solved.

In this chapter, vector control for an induction machine is implemented using the MBD MiL scenario for its implementation in a microcontroller using the toolbox Simscape Electrical™ from Simulink.

The basic system specification consists of a 5 HP induction machine which should operate in a speed range from 0 to 3700 RPM in both directions of rotation and with speed sensing. On the other hand, for automotive application, the dynamic of EV is modeled in order to test the vector control with a large induction machine of 109 HP of power during a drive cycle.

11.2 Control Loop Analysis

The basic premise of vector control in AC electrical machines is to have an instantaneous and independent control of torque and flux as discussed previously. The response of the control dynamics to the transients as in changes of speed commands or load changes is defined fundamentally by the bandwidth of the speed, torque, and flux controllers. In Sect. 10.2.1, it was shown how the gain of the continuous-time speed PI controller was defined by the inertia of the system, the desired bandwidth, and the desired damping coefficient being also valid for AC machines. On the other hand, like DC machine current loop analysis of Sect. 10.2.2, if the bandwidth of the controllers is set as ω_c, then the gains of the continuous timer current PI controller for induction machine can be calculated (11.1) assuming a balanced three-phase circuit and synchronously rotating reference frame as depicted in Fig. 11.1 (Sul 2011).

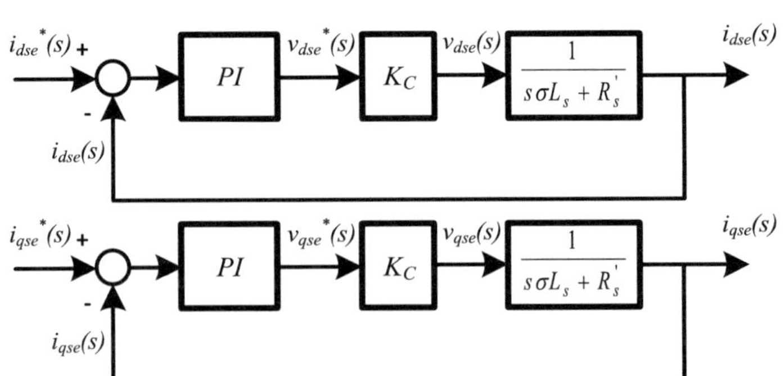

Fig. 11.1 Induction machine modeled with a simplified RL circuit with a PI current controller in synchronously rotating reference frame for every component

It is possible to observe in (11.1) that the gains of PI controller are related to the induction machine parameters as DC machine.

$$K_p = \frac{\omega_c \sigma L_s}{K_c}$$
$$\tau_i = \frac{K_p K_c}{R'_s \omega_c} \tag{11.1}$$

where

$$R'_s = R_s + R_r \left(\frac{L_m}{L_r} \right)^2 \tag{11.2}$$

If the power stage constant K_c is approximated to unity, for the same current bandwidth, the proportional gain parameter is proportional to the σL_s factor. It means that for larges machines (>10 HP), the σL_s factor is smaller, then the gain K_P should be lower also. Something similar occurs with an integral time constant, where for large machines the R'_s factor decreases and then $K_I = K_P/\tau_i$ tends to decrease but to depend on the gain value of K_P. Table 11.1 describes the parameters of two different induction machines. The first induction machine is a general-purpose industrial induction machine of 5 HP of power, while the second machine is a traction induction machine for EV of 109 HP of power.

The gain of the PI controllers, according to Eqs. 11.1 and 11.2, for both induction machines is represented in Table 11.2 when bandwidth is set to 1000 Hz.

Table 11.1 5 HP general-purpose and 109 HP traction induction machine parameters

IM parameters	Symbol	5 HP	109 HP	Unit
Rated voltage line-to-line	V	200	375	V
Leakage stator inductance	L_{ls}	1.62	0.068	mH
Leakage rotor inductance	L_{lr}	2.48	0.067	mH
Magnetizing inductance	L_m	45.52	0.91	mH
Stator resistance	R_s	0.3825	0.016	Ohms
Rotor resistance	R_r	0.2725	0.021	Ohms
Rotor inertia	J	0.019	0.025	$kg \cdot m^2$
Damping coefficient	B	0	0	$kg \cdot m^2/s$

Table 11.2 Gains of the current PI controllers

PI controllers	Symbol	Value
Current proportional part 5 HP	K_{Pc}	25.13
Current integral part 5 HP	τ_{Ic}/K_{Ic}	0.00637/3942.9
Current proportional part 110 HP	K_{Pc}	0.8193
Current integral part 110 HP	τ_{Ic}/K_{Ic}	0.00381/214.8

In vector control, the PI controllers are a part of the control loop task which has to be discretized for digital control. In order to accomplish a defined bandwidth, the execution control loop task and the sampling time of the variables, needed for the control, should be selected carefully. For example, in the indirect vector sensor control, the currents components i_{ds} and i_{qs} should be sampled and updated at least at the speed that these control loops are executed to guarantee a minimum bandwidth of both controllers. In a four-quadrant chopper or full-bridge power stage, if the measure is sampled at peak or valley of carrier waveform, the maximum bandwidth for a PI regulator can be calculated as *fs*/21, while if the measure is sampled at peak and valley of carrier waveform, the maximum bandwidth is approximately *fs*/10 (Sul 2011). In the example of Sect. 10.7.1, the sampling frequency *fs* and PWM was 10 kHz, where the current control loop was executed every 100 µs (1/*fs*). As mentioned above, the bandwidth of the digital PI controller for the first case is 476 Hz and for the second one 1 kHz. As it is possible to observe the difference is more than double where the second one offers a faster dynamic response to the load change or speed command than the first. As it can be seen, in these conditions, if the gain current PI controller is calculated according to (11.1) with a bandwidth of 1000 Hz, sampling in the peak or valley, it will not be fulfilled. However, sampling at peak and valley of carrier waveform, the bandwidth of the control regulation will be performed. The examples of vector control of this chapter are used the valley sampling with sampling time and current control loop of 100 µs, where the maximum bandwidth of the digital PI controller is 476 Hz as commented before.

On the other hand, since machine control is usually a cascade control system, the internal control loops that correspond to the torque and flux control loops must run faster than the external speed loop as seen in Chap. 2. The rotor time constant as T_r becomes, in general, increasingly more significant as the power of the machine increases. The rotor and load inertia are also more significant for larger machines, where the speed dynamic response is in general lower than for smaller machines so that it is possible to limit the speed regulation bandwidth. One criterion that offers excellent results is to execute the internal loop (current and flux controllers) at least ten times faster than the external loop (speed controller).

11.3 Rapid Prototype Simulation Without Power Plant

In Sect. 10.7.1, a four-pole, 220 V, 5 HP induction machine was simulated in an open loop. However, in this section, the indirect closed-loop vector control technique, with the same machine, is simulated.

The indirect vector control was based on imposing the required slip on the machine to force the orientation of the rotor field. In the literature, it is possible to find it as an indirect rotor field orientation (IRFO). Figure 11.2 shows the schematic of the indirect vector control with speed sensor and flux weakening without using any power plant to rapid prototyping the control. It is important to mention that in this model, there are no inverter or PWM-type signals.

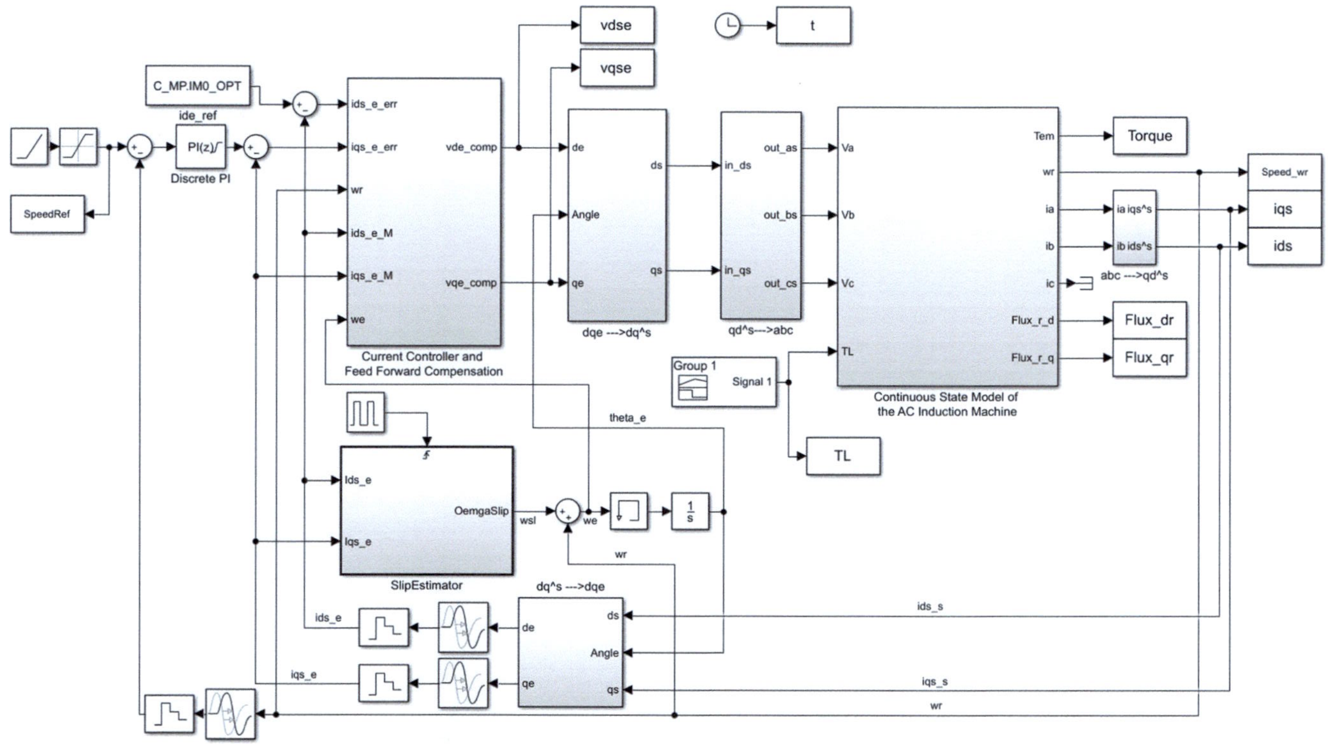

Fig. 11.2 Indirect sensored vector control for IM with flux weakening, feedforward compensation, slip estimator, and speed controller based in a Fuzzy + PI structure

The speed control is made with a non-linear controller based on the Fuzzy + PI controller as seen in Sect. 2.6.4, while the *dq* current controllers are conventional PI regulators with anti-windup. The feedforward compensator is also incorporated to avoid the cross-coupling effect present in the induction machine. The flux weakening block consists of a PI controller that regulates the amount of the *d* current component as a function of the voltage supplied to the machine as discussed in Sect. 10.5.3. That is, as the supply voltage of the machine approaches the maximum voltage $V_{\max}$ minus an offset equal (10 V in this case), the magnetic flux is weakened with the decrease of the current component *d*. The value of the offset can be fine-tuned in the simulation and experimentally. The voltage of the machine is constructed from the module of the components v_{dse} and v_{qse} as represented. Below $V_{\max}$ and its corresponding offset, the *d* current component takes the optimum and constant value defined by *IM0_OPT*. The sampling time T_s is 100 µs. Speed control and flux weakening run 50 times slower than T_s, that is, every 5 ms. However, the current regulators and the slip estimator are executed every T_s. The slip estimator is based on (10.128). Finally, the magnetizing current of the i_{Mr} rotor is necessary for the flux weakening and to get a better approximation of the slip estimator. It is related to the stator current of the *d* axis from the dynamic Eq. 11.3 (Vas 1998).

$$T_r \frac{\mathrm{d}i_{Mre}}{\mathrm{d}t} + i_{Mre} = i_{dse} \tag{11.3}$$

where T_r was the rotor time constant of the induction machine (L_r/R_r). It is possible to observe that Eq. 11.3 has the form of a first-order delay element with a unity gain and time constant T_r. That is, the magnetizing current of the rotor is the result of filtering the I_{dse} current measured in the stator.

The simulation test results of the indirect sensored vector control for the 5 HP induction machine are shown in Figs. 11.3, 11.4, and 11.5. The response to a mechanical speed ramp reference of 1000 RPM with a load of 10 Nm is shown in Fig. 11.3. It can be seen that at start-up, at the time instant $t = 0.2$ s, the reference speed increases, and at $t = 0.5$ s, the reference speed well matches the measured speed. During acceleration, the torque produced by the machine takes a peak value of approximately 57 Nm with a reference I_{qse} current of 21 A peak. The phase current A reaches a maximum value of 25 A as can also be seen. As the rotor speed of the induction machine approaches the maximum reference speed, the reference current I_{qse} becomes smaller to reduce the developed electromagnetic torque. The machine speed has an overshoot of 5%, i.e., 50 RPM. The current reference I_{dse} is set to its optimum value of *IM0_OPT*, i.e., 14 A peak. At the time instant $t = 3$ s, a disturbance load torque from 10 to 20 Nm (50–100% load torque) occurs to observe the effect it produces on the speed. As can be seen, the speed of the machine undergoes a slight alteration thanks to the high bandwidth of the current controller's and the non-linear Fuzzy + PI regulator of speed control.

The estimated slip is shown in Fig. 11.2a. In the acceleration, the slip takes a more significant value of up to 22 rad/s (3.5 Hz) but decreases rapidly at a constant speed. When the load is 20 Nm, the slip takes a value of 2.5 rad/s. It is possible to observe

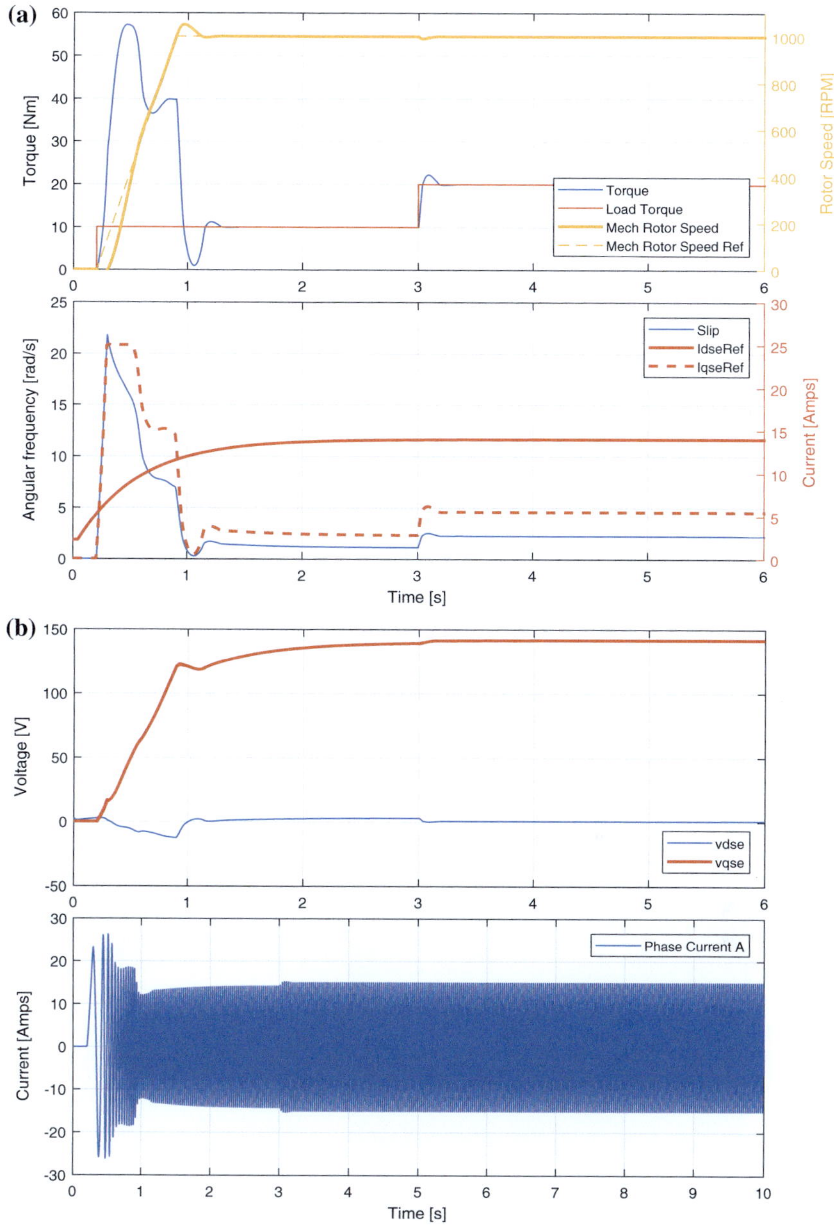

Fig. 11.3 Run-up speed till 1000 RPM with 10 Nm of load torque till $t = 3$ s, where load torque changes suddenly to 20 Nm. **a** Reference speed, rotor speed, torque developed, load torque, slip, and current reference components i_{dse} and i_{qse}. **b** Voltage components v_{dse} and v_{qse}, and phase current A. Simulation fixed-step, minimum step auto, and automatic solver selection

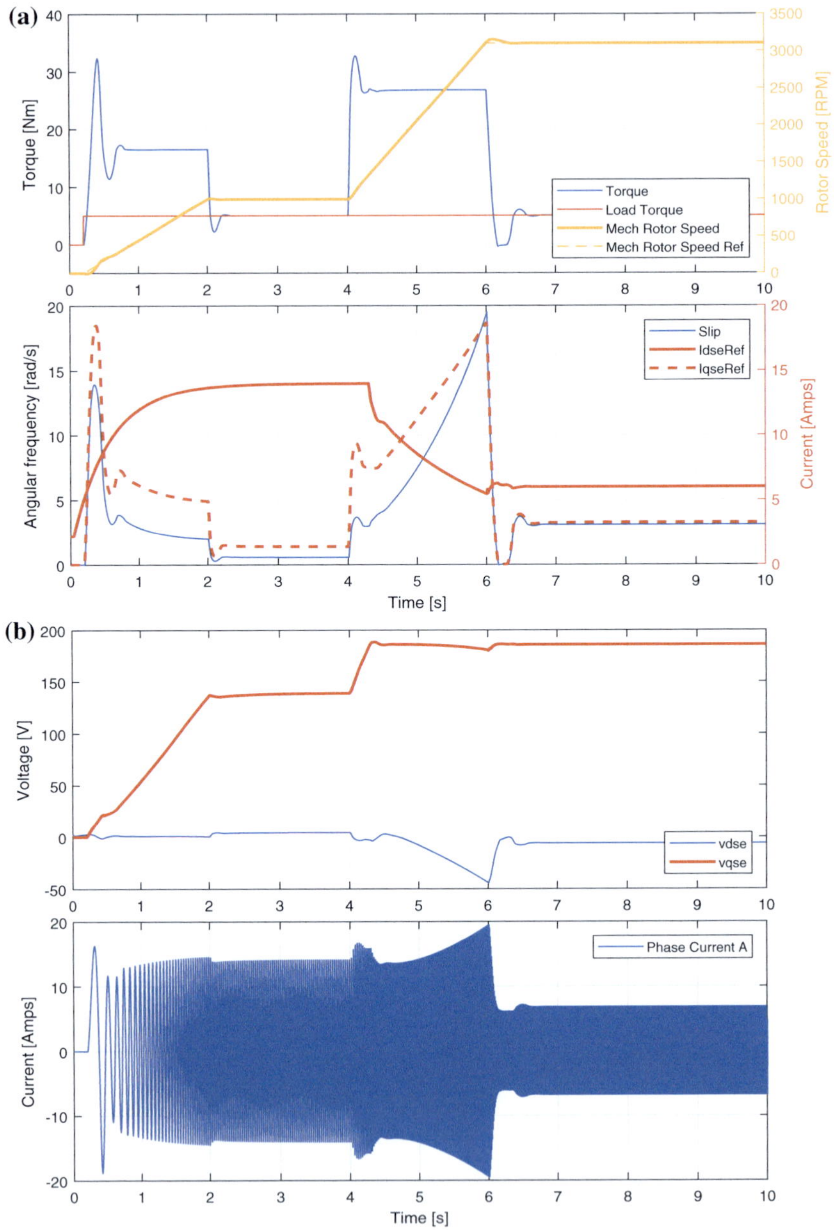

Fig. 11.4 Run-up speed till 1000 RPM with 5 Nm of load torque till $t = 4$ s. At $t = 4$ s speed reference increase till 3103 RPM to see the flux weakening behavior. **a** Reference speed, rotor speed, torque developed, load torque, slip, and current reference components i_{dse} and i_{qse}. **b** Voltage components v_{dse} and v_{qse}, and phase current A. Simulation fixed-step, minimum step auto, and automatic solver selection

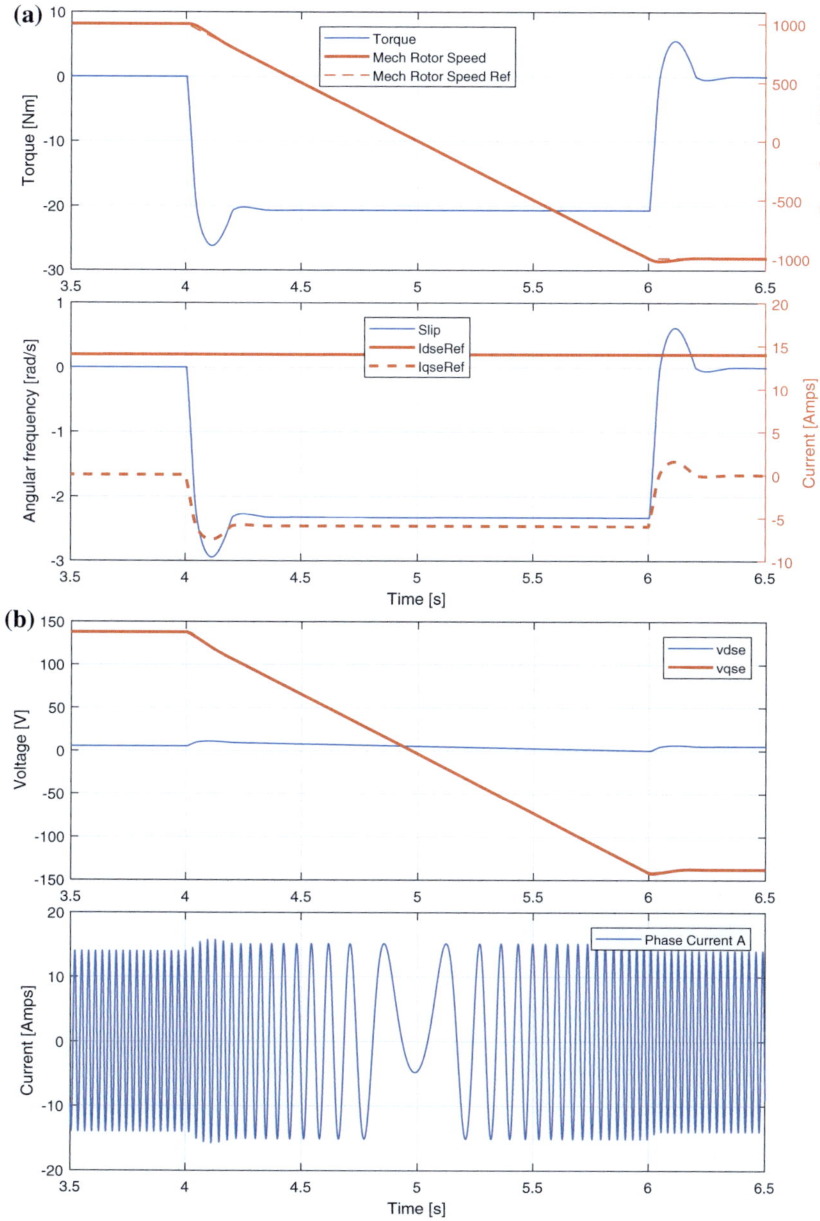

Fig. 11.5 Speed reversal under no-load from 1000 to −1000 RPM. **a** Reference speed, rotor speed, torque developed, slip, and current reference components i_{dse} and i_{qse}. **b** Voltage components v_{dse} and v_{qse}, and phase current A. Simulation fixed-step, minimum step auto, and automatic solver selection

in Fig. 11.2b that the module of the components v_{dse} and v_{qse} takes an approximate value of 140 V from the instant $t = 4$ s. The maximum voltage $V_{\max}$ is set to 195.95 V since the DC-link voltage is approximately 340 V for a linear SVPWM modulation.

In Fig. 11.4, the speed control performance during the flux weakening is demonstrated. The simulation is carried out at speed higher than the machine base speed as can be observed. It can be seen that at time instant $t = 4$ s, the speed is increased from 1000 to 3103 RPM with a load torque of 5 Nm. It is possible to observe that during the acceleration, the reference current i_{qseRef} increases to increase the developed torque. Meanwhile, when the machine voltage reaches its maximum value $V_{\max}$, the current reference i_{dseRef} has to decrease to reduce the magnetic field as represented in Fig. 11.4a.

In Fig. 11.4a, when the rotor speed reaches the speed of 3103 RPM, the reference current i_{qseRef} decreases rapidly, while the current reference i_{dseRef} takes the necessary value to maintain the nominal voltage of the machine. On the other hand, it is possible to observe that the module of the components v_{dse} and v_{qse} is very close to the maximum voltage $V_{\max}$ above time instant equal to 4.2 s.

Finally, Fig. 11.5 shows the change of the rotation direction from 1000 to -1000 RPM when the machine has no load. It is possible to observe how the machine speed well matches the speed reference command during deceleration with a developed negative torque.

As expected, while the machine slows down, the torque current component q is negative and consequently the slip too. It happens until it reaches the steady-state at the time instant $t = 6.2$ s approximately, where the torque current and the slip take the value 0 when there is no load. In Fig. 11.5b, it is possible to observe the change of sign of the voltage v_{qse} as well as the direction of the phase current A at the time instant $t = 5$ s approximately.

11.4 Software Architecture Design

In the software architecture design, there are different options. In the case of the AUTOSAR architecture, it offers two possible options. The first consists of performing all the logical part of the control of the induction machine and VSI in the application layer, while the second option consists in carrying out only a part in the application layer, reserving the most critical tasks (a critical period between 40 and 100 μs) in the complex device drivers (CDD). Both offer their respective advantages/disadvantages. For example, if the most critical tasks are managed in the CDD, the RTE is not so saturated because the information to manage is a non-critical update. However, if the critical tasks are performed in the application layer, the RTE could become saturated due to its complexity. In any of the cases, Figs. 11.6 and 11.7 in the last layer are shown the necessary microcontroller peripherals for the vector control implementation by using a VSI . The peripherals are a timer with different channels for the management of the PWM outputs and digital inputs for the capture of the duty cycles, an A/D converter, and an input/output interface of the ports for the activation of the pre-charge relay.

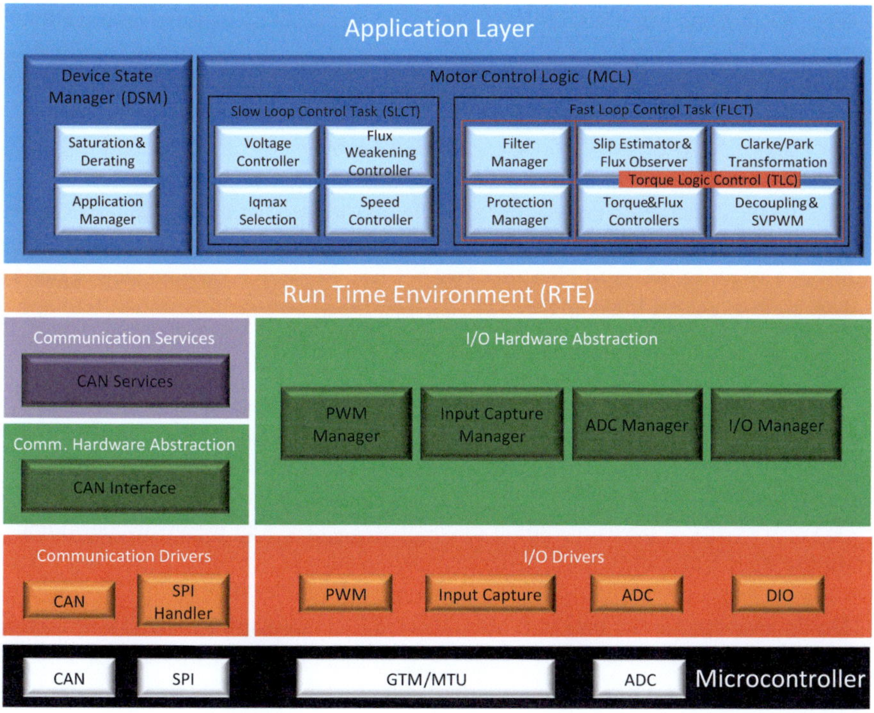

Fig. 11.6 AUTOSAR classic platform architecture applied to machine control with a VSI . The critical task are made in application layer

As it is possible to observe, Fig. 11.6 shows the AUTOSAR option, where all the control logic is in the application layer. At the MCAL level, without taking into account, the CAN interface, there is the driver for the management of the PWM, the driver of the input capture, the driver of the A/D converter, and the driver for direct activation/deactivation of I/O inputs and outputs. At the ECUAL level, the PWM manager manages the PWM to correctly generate the duty cycles with the deadtime and the set point frequency based on the SVPWM that the application layer generates. The input capture manager is responsible for managing the duty cycles captured by the timer. Then, in the application layer is processed the reconstruction of the voltage measured at the machine terminals.

The ADC manager is responsible for managing the measurements of the different A/D channels available by applying priorities and verifying if the peripheral is free or busy. For example, The A/D priority for the measurement of the slow variation signals such as temperature through an NTC should be lower than for the phase current and voltage measurements. The A/D converter is configured so that, if the A/D conversion is completed for measuring the phase currents of the machine and the DC-link voltage, an ISR interrupt is released.

The I/O manager manages the activation/deactivation of the corresponding outputs and inputs. The I/O Hardware Abstraction Layer communicates with the application layer using the RTE. The RTE implements the virtual functional bus (VFB) in a specific ECU. The VFB is the sum of the communication mechanisms and the interfaces of the basic software provided by AUTOSAR in an abstract and technologically independent level as it was seen in Chap. 1.

Finally, the application layer is responsible for the management of all the machine and inverter control. As seen in Sect. 1.4.2, the application layer can be formed by different SWCs. In this case, there are two SWCs, the device state manager (DSM), and the motor control logic engine (MCL). The DSM has the functionality to apply the necessary saturations and deratings in case the inverter or the machine must limit its performance due to an excess of temperature, current, or voltage. It is also responsible for monitoring the errors that have occurred and for managing the DC-link pre-charge. Its periodicity of execution is relatively slow, between 5 and 100 ms. On the other hand, the MCL manages the logical control of the machine and is divided into two modules, the slow-loop control task (SLCT) and the fast-loop control task (FLCT). The SLCT performs some cyclical tasks at a relatively low periodicity, while the FLCT executes some tasks that are also cyclical, but at a relatively high periodicity, precisely at the switching frequency of the PWM.

On the other hand, the option is also AUTOSAR, but using the CDD is shown in Fig. 11.7. In this case, the FLCT is executed in the CDD, leaving the SLCT in the application as shown in Fig. 11.7.

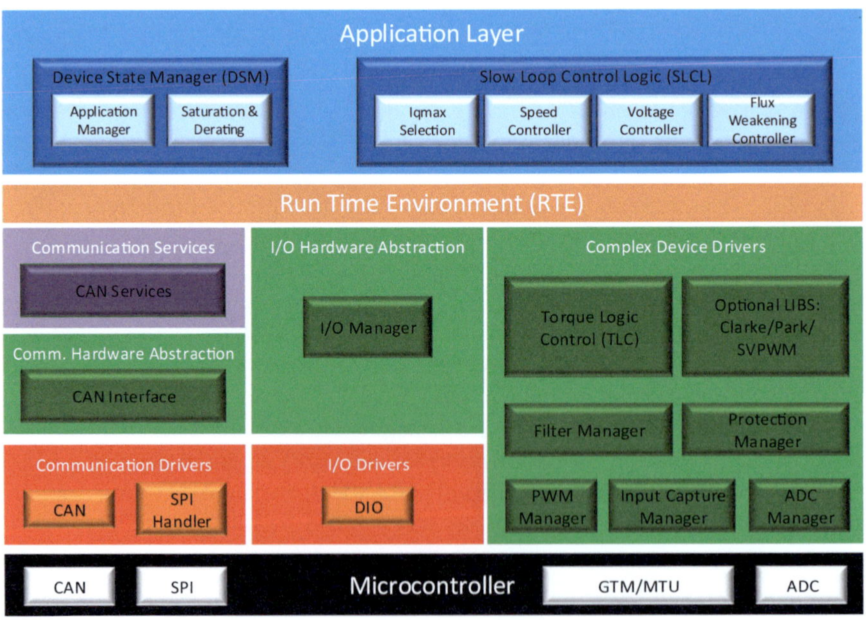

Fig. 11.7 AUTOSAR classic platform architecture where critical tasks for machine control with a VSI are made in CDD

As was seen in Chap. 1, the CDD has direct access to resources for critical applications, where the time-critical current loops of the electric machine control can be handled in an interrupt context. Also, the CDD is not provided by AUTOSAR, and then, it is possible to describe similar modules shown with the previous option. Some of the modules in the application layer are now in the CDD. Some microcontroller manufacturers offer libraries dedicated to electric machine control where the most used standard libraries are defined such as the Clarke; Park transforms, including the SVPWM modulation algorithm, where their execution efficiency is optimized. These libraries are usually below the RTE so that the CDD can use them efficiently.

It is important to comment that if the architecture does not require it to be AUTOSAR, even if it is an automotive ECU, or if the ECU is for domestic or industrial use, the weight and complexity of the modules and the RTE disappear in such way that it is more feasible to perform all the control logic in the application layer.

11.5 MCL SWC Design

Once the inverter and the electric machine are modeled and verified in the simulation environment, the next step is to develop the different software components necessary for their control. Previously two SWCs were seen, the device state manager (DSM) and the motor control logic (MCL). The MCL was divided into two modules, the slow-loop control task (SLCT) and the fast-loop control task (FLCT). Regardless of the chosen architecture, both can be modeled using the model-based design technique. In this section, only the MCL SWC will be treated.

The Stateflow® diagram of Simulink® of the MCL consisting of a scheduler, voltage, and current filtering, the SLCT, and the FLCT is shown in Fig. 11.8.

The MCL SWC is divided into two modules, the SLCT and the FLCT as seen above. The SLCT performs tasks or subroutines at a slower periodicity such as the DC-link voltage controller, the flux weakening controller, and the speed controller. In addition, it determines the maximum q current component according to the circle constraint depending on the operation of the machine. The FLCT performs other critical tasks such as protections, filtering, and torque logic control (TLC). Here is where the Clarke transformation, rotation transformation, the slip estimator, the flux observer, the torque and flux controllers, the decoupling mechanism, and the SVPWM are implemented. These critical tasks must be executed in synchronization with the PWM and must last a total of less than the PWM period. It is recommended not to exceed 50% of the time of the PWM period since the microcontroller has to perform other tasks, such as the DSM, or a possible CAN communication. For example, for a PWM frequency of 10 kHz and a cycle-to-cycle control as the vector control, the sum of time it takes to execute all critical tasks should be a maximum of 50 μs. It supposes an occupation of the CPU of 50% so that another 50% of CPU is free for other tasks.

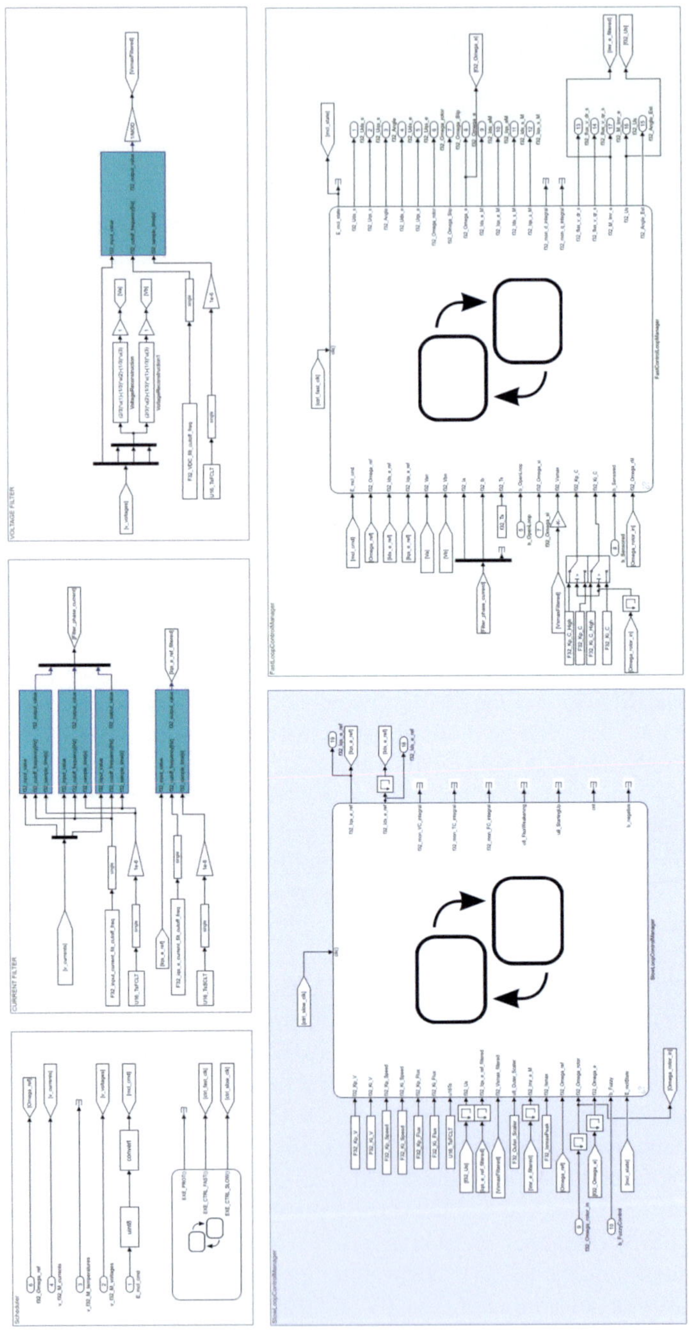

Fig. 11.8 Stateflow block diagram of MCL, with the SLCT, FLCT, scheduler, current, and voltage filters

11.5.1 Slow Control Loop Task

The SCLT module in Simulink Stateflow® is shown in Fig. 11.9, while the content is shown in Fig. 11.10.

It is possible to observe the four subroutines, voltage controller, flux controller, maximum torque component selection, and the speed controller, which are sequentially called one after the other. The order of call is designated by the number that is indicated. If *E_mclState* takes the value of *E_MCL_CTRL_MOTOR_ON*, it will go to the SLCM state.

The voltage controller consists of a PI controller that reduces the amount of magnetizing reference current of the rotor as the speed of the machine increases above the rated speed, thus controlling the machine voltage measured in the terminals is the rated voltage as seen in Sect. 11.3. The flux controller also consists of a PI controller that controls the *d* component of the current as a function of the error between the

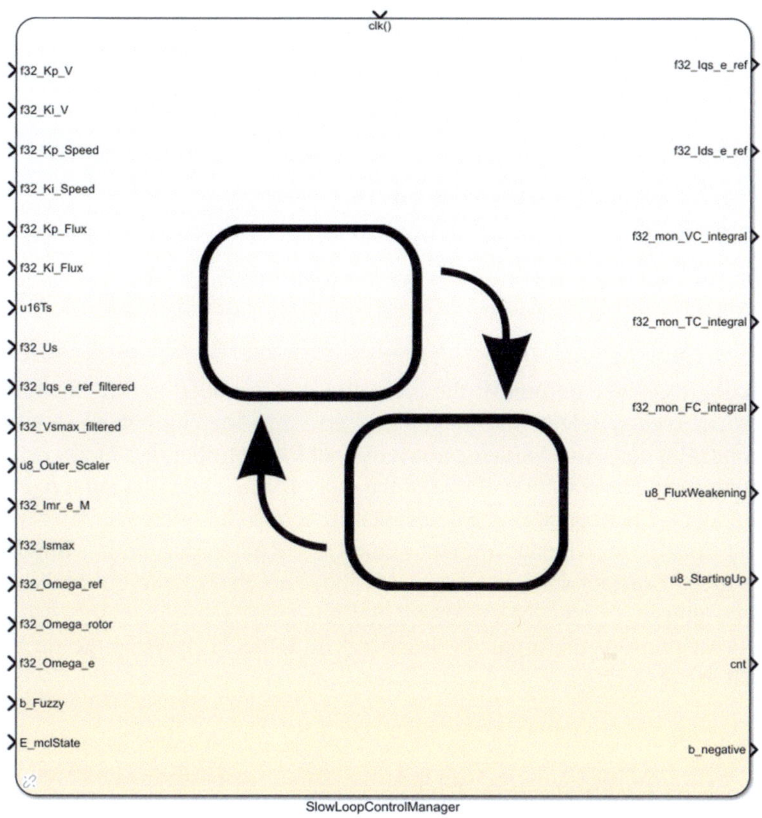

Fig. 11.9 Slow control loop manager which executes the slow outer loops as speed, flux weakening, and DC-link voltage control

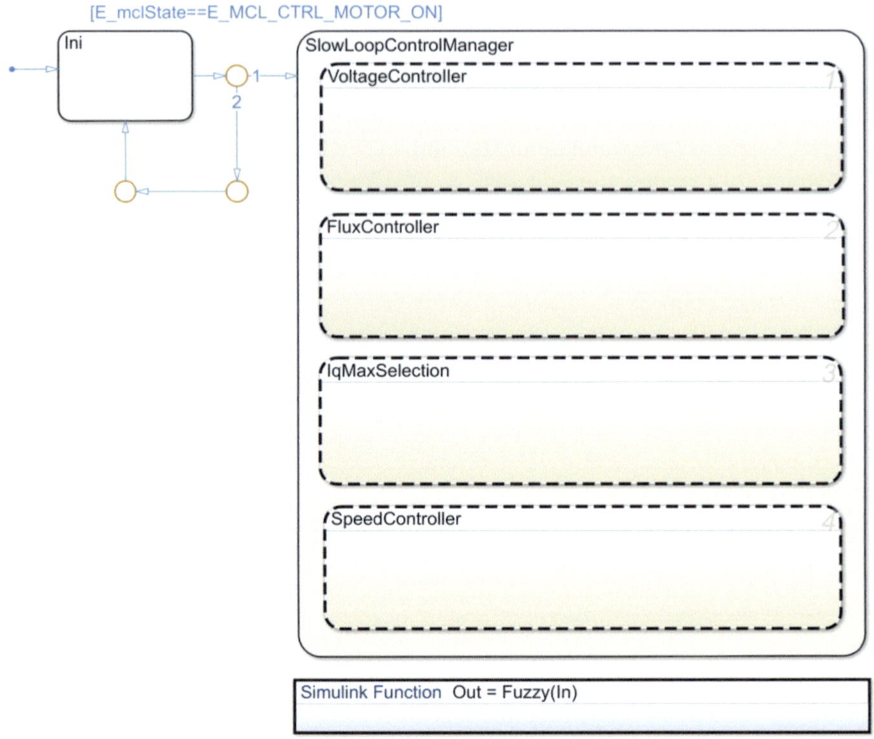

Fig. 11.10 Inside of slow control loop manager. The main subroutines are the voltage controller, the flux controller, the maximum torque component selection, and the speed controller

magnetizing reference current of the rotor and that provided by the measurement. The *IqMaxSelection* selects the maximum current of the component *q*, while the speed controller consists of either a conventional PI controller or a Fuzzy + PI nonlinear controller. The selection of one or the other is made with the input *b_Fuzzy* of the component. In either case, the controller compares the reference speed with the measured speed to regulate the amount of the current *q* component. It is important to mention that all current variables discussed above are in the synchronously rotating reference frame.

The description of the inputs is described in Table 11.3, while the outputs are described in Table 11.4.

11.5.2 Fast Control Loop Task

The FCLT module consists of three sub-modules: the filtering sub-module, the protection sub-module, and the TLC. The filtering sub-module applies a series of digital

Table 11.3 Inputs signals of the SLCM

Input	Description	Values
clk	Periodic clock to define the rated frequency which should be called the module	{0–100 ms}
f32_Kp_V	Proportional constant parameter of voltage control loop	{0–100,000}
f32_Ki_V	Integral constant parameter of voltage control loop	{0–100,000}
f32_Kp_Speed	Proportional constant parameter of speed control loop	{0–100,000}
f32_Ki_Speed	The integral constant parameter of the speed control loop	{0–100,000}
f32_Kp_Flux	Proportional constant parameter of flux control loop	{0–100,000}
f32_Ki_Flux	Integral constant parameter of flux control loop	{{0–100,000}
u16Ts	Sample time in μs	{1–500}
f32_Us	Voltage in machine terminals	{0–500 V}
f32_Iqs_e_ref_filtered	Reference *q* current component filtered in synchronous rotating reference frame	{−30 to 30 A}
f32_Vsmax_filterd	Vsmax measured and filtered in machine terminals for linear modulation	{0–500 V}
u8_Outer_Scaler	Scaler for outer loops (speed control loop)	{1–100}

(continued)

Table 11.3 (continued)

Input	Description	Values
f32_Imr_e_M	Rotor magnetizing current filtered	{−30 to 30 A}
f32_Ismax	Maximum current for VSI (peak)	{−30 to 30 A}
f32_Omega_ref	Reference speed in rad/s	{−5000 to 5000 rad/s}
f32_Omega_rotor	Machine rotor speed in rad/s	{−5000 to 5000 rad/s}
f32_Omega_e	Synchronous speed in rad/s	{−5000 to 5000 rad/s}
b_Fuzzy	To enable Fuzzy + PI controller	{0–1}
E_mclState	MCL state	MOTOR_ON, MOTOR_OFF, MOTOR_FAULT

Table 11.4 Outputs signals of the SLCM

Output	Description	Values
f32_Iqs_e_ref	Reference *q* current component in synchronous rotating reference frame	{−30 to 30 A}
f32_Ids_e_ref	Reference *d* current component in synchronous rotating reference frame	{−30 to 30 A}
f32_mon_VC_integral	Integral part of voltage controller for debugging purpose	{−30 to 0 A}
f32_mon_TC_integral	Integral part of torque controller for debugging purpose	{−30 to 30 A}
f32_mon_FC_integral	Integral part of flux controller for debugging purpose	{−30 to 30 A}
U8_FluxWeakening	To indicate is the machine is in flux weakening	{0–1}
U8_StartUp	To indicate is the machine is starting	{0–1}
cnt	Signal for debugging purpose	–
b_negative	Signal for debugging purpose	–

filters to the inputs to minimize the noise captured by the A/D converter. To avoid excessive delay and not to penalize the control dynamics, the cut-off frequency should not be too low. It is important to mention that it is mandatory to implement a hardware low-pass filter for the voltage and current measurements. The cut-off frequency must be higher than 10–15 times the PWM frequency to obtain the highest performance cycle-to-cycle control. The protection module serves mainly to protect the VSI and the machine, as discussed in Sects. 8.7 and 8.8. It detects currents, overvoltage, undervoltage, overtemperature and if necessary stops the machine to pass to a safe state. Lastly, the TLC SWC is responsible for controlling the required torque, which is executed in a fast loop every new acquisition of the voltage and current terminals of the machine. The execution is synchronized with the PWM. Every sampling time with actual acquisitions for voltage and current, executes the critical tasks as Clarke and rotation transformation, the two PI loops, one for torque and the other for the flux, the flux observed, the slip observer, the feedforward compensation, and update the new voltages to set the new PWM duty cycle (SVPWM). Figure 11.11 shows the Stateflow® model of the FCLT SWC.

Figure 11.12 shows the content of the FCLM where two main subroutines are observed, the DC-link voltage monitoring and the TLC unit. It also shows the different MATLAB functions required for the TLC. These are the flux estimator, the Clarke and Park (rotation) transformations, its inverses, and the slip estimator.

Table 11.5 describes the inputs, while the outputs of the FCLM are described in Table 11.6.

11.5.3 MCL Unit Test

The SWC MCL test can be performed on the Simulink® environment. The test consists of exciting the inputs and seeing how the outputs behave compared to what is expected. The designer of the module is the one who knows how it should behave so that it is usually the same person who performs the test. Different test cases can be considered to try to test most of the possible conditions and find the most errors in this phase. The more test cases are defined, the better tested the model will be. However, the number of test cases depends on the complexity of the model.

On the other hand, it is also possible to use the Simulink Coverage™ tool that performs coverage analysis of models and codes that measure the integrity of the tests in models and generated codes. Apply standard industry metrics such as decision, condition, modified condition/decision coverage, and coverage of relational limits to evaluate the effectiveness of model simulation tests. It also produces interactive reports that show how much the model has been exercised.

Before the MiL test of the following section, it is advisable to test the SCLT and FCLT modules separately since it facilitates the analysis of the results and allows finding possible errors more easily. Figure 11.13 shows schematically the test carried out on the SCLT module, where the excitation is carried out using the signal builder.

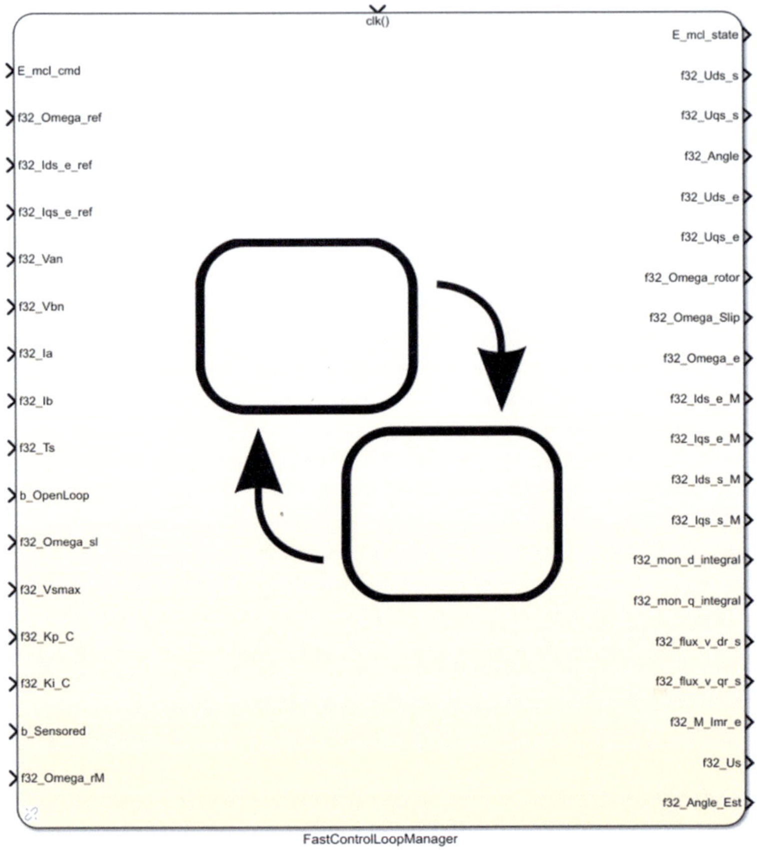

Fig. 11.11 FCLT Stateflow implementation which is executed every sampling time

It is important to mention that in this section, only one test case is shown as an example to give an idea of how to apply the unit test in the models.

If the SLCM model is applied the test case 1 of Fig. 11.14, the result of the simulation obtained in the outputs of the SLCM is shown in Fig. 11.15.

As can be seen in this test, it is possible to verify the functioning of the SLCM where the results must be corroborated with the results expected by the designer.

In particular, in the test case 1, it is possible to see the operation of the Fuzzy + PI controller where the q component of the current varies according to the reference speed and the current speed. In the same way, it is possible to observe at what point the flux weakening is activated and to see how the current component d varies. If the parameters of the controllers are changed, the behavior of both current components also changes.

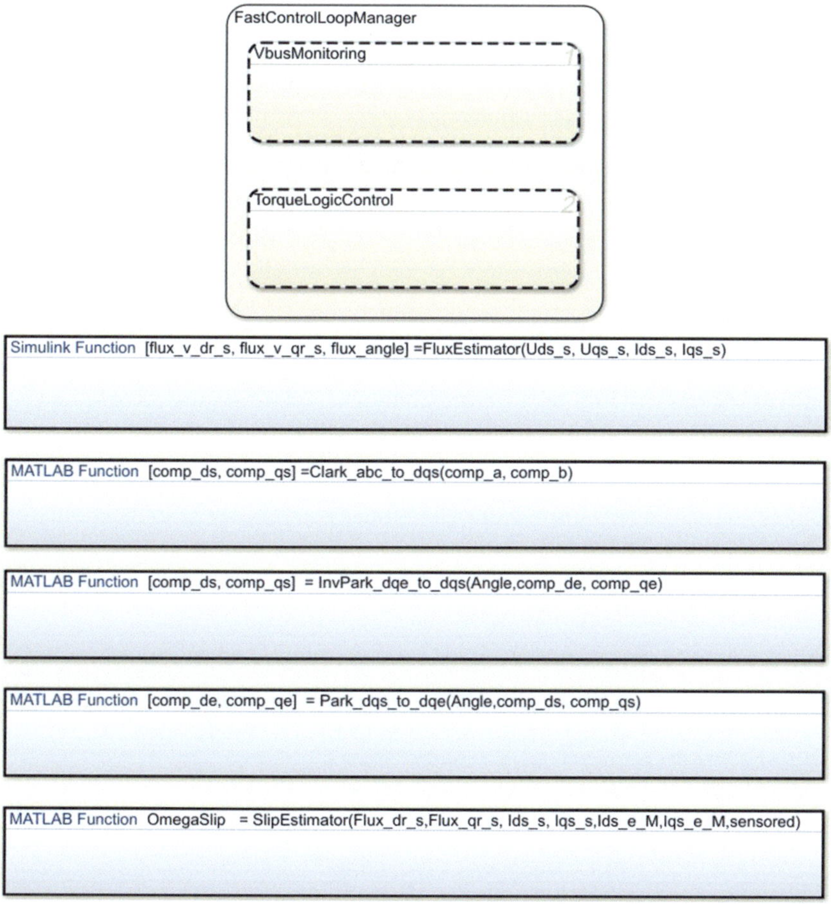

Fig. 11.12 Inside of FLCM. The main subroutines are the DC-link monitor and the TLC. The necessary MATLAB functions are also shown

11.6 Model-in-the-Loop Test (MiL)

In this section, some representative results regarding sensorless vector control of induction machine simulation in MiL scenario are presented and analyzed for a four-pole, 220 V, 5 HP induction machine.

The objective of the MiL is to completely simulate the MCL SWC by using the models necessary for its accurate testing. It requires an inverter model, a model of the induction machine, an A/D converter model, a model to generate the SVPWM, and the microcontroller timer model, these last two seen in Chaps. 7 and 9, respectively. This simulation is already much more realistic than the view in Sect. 11.3 since it takes into account a plant based on a VSI , a signal acquisition through a simulated A/D converter, a PWM generation, and a more precise model of the induction

Table 11.5 Input signals of the FLCM

Input	Description	Values
clk	Periodic clock to define the rated frequency which should be called the module	{0-100 ms}
E_mcl_cmd	MCL command	NO_CONTROL MOTOR_ON
f32_Omega_ref	Reference speed in rad/s	{−5000 to 5000 rad/s}
f32_Ids_e_ref	Reference *q* current component in synchronous rotating reference frame	{−30 to 30 A}
f32_Iqs_e_ref	Reference d current component in synchronous rotating reference frame	{−30 to 30 A}
f32_Van	Voltage measured of Phase A respect to neutral	{−500 to 500 V}
f32_Vbn	Voltage measured of Phase A respect to neutral	{−500 to 500 V}
f32_Ia	Phase A current measured	{−30 to 30 A}
f32_Ib	Phase A current measured	{−30 to 30 A}
f32_Ts	Sample time of current loops in seconds	{1-500 μs}
b_OpenLoop	To set open-loop control	{0-1}
f32_Omega_sl	Slip in rad/s	{−5000 to 5000 rad/s}
f32_Vsmax	Maximum stator voltage allowed	{0-500 V}
f32_Kp_C	Proportional constant parameter of current control loop	{0-100,000}
f32_Ki_C	Integral constant parameter of current control loop	{0-100,000}
b_Sensored	To set sensored or sensorless	{0-1}
f32_Omega_rM	Rotor speed measured in rad/s	{−5000 to 5000 rad/s}

machine. The effects of PWM switching, delays, noise, etc. can be analyzed with this simulation. These effects directly affect the stabilization and performance of the control so that the tuning of the different controllers must be done in small variations. The filtering of the signals as the measurement of voltage and current in the machine terminals is essential to mitigate the effect of the PWM, although it implies a small

Table 11.6 Output signals of the FLCM

Output	Description	Values
E_mcl_state	MCL state	NO_CONTROL MOTOR_ON
f32_Uds_s	Measured *d* voltage component in the stationary reference frame	{−500 to 500 V}
f32_Uqs_s	Measured *q* voltage component in the stationary reference frame	{−500 to 500 V}
f32_Angle	Open-loop angle	{−π to π}
f32_Uds_e	Measured *d* voltage component in the synchronously rotating reference frame	{−500 to 500 V}
f32_Uqs_e	Measured *q* voltage component in the synchronously rotating reference frame	{−500 to 500 V}
f32_Omega_rotor	Estimated rotor speed in rad/s	{−5000 to 5000 rad/s}
f32_Omega_slip	Estimated slip in rad/s	{−5000 to 5000 rad/s}
f32_Omega_e	Synchronous speed in rad/s	{−5000 to 5000 rad/s}
f32_Ids_e_M	Measured *d* current component in the synchronously rotating reference frame	{−30 to 30 A}
f32_Iqs_e_M	Measured *q* current component in the synchronous rotating reference frame	{−30 to 30 A}
f32_Ids_s_M	Measured *d* current component in the stationary reference frame	{−30 to 30 A}

(continued)

Table 11.6 (continued)

Output	Description	Values
f32_Iqs_s_M	Measured *q* current component in the stationary reference frame	{−30 to 30 A}
f32_mon_d_integral	Integral part of current *d* controller for debugging purpose	{−500 to 500 V}
f32_mon_q_integral	Integral part of current *q* controller for debugging purpose	{−500 to 500 V}
f32_flux_v_dr_s	Estimated rotor flux *d* component in the stationary reference frame	{−10 to 10 Wb}
f32_flux_v_qr_s	Estimated rotor flux *q* component in the stationary reference frame	{−10 to 10 Wb}
f32_M_Imr_e	Rotor magnetizing current filtered	{−30 to 30 A}
f32_Us	Voltage in machine terminals	{0–500 V}
f32_Angle_Est	Estimated angle of the rotor flux	{−π to π}

delay in the signals. The inverter voltage source corresponds to a single-phase AC source of 240 V and 50 Hz, which is rectified and filtered by an inductance and a DC-link capacitor. Figure 11.16 shows the schematic of the MiL where all the models discussed above can be seen.

The power plant or VSI is made with elements of the Simscape Electrical™ toolbox from Simulink®. Using the SimCoupler tool, it is possible to use a more precise plant made with PSIM elements. When the MiL test is thoroughly tested, it is possible to proceed to generate the MCL model code for later use in a microcontroller or DSP.

11.6.1 Test Below Nominal Speed

The purpose of this test is to verify the behavior of the vector control and the VSI when the machine speed is below the nominal speed, that is, in the constant torque

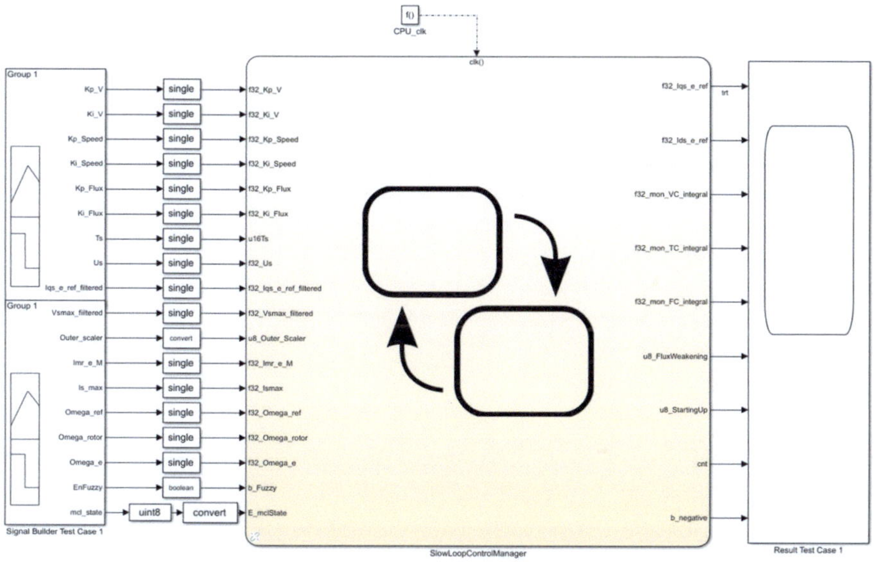

Fig. 11.13 Test case applied in the SLCM where the inputs are built with the signal builder tool of simulink and the results obtained for this test case are represented in the scope

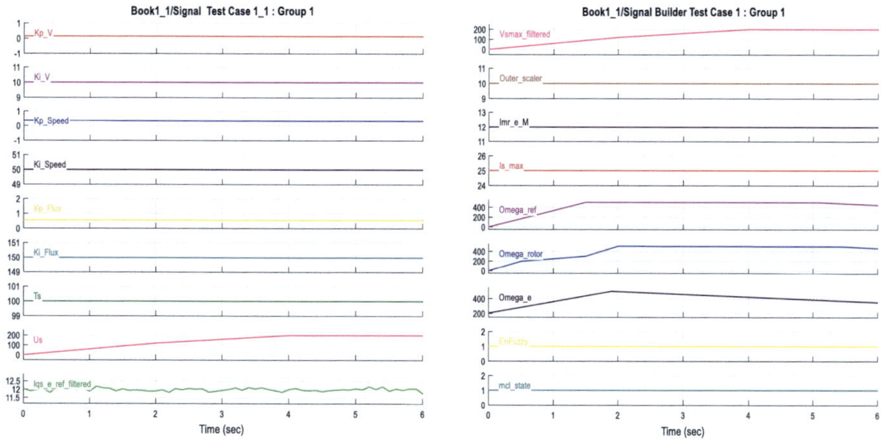

Fig. 11.14 Input signals for SLCM. Test case 1

region. In this region, a disturbance load torque in step manner is applied. The DC-link capacitor of the VSI is set to 2.2 mF.

The results of the simulation are shown in Figs. 11.17, 11.18, and 11.19. The reference speed reaches a final value of 1000 RPM at time instant $t = 0.7$ s. During this acceleration, the load torque kept constant with a value of 5 Nm. At the instant $t = 1.5$ s where the machine is in steady-state, a disturbance load torque of 20

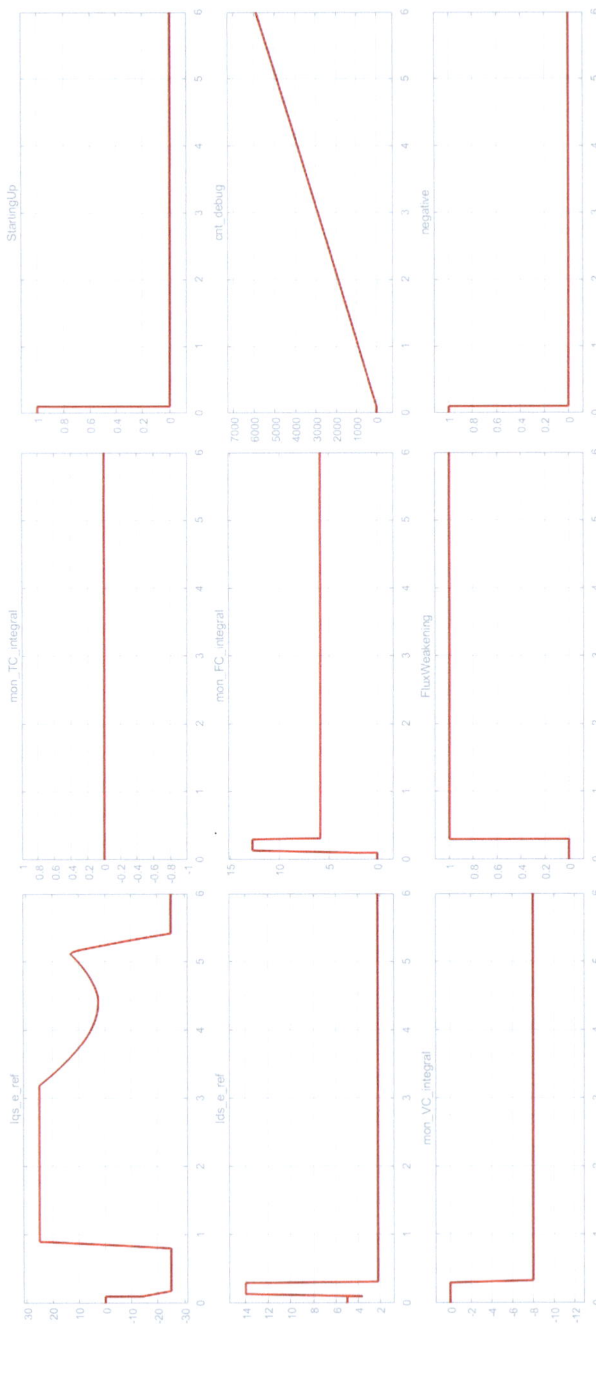

Fig. 11.15 Output signal simulation result of SLCM when test case 1 is applied

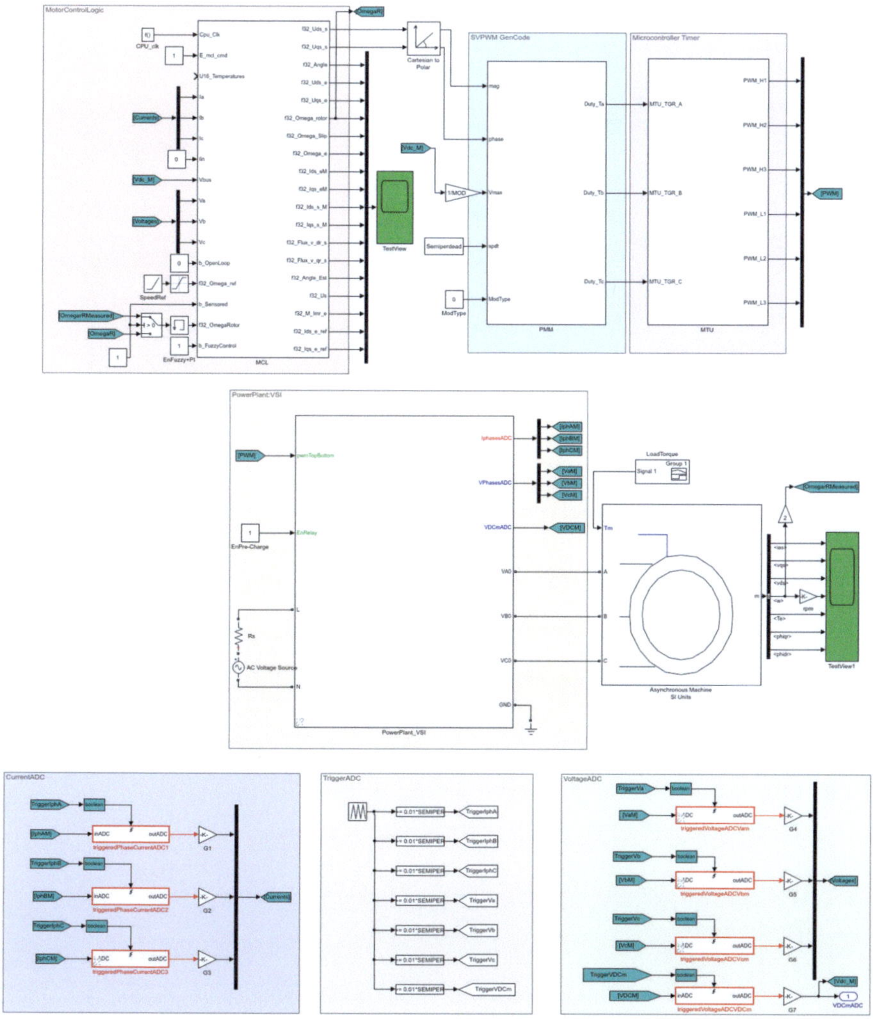

Fig. 11.16 MiL with all the necessary models to test the MCL with a more realistic simulation

Nm (100%) is applied to verify how the vector control responds to this load step change. Figure 11.16a shows machine speed, developed torque, load torque, and reference speed. It also shows the reference currents of the components *d* and *q* in the synchronously rotating reference frame and its corresponding measured values. It is possible to observe that the current components *dqe* measured well match their corresponding references, although it has some ripple component. It is most likely caused by dead zones due to the non-linearity of the VSI . The dead zones occur when any of the phase current crosses zero, and its frequency is six times the fundamental frequency.

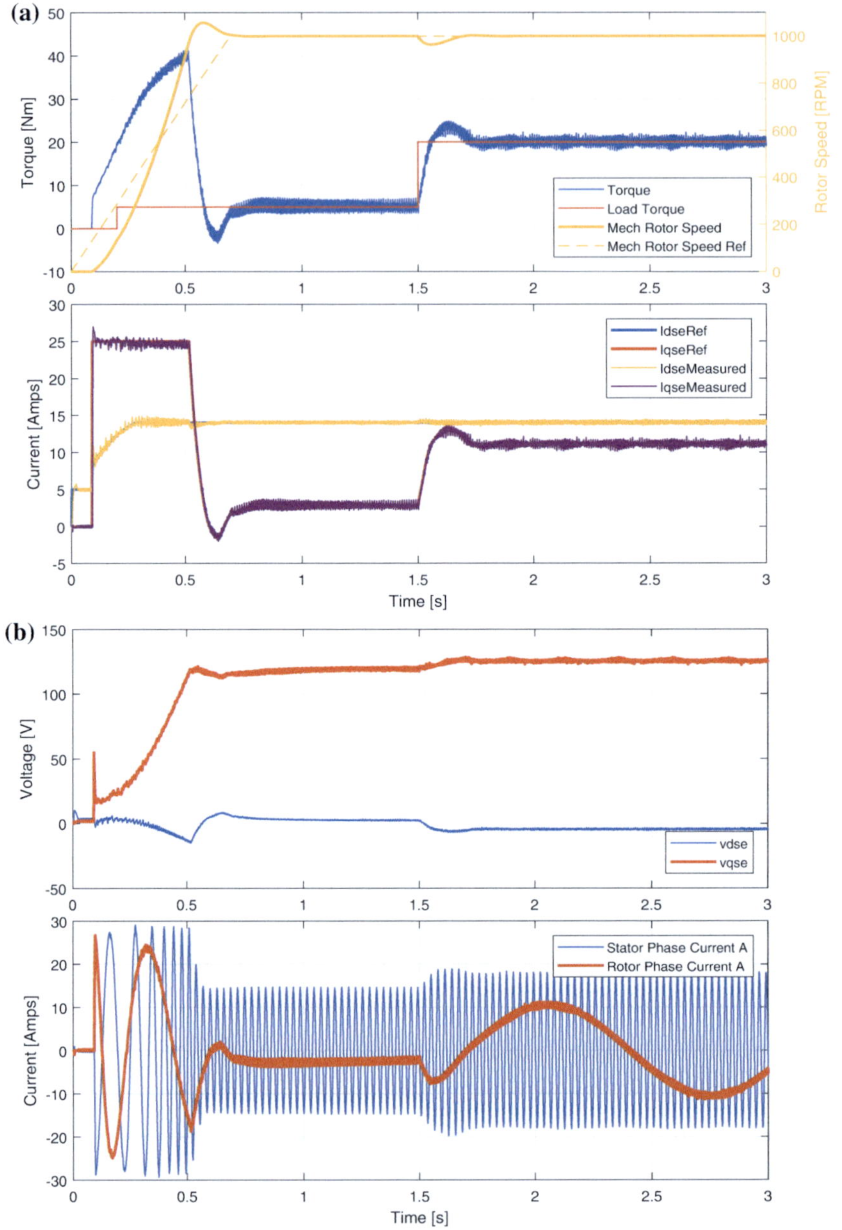

Fig. 11.17 Run-up speed till 1000 RPM with 5 Nm of load torque till $t = 1.5$ s. At $t = 1.5$ s, load torque is increased to 20 Nm. **a** Reference speed, rotor speed, torque developed, load torque, reference current components i_{dseRef}, i_{qseRef}, and measured i_{dse}, i_{qse}. **b** Voltage components v_{dse} and v_{qse}, stator and rotor phase current A. Simulation fixed-step 1 µs and automatic solver selection

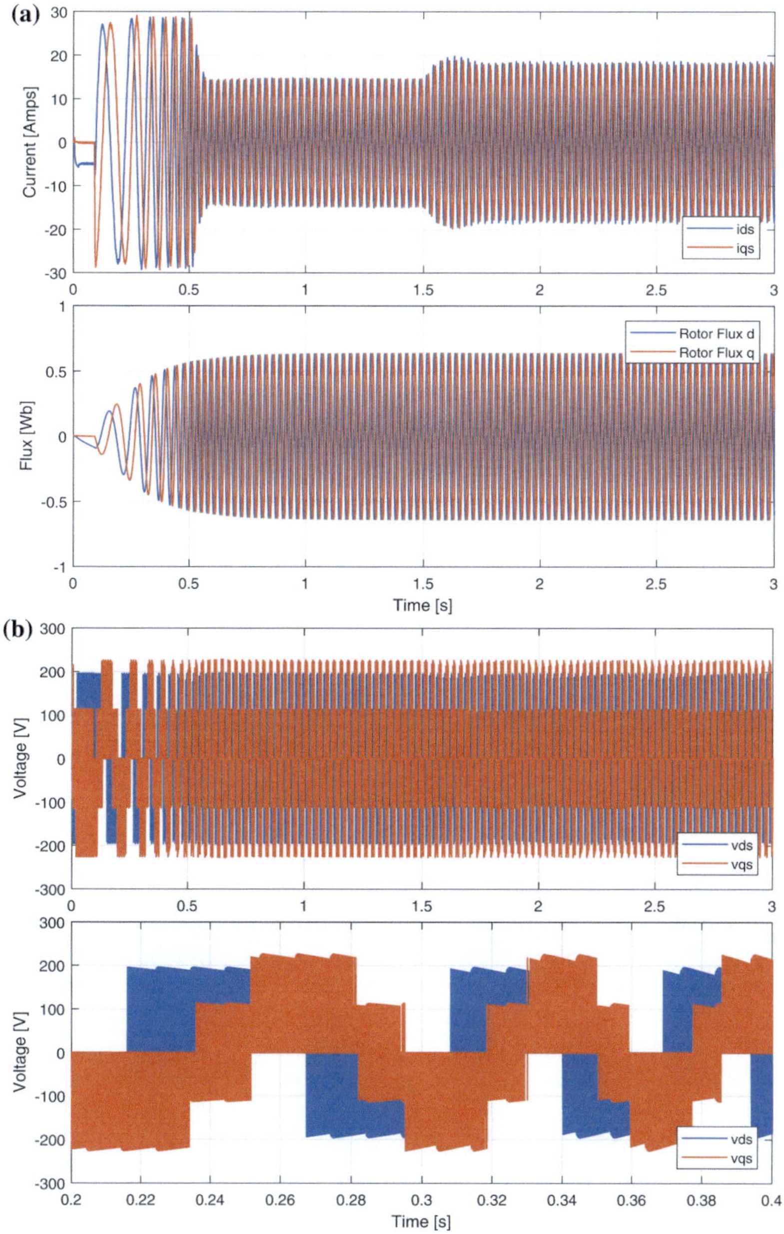

Fig. 11.18 Run-up speed till 1000 RPM with 5 Nm of load torque till $t = 1.5$ s. At $t = 1.5$ s, load torque is increased to 20 Nm. **a** Current components i_{ds}, i_{qs}, and rotor flux dqs components. **b** Voltage components v_{ds} and v_{qs} with zoom detail. Simulation fixed-step 1 μs and automatic solver selection

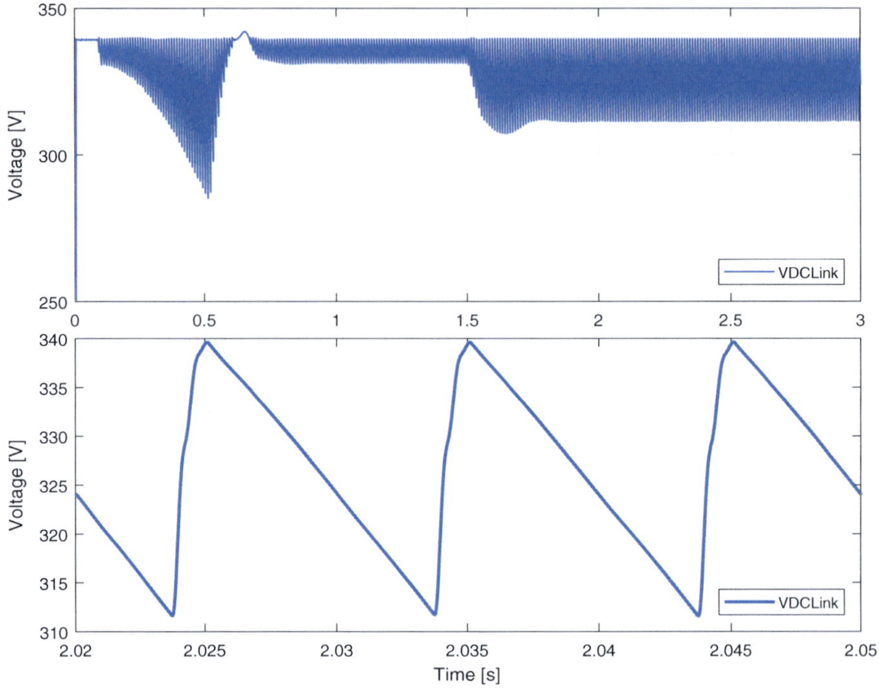

Fig. 11.19 Run-up speed till 1000 RPM with 5 Nm of load torque till $t = 1.5$ s. At $t = 1.5$ s, load torque is increased to 20 Nm. DC-link voltage with zoom detail. Simulation fixed-step 1 μs and automatic solver selection

The voltage components dq in the synchronously rotating reference frame with the stator and rotor currents for phase A are shown in Fig. 11.17b. At the start-up, the rotor current frequency reaches the maximum value, and when the machine reaches the steady-state at time instant $t = 0.75$ s, the slip is close to zero since the rotor frequency is also zero. When the disturbance load torque step is applied, the slip is increased, and the frequency of the rotor current of phase A reaches the value of 0.7 Hz approximately. In Fig. 11.18a, the current and rotor flux components are shown in the stationary reference frame. The voltage components measured in the machine terminals in the stationary reference frame are shown in Fig. 11.8b. These last signals are PWM signals, and it is possible to observe in the zoom part of Fig. 11.8b. The crest shape follows the DC-link voltage ripple shape.

Finally, the DC-link voltage is shown in Fig. 11.19. The maximum ripple it reaches during acceleration is approximately 50 Vpp. In the steady-state, the ripple is approximately 28 Vpp. It is possible to observe that when the speed of the machine exceeds the reference speed, the developed torque becomes negative at the instant of time $t \approx 0.6$ s, regenerating a little energy in the DC-link voltage.

11.6.2 Test Above Nominal Speed

In a four-pole induction machine, to reach the requirement of 3700 RPM as maximum speed, the flux weakening should be applied. The purpose of this test is to verify the flux weakening above the nominal speed and the stabilization for two DC-link capacitors options: one DC-link capacitor of 2.2 mF and another of 4.4 mF. The ramp of the speed set point applied in this test reaches its maximum value of 3700 RPM in 4 s. The load torque is kept constant and value of 4 Nm.

The results obtained, as described above, are shown in Figs. 11.20, 11.21, 11.22, 11.23 and 11.24. The reference speed, the speed of the machine, the torque developed, and the components of the reference *dq* currents and its corresponding measurements in the synchronously rotating reference frame are shown in Fig. 11.20 for both DC-link capacitors. In this figure, the flux weakening can be observed from a speed of approximately 1400 RPM where the current reference i_{dse} begins to decrease when the voltage of the machine is close to the nominal voltage. In the case of DC-link capacitor of 2.2 mF, the ripple of the current components *dqe* has more significant amplitude than for the DC-link capacitor of 4.4 mF. The main reason for this increment in the amplitude of the oscillation is the DC-link voltage ripple. As commented before, the ripple frequency component is six times the fundamental frequency. Note that the only difference between both tests is the DC-link capacitor. On the other hand, practically, throughout the acceleration, the torque developed by the machine remains constant at a value between 12 and 14 Nm.

Figure 11.21 shows the voltage components *dq* in the synchronously rotating reference frame with the stator and rotor current for phase A.

The currents and the rotor flux in the stationary reference frame are shown in Fig. 11.22. It is possible to observe the beginning of the flux weakening at approximate time $t = 1.4$ s, at a speed of approximately 1400 RPM as previously mentioned. As can be seen, the reduction of the flux is inversely proportional to the speed of the machine. For the DC-link capacitor of 4.4 mF, the *dqs* currents are more stable while in the rotor flux, no significant difference is observed.

Figure 11.23 shows the DC-link voltage ripple for both 2.2 mF and 4.4 mF respectively. In Fig. 11.23a, during the acceleration of the induction machine, the ripple reaches its maximum value at $t = 3.5$ s of about 60 Vpp, while in Fig. 11.23b, it is 33 Vpp at the time instant $t = 3.7$ s. In a steady-state, as can be seen, the ripple is approximately 22 Vpp and 11 Vpp, respectively, and with a frequency equal to 100 Hz.

When the machine speed exceeds the reference speed, a little regeneration of energy occurs at time $t = 4.25$ s since the DC-link voltage increases to about 368 V, as shown in Fig. 11.23a. However, when the DC-link capacity is 4.4 mF, this regeneration of energy does not occur as can be seen, because the measured current i_{qse} and the developed torque, according to Fig. 11.20b, do not take a negative value. This regenerative braking, although the current reference i_{qseRef} takes as minimum value zero, due to the high ripple effect that affects the voltage and phases current, as shown in Fig. 11.24, makes it unable to control the current i_{qse} properly taking

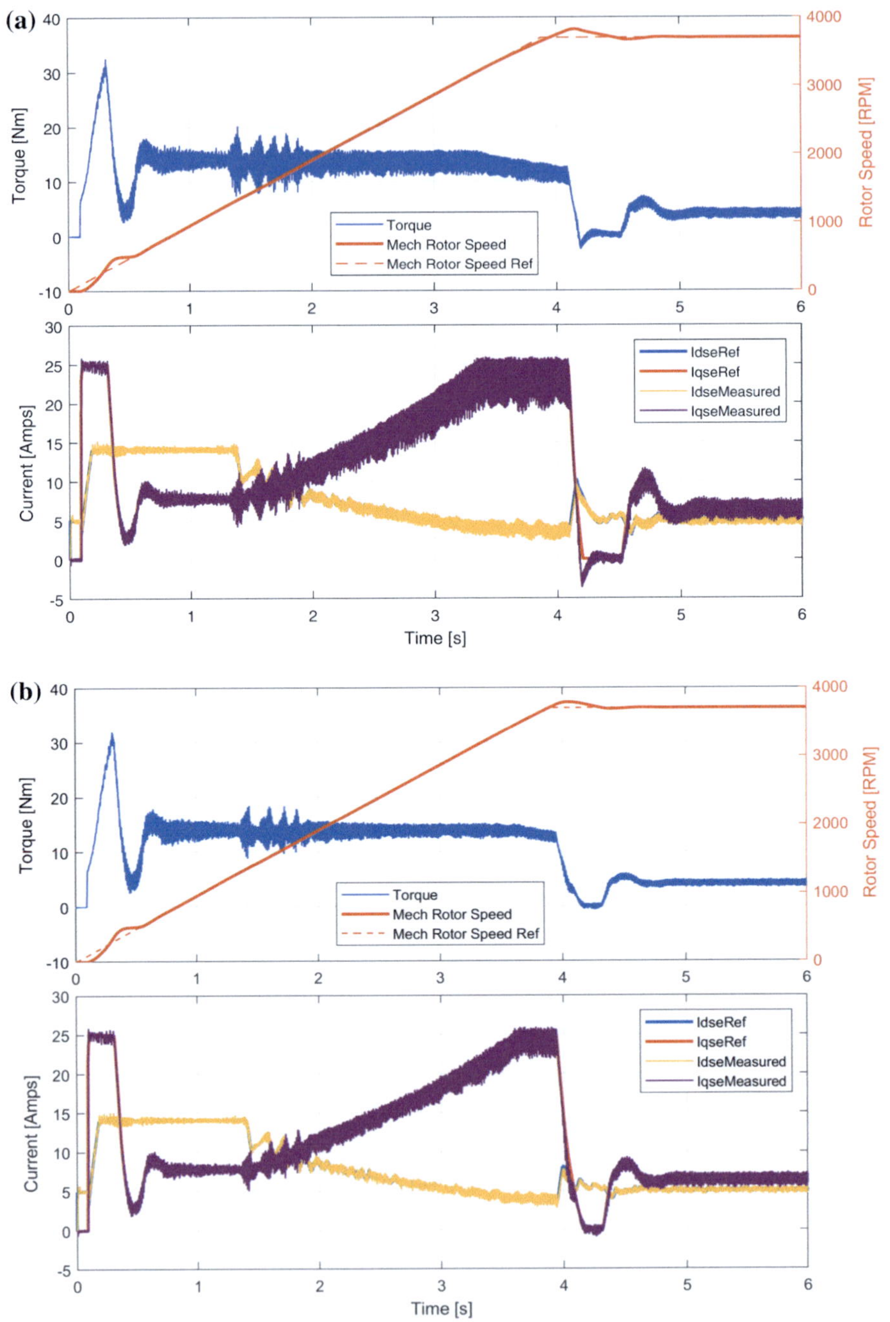

Fig. 11.20 Run-up speed till 3700 RPM with 4 Nm of load torque. Reference speed, rotor speed, torque developed, reference current components i_{dseRef}, i_{qseRef}, and measured i_{dse}, i_{qse}. **a** DC-link capacitor 2.2 mF. **b** DC-link capacitor 4.4 mF. Simulation fixed-step 1 μs and automatic solver selection

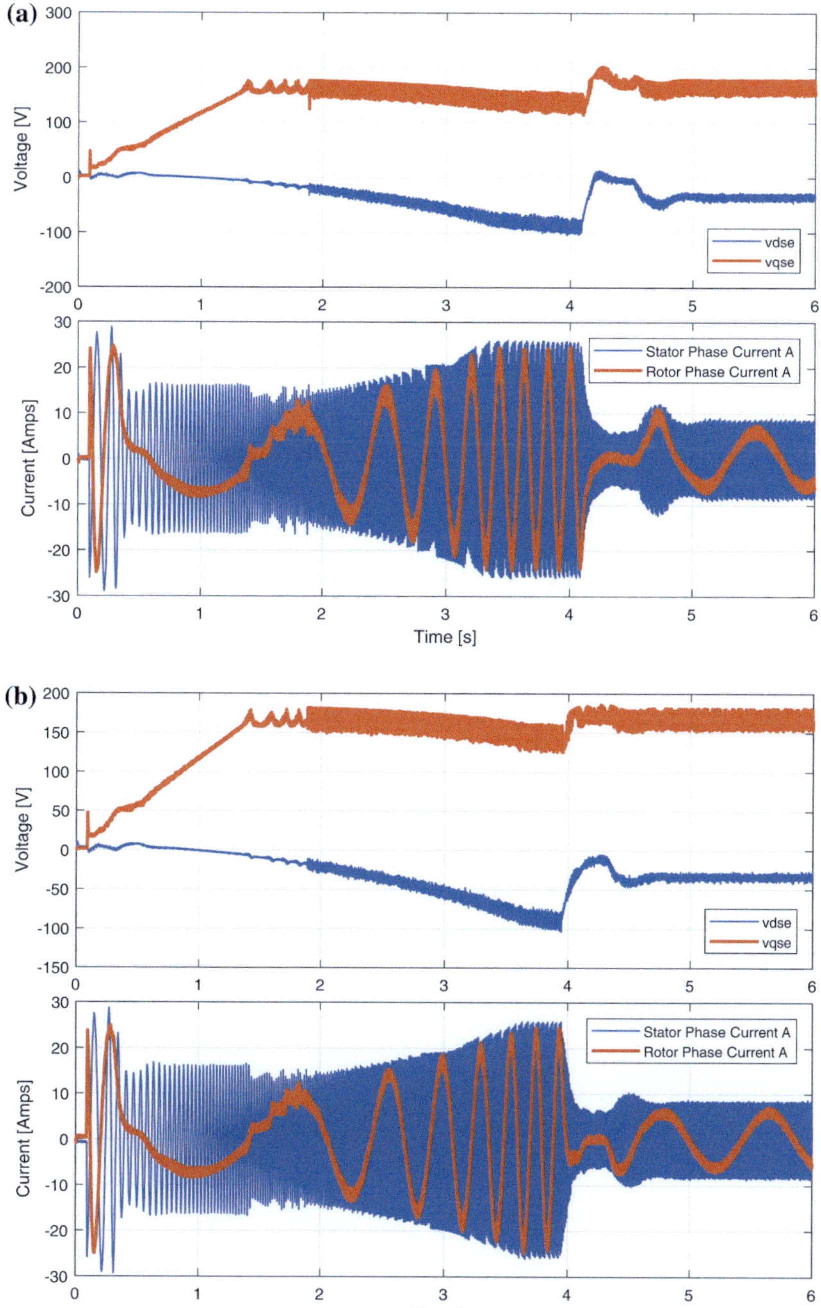

Fig. 11.21 Run-up speed till 3700 RPM with 4 Nm of load torque. Voltage components v_{dse} and v_{qse}, stator, and rotor phase current A. **a** DC-link capacitor 2.2 mF. **b** DC-link capacitor 4.4 mF. Simulation fixed-step 1 μs and automatic solver selection

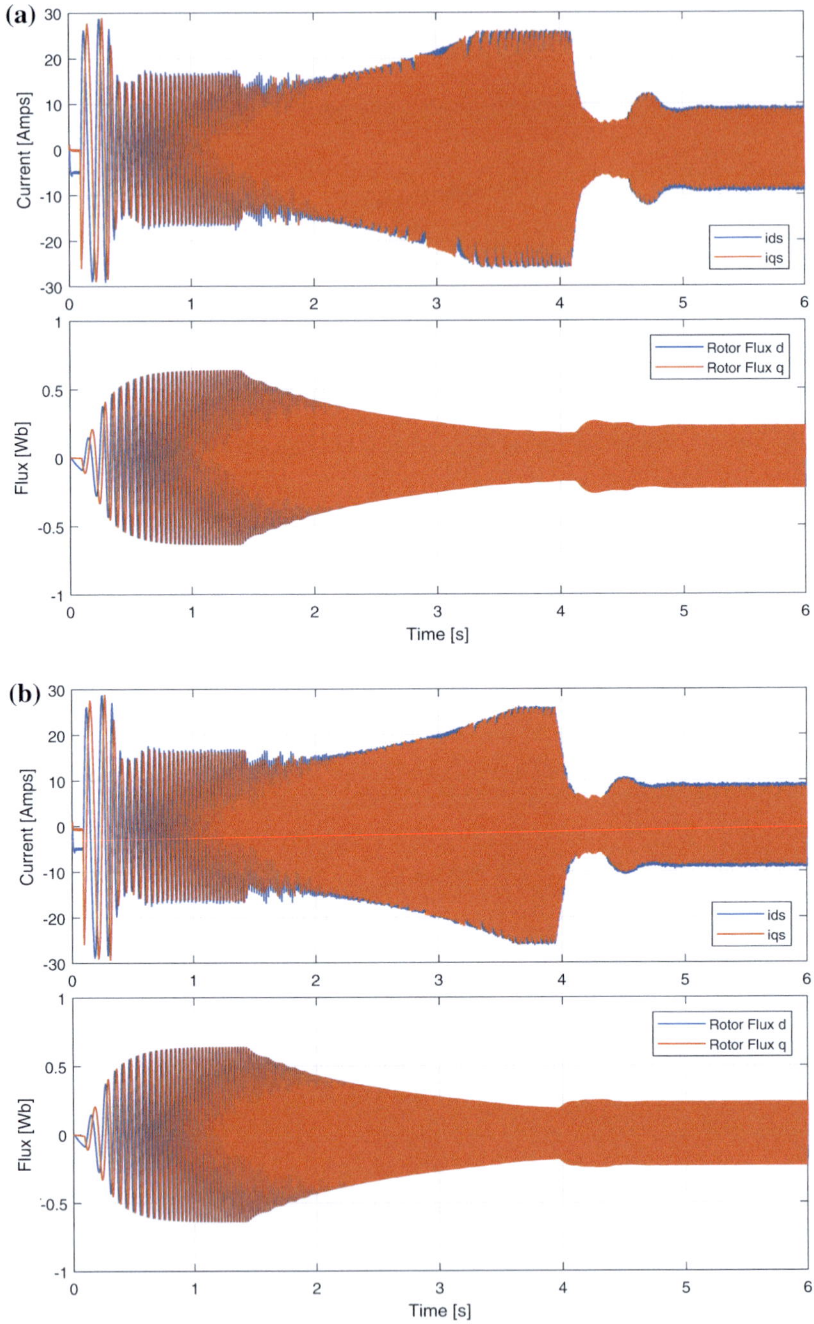

Fig. 11.22 Run-up speed till 3700 RPM with 4 Nm of load torque. Current components i_{ds}, i_{qs}, and rotor flux dqs components. **a** DC-link capacitor 2.2 mF. **b** DC-link capacitor 4.4 mF. Simulation fixed-step 1 μs and automatic solver selection

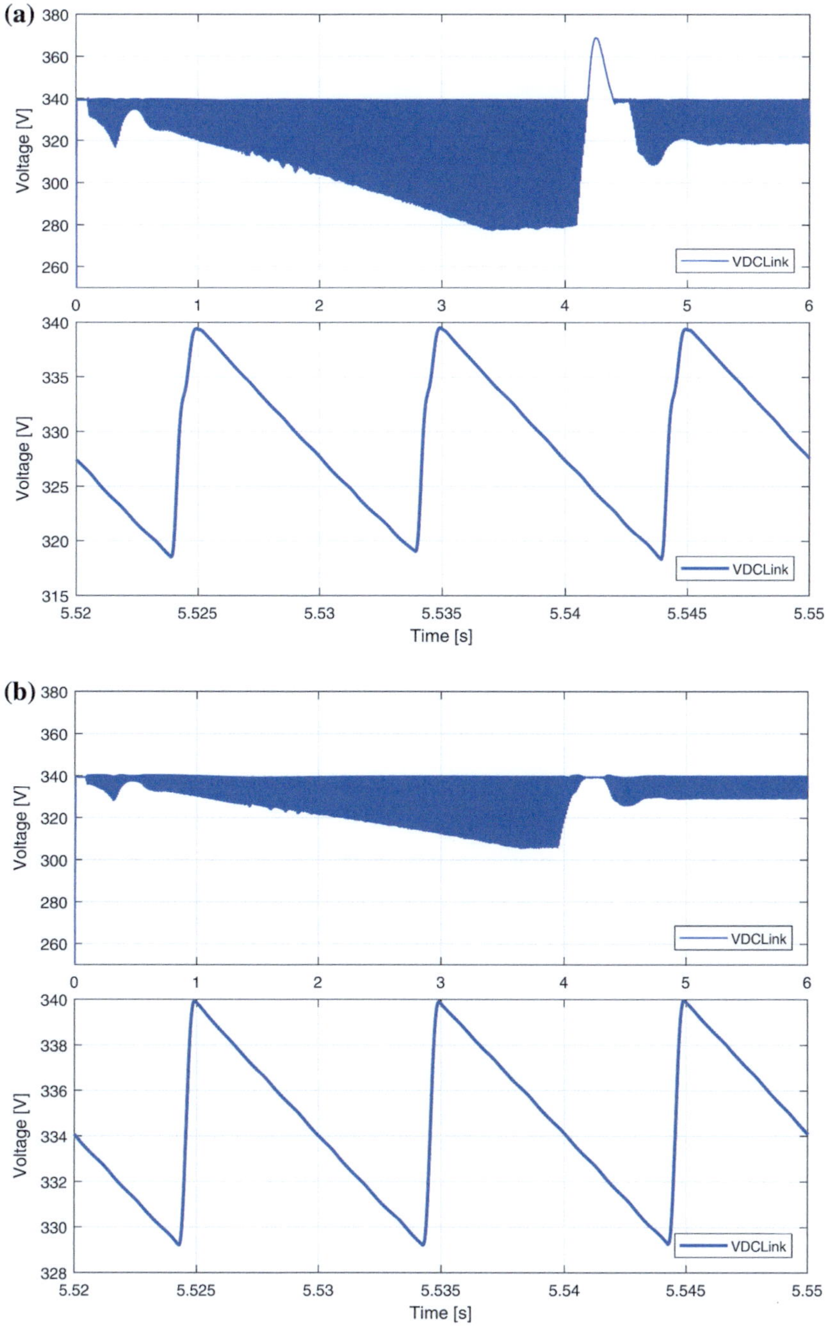

Fig. 11.23 Run-up speed till 3700 RPM with 4 Nm of load torque. DC-link voltage with zoom detail. **a** DC-link capacitor 2.2 mF. **b** DC-link capacitor 4.4 mF. Simulation fixed-step 1 μs and automatic solver selection

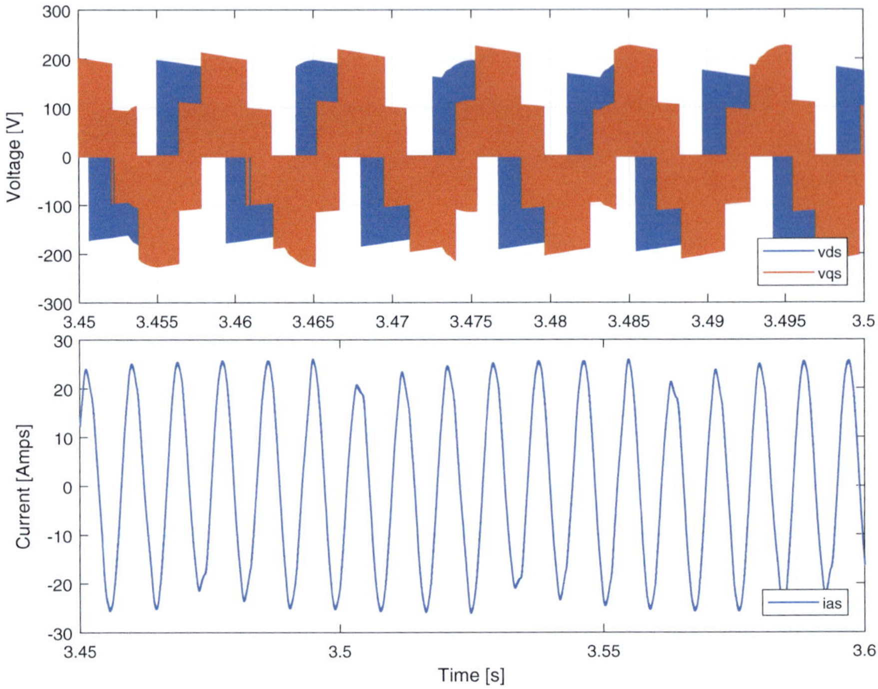

Fig. 11.24 Voltage components v_{ds} and v_{qs} and stator phase current A detail. DC-link capacitor 2.2 mF. It is possible to observe how the current and voltage in the machine terminals are affected by the ripple of the DC-link voltage. Simulation fixed-step 1 μs and automatic solver selection

a negative value. By performing a finer adjustment of the current controllers, it is possible to reduce this regeneration of energy. One of the objectives of these tests was also to show how the DC-link capacitors affect the stabilization of the vector control with an identical adjustment of the parameters of the different controllers.

11.7 Application in Electrical Vehicle

The restrictions of CO_2 emissions in the automotive sector in recent years is allowing the growth of demand for electric vehicles (EVs). They are considered the most viable solution to help protect the environment achieving an improvement of energy efficiency for transportation. Vehicles require certain performance for different driving conditions such as a high torque to start at zero speed, an acceptable acceleration and deceleration capacity, a capacity to climb slopes with heavy load, a braking capacity, and high power for highway cruising speed (Wang et al. 2013). The combustion engine can meet all these requirements, although with a very low efficiency compared to the one offered today by an electric traction machine. In combustion engine vehicles, the accelerator pedal gives a torque reference for the engine while in AC machines, the torque control can be achieved by a vector control as discussed in Sect. 10.3. In EV, electric traction machines are the key component for propulsion which transforms the electrical energy stored in a battery into mechanical energy in a very efficient way. The characteristics offered by an internal combustion engine are far outweighed by electric traction machines, because of their more efficient operating principle and because they are essentially designed to produce maximum torque or power, that is, several times the nominal values. The requirements demanded by the automotive sector of electric traction machines are usually high torque and power density, wide speed range with overload capacity, high efficiency, regenerative braking capacity, high reliability, low noise level, and a reasonable cost. The efficiency and the power factor of electric machines play a significant role since a worse efficiency increases the capacity requirements of the battery, and a lower power factor means that the inverter must be designed with a higher capacity of reactive power. The PMSM usually offers a better power factor, a higher torque density, a higher power density, and a better efficiency than an induction machine in a wider speed region. Recently, in PMSM, dual rotors were proposed to increase the power density with a drawback of the cogging torque increment. However, in order to achieve overload capability in the speed range (Pellegrino et al. 2012), a more secure back-EMF from an inverter failure and a lower sensitivity to the temperature of the permanent magnet, the rotor of the PMSM must be designed with multiple flux barriers to have a high saliency, where its industrial manufacturing is usually complicated. Also, the permanent magnet has a risk of demagnetization, and the high and changing price of the materials used in the permanent magnets, composed of rare earth, make the PMSM have a higher cost than an induction machine. For example, a PMSM for EV 150 kW can contain an equivalent weight of permanent magnets of up to 2 kg. On the other hand, permanent magnet assisted synchronous reluctance machine (PMASynRM) can be an excellent candidate since the number of permanent magnets is less than in a PMSM, and ferrite magnet or rare earth can be used in the rotor core. However, the danger of demagnetization still exists, and the rotor must be well designed to optimize the reluctant torque and compensate for the loss of electromagnetic torque due to the lower number of permanent magnets. A variant of the synchronous machine used in EV is the synchronous machine with rotor wound and hybrid excitation by a DC voltage. This machine which has no permanent magnets offers an improvement of the efficiency concerning the PMSM in the flux weakening region. On the contrary, it adds losses of the copper in the excitation winding nonexistent in the PMSM, as well as a greater degree of complexity in control between armature current and excitation current to achieve high torque and high efficiency throughout the speed range. Also, these usually require liquid cooling in the rotor to contain the temperature of the wound rotor. Another option without permanent magnets is the induction machine that is also used in EV. The induction machine is naturally deexcited under the condition of inverter fault, which is an important safety reason to be used among EV manufacturers. The copper die-cast rotor cage manufacturing technology increases the costs of the induction machine due to the high melting point of copper, which in the past made the life of the die significantly shorter. Recent research has improved the technology of copper die-casting manufacturing being today very competitive with the traditional squirrel-cage die-cast aluminum. The electrical properties of copper with respect to aluminum, allow reducing the equivalent resistance of the rotor, thus increasing its average efficiency. Besides, other improvements such as the optimal design of the geometry of the rotor slots have also achieved better efficiency and an improvement of the power factor of the induction machine. If also it is added that the induction machine offers better performance at higher speeds between 7000 and 12,000 RPM, it is possible to affirm that its average efficiency is very close to that of a PMSM. In any case, for the same electrical power of an EV, in the case of using an induction machine, a higher capacity battery could be necessary to obtain the same benefits in terms of the state of charge (SOC) of the battery for the same driving cycle as with a PMSM.

On the other hand, some EV manufacturers tend to adopt a dual machine topology, one for each axis of the same type of AC machine. It is possible to think that this topology would have the possibility of installing an induction machine on the front axle and a PMSM on the rear axle or vice versa to have the advantages of both machines. The induction machine offers more optimal performances at higher speeds, where the required torque is lower, while the PMSM works well at low and medium speeds where torque demand is higher. Figure 11.25 illustrates the Audi e-tron 55 quattro model which uses rear and front induction machines with a liquid-cooled lithium-ion battery of 95 kWh.

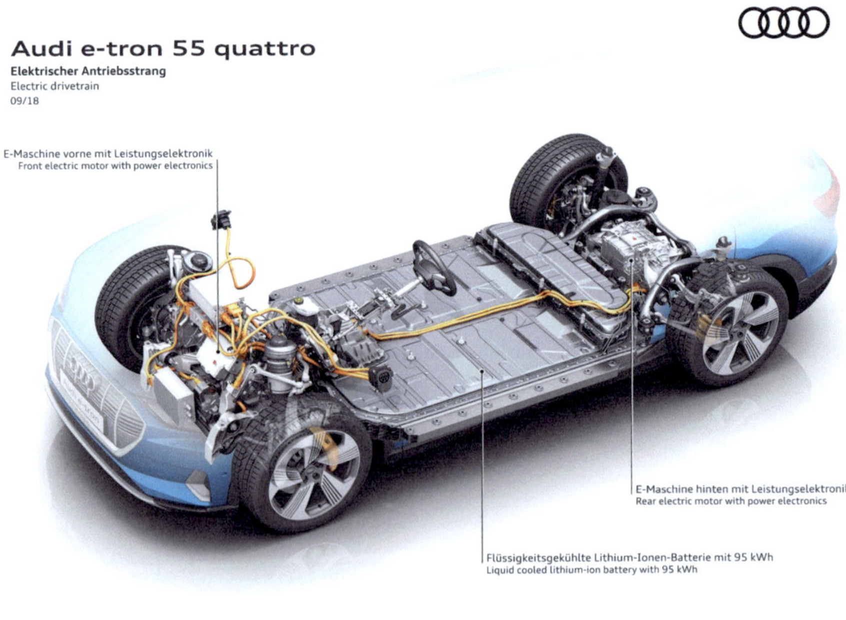

Fig. 11.25 Audi e-tron 55 quattro with two electric machines. Copyright: AUDI AG

The 265 kW model can develop a peak electric power of 265 kW with an electromagnetic torque of 561 Nm.

The control strategy for EV is similar to that seen in Sect. 11.3, that is, a sensored vector control where high efficiency must be maintained for as a wide range of speed as possible. It can be achieved using the strategy of maximum torque control per ampere (MTPA) for low speeds to maximize torque, while at higher speeds the approach is the flux weakening.

In this section, is presented a simulation using Simscape™ components of a twin-axle drive EV with a traction induction machine of 110 HP of power, one per axle. Thus, there is one VSI per machine which is supplied both by lithium battery. The same architecture approach of previous Sect. 11.5 will be used to control the induction machine and the VSI .

11.7.1 Vehicle Movement Simulation

In EVs, the electric traction machine is connected to the wheels of the vehicle through a single or differential gear of one or several speeds. The simple gear represents a gearbox that restricts the two axes of the connected transmission line to rotate together with a specified fixed ratio. The torque transferred and the power is reduced by the friction between the surfaces of the teeth in the gears, characterized by the efficiency η, and viscous coupling parameterized by the coefficient of viscous friction μ. However, the differential gear represents a gear mechanism that allows the two output shafts to rotate at different speeds. Differentials are standard in cars where they allow different wheels to turn at different speeds during a curve. The differential also has efficiency and a viscous friction coefficient as the simple gear. For the simulation of Sect. 11.7.2, single gear and speed will be used.

Figure 11.26 shows the different forces that act in a vehicle.

According to Newton's second law, the translational velocity of the vehicle is

$$m_{EV} \cdot a = F_t - F_g - F_w + F_r \tag{11.4}$$

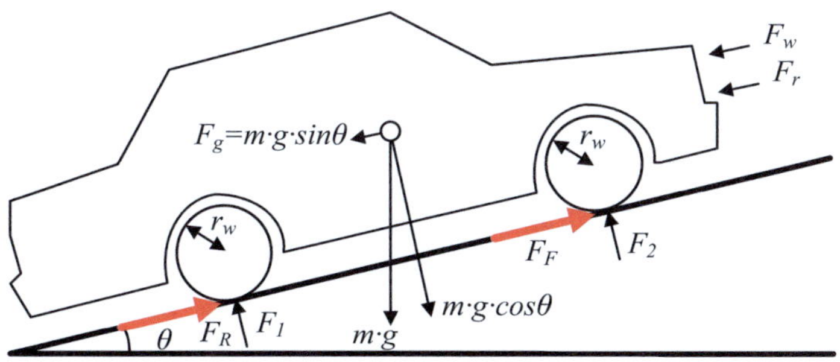

Fig. 11.26 Different forces acting on a vehicle driving uphill

where traction force $F_t = (F_F + F_R)$ is the propelling force of the electric vehicle, the F_g is the grading force due to the gradient of the road, F_r is the rolling resistance force on the wheels, and F_w is aerodynamic drag friction force.

The traction force can be expressed as a function of the radius and the traction torque T_t acting on the center of the wheels. For example, the rear wheel is:

$$F_R = \frac{T_R}{r_w} \tag{11.5}$$

when a vehicle driving on the road, the resistance or load torque which sees the shaft of one of the electric machines is (Ehsani et al. 2010)

$$T_L = \frac{1}{2}\frac{r_w}{N}\left(F_g + F_r + F_w + m_{EV} \cdot \frac{r_w}{N}\frac{\mathrm{d}\omega_m}{\mathrm{d}t}\right) \tag{11.6}$$

The load torque can also be expressed as a function of the developed torque of the electric machine

$$T_L = (T_e - T_\mu)\eta_{gear} \tag{11.7}$$

where T_u is the friction of the bearings in an electric machine, and η_{gear} is the efficiency of the gear.

Equation 11.6 can be rewritten as

$$T_L = \frac{1}{2}\frac{r_w}{N}(F_g + F_r + F_w) + \frac{1}{2}J_{EV} \cdot \frac{1}{N^2}\frac{\mathrm{d}\omega_m}{\mathrm{d}t} \tag{11.8}$$

where inertia J_{EV} of the vehicle can be calculated as

$$J_{EV} = m_{EV} \cdot r_w^2 \tag{11.9}$$

The aerodynamic drag F_w is a function of vehicle speed v in m/s, vehicle frontal area A_f, the aerodynamic drag coefficient C_D, the wind speed v_w on the vehicle's moving direction, and air density ρ. Aerodynamic drag can be expressed as

$$F_w = \frac{1}{2}\rho \cdot C_D A_f \cdot (v + v_w)^2 \tag{11.10}$$

The grading force F_g is a function of the vehicle mass, gravity, and the angle θ of the ramp, while the rolling resistance force F_r is a function of the same variables except for the vehicle mass and with the rolling resistance coefficient f_r. The rolling resistance coefficient f_r is a function of the tire material, the pressure, and the road surface. Lastly, the acceleration resistance force F_a is a function of the inertia of the vehicle J_{EV} and the acceleration $\mathrm{d}\omega_m/\mathrm{d}t$. Equation 11.11 is expressed the load torque with all the above variables and constants:

$$T_L(t) = \frac{1}{2}\frac{r_w}{N}\left(m_{EV}\cdot g\cdot \sin(\theta) + m_{EV}\cdot f_r\cdot g\cdot \cos(\theta) + \frac{1}{2}\rho\cdot C_D A_f\cdot (v+v_w)^2\right) + \frac{1}{2}J_{veh}\frac{1}{N^2}\frac{d\omega}{dt} \tag{11.11}$$

The angle of the road and the wind speed are driving conditions, which affect the dynamics of the electric vehicle, as seen in (11.4).

On the other hand, the power of the shaft of the electric traction machine is expressed as

$$P_{EM} = \left(T_e - T_\mu\right)\omega_m \tag{11.12}$$

Then, the power transferred to the wheels is:

$$P_{wheel} = P_{EM}\cdot \eta_{gear} \tag{11.13}$$

Figure 11.27 shows the simplified modeling of EV dynamics according to Eq. (11.11) performed with essential components of Simulink®. The model has three inputs and two outputs. The inputs are the speed of the electric machine, the wind speed, and the road angle. The outputs are the load torque that supposes for the electrical machine and the mechanical power.

An alternative to the previous model is to use the Simscape™ physical modeling components, as shown in Fig. 11.28. It is possible to observe for each of the axes, the

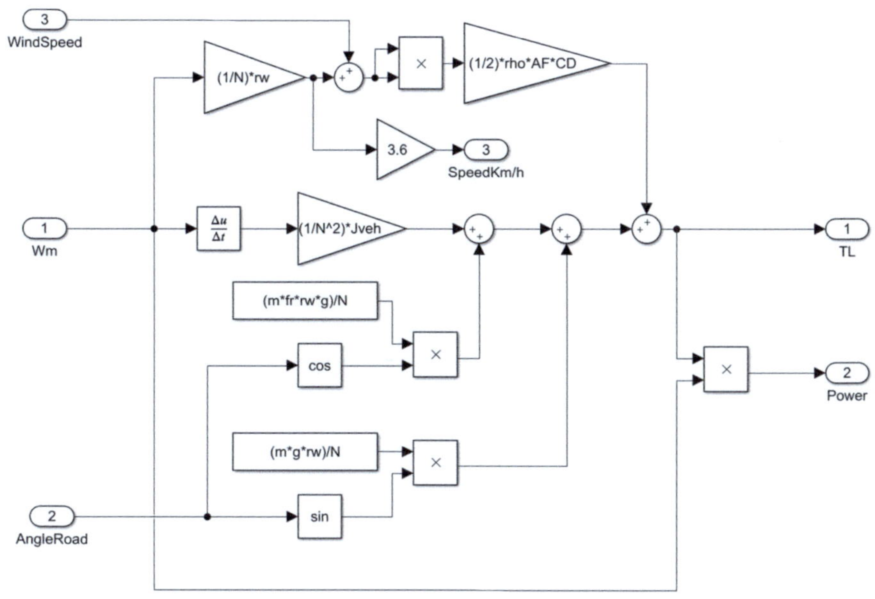

Fig. 11.27 Simulink block diagram of the vehicle dynamic with single-axle drive propulsion

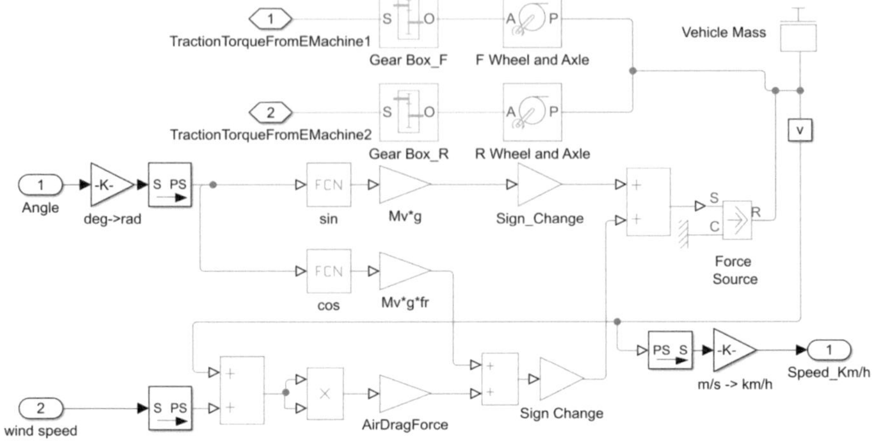

Fig. 11.28 Simscape^{TM} model vehicle dynamic with twin-axle drive propulsions

model of the gearbox, an ideal converter between the mechanical rotational torque to mechanical translation force, a component of a mechanical translation mass (EV), and a force source composed by the rolling resistance force, the aerodynamic drag force, and the grading force.

11.7.2 Vehicle Speed Control Simulation

In this section, it is shown drive cycle simulation of a simplified twin-axle drive EV propelled with an induction machine of 110 HP, one per axle. The EV is equipped with a high-voltage Li-ion battery of 70 kWh. The optimization of the magnetizing current at all operating points of the induction machine is mandatory to maximize the autonomy of the EV in a drive cycle. It is possible to map at each torque and speed, the magnetizing current to provide the best efficiency, minimizing the total loss of the VSI and IM. In the present simulation, the base magnetizing current is applied during the constant torque region, and in the flux weakening region, the magnetizing current is a function of rotor speed and IM voltage.

On the other hand, in the modeling of the dynamics of the vehicle (drive train model) the dynamics of the tires, the skid between the wheel and the road are neglected, and efficiency in the transmission gear η is considered of one. The N ratio between the gear input and output and the wheel radius are carefully selected according to the power and performance of the vehicle. It is possible to consider a ratio N of $10:1$, a wheel radius r_w of 0.35 cm, and a vehicle weight m_{EV} of 1100 kg. It is possible to make a curve that relates the sum of the torques of the two induction machines and their speed in RPM together with speed in km/h of the electric

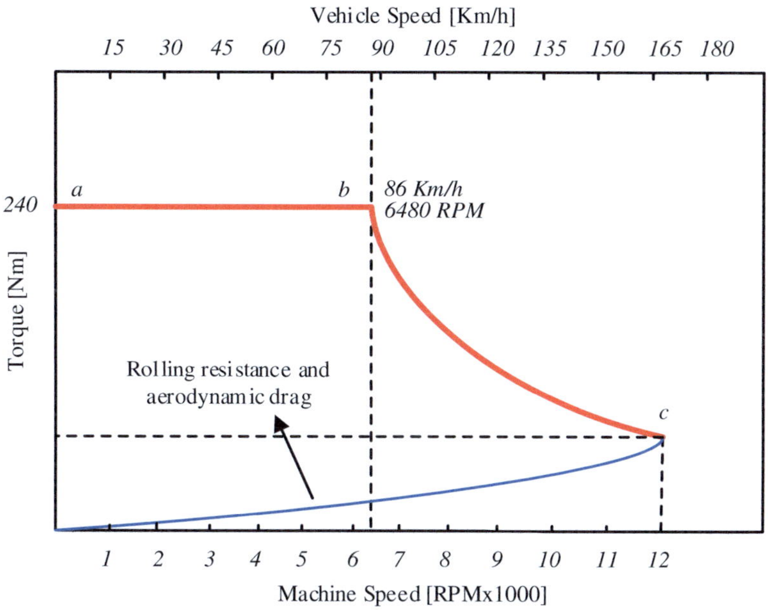

Fig. 11.29 Electromagnetic torque effort from 2 × 110 HP (2 × 82 kW) traction induction machines with single-gear transmission versus vehicle and machine speed. The load torque is counteracting the electromagnetic torque

vehicle as shown in Fig. 11.29. The load curve due to the rolling resistance and the aerodynamic drag is also shown.

The region *a*–*b* is the constant torque region, and from point *b* the constant power region begins. At point *b*, the flux weakening of the induction machine begins, and it is just when the mechanical power reaches its maximum value of 81.43 kW per machine. At this point, the speed of the machine is 6480 RPM, and the speed of the EV corresponds to 86 km/h. On the other hand, point *c* indicates the maximum speed that the vehicle will reach based on the rolling resistance and the aerodynamic drag force. Table 11.7 shows the constants used in the following simulations.

Table 11.7 Constant values used in the simulation of the EV

Constant	Description	Value	Unit
η_{gear}	Gear efficiency	1	–
N	Gear ratio	10	–
A_f	Vehicle frontal area	1.78	m^2
C_D	Aerodynamic drag coefficient	0.24	–
ρ	Air density	1.225	kg/m^3
r_w	Radius of wheels	0.35	m
m_{EV}	Mass of vehicle	1100	kg
f_r	Friction coefficient	0.0056	–

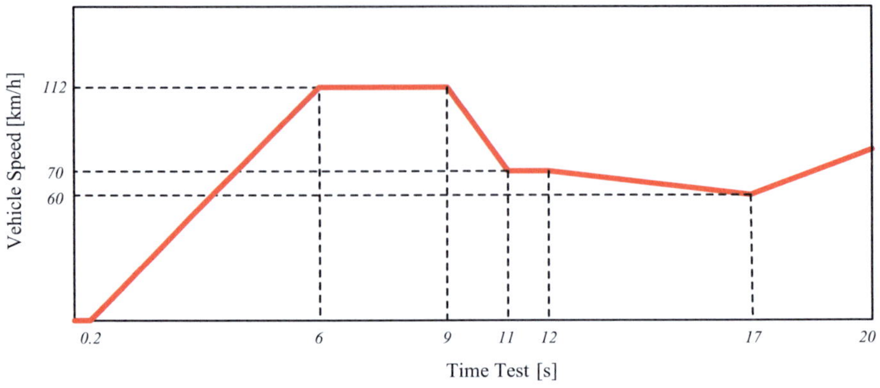

Fig. 11.30 Drive cycle used in the simulation

The drive cycle used for the evaluation of the machine control simulation is depicted in Fig. 11.30. The length of the drive cycle is 20 s and stretches over about 0.8 km, where the maximum speed is 111.5 km/h. It corresponds to a very short inter-urban drive cycle.

Applying different standard drive cycles is possible. For example, the EPA urban dynamometer driving schedule (UDDS) which represents city driving conditions with a length of 1369 s and 11.92 km, or the UN/ECE extra-urban driving cycle depicted in Fig. 11.31.

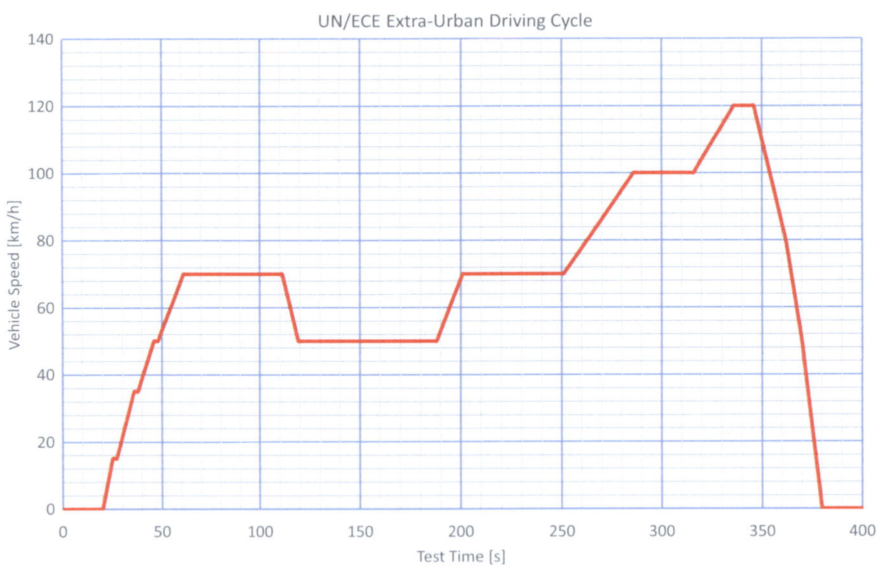

Fig. 11.31 UN/ECE extra-urban driving cycle

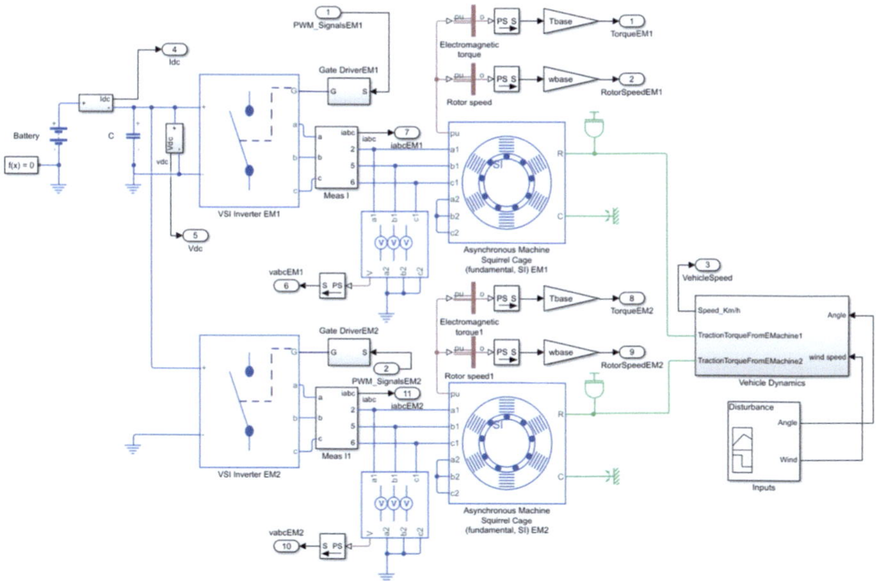

Fig. 11.32 Simscape™ model of twin-axle drive in an EV composed by two induction machines, two VSI , a lithium battery, and the EV dynamic model

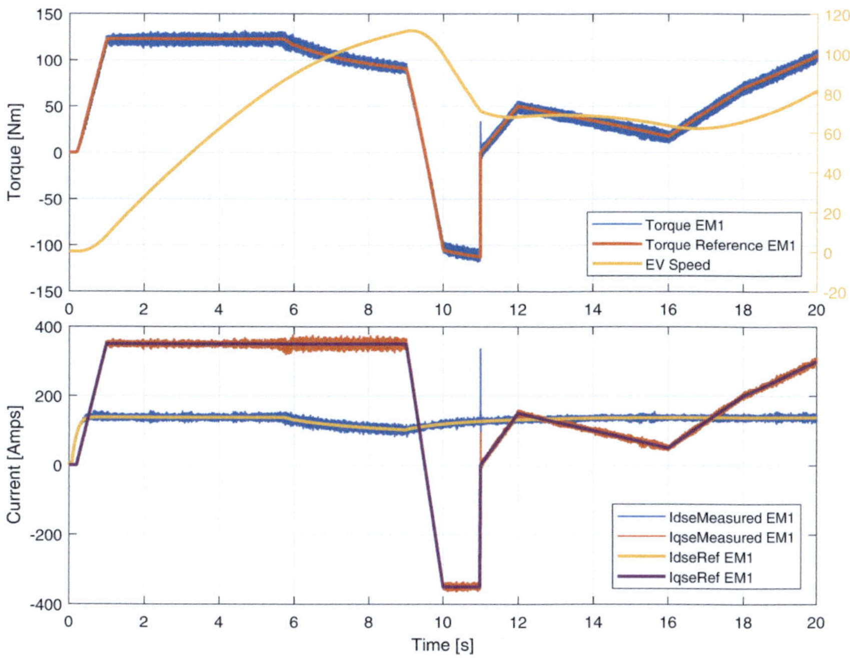

Fig. 11.33 Drive cycle simulation results. It is shown the torque reference of EM1, EV speed in km/h, developed torque of EM1, reference current components i_{dseRef}, i_{qseRef}, and measured i_{dse}, i_{qse} for EM1. Simulation fixed-step 10 ns and automatic solver selection

Figure 11.32 shows the Simscape™ components of the model of the two induction machines, EM1 and EM2, the two VSI , the lithium battery, and the EV dynamics seen in the previous section. It is important to mention that both inverters and/or controls are independent and that the idea of this mechanical composition is to distribute the power in the two axes of the EV. The total horsepower of the present EV is 220 HP at a speed of 6480 RPM of the electric machine.

The results obtained are described in Figs. 11.33 and 11.34. As shown in Fig. 11.33, at the instant time $t = 0.2$ s, the torque of the EM1 begins to increase to its maximum value of approximately 120 Nm, where it remains constant up to the instant time $t = 5.7$ s. At this point, the flux weakening of both induction machines EM1 and EM2 begins. The speed of the EV continues to rise but with a lower acceleration due mainly to the lower torque, and to a lesser degree, to the increase in the rolling resistance and aerodynamic drag. As can be seen, the EV is capable of reaching the speed of 100 km/h in approximately 6.8 s. At the instant time $t = 9$ s, the EV is braked using only the regenerative braking of both induction machines to

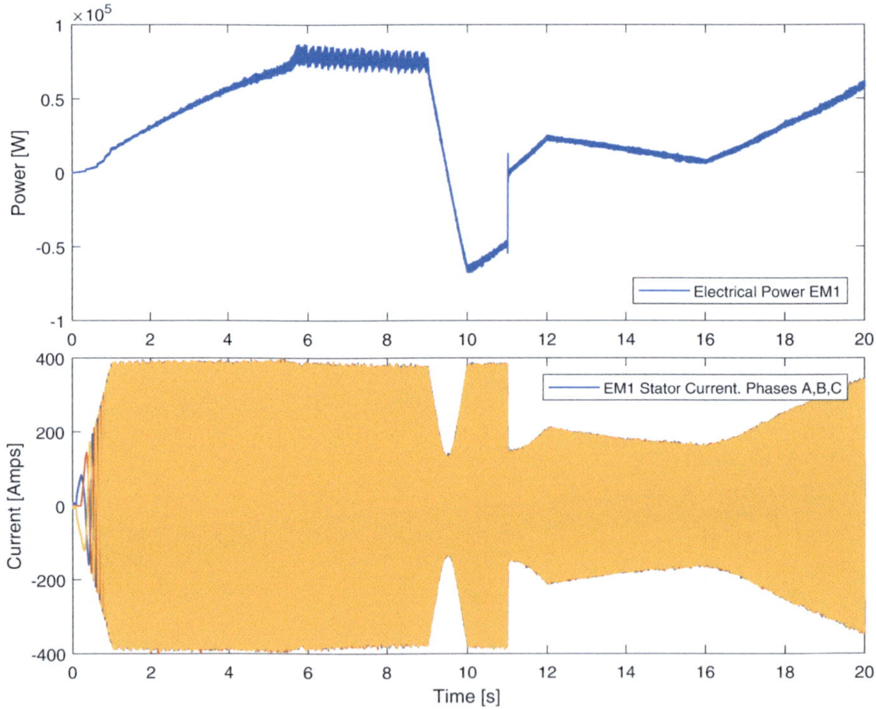

Fig. 11.34 Electric power and phase current a, b, c of EM1 during the drive cycle. Simulation fixed-step 10 ns and automatic solver selection

see how the battery accumulates energy up to the instant of time $t = 11$ s. From this moment, it accelerates again to a lower acceleration according to the previous drive cycle.

The reference current components *dqe* and its corresponding measurements in the synchronously rotating reference frame are also shown. The maximum current for torque reference is 350 A, while for flux component *d* is 140 A approximately. It is possible to observe that the measured values follow exactly the reference values with a certain ripple, most likely caused by dead zones of non-linearity of the VSI .

Figure 11.34 shows the electrical power and phase currents of the EM1 machine during the drive cycle. During the acceleration, it is possible to observe how the power increases in a linear form, while the torque developed by the machine EM1 is constant. In the flux weakening, the power remains approximately constant, and during the regenerative braking, the power is negative with what energy is returned to the battery.

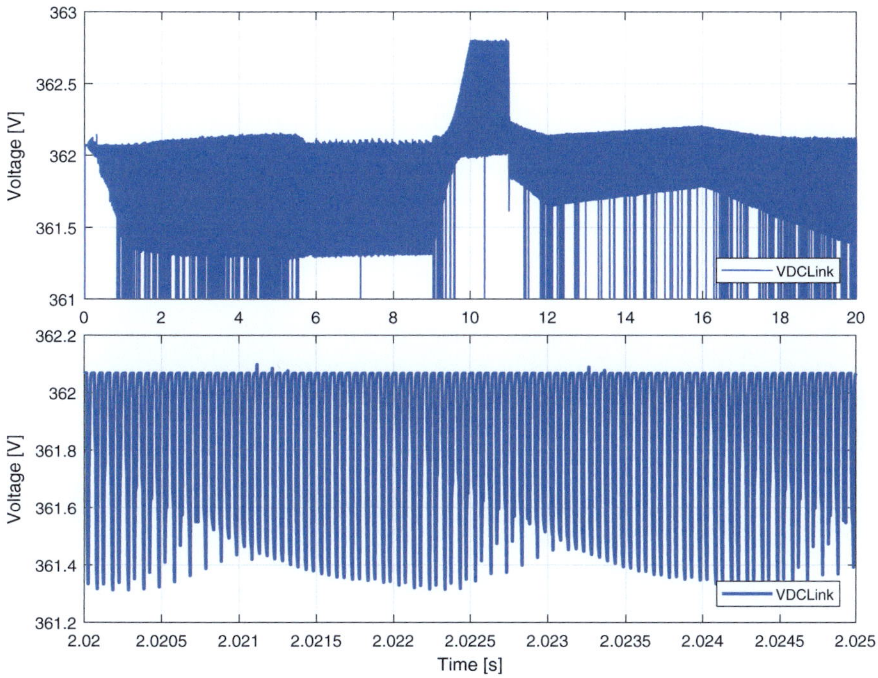

Fig. 11.35 DC voltage battery during the drive cycle. The voltage of the battery drops during positive power, while the voltage gets higher values during regeneration braking. It is shown a detailed zoom of voltage battery to see the time instants where the VSI gets energy from the battery. Simulation fixed-step 10 ns and automatic solver selection

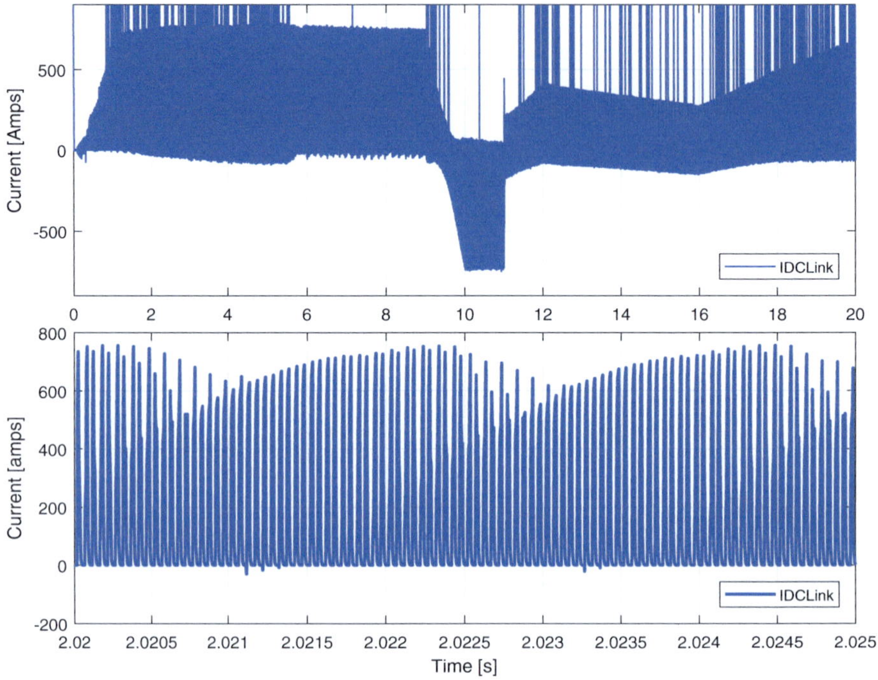

Fig. 11.36 Current of the battery during the drive cycle. The current of the battery is positive during positive power, while it is negative during regeneration braking. It is shown a detailed zoom of the current battery to see the time instants where the VSI gets energy from the battery. Simulation fixed-step 10 ns and automatic solver selection

The voltage and current of the battery during the drive cycle are shown in Figs. 11.35 and 11.36. It is possible to observe how the battery voltage decreases when the power is positive and how it increases when the power is negative. The detail is also shown by zooming in on the instant times where the VSI demands power from the battery.

Finally, Fig. 11.37 shows the stator currents of each of the phases during start-up and at maximum power.

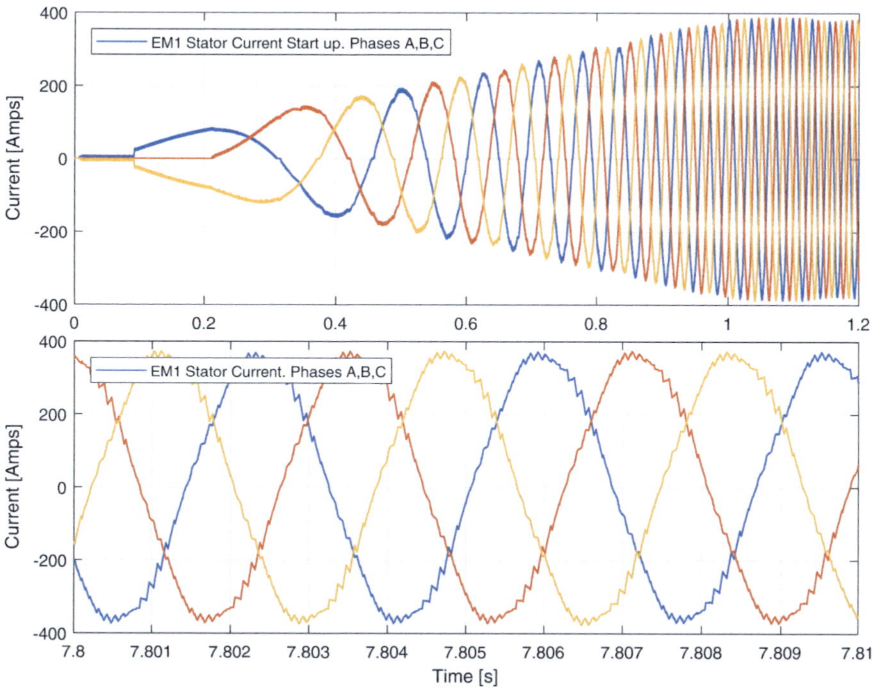

Fig. 11.37 Starting-up stator phase current and full zoom in maximum power for EM1. Simulation fixed-step 10 ns and automatic solver selection

11.8 Application in Propeller Aircraft

Recently, the electrified aircraft propulsion (EAP) uses propulsors such as propeller or fans driven by electric machines. Some companies, and in especial, NASA are making significant progress toward establishing the viability of EAP through a combination of aircraft conceptual design studies. The investigated electric machines are induction machines, PMSM, and wound field machine where the goal is to achieve a power density 2–3 times state of the art for machines in the MW power range. Different research centers have developed prototypes with a power density between 13 and 16 kW/kg with higher efficiencies than 96%. It means that for 1 MW machine, the weight is only 62.5 kg for the best case. The power converter (inverter) developed for this high-power machines meet with the complex requirements of efficiencies of 99% and power densities higher than 18 kW/kg thanks to the silicon carbide, and gallium nitride power switches with conventional liquid cooling, or higher power densities above 20 kW/kg with and cryogenically cooling system. In the best case, the weight of the inverter is around 50 kg. It is important to comment that the validation of the inverters is a complicated task because of the altitude operation.

In case of propeller aircraft, when a conventional internal combustion engine is used, the energy conversion efficiency is around 42%, while for full electric, it is around 90%. In full-electric propeller aircraft, the energy storage comes from a battery which usually has an efficiency of 95% and the present lithium–air batteries the current energy density is 363 Wh/kg (Voskuijl et al. 2018). The efficiency of the inverter and the electric machine typically are 97% for both.

Some full-electric propeller aircraft has been successful in the recent past covering different flight distance but seems reserved for 1–4 passengers. The typical characteristic of one passenger aircraft consists of a maximum take-off weight (MTOW) of 600 kg and wingspan around 8.0 m powered with 75 kW electric machine and battery of 10 kWh. The propeller speed is around 2500 RPM, and the developed torque is close to 286 Nm. In some altitude conditions, they achieve an airspeed of 100 KIAS (Knots-Indicated Air Speed), i.e., around 185 km/h. Figure 11.38 represents a fully electric aircraft propulsion propeller from Siemens which develops a maximum power of 85 kW (Siemens 2016).

The application in larger aircraft is one of the upcoming steps in the electric propulsion systems. As it happens in the automotive market, the pure-electric propulsion concepts have disadvantages such as autonomy due to the current energy density of the batteries. Hence, since the full-electric aircraft is still complicated for commercial, in the meantime, hybrid-electric architecture concepts are developing (Voskuijl et al. 2018) as it happens in the automotive market. For example, a propulsion architecture based on serial-hybrid-electric propulsion uses fuel energy to be converted into electric power through an internal combustion engine connected to a generator. In this configuration, the generator can supply energy to the inverter and/or charge the battery. Thus, the inverter can be supplied by the electric generator or through

Fig. 11.38 Full-electric aircraft propulsion propeller. Copyright: Siemens

the battery. The main advantages of this configuration are the reduction of CO_2 and noise emissions significantly. Then, a silent take-off and landing are possible.

In electric propeller aircraft, the electric propulsion machine is usually connected directly to the propeller. However, a gear of one speed can be used depending on the specification of the machine as the following is discussed.

The 110 HP induction machine discussed in the previous section develops an electromagnetic torque of 120 Nm during the constant torque region, i.e., till 6480 RPM rotor speed. The electric propeller usually operates in a lower speed range a needs higher torque. With a gear box with ratio 2 : 1, the new torque and propeller speeds are 240 Nm and 3240 RPM, respectively. Now, the conditions are ideal for one passenger electric propeller aircraft where the maximum electric power of the machine is 81.43 kW.

The detail of how a propeller generates thrust is a complex task (Hartman and Biermann 1937), but here only some fundamentals concepts will be defined. Figure 11.39 shows an electric propulsor propeller with four blades of a Siemens electric aircraft. As can be observed, the blade has a twisted airfoil of irregular planform.

Some propeller aircraft allow for the propeller to be pitched in flight (Teeuwen 2017). The geometric pitch angle β of the blades is changed to adjust the propeller blade to the varying free stream speed of the aircraft. This change in pitch angle

Fig. 11.39 Electric propulsion propeller. Copyright: Siemens

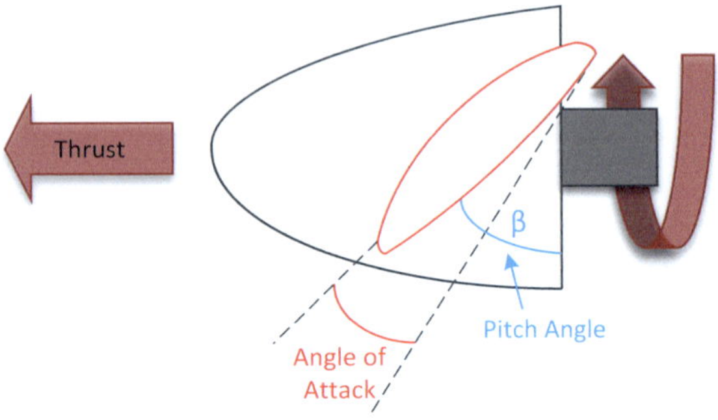

Fig. 11.40 Lateral view of a propeller with an adjustable pitch angle

ensures an optimal angle to the relative velocity on the blade. Figure 11.40 illustrates the pitch angle in a lateral view.

Usually, there is a misunderstanding between the pitch and the geometric pitch angle β concepts. The pitch is defined as the distance usually in inches traveled forward in one revolution (e.g., 5 in./revolution) if there were no slippage (McCormick 1995) Thus, geometric or theoretical pitch is based on no slippage. Then, pitch p is defined as

$$p = 2\pi r \tan\ \beta \tag{11.14}$$

where r is the radius of the propeller. Sometimes, it is defined as the pitch–diameter ratio p/D_P as

$$p/D_P = \pi \frac{r}{R} \tan\ \beta \tag{11.15}$$

where r/R is the relative radius of the blade section. As commented early, the blade of a typical propeller consists of a twisted airfoil of irregular planform. However, a constant pitch propeller, the pitch does not vary with radius, i.e., r/R equals unity.

The thrust of the propeller propulsion in N can be calculated according to

$$T = C_T \rho n^2 D_p^4 \tag{11.16}$$

where C_T is the non-dimensional propeller thrust coefficient, n is de propeller speed in revolutions per second, D_P is the propeller diameter, and ρ is the air density. The power required by the propeller can be calculated as

$$P_p = C_P \rho n^3 D_p^5 \tag{11.17}$$

where C_P is the non-dimensional power coefficient. The power increases as the fifth power of diameter and the cubic power of the revolution rate.

Both coefficients C_T and C_P are fully characterized by the advance ratio J, and the pitch-diameter ratio of the propeller p/D_P.

The advance ratio J is defined by:

$$J = \frac{V}{nD_p} \quad (11.18)$$

where V is the axial or forward speed of the propeller.

Maximum thrust is achieved at the lowest advance ratio since C_T takes higher values.

From parameters data (Jeracki et al. 1981) of Table 11.8, it is possible to approximate the power required by a three-blade propeller Clark Y airfoil during take-off, climb, and cruise speed as following.

For take-off the advance ratio J is 41/(51.07 × 1.524) = 0.526.

By using the advance ratio value, and the wind tunnel test results of (Jeracki et al. 1981), the thrust coefficient C_P takes a value of 0.192. Then, the power required by the propeller in take-off can be calculated as:

$$P_{Pto} = C_P \rho n^3 D_p^5 = 0.192 \cdot 1.225 \cdot 51.07^3 \cdot 1.524^5 = 257548.91\,(W) \quad (11.19)$$

In case of climb conditions, the advance ratio J is 78/(46.45 × 1.524) = 1.101. Similarly, the thrust coefficient C_P takes a value of 0.204. The power required by the propeller in climb conditions can be calculated as:

$$P_{Pcl} = C_P \rho n^3 D_p^5 = 0.204 \cdot 1.225 \cdot 46.45^3 \cdot 1.524^5 = 205896.16\,(W) \quad (11.20)$$

Lastly, for cruise speed, the advance ratio J is 121/(35.58 × 1.524) = 2.23. By using the advance ratio value, from a power coefficient graph of the propeller (Jeracki et al. 1981) it is possible to find a thrust coefficient C_P of 0.304 for a geometric pitch angle β equal to 48°. Then, power required by the propeller in cruise speed can be calculated as:

$$P_{Pcs} = C_P \rho n^3 D_p^5 = 0.304 \cdot 1.225 \cdot 35.58^3 \cdot 1.524^5 = 137896\,(W) \quad (11.21)$$

Note that the air density taken corresponds to ground (0 m altitude, 1.225 kg/m^3), which is lower when altitude is incremented. For example, at 8000 m of altitude, the air density is 0.525168 kg/m^3.

Since the mechanical power of the induction machine was 81.43 kW (31.61%), in worse case (take-off), it is needed an additional power of 176.12 kW (68.39%).

Table 11.8 Low-speed propeller operation conditions (Jeracki et al. 1981)

Description	Take-off	Climb	Cruise	Units
Forward speed (V)	41	78	121	m/s
Propeller speed (n)	3064-/51.07	2787-/46.45	2135-/35.58	RPM-/rev/second
Propeller diameter (D_P)	1.524	1.524	1.524	m

This power can be supplied by and turboshaft engine in a parallel hybrid architecture where both power powers are coupled to the propeller. A typical design which has 34% electric shaft power requires 28% less mission fuel at the expense of a larger aircraft in terms of weight and wing area (Voskuijl et al. 2018).

With this parallel hybrid architecture, it is possible to get a cruise speed of 235 KIAS for an 800–1100 kg aircraft with a lithium battery of 20 kWh. The battery expected weight is 55 kg according to the battery energy density of 363 Wh/kg.

In this chapter, the field-oriented control of low-power and medium-power induction machines has been seen by using the MBD techniques. As discussed, the applications were for a general-purpose industrial market such as pumps, fans, compressors, for automotive market as a traction machine in EVs, and aircraft market as electric propulsion. The same approach can be applied for other applications with induction machines, permanent magnet machines, wound field, and synchronous reluctance machines. For example, drones usually use high-efficiency PMSM where the speed precision and wide speed range are required for proper control. Thus, field-oriented control is mandatory.

References

Ehsani M, Gao Y, Emadi A (2010) Modern electric, hybrid electric, and fuel cell vehicles fundamentals, theory, and design. CRC Press, Taylor & Francis Group, 2 editions

Hartman EP, Biermann D (1937) The aerodynamic characteristics of full-scale propellers having 2, 3, and 4 blades of Clark Y and R.A.F. 6 aifoil sections, NACA report 640

Jeracki RJ, Mitchell GA (1981) Low and high speed propellers for general aviation—performance potential and recent wind tunnel test results. Lewis Research Center. Cleveland, Ohio

McCormick BW (1995) Aerodynamics, aeronautics, and flight mechanics, 2nd edn. Wiley

Pellegrino G, Vagati A, Guglielmi P, Boazzo B (2012) Performance comparison between surface-mounted and interior PM motor drives for electric vehicle application. IEEE Trans Ind Electron 59(2):803–811

RUAG Aerospace Services GmbH, Dornier 228 Advanced Commuter (AC) Facts & Figures. https://dornier228.ruag.com. Accessed 02 Aug 2017

Siemens AG (2016) Aerobatic airplane "Extra 330LE". https://www.siemens.com/press/pool/de/events/2016/corporate/2016-12-innovation/inno2016-aerobaticairplane-e.pdf. Accessed 31 Aug 2017

Sul S-K (2011) Control of electric machine drive systems. Wiley-IEEE Press

Teeuwen YAP Propeller design for conceptual turboprop aircraft

Vas P (1998) Sensorless vector and direct torque control. University Press, Oxford

Voskuijl M, van Bogaert J, Rao AG (2018) Analysis and design of hybrid electric regional turboprop aircraft. CEAS Aeronaut J 9:15. https://doi.org/10.1007/s13272-017-0272-1

Wang J, Yuan X, Kais A (2013) Design optimization of a surface-mounted permanent-magnet motor with concentrated windings for electric vehicle applications. IEEE Trans Veh Technol 62:1053–1064

Chapter 12
Appendices

12.1 Real-Time Implementation: PiL Testing

For faster development, the hardware composed by the machine, inverter, and controller can be replaced by a precise real-time model for emulation. It is true that control systems increasingly use their verification and development through a digitized plant on an FPGA (field-programmable gate array) platform so that control algorithms can be evaluated without the need for real hardware, in this case an inverter and an electric machine (Tavana and Dinavahi 2015). However, real-time simulation of electric machine models and the VSI can be especially complicated due to the rapid nature of the dynamics, that is, reduced time constants, especially on very low-power machines. The switching of PWM signals of up to tens of kHz requires sampling rates of the order of several MHz to obtain reasonable accuracy, for example, to model the ripple produced by the PWM in the inductance of the machine. That is why FPGAs are the ideal platform for complex real-time simulations due to their ability to process data in parallel allowing sampling and execution rates up to the MHz range. An FPGA is a reconfigurable digital logic platform, which allows the execution of millions of operations in parallel. As cited in Le-Huy et al. (2006), research has advanced considerably in real-time modeling and simulation of different power systems that use FPGA as computational devices. The control algorithm designed in this case for the control of an electric machine is loaded on a target where it will be tested with the FPGA that models the VSI and the machine. Different test cases can be verified quickly by advancing many of the possible problems such as control stability. On the other hand, if VSI is avoided and only the machine model should be running in real time, the DSP platform instead of FPGA can be good and cheaper option.

© Springer Nature Switzerland AG 2020

R. Molina Llorente, *Practical Control of Electric Machines*,
Advances in Industrial Control, https://doi.org/10.1007/978-3-030-34758-1_12

The discrete model of an AC machine, such as the one seen for the PMASynRM in Sect. 5.4.2.2.2, can be loaded into a DSP to test the control algorithm on the actual ECU microcontroller in real time. The discrete model can be refined by adding thermal effects, non-linearity, saturation effect, and demagnetization of the permanent magnet, which add complexity in the model that requires more computation for real-time emulation, but in this case, it will be ignored. Figure 12.1 shows the block diagram of a possible system that emulates the PMASynRM in a DSP.

As can be seen, the analog output signals of interest of the machine model are the *dqe* current components such as i_{dse} and i_{qse} currents, together with the instantaneous rotor speed ω_r. As discussed in Chap. 10, these output variables permit closing the current and speed control loops. It is important to comment that these signals should be converted to analog signals through an D/A converter as figure shows. In this case, the DSP has an internal D/A converter which converts the machine model discrete signals at sampling time rate. On the other hand, the input signals are the components of the voltage V_{dse} and V_{qse}. In this case, they must be converted to digital signals through an A/D converter as observed in the figure. The implementation of the machine's discrete code begins with the Embedded Coder® code generation tool, which is a Simulink tool. The MISRA® compliant C code is generated with this tool for implementation in the DPS or microcontrollers. During the configuration phase of the code generation, it is possible to select different families of microcontrollers and DSPs, as well as speed or space optimization. At the end of the code generation, a report is automatically generated indicating, among other things, metrics of code size, lines of code, estimation of the necessary stack size, occupation of variables in RAM, etc. The generated code is ready to be used in the DSP without requiring any post-processing. Two *.*c* and *.*h* files are typically generated. The *.*c* file contains the main function of the machine model, while the *.*h* file contains the declaration of variables and structures necessary for the main function. Typically, the generated constants, if not indicated otherwise, are real constants. The DSP contains a floating-point unit that can process real numbers efficiently.

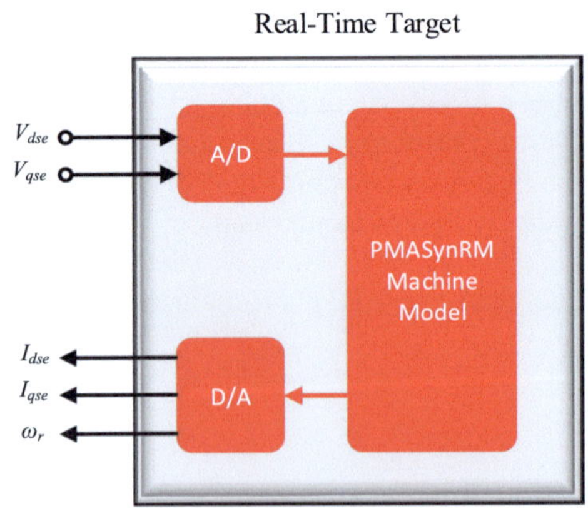

Fig. 12.1 PMASynRM emulation in a DSP with A/D and D/A converters

The generated C code of the PMASynRM for a sampling time of 4 ms is shown below. The constants that can be observed correspond to the constants k_1 to k_{10} (Sect. 5.4.2.2.2) which are real variables. As observed, there is a main function called "Discrete_step" which implements Eqs. (5.105), (5.106), and (5.109). The function should be called every sampling time T equal to 4 ms. For other sampling time rate, e.g., 40 μs, only the constants that depend on sampling time change, so it is easy to generate the code for $T = 40$ μs.

```
/* Model step function T=4ms */
void Discrete_step(void)
{
  real_T rtb_Delay5;
  real_T rtb_Sum4;
  real_T rtb_Sum6;
  real_T rtb_Sum6_tmp;
  /* Delay: '<S1>/Delay5' */
  rtb_Delay5 = rtDW.Delay5_DSTATE;
  rtb_Sum4 = ((0.970333745364647 * rtDW.Delay_DSTATE + 0.12360939431396786 *
              rtU.vde) + rtb_Delay5 * rtDW.Delay3_DSTATE *
             0.00030902348578491968) - 0.0 * rtb_Delay5;
  /* Outport: '<Root>/ide' */
  rtY.ide = rtb_Sum4;
  rtb_Delay5 = (0.7225433526011561 * rtDW.Delay1_DSTATE - rtDW.Delay4_DSTATE *
               rtb_Delay5 * 0.036300578034682084) + 1.1560693641618498 *
    rtU.vqe;
   /* Outport: '<Root>/iqe' */
   rtY.iqe = rtb_Delay5;
   rtb_Sum6_ tmp = rtb_Delay5 * rtb_Sum4 * 0.0289 + 0.0 * rtb_Sum4;
   rtb_Sum6 = (rtb_Sum6_ tmp * 0.8 - 0.26666666666666666 * rtU.TL) +
    rtDW.Delay2_DSTATE;
   /* Outport: '<Root>/wr' */
   rtY.wr = rtb_Sum6;
   rtY.Tem = rtb_Sum6_ tmp * 3.0;
   /* Update for Delay: '<S1>/Delay' */
   rtDW.Delay_DSTATE = rtb_Sum4;
   /* Update for Delay: '<S1>/Delay5' */
   rtDW.Delay5_DSTATE = rtb_Sum6;
   /* Update for Delay: '<S1>/Delay3' */
   rtDW.Delay3_DSTATE = rtb_Delay5;
   /* Update for Delay: '<S1>/Delay1' */
   rtDW.Delay1_DSTATE = rtb_Delay5;
   /* Update for Delay: '<S1>/Delay4' */
   rtDW.Delay4_DSTATE = rtb_Sum4;
   /* Update for Delay: '<S1>/Delay2' */
   rtDW.Delay2_DSTATE = rtb_Sum6;
}
```

The next step is to integrate the files generated in the DSP compilation project so that they are part of the compilation chain. Once this step is completed, the model can be executed in the DSP. The experimental result concludes that the time it takes to execute in real time of the discrete model of the PMASynRM is 4.95 μs. If the sampling time is 40 μs, the discrete model assumes a CPU usage of 12.38%. It should be noted that the processing time can be reduced even further by using an FPGA because of its excellent computing power as discussed early. For this case, the code generation options for the desired type of FPGA would have to be reconfigured.

The real-time results of the model running in the DSP are shown in Fig. 12.2 for a sampling time of 40 μs and 4 ms. The model is excited in open loop with a voltage $V_{dse} = 2\ V$, $V_{qse} = 15\ V$, and a load torque of 5 Nm. The torque and mechanical speed of the machine are represented. As it can see, the dynamics of the response are different for each of the sampling times. In the case of sampling time $T = 40\ \mu s$, its behavior is much more similar to that achieved with a continuous model, but on the contrary, the necessary computing power is higher. The overshoot is higher for the higher sampling time, making its dynamics slower compared to shorter sampling time. In other words, the stability depends strongly on the sampling time, and for smaller sampling times the higher the stability area and the higher the usable speed range. Therefore, there is an optimal compromise between sampling time, stability, and computing power that should be found.

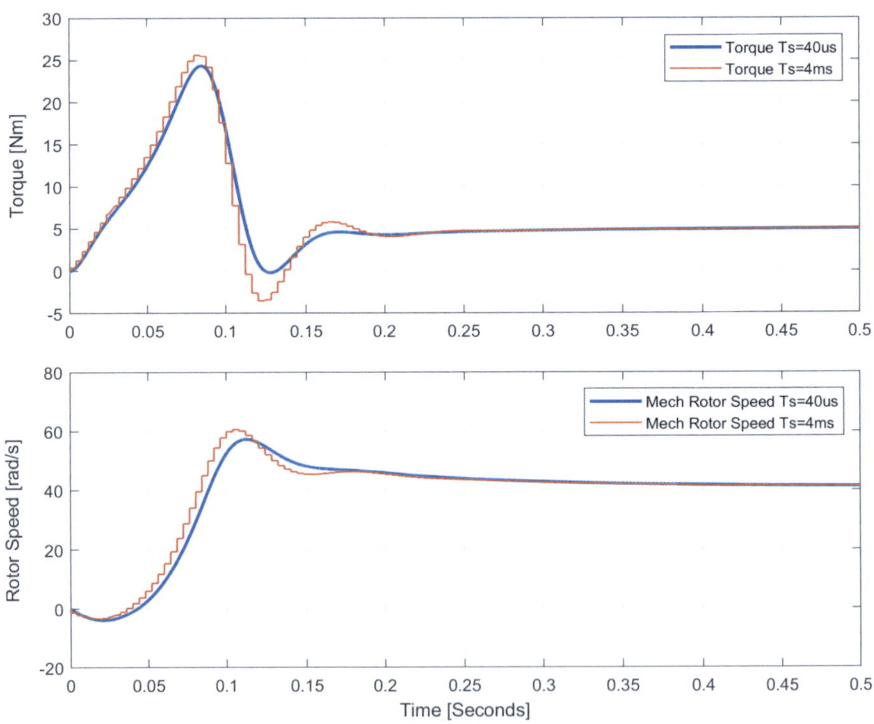

Fig. 12.2 Real-time output result for run-up speed and torque, of loaded PMASynRM at different sampling time in open loop. Load torque is 5 Nm. DSP clock frequency: 100 MHz. Data storage every Ts: 40 μs or 4 ms, depending on the sampling time

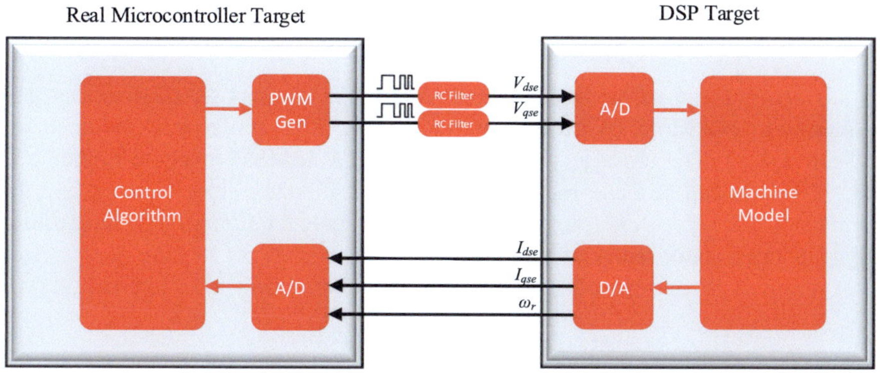

Fig. 12.3 Real microcontroller target connected to the DSP target, which emulates an AC machine. The real microcontroller is evaluated, then the processor is involved in the loop (PiL testing)

Figure 12.3 represents the generic system block diagram for the control algorithm evaluation on a real microcontroller through the AC machine emulation without inverter model. The AC machine model could be the one studied previously, while the control algorithm could be directly designed in the MBD phase.

In the absence of an inverter (there is no PWM), it is not necessary to sample in orders of magnitude of MHz to achieve a good approximation to the real model, but with orders of magnitude of several kHz, it is sufficient for most electrical machines. The embedded controller can be a standard evaluation microcontroller target (PiL), or much more interesting, the real hardware of the VSI to design (HiL).

The strategy of the control algorithm may be different, such as a field orientation control or a voltage–frequency control. Regardless of the chosen control strategy, the control loops can be tested in real time without the need for the real machine as previously mentioned.

At this stage, it is not necessary that the entire software architecture is fully implemented in the microcontroller, although it will always be better to perform tests with reasonably advanced architecture. In this phase, the control loops can be verified in real time without the need for the real machine as previously mentioned.

Stability tests can be performed under different machine operating conditions that will help to fine-tune the most optimal control in the control algorithm. This algorithm is already the code that will be part of the real platform so that it can be assured successful results, although small adjustments must be made to compensate among other factors, delays, and noise in A/D conversions, as well as the inverter function transfer. Moreover, an approximate measure of the CPU load, memory, and stack used can be accomplished in this stage, so that the performance of the basic microcontroller can be more adequately sized.

12.2 55 kW IPMSM Simulation Results

In this section, the motor and generator mode simulations detailed results performed with FluxMotor™ are presented according to a similar machine analyzed in Sect. 4.4.2.1. The cross-sectional view of 55 kW IPMSM machine is shown in Fig. 12.4a. As can be observed, the machine has 48 stator slots, 8 poles, and the permanent magnets of the rotor are mounted in *V*-pole configuration. The machine has an external stator radius of 134.5 mm and a stack length of 84 mm. The external rotor radius is 80.235 mm. The stator and rotor are made with a lamination type

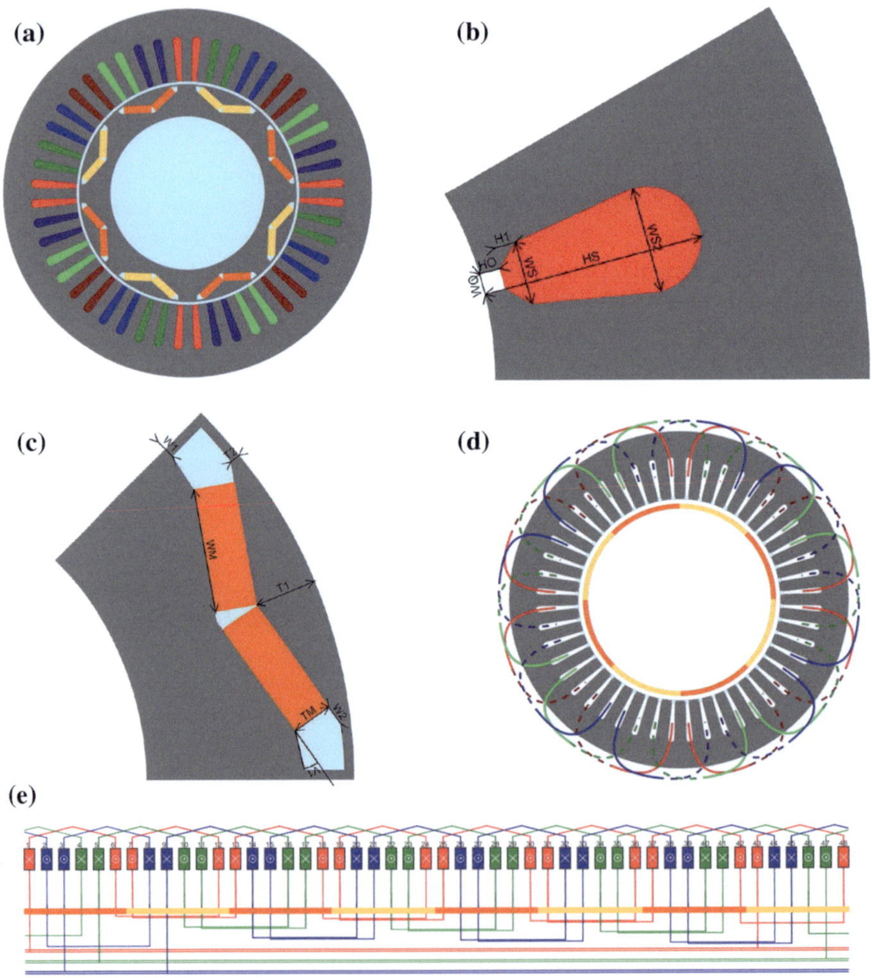

Fig. 12.4 **a** Cross-sectional view of the IPMSM machine. **b** Cross-sectional view of stator slot. **c** Cross-sectional view of magnet pole. **d** Winding configuration. **e** Winding layout. Image developed using FluxMotor™ provided courtesy of Altair Engineering, Inc.

M330-35A, that is, electrical steel stacks of 0.35 mm of thickness. The main characteristics of this material are high permeability to let higher amount of flux through the core, low electrical conductivity to reduce the eddy currents and hence the Foucault losses, and a mass density of 7650 kg/m^3. The rotor permanent magnets type is neodymium iron boron (NdFeB), and the dimensions are 18.9 × 6.5 × 84 mm, with a mass density of 7500 kg/m^3. The PM has a magnetic flux remanence B_r equal to 1.23 T and coercivity field strength H_c equal to 1400 kA/m at 20 °C. The total mass of the machine is 32.137 kg, and the total mass of magnets is 1.238 kg. Figure 12.4 represents some of the main characteristics of the IPMSM machine.

Tables 12.1 and 12.2 show the slot and the magnet dimension details, respectively.

Table 12.1 Slot measurement details

Description	Name	Value	Unit
Slot height	HS	33.5	mm
Slot width	WS2	5.8	mm
Intermediary height of the slot	H1	0.53	mm
Intermediary width of the slot	WS1	3.19	mm
Height of slot opening	HO	1.02	mm
Width of slot opening	WO	1.93	mm

Table 12.2 Magnet measurement details

Description	Name	Value	Unit
Magnet thickness	TM	6.5	mm
Magnet width	WM	18.9	mm
Height of rotor pole cap	T1	10.12	mm
Bridge thickness	T2	1.48	mm
Q-axis width	W1	13.15	mm
Window width	W2	0.1	mm
Window angle	V1	0.0	Deg.

12.2.1 Static Simulation

The following figures show the static simulation detail results performed with FluxMotor™, where *dq* flux linkage and *dq* inductances are represented in the *dq* current plane (Figs. 12.5, 12.6, 12.7 and 12.8).

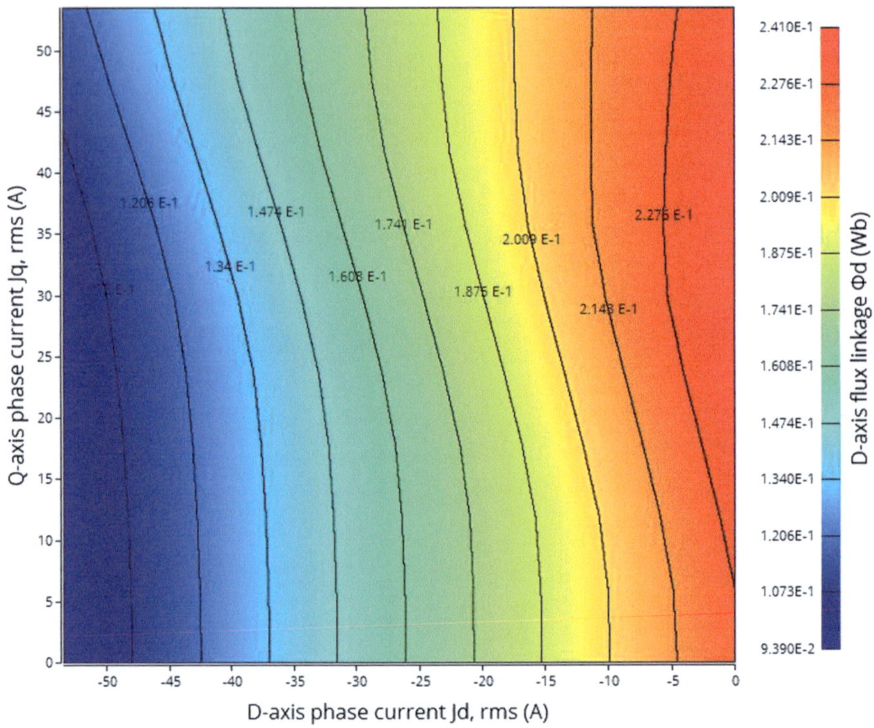

Fig. 12.5 *D*-axis flux linkage in the *dq* current plane. Image developed using FluxMotor™ provided courtesy of Altair Engineering, Inc.

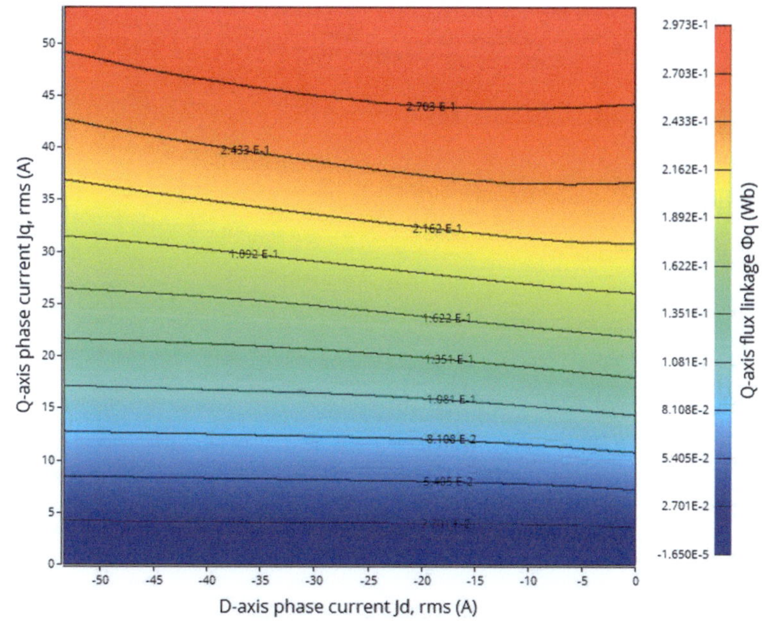

Fig. 12.6 *Q*-axis flux linkage in the *dq* current plane. Image developed using FluxMotor™ provided courtesy of Altair Engineering, Inc.

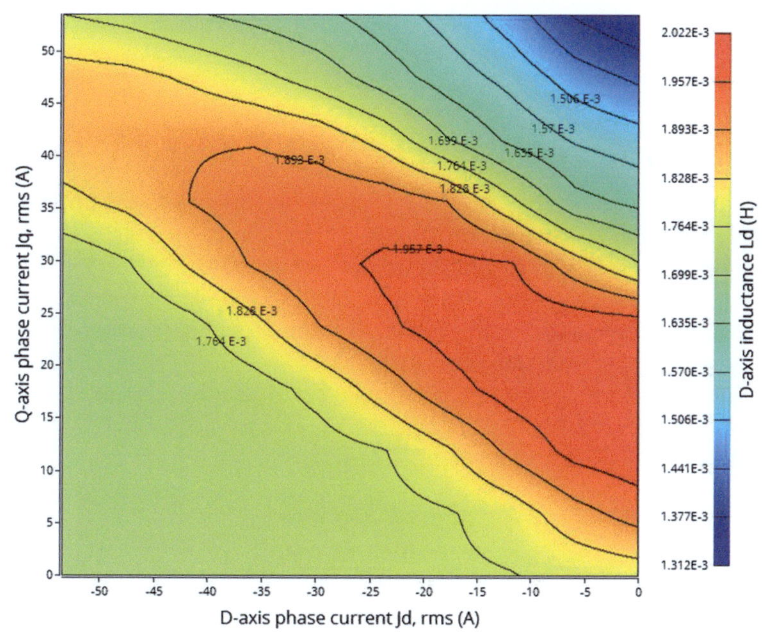

Fig. 12.7 *D*-axis inductance in the *dq* current plane. Image developed using FluxMotor™ provided courtesy of Altair Engineering, Inc.

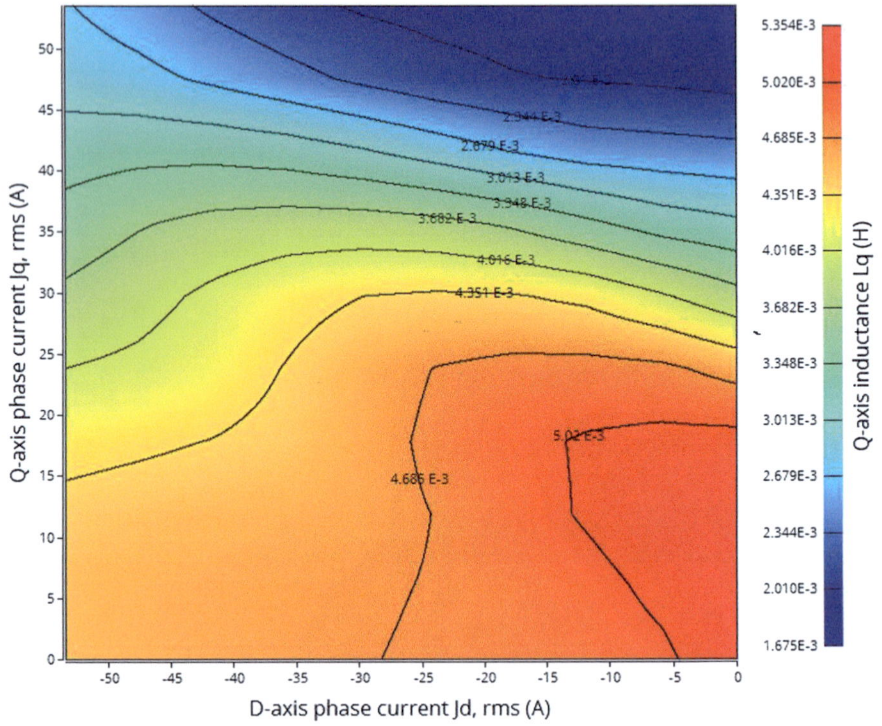

Fig. 12.8 Q-axis inductance in the dq current plane. Image developed using FluxMotor™ provided courtesy of Altair Engineering, Inc.

12.2.2 Motor Mode

The simulation results are performed with a three-phase inverter with a voltage constraint of 450 V (DC-link voltage) and 176.5 A RMS for current. Thus, the maximum line-to-line voltage is 318.198 V RMS, which means a phase voltage of 183.71 V RMS. The control strategy is the maximum torque per voltage (MTPV), and the speed range is from 0 to 6000 RPM.

The following figures show the simulation detail results performed with FluxMotor™ (Figs. 12.9, 12.10, 12.11, 12.12, 12.13, 12.14, 12.15, 12.16, 12.17, 12.18, 12.19, 12.20, 12.21, 12.22, 12.23 and 12.24).

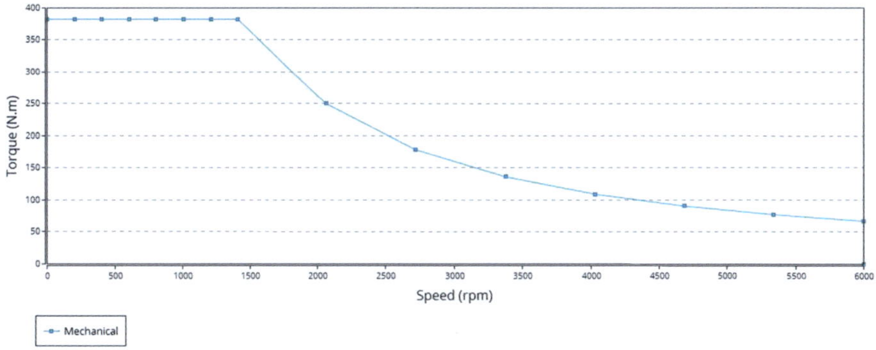

Fig. 12.9 Mechanical torque versus speed. Image developed using FluxMotor™ provided courtesy of Altair Engineering, Inc.

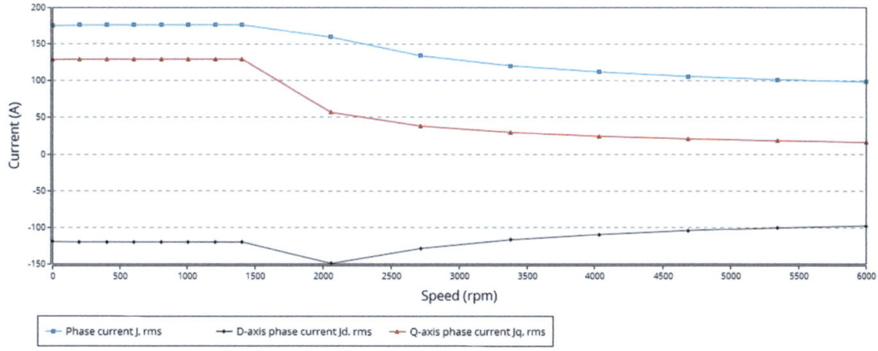

Fig. 12.10 Phase current and *dqe* currents versus speed. Image developed using FluxMotor™ provided courtesy of Altair Engineering, Inc.

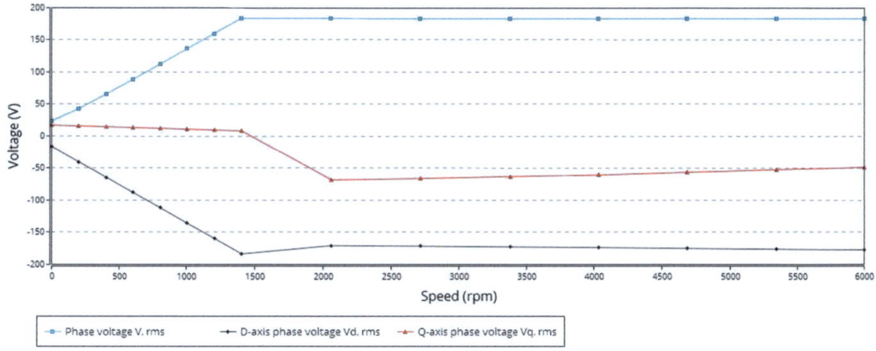

Fig. 12.11 Phase voltage and *dqe* voltage versus speed. Image developed using FluxMotor™ provided courtesy of Altair Engineering, Inc.

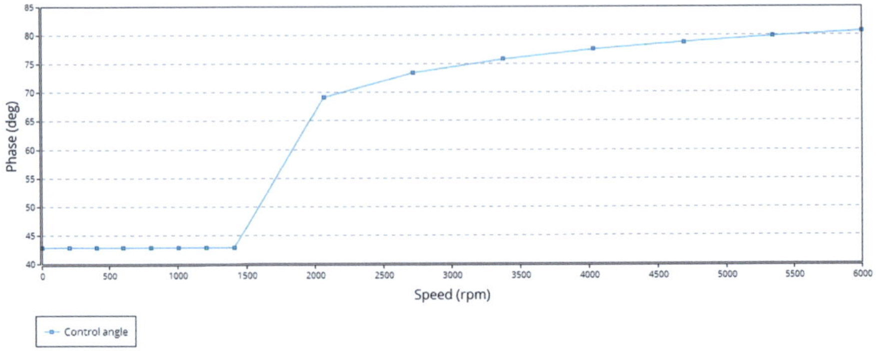

Fig. 12.12 Control angle versus speed. Image developed using FluxMotor™ provided courtesy of Altair Engineering, Inc.

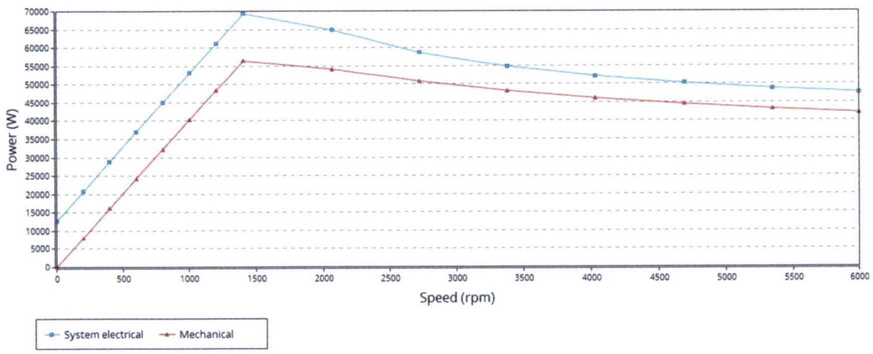

Fig. 12.13 Electrical and mechanical power versus speed. Image developed using FluxMotor™ provided courtesy of Altair Engineering, Inc.

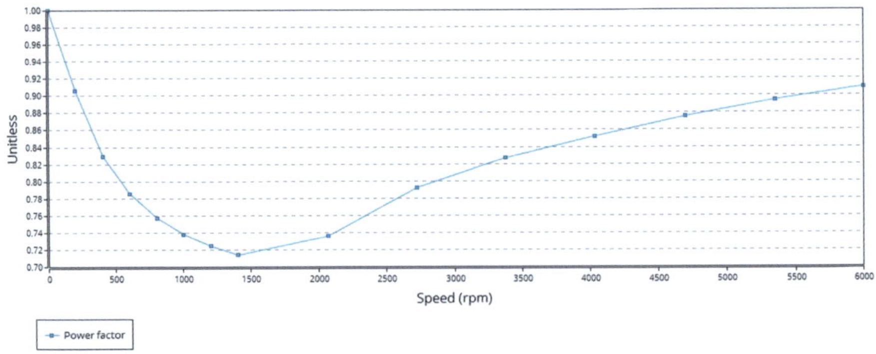

Fig. 12.14 Power factor versus speed. Image developed using FluxMotor™ provided courtesy of Altair Engineering, Inc.

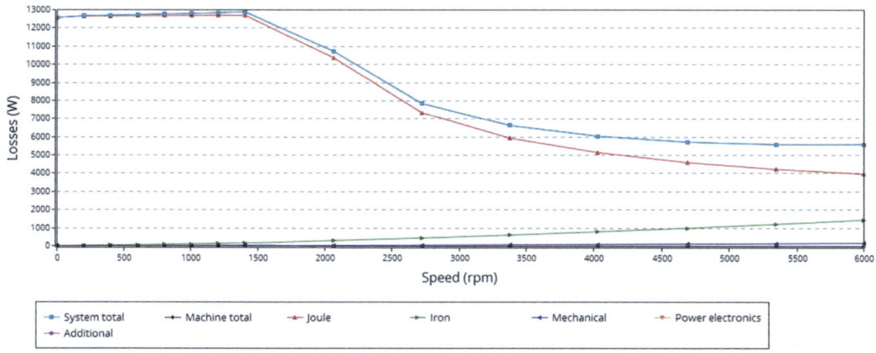

Fig. 12.15 Losses versus speed. Image developed using FluxMotor™ provided courtesy of Altair Engineering, Inc.

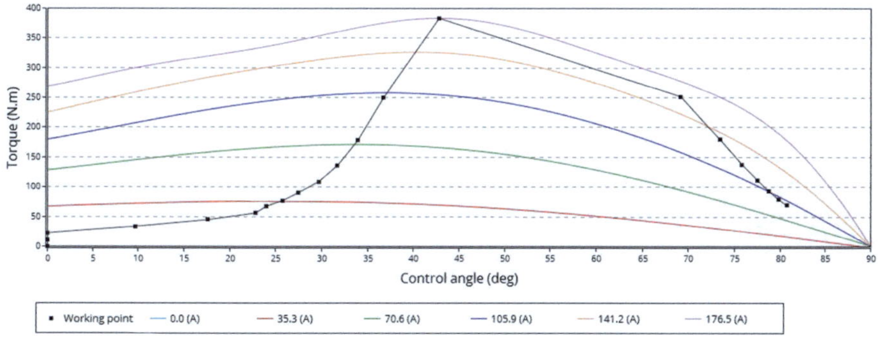

Fig. 12.16 Electromagnetic torque versus current and control angle. Image developed using FluxMotor™ provided courtesy of Altair Engineering, Inc.

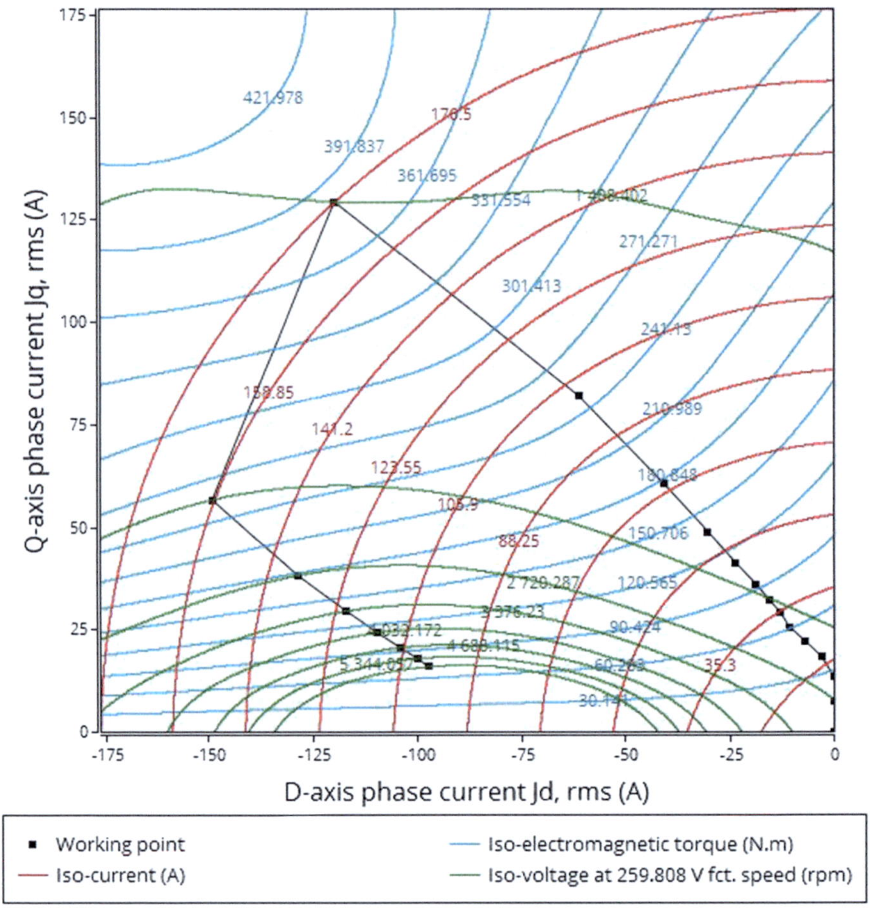

Fig. 12.17 DQ current plane. Image developed using FluxMotor™ provided courtesy of Altair Engineering, Inc.

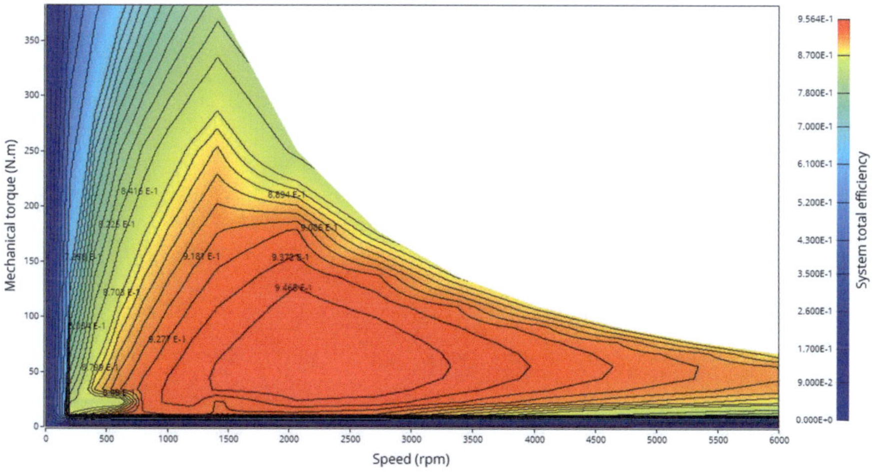

Fig. 12.18 Efficiency in the torque–speed area. Image developed using FluxMotor™ provided courtesy of Altair Engineering, Inc.

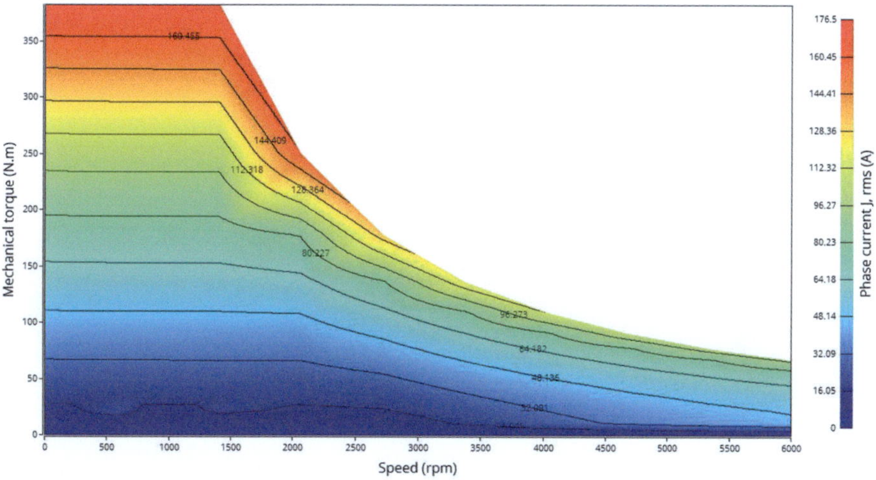

Fig. 12.19 Current in torque–speed area. Image developed using FluxMotor™ provided courtesy of Altair Engineering, Inc.

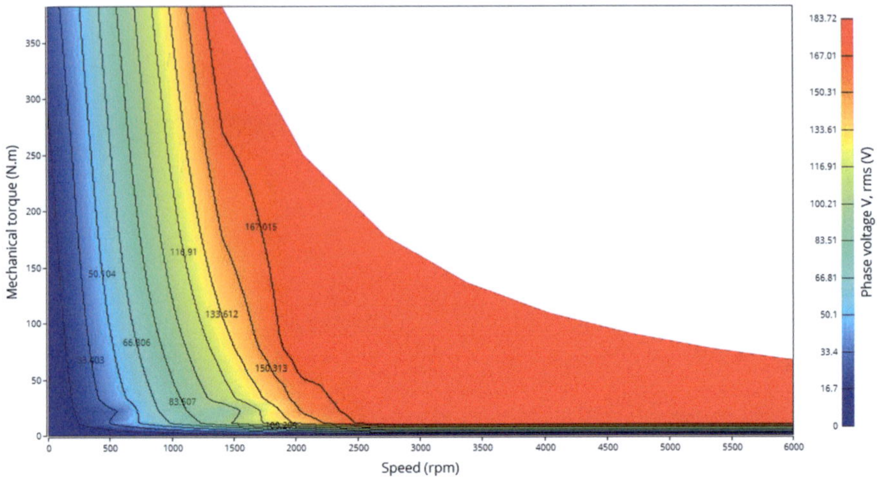

Fig. 12.20 Voltage in torque–speed area. Image developed using FluxMotor™ provided courtesy of Altair Engineering, Inc.

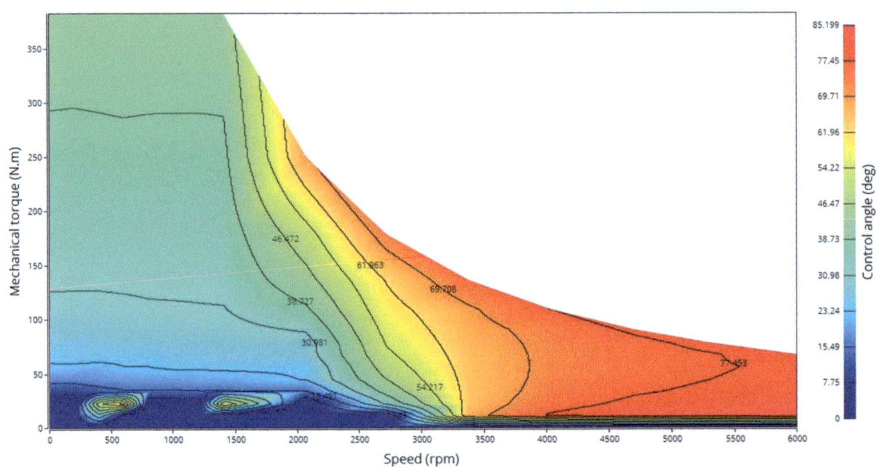

Fig. 12.21 Control angle in the torque–speed area. Image developed using FluxMotor™ provided courtesy of Altair Engineering, Inc.

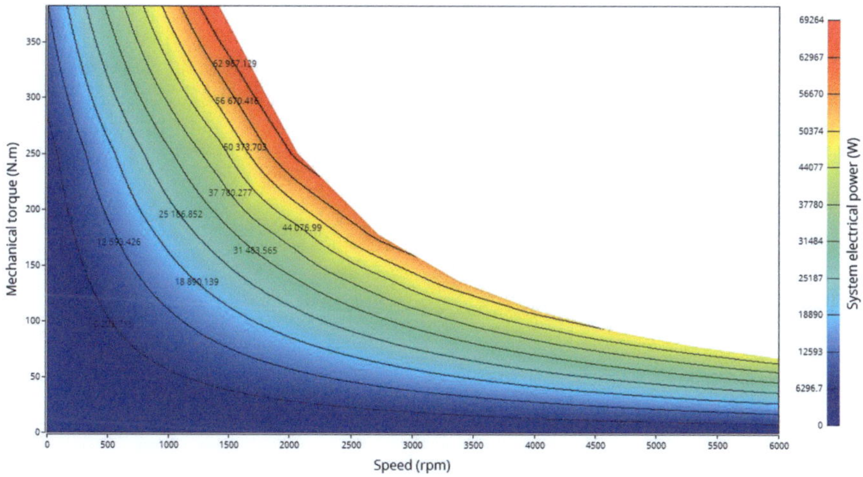

Fig. 12.22 Electrical power in the torque–speed area. Image developed using FluxMotor™ provided courtesy of Altair Engineering, Inc.

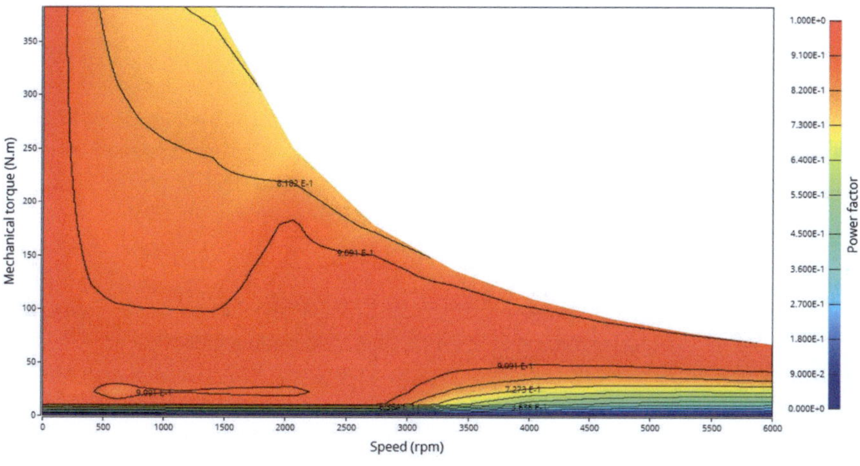

Fig. 12.23 Power factor in the torque–speed area. Image developed using FluxMotor™ provided courtesy of Altair Engineering, Inc.

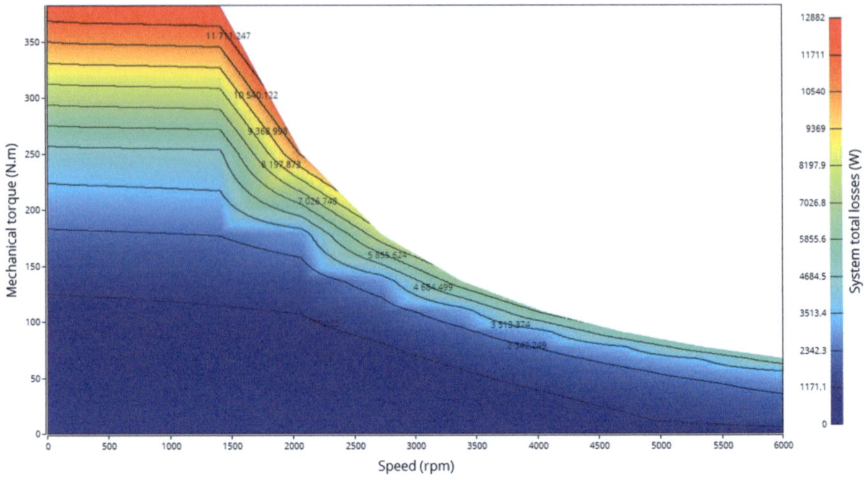

Fig. 12.24 Losses in torque–speed area. Image developed using FluxMotor™ provided courtesy of Altair Engineering, Inc.

12.2.3 Generator Mode

The control strategy in generator mode is set to maximum torque per ampere (MTPA), and the speed range is from 0 to 3000 RPM. The simulations result for generator mode in the corner speed command of 2262.461 RPM sets and electric power generated of 35,165 VA with a power factor of 0.879 and an efficiency of 97.09%. The torque is 134.331 Nm. The line-to-line voltage is 380.606 V RMS, and the phase current is 53.343 A RMS.

For maximum speed command of 3226.272 RPM, the power generated is 4429.149 VA with a power factor of 0.9989. The efficiency and torque are 97.728% and 13.55 Nm, respectively. The line-to-line voltage is 379.573 V RMS, and the phase current is 6.737 A RMS.

The following figures show the simulation detail results performed with FluxMotor™ (Figs. 12.25, 12.26, 12.27, 12.28, 12.29, 12.30, 12.31, 12.32, 12.33, 12.34, 12.35, 12.36, 12.37, 12.38, 12.39, and 12.40).

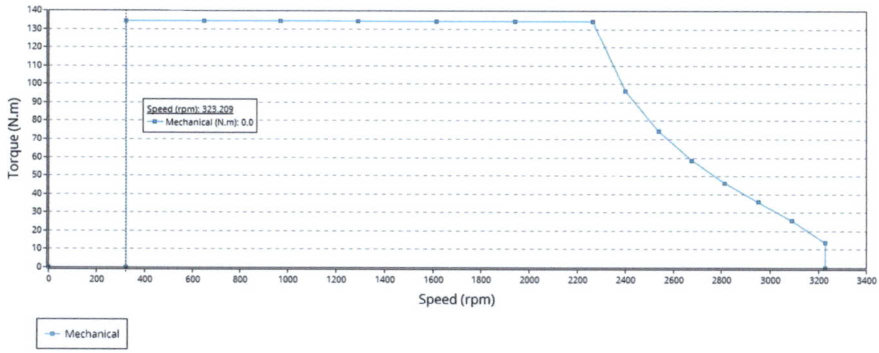

Fig. 12.25 Mechanical torque versus speed. Image developed using FluxMotor™ provided courtesy of Altair Engineering, Inc.

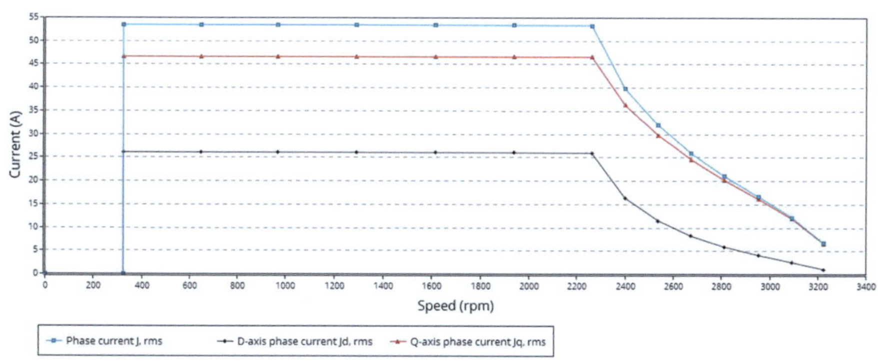

Fig. 12.26 Phase current and *dqe* current versus speed. Image developed using FluxMotor™ provided courtesy of Altair Engineering, Inc.

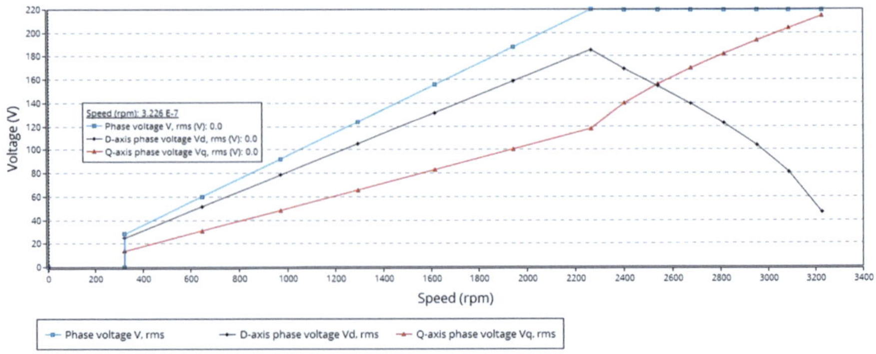

Fig. 12.27 Phase voltage and *dqe* voltage versus speed. Image developed using FluxMotor™ provided courtesy of Altair Engineering, Inc.

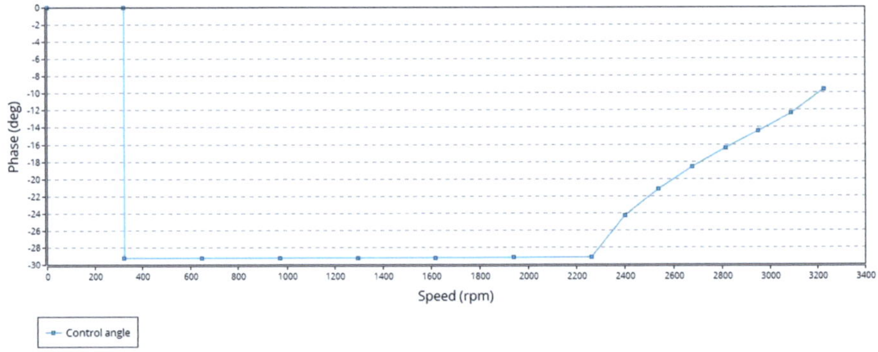

Fig. 12.28 Control angle versus speed. Image developed using FluxMotor™ provided courtesy of Altair Engineering, Inc.

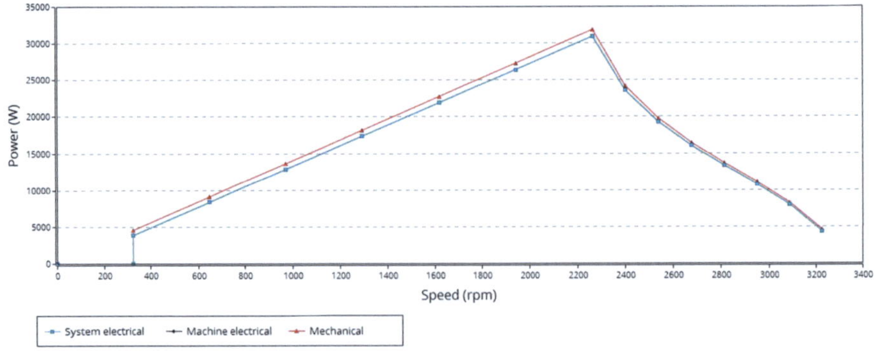

Fig. 12.29 Electrical and mechanical power versus speed. Image developed using FluxMotor™ provided courtesy of Altair Engineering, Inc.

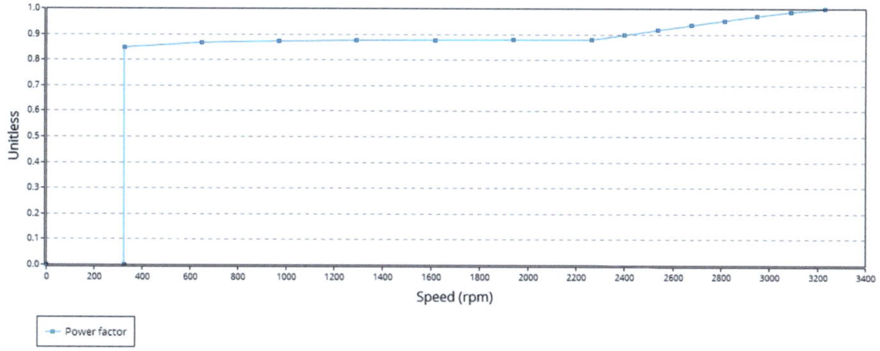

Fig. 12.30 Power factor versus speed. Image developed using FluxMotor™ provided courtesy of Altair Engineering, Inc.

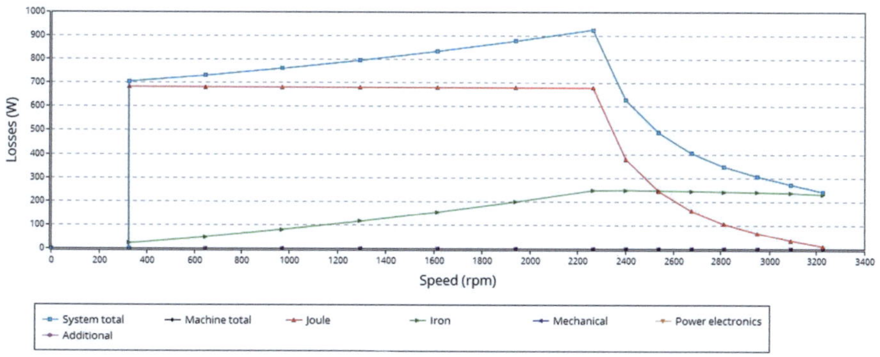

Fig. 12.31 Losses versus speed. Image developed using FluxMotor™ provided courtesy of Altair Engineering, Inc.

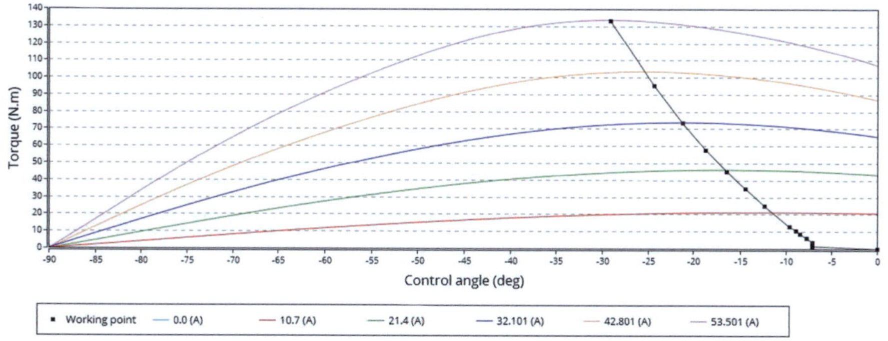

Fig. 12.32 Electromagnetic torque versus current and control angle. Image developed using FluxMotor™ provided courtesy of Altair Engineering, Inc.

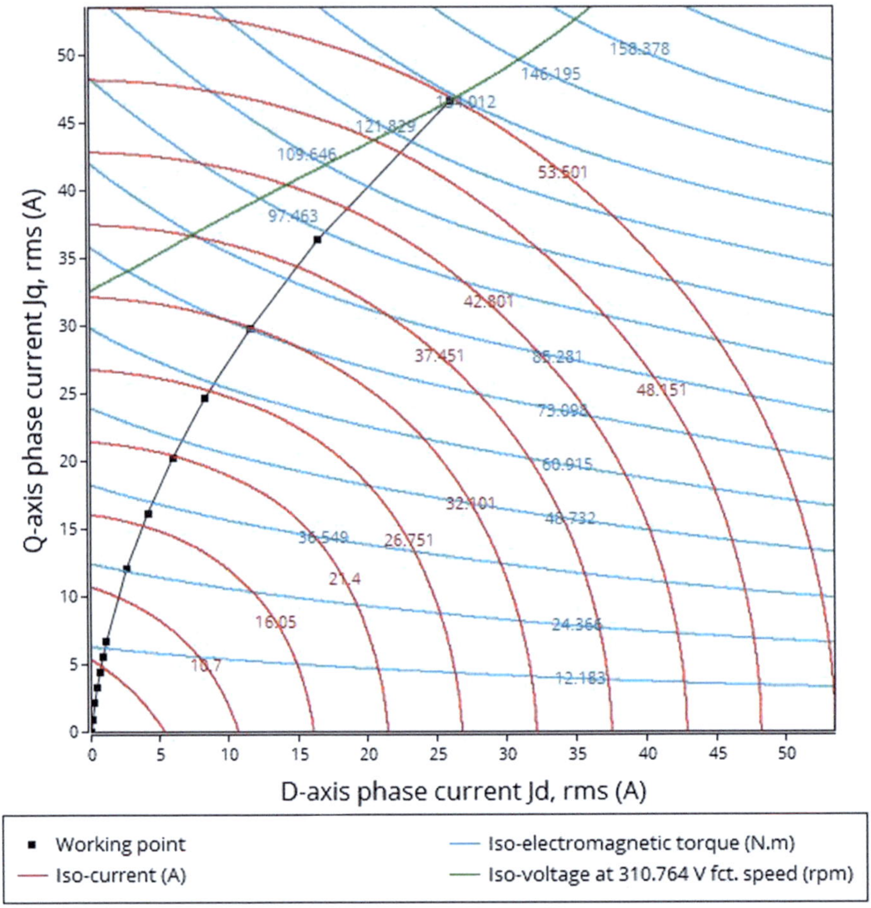

Fig. 12.33 DQ current plane. Image developed using FluxMotor™ provided courtesy of Altair Engineering, Inc.

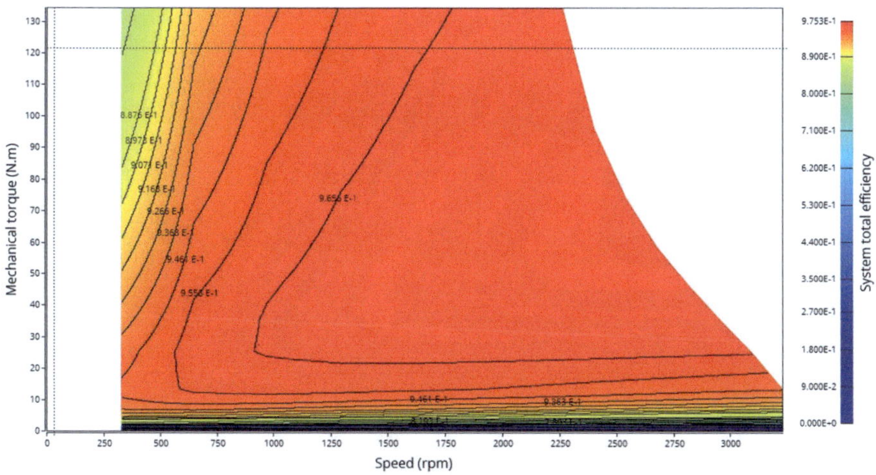

Fig. 12.34 Efficiency in the torque–speed area. Image developed using FluxMotor™ provided courtesy of Altair Engineering, Inc.

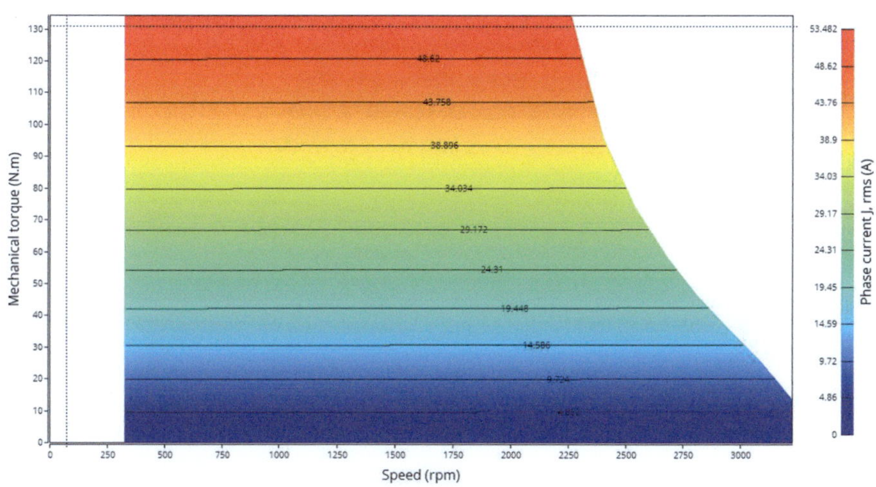

Fig. 12.35 Current in torque–speed area. Image developed using FluxMotor™ provided courtesy of Altair Engineering, Inc.

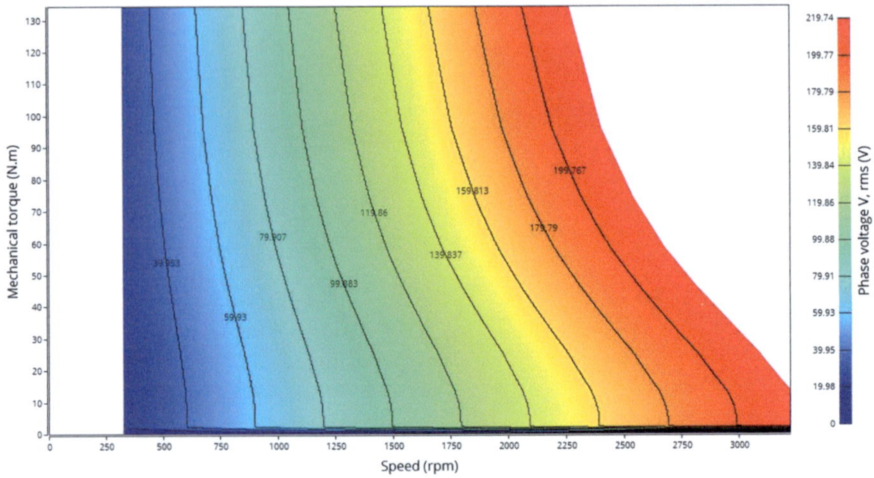

Fig. 12.36 Voltage in torque–speed area. Image developed using FluxMotor™ provided courtesy of Altair Engineering, Inc.

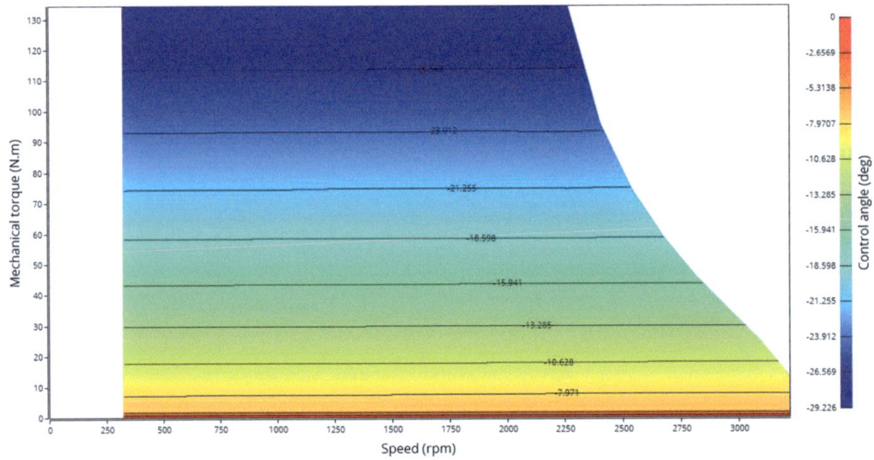

Fig. 12.37 Control angle in the torque–speed area. Image developed using FluxMotor™ provided courtesy of Altair Engineering, Inc.

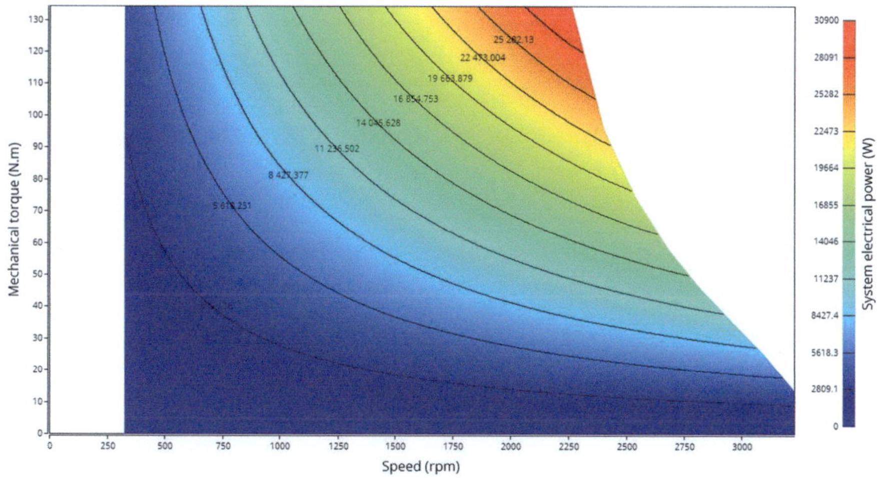

Fig. 12.38 Power in torque–speed area. Image developed using FluxMotor™ provided courtesy of Altair Engineering, Inc.

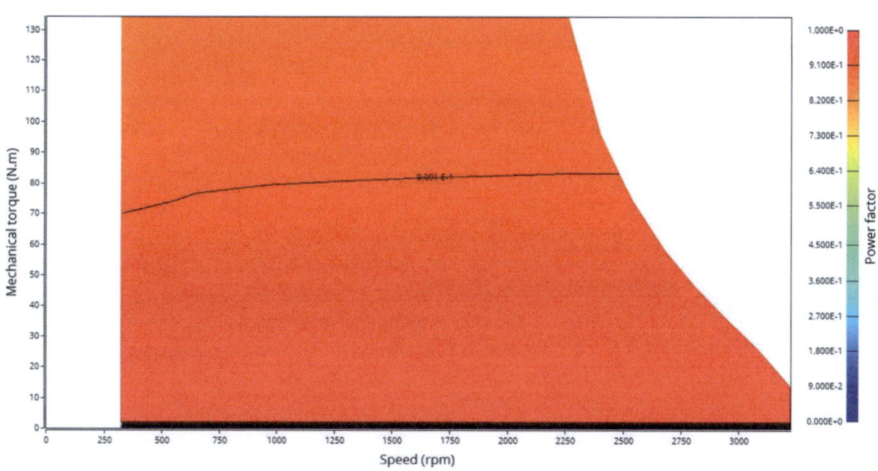

Fig. 12.39 Power factor in the torque–speed area. Image developed using FluxMotor™ provided courtesy of Altair Engineering, Inc.

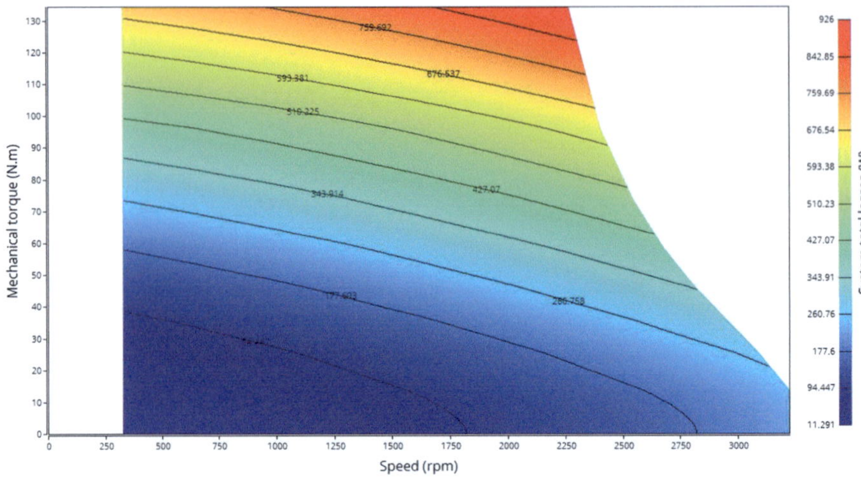

Fig. 12.40 Losses in torque–speed area. Image developed using FluxMotor™ provided courtesy of Altair Engineering, Inc.

References

Le-Huy P, Guerette S, Dessaint LA, Le-Huy H (2006) Real-time simulation of power electronics in power systems using an FPGA. In: Canadian conference on electrical and computer engineering, May 2006, pp 873–877

Tavana NR, Dinavahi V (2015) A general framework for FPGA-based real-time emulation of electrical machines for HIL applications. IEEE Trans Ind Electron 62(4):2041–2053

Index

© Springer Nature Switzerland AG 2020

R. Molina Llorente, *Practical Control of Electric Machines*,
Advances in Industrial Control, https://doi.org/10.1007/978-3-030-34758-1

D

E

Q

R

MIX
Papier aus verantwortungsvollen Quellen
Paper from responsible sources
FSC® C105338

If you have any concerns about our products, you can contact us on
ProductSafety@springernature.com

In case Publisher is established outside the EU, the EU authorized representative is:
Springer Nature Customer Service Center GmbH
Europaplatz 3, 69115 Heidelberg, Germany

Printed by Libri Plureos GmbH
in Hamburg, Germany